YERSINIA

MOLECULAR AND CELLULAR BIOLOGY

Edited by:

Elisabeth Carniel
Institut Pasteur, Paris, France

and

B. Joseph Hinnebusch
National Institutes of Health, Hamilton, USA

horizon bioscience

Horizon Bioscience
32 Hewitts Lane
Wymondham
Norfolk NR18 0JA
U.K.

www.horizonbioscience.com

British Library Cataloguing-in-Publication Data

A catalogue record for this book is available from the British Library

ISBN: 1-904933-06-8

Printed in Great Britain by Antony Rowe Ltd, Chippenham, Wiltshire

Contents

Section IIIA. Molecular Biology of *Yersinia* Chromosome-Encoded Factors

Section IIIB. Molecular Biology of *Yersinia* Plasmid-Encoded Factors

Contributors

Jun Abe*
Department of Allergy and Immunology
National Research Institute for Child Health and Development
Tokyo
Japan

Mark Achtman*
Max-Planck Institut für Infektionsbiologie
Dept. of Molecular Biology
Schumannstrasse 21/22
10117 Berlin
Germany
achtman@mpiib-berlin.mpg.de

Steven Atkinson
Institute of Infection, Immunity and Inflammation
Centre for Biomolecular Sciences
University of Nottingham
Nottingham
NG7 2RD
UK

Ingo Birger Autenrieth*
Institut für Med Mikrobiologie
Universitatsklinikum
Elfriede-Aulhorn-Strasse 6
Tubingen
D-72076
Germany

Penelope Barnes
Div. Allergy and Infectious Diseases
University of Washington
Seattle
WA
USA

Erwin Bohn
Institut für Med Mikrobiologie
Universitatsklinikum
Elfriede-Aulhorn-Strasse 6
Tubingen
D-72076
Germany

Miguel Cámara
Institute of Infection, Immunity and Inflammation
Centre for Biomolecular Sciences
University of Nottingham
Nottingham
NG7 2RD
UK

Elisabeth Carniel*
Yersinia Research Unit
Institut Pasteur
28 rue du Dr. Roux
75724 Paris cedex 15
France

François Collyn
Inserm E0364
Université de Lille II (Faculté de Médecine Henri Warembourg)
Département de Pathogenèse des Maladies Infectieuses
Institut de Biologie de Lille
1 rue du Professeur Calmette
F-59021 Lille Cedex
France

Guy R. Cornelis*
Molekulare Mikrobiologie
Biozentrum der Universität Basel
Klingelbergstr. 50-70
CH - 4056 Basel
Switzerland

Andrew J. Darwin*
Department of Microbiology MSB 228
New York University School of Medicine
550 First Avenue
New York
NY 10016
USA

Jacqueline D. Fetherston
Department of Microbiology Immunology
and Molecular Genetics
MS415 Medical Center
University of Kentucky
Lexington
KY 40536-0298
USA

Laura J. Fooks
School of Animal and Microbial Sciences
University of Reading
UK

B. Joseph Hinnebusch*
Laboratory of Human Bacterial Pathogenesis
Rocky Mountain Laboratories
National Institute of Allergy and Infectious Diseases
National Institutes of Health
Hamilton
MT 59840
USA

Ralph R. Isberg*
Department of Molecular Biology and Microbiology
Tufts University School of Medicine
150 Harrison Ave.
Boston
MA 02111
USA

Stefan D. Knight
Uppsala Biomedical Center
Swedish University of Agricultural Sciences
Uppsala
Sweden

Timo K. Korhonen*
Division of General Microbiology
Department of Biosciences
FIN-0014 University of Helsinki
Finland

Maini Kukkonen
Division of General Microbiology
Department of Biosciences
FIN-0014 University of Helsinki
Finland

Päivi Kyllönen
Division of General Microbiology
Department of Biosciences
FIN-0014 University of Helsinki
Finland

Kaarina Lähteenmäki
Division of General Microbiology
Department of Biosciences
FIN-0014 University of Helsinki
Finland

Hannu Lang
Division of General Microbiology
Department of Biosciences
FIN-0014 University of Helsinki
Finland

Biliana Lesic
Yersinia Research Unit
Institut Pasteur
28 rue du Dr. Roux
75724 Paris cedex 15
France

Luther E. Lindler*
National Biodefense Analysis and Countermeasures Center
Science and Technology Directorate
Department of Homeland Security
Washington
DC 20528
USA

Sheila MacIntyre*
School of Animal and Microbial Sciences
University of Reading
UK

Michaël Marceau*
Inserm E0364
Université de Lille II (Faculté de Médecine Henri Warembourg)
Département de Pathogenèse des Maladies Infectieuses
Institut de Biologie de Lille
1 rue du Professeur Calmette
F-59021 Lille Cedex
France

Marie-Noëlle Marenne
Christian de Duve Institute of Cellular Pathology and Faculté de Médecine
Université de Louvain
B-1200 Brussels
Belgium

Luís Jaime Mota
Molekulare Mikrobiologie
Biozentrum der Universität Basel
Klingelbergstr. 50-70
CH - 4056 Basel
Switzerland

Julian Parkhill
The Sanger Institute
Wellcome Trust Genome Campus
Hinxton
Cambridge
CB10 1SA
UK

Robert D. Perry*
Department of Microbiology
Immunology
and Molecular Genetics
MS415 Medical Center
University of Kentucky
Lexington
KY 40536-0298
USA

Michel Simonet*
Inserm E0364
Université de Lille II (Faculté de Médecine Henri Warembourg)
Département de Pathogenèse des Maladies Infectieuses
Institut de Biologie de Lille
1 rue du Professeur Calmette
F-59021 Lille Cedex
France

Mikael Skurnik*
Department of Bacteriology and Immunology
Haartman Institute
University of Helsinki
and Helsinki University Central Hospital Laboratory Diagnostics
PO Box 63
00014 University of Helsinki
Helsinki
Finland

R. Elizabeth Sockett
Institute of Genetics
University of Nottingham
Queen's Medical Centre
Nottingham
NG7 2UH
UK

G. B. Smirnov*
The Gamaleya Research Institute
for Epidemiology and Microbiology
RAMS
123098 Moscow
Russian Federation

Marjo Suomalainen
Division of General Microbiology
Department of Biosciences
FIN-0014 University of Helsinki
Finland

Nicholas R. Thomson*
The Sanger Institute
Wellcome Trust Genome Campus
Hinxton
Cambridge
CB10 1SA
UK

Ritva Virkola
Division of General Microbiology
Department of Biosciences
FIN-0014 University of Helsinki
Finland

Paul Williams*
Institute of Infection, Immunity and Inflammation
Centre for Biomolecular Sciences
University of Nottingham
Nottingham
NG7 2RD
UK

Ka-Wing Wong
Department of Molecular Biology and Microbiology
Tufts University School of Medicine
150 Harrison Ave.
Boston
MA 02111
USA

Glenn M. Young*
University of California
One Shield Avenue
FS and T
217 Cruess Hall
Davis
CA 95616
USA

* Corresponding author

Preface

The genus *Yersinia* occupies a prominent place in the history of microbiology and medicine. *Yersinia pestis* has claimed millions of lives in periodic pandemics; perhaps no other pathogen has influenced human history and civilization so dramatically. In recent years, the three pathogenic *Yersinia* species, *Y. pestis*, *Y. pseudotuberculosis*, and *Y. enterocolitica*, have provided highly insightful models to study many of the key aspects of microbial pathogenesis, including adherence to host cells, invasion of host tissues, the regulation and controlled secretion of virulence factors, in vivo acquisition of iron, and evasion of the host immune system. This book presents a comprehensive and comparative update on these and other elements of the host-parasite relationships of the Yersiniae.

It is clear that this book appears at a watershed moment in the history of *Yersinia* research. The genome sequences of at least one strain of all three species are available, and the genomes of additional strains will soon follow. The first chapter is a review of *Y. pestis* genomics, but the influence of this new information is clear in many of the chapters. Preliminary comparative genomics analyses have provided insight into the evolution of *Yersinia*, but have perhaps raised still more questions about how genetic differences translate into different pathogenesis phenotypes. Many of these questions, and the experimental framework from which the answers will come, will be apparent to the readers of this book. New knowledge on the molecular pathogenesis of *Yersinia* has accumulated at an impressive rate, and there is little doubt that these bacteria will continue to hold a prominent position in the future.

Thus, this book provides a timely summation of the molecular and cellular biology of the three pathogenic *Yersinia* species at the beginning of a promising new era. For this, we would like to express our thanks to each of our fellow authors, and to Annette Griffin and her staff at Horizon Scientific Press.

Elisabeth Carniel
B. Joseph Hinnebusch

October 2004

Cover image

Chain-forming yersiniae and phagocytes on the surface of the destroyed follicle-associated epithelium (see page 174).

Books of Related Interest

SAGE: Current Technologies and Applications	2005
Microbial Toxins: Molecular and Cellular Biology	2005
Vaccines: Frontiers in Design and Development	2005
Antimicrobial Peptides in Human Health and Disease	2005
Campylobacter: Molecular and Cellular Biology	2005
The Microbe-Host Interface in Respiratory Tract Infections	2005
Malaria Parasites: Genomes and Molecular Biology	2004
Pathogenic Fungi: Structural Biology and Taxonomy	2004
Pathogenic Fungi: Host Interactions and Emerging Strategies for Control	2004
Strict and Facultative Anaerobes: Medical and Environmental Aspects	2004
Brucella: Molecular and Cellular Biology	2004
Yersinia: Molecular and Cellular Biology	2004
Bacterial Spore Formers: Probiotics and Emerging Applications	2004
Foot and Mouth Disease: Current Perspectives	2004
Sumoylation: Molecular Biology and Biochemistry	2004
DNA Amplification: Current Technologies and Applications	2004
Prions and Prion Diseases: Current Perspectives	2004
Real-Time PCR: An Essential Guide	2004
Protein Expression Technologies: Current Status and Future Trends	2004
Computational Genomics: Theory and Application	2004
The Internet for Cell and Molecular Biologists (2nd Edition)	2004
Tuberculosis: The Microbe Host Interface	2004
Metabolic Engineering in the Post Genomic Era	2004
Peptide Nucleic Acids: Protocols and Applications (2nd Edition)	2004
Ebola and Marburg Viruses: Molecular and Cellular Biology	2004
MRSA: Current Perspectives	2003
Genome Mapping and Sequencing	2003
Bioremediation: A Critical Review	2003
Frontiers in Computational Genomics	2003
Transgenic Plants: Current Innovations and Future Trends	2003
Bioinformatics and Genomes: Current Perspectives	2003
Vaccine Delivery Strategies	2003
Multiple Drug Resistant Bacteria: Emerging Strategies	2003
Regulatory Networks in Prokaryotes	2003
Genomics of GC-Rich Gram-Positive Bacteria	2002
Genomic Technologies: Present and Future	2002
Probiotics and Prebiotics: Where are We Going?	2002

Full details of all these books at: www.horizonbioscience.com

Chapter 1

The *Yersinia pestis* Chromosome

Nicholas R. Thomson and Julian Parkhill

ABSTRACT

The whole genome sequences of two *Y. pestis* biovars, Mediaevalis and Orientalis, have recently been published. Each isolate sequenced is thought to represent the predominant strain from different plague pandemics: The Black Death and Modern Plague. Comparison of the two sequenced strains and other members of the *Enterobacteriaceae* reveal that *Y. pestis* exhibits many of the characteristics of an organism that has undergone a recent dramatic change in lifestyle. This chapter will focus on the global architecture of the *Y. pestis* chromosome as well as looking at the fine detail contained within the genome that gives some insights into how this pathogen has developed into such a potent foe.

INTRODUCTION

Since the causative agent of plague was first identified as a Gram-negative bacterium by Alexander Yersin in 1894, our understanding of the epidemiology, genetics and evolution of *Yersinia pestis* has come a long way. *Y. pestis* is thought to have been responsible for three human pandemics; the Justinian plague (6th to 8th centuries), The Black Death (14th to 19th centuries) and the current pandemic or Modern plague (19th century to present day). The Black Death alone is estimated to have claimed one third of the European population. A wealth of paintings and detailed historical writings have recorded the dramatic impact that plague has had on the development of modern civilization. This historical data combined with phenotypic analysis of strains isolated from foci known to have been affected in different pandemics has allowed isolates of *Y. pestis* to be split into three biovars: Biovar Antiqua is thought to be representative of strains that were responsible for the Justinian plague, similarly biovar Medievalis is thought to be representative of those responsible for The Black Death and biovar Orientalis is responsible for the current pandemic (Devignat, 1951).

The current pandemic of plague is widespread. Originating in Hong Kong in 1894, it has subsequently spread over five continents, with foci in Africa, Asia and North and South America. The reasons for the rise and fall of the human plague pandemics in the pre-antibiotic era remain obscure. More recently it is thought that improvements in public health and sanitation, which acts to limit human contact with animal reservoirs (and therefore infected fleas), as well as the development of both efficacious antibiotic and vaccine regimes has dramatically reduced the impact of the current pandemic on world health (Perry and Fetherston, 1997; Titball and Williamson, 2001). The WHO figure for the number of instances of plague in 1999 was 2,603 cases, which approximates to the current yearly average (World Health Organisation figures see: www.who.int/csr/disease/plague/impact/en/). However, there are still ongoing epidemics such as those in Surat, India (1994), Madagascar and the more recent outbreak in Algeria (2003) all of which serve to remind us that *Y. pestis* remains a potential threat (see Titball and Williamson, 2001 and references therein; World Health Organisation, Communicable Disease Surveillance and Response (CSR) http://www.who.int/csr/don/en/).

Yersinia pestis is primarily a rodent pathogen, usually transmitted subcutaneously to humans by the bite of an infected flea (principally *Xenopsylla cheopis*), but also transmitted by the airborne route during epidemics of pneumonic plague. Significantly, *Y. pestis* is very closely related to the gastrointestinal pathogen *Yersinia pseudotuberculosis* and it has been proposed that *Y. pestis* is a clone that evolved from *Y. pseudotuberculosis* as recently as 1,500-20,000 years ago (Achtman et al., 1999). Thus *Y. pestis* appears to be a species that has rapidly adapted from being a mammalian enteropathogen, spread via the fecal-oral route of infection and widely found in the environment, to an obligate blood-borne pathogen of mammals, which can also parasitize insects and utilize them as vectors for onward dissemination of infection (Perry and Fetherston, 1997; Achtman et al., 1999).

Two representatives of *Y. pestis* have recently been sequenced. The first to be published was *Y. pestis* biovar Orientalis strain CO92 (CO92), which was isolated from a fatal human case of primary pneumonic plague contracted from an infected cat in Colorado in 1992 (Parkhill et al., 2001b). The second *Y. pestis* genome sequence was from biovar Mediaevalis strain KIM10+ (KIM), a genetically amenable laboratory strain (Deng et al., 2002). The genome sequence of *Y. pestis* revealed features characteristic of a bacterium that has undergone a dramatic change in lifestyle. These features include the presence of many pseudogenes, a large population of insertion sequences and a significant level of genomic rearrangement. It was also apparent from the genome that gene acquisition had played a major role in the development of this pathogen; *Y. pestis* has multiple chromosomal loci exhibiting many of the characteristics of horizontally acquired DNA. These regions include previously characterized loci, such as the high pathogenicity island (see below), as well as an array of novel regions encoding, for example, possible insect toxins and proteins similar to insect viral enhancins. Many of these features will be discussed here using specific examples to illustrate points. For a more exhaustive analysis of the CO92 and KIM genome sequences see (Parkhill et al., 2001b and Deng et al., 2002). The contribution of the plasmids to the phenotype of the organism will be discussed in Chapter 3 of this book.

GENOMIC ARCHITECTURE OF THE *Y. PESTIS* CHROMOSOME

As anticipated, the genomes of *Y. pestis* CO92 and KIM were found to be very similar in size consisting of a 4.65 Mb and 4.60 Mb chromosome, respectively. One of the most striking features visible from looking at the global composition of the *Y. pestis* CO92 genome is the large anomalies in GC-bias (Figure 1A). Base compositional asymmetry is seen in most eubacterial genome sequences (Lobry, 1996; Lobry and Sueoka, 2002). This manifests itself as a small but detectable bias towards G nucleotides on the leading strand of the bi-directional replication fork, a phenomenon which is thought to be due to differential mutational biases on the leading and lagging strands during replication. This can be clearly seen in the *Escherichia coli* GC-bias plot (Figure 1B), a phenomenon which is often used to identify the origin and terminus of replication. Anomalies in this plot can be caused by the acquisition of DNA (such as prophage or integrative plasmids) or by the inversion or translocation of blocks of DNA within the genome. Since there is a constant pressure to maintain this polarization of the replichores (Lobry and Louarn, 2003), the

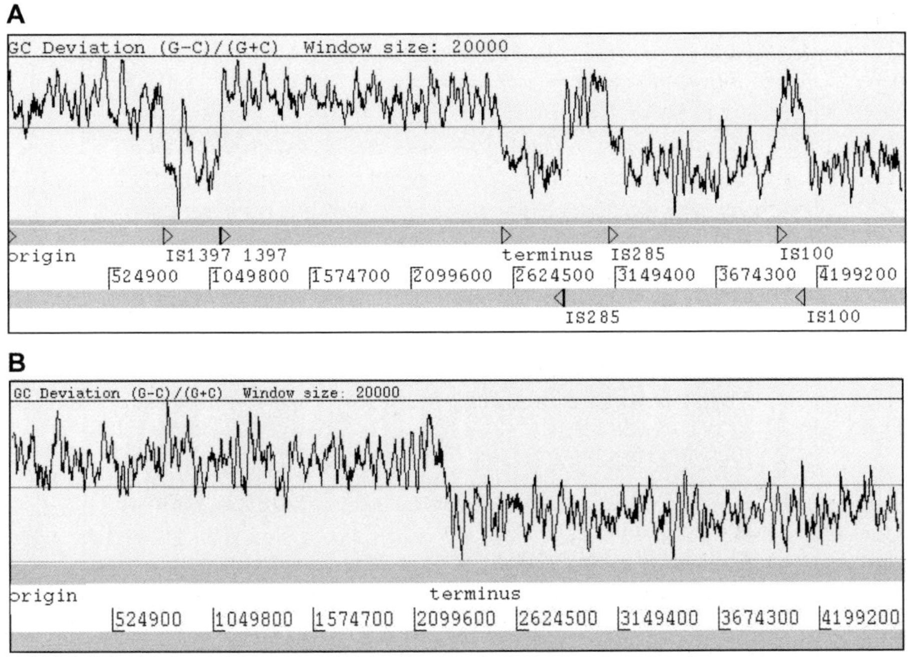

Figure 1. Global GC deviation plot for the genomes of *Y. pestis* CO92 (A) and *E. coli* K12 (B). The GC deviation plots show the asymmetrical substitution patterns for the G and C nucleotides on the upper strand of the genomic DNA using the formula (G-C)/(G+C). The grey bars below the GC deviation plot represent the forward and reverse strands of DNA and the figures refer to the base position within the genome. The GC deviation plot of *Y. pestis* CO92 (A) shows three anomalous regions bordered by IS elements (IS*285*, IS*100* and IS*1397* elements; open black triangles). The position of the origin and terminus of replication are also marked for both genomes.

majority of the loci showing an atypical GC bias are likely to be the result of relatively recent events. Three anomalies are visible in a G/C bias plot of CO92 (Figure 1A). These regions are bounded by IS elements, suggesting that they could be the result of recent recombination between these perfect repeats. Indeed, PCR studies indicated that these regions are capable of inverting during the growth of a standard bacterial culture. Similar observations were also made in other strains of *Y. pestis* biovar Orientalis, GB and MP6 (Parkhill et al., 2001b). Direct genome comparisons of CO92 and KIM extended these observations of genome fluidity in *Y. pestis,* showing that there were many more instances of intra-chromosomal recombination that differentiated these two biovars (Figure 2).

Deng et al. (2002) mapped the recombinations that differentiated CO92 from KIM and proposed a detailed model to describe the minimum number of recombination events that would explain the current relative arrangement of the genomes. Three regions within the KIM genome were seen to have undergone multiple rearrangements. One of these regions, multiple-inversion region I, spans the origin and carries five of the seven KIM rRNA operons (Figure 2). Deng et al. (2002) show how rearrangement of this region might have brought two rRNA operons into close proximity to form a tandem array. Subsequently, in CO92 this it thought to have resulted in the deletion of one of these rRNA operons by homologous recombination, explaining why CO92 has only 6 rRNA operons compared

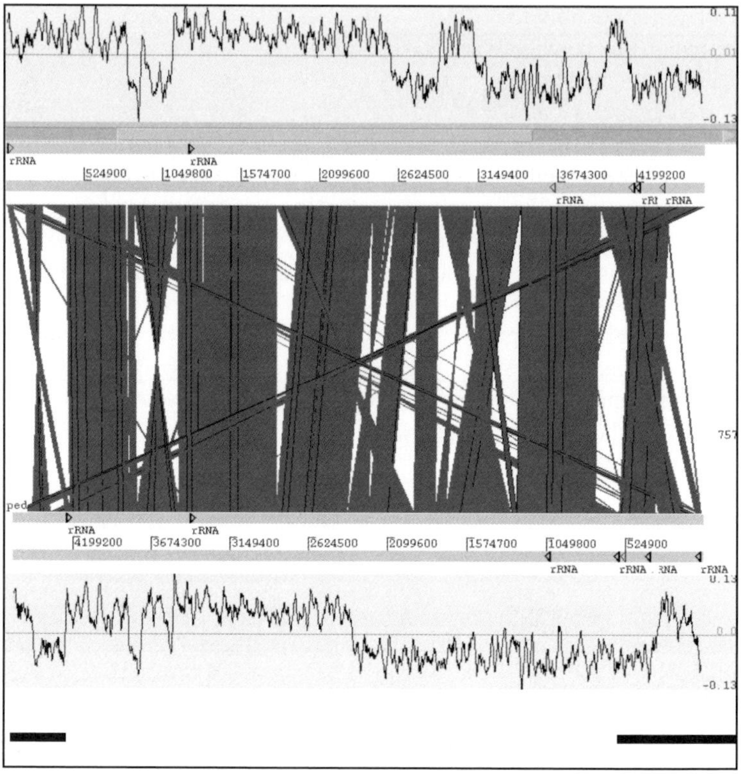

Multiple inversion region 1 (Deng et al, 2002).

Figure 2. Global comparison between *Y. pestis* CO92 and *Y. pestis* KIM. The figure shows DNA:DNA matches (computed using BLASTN and displayed using ACT http://www.sanger.ac.uk/Software/ACT) between the two published *Y. pestis* genomes. The grey bars between the genomes represent individual BLASTN matches. Some of the shorter and weaker BLASTN matches have been removed to show the overall structure of the comparison. The GC deviation plot is shown for both genomes, for an explanation see the legend for figure 1. The approximate location of the multiple invertion region identified by Deng *et al.* (Deng et al., 2002) is shown as a solid black line under the *Y. pestis* KIM genome. The positions of all the rRNA operons are marked (open black triangles) for both genomes.

with the 7 possessed by KIM and most other members of the *Enterobacteriaceae* (see above; Rakin and Heesemann, 1995). This model also shows how by bringing the rRNA operons together the loss of one of these operons could have occurred without the deletion of large sections of the genome, and therefore other possible essential genes held within.

Whole genome comparisons of *Y. pestis* with more distantly related members of the *Enterobacteriaceae* shows that the majority of the genomic rearrangements observed have occurred following symmetrical recombination around the origin. Such reciprocal inversions around the origin and terminus of replication are the most common form of chromosomal rearrangement seen when comparing related bacteria (Eisen et al., 2000; Tillier and Collins, 2000). This may reflect the fact that because DNA close to the replication forks are the only regions unpackaged during replication they are the most likely sites for recombination to occur (Tillier and Collins, 2000; Deng et al., 2002).

Alternatively, it is possible that other recombinational events occur, but are selectively disadvantageous (Mackiewicz et al., 2001). Either way, the effect of this type of reciprocal recombination is to maintain the gene order, orientation and distance with respect to the origin, as well as preserving the size of the replichores and the G/C bias of the leading strand.

There are several hypotheses that attempt to explain why this conservation about the origin could be important. Once such theory relies on the fact that because chromosomal replication occurs substantially in advance of cell division in rapidly-growing bacteria, genes closest to the origin are present in more copies per cell than those closer to the terminus. It is likely that this gene-dosage effect will have an effect on specific expression levels in the cell, and the organism may well have adapted to exploit this. It has also been proposed that since genes are predominantly transcribed in the same direction as the movement of the replication fork, preserving gene order and orientation may act to avoid collisions between the replication and transcription machinery.

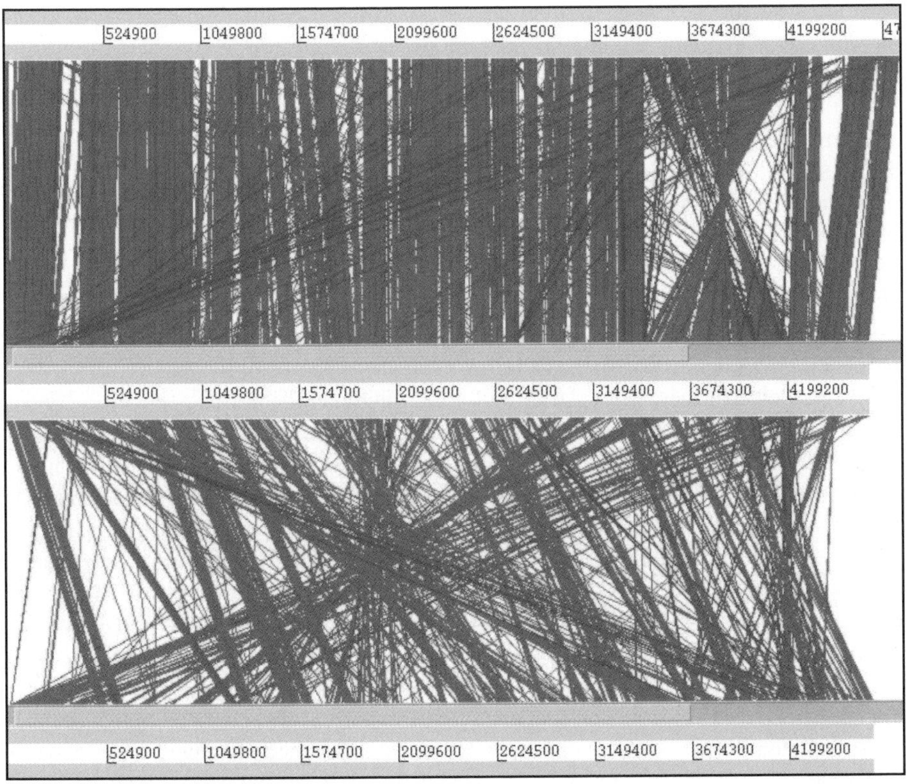

Figure 3. Global comparison between *S.* Typhi CT18, *E. coli* K12 and *Y. pestis* CO92. The figure shows DNA: DNA matches (computed using BLASTN and displayed using ACT http://www.sanger.ac.uk/Software/ACT). The genomes are, from the top down; *S.* Typhi CT18, *E. coli* K12 and *Y. pestis* CO92. The grey bars between the genomes represent individual BLASTN matches. Some of the shorter and weaker BLASTN matches have been removed to show the overall structure of the comparison.

A direct comparison of the *Y. pestis*, *E. coli* K12 and *S.* Typhi genomes (Figure 3) reveals that much of the co-linearity preserved between other enterics, for example *Salmonella* and *E. coli,* has been lost. However, this may not be surprising considering that *E. coli* and *Y. pestis* are estimated to be separated by as much as 500 million years of evolution, compared to only 110 million years between *E. coli* from *Salmonella* (Doolittle et al., 1996; Achtman et al., 1999; Deng et al., 2002). However, if instead of looking at the overall co-linearity of the two genomes, the distances of orthologous genes from the origins in *E. coli* and *Y. pestis* are compared, then there is a surprisingly high level of conservation, suggesting that most of the observed rearrangement has maintained these distances and were therefore reciprocal around the origin and terminus. Based on these data it was estimated that almost 50% of the core genes of KIM had been subject to inter-replichore inversions during the evolution of the species (Deng et al., 2002).

Observations relating to the ability of the *Y. pestis* genome to rearrange and the variations in the number of rRNA operons have been noted previously. It has been shown by pulse-field gel electrophoresis and ribotyping that although the majority of Orientalis isolates from around the world relate to ribotype B there are generally a detectable number of bacteria displaying variant ribotypes present within that population (Lucier and Brubaker, 1992; Guiyoule et al., 1994; Rakin et al., 1995). In foci, such as Madagascar, it appears that once the Orientalis clone became established a significant level of chromosomal rearrangements occurred leading to the emergence of new strains which have persisted within that population (Guiyoule et al., 1994). There is some evidence to suggest that this type of rearrangement has led to the generation of new *Y. pestis* variants which are better able to survive and are potentially more pathogenic (Guiyoule et al., 1997). However, the precise effects of these rearrangements on the biology and pathogenicity of *Y. pestis* remains equivocal.

GENOME CONTENTS

The genome contents of KIM and CO92 are also highly similar, with up to 95% of the sequence being common to both. Detailed analysis of the coding capacity of KIM and CO92 revealed that there were 4,198 and 4,012 predicted genes, respectively (differing by 186). Of the total genes predicted for KIM, 3672 were identical to those assigned for CO92. Of the remaining 526 KIM-specific genes most were of unknown function and 318 were less than 100 aa. Considering how closely conserved the genomes of KIM and CO92 are the majority of the uniquely predicted genes for both KIM and CO92 are more likely to reflect the inherent uncertainties in gene prediction methods rather than truly representing unique gene sets.

However, there are significant regions which differ between CO92 and KIM. KIM carries an additional rRNA operon (as mentioned above) and the CO92 genome is ~50 kb larger than that of KIM. The additional CO92 DNA is essentially made up of an expanded population of insertion sequence (IS) elements (140 vs. 122 complete or partial IS elements for CO92 and KIM, respectively) and an integrated prophage, which is absent from KIM (Figure 4). This prophage is ~9 kb in length and displays small regions of similarity with other enterobacterial phage. It is predicted to carry 12 genes which are all thought to be involved with phage replication and growth (Figure 4). No genes involved in lysogenic conversion could be identified.

The other large-scale variation between CO92 and KIM concerns one of the two flagellar operons encoded by both biovars. Analysis of the KIM genome revealed that a

Yersinia pestis CO92

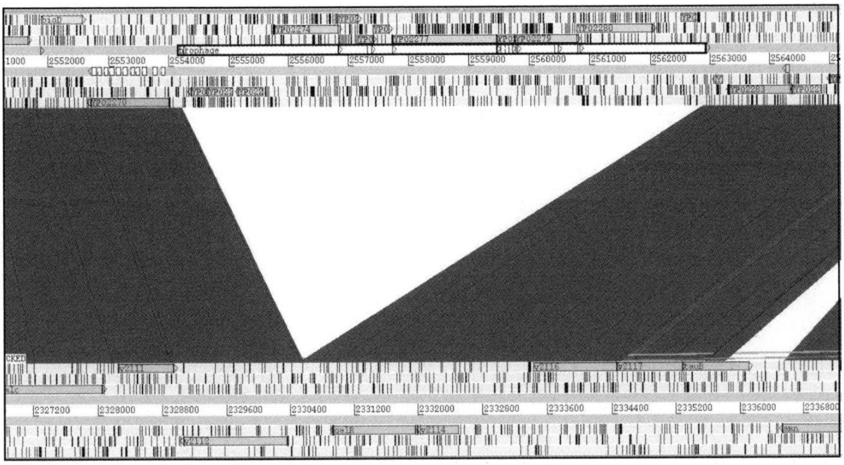

Yersinia pestis KIM

Figure 4. The *Y. pestis* CO92 unique prophage. The figure shows DNA:DNA matches (computed using BLASTN and displayed using ACT) between *Y. pestis* CO92 (top) and *Y. pestis* KIM (bottom). The full extent of the prophage is marked as white box on the upper genome. Coding sequences are marked as differently shaded boxes displaying their systematic or functional gene names.

large number of the flagella genes present in CO92 have been deleted from KIM. These include the multiple flagellin genes *flaA1, flaA2* and *flaA3,* as well as many others involved in flagella biogenesis and chemotaxis (Figure 5).

PATHOGENICITY ISLANDS/LATERALLY ACQUIRED DNA

It is a phenomenon of many enteric pathogens that the genome sequence is composed of a conserved backbone or core sequence, which encodes essentially 'house keeping' functions. Interdispersed within this core sequence are regions ranging from ~5-150 kb which share many common features including; being inserted adjacent to stable RNA genes; displaying an atypical G+C content and encoding a preponderance of virulence-related, transposase or integrase-like proteins. These regions can be seen to be of limited phylogenetic distribution and in some cases unstable and/or self-mobilisable (Hacker et al., 1997). These regions are generally referred to as islands, pathogenicity islands (PAI) or strain specific loops (we shall refer to them jointly as PAI).

The best characterised *Yersinia* PAI is the *Y. pestis* high pathogenicity island (HPI; (Buchrieser et al., 1998); discussed in more detail in chapter 14 of this book) which directs the production and uptake of the siderophore yersiniabactin. HPI-like elements are widely distributed in enterobacteria including *E. coli, Klebsiella, Enterobacter* and *Citrobacter* spp (Bach et al., 2000; Schubert et al., 2000; Hayashi et al., 2001) and like many prophage, HPI can be found adjacent to several tRNA genes. tRNA genes are common sites for bacteriophage integration into genomes and the movement of PAIs, such as HPI, is thought to have much in common with the method of bacteriophage integration (Rakin et al., 2001).

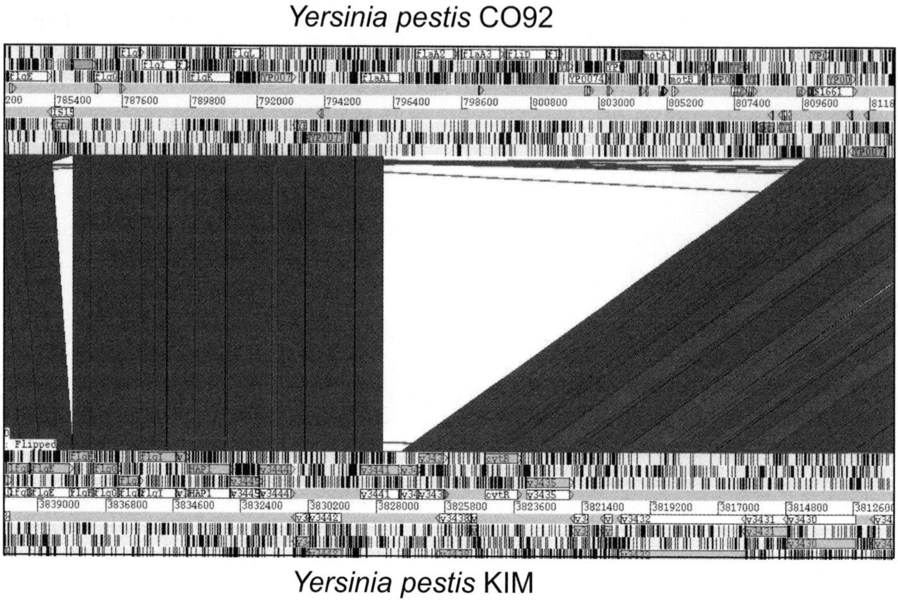

Figure 5. Mutations and deletions present in the flagella operons of *Y. pestis* CO92 and *Y. pestis* KIM. The figure shows DNA:DNA matches (computed using BLASTN and displayed using ACT) between *Y. pestis* CO92 (top) and *Y. pestis* KIM (bottom). The region encoding the flagellins and some of the chemotaxis genes can be seen to be absent from *Y. pestis* KIM. Coding sequences are marked as differently shaded boxes displaying their systematic or functional gene names. Gene *flgF* (top left) has been interrupted by an IS1541 element in CO92, but remains intact in KIM (bottom left).

A large number of additional regions displaying many of the characteristics of such PAIs were also identified within the *Y. pestis* chromosome. These included some encoding genes involved in uptake of, for example, iron or sugars, as well as those encoding toxins and potential virulence determinants. One such island, denoted HPI-2, showed a significant level of sequence similarity, and some synteny, with HPI. The protein products predicted to be encoded by HPI-2 include two possible non-ribosomal peptide synthases, similar to those encoded by HPI. In addition HPI-2 was also shown to encode siderophore regulatory and uptake proteins, but no apparent siderophore receptor (Parkhill et al., 2001b).

Other islands within the *Y. pestis* genome include those that appear to have been derived from pathogens of insects. One such locus is predicted to encode homologues of the high molecular weight insecticidal toxin complexes (Tcs) from *Photorhabdus luminescens*, *Serratia entomophila* and *Xenorhabdus nematophilus* (Waterfield et al., 2001). These toxins have been shown to be complexes of the products of three different gene families: *tcaB*/*tcdA*, *tcaC*/*tcaB*, and *tccC*. Analysis of the histopathological effects of purified Tca toxin on an insect midgut has shown that it causes the lining epithelium to bleb into the midgut lumen and eventually to disintegrate. A similar observation was also made if the toxin is introduced into the insect via the hemocoel (Blackburn et al., 1998). In *Y. pestis,* three adjacent genes encoding homologues of *P. luminescens* TcaA, TcaB and TcaC were identified, separated from nearby homologues of TccC by phage-like genes. However, although the *Y. pestis tcaA* gene was intact, *tcaB* contained a frameshift mutation and *tcaC* possessed an internal deletion. It is not clear whether these genes are remnants from a previous pathogenic association with an insect host or are required currently. Indeed it has been shown that orthologues of the *Y. pestis tca* insecticidal toxin genes were also present within a strain of *Y. pseudotuberculosis* (Parkhill et al., 2001b). If these genes were in fact important prior to the split with *Y. pseudotuberculosis* it may be that the disruptions of these genes might be necessary for the new lifestyle of *Y. pestis*, which persists in the flea gut for relatively long periods of time.

Other insect pathogen-related islands include a low G+C region encoding a protein similar to viral enhancins. This island is inserted alongside a tRNA gene and carries several transposase fragments. Like the Tca toxin, viral enhancins attack the insect midgut. The peritrophic membrane, a non-cellular matrix composed of chitin, proteins and glycoproteins, lines the insect midgut and is thought to act as a barrier for pathogenic microorganisms (Wang and Granados, 1997). The proteolytic activity of enhancins has been shown to degrade the peritrophic membrane and allow the escape of viral pathogens from the gut into the deeper tissues. Like the genes encoding the Tca toxin it is not clear whether these genes are vestiges of a former lifestyle or could be involved in the colonisation of the flea.

As described in chapter 16 of this book, the *Y. pestis* type III secretion system located on the *Yersinia* virulence plasmid (pCD1) has been the subject of extensive research. This secretion system is highly important for virulence through transport of a range of effector proteins (Yops; Yersinia outer proteins; Brubaker, 1991; Perry and Fetherston, 1997), which act to subvert the functioning of host cells. An additional, chromosomally-located type-III secretion system, distinct from both those located on pCD1 and from the chromosomally encoded type III system of *Y. enterocolitica*, (Haller et al., 2000) was found on the *Y. pestis* chromosome. This novel system displayed a remarkable degree of sequence similarity and synteny with the type III secretion system located on the *Salmonella* pathogenicity Island 2 (SPI-2; Shea et al., 1996; Figure 6). *Salmonella enterica* serovars possess two type III secretion systems located on pathogenicity islands called SPI-1 and SPI-2. The gene products encoded by SPI-1 and SPI-2 have been shown to be important for different stages of the infection process (Mills et al., 1995; Galan, 1996; Ochman et al., 1996; Shea et al., 1996). Both of these islands carry type III secretion systems as well as associated secreted protein effectors. SPI-1 confers the ability to invade epithelial cells on all salmonellae. SPI-2 has been shown to be important for various aspects of the systemic infection, allowing *Salmonella* to spread from the intestinal tissue into the blood and eventually to infect, and survive within, macrophages of the liver and spleen (reviewed in Kingsley and Baumler, 2002). It is possible that a similar approach is employed by *Y. pestis* for different stages of the infection process, or that these systems may allow interactions with different hosts.

Other potential islands included loci encoding fimbriae and adhesins. Five out of the eight operons encoding these proteins were found to be flanked by genes encoding transposases or integrases and had several other features consistent with them being of foreign origin. Such extremely high redundancy of fimbria-related genes is also found in other bacterial pathogens, such as *E. coli* and *S.* Typhi (Townsend et al., 2001). A large arsenal of independent gene clusters specialized in production of different fimbriae and adhesins is proposed to be beneficial in evading host immune response, or interacting with a variety of host cells and receptors.

Unlike the genomes of many other enterics, bacteriophage do not make a substantial contribution to the total laterally acquired DNA held within this genome (Boyd and

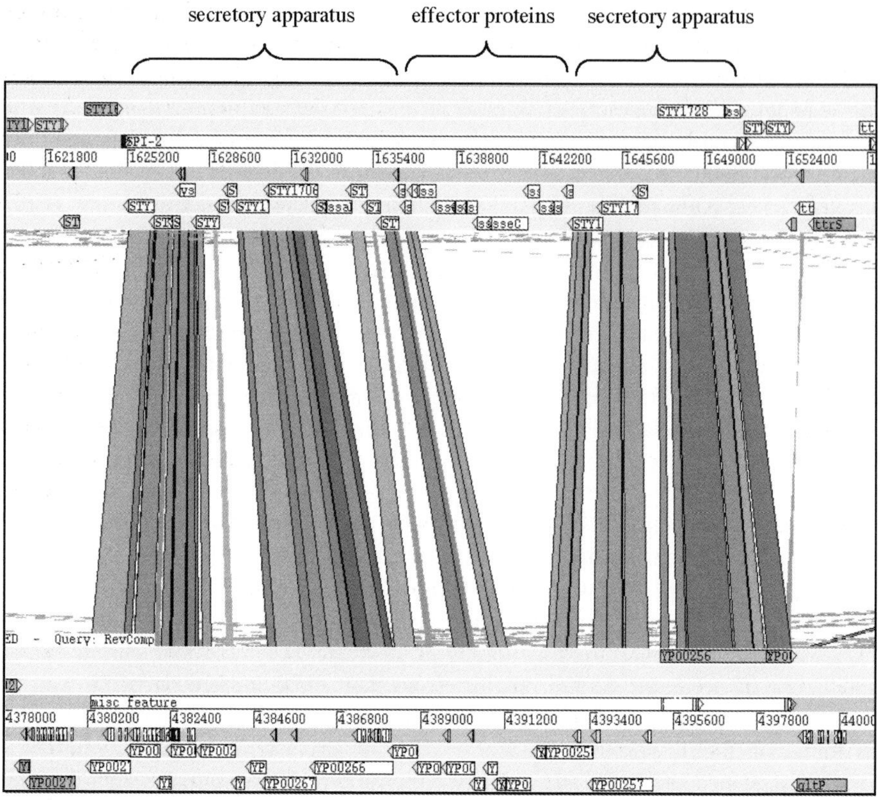

Figure 6. Global comparison between the *S.* Typhi CT18 and *Y. pestis* CO92 type III secretions systems. The figure shows DNA:DNA matches (computed using BLASTN and displayed using ACT http://www.sanger.ac.uk/Software/ACT). The grey bars between the genomes represent individual BLASTN matches. Some of the shorter and weaker BLASTN matches have been removed to show the overall structure of the comparison. The regions in SPI-2 encoding the secretory apparatus and the type III secreted effector proteins have been marked. The 'misc_feature' shown for *Y. pestis* CO92 delimits the low G+C region carrying the proposed type III secretion system.

Brussow, 2002). Deng et al. (2002) identified six regions in KIM that resembled integrated prophage or prophage-remnants. All of these regions are also present in CO92. As mentioned above CO92 also carries an additional phage not found in KIM. Of the remaining prophage the largest, and most complete, is ~40 kb and shows a significant level of similarity to bacteriophage Lambda. However, the sequence of this phage is interrupted by multiple IS elements and so it is unlikely to be active. Similarly to the 9 kb prophage, unique to CO92, this phage does not appear to carry genes which would impinge on the virulence of the host, such as toxins and type III secreted effectors, that are often found in other enteric bacteriophage (Boyd and Brussow, 2002).

GENE DECAY, INSERTION AND DISRUPTION

Perhaps one of the most surprising observations to be made from looking at *Y. pestis* was not the gross differences between the CO92 and KIM genomes, and/or other enterics, but the large number of pseudogenes present in both genomes. The significance of pseudogenes has been the cause of much debate, especially against the background of the common perception that bacterial genomes are highly efficient systems that do not carry DNA that no-longer confers a selective advantage. However, some genomes such as *Mycobacterium leprae* (Cole et al., 2001) and certain members of the *Rickettsia* (Andersson et al., 1998) have been found to carry large numbers of defunct genes or gene-remnants. Similarly the *Y. pestis* genome sequence was also found to carry a large number of pseudogenes; >140 representing ~4 % of the total gene complement. Most of these pseudogenes had been disrupted by the insertion of IS elements (51) or carried frame-shift mutations (58). Of the remaining 40 pseudogenes, 32 were the result of deletion events with the remainder having in-frame stop codons. The number of pseudogenes identified so far is likely to be an underestimate. Identification of disrupted genes in the absence of strong database matches, or, a highly related genome sequence with which to compare, is a problematical task. This necessitates that only those genes with obvious in-frame stop codons, frameshifts, deletions or IS element insertions can be described as being inactive with any reasonable level of confidence. Consequently, genes with mutations affecting the promoter regions or mis-sense coding changes are not generally detected. Similarly, novel genes lacking significant similarity to anything currently in the databases or displaying atypical coding characteristics can also be mistakenly defined as two or more genes rather than a single gene with a frameshift mutation. Close comparisons of CO92 and KIM with *Y. pseudotuberculosis* will significantly aid this task.

As mentioned above, IS insertions were responsible for almost half of the total pseudogenes reported in *Y. pestis*. The copy number of individual IS elements in *Y. pestis* is high, contributing 3.7% of the total genes in the genome. This exceeds that described for most other enteric genomes, the most notable exception to this being the shigellae (Jin et al., 2002). Four types of IS elements were identified in both of the sequenced *Y. pestis* biovars. As mentioned previously CO92 was found to have significantly more IS elements than KIM and an estimated 10 fold more than *Y. pseudotuberculosis* (Odaert et al., 1996; McDonough and Hare, 1997).

Interestingly, of the 58 genes with frameshift mutations, 21 occurred within homopolymeric tracts. It is known that some mucosal pathogens use slipped-strand mispairing of repeat sequences during replication to switch surface-expressed antigens on or off *in vitro* and *in vivo* (Henderson et al., 1999). A similar process has been demonstrated in *Y. pestis* (Sebbane et al., 2001); the organism is characteristically urease negative but activity can be restored *in vitro* by the spontaneous deletion of a single base pair in a homopolymeric tract in *ureD*. This type of reversible mutation would reduce the metabolic burden of producing proteins unnecessary to *Y. pestis* in its new flea/mammal life-cycle yet still allow the potential to express these should a subsequent need arise.

Parallels can be drawn for both the level of IS expansion and the loss of gene function observed in *Y. pestis* and other enteric pathogens, such as *S.* Typhi and *Shigella flexneri*, (Parkhill et al., 2001a; Jin et al., 2002; Wei et al., 2003) which like *Y. pestis* have undergone recent dramatic changes in lifestyle. It is apparent from these genomes that this process is often accompanied with the loss of genes required solely for their former lifestyle, as discussed below.

PHENOTYPIC STREAMLINING

All the pseudogenes observed in *Y. pestis* fell into a broad spectrum of functional classes, the largest of which contained genes associated with pathogenicity and/or host interaction.

Well characterised examples of these are the *Y. pestis yadA* and *inv* genes. Inactivated in *Y. pestis*, the products of these genes encode an adhesin and an invasin which are important for adherence to surfaces of the gut, and invasion of the cells lining it, in other *Yersinia* (Pepe and Miller, 1993; reviewed by El Tahir and Skurnik, 2001). Complementation of the YadA phenotype in *Y. pestis* has been shown to result in a significant decrease in virulence by the subcutaneous infection route (Rosqvist et al., 1988). This phenomenon has also been observed in other enteric pathogens, such as *E. coli* and the shigellae (Maurelli et al., 1998). In addition to *yadA* and *inv* many other pseudogenes were also predicted to encode other adhesin-like molecules which may also have been important for enteropathogenicity.

The second largest grouping of *Y. pestis* pseudogenes (26%) was predicted to encode surface exposed or exported proteins. Five pseudogenes were found to be lipopolysaccharide (LPS) biosynthesis genes. The LPS O-side chain plays a key role in the virulence of a range of pathogens including *Y. enterocolitica*, where it is known to mediate resistance to complement-mediated and phagocyte killing (Darwin and Miller, 1999). It might also play a role in survival in the gut by protecting the bacterium from bile salts (Nesper et al., 2001). As a consequence of these mutations within the O-antigen biosynthesis cluster *Y. pestis* produces a rough LPS, lacking an O-antigen (Skurnik et al., 2000; Prior et al., 2001). It is not clear what advantage, if any, this phenotype may confer to *Y. pestis*.

Motility is required for efficient host cell invasion by *Y. enterocolitica* (Young et al., 2000) and *Y. pseudotuberculosis*; *Y. pestis* strains are uniformly non-motile but CO92 was found to possess a total of more than 80 motility-related genes (2 % of the total gene set), including two separate clusters of flagella genes as well as a chemotaxis operon. Consistent with the failure to observe motility in *Y. pestis* the flagellar and chemotaxis gene clusters contains several mutations. The disrupted genes within the flagella operon of CO92 include *flhD* (part of the master regulator), *flhA* (required for flagella rod formation) and *flgF* (flagella rod) (Figure 5). These multiple mutations together within the flagella regulation and biosynthetic operons make it highly unlikely that this phenotype would or could revert. As mentioned above there are differences between the flagella and chemotaxis genes of KIM and CO92. One of the KIM flagella operons has been subject to a deletion event and the *flgF*, which is disrupted by an IS insertion in CO92, remains intact in KIM (Figure 5).

Mutations in energy metabolism and central and intermediary metabolism are underrepresented in *Y. pestis*. One of the phenotypes that differentiates *Y. pestis* biovars Orientalis and Mediaevalis is the ability to ferment glycerol. KIM is able to ferment glycerol and has a full complement of glycerol utilization genes. CO92 (Mediaevalis), however is unable to ferment glycerol. Whilst the majority of the *glp* regulon genes remain intact in CO92, *glpK* (glycerol kinase) and *glpX* (part of the *glp* regulon, but of unknown function) have been subject to a deletion event resulting in the loss of their 3' and 5' termini, respectively. In addition, *glpD* (aerobic glycerol-3-phosphate dehydrogenase) has also been subject to a small deletion leading to a frameshift in the sequence. Other pseudogenes in this class include mutations in genes involved in amino acid biosynthetic genes such as *pheA*, which is consistent with previous observations that *Y. pestis* CO92 strains require one or more of the amino acids for growth *in vitro* (Perry and Fetherston, 1997).

Many other mutations were found in uptake and transport systems. These mutations may indicate that compared to life as a gastrointestinal pathogen, there were fewer or possibly markedly different types of nutrients available to *Y. pestis* within the flea or its new niche within the mammalian host. Considering how important the ability to sequester iron is for many human and animal pathogens, it was surprising to find that three genes possibly involved in iron uptake were inactivated. One such example was *iucA*, a gene essential for the production of aerobactin. It is likely that to compensate for this loss other potentially novel iron uptake genes are present within the genome.

Other differences between KIM and CO92 relate to the in-frame opal stop codon present in the formate dehydrogenase gene *fdoG*. In *E. coli* this stop codon is translated into selenocysteine by the gene products of *selA*, *selB* and *selD* along with the *selC*-tRNA. There are other *E. coli* genes, such as *fdhF,* for which this is also true. All of the genes required for selenocysteine incorporation are present in KIM, but in CO92 the *selB* gene is disrupted by a frameshift and so, as has been observed in other systems, it is assumed that FdoG is non-functional in CO92 (Tormay et al., 1996; Rother et al., 2003). Moreover, the selenocysteine codons of the *fdhF* genes in KIM and CO92 are replaced by cysteine codons, which should preserve function and may be a consequence of environmental adaptation (Deng et al., 2002).

Other biovar differences include the *rscBAC* genes, required for systemic invasion in *Y. enterocolitica* (Nelson et al., 2001), they are intact in CO92; however, *rscA* is disrupted by a frameshift mutation in KIM. What is clear is that like many other enterics the evolution of *Y. pestis* has involved significant gene loss as well as gain and that these pseudogenes appear to be relics of a former lifestyle which can be readily detected from whole genome analysis.

DISCUSSION

Sequencing of the *Y. pestis* genome has revealed a pathogen that has undergone recent rapid genetic flux. The fine detail of the genome has shown that, like many other enterics, the evolution of *Y. pestis* has not simply been a process of acquisition and deletion of large islands of DNA, but that, concomitant with these gross changes in the chromosome, there has been a more subtle process of gene decay involving mainly single genes.

These wide ranging genetic changes are thought to have been a consequence of a recent speciation event, accompanying a change in pathogenic niche. Many of the genes which are thought to be of foreign origin may have been acquired incrementally as an antecedent to speciation. The genome contents suggest that the ancestor of *Y. pestis* had adapted to colonise the gut of both mammalian and insect hosts. At some stage these adaptations converged and *Y. pestis* was able to spread directly from the flea gut to the mammalian host via subcutaneous injection. During, or immediately following this predicted evolutionary bottleneck it is likely that genes essential for the colonization of the gut, or those deleterious for life in this new niche, became redundant and were lost through the accumulation of mutations and disruption by the expansion of IS elements. These vestiges of the former lifestyle can be clearly identified in the genome. It is also evident from comparisons of the two sequenced *Y. pestis* biovars, and from previous studies, that the *Y. pestis* genome is still undergoing rearrangement and change. Comparison with the *Y. pseudotuberculosis* genome sequence should allow for a more complete analysis of the changes that occurred preceding the emergence of this new pathogen.

REFERENCES

Achtman, M., Zurth, K., Morelli, G., Torrea, G., Guiyoule, A. and Carniel, E. 1999. *Yersinia pestis*, the cause of plague, is a recently emerged clone of *Yersinia pseudotuberculosis*. Proc. Natl. Acad. Sci. USA. 96: 14043-8.

Andersson, S.G., Zomorodipour, A., Andersson, J.O., Sicheritz-Ponten, T., Alsmark, U.C., Podowski, R.M., Naslund, A.K., Eriksson, A.S., Winkler, H.H. and Kurland, C.G. 1998. The genome sequence of *Rickettsia prowazekii* and the origin of mitochondria. Nature. 396: 133-40.

Bach, S., de Almeida, A. and Carniel, E. 2000. The *Yersinia* high-pathogenicity island is present in different members of the family *Enterobacteriaceae*. FEMS Microbiol. Lett. 183: 289-94.

Blackburn, M., Golubeva, E., Bowen, D. and Ffrench-Constant, R.H. 1998. A novel insecticidal toxin from *Photorhabdus luminescens*, toxin complex a (Tca), and its histopathological effects on the midgut of *Manduca sexta*. Appl. Environ. Microbiol. 64: 3036-41.

Blattner, F.R., Plunkett, G., Bloch, C.A., Perna, N.T., Burland, V., Riley, M., Collado-Vides, J., Glasner, J.D., Rode, C.K., Mayhew, G.F., Gregor, J., Davis, N.W., Kirkpatrick, H.A., Goeden, M.A., Rose, D.J., Mau, B. and Shao, Y. 1997. The complete genome sequence of *Escherichia coli* K-12. Science. 277: 1453-74.

Boyd, E.F. and Brussow, H. 2002. Common themes among bacteriophage-encoded virulence factors and diversity among the bacteriophages involved. Trends Microbiol. 10: 521-9.

Brubaker, R.R. 1991. Factors promoting acute and chronic diseases caused by yersiniae. Clin. Microbiol. Rev. 4: 309-24.

Buchrieser, C., Prentice, M. and Carniel, E. 1998. The 102-kilobase unstable region of *Yersinia pestis* comprises a high- pathogenicity island linked to a pigmentation segment which undergoes internal rearrangement. J. Bacteriol. 180: 2321-9.

Cole, S.T., Eiglmeier, K., Parkhill, J., James, K.D., Thomson, N.R., Wheeler, P.R., Honore, N., Garnier, T., Churcher, C., Harris, D., Mungall, K., Basham, D., Brown, D., Chillingworth, T., Connor, R., Davies, R.M., Devlin, K., Duthoy, S., Feltwell, T., Fraser, A., Hamlin, N., Holroyd, S., Hornsby, T., Jagels, K., Lacroix, C., Maclean, J., Moule, S., Murphy, L., Oliver, K., Quail, M.A., Rajandream, M.A., Rutherford, K.M., Rutter, S., Seeger, K., Simon, S., Simmonds, M., Skelton, J., Squares, R., Squares, S., Stevens, K., Taylor, K., Whitehead, S., Woodward, J.R. and Barrell, B.G. 2001. Massive gene decay in the leprosy bacillus. Nature. 409: 1007-11.

Darwin, A.J. and Miller, V.L. 1999. Identification of *Yersinia enterocolitica* genes affecting survival in an animal host using signature-tagged transposon mutagenesis. Mol. Microbiol. 32: 51-62.

Deng, W., Burland, V., Plunkett, G., 3rd, Boutin, A., Mayhew, G.F., Liss, P., Perna, N.T., Rose, D.J., Mau, B., Zhou, S., Schwartz, D.C., Fetherston, J.D., Lindler, L.E., Brubaker, R.R., Plano, G.V., Straley, S.C., McDonough, K.A., Nilles, M.L., Matson, J.S., Blattner, F.R. and Perry, R.D. 2002. Genome sequence of *Yersinia pestis* KIM. J. Bacteriol. 184: 4601-11.

Devignat, R. 1951. Variétés de l'espèce *Pasteurella pestis*. Nouvelle hypothèse. Bull. Org. Mond. Sante. 4: 247-263.

Doolittle, R.F., Feng, D.F., Tsang, S., Cho, G. and Little, E. 1996. Determining divergence times of the major kingdoms of living organisms with a protein clock. Science. 271: 470-7.

Eisen, J.A., Heidelberg, J.F., White, O. and Salzberg, S.L. 2000. Evidence for symmetric chromosomal inversions around the replication origin in bacteria. Genome Biol. 1.

El Tahir, Y. and Skurnik, M. 2001. YadA, the multifaceted *Yersinia* adhesin. Int. J. Med. Microbiol. 291: 209-18.

Galan, J.E. 1996. Molecular genetic bases of *Salmonella* entry into host cells. Mol. Microbiol. 20: 263-71.

Guiyoule, A., Grimont, F., Iteman, I., Grimont, P.A., Lefevre, M. and Carniel, E. 1994. Plague pandemics investigated by ribotyping of *Yersinia pestis* strains. J. Clin. Microbiol. 32: 634-41.

Guiyoule, A., Rasoamanana, B., Buchrieser, C., Michel, P., Chanteau, S. and Carniel, E. 1997. Recent emergence of new variants of *Yersinia pestis* in Madagascar. J. Clin. Microbiol. 35: 2826-33.

Hacker, J., Blum-Oehler, G., Muhldorfer, I. and Tschape, H. 1997. Pathogenicity islands of virulent bacteria: structure, function and impact on microbial evolution. Mol. Microbiol. 23: 1089-97.

Haller, J.C., Carlson, S., Pederson, K.J. and Pierson, D.E. 2000. A chromosomally encoded type III secretion pathway in *Yersinia enterocolitica* is important in virulence. Mol. Microbiol. 36: 1436-46.

Hayashi, T., Makino, K., Ohnishi, M., Kurokawa, K., Ishii, K., Yokoyama, K., Han, C.G., Ohtsubo, E., Nakayama, K., Murata, T., Tanaka, M., Tobe, T., Iida, T., Takami, H., Honda, T., Sasakawa, C., Ogasawara, N., Yasunaga, T., Kuhara, S., Shiba, T., Hattori, M. and Shinagawa, H. 2001. Complete genome sequence of enterohemorrhagic *Escherichia coli* O157:H7 and genomic comparison with a laboratory strain K-12. DNA Res. 8: 11-22.

Henderson, I.R., Owen, P. and Nataro, J.P. 1999. Molecular switches--the ON and OFF of bacterial phase variation. Mol. Microbiol. 33: 919-932.

Jin, Q., Yuan, Z., Xu, J., Wang, Y., Shen, Y., Lu, W., Wang, J., Liu, H., Yang, J., Yang, F., Zhang, X., Zhang, J., Yang, G., Wu, H., Qu, D., Dong, J., Sun, L., Xue, Y., Zhao, A., Gao, Y., Zhu, J., Kan, B., Ding, K., Chen, S., Cheng, H., Yao, Z., He, B., Chen, R., Ma, D., Qiang, B., Wen, Y., Hou, Y. and Yu, J. 2002. Genome sequence of *Shigella flexneri* 2a: insights into pathogenicity through comparison with genomes of *Escherichia coli* K12 and O157. Nucleic Acids Res. 30: 4432-41.

Kingsley, R.A. and Baumler, A.J. 2002. Pathogenicity islands and host adaptation of *Salmonella* serovars. Curr. Top. Microbiol. Immunol. 264: 67-87.

Lobry, J.R. 1996. Asymmetric substitution patterns in the two DNA strands of bacteria. Mol. Biol. Evol. 13: 660-665.

Lobry, J.R. and Louarn, J.M. 2003. Polarisation of prokaryotic chromosomes. Curr. Opin. Microbiol. 6: 101-108.

Lobry, J.R. and Sueoka, N. 2002. Asymmetric directional mutation pressures in bacteria. Genome Biol. 3: RESEARCH0058.

Lucier, T.S. and Brubaker, R.R. 1992. Determination of genome size, macrorestriction pattern polymorphism, and nonpigmentation-specific deletion in *Yersinia pestis* by pulsed-field gel electrophoresis. J. Bacteriol. 174: 2078-2086.

Mackiewicz, P., Mackiewicz, D., Kowalczuk, M. and Cebrat, S. 2001. Flip-flop around the origin and terminus of replication in prokaryotic genomes. Genome Biol. 2: INTERACTIONS1004.

Maurelli, A.T., Fernandez, R.E., Bloch, C.A., Rode, C.K. and Fasano, A. 1998. "Black holes" and bacterial pathogenicity: a large genomic deletion that enhances the virulence of *Shigella* spp. and enteroinvasive *Escherichia coli*. Proc. Natl. Acad. Sci. USA. 95: 3943-3948.

McDonough, K.A. and Hare, J.M. 1997. Homology with a repeated *Yersinia pestis* DNA sequence IS100 correlates with pesticin sensitivity in *Yersinia pseudotuberculosis*. J. Bacteriol. 179: 2081-2085.

Mills, D.M., Bajaj, V. and Lee, C.A. 1995. A 40 kb chromosomal fragment encoding *Salmonella typhimurium* invasion genes is absent from the corresponding region of the *Escherichia coli* K- 12 chromosome. Mol. Microbiol. 15: 749-759.

Nelson, K.M., Young, G.M. and Miller, V.L. 2001. Identification of a locus involved in systemic dissemination of *Yersinia enterocolitica*. Infect. Immun. 69: 6201-6208.

Nesper, J., Lauriano, C.M., Klose, K.E., Kapfhammer, D., Kraiss, A. and Reidl, J. 2001. Characterization of *Vibrio cholerae* O1 El tor *galU* and *galE* mutants: influence on lipopolysaccharide structure, colonization, and biofilm formation. Infect. Immun. 69: 435-445.

Ochman, H., Soncini, F.C., Solomon, F. and Groisman, E.A. 1996. Identification of a pathogenicity island required for *Salmonella* survival in host cells. Proc. Natl. Acad. Sci. USA. 93: 7800-7804.

Odaert, M., Berche, P. and Simonet, M. 1996. Molecular typing of *Yersinia pseudotuberculosis* by using an IS*200*-like element. J. Clin. Microbiol. 34: 2231-2235.

Parkhill, J., Dougan, G., James, K.D., Thomson, N.R., Pickard, D., Wain, J., Churcher, C., Mungall, K.L., Bentley, S.D., Holden, M.T., Sebaihia, M., Baker, S., Basham, D., Brooks, K., Chillingworth, T., Connerton, P., Cronin, A., Davis, P., Davies, R.M., Dowd, L., White, N., Farrar, J., Feltwell, T., Hamlin, N., Haque, A., Hien, T.T., Holroyd, S., Jagels, K., Krogh, A., Larsen, T.S., Leather, S., Moule, S., O'Gaora, P., Parry, C., Quail, M., Rutherford, K., Simmonds, M., Skelton, J., Stevens, K., Whitehead, S. and Barrell, B.G. 2001a. Complete genome sequence of a multiple drug resistant *Salmonella enterica* serovar Typhi CT18. Nature. 413: 848-852.

Parkhill, J., Wren, B.W., Thomson, N.R., Titball, R.W., Holden, M.T., Prentice, M.B., Sebaihia, M., James, K.D., Churcher, C., Mungall, K.L., Baker, S., Basham, D., Bentley, S.D., Brooks, K., Cerdeno-Tarraga, A.M., Chillingworth, T., Cronin, A., Davies, R.M., Davis, P., Dougan, G., Feltwell, T., Hamlin, N., Holroyd, S., Jagels, K., Karlyshev, A.V., Leather, S., Moule, S., Oyston, P.C., Quail, M., Rutherford, K., Simmonds, M., Skelton, J., Stevens, K., Whitehead, S. and Barrell, B.G. 2001b. Genome sequence of *Yersinia pestis*, the causative agent of plague. Nature. 413: 523-527.

Pepe, J.C. and Miller, V.L. 1993. *Yersinia enterocolitica* invasin: a primary role in the initiation of infection. Proc. Natl. Acad. Sci. USA. 90: 6473-6477.

Perry, R.D. and Fetherston, J.D. 1997. *Yersinia pestis*--etiologic agent of plague. Clin. Microbiol. Rev. 10: 35-66.

Prior, J.L., Parkhill, J., Hitchen, P.G., Mungall, K.L., Stevens, K., Morris, H.R., Reason, A.J., Oyston, P.C., Dell, A., Wren, B.W. and Titball, R.W. 2001. The failure of different strains of *Yersinia pestis* to produce lipopolysaccharide O-antigen under different growth conditions is due to mutations in the O-antigen gene cluster. FEMS Microbiol. Lett. 197: 229-233.

Rakin, A. and Heesemann, J. 1995. The established *Yersinia pestis* biovars are characterized by typical patterns of I-CeuI restriction fragment length polymorphism. Mol. Gen. Mikrobiol. Virusol. 26-29.

Rakin, A., Noelting, C., Schropp, P. and Heesemann, J. 2001. Integrative module of the high-pathogenicity island of *Yersinia*. Mol. Microbiol. 39: 407-415.

Rakin, A., Urbitsch, P. and Heesemann, J. 1995. Evidence for two evolutionary lineages of highly pathogenic *Yersinia* species. J. Bacteriol. 177: 2292-2298.

Rosqvist, R., Skurnik, M. and Wolf-Watz, H. 1988. Increased virulence of *Yersinia pseudotuberculosis* by two independent mutations. Nature. 334: 522-524.

Rother, M., Mathes, I., Lottspeich, F. and Bock, A. 2003. Inactivation of the *selB* gene in *Methanococcus maripaludis*: effect on synthesis of selenoproteins and their sulfur-containing homologs. J. Bacteriol. 185: 107-114.

Schubert, S., Cuenca, S., Fischer, D. and Heesemann, J. 2000. High-pathogenicity island of *Yersinia pestis* in enterobacteriaceae isolated from blood cultures and urine samples: prevalence and functional expression. J. Infect. Dis. 182: 1268-1271.

Sebbane, F., Devalckenaere, A., Foulon, J., Carniel, E. and Simonet, M. 2001. Silencing and reactivation of urease in *Yersinia pestis* is determined by one G residue at a specific position in the *ureD* gene. Infect. Immun. 69: 170-176.

Shea, J.E., Hensel, M., Gleeson, C. and Holden, D.W. 1996. Identification of a virulence locus encoding a second type III secretion system in *Salmonella typhimurium*. Proc. Natl. Acad. Sci. USA. 93: 2593-2597.

Skurnik, M., Peippo, A. and Ervela, E. 2000. Characterization of the O-antigen gene clusters of *Yersinia pseudotuberculosis* and the cryptic O-antigen gene cluster of *Yersinia pestis* shows that the plague bacillus is most closely related to and has evolved from *Y. pseudotuberculosis* serotype O:1b. Mol. Microbiol. 37: 316-330.

Tillier, E.R. and Collins, R.A. 2000. Genome rearrangement by replication-directed translocation. Nat Genet. 26: 195-197.

Titball, R.W. and Williamson, E.D. 2001. Vaccination against bubonic and pneumonic plague. Vaccine. 19: 4175-4184.

Tormay, P., Sawers, A. and Bock, A. 1996. Role of stoichiometry between mRNA, translation factor SelB and selenocysteyl-tRNA in selenoprotein synthesis. Mol. Microbiol. 21: 1253-1259.

Townsend, S.M., Kramer, N.E., Edwards, R., Baker, S., Hamlin, N., Simmonds, M., Stevens, K., Maloy, S., Parkhill, J., Dougan, G. and Baumler, A.J. 2001. *Salmonella enterica* serovar Typhi possesses a unique repertoire of fimbrial gene sequences. Infect. Immun. 69: 2894-2901.

Wang, P. and Granados, R.R. 1997. An intestinal mucin is the target substrate for a baculovirus enhancin. Proc. Natl. Acad. Sci. USA. 94: 6977-6982.

Waterfield, N.R., Bowen, D.J., Fetherston, J.D., Perry, R.D. and ffrench-Constant, R.H. 2001. The tc genes of *Photorhabdus*: a growing family. Trends Microbiol. 9: 185-191.

Wei, J., Goldberg, M.B., Burland, V., Venkatesan, M.M., Deng, W., Fournier, G., Mayhew, G.F., Plunkett, G., 3rd, Rose, D.J., Darling, A., Mau, B., Perna, N.T., Payne, S.M., Runyen-Janecky, L.J., Zhou, S., Schwartz, D.C. and Blattner, F.R. 2003. Complete genome sequence and comparative genomics of *Shigella flexneri* serotype 2a strain 2457T. Infect. Immun. 71: 2775-2786.

Young, G.M., Badger, J.L. and Miller, V.L. 2000. Motility is required to initiate host cell invasion by *Yersinia enterocolitica*. Infect. Immun. 68: 4323-4326.

Chapter 2

Age, Descent and Genetic Diversity within *Yersinia pestis*

Mark Achtman

ABSTRACT

This article was originally intended as a review of the literature on the population genetics of *Yersinia pestis* but, unfortunately, there is essentially no such literature to review. Therefore, in lieu of a literature review, I shall summarize molecular evidence from which the source and age of these bacteria can be deduced. I shall attempt to also explain some of the population genetic concepts on the basis of which such data can be interpreted.

TAXONOMY

Reliable taxonomic relationships can provide a solid basis for deducing relationships and evolutionary history. Yet, when I began working with *Y. pestis* in 1998, the relationships between *Y. pestis* and *Yersinia pseudotuberculosis* were not obvious from the general literature, nor was it clear which of these closely related organisms was more ancient (Brossollet and Mollaret, 1994). 25 years ago, DNA-DNA hybridizations between multiple isolates of *Y. pestis* or *Y. pseudotuberculosis* and 3 labelled DNA preparation from either *Y. pestis* or *Y. pseudotuberculosis* were unable to distinguish between the bacteria from these two "species" (Bercovier et al., 1980). These analyses included 13 *Y. pestis* isolates of the biovars Orientalis, Medievalis and Antiqua, but did not include pestoides isolates (see below). Although Bercovier *et al.* recommended that taxonomists refer to both *Y. pestis* and *Y. pseudotuberculosis* as subspecies of *Y. pseudotuberculosis* (Bercovier et al., 1980), this recommendation has met with little or no compliance, possibly because scientists working with *Y. pestis* are so focussed on its pathogenicity, which is not properly a taxonomic criterion. (For similar reasons, a unique genus designation, *Shigella*, continues to be used for some *Escherichia coli* that cause epidemic dysentery (Lan and Reeves, 2002)). This situation may be convenient for medically oriented scientists, but can be very confusing for population geneticists.

rRNA SEQUENCES

Although DNA-DNA hybridization remains the gold standard for the differentiation of bacterial species, the technically simpler alternative of examining sequence differences between ribosomal RNAs (rRNA) is being used more and more frequently for taxonomic purposes. It might therefore be expected that rRNA sequences could shed light on the relationships between *Y. pestis* and *Y. pseudotuberculosis*. No sequence differences were detected between the 16S rRNA of a serotype O:3 isolate of *Y. pseudotuberculosis* and a biovar Orientalis isolate of *Y. pestis* (Trebesius et al., 1998) and the only difference detected within a 500 bp fragment of 23S rRNA was a polymorphic (C or U) nucleotide at position 1534 in *Y. pseudotuberculosis versus* a uniform nucleotide (C) in *Y. pestis*. Identical or nearly identical rRNA sequences provide further support for assigning both *Y. pestis* and *Y. pseudotuberculosis* to a single species. To my knowledge, no other bacteria are known that have distinct species designations despite such similar rRNA sequences!

Y. PESTIS IS A CLONE OF *Y. PSEUDOTUBERCULOSIS*

Neutral sequence diversity can be used to define population structure and relatedness between populations. Therefore, sequence variation was determined within *Y. pestis* and *Y. pseudotuberculosis* for 2.5 kb of sequences from five housekeeping genes and one gene involved in lipopolysaccharide synthesis (Achtman et al., 1999). No sequence variation was observed in 36 *Y. pestis* isolates of biovars Medievalis, Orientalis or Antiqua, but two to five alleles were found at each of these loci for 12 isolates of *Y. pseudotuberculosis*. Furthermore, each of the alleles from *Y. pestis* was identical (3 loci) or very similar to an allele in *Y. pseudotuberculosis*. These results show that *Y. pestis* is highly uniform and corresponds to a clone within the more diverse and more ancient species of *Y. pseudotuberculosis*. *Y. pestis* contains two unique virulence plasmids encoding a plasminogen activator (Hinnebusch et al., 1998) and a phospholipase D homologue (Hinnebusch et al., 2002b), respectively, that are lacking in *Y. pseudotuberculosis*. It is conceivable that the acquisition of these two plasmids, possibly by conjugation during mixed infections of fleas (Hinnebusch et al., 2002a), may have sufficed to endow a novel form of transmission upon *Y. pseudotuberculosis*.

Y. pestis expresses a rough lipopolysaccharide (LPS) whereas *Y. pseudotuberculosis* expresses a smooth, long-chained LPS that has been used as the basis for serological typing. Both *Y. pseudotuberculosis* and *Y. pestis* contain a 20 kb O-antigen gene cluster but multiple mutations in this region result in the inability of *Y. pestis* to synthesize smooth LPS (Skurnik et al., 2000). Hybridization analyses showed that most *Y. pseudotuberculosis* O serotypes possess a different complement of genes in this region than does *Y. pestis* but this region is almost identical between *Y. pestis* and an O1:b strain of *Y. pseudotuberculosis*. As a result, Skurnik *et al.* (2000) proposed that *Y. pestis* descended from an O1:b strain and the genome sequence of the serotype I strain, IP32953, of *Y. pseudotuberculosis* has been completed (http://greengenes.llnl.gov/bbrp/html/microbe.html).

A similarity within the O-antigen region between two isolates indicates either that they are related through descent from a common ancestor, which in this case probably expressed the O:1b serotype, or that an ancestor of at least one of the modern isolates acquired these genes by horizontal genetic exchange. If they were descended from a common ancestor, other genes should also be very similar. However, the housekeeping genes tested by Achtman *et al.* (1999) were more similar between *Y. pestis* and serotype IV (3/6 loci identical) than between *Y. pestis* and serotype I (strains IP32949, 2/6 loci; IP32953: 1/6; IP 32790: 1/5). (Note that the strain whose genome has been completed, IP32953, is even less similar to *Y. pestis* than are other serotype I isolates, as also confirmed by a blast search of the genome.) Thus, it is not clear at this moment, whether strain IP32953 used for the genome analysis is as closely related to *Y. pestis* as the serotype O:1b strain Pa3606 that was tested by Skurnik *et al.* Similarly, until the corresponding housekeeping genes of Pa3606 have been sequenced, it will not be clear whether that strain is particularly similar to *Y. pestis* at regions outside of the O-antigen cluster. At the moment, the bold statement that *Y. pestis* has evolved from an O:1b strain still remains to be proven.

MICROEVOLUTION OF rRNA OPERONS

Further information on the relatedness of *Y. pestis* and *Y. pseudotuberculosis* might be achieved by comparing multiple genome sequences from each of these "species" but the annotated genome of *Y. pseudotuberculosis* IP32953 has not yet been published. As a substitute, I performed an analysis of the similarities between 16S and 23S rRNA sequences from the currently unpublished genome of IP32953 (http://greengenes.llnl.gov/bbrp/html/microbe.html) with homologous sequences from the two published *Y. pestis* genomes of CO92 and KIM (Parkhill et al., 2001; Deng et al., 2002). At the time of analysis (early January, 2004), the ends of the 16S sequence were incorrectly annotated in both *Y. pestis* genomes and the correct start and end nucleotides were determined by comparison to the 16S sequence from *E. coli* (GenBank J01695, bases 1518..3059; pdb 1P6G). The results (Figure 1) indicate that the relationship of rRNA between species is somewhat more complicated than indicated above. Firstly, the Medievalis (KIM) and the *Y. pseudotuberculosis* genomes contain seven operons with rRNA sequences while the Orientalis genome (CO92) contains only six, in agreement with conclusions based on digestion with I-*Ceu*I, that biovars Medievalis (strain KIM) or Antiqua (4 isolates) possess seven rRNA operons while Orientalis (4 isolates) possesses only six (Rakin and Heesemann, 1995). Secondly, relatively large differences exist within CO92 or

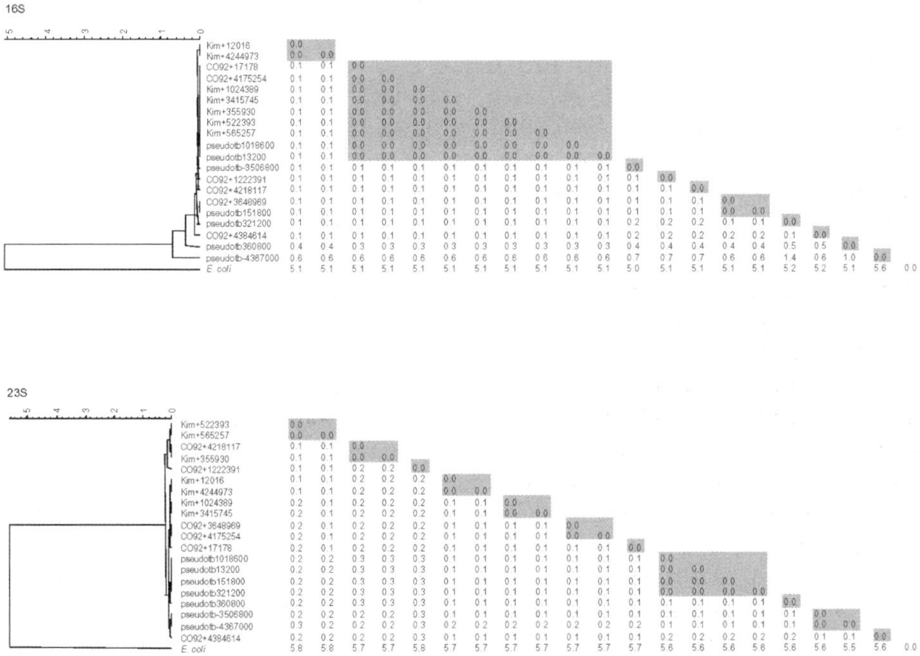

Figure 1. Sequence diversity of rRNA operons in *Y. pestis* and *Y. pseudotuberculosis*. The complete 16S and 23S rRNA genes were extracted from the genomes of *Y. pestis* (strains CO92 and KIM) and *Y. pseudotuberculosis* (strain IP32953, abbreviated as pseudotb) as described in the text and aligned with reference sequences from *E. coli*. UPGMA (unpaired pair group with arithmetic means) dendrograms of relatedness are shown at the left and difference matrices are shown at the right. Blocks of identity are indicated by shadowing. Top: 16S rRNA. Bottom: 23S rRNA. The different operons encoding each of these genes are indicated by numbers after the strain designation that indicate their genomic coordinates.

IP32953 between some 16S genes from different operons at the same time as other 16S sequences from CO92 (2 operons), KIM (5 operons) and IP32953 (2 operons) are identical (Figure 1). Four rRNA operons in IP32953 contained identical 23S sequences but most other 23S sequences were unique or identical to only one other sequence in the same genome. Sequences from IP32953 clustered together, as did sequences from CO92 or KIM. However, one 23S sequence from CO92 was most closely related to sequences from IP32953, and a second was most closely related to a sequence in KIM.

These results provide further support for the assignment of *Y. pestis* and *Y. pseudotuberculosis* to a single species. The observation that rRNA sequences in different operons are identical can most simply be explained by intragenomic gene conversion (Abdulkarim and Hughes, 1996), wherein a sequence at one genomic location is replaced by a homologous sequence from a second location on the same genome. Such gene conversion events might have possibly occurred well before the emergence of *Y. pestis*. Similarly, the observation that particular rRNA sequences differ more between two isolates of *Y. pestis* than do others between *Y. pestis* and *Y. pseudotuberculosis* might indicate that ancestral *Y. pestis* contained diverse sequences at different operons, which then converged to greater uniformity in some isolates by gene conversion. However, the results are also reminiscent of the *Mhc* loci in humans, among which certain alleles are more closely related to homologues alleles in other higher primates than they are to other human *Mhc* alleles (Klein et al., 1990). The *Mhc* data are interpreted to indicate that the smallest human population that ever existed during speciation contained at least 10 individuals and probably more than 10,000, i.e. that sequence diversity was never totally removed during the speciation bottleneck (see chapter 11 in (Klein and Takahata, 2002) for a general discussion of the effects of bottlenecks and population size during recent speciation). If a similar explanation applied to *Y. pestis*, then modern isolates could be the descendents of a diverse population that underwent speciation rather than having all been derived from a single cell.

AGE OF *Y. PESTIS* AS INFERRED FROM GENETIC DIVERSITY

Microbiologists search for diversity in order to be able to distinguish between isolates ("typing"). And diversity within *Y. pestis* has been demonstrated by a variety of properties, ranging from nutritional requirements and fermentative capacities (Devignat, 1951; Anisimov et al., 2004), the number of rRNA operons (Rakin and Heesemann, 1995), variable intergenic regions within rRNA operons as indicated by restriction-fragment length polymorphism (RFLP) (Guiyoule et al., 1994) through to variable insertion sites for the IS*100* insertion element (Achtman et al., 1999; Motin et al., 2002). A comparison of the genomes of KIM and CO92 showed numerous rearrangements mediated by insertion elements and only 3,672 of the 4,198 coding sequences (CDS) are strongly homologous (Deng et al., 2002). However, many differences between the genomes seem to represent gene loss or supernumerary DNA ("islands") (Radnedge et al., 2002) and no algorithms are available for deducing age from these forms of genetic diversity.

Population geneticists often approach genetic diversity in a different manner from microbiologists. Neutral genetic diversity is a measure of elapsed time since a common ancestor, which can be approximated by established algorithms from the frequency of synonymous substitutions (nucleotide changes at the first and third codon position that do not result in amino acid changes). The lack of sequence variation from over 22,000 synonymous sites in six apparently neutral genes was therefore used to estimate that at most 1,000-20,000 years had elapsed since the evolution of *Y. pestis* (Achtman et al., 1999). In other experiments, no synonymous substitutions were detected within the 36 kb Ybt region upon comparing strains 6/69M (Orientalis) and KIM (Medievalis) or within the 6 kb *hms* region upon comparing 6/69M, KIM and CO92 (Buchrieser et al., 1999). (Two non-synonymous sites were detected but these are less reliable for age calculations since they might reflect selection).

In the absence of published, genome-wide estimates of the synonymous diversity within conserved CDS in *Y. pestis*, I have now calculating this value for homologous CDS that are common to the CO92 and KIM genomes. Julian Parkhill and Nicholas R. Thomson from the Sanger Centre, Cambridge, identified 3,237 orthologues between CO92 and KIM by pairwise alignment after blast searches, of which 67 putatively contained synonymous single nucleotide polymorphisms (sSNPs) (pers. comm.). Only 40 of these orthologues were investigated further because the other 37 contain repetitive DNA. Four more putative sSNPs were excluded because sequencing the corresponding stretches from both CO92 and KIM showed that they reflected sequencing errors (man. in prep.), leaving 38 sSNPs (two orthologues each contained 2 sSNPs) from 3,235,075 bp containing 767,352 potential synonymous sites in strain CO92. The synonymous distance between KIM and CO92 (D_S) is then $38/767{,}352 = 5 \times 10^{-5}$.

The age since a common ancestor of two organisms can be calculated from the synonymous distance and the mutation rate, μ, by the formula: Age = $D_S/(2 \cdot \mu)$ (Page and Holmes, 1998). The factor 2 in the denominator accounts for the fact that D_S measures the current distance between the lineages after the independent accumulation of mutations along both branches. Previous calculations of age based on the lack of diversity in 6 gene fragments (Achtman et al., 1999) used an incorrect mutation rate, and it seems worthwhile to repeat these calculations here in order to compare them with estimates based on the newer data.

When no mutations are found, an upper bound for the age can be estimated at the 50% or 95% confidence limit on the basis of the Poisson distribution by calculating D_S for 0.693 or 2.996 instead of 0 mutations, respectively (Rich et al., 1998). However, the mutation rate for *Y. pestis* is unknown. In the laboratory, the mutation rate is approximately the same between *Y. pestis* and *E. coli*, and therefore a first approximation would be to use the mutation rate for *E. coli* of 3×10^{-8} per year (Guttman and Dykhuizen, 1994), which is based on the frequency of laboratory mutations and the assumption that *E. coli* undergoes 300 generations per year in nature. This estimate is almost certainly inaccurate, because the extrapolation of short term rates to long-term processes ignores the loss of mutations due to genetic drift, whose magnitude depends in turn on population size and bottlenecks (Ohta, 1996). A second estimate of the mutation rate for *E. coli* of 6.7×10^{-9} (Whittam, 1996) is based on the value of 0.94 for D_S between *E. coli* and *Salmonella typhimurium* (Sharp, 1991), which have undergone separate evolution for 140 million years (Ochman and Wilson, 1987). However, a twofold error crept into this estimate: if these two species descended from a common ancestor that existed 140 Myr ago, the elapsed evolutionary time between them is 280 Myr, leading to a mutation rate of 3.4×10^{-9}.

The molecular clock hypothesis for neutral or nearly neutral mutations has been largely substantiated after extensive debate for eukaryotic organisms (see standard textbooks such as Chapter 7 in Page and Holmes (Page and Holmes, 1998)), but has not been extensively tested for bacterial DNA. Sequence diversity in 16S rRNA does accumulate at a fairly uniform clock rate in many bacterial species (Ochman and Wilson, 1987), but

Table 1. Age calculations based on synonymous polymorphisms*.

Sequences	Polymorphisms	Potential synonymous sites	Age (years) (Whittam) ($\mu = 3.4 \times 10^{-9}$)¶	Age (years) (Dykhuizen) ($\mu = 3 \times 10^{-8}$)¶
hms	0	3,000	<35,000-150,000	<4,000-17,000
Ybt	0	8,500	<12,000-52,000	<1,500-6,000
6 genes	0	22,000	<9,000-40,000	<1,000-4,500
Genomic	38	767,352	7,282	825

**hms* and Ybt refer to the data of Buchrieser *et al.*, 1999, which were obtained from a comparison of 3 and 2 isolates, respectively. 6 genes refers to the 2.5 kb from six genes that was sequenced from 36 isolates (Achtman et al., 1999). Genomic refers to the genomic comparisons between homologous CDS in CO92 and KIM that are described in the text. The number of potential synonymous sites was calculated for the genomic data for each nucleotide using the simplified rules described by Weng-Hsiu Li (Li et al., 1985), where each nucleotide in 2-fold or 3-fold divergent sites counts as 0.33 and 4-fold divergent sites count as 1.0. No correction was made for forward and backward mutations, which should be exceedingly rare at such low levels of sequence diversity. The resulting frequency of potential synonymous sites (0.237) was then applied to the 6.3 kb *hms* region and the 35.8 Ybt region. The number of potential synonymous sites within the six genes was taken from published data (Achtman et al., 1999), where it was calculated using the program DnaSP (Rozas and Rozas, 1999).

¶Age was calculated as $D_S/(2 \cdot \mu)$ for *hms*, Ybt and genomic comparisons. The factor of 2 is needed because the diversity between isolates has occurred during the combined evolution along both branches. D_S was calculated as synonymous sites observed in all pairwise comparisons / (length of sequence from each isolate · (n · n-1)/2), where n is the number of isolates. Note that (n · n-1)/2) is 1 for a comparison between two isolates and 3 for a comparison among three isolates. However, this calculation is inappropriate for a comparison of 36 identical isolates and would yield an artificially low value of D_S. Therefore, for the 6 genes, the distance calculated was the theoretical distance from a central node from which all radiated in a star phylogeny without acquiring any sSNPs and corresponds to observed synonymous sites/(length of sequence from each isolate · n). Since the distance from the source is measured directly, the factor of 2 is not needed and Age = distance/μ. Because 0 synonymous sites were detected for *hms*, Ybt or the 6 genes, 0 was replaced by 0.693 and 2.996, which correspond to the 50% and 95% upper confidence estimates according to the Poisson distribution (Rich et al., 1998) to yield a range of values.

recent analyses have shown that the ratio of synonymous substitution in protein-coding genes to substitution in 16S rRNA differs between different bacterial species by up to 10-fold (Ochman et al., 1999). Thus, although it is the best that is currently possible, age calculations for *Y. pestis* based on an *E. coli* clock rate should be considered as tentative and demand confirmation by independent methods. Given this proviso, I have recalculated these age estimates on the basis of the data summarized above (Table 1).

The maximal age calculations based on a lack of detection of sSNPs diminish with the number of potential synonymous sites that are examined. If the calculations based on Dykhuizen's estimate are ignored, which seems reasonable based on the analysis above, the current calculations for the 6 genes examined by Achtman *et al.*, 1999 indicate that *Y. pestis* evolved within the last 9,000-40,000 years while the lower numbers of potential synonymous sites examined by Buchrieser *et al.*, 1999 result in greater values. The genomic comparisons indicate that Orientalis split from Medievalis 7,000 years ago (Table 1). More accurate estimates would require measurements of the mutation rate for *Y. pestis* in nature and testing even more potential synonymous sites from multiple isolates.

BIOVARS AND HISTORY

Can we deduce features of the history of *Y. pestis* by examining the properties of the biovars, which differ in nutritional markers (Table 2)? Based on historical records

Table 2. Phenotypes of biovars within *Y. pestis*.

Biovar	Glycerol fermentation	Nitrate reduction	Fermentation of rhamnose, melibiose
pestoides	+	+/-	+,+
Antiqua	+	+	-,-
Medievalis	+	-	-,-
Orientalis	-	+	-,-

+/-, some isolates are positive and others are negative.

(Devignat, 1951), it seems fairly certain that the global distribution of biovar Orientalis reflects global expansion since the late 1890's via marine shipping from Hong Kong. This would infer that these bacteria have undergone a strong, recent bottleneck and should have been purified of much of their long-standing diversity. In other words, Orientalis isolates from the US or from Africa would be expected to be extremely homogeneous because their current diversity has accumulated over slightly more than 100 years. However, the extent of diversity of Orientalis in East Asia is unknown, as is its history of epidemic spread and geographical location prior to the 20th century.

Even less is known about the other biovars: the few Medievalis isolates that have been investigated stem from southern Eurasia while the few Antiqua isolates come from Africa or the near East. The biovar named "pestoides" is even more poorly defined, partially because most investigations of these bacteria were published decades ago in Russian and Chinese (Anisimov et al., 2004). These publications indicate that various rodent species in large areas of the former Soviet Union and China are infected by diverse bacteria of uncertain parentage. Recent molecular work from the U.S.A. has largely focused on a very small, and somewhat obscure, collection of pestoides isolates from Russia that are within the possession of the U.S. Army Medical Research Institute of Infectious Disease. Analyses on these few isolates need to be confirmed with studies based on more representative collections that reflect the diversity in Russia and China.

The molecular basis for the nutritional differences between the biovars has not yet been completely resolved. The inability to ferment glycerol by biovar Orientalis is associated with a 93 bp deletion in the *glpD* gene (Motin et al., 2002). We have confirmed the association of this deletion with biovar Orientalis (unpublished data) but direct evidence is still lacking that this mutation inactivates the ability to ferment glycerol. The inability of biovars Medievalis and some pestoides to reduce nitrate does not reflect a genetic relatedness because we have identified a mutation that is specific to Medievalis which inactivates nitrate reduction while pestoides isolates are wild-type at this site (unpublished data). The molecular basis for the inability of biovars Antiqua, Medievalis and Orientalis to ferment rhamnose and melibiose remain unknown.

Based on their geographical associations, Devignat suggested the very striking concept that the primary biovars are each responsible for a different wave of plague epidemics (Devignat, 1951). Orientalis is clearly responsible for the ongoing third pandemic (1894-present). However, the only basis for the assignment of Justinian's plague (541-767) to Antiqua and the "Black Death" (1346-1850) to Medievalis is wishful thinking plus the observation that these biovars are currently isolated in the same areas where those epidemics are thought to have arisen. In fact, even the widespread belief that Justinian's plague and the black death were to a large degree caused by *Y. pestis* has been subjected to criticism for at least 120 years on the basis of incompatible historical and modern epidemiological patterns (Hirsch, 1881; Cohn, Jr., 2002). Aside from similarities (and some remarkable differences) between the pathologies reported from historical epidemics and modern disease, the primary ground to believe that *Y. pestis* has caused historically important human disease are reports on ancient DNA. *Y. pestis*-specific PCR products were amplified from the tooth pulp of several corpses from mass graves dating to the 14th and 16th centuries (Drancourt et al., 1998; Raoult et al., 2000). These results await confirmation from independent sources, which have failed abysmally until now (Gilbert et al., 2004) and historians remain unconvinced by the published evidence (Karlsson, 1996; Twigg, 2003; Wood and DeWitte-Avina, 2003). If such results with ancient DNA could be confirmed and indicated which biovar was responsible for ancient disease, it might be possible to reconstruct the history of epidemic spread of *Y. pestis* (at least in humans). Until then, the only tools that are available are genetic signatures in the genomes of modern isolates.

DO THE BIOVARS CORRESPOND TO GENETICALLY RELATED GROUPS OF ISOLATES?

How should genetic diversity be investigated in a "species" that is as uniform as *Y. pestis*? And does each of the biovars contain only isolates that share common descent from a common ancestor? Initial analyses using RFLP of ribosomal operons (ribotyping) showed that many Orientalis isolates share one common ribotype and all except one Medievalis isolate share a second ribotype (Guiyoule et al., 1994). However, each biovar does include multiple ribotypes and a few ribotypes were present in two biovars. If biovars represented relatively uniform groupings, as suggested below, the latter observation demands that apparently indistinguishable ribotypes have arisen on multiple occasions. In subsequent analyses, three novel ribotypes were found during an extensive investigation of numerous Orientalis isolates from Madagascar (Guiyoule et al., 1997), suggesting that novel ribotypes can arise over a period of decades.

RFLP of the locations of the IS*100* insertion element yielded a clear distinction between the biovars (Achtman et al., 1999). An average of 30 well-resolved *Eco*RI fragments in which IS*100* had integrated were found for 33 Orientalis isolates while an average of 20 was found for 5 Medievalis and 11 Antiqua isolates. Although each isolate possessed a unique banding pattern, the patterns clustered into biovar-specific groups (Achtman et al., 1999). All the Medievalis isolates tested had been assigned the O ribotype, as had two of the Antiqua isolates (E. Carniel, pers. comm.), but the IS*100* RFLP clustering clearly separated these isolates into the appropriate biovar-specific group. Either the presence of the O ribotype in biovars Medievalis and Antiqua reflects a common ancestor for these two biovars or biovar O has arisen on multiple occasions (homoplasy), which seems more likely given the evidence for rapid evolution of ribotypes in Madagascar.

RFLP, like all fingerprinting methods, is difficult to reproduce in different laboratories and yields uncertain data because multiple bands can co-migrate and the pattern of band migration can vary from day to day. Binary methods based on the presence/absence of individual insertions are more reproducible and easier to analyse. A binary PCR-based method based on the insertion of IS*100* in 16 loci was applied to 73 isolates of the biovars pestoides, Antiqua, Medievalis, Orientalis plus 2 atypical strains (Motin et al., 2002). The results confirmed that the insertion of IS*100* correlates well with biovar (if pestoides are thought of a separate biovar), with one major and four minor exceptions.

The major exception is that biovar Antiqua contains (at least) two groups of unrelated organisms, one isolated in East Asia and most closely related to Medievalis, and the second isolated in Africa. One minor exception consisted of biovar Antiqua strain Nicholisk 51, which possessed the same IS*100* pattern as Orientalis isolates. Nicholisk 51 also possesses two nucleotides in the *glpD* gene that are typical of Orientalis (Motin et al., 2002) and the same difference region (DRF) pattern as Orientalis isolates (Radnedge et al., 2002) (see below). Motin *et al.*, interpreted Nicholisk 51 as representing an example of horizontal genetic exchange that replaced the 69 bp deletion within *glpD* that is characteristic of Orientalis strains by import from a $glpD^{+}$ strain (Motin et al., 2002). In the absence of any serious evidence for horizontal genetic exchange of chromosomal genes in *Y. pestis*, it seems equally likely to me that Nicholisk 51, which was isolated in Manchuria, may represent a sub-branch of Orientalis that separated before the *glpD* deletion arose.

A second exception, pestoides J, has IS*100* and DRF patterns that are typical of Medievalis isolates but cannot ferment glycerol. However, pestoides J is properly considered as an atypical isolate that neither reduces nitrate nor ferments glycerol (Radnedge et al., 2002). This strain might be a member of the Medievalis biovar with a mutation in a gene important for glycerol fermentation that was assigned to the pestoides biovar because it was isolated from the former Soviet Union. Thirdly, a Japanese Antiqua isolate, Yokohama, has the same IS*100* profile as do Medievalis isolates from Kurdistan (Motin et al., 2002). This observation is difficult to interpret because Yokohama was found to possess the same DRF pattern as other Antiqua isolates (Radnedge et al., 2002). And finally, strain Angola groups with pestoides, although it is an Antiqua that was isolated in Africa. Both DRF analysis and VNTR analysis (Klevytska et al., 2001) (see below) also group Angola with pestoides, suggesting that this group of organisms is not restricted to Central Asia.

These four minor exceptions may all reflect a variety of secondary genetic events and I conclude that except for Antiqua, biovars contain groups of related organisms. Unfortunately, the IS*100* data do not permit deducing a parsimonious evolutionary tree of how the biovars evolved: the loci were defined exclusively on IS*100* insertions that are present in CO92 and give no information on additional insertions that might have happened during the evolution of the other biovars. The tree that was presented (Motin et al., 2002) is based on a similarity matrix of how many PCR reactions were successful with each isolate and is inappropriate for reconstructing evolutionary pathways.

An independent analysis of the same isolates that were tested by Motin *et al.*, used six differentially present chromosomal regions (DRF) that were identified by subtractive hybridization (Radnedge et al., 2002). These results also assigned most Orientalis isolates to one group and most Medievalis isolates to a second group. Numerous DRF patterns were found within Antiqua and pestoides and their groupings remain unclear. Otherwise, the results were congruent with those described above, except that DRF analysis indicated different relationships for strains Yokohama and A16 (both Antiqua) and for 366 and Nicholisk 41 (both Medievalis) than did the PCR-based IS*100* analysis of Motin *et al.* (Motin et al., 2002). Given that only six markers are involved in this analysis and that several markers were identified from experiments wherein closely related strains were used as driver and tester, I am uncertain how much importance these apparent exceptions deserve.

In still another publication, 11 geographically distinct isolates were investigated by multilocus VNTR analysis using 42 VNTRs (stretches of variable numbers of tandem repeats) (Klevytska et al., 2001). Once more, Orientalis isolates formed one cluster, Medievalis isolates formed a second cluster and pestoides were highly different from the others. VNTR is clearly capable of distinguishing very closely related isolates (Girard et al., 2004), indicating that these groupings are highly significant.

All these analyses indicate that Orientalis is uniform and that most Medievalis strains are similar. But there is apparent diversity within Antiqua and pestoides that warrants closer investigation. And the great diversity expected to be present among isolates from rodents in central Asia and China (Anisimov et al., 2004) has not yet been properly addressed by any molecular analysis. We also lack a convincing evolutionary model for the descent of the biovars and their subsequent differentiation after *Y. pestis* diverged from *Y. pseudotuberculosis*.

CONCLUDING REMARKS

The new approach of DNA-sequence based population genetic analyses has revealed fascinating features about the biology of several bacterial pathogens (Spratt et al., 2002; Achtman, 2002), but has not yet been extensively applied to *Yersinia pestis*. Most previous investigations of *Y. pestis* were motivated by its supposed historical association with large epidemics of fatal disease in humans and by its high lethality in experimental animals. In contrast, a population genetic approach to these bacteria (and those causing other zoonoses) would concentrate on aspects related to the persistence and spread in their natural hosts, namely wild rodents. The global natural genetic diversity of *Y. pestis* in rodents is largely unexplored and such experiments are becoming even more difficult because current hysteria about the potential use of these bacteria for bioterrorism greatly hinders the facile international exchange of isolates. Given the lack of extensive population genetic analysis, this review has attempted to focus on open questions and concepts, rather than providing an extensive summary of the (largely non-existent) literature.

Current knowledge shows that *Y. pestis* is a clone of *Y. pseudotuberculosis*, which evolved approximately 20,000 years ago according to the best current estimates. During this time, it has spread globally and established itself as an endemic infection within many rodents and other mammals and has, occasionally, inadvertently caused epidemics in humans. Over the extremely short period of time that it has existed as a distinct group of organisms, *Y. pestis* has diverged into several, distinct groups of closely related organisms, several of which possess distinguishing nutritional properties that have resulted in biovar designations. However, the true global diversity of these microbes will first become apparent after extensive sampling of endemic populations, particularly in Central Asia. Furthermore, such isolates will need to be investigated with a common suite of molecular techniques whose discriminatory power is sufficient to elucidate evolutionary patterns within highly homogeneous bacteria, and which yield numerical data that can be readily exchanged by the internet. Such techniques have not yet been published for *Y. pestis*, but can potentially be implemented in the near future due to the exploratory work that is summarized here. A similar situation for other bacterial species existed less than 6 years ago and was solved by the implementation of global, public databases for multilocus sequence typing (Maiden et al., 1998; Feil and Spratt, 2001) (see http://www.mlst.net). Once such techniques are available for *Y. pestis*, they could result in a global resource for the community interested in the natural history and evolution of this fascinating group of organisms.

ACKNOWLEDGEMENTS

This review would not have been possible without the support and help of the following colleagues, to all of whom I hereby express my gratitude. Analyses of the genomic diversity of *Y. pestis* were performed in close collaboration with Giovanna Morelli (MPI für Infektionsbiologie, Berlin) and Elisabeth Carniel (Institut Pasteur, Paris) and supported by grants from the Deutsche Forschungsgemeinschaft (Ac 36/9-1 through Ac 36/9-3). My knowledge about the pestoides biovar is based on communications from Luther E. Lindler (WRAIR, MD) and Paul Keim (Northern Arizona University, Flagstaff, Arizona). If the calculations of age are correct, then it is largely as a result of extensive discussions with Edward Holmes (Oxford University); if not, it is my own fault for not understanding him. Genomic comparisons at synonymous sites were performed by Nicholas R. Thomson and Julian Parkhill (Sanger Centre, Cambridge). This review would never have been published without the almost inexhaustible patience of Mme La Peste.

REFERENCES

Abdulkarim, F., and Hughes, D. 1996. Homologous recombination between the *tuf* genes of *Salmonella typhimurium*. J. Mol. Biol. 260: 506-522.

Achtman, M. 2002. A phylogenetic perspective on molecular epidemiology In:Molecular Medical Microbiology.M.Sussman ed. Academic Press, London, pp. 485-509.

Achtman, M., Zurth, K., Morelli, G., Torrea, G., Guiyoule, A., and Carniel, E. 1999. *Yersinia pestis*, the cause of plague, is a recently emerged clone of *Yersinia pseudotuberculosis*. Proc. Natl. Acad. Sci. USA. 96: 14043-14048.

Anisimov,A.P., Lindler,L.E., and Pier,G.B. 2004. Intraspecific diversity of *Yersinia pestis*. Clin. Microbiol. Rev. in press.

Bercovier, H., Mollaret, H. H., Alonso, J. M., Brault, J., Fanning, G. R., Steigerwalt, A. G., and Brenner, D. J. 1980. Intra- and interspecies relatedness of *Yersinia pestis* by DNA hybridization and its relationship to *Yersinia pseudotuberculosis*. Curr. Microbiol. 4: 225-229.

Brossollet, J., and Mollaret, H. 1994. Pourquoi la peste? Le rat, la puce et le bubon. Gallimard, Paris, France, p. 1-160.

Buchrieser, C., Rusniok, C., Franguel, L., Couve, E., Billault, A., Kunst, F., Carniel, E., and Glaser, P. 1999. The 102 kb pgm locus of *Yersinia pestis*: sequence analaysis and comparison of selected regions among different *Yersinia pestis* and *Yersinia pseudotuberculosis* strains. Infect. Immun. 67: 4851-4861.

Cohn, S. K., Jr. 2002. The Black Death Transformed: Disease and culture in early Renaissance Europe. Arnold, London, p. 1-318.

Deng, W., Burland, V., Plunkett III, G., Boutin, A., Mayhew, G. F., Liss, P., Perna, N. T., Rose, D. J., Mau, B., Zhou, S., Schwartz, D. C., Fetherston, J. D., Lindner, B., Brubaker, R. R., Plano, G. V., Straley, S. C., McDonough, K. A., Nilles, M. L., Matson, J. S., Blattner, F. R., and Perry, R. D. 2002. Genome sequence of *Yersinia pestis* KIM. J. Bacteriol. 184: 000.

Devignat, R. 1951. Variétés de l'espèce *Pasteurella pestis*. Nouvelle hypothèse. Bull. World Hlth. Org. 4: 247-263.

Drancourt, M., Aboudharam, G., Signoli, M., Dutour, O., and Raoult, D. 1998. Detection of 400-year-old *Yersinia pestis* DNA in human dental pulp: an approach to the diagnosis of ancient septicemia. Proc. Natl. Acad. Sci. USA. 95: 12637-12640.

Feil, E. J., and Spratt, B. G. 2001. Recombination and the population structures of bacterial pathogens. Annu. Rev. Microbiol. 55: 561-590.

Gilbert, M. T., Cuccui, J., White, W., Lynnerup, N., Titball, R. W., Cooper, A., and Prentice, M. B. 2004. Absence of *Yersinia pestis*-specific DNA in human teeth from five European excavations of putative plague victims. Microbiology. 150: 341-354.

Girard, J.M., Wagner, D.M., Vogler, A.J., Keys, C., Allender, C.J., Drickamer, L.C., and Keim, P. 2004. Differential plague transmission dynamics determine *Yersinia pestis* population genetic structure at local, regional and global scales. Proc. Natl. Acad. Sci. USA.11: 8408-8413.

Guiyoule, A., Grimont, F., Iteman, I., Grimont, P. A. D., Lefèvre, M., and Carniel, E. 1994. Plague pandemics investigated by ribotyping of *Yersinia pestis* strains. J. Clin. Microbiol. 32: 634-641.

Guiyoule, A., Rasoamanana, B., Buchrieser, C., Michel, P., Chanteau, S., and Carniel, E. 1997. Recent emergence of new variants of *Yersinia pestis* in Madagascar. J. Clin. Microbiol. 35: 2826-2833.

Guttman, D. S., and Dykhuizen, D. E. 1994. Clonal divergence in *Escherichia coli* as a result of recombination, not mutation. Science 266: 1380-1383.

Hinnebusch, B. J., Fischer, E. R., and Schwan, T. G. 1998. Evaluation of the role of *Yersinia pestis* plasminogen activator and other plasmid-encoded factors in temperature-dependent blockage of the flea. J. Infect. Dis. 178: 1406-1415.

Hinnebusch, B. J., Rosso, M. L., Schwan, T. G., and Carniel, E. 2002a. High-frequency conjugative transfer of antibiotic resistance genes to *Yersinia pestis* in the flea midgut. Mol. Microbiol. 46: 349-354.

Hinnebusch, B. J., Rudolph, A. E., Cherepanov, P., Dixon, J. E., Schwan, T. G., and Forsberg, A. 2002b. Role of *Yersinia murine* toxin in survival of *Yersinia pestis* in the midgut of the flea vector. Science 296: 733-735.

Hirsch, A. 1881. Beulenpest In:Handbuch Der Historisch-Geographischen Pathologie. Volumen I. Verlag von Ferdinand Enke, Stuttgart, pp. 349-184.

Karlsson, G. 1996. Plague without rats: the case of fifteenth-century Iceland. J. Medieval History 22: 263-284.

Klein, J., Gutknecht, J., and Fischer, N. 1990. The major histocompatibility complex and human evolution. Trends Genet. 6: 7-11.

Klein, J., and Takahata, N. 2002. Where do we come from? Springer-Verlag, Berlin, p. 1-462.

Klevytska, A. M., Price, L. B., Schupp, J. M., Worsham, P. L., Wong, J., and Keim, P. 2001. Identification and characterization of variable-number tandem repeats in the *Yersinia pestis* genome. J. Clin. Microbiol. 39: 3179-3185.

Lan, R., and Reeves, P. R. 2002. *Escherichia coli* in disguise: molecular origins of *Shigella*. Microbes. Infect. 4: 1125-1132.

Li, W.-H., Wu, C.-I., and Luo, C.-C. 1985. A new method for estimating synonymous and nonsynonymous rates of nucleotide substitution considering the relative likelihood of nucleotide and codon changes. Mol. Biol. Evol. 2: 150-174.

Maiden, M. C. J., Bygraves, J. A., Feil, E., Morelli, G., Russell, J. E., Urwin, R., Zhang, Q., Zhou, J., Zurth, K., Caugant, D. A., Feavers, I. M., Achtman, M., and Spratt, B. G. 1998. Multilocus sequence typing: a portable approach to the identification of clones within populations of pathogenic microorganisms. Proc. Natl. Acad. Sci. USA. 95: 3140-3145.

Motin, V. L., Georgescu, A. M., Elliott, J. M., Hu, P., Worsham, P. L., Ott, L. L., Slezak, T. R., Sokhansanj, B. A., Regala, W. M., Brubaker, R. R., and Garcia, E. 2002. Genetic variability of *Yersinia pestis* isolates as predicted by PCR- based IS100 genotyping and analysis of structural genes encoding glycerol-3-phosphate dehydrogenase (glpD). J. Bacteriol. 184: 1019-1027.

Ochman, H., Elwyn, S., and Moran, N. A. 1999. Calibrating bacterial evolution. Proc. Natl. Acad. Sci. USA. 96: 12638-12643.

Ochman, H., and Wilson, A. C. 1987. Evolution in bacteria: Evidence for a universal substitution rate in cellular genomes. J. Mol. Evol. 26: 74-86.

Ohta, T. 1996. The current significance and standing of neutral and nearly neutral theories. BioEssays 18: 673-677.

Page, R. D. M., and Holmes, E. C. 1998. Molecular evolution. A phylogenetic approach. Blackwell Science Ltd, Oxford, UK, p. 1-346.

Parkhill, J., Wren, B. W., Thomson, N. R., Titball, R. W., Holden, M. T., Prentice, M. B., Sebaihia, M., James, K. D., Churcher, C., Mungall, K. L., Baker, S., Basham, D., Bentley, S. D., Brooks, K., Cerdeno-Tarraga, A. M., Chillingworth, T., Cronin, A., Davies, R. M., Davis, P., Dougan, G., Feltwell, T., Hamlin, N., Holroyd, S., Jagels, K., Karlyshev, A. V., Leather, S., Moule, S., Oyston, P. C., Quail, M., Rutherford, K., Simmonds, M., Skelton, J., Stevens, K., Whitehead, S., and Barrell, B. G. 2001. Genome sequence of *Yersinia pestis*, the causative agent of plague. Nature. 413: 523-527.

Radnedge, L., Agron, P. G., Worsham, P. L., and Andersen, G. L. 2002. Genome plasticity in *Yersinia pestis*. Microbiology. 148: 1687-1698.

Rakin, A., and Heesemann, J. 1995. The established *Yersinia pestis* biovars are characterized by typical patterns of I-CeuI restriction fragment length polymorphism. Mol. Genet. Microbiol. Virusol. 3: 26-29.

Raoult, D., Aboudharam, G., Crubezy, E., Larrouy, G., Ludes, B., and Drancourt, M. 2000. Molecular identification by "suicide PCR" of *Yersinia pestis* as the agent of Medieval Black Death. Proc. Natl. Acad. Sci. USA. 97: 12800-12803.

Rich, S. M., Licht, M. C., Hudson, R. R., and Ayala, F. J. 1998. Malaria's Eve: evidence of a recent population bottleneck throughout the world populations of *Plasmodium falciparum*. Proc. Natl. Acad. Sci. USA. 95: 4425-4430.

Rozas, J., and Rozas, R. 1999. DnaSP version 3.0: an integrated program for molecular population genetics and molecular evolution analysis. Bioinformatics 15: 174-175.

Sharp, P. M. 1991. Determinants of DNA sequence divergence between *Escherichia coli* and *Salmonella typhimurium*: codon usage, map position, and concerted evolution. J. Mol. Evol. 33: 23-33.

Skurnik, M., Peippo, A., and Ervela, E. 2000. Characterization of the O-antigen gene clusters of *Yersinia pseudotuberculosis* and the cryptic O-antigen gene cluster of *Yersinia pestis* shows that the plague bacillus is most closely related to and has evolved from *Y. pseudotuberculosis* serotype O:1b. Mol. Microbiol. 37: 316-330.

Spratt, B. G., Feil, E., and Smith, N. H. 2002. Population genetics of bacterial pathogens In:Molecular Medical Microbiology.M.Sussman ed. Academic Press, London, pp. 445-484.

Trebesius, K., Harmsen, D., Rakin, A., Schmelz, J., and Heesemann, J. 1998. Development of rRNA-targeted PCR and in situ hybridization with fluorescently labelled oligonucleotides for detection of *Yersinia* species. J. Clin. Microbiol. 36: 2557-2564.

Twigg, G. 2003. The Black Death and DNA. Lancet Infect. Dis. 3: 11.

Whittam, T. S. 1996. Genetic variation and evolutionary processes in natural populations of *Escherichia coli* In: *Escherichia coli* and *Salmonella*. R. Curtiss III, J.L. Ingraham, E.C.C. Lin, K.B. Low, B. Magasanik, W.S. Reznikoff, M. Riley, M. Schaechter, and H.E. Umbarger eds. ASM Press, Washington,D.C., pp. 2708-2720.

Wood, J., and DeWitte-Avina, S. 2003. Was the Black Death yersinial plague? Lancet Infect. Dis. 3: 327-328.

Chapter 3

The *Yersinia pestis*-Specific Plasmids pFra and pPla

Luther E. Lindler

ABSTRACT

In addition to the common *Yersinia* virulence plasmid that is ~ 70-kb, typical strains of *Yersinia pestis* harbor two unique plasmids. These two plasmids are generally referred to as the "Murine Toxin" plasmid, pFra, and the "pesticin" plasmid, pPla. The *Y. pestis*-specific plasmids are approximately 100 and 9.6-kb in size, respectively. Both of these molecules encode DNA sequences similar to those found on plasmids from Gram-negative enteric organisms with which *Y. pestis* may have exchanged genetic material. Thus, these plasmids contain the molecular fingerprints left behind during the evolution and emergence of *Y. pestis* as a unique pathogen. Furthermore, the detailed analysis of pFra at the genomic-level revealed how virulence plasmids can be assembled in nature as a mosaic of multiple genetic elements. Taken together, the study of these *Y. pestis*-specific plasmids has yielded valuable lessons in how serial coalescence of multiple genetic elements in a bacterium can contribute to the emergence of new bacterial diseases. This chapter presents the salient features of *Y. pestis* pathogenesis in relation to the factors encoded by the pFra and pPla plasmids as well as the genetics of the virulence factors encoded by these extrachromosomal elements.

INTRODUCTION

Typical virulent isolates of *Yersinia pestis* harbor three plasmids. One of these plasmids is ~70 kb in size and is also found in the other pathogenic *Yersinia* species, *Yersinia pseudotuberculosis* and *Yersinia enterocolitica*. This molecule is commonly referred to as pLcr (or pYV) and is discussed in detail in Chapter 16. The two other plasmids found in typical *Y. pestis* strains are specific for this species. The first is the ~100 kb plasmid generically designated pFra and the second is the ~9.6 kb element pPla. Both of these plasmids encode proteins unique to *Y. pestis* and factors necessary for full virulence of the organism. The fact that these molecules are *Y. pestis*-specific suggested that they might be involved in the highly invasive nature of the organism compared to the enteropathogenic *Yersinia* spp. (Brubaker, 1991). Genetic and virulence characterization of some of the genes encoded by these plasmids as well as virulence studies of natural isolates lacking pPla have demonstrated that virulence of this organism is not so simple. Certainly the acquisition of these plasmids was a major leap in the evolution of the organism we now know as *Y. pestis*, however these elements do not completely account for the major increase in virulence of this species.

This chapter will focus on the putative virulence elements encoded by pFra and pPla as well as on their genetic architecture. Several versions (from different strains as well as isolates) of each of these plasmids have been completely sequenced as part of genomics efforts aimed at understanding the pathogenesis of *Y. pestis* infection and the evolution of this virulent organism from a relatively low-level pathogen like *Y. pseudotuberculosis*. Some basic properties and references for complete DNA sequences are listed in Table 1. These genomic efforts have allowed a molecular examination of both pFra and pPla in order to identify new potential virulence factors, compare molecules in an attempt to understand intra- and interspecies evolution and gain a better understanding of the life cycle of *Y. pestis*. This chapter will relate the genomic studies of pFra and pPla to virulence properties of *Y. pestis*.

COMPARATIVE VIRULENCE

To better understand the roles that *Y. pestis*-specific plasmids might play in the pathogenesis of plague infection, a brief comparison of the pathogenesis with *Y. pseudotuberculosis* is in order. The DNA relationship between Y. *pestis* and *Y. pseudotuberculosis* chromosomes is greater than eighty percent as measured by DNA::DNA hybridization analysis (Moore and Brubaker, 1975, Bercovier et al. 1980). However, these two species cause vastly different diseases although both have an affinity for lymphoid tissues. The major difference in virulence between *Y. pestis* and *Y. pseudotuberculosis* in rodent experimental infection models is the ability of the former to cause disease from peripheral sites of inoculation and the fact that this species' transmission involves a flea vector. *Y. pestis* can initiate infection following the subcutaneous injection of less than 10 CFU in the mouse (Brubaker, 1983; Welkos et al., 1995; Welkos et al., 1997). In contrast, *Y. pseudotuberculosis* has an LD50 of ~10^4 when delivered subcutaneously but is of similar virulence to *Y. pestis* when delivered intravenously (Brubaker, 1983). Given the fact that both of these species harbor the ~70 kb virulence plasmid pLcr and the high level of DNA homology between them, these virulence studies have suggested that the *Y. pestis*-specific plasmids may at least partially explain the increased invasiveness of this species from peripheral routes.

As indicated above, the typical infection caused by *Y. pestis* is initiated through the bite of an infected flea. For this reason, unlike many *Y. pseudotuberculosis* infections, bubonic or pneumonic plague can occur in otherwise healthy individuals. Following the initial insult with infected material delivered by the flea, the bacteria would be expected to be taken up by dendritic cells and other professional phagocytes. *Y. pestis* is well-

Table 1. Properties of the *Y. pestis*-specific plasmids pFra and pPla.

Plasmid	Properties	Accession # (strain and size)	References for Complete Sequence
pFra	Large plasmid encoding the fraction 1 (F1) capsular protein and the murine toxin (Ymt); Known to be highly variable in size among some natural isolates; Has been shown to integrate into the chromosome of the organism; Contains two complete copies of IS*100* and a single copy of IS*285* and IS*1541* as well as many remnants of mobile elements; RepFIB replicon and ParABS partitioning	AF053947 (KIM 100984 bp) AF074611 (KIM 100990 bp) AL117211 (CO92 96210 bp)	Hu et al., 1998 Lindler et al., 1998 Prentice et al., 2001
pPla	Small plasmid encoding the plasminogen activator, pesticin and pesticin immunity proteins; Single copy of IS*100*; Encodes a ColE1-like replicon	AF053945 (KIM 9610 bp) AL109969 (CO92 9612 bp)	Hu et al., 1998 Prentice et al., 2001

known to be able to survive macrophage phagocytosis (Cavanaugh and Randall, 1959; Straley and Harmon, 1984a; Straley and Harmon, 1984b; Charnetzky and Shuford, 1985; Goguen et al., 1986). At least some virulence genes have been shown to be activated in this intracellular environment (Lindler and Tall, 1993; Makoveichuk et al., 2003). Many other virulence determinants are induced by growth at mammalian body temperature (Straley and Bowmer, 1986; Perry and Fetherston, 1997) but their induction inside macrophages has yet to be demonstrated. The role that the interaction of pathogenic *Yersinia* with host macrophages plays in the infectious process is an area of debate within the research community. This has been precipitated by the demonstration that many virulence factors upregulated at 37°C have now been shown to be involved in blocking bacterial phagocytosis (Perry and Fetherston, 1997; Cornelis et al., 1998; Du et al., 2002). Regardless of the need to interact with macrophages, the organism migrates to the regional lymph nodes where it proliferates and stimulates inflammation. This region is then referred to as a bubo. Alternatively, septicemic plague occurs where the organism can be isolated from the blood in the absence of a developed bubo. The organism gains access to the liver and spleen where it reaches very high numbers and eventually causes a fulminating bacteremia allowing infection of fleas that feed on the diseased animal (Perry and Fetherston, 1997). The infected fleas then transmit the organism to other disease-free hosts. In some cases, bubonic plague may result in pneumonic plague by colonization of the lungs of the infected animal. This then allows spread of the organism by flea-independent aerosolized respiratory droplets and results in a fulminating pneumonia and rapid death. In all forms of the disease, death occurs by multiple organ failure and disseminated intravascular coagulation (DIC) (Butler, 1983).

The infection caused by *Y. pseudotuberculosis* results in a much milder and usually self-limiting disease (Butler, 1983). The infection occurs by the oral route so that the organism invades its host through the gastrointestinal mucosa and causes a gastroenteritis. *Y. pseudotuberculosis* typically targets the mesenteric lymph nodes but does not typically infect deeper tissues such as the liver and spleen.

Taken together, the description of *Y. pestis* pathogenesis and transmission indicates that this bacterium must synthesize certain factors that *Y. pseudotuberculosis* does not. Besides the factors needed to allow *Y. pestis* to be transmitted by fleas, *Y. pestis* also must produce virulence factors that allow rapid proliferation of the organism in deep tissues and precipitate the development of DIC. I will present the factors that are encoded by the pPla and pFra plasmids as they relate to the difference in the pathogenic properties of *Y. pestis* and *Y. pseudotuberculosis* in the sections below.

THE "MURINE TOXIN" PLASMID

The Murine Toxin plasmid (pFra) is the largest of the extrachromosomal elements harbored by *Y. pestis*. Typical strains harbor a molecule that is ~100 kb; however, the plasmid can vary in size from ~90 to 288 kb, especially in *Y. pestis* strains with atypical properties (Filippov et al., 1990). Atypical strains of *Y. pestis* isolated in states of the Former Soviet Union are defined by changes in the biochemical properties, plasmid profile and virulence in various hosts (Anisimov et al, 2004). Although the mechanism of variability has not been determined, it may be due to the activity of insertion sequence (IS) elements or the acquisition of new genetic material from natural sources (Parkhill et al., 2001; Hinnebusch et al., 2002a). This plasmid has been shown to integrate into the chromosome of both laboratory (Protsenko et al., 1991) and natural (Cavalcanti et al., 2002) *Y. pestis* isolates.

The largest plasmid of *Y. pestis* was generally referred to as the "cryptic plasmid" until it was discovered that both the Murine Toxin (Ymt) and fraction one (F1) capsular antigens were encoded by this plasmid (Protsenko et al., 1983). Two different designations have been used for this plasmid. In strain KIM, it has been specifically designated pMT1 (Perry and Fetherston, 1997). The largest *Y. pestis* plasmid has also been referred to generically as pFra. However, in some cases this has been used as a specific name as well (Prentice et al., 2001). Here, I will use the generic designation of pFra to refer to this molecule and note specific differences in isolates or strains where appropriate.

pFra AND VIRULENCE

As indicated above, pFra was initially thought to be important for the increased ability of *Y. pestis* to colonize host deep tissues from peripheral sites. The experimental evidence for this role in *Y. pestis* pathogenesis is not so clear. The animal model used has been shown to influence the amount of attenuation seen in pFra-negative strains. When isogenic strains of *Y. pestis* KIM devoid of pFra were examined in mice, no effect on virulence was detected (Brubaker, 1983). In contrast, complementary studies in guinea pigs using the same strain showed that loss of pFra resulted in an approximately ten thousand-fold decrease in virulence of *Y. pestis*. The influence that pFra has on virulence in guinea pigs has also been shown to vary according to the strain tested. *Y. pestis* strains 231 (biovar Antiqua) and 358 (biovar Medievalis) that were rendered pFra negative have both been found to have comparable virulence in guinea pigs with the wild type isogenic strains (Kutyrev et al., 1989; Drozdov et al., 1995). In contrast to the overall virulence of the organism as indicated by LD50 values, it is clear from all studies that the presence of pFra does influence the rapidity of death of the infected animal (Kutyrev et al., 1989; Drozdov et al., 1995; Welkos et al., 1995). Taken together, these results suggest that pFra has little effect on overall virulence of the organism in laboratory animal models. This must be reconciled with the fact that pFra is harbored at a high rate in typical strains isolated from natural foci indicating that there is selective pressure to maintain the plasmid in the wild. Thus, there may be a relationship between the extreme virulence of the organism and natural maintenance of endemic plague. Alternatively, there may be yet undefined factors similar to Ymt that are encoded by pFra and are necessary for natural transmission (Hinnebusch et al., 2002b; and see also Chapter 4).

There are two specific factors encoded by *Y. pestis* pFra that have at least been associated with properties of the disease caused by the organism. The first is the F1 capsular protein antigen. F1 has been shown to be antiphagocytic when produced by the organism. Recently, Du et al. (2002) demonstrated that F1-negative strains were phagocytosed ~25-30 percent more efficiently than F1-positive isogenic strains. The ability of F1 producing bacteria to remain extracellular was not influenced by prior opsonization. Given that F1-negative strains are virulent in mice (Drozdov et al., 1995; Welkos et al., 1995) and in monkeys (Davis et al., 1996), other factors encoded by *Y. pestis* may account or compensate for the loss of F1 in antiphagocytosis. One group of known factors that plays a major role in blocking uptake of *Y. pestis* into phagocytes is encoded by the "Yop" plasmid pLcr (Cornelis et al., 1998; Du et al., 2002), although there may be other factors yet to be defined.

The second factor encoded by pFra that has been characterized is the Ymt. At one time Ymt was thought to play an important role in pathogenesis because of the high toxicity it has for mice (Montie, 1981), hence its name. Virulence data did not support any role for the toxin in the mammalian host (Du et al., 1995; Hinnebusch et al., 2000). It has recently been shown that Ymt is actually a phospholipase involved in the survival of *Y. pestis* in the midgut of the flea vector (Hinnebusch et al., 2002b; and Chapter 4 of this book). Accordingly, Ymt is involved in the ability of *Y. pestis* to be transmitted by the insect vector. This activity would explain the selective pressure for the maintenance of pFra in the wild without having obvious effects in the virulence models discussed above. The fact that pFra-negative strains would not be transferred by the flea vector in the wild does not explain the reduced time to death observed in experimental animals. Thus, there are probably other as yet undefined factors encoded by *Y. pestis* pFra that may function in the mammalian host.

COMPLETE DNA SEQUENCES OF pFra

As indicated in Table 1, three versions of pFra have been sequenced. The first to be reported was from *Y. pestis* KIM (Hu et al., 1998) and was obtained from the Brubaker laboratory at Michigan State University. This sequencing group identified 81 potential open reading frames (ORFs). The second sequence to be reported a month later (Lindler et al., 1998) was also strain KIM but was obtained from the Perry laboratory at the University of Kentucky. The second group performed a more extensive analysis of the plasmid sequence and identified 115 potential coding regions within the pMT1 sequence. The differences in the number of ORFs identified by the two groups can be accounted for by differences in the analysis software used, the sizes of the coding regions considered as significant and subjective matters such as the decision to report sequences predicted to encode remnants of proteins as ORFs. *Y. pestis* KIM is of the Medievalis biotype. Sequencing of these two pMT1 molecules has allowed a comparison of the same plasmid from different sources, i.e. isolates of the same strain. The third pFra sequence to be reported was that of *Y. pestis* CO92 of the Orientalis biotype (Prentice et al., 2001). The CO92 pFra sequence was predicted to encode 103 ORFs. The reason for the size difference between the pMT1 plasmids and pFra from CO92 (Table 1) is a deletion near the F1 operon potentially involving IS elements (Prentice et al., 2001). This deletion in the CO92 pFra is the only major sequence difference noted between all of the pFra molecules that have been sequenced (Prentice et al., 2001). Most other differences were conservative single nucleotide changes or were in non-coding regions of the plasmid. In terms of comparative genomics, the number of polymorphisms noted for the two sequenced pMT1 molecules from strain KIM were as frequent as differences between one of the KIM pMT1 molecules and CO92 pFra (Prentice et al., 2001). A 63-bp deletion was noted between two 12-bp direct repeat sequences that resulted in the fusion of two pMT1 ORFs to form one coding region in CO92 pFra.

The following sections will describe the features discovered from these sequencing projects and, where appropriate, the differences noted between isolates and strains.

pFra HOMOLOGY WITH A *SALMONELLA* CRYPTIC PLASMID

From an evolutionary and emerging disease viewpoint, one of the remarkable discoveries made from the complete genome sequences of pFra was the fact that approximately half the plasmid is >90% homologous at the DNA level to the *Salmonella enterica* serovar Typhi cryptic plasmid, pHCM2 (Prentice et al., 2001). This finding was striking given the fact that *Salmonella* is an enteric pathogen and *Y. pestis* is an insect-borne disease. Interestingly, *Y. pestis* does have a very close relative in *Y. pseudotuberculosis* that is an enteropathogen as discussed above. It is easy to imagine how *Y. pestis* might have evolved from a *Y. pseudotuberculosis*-like organism. However, *Y. pestis* "lives" a relatively closed lifestyle, either in the flea or in the mammalian host. Therefore, the opportunity for genetic transfer would be expected to be limited. A possible environment for transfer of plasmid DNA was revealed when it was recently demonstrated that *Y. pestis* could serve as a recipient for a resistance transfer factor from *Escherichia coli* when fleas were co-infected with the two bacteria (Hinnebusch et al., 2002a). A similar mechanism could well explain the acquisition of antibiotic resistance by strains of *Y. pestis* circulating in Madagascar (Galimand et al., 1997; Guiyoule et al., 2001). Thus, the potential mechanism that may have contributed to the emergence of *Y. pestis* as a pathogen may still be involved in the development of new more dangerous strains of the pathogen.

Although it cannot be ascertained where pFra may have originated simply from the homology with pHCM2, some features that are common between the two plasmids may shed some light on this subject. The regions of homology between these two plasmids are concentrated between 0 - 40 kb, and 85 - 100 kb on pFra as shown in Figure 1. The origin of replication is common to both plasmids. The origin of replication is thought to be the most fundamental defining unit of a plasmid (Osborn et al., 2000). The fact that pHCM2 and pFra share the same origin of replication clearly suggests that they share a common ancestral molecule rather than just common DNA segments acquired by chance. This fact does not, however, clarify the source or direction of transfer of pFra. The ancestor of pFra may have come directly from pHCM2 or vioe versa, or possibly from a third molecule. A second common feature shared by pFra and pHCM2 that is easily recognizable is the presence of a remnant of a lambdoid phage (Figure 1). The pFra molecule encodes a contiguous ~5 kb larger segment of the phage remnant than does the *Salmonella* plasmid suggesting that the *Yersinia* plasmid may have arisen first. The phage remnant stops at a copy of IS*100* that is not present on pHCM2. In fact, none of the IS elements present on pFra are shared with the *Salmonella* plasmid. The final region with recognizable features that is common to the two plasmids is an ~10 kb region that encodes a phage integrase, two genes involved in vitamin B12 synthesis (*cobT cobS*) and a DNA Polymerase III-like protein (Figure 1). All of these features demonstrate phage-mediated genetic exchange

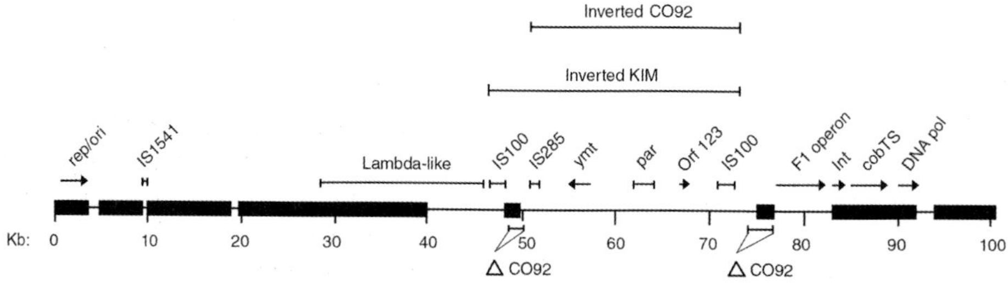

Figure 1. Map of *Y. pestis* KIM pFra. This map was derived from GenBank accession number AF074611. Darkened and wide areas on the map represent areas shared by the *S. enterica* serovar Typhi pHCM2 plasmid at greater than 94% DNA identity. Deletions noted in the *Y. pestis* CO92 sequence relative to the strain KIM sequence are shown below the map. Major features and ORFs are shown above the map. The direction of transcription is indicated by the arrows. The regions between the two copies of IS*100* that were inverted in the two KIM sequences is labeled. The region within the strain CO92 pFra sequence between one of the IS*100* and IS*285* elements with inverted gene order is also shown. Abbreviations: rep/ori, origin of replication; ymt, Murine Toxin gene; par, partitioning locus; ORF123, adenine-specific DNA methylase gene; int, phage integrase gene; DNA pol, DNA Polymerase III-like gene.

Table 2. Genes identified on pMT1 (AF074611) that are associated with DNA metabolism.

ORF[a]	Comments
Y1095 (ORF2)	Bacteriophage T3 DNA ligase
Y1101 (ORF12)	Phage integrase similar to *Vibrio cholerae*
Y1106 (ORF16)	*E. coli* DNA Polymerase III
Y1111(ORF23)	Remnant of DNA Polymerase I similar to *Streptococcus*
Y1113 (ORF26)	RecA of *Bacteroides*
RepA (ORF34)	Similar to *Salmonella enterica* serovar Cholerasuis plasmid replication protein
Y1006 (ORF41)	Bacteriophage T4 exonuclease g47
Y1008 (ORF43)	Bacteriophage T4 exonuclease g46
Y1023 (ORF60)	Plasmid partitioning (ParB) *Vibrio*
Y1060 (ORF99)	Remnant of UvrC-like protein
Y1063 (ORF102)	Probable transposase
Y1064 (ORF103)	Remnant of transposase
Y1065) (ORF103a)	Remnant of IS*600* transposase
Y1068 (ORF106)	Remnant of bacterial reverse transcriptase
(ORF106a)	Remnant of IS*801* transposase
Y1070 (ORF108)	Probable membrane endonuclease
Y1073 (ORF111)	Site specific recombinase/resolvase *Shewanella oneidensis*
ParA (ORF113)	Similar to the partitioning protein A of bacteriophage P7
ParB (ORF114)	Similar to the partitioning protein B of bacteriophage P7
Y1085 (ORF123)	Bacteriophage P1 methylase
Y1090 (ORF128)	Plasmid anti-restriction protein from *E. coli*
Y1092 (ORF135)	Probable chromosome partitioning protein

[a]ORF designations in parenthesis match those reported previously (Lindler et al., 1998). The list does not include known IS elements. Underlined ORFs are also present in *Salmonella* pHCM2.

that includes movement of putative chromosomal genes (*cob* and *polIII*) onto a common precursor plasmid that now is harbored by *Y. pestis* and *Salmonella*. Regions that are not in common between pFra and pHCM2 carry the *Y. pestis*-specific genes *ymt* and the *caf1* locus (Figure 1), demonstrating the acquisition of unique properties that contribute to the life cycle of this flea-borne pathogen.

GENETIC ARCHITECTURE

The *Y. pestis* murine toxin plasmid is a patchwork assembled from many mobile genetic elements. Besides the ones discussed in the previous section, many other phage and plasmid-associated ORFs make up this molecule. In many cases, the genes from these elements that are involved in DNA metabolism, such as replication, DNA repair or integration, are still present on pFra. A list of genes involved in DNA reactions that were identified on pMT1 (Lindler et al., 1998) is presented in Table 2. One can see from this list that many mobile elements have contributed to the mosaic nature of this plasmid. In many cases, only a portion of the functional group of genes remains. For example, ORF60 and ORF135 encode factors homologous to partitioning proteins, but no other accessory proteins or *cis*-acting elements necessary for DNA partitioning are linked to these putative coding regions. Similarly, several regions encode proteins homologous to specific bacteriophage proteins. These regions are dispersed throughout the pFra sequence.

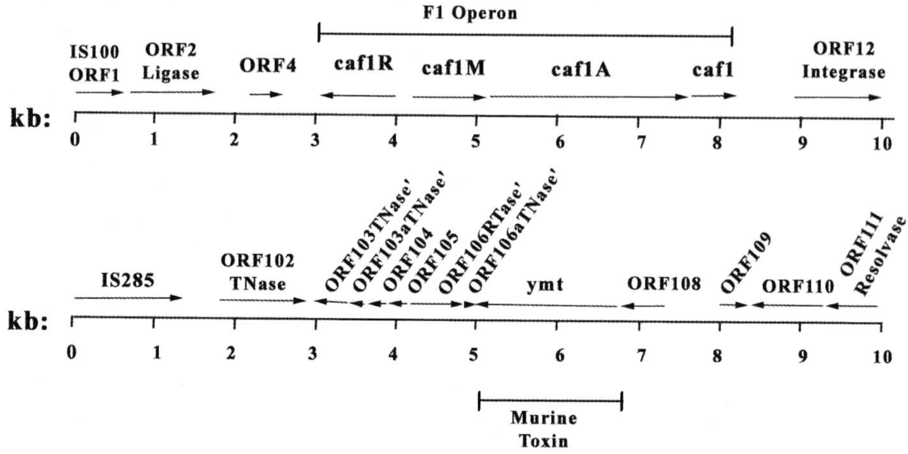

Figure 2. Genes linked to known virulence-associated loci of pFra. The top panel shows the F1 capsular locus and surrounding ORFs. Phage-associated ligase and integrase are located on either side of the F1 gene cluster. The bottom panel shows the Murine Toxin gene (*ymt*) and surrounding ORFs. This region is marked with many IS-associated genes including remnants designated with a "prime" at the end of the gene name. This region also includes a plasmid associated resolvase (ORF111). The identified transposase (TNase) and reverse transcriptase (RTase) are shown.

The general features described in the above paragraph pertain to the overall properties of the plasmid. The activity of mobile genetic elements is especially concentrated near the two known virulence-associated factors encoded by pFra (Lindler et al., 1998) as shown in Figure 2. Specifically, the region flanking the *ymt* locus contains remnants of multiple elements (Figure 2, lower panel). These include short remnants of IS*600*, IS*801* and a bacterial retron element downstream of *ymt* itself. Upstream of *ymt* (Table 2) there is a remnant of a partitioning locus including the resolvase (ORF111) as well as a plasmid-encoded endonuclease (ORF108). The region of pFra that flanks the F1 capsular locus also contains elements associated with bacterial mobile DNA (Figure 2, upper part panel). A bacteriophage T3-like ligase (ORF2) and a phage-associated ligase (ORF12) are located on either side of the *caf1* operon (Table 2). In support of the concept that many virulence factors such as pathogenicity islands or toxins are acquired by horizontal gene transfer (Kaper and Hacker, 1999), both the *ymt* and *caf1* regions have significantly lower guanine plus cytosine ratios than the rest of pFra (Lindler et al., 1998). The elements near these two regions are not present on the *Salmonella* plasmid pHCM2. Thus, these *Y. pestis*-unique regions were assembled through multiple mobile genetic events, added to the plasmid profile and contributed to the emergence of bubonic/pneumonic plague as a disease.

The operon involved in the synthesis of the F1 capsule warrants some discussion in relation to molecular architecture of pFra. Although the F1 capsule has never been shown to be involved in adherence to host cells, the genes involved in synthesis of this protein show a high degree of identity to genes involved in pilus secretion and assembly (Galyov et al., 1990; Galyov et al., 1991; Karlyshev et al., 1992). These include two genes coding for a membrane anchor (*caf1A*) and a chaperone protein (*caf1M*) that are both upstream of *caf1* (the F1 structural gene) in an operon. Finally, the gene encoding the transcriptional regulator, *caf1R*, is transcribed from a divergent promoter. Caf1R is a member of the

AraC-family of regulatory proteins. For more details about the F1 system see Chapter 18. Thus, the organization of the F1 operon is similar to pilus biosynthetic operons and is situated within the larger mosaic of the Murine Toxin plasmid.

The most obvious group of genes on pFra that resembles a known mobile element capable of gene transfer is similar to phage lambda (Figure 1). The contiguous cluster of genes is ~17.7 kb on the pFra molecule and includes coding regions for tail proteins including those involved in tail assembly. These genes are in the same order as on phage lambda itself. There appears to have been a duplication of the lambda V gene such that a second copy, designated ORF74a, is located outside the main cluster of lambda-like genes. This protein has a significant amount of amino acid similarity to several proteins predicted to be involved in cell adhesion and therefore may be an example of a virulence factor derived from a phage protein, although this remains to be tested. Five other predicted proteins that might play a role in *Y. pestis* pathogenesis were also identified as part of the genomic sequencing effort (Lindler et al., 1998). A portion of the cluster of lambda-like genes is also found on the *Salmonella* plasmid pHCM2, including ORF74a.

The murine toxin gene and the partitioning locus of pMT1 lie between two copies of IS*100* in *Y. pestis* KIM (Figure 1). This region was found to be exactly inverted when the sequences derived from the two different KIM isolates were examined (Hu et al., 1998; Lindler et al., 1998). These copies of IS*100* were flanked by 5 bp direct repeats in the pMT1 sequence derived from the Brubaker laboratory source (Hu et al., 1998) but not in the sequence derived from the Perry laboratory (Lindler et al., 1998). The reason for this is unclear. Part of this region is also inverted in *Y. pestis* CO92 pFra relative to one of the KIM sequences (Prentice et al., 2001). However, this is not a simple inversion between the two copies of IS*100*, but rather an inversion of gene order between one copy of this IS and the single copy of IS*285* as shown in Figure 1. There has also been movement of one copy of IS*100* in the pFra molecule harbored by strain CO92 such that the DNA Polymerase III gene is interrupted. These illegitimate recombination events have also generated a deletion of DNA in pFra that appears to be specific to the Orientalis biotype (Prentice et al., 2001). Taken together, these observations demonstrate the fluid nature of pFra within strains and biotypes of *Y. pestis* and is consistent with the general genomic fluidity noted for this organism (Parkhill et al., 2001; Deng et al., 2002).

REPLICATION AND PARTITIONING FUNCTIONS

The pFra replication and partitioning regions were identified within the pMT1 sequence based on protein and DNA sequence homologies with systems previously defined in other plasmids (Lindler et al., 1998). The predicted replication initiation protein, RepA, displayed the highest similarity with the *repA* gene product of the RepFIB replication system. Both upstream and downstream of the pMT1 *repA* are located 19-bp direct and inverted repeated DNA sequences similar to other iteron-based replication control elements (DelSolar et al., 1998). These repeated sequences share the highest level of similarity with the RepHI1B replicons. Thus, the origin of replication of pFra may represent a hybrid or diverged element with characteristics of two different systems; the RepFIB and RepHI1B origins. Hybrid origins of replication may promote plasmid compatibility and aid maintenance of the molecule in nature (Thomas, 2000). Other DNA sequence elements near the *repA* protein coding region include two AT-rich regions, DnaA boxes and Dam methylation sites. This region was designated as the origin of replication for pFra since it was the only one identified with all of the necessary coding sequences that included a replication initiation protein and *cis*-acting DNA sequences known to be necessary for plasmid propagation. A similar region was also found on the *S. enterica* serovar Typhi plasmid (Prentice et al., 2001) as shown in Figure 1.

The partitioning region of pFra is most similar to the *parABS* sequence of bacteriophage P1 and P7 (Hu et al., 1998; Lindler et al., 1998; Youngren et al., 2000). The highest level of identity was with the ParA (90%) and ParB (67%) proteins of P7. The *parS* site is a *cis*-acting element involved in ParB binding during partitioning (Davis and Austin, 1988). The pMT1 molecule encoded all of the hexamer and heptamer boxes with the proper spacing to potentially function as a *parS* sequence (Lindler et al., 1998). Youngren et al. (2000) showed that the *parABS* pMT1 sequences could function in plasmid partitioning. These investigators also demonstrated that the P1 and P7 sequences could not substitute for the pMT1 sequences, nor could the pMT1 *parABS* sequences be used by P1 or P7 to support replication. Partitioning of either plasmid was also not inhibited by the presence of the other partitioning loci, i.e. the partitioning systems of P1 and P7 were compatible with that of pMT1. Accordingly, although the levels of protein and DNA similarities are high between P1, P7 and pMT1, the sequences have diverged enough so that the partitioning systems were compatible.

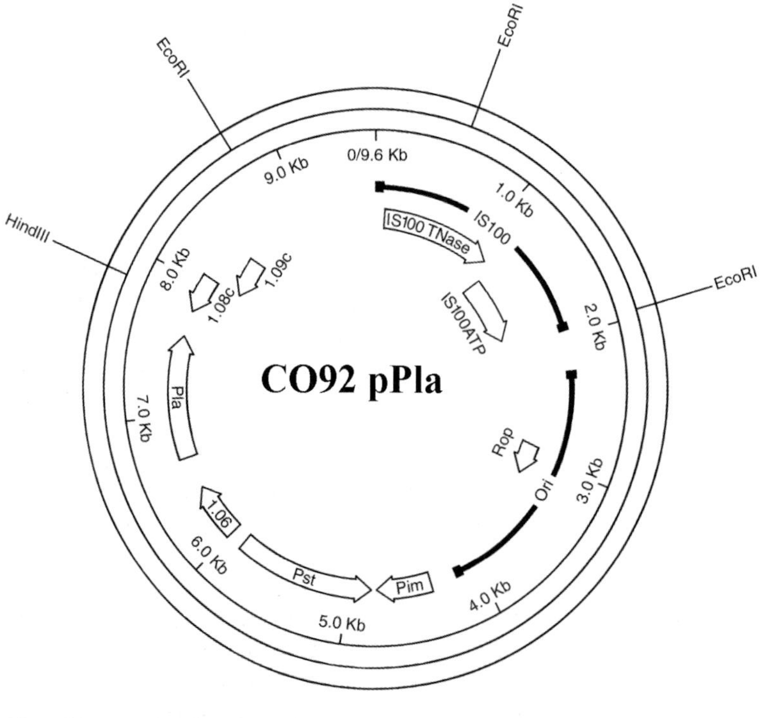

Figure 3. Map of *Y. pestis* CO92 pPla. Individual genes are indicated by the placement and direction of the arrows. The origin of replication (*ori*) that is similar to ColE1 is indicated inside the map. Genes for replication regulatory protein (*rop*), pesticin (*pst*), pesticin immunity (*pim*), plasminogen activator (*pla*), IS*100* transposase (TNase)/ATP binding protein as well as putative ORFs are labeled. Restriction sites for *Eco*RI and *Hin*dIII are shown outside the circular map.

THE "PESTICIN PLASMID"

The smallest plasmid harbored by typical strains of *Y. pestis* is generically referred to as the "pesticin plasmid", or pPla for plasminogen activator. The plasmid harbored by *Y. pestis* KIM has been designated pPCP1 (Sodeinde and Goguen, 1988). The pPla molecule is typically ~9.6-kb in size, although some strains within the United States harbor a dimer of this element (Chu et al., 1998). The complete sequence and analysis of pPla from *Y. pestis* strains KIM (Hu et al., 1998) and CO92 (Prentice et al., 2001) have been reported. A map of *Y. pestis* CO92 pPla is shown in Figure 3. I will present information on the functions and contribution to virulence of the pPla-encoded factors, and the complete sequence analysis of this plasmid in the following sections of this chapter. For detailed information on the activity of Pla see Chapter 17.

ROLE IN PATHOGENESIS AND VIRULENCE

As indicated above for the pFra plasmid, the fact that the small 9.6 kb pPla molecule appeared to be *Y. pestis*-specific suggested that it could be involved in causing severe disease by this pathogen (Brubaker, 1983; Brubaker, 1991). *Y. pestis* bacteriocin (pesticin; Pst) and fibinolysin/coagulase activities were shown to be associated with each other and thought to be located on a plasmid (Brubaker et al., 1965; Beesley et al., 1967). In 1973, it was shown that Pst and fibinolysin activities could be mobilized from one *Y. pestis* strain to another, suggesting that these activities were indeed encoded on a plasmid (Koltsova et al., 1973). Finally, the plasmid encoding the Pst activity was identified by isolation of isogenic Pst-negative strains and analysis of their plasmid content (Ben-Gurion and Shafferman, 1981; Ferber and Brubaker, 1981).

Brubaker et al. (1965) demonstrated that a pPla-negative strain of *Y. pestis* was attenuated by the intraperitoneal and subcutaneous routes of infection in mice. In contrast, this strain was found to be near wild type in virulence by the intravenous route. Thus, lethality of pPla-cured strains of *Y. pestis* introduced intravenously was similar to that of *Y. pseudotuberculosis* in mice, supporting the idea that this plasmid aided the former organism in invading the host from peripheral sites. Later, Sodeinde et al. (1992) studied defined plasminogen activator (*pla*) mutants in mice. The plasminogen activator activity was also referred to as the fibrinolysin/coagulase activity (McDonough et al., 1988; Sodeinde and Goguen, 1988). These mutants had a one million-fold reduction of virulence by the subcutaneous route of infection in mice and were found to induce a higher inflammatory response near the site of infection. *Y. pestis* KIM *pla*-negative strains were unable to reach the liver and spleen in numbers comparable to the that of the isogenic wild type parent strain (Sodeinde and Goguen, 1988). Similar observations have also been made for strain CO92 (Welkos et al., 1997). In contrast, some strains of *Y. pestis* do not require pPla for virulence. *Y. pestis* strains 358 (biotype Medievalis) and Pestoides F (biovar Antiqua) lacking pPla displayed subcutaneous LD50 values on the order of 3-4 CFU in mice (Welkos et al., 1997). Strain 358, which lacks pPla, has also been shown to be fully virulent in guinea pigs (Samoilova et al., 1996). Taken together, these observations suggest that Pla is necessary for the virulence of some strains of *Y. pestis* but that others may have factors that can compensate for this activity. It is clear that in strains requiring *pla* for full virulence, this activity is needed for deep tissue invasion. The addition of Pla activity to *Y. pseudotuberculosis* does not render that organism able to invade its host from peripheral routes, however (Kutyrev et al., 1999). Accordingly, there are other factors that contribute to the higher virulence of *Y. pestis* besides Pla. Furthermore, there appear to be some strains of *Y. pestis* that do not require Pla activity at all for virulence, further expanding the number of virulence factors involved in the high pathogenicity seen in this species.

Pla ACTIVITIES

PLASMINOGEN ACTIVATOR

Y. pestis Pla has been shown to cleave human plasminogen at the same place as human urokinase, a major plasminogen activator in the host, to yield active plasmin (Sodeinde et al., 1992). These studies confirmed that the major activity of Pla was as a plasminogen activator, as suggested years earlier by Beesley et al. (1967). The *Y. pestis* Pla protein is one of the only bacterial plasminogen activators that has been studied using an *in vivo* animal model (Sodeinde et al., 1992; Lahteenmaki et al., 2001). The observation that *pla*-negative mutants of *Y. pestis* cannot invade the liver and spleen from peripheral sites, along with the *in vitro* enzymatic activity of the protein suggest that this virulence factor may promote disease by degradation of the extracellular matrix, thus promoting migration within the host. Plasminogen activation to plasmin may also interfere with confinement of the bacteria by decreasing fibrin deposition at the site of infection. Pla has also been shown to cleave complement factor C3 (Sodeinde et al., 1992). This activity could reduce the chemoattractant signals for polymorphonuclear leukocytes elicited by the alternate complement pathway following infection, and explain the lack of these immune cells near the site of infection by *pla*-positive *Y. pestis* (Sodeinde et al., 1992). However, Pla is not necessary for complement resistance by this organism. These observations demonstrate that Pla may play a role in the ability of *Y. pestis* to establish infection from distal sites and to then migrate to deeper tissues and organs.

COAGULASE

Besides plasminogen activator activity, *Y. pestis* Pla also displays coagulase activity (Sodeinde and Goguen, 1988; McDonough and Falkow, 1989). Pla coagulase enzyme activity is maximal when incubated at 26°C, whereas plasminogen activator activity is maximal at 37°C. This suggested that Pla might have a function in the flea portion of the life cycle. In fact, *pla*-positive *Y. pestis* were found to cause an ~two-fold higher death rate in fleas compared to their mutant counterparts (McDonough et al., 1993). The increased mortality of fleas infected with Pla expressing bacteria was not due to increased bacterial load. The mechanism for the enhanced ability of Pla-expressing bacteria to cause the death of infected fleas could not be determined. However, from these studies it is clear that Pla activity does have an adverse effect on the plague bacillus-flea interaction and these data support the notion that coagulase activity might be involved as well (Cavanaugh, 1971). pPla was hypothesized to play a role in producing the foregut blockage in the flea vector that precedes transmission. However, *Y. pestis* strains cured of pPla infected and blocked fleas normally. Thus, procoagulant ability of Pla did not mediate blockage, nor did its ability to induce fibrinolysis at >28°C account for failure to block at elevated temperatures (Hinnebusch et al., 1998).

PROTEASE

Y. pestis Pla has been found to have other activities that might be related to virulence in the mammalian host. One of them is the proteolytic cleavage (Sodeinde et al., 1988) of the pYV-encoded virulence factors known as the Yops (*Yersinia* Outer Proteins). The role that the degradation of these proteins might play in pathogenesis is unknown although it can be speculated that destruction of excessed Yops might aid the organism in immune avoidance during deep tissue invasion.

ADHESIN-INVASIN

An association between Pla activity and the ability of *Y. pestis* to attach to the mammalian extracellular matrix has been observed (Lahteenmaki et al., 1998). Bacterial adherence to the extracellular matrix may generally promote the colonization of host tissues by pathogens, especially at wounds. Furthermore, pPla is essential for highly efficient invasion in some strains of *Y. pestis*. It encodes an invasion-promoting factor and not just an adhesin, because *Y. pestis* lacking this plasmid still adheres to HeLa cells (Cowan et al., 2000).

DNA SEQUENCE AND ANALYSIS

Sodeinde and Goguen (1988) first presented genetic evidence that a single gene encoded both the plasminogen activator and coagulase activities within the pPCP1 sequence. This was accomplished through a combination of transposon mutagenesis, cloning, and protein expression analysis. They also showed that pPCP1 replication was PolA-dependent as is characteristic of ColE1-like replicons. One year later, two groups reported the sequence of the gene encoding Pla (McDonough and Falkow, 1989; Sodeinde and Goguen, 1989). *Y. pestis* Pla was found to have sequence similarity with other enteric outer membrane proteins, specifically OmpT and Gene E (Sodeinde and Goguen, 1989). Pla is now known to be part of the omptin family of proteases that includes eight other proteins (MEROPS protease database at: http://www.merops.ac.uk/). The discussion and relevance of these similarities can be found in Chapter 17.

Two complete sequences of the *Y. pestis* pPla molecule have been reported. One sequence was derived from pPCP1 of strain KIM (Hu et al., 1998) and the other was derived for the strain CO92 plasmid (Prentice et al., 2001). A complete map of the CO92 plasmid is shown in Figure 3. The CO92 pPla molecule is predicted to encode nine ORFs. These include the genes encoding Pla, pesticin (Pst), pesticin immunity protein (Pim), a putative replication regulatory protein (Rop), IS*100*-associated protein, and three hypothetical proteins (Figure 3). The Pst and Pim proteins were predicted to be transcribed in a convergent manner and are linked on the pPla sequence. A 1252-bp region of pPla was found to be similar to the ColE1 origin of replication (Hu et al., 1998; Prentice et al., 2001), in agreement with the finding that replication of this plasmid is dependent on a functional *polA* (Sodeinde and Goguen, 1988).

The *Y. pestis* KIM and CO92 sequences were found to be almost identical. There were two single nucleotide differences and a two bp insertion in the CO92 sequence compared to the KIM sequence (Prentice et al., 2001). However, all of these changes were in polymeric nucleotide tracts that occur between coding regions and therefore did not change the predicted protein sequences. One interesting feature near the single copy of IS*100* was noted on pPla. The insertion of this element included the typical five bp direct repeat on either side of the IS. On one side of the insertion a five bp sequence was identical to one of the IS*100* insertion sites in pFra, suggesting interplasmid recombination before the divergence of the Medievalis (strain KIM) and Orientalis (strain CO92) biotypes.

CONCLUSION

Plasmids play an important part in the pathogenesis of *Yersinia* infection in general and in the ability to cause a unique disease by *Y. pestis*. Genomic sequencing efforts have revealed that there was likely an association between the progenitor of *Y. pestis* and insects (Parkhill et al., 2001). The addition of plasmids to this ancestral "platform" created a pathogen that could then infect an insect vector and use this vector to propagate by transmission to a warm-blooded host. The pFra molecule encodes proteins that certainly promote the transmission of *Y. pestis* by fleas (Ymt) as well as protect the organism from destruction by professional phagocytes in the mammalian host (F1). Similarly, pPla encodes a protein that has functions in both the vertebrate and invertebrate hosts. The understanding of genes carried by these two *Y. pestis*-specific plasmids has added to our understanding of the emergence of this important disease. Future endeavors to identify and characterize the function of other genes that are unique to *Y. pestis* as well as to individual groups within this species in general will certainly yield invaluable lessons in the field of pathogenesis of infectious disease as well as emerging diseases.

ACKNOWLEDGEMENTS

The author would like to thank all of the laboratory workers in his lab who have been engaged in Biodefense research for the past five years. I appreciate the excellent graphics work of Lee Collins and Patricia Stroy. The support of the US Army Medical Chemical and Biological Defense program is gratefully acknowledged. The views presented here do not reflect those of the US Army, of the Department of Defense or Department of Homeland Security.

REFERENCES

Anisimov, A.P., Lindler, L.E. and Pier, G. B., 2004. Interspecific diversity of *Yersinia pestis*. Clinical Microbiol. Rev. 17: 434-464.

Bercovier, H., Mollaret, H.H., Alonso, J.M., Brault, J., Fanning, G.R., Steigerwalt, A., and Brenner, D.J., 1980. Intra- and Interspecies relatedness of *Yersinia pestis* by DNA hybridization and its relationship to *Yersinia pseudotuberculosis*. Curr. Microbiol. 4: 225-229.

Beesley, E.D., Brubaker, R.R., Janssen, W.A., and Surgalla, M.J. 1967. Pesticins III. Expression of coagulase and mechanisms of fibrinolysis. J. Bacteriol. 94: 19-26.

Ben-Gurion, R. and Shafferman, A. 1981. Essential virulence determinants of different *Yersinia* species are carried on a common plasmid. Plasmid. 5: 183-187.

Brubaker, R.R. 1983. The Vwa+ virulence factor of yersiniae: the molecular basis of the attendant nutritional requirement for Ca++. Rev. Infect. Dis. 5: S748-S758.

Brubaker, R.R. 1991. Factors promoting acute and chronic diseases caused by yersiniae. Clin. Microbiol. Rev. 4: 309-24.

Brubaker, R.R., Beesley, E.D., and Surgalla, M.J. 1965. *Pasteurella pestis*: role of pesticin I and iron in experimental plague. Science. 149: 422-424.

Butler, T. 1983. Plague and other *Yersinia* infections. Plenum Medical Book Co., New York.

Cavalcanti, Y.V., Leal, N.C., and De Almeida, A.M. 2002. Typing of *Yersinia pestis* isolates from the state of Ceara, Brazil. Lett. Appl. Microbiol. 35: 543-547.

Cavanaugh, D.C. 1971. Specific effect of temperature upon transmssion of the plague bacillus by the Oriental rat flea *Xenopsylla cheopis*. Am. J. Trop. Med. Hyg. 20: 264-273.

Cavanaugh, D.C., and Randall, R. 1959. The role of multiplication of *Pasteurella pestis* in mononuclear phagocytes in the pathogenesis of flea-borne plague. J. Immunol. 83: 348-363.

Charnetzky, W.T., and Shuford, W.W. 1985. Survival and growth of *Yersinia pestis* within macrophages and an effect of the loss of the 47-megadalton plasmid on growth in macrophages. Infect. Immun. 47: 234-241.

Chu, M.C., Dong, X.Q., Zhou, X., and Garon, C.F. 1998. A cryptic 19-kilobase plasmid associated with U.S. isolates of *Yersinia pestis*: a dimer of the 9.5-kilobase plasmid. Am. J. Trop. Med. Hyg. 59: 679-686.

Cornelis, G.R., Boland, A., Boyd, A.P., Geuijen, C., Iriarte, M., Neyt, C., Sory, M.P., and Stainier, I. 1998. The virulence plasmid of *Yersinia*, an antihost genome. Microbiol. Mol. Biol. Rev. 62: 1315-1352.

Cowan, C., Jones, H.A., Kaya, Y.H., Perry, R.D., and Straley, S.C. 2000. Invasion of epithelial cells by *Yersinia pestis*: Evidence for a *Y. pestis*-specific invasin. Infect. Immun. 68: 4523-4530.

Davis, K.J., Fritz, D.L., Pitt, M.L., Welkos, S.L., Worsham, P.L., and Friedlander, A.M. 1996. Pathology of experimental pneumonic plague produced by fraction 1- positive and fraction 1-negative *Yersinia pestis* in African green monkeys (*Cercopithecus aethiops*). Arch. Pathol. Lab. Med. 120: 156-163.

Davis, M.A., and Austin, S.J. 1988. Recognition of the P1 plasmid centromere analog involves binding of the ParB protein and is modified by a specific host factor. EMBO J. 7: 1881-1888.

DelSolar, G., Giraldo, R., Ruiz-Echevarria, M.J., Espinosa, M., and Diaz-Orejas, R. 1998. Replication and control of circular bacterial plasmids. Micribiol. Mol. Biol. Rev. 62: 434-464.

Deng, W., Burland, V., Plunkett, G., Boutin, A., Mayhew, G.F., Liss, P., Perna, N.T., Rose, D.J., Mau, B., Schwartz, D.C., Zhou, S., Fetherston, J.D., Lindler, L.E., Brubaker, R.R., Plano, G.V., Straley, S.C., McDonough, K.A., Nilles, M.L., Matson, J.S., Blattner, F.R., and Perry, R.D. 2002. Genome sequence of *Yersinia pestis* KIM. J. Bacteriol. 184: 4601-4611.

Drozdov, I.G., Anisimov, A.P., Samoilova, S.V., Yezhov, I.N., Yeremin, S.A., Karlyshev, A.V., Krasilnikova, V.M., and Kravchenko, V.I. 1995. Virulent non-capsulate *Yersinia pestis* variants constructed by insertion mutagenesis. J. Med. Microbiol. 42: 264-268.

Du, Y., Galyov, E., and Forsberg, A. 1995. Genetic analysis of virulence determinants unique to *Yersinia pestis*. Contrib. Microbiol. Immunol. 13: 321-324.

Du, Y., Rosqvist, R., and Forsberg, A. 2002. Role of Fraction 1 Antigen of *Yersinia pestis* in Inhibition of Phagocytosis. Infect. Immun. 70: 1453-1460.

Ferber, D.M., and Brubaker, R.R. 1981. Plasmids in *Yersinia pestis*. Infect. Immun. 31: 839-841.

Filippov, A.A., Solodovnikov, N.S., Kookleva, L.M., and Protsenko, O.A. 1990. Plasmid content in *Yersinia pestis* strains of different origin. FEMS Microbiol. Lett. 55: 45-48.

Galimand, M., Guiyoule, A., Gerbaud, G., Rasoamanana, B., Chanteau, S., Carniel, E., and Courvalin, P. 1997. Multidrug resistance in *Yersinia pestis* mediated by a transferable plasmid. N. Engl. J. Med. 337: 677-680.

Galyov, E.E., Karlishev, A.V., Chernovskaya, T.V., Dolgikh, D.A., Smirnov, O., Volkovoy, K.I., Abramov, V.M., and Zav'yalov, V.P. 1991. Expression of the envelope antigen F1 of *Yersinia pestis* is mediated by the product of caf1M gene having homology with the chaperone protein PapD of *Escherichia coli*. FEBS Lett. 286: 79-82.

Galyov, E.E., Smirnov, O., Karlishev, A.V., Volkovoy, K.I., Denesyuk, A.I., Nazimov, I.V., Rubtsov, K.S., Abramov, V.M., Dalvadyanz, S.M., and Zav'yalov, V.P. 1990. Nucleotide sequence of the *Yersinia pestis* gene encoding F1 antigen and the primary structure of the protein. Putative T and B cell epitopes. FEBS Lett. 277: 230-232.

Goguen, J.D., Walker, W.S., Hatch, T.P., and Yother, J. 1986. Plasmid-determined cytotoxicity in *Yersinia pestis* and *Yersinia pseudotuberculosis*. Infect. Immun. 51: 788-794.

Guiyoule, A., Gerbaud, G., Buchrieser, C., Galimand, M., Rahalison, L., Chanteau, S., Courvalin, P., and Carniel, E. 2001. Transferable plasmid-mediated resistance to streptomycin in a clinical isolate of *Yersinia pestis*. Emerg. Infect. Dis. 7: 43-48.

Hinnebusch, B.J., Fischer, E.R., and Schwan, T.G. 1998. Evaluation of the role of the *Yersinia pestis* plasminogen activator and other plasmid-encoded factors in temperature-dependent blockage of the flea. J. Infect. Dis. 178: 1406-1415.

Hinnebusch, B.J., Rosso, M.L., Schwan, T.G., and Carniel, E. 2002a. High-frequency conjugative transfer of antibiotic resistance genes to *Yersinia pestis* in the flea midgut. Mol. Microbiol. 46: 349-354.

Hinnebusch, B.J., Rudolph, A.E., Cherepanov, P., Dixon, J.E., Schwan, T.G., and Forsberg, A. 2002b. Role of *Yersinia* murine toxin in survival of *Yersinia pestis* in the midgut of the flea vector. Science. 296: 733-735.

Hinnebusch, J., Cherepanov, P., Du, Y., Rudolph, A., Dixon, J.D., Schwan, T., and Forsberg, A. 2000. Murine toxin of *Yersinia pestis* shows phospholipase D activity but is not required for virulence in mice. Int. J. Med. Microbiol. 290: 483-487.

Hu, P., Elliott, J., McCready, P., Skowronski, E., Garnes, J., Kobayashi, A., Brubaker, R.R., and Garcia, E. 1998. Structural organization of virulence-associated plasmids of *Yersinia pestis*. J. Bacteriol. 180: 5192-5202.

Kaper, J.B., and Hacker, J. 1999. Pathogenicity islands and other mobile virulence elements. American Society for Microbiology press, Washington, DC.

Karlyshev, A.V., Galyov, E.E., Abramov, V.M., and Zav'yalov, V.P. 1992. Caf1R gene and its role in the regulation of capsule formation of *Y. pestis*. FEBS Lett. 305: 37-40.

Koltsova, E.G., Suchkov, Y.G., and Legedeva, S.A. 1973. Transmission of a bacteriocinogenic factor in *Pasteurella pestis*. Sov. Genet. 7: 507-510.

Kutyrev, V., Mehigh, R.J., Motin, V.L., Pokrovskaya, M.S., Smirnov, G.B., and Brubaker, R.R. 1999. Expression of the plague plasminogen activator in *Yersinia pseudotuberculosis* and *Escherichia coli*. Infect. Immun. 67: 1359-1367.

Kutyrev, V.V., Filippov, A.A., Shavina, N., and Protsenko, O.A. 1989. Genetic analysis and simulation of the virulence of *Yersinia pestis*. Mol. Gen. Mikrobiol. Virusol. 42-47.

Lahteenmaki, K., Kuusela, P., and Korhonen, T.K. 2001. Bacterial plasminogen activators and receptors. FEMS Microbiol. Rev. 25: 531-552.

Lahteenmaki, K., Virkola, R., Saren, A., Emody, L., and Korhonen, T.K. 1998. Expression of plasminogen activator Pla of *Yersinia pestis* enhances bacterial attachment to the mammalian extracellular matrix. Infect. Immun. 66: 5755-5762.

Lindler, L.E., Plano, G.V., Burland, V., Mayhew, G.F., and Blattner, F.R. 1998. Complete DNA sequence and detailed analysis of the *Yersinia pestis* KIM5 plasmid encoding murine toxin and capsular antigen. Infect. Immun. 66: 5731-5742.

Lindler, L.E., and Tall, B.D. 1993. *Yersinia pestis* pH 6 antigen forms fimbriae and is induced by intracellular association with macrophages. Mol. Microbiol. 8: 311-324.

Makoveichuk, E., Cherepanov, P., Lundberg, S., Forsberg, A., and Olivecrona, G. 2003. pH6 antigen of *Yersinia pestis* interacts with plasma lipoproteins and cell membranes. J. Lipid Res. 44: 320-330.

McDonough, K.A., Barnes, A.M., Quan, T.J., Montenieri, J., and Falkow, S. 1993. Mutation in the *pla* gene of *Yersinia pestis* alters the course of the plague bacillus-flea (*Siphonaptera: Ceratophyllidae*) interaction. J. Med. Entomol. 30: 772-780.

McDonough, K.A., and Falkow, S. 1989. A *Yersinia pestis*-specific DNA fragment encodes temperature-dependent coagulase and fibrinolysin-associated phenotypes. Mol. Microbiol. 3: 767-775.

McDonough, K.A., Schwan, T.G., Thomas, R.E., and Falkow, S. 1988. Identification of a *Yersinia pestis*-specific DNA probe with potential for use in plague surveillance. J. Clin. Microbiol. 26: 2515-2519.

Montie, T.C. 1981. Properties and pharmacological action of plague murine toxin. Pharmacol. Ther. 12: 491-499.

Moore, R.L., and Brubaker, R.R. 1975. Hybridization and deoxyribonucleotide sequences of *Yersinia enterocolitica* and other selected members of *Enterobacteriaceae*. Int. J. Syst. Bacteriol. 25: 336-339.

Osborn, M., Bron, S., Firth, N., Holsappel, S., Huddleston, A., Kiewiet, R., Meijer, W., Seegers, J., Skurray, R., Terpstra, P., Thomas, C.M., Thorsted, P., Tietze, E., and Turner, S.L. 2000. The evolution of bacterial pasmids. In: The Horizontal Gene Pool. Thomas, C.M. (Ed.), Harwood Academic Publishers, Amsterdam, Netherlands, p. 301-361.

Parkhill, J., Wren, B.W., Thomson, N.R., Titball, R.W., Holden, M.T., Prentice, M.B., Sebaihia, M., James, K.D., Churcher, C., Mungall, K.L., Baker, S., Basham, D., Bentley, S.D., Brooks, K., Cerdeno-Tarraga, A.M., Chillingworth, T., Cronin, A., Davies, R.M., Davis, P., Dougan, G., Feltwell, T., Hamlin, N., Holroyd, S., Jagels, K., Karlyshev, A.V., Leather, S., Moule, S., Oyston, P.C., Quail, M., Rutherford, K., Simmonds, M., Skelton, J., Stevens, K., Whitehead, S., and Barrell, B.G. 2001. Genome sequence of *Yersinia pestis*, the causative agent of plague. Nature. 413: 523-527.

Perry, R.D., and Fetherston, J.D. 1997. *Yersinia pestis*-etiologic agent of plague. Clin. Microbiol. Rev. 10: 35-66.

Prentice, M.B., James, K.D., Parkhill, J., Baker, S.G., Stevens, K., Simmonds, M.N., Mungall, K.L., Churcher, C., Oyston, P.C., Titball, R.W., Wren, B.W., Wain, J., Pickard, D., Hien, T.T., Farrar, J.J., and Dougan, G. 2001. *Yersinia pestis* pFra shows biovar-specific differences and recent common ancestry with a *Salmonella enterica* serovar Typhi plasmid. J. Bacteriol. 183: 2586-2594.

Protsenko, O.A., Anisimov, P.I., Mozharov, O.T., Konnov, N.P., and Popov, I.A. 1983. Detection and characterization of the plasmids of the plague microbe which determine the synthesis of pesticin I, fraction I antigen and "mouse" toxin exotoxin. Genetika. 19: 1081-1090.

Protsenko, O.A., Filippov, A.A., and Kutyrev, V.V. 1991. Integration of the plasmid encoding the synthesis of capsular antigen and murine toxin into *Yersinia pestis* chromosome. Microb. Pathog. 11: 123-128.

Samoilova, S.V., Samoilova, L.V., Yezhov, I.N., Drozdov, I.G., and Anisimov, A.P. 1996. Virulence of pPst+ and pPst- strains of *Yersinia pestis* for guinea-pigs. J. Med. Microbiol. 45: 440-444.

Sodeinde, O.A., and Goguen, J.D. 1988. Genetic analysis of the 9.5-kilobase virulence plasmid of *Yersinia pestis*. Infect. Immun. 56: 2743-2748.

Sodeinde, O.A., and Goguen, J.D. 1989. Nucleotide sequence of the plasminogen activator gene of *Yersinia pestis*: relationship to *ompT* of *Escherichia coli* and gene *E* of *Salmonella typhimurium*. Infect. Immun. 57: 1517-1523.

Sodeinde, O.A., Sample, A.K., Brubaker, R.R., and Goguen, J.D. 1988. Plasminogen activator/coagulase gene of *Yersinia pestis* is responsible for degradation of plasmid-encoded outer membrane proteins. Infect. Immun. 56: 2749-2752.

Sodeinde, O.A., Subrahmanyam, Y.V., Stark, K., Quan, T., Bao, Y., and Goguen, J.D. 1992. A surface protease and the invasive character of plague. Science. 258: 1004-1007.

Straley, S.C., and Bowmer, W.S. 1986. Virulence genes regulated at the transcriptional level by Ca2+ in *Yersinia pestis* include structural genes for outer membrane proteins. Infect. Immun. 51: 445-454.

Straley, S.C., and Harmon, P.A. 1984a. Growth in mouse peritoneal macrophages of *Yersinia pestis* lacking established virulence determinants. Infect. Immun. 45: 649-654.

Straley, S.C., and Harmon, P.A. 1984b. *Yersinia pestis* grows within phagolysosomes in mouse peritoneal macrophages. Infect. Immun. 45: 655-659.

Thomas, C.M. 2000. Paradigms of plasmid organization. Mol. Microbiol. 37: 485-491.

Welkos, S.L., Davis, K.M., Pitt, L.M., Worsham, P.L., and Freidlander, A.M. 1995. Studies on the contribution of the F1 capsule-associated plasmid pFra to the virulence of *Yersinia pestis*. Contrib. Microbiol. Immunol. 13: 299-305.

Welkos, S.L., Friedlander, A.M., and Davis, K.J. 1997. Studies on the role of plasminogen activator in systemic infection by virulent *Yersinia pestis* strain C092. Microbiol. Pathog. 23: 211-223.

Youngren, B., Radnedge, L., Hu, P., Garcia, E., and Austin, S. 2000. A plasmid partition system of the P1-P7par family from the pMT1 virulence plasmid of *Yersinia pestis*. J. Bacteriol. 182: 3924-3928.

Chapter 4

The Evolution of Flea-Borne Transmission in *Yersinia pestis*

B. Joseph Hinnebusch

ABSTRACT

Transmission by fleabite is a recent evolutionary adaptation that distinguishes *Yersinia pestis*, the agent of plague, from *Yersinia pseudotuberculosis* and all other enteric bacteria. The very close genetic relationship between *Y. pestis* and *Y. pseudotuberculosis* indicates that just a few discrete genetic changes were sufficient to give rise to flea-borne transmission. *Y. pestis* exhibits a distinct infection phenotype in its flea vector, and a transmissible infection depends on genes that are specifically required in the flea, but not the mammal. Transmission factors identified to date suggest that the rapid evolutionary transition of *Y. pestis* to flea-borne transmission within the last 1,500 to 20,000 years involved at least three steps: acquisition of the two *Y. pestis*-specific plasmids by horizontal gene transfer; and recruitment of endogenous chromosomal genes for new functions. Perhaps reflective of the recent adaptation, transmission of *Y. pestis* by fleas is inefficient, and this likely imposed selective pressure favoring the evolution of increased virulence in this pathogen.

INTRODUCTION

Pathogenic bacteria must overcome several physiological and immunological challenges to successfully infect even a single type of host, such as a mammal. It is remarkable, then, that bacteria transmitted by blood-feeding arthropods are capable of infecting two very different hosts during their life cycle: an invertebrate (usually an insect or tick) and a mammal. As if this were not enough of a challenge, it is not sufficient that an arthropod-borne bacterium successfully infect both vector and host. It must establish a transmissible infection in both; that is, it must infect the vector in such a way as to be transmitted during a blood meal, and it must infect the mammal in a way that allows uptake by a blood-feeding arthropod. This feat of evolution has occurred relatively rarely, but nonetheless arthropod-borne transmission has developed independently in a phylogenetically diverse group of microorganisms, including the rickettsiae, spirochetes in the genus *Borrelia*, and the Gram-negative bacteria.

Compared to the ancient relationship of rickettsiae and spirochetes with arthropods, the vector relationship between *Y. pestis* and fleas is new. As reviewed in Chapter 2, population genetics evidence indicates that *Y. pestis* is a clonal variant of *Y. pseudotuberculosis* that diverged only within the last 1,500 to 20,000 years (Achtman et al., 1999). Presumably, the change from the food- and water-borne transmission of the *Y. pseudotuberculosis* ancestor to the flea-borne transmission of *Y. pestis* occurred during this evolutionarily short period of time. The monophyletic relationship of these two sister-species implies that the genetic changes that underlie the ability of *Y. pestis* to use the flea for its transmission vector are relatively few and discrete. Therefore, the *Y. pseudotuberculosis-Y. pestis* species complex provides an interesting case study in the evolution of arthropod-borne transmission. Some of the genetic changes that led to flea-borne transmission have been identified using the rat flea *Xenopsylla cheopis* as model organism, and an evolutionary pathway can now be surmised. Reliance on the flea for transmission also imposed new selective pressures on *Y. pestis* that help explain the evolution of increased virulence in this pathogen.

Y. PESTIS – FLEA INTERACTIONS

There are an estimated 2,500 species and subspecies of fleas that constitute 220 genera and 15 families in the insect order Siphonaptera (Lewis, 1998). Of these, approximately 80 species, associated with some 200 species of wild rodents, have been found to be infected with *Y. pestis* in nature, or to be susceptible to experimental infection (Pollitzer, 1954). Accordingly, the ecology of plague is extremely complex, involving many different rodent-flea cycles. Different species of fleas vary greatly in their ability to transmit *Y. pestis*, at least under laboratory conditions; some species, such as the common cat flea *Ctenocephalides felis*, are incapable of transmission. Wheeler and Douglas (1945) and Burroughs (1947) developed a mathematical model to estimate the vector efficiency of different flea species that took into account i) infection potential (the percentage of fleas becoming infected after feeding on a septicemic animal); ii) vector potential (the percentage of infected fleas which become infective or blocked, i.e., develop a transmissible infection as described below); and iii) the transmission potential (the average number of successful transmissions per flea). Kartman (1957) later added two more factors: iv) the life span of infective (blocked) fleas; and v) the field prevalence index (the average number of fleas per species per rodent or rodent nest). By these measures an experimental vector efficiency could be calculated for

Table 1. Flea-specific factors which might affect the ability of *Y. pestis* to produce a transmissible infection.
Flea anatomy and physiology
Midgut
pH, redox potential, osmolarity, etc.
biochemical composition of host blood
digestive enzymes, digestive byproducts
endogenous microbial flora
frequency of feeding and defecation
insect immunity components
Proventriculus
size (volume)
number, density, length, and shape of spines
rate of opening and closing during feeding
hydrodynamic forces generated during feeding
insect immunity components
Ecology
ambient temperature
flea life span after infection

different fleas (Wheeler and Douglas, 1945; Burroughs, 1947; Kartman, 1957; Kartman and Prince, 1956). These comparisons have sometimes been intriguing and enigmatic. For example, the rat flea *Xenopsylla cheopis* has most frequently been identified as the most efficient vector, yet the closely related *Xenopsylla astia* is a poor vector (Hirst, 1923). The physiological mechanisms that account for differences in vector efficiency among different flea species are not known, but some possible factors are described in the following sections and listed in Table 1.

A high degree of vector specificity is characteristic of many arthropod-borne agents. For example, human malaria is transmitted by anopheline but not culicine mosquitoes, different subspecies of *Leishmania* are transmitted by different sandfly species, and the closely related North American species of *Borrelia* spirochetes that cause relapsing fever are each transmitted by a different species of *Ornithodoros* tick (Sacks and Kamhawi, 2001; Barbour and Hayes, 1986). Whether the same co-evolutionary process is occurring in *Y. pestis* remains to be demonstrated, but Russian researchers have proposed that, at least for some natural plague cycles, discrete triads of flea species, rodent, and subspecies or strain of *Y. pestis* have co-evolved (Anisimov et al., 2004).

THE FLEA GUT ENVIRONMENT

Y. pestis infection of the flea is confined to the digestive tract, which is depicted in Figure 1. Storage, digestion, and absorption of the blood meal all occur in the simple midgut made of a single layer of columnar epithelial cells and associated basement membrane. The proventriculus, a valve at the base of the esophagus that guards the entrance to the midgut,

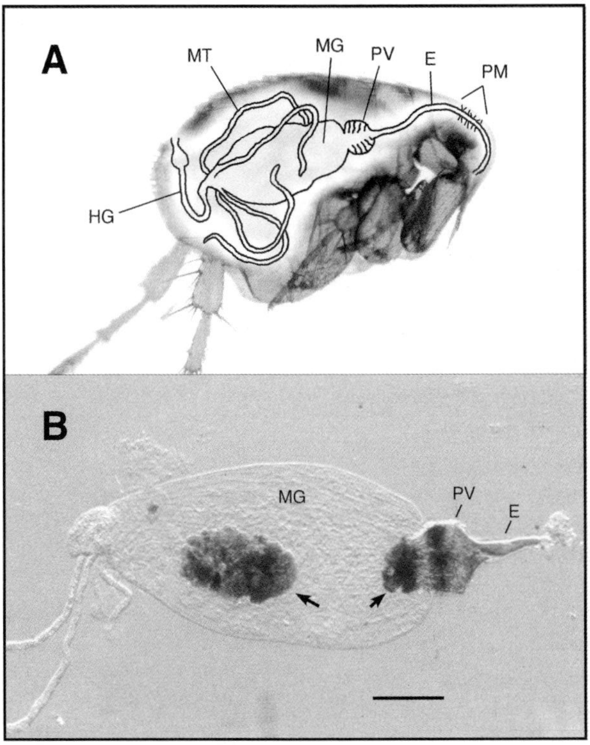

Figure 1. (A) Digestive tract anatomy of fleas. E = esophagus; PV = proventriculus; MG = midgut; HG = hindgut. The muscles that pump blood into the midgut (PM) and the malpighian tubules (MT) are also indicated. (B) Digestive tract dissected from a blocked *X. cheopis* flea. Arrows indicate the large aggregate of *Y. pestis* that fills and blocks the proventriculus and an independent bacterial aggregate in the midgut. The *Y. pestis* aggregates are surrounded by a dark colored extracellular matrix. Bar = 0.25 mm.

is central to the transmission mechanism. The interior of the proventriculus is arrayed with densely packed rows of inward-directing spines, which are coated with an acellular layer of cuticle, the same material that makes up the insect exoskeleton (Figure 2). In *X. cheopis*, there are a total of 264 proventricular spines in the male and 450 in the female (Munshi, 1960). The proventricular valve is normally tightly closed by layers of surrounding muscle. During feeding periods, however, the proventricular muscles rhythmically open and close the valve in concert with a series of three sets of pump muscles located in the flea's head to propel blood into the midgut and to keep it from leaking back out. Fleas usually live on or in close association with their hosts and take small but frequent (every few days) blood meals. Digestion of the blood meal begins quickly, resulting in hemolysis and liquefaction of ingested blood cells by six hours (Vaughan and Azad, 1993). During the next two to three days, the blood meal digest is brown-colored, viscous, and contains many large and small lipid droplets, but is eventually processed to a compact dark residue. Fleas defecate partially digested portions of their blood meals, which are used as a food source by flea larvae. Unlike other blood-feeding arthropods, fleas do not secrete a chitinous peritrophic membrane around the blood meal.

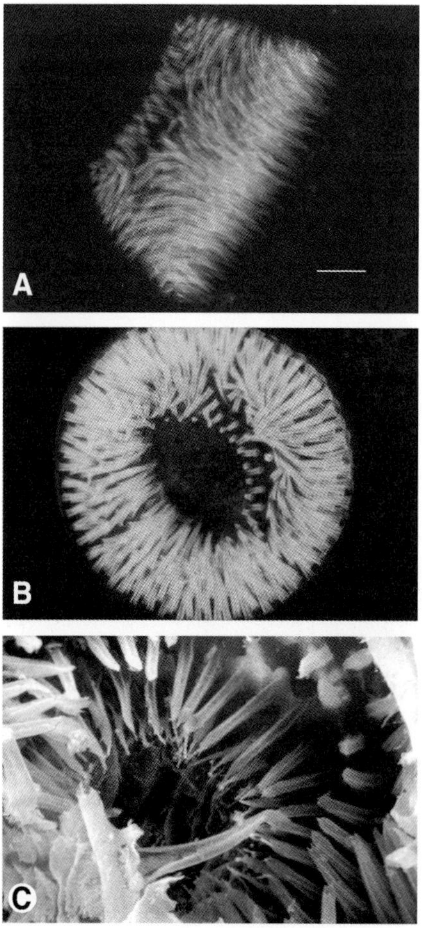

Figure 2. Proventricular spines of *Xenopsylla cheopis*. Side (A) and front (B) views of an uninfected proventriculus viewed by fluorescence microscopy. The proventricular spines are covered with cuticle, which is autofluorescent. (C) Scanning electron microscopy of the interior of an uninfected proventriculus. Bar = 5 μm.

Few details are known about flea gut physiology and associated environmental conditions in the digestive tract. A probable midgut pH of 6 to 7 has been cited (Wigglesworth, 1972), but other basic parameters such as osmotic pressure and redox potential are unknown. The biochemical composition may initially reflect that of hemolyzed blood, but is subject to rapid change due to selective absorption of certain nutrients, ions, and water. Insect midgut epithelium secretes a variety of digestive enzymes that are similar to those of vertebrates, including trypsin, chymotrypsin, amino- and carboxypeptidases, cathepsins, lysozymes, glycosidases, and lipases (Terra and Ferreira, 1994). Mammalian blood is composed principally of protein and lipid, and lipids are a major energy source for hematophagous arthropods. Lipids are relatively insoluble in water, and the mechanism of their solubilization and absorption in fleas is unknown.

It is in this active digestive milieu that *Y. pestis* lives in the flea. These conditions must be relatively hostile and refractory to colonization, because fleas have rather limited normal digestive tract flora, and few pathogens are transmitted by fleas (Savalev et al., 1978; Beard et al., 1990). Besides *Y. pestis*, flea-borne pathogens include *Bartonella henselae*, the agent of cat-scratch disease and bacillary angiomatosis, and *Rickettsia typhi* and *Rickettsia felis*, members of the typhus group (Chomel et al., 1996; Azad et al., 1997). The Gram-negative bacterium *Francisella tularensis* is also associated with fleas, although the importance of flea vectors in the overall ecology of tularemia is unclear (Hopla, 1974). Fleas have also been implicated in transmission of the poxvirus that causes myxomatosis in rabbits (Chapple and Lewis, 1965). Vaughan and Azad (1993) have hypothesized that the rapid digestive process of fleas and lice is not conducive to the development of eukaryotic parasites, but that it can be better tolerated by prokaryotes.

The midgut epithelium of mosquitoes and the blood sucking fly *Stomoxys calcitrans* has been shown to be an immune-competent tissue, and the presence of bacteria in the blood meal of these insects induces the secretion of antimicrobial peptides into the gut lumen (Dimopoulos et al., 1997; Lehane et al., 1997). Whether this occurs in fleas is unknown. At any rate, *Y. pestis* appears to be inherently resistant to the flea immune response (Hinnebusch et al., 1996; and unpublished data).

BIOLOGICAL TRANSMISSION OF *Y. PESTIS* BY FLEAS: THE PROVENTRICULAR BLOCKAGE MODEL

The general mechanism of *Y. pestis* transmission by fleas was described by Bacot and Martin (1914). They observed that in infected *X. cheopis* and *Ceratophyllus fasciatus* fleas, solid masses of *Y. pestis* could fill the lumen of the proventriculus and obstruct the flow of blood (Figure 1B). These fleas made persistent, vigorous attempts to feed, but could not pump blood past the blocked proventriculus. Instead, the esophagus became distended with blood which was then partially refluxed back into the bite site due to contraction and relaxation of the cibarial and pharyngeal pumps. This phenomenon correlated with efficient transmission. Based on their observations, Bacot and Martin proposed the proventricular blockage-regurgitation mechanism for plague transmission. According to this model, blockage of the proventriculus by a mass of *Y. pestis* precedes transmission, and *Y. pestis* is conveyed to the bite site by regurgitation when blocked fleas attempted to feed. Bacot (1915) later amended this model by proposing that fleas with only partial obstruction of the proventriculus were actually better transmitters. Partially blocked fleas can still pump blood into the midgut through an open channel in the proventriculus. The bacterial growth prevents complete closing of the valve, however, so that blood mixed with *Y. pestis* from the midgut is able to flow back out the proventriculus into the bite site. This can happen because during flea feeding the pumping action is not continuous, but stops for short intervals. Transmission as a result of partial blockage was an important addendum to the model because complete proventricular blockage does not readily develop in some flea species that are good vectors of plague (Burroughs, 1947; Pollitzer, 1954). According to the Bacot model, complete blockage of the proventriculus is not necessary for efficient transmission; partial interference with its valvular function is sufficient.

MECHANICAL TRANSMISSION OF *Y. PESTIS* BY FLEAS

Although several investigators have established that biological transmission (requiring *Y. pestis* growth in the digestive tract to produce a proventricular infection) is the only reliable means of transmission (Burroughs, 1947; Pollitzer, 1954), there is evidence that mechanical transmission may also play a role in the ecology of plague. For mechanical transmission, infection of the vector is not necessary. It is only necessary that septicemia levels are high and that *Y. pestis* survive on the blood-stained mouthparts of fleas between consecutive feedings. For example, *X. cheopis* and the wild rodent flea *Malaraeus telchinum* allowed to feed *en masse* on uninfected mice one day after feeding on a highly septicemic mouse consistently transmitted the disease (Burroughs, 1947). Since that time interval is too short for proventricular infection to develop, transmission presumably occurred by mechanical transference of bacteria on contaminated mouthparts. The phenomenon of mechanical or mass transmission provides a potential mechanism for fleas that do not develop proventricular blockage readily, such as *M. telchinum* and the human flea *Pulex irritans*, to transmit *Y. pestis* during epidemics. Human to human transmission via *P. irritans* has been hypothesized to have contributed to the plague pandemics of medieval Europe (Beaucournu, 1999). Because mechanical transmission does not rely on specific interactions with the vector, it will not be considered further here.

Y. PESTIS TRANSMISSION FACTORS

A central hypothesis, now substantiated by experimental evidence, is that bacteria that cycle between a mammal and an arthropod express distinct subsets of genes in their two hosts. Genes specifically required to infect the vertebrate host are referred to as virulence factors, and the analogous genes required to produce a transmissible infection in the arthropod vector have been termed transmission factors (Hinnebusch et al., 1996; Paskewitz, 1997). Many virulence factor genes of *Y. pestis* that are required to infect and cause disease in the mammal have been identified and studied. In contrast, the genetic factors required in the insect host have been relatively neglected. Nevertheless, some of the genetic factors of *Y. pestis* that are specifically involved in flea-borne transmission have been identified (Table 2).

THE *YERSINIA* MURINE TOXIN: A PHOSPHOLIPASE D REQUIRED FOR FLEA GUT COLONIZATION

The *Yersinia* murine toxin (Ymt) was described in the 1950s as a protein fraction of *Y. pestis* that was toxic to mice and rats (Ajl et al., 1955), so it has universally been considered to be a virulence factor. Brown and Montie (1977) presented evidence that Ymt is a β-adrenergic receptor antagonist, blocking epinephrine-induced mobilization of glucose and fatty acids. In mice, Ymt causes circulatory failure due to vascular collapse,

Table 2. *Y. pestis* genes important for flea-borne transmission.

Gene	Location	Present in: *Y. pestis*	Present in: *Y. pstb*	Function in *Y. pestis* and role in transmission
ymt	pFra plasmid	+	–	Phospholipase D, survival in flea midgut
hmsHFSR, T	chromosome	+	+	Extracellular matrix synthesis, biofilm formation, infection and blockage of the proventriculus
pla	pPst plasmid	+	–	Plasminogen activator, dissemination from fleabite site

resulting in death in ten hours with an LD_{50} of 0.2 to 3.7 μg (Schär and Meyer, 1956). Ymt is not toxic to guinea pigs, rabbits, dogs, or primates even in enormous doses, however (Montie and Ajl, 1970). Murine toxin was described as a cell-associated protein that is only released upon bacterial death, and, correspondingly, its effects are seen only in the late stages of septicemic murine plague, when the animal is already succumbing to the disease (Montie and Ajl, 1970). Interestingly, recombinant Ymt protein produced in and purified from *Escherichia coli* is nontoxic for mice, whereas native Ymt purified from *Y. pestis* is toxic (Hinnebusch et al., 2000; and unpublished data). The explanation for this is not known, but Walker (1967) suggested that synergism between Ymt, endotoxin, and possibly other *Y. pestis* factors was responsible for murine toxicity.

Sequence analysis showed that the *Y. pestis ymt* gene mapped to the 100-kb pFra plasmid and encodes a 61-kDa protein that is a member of a newly described family of phospholipase D enzymes found in all kingdoms of life: animals, plants, fungi, and eukaryotic viruses as well as bacteria (Cherepanov et al., 1991; Ponting and Kerr, 1996). All members of this PLD family have two copies of a signature HKD ($HxKx_4Dx_6GG/S$) motif, which come together to form the catalytic site for binding and hydrolysis of the phosphodiester bond (Stuckey and Dixon, 1999). The *Y. pestis* Ymt has classic PLD activity, as shown by its ability to cleave the polar head group from phosphatidylcholine, phosphati dylethanolamine, and other phospholipids. Ymt is also capable of transphosphatidlyation of phospholipid with an alcohol acceptor, a second characteristic PLD reaction (Rudolph et al., 1999). Because of this proven biochemistry, it has been proposed that the *Y. pestis ymt* gene should be renamed *pldA* (Carniel, 2003).

Despite the known toxic effects of murine toxin, a *ymt* deletion mutant of *Y. pestis* was essentially fully virulent for mice (Drozdov et al., 1995; Du et al., 1995; Hinnebusch et al., 2000). Thus, Ymt is not required for morbidity or mortality, even in mice, but only adds insult to injury. This likely reflects the fact that Ymt is not a classic exotoxin, but is a cytoplasmic enzyme that is only released upon bacterial cell death and lysis. The full virulence of Ymt^- *Y. pestis*, and other results showing that *ymt* expression is downregulated at 37°C (Du et al., 1995), suggested that the principle biological function of this PLD is not as a virulence factor.

A role for Ymt in transmission was first indicated by a study evaluating the fate of plasmid-cured *Y. pestis* strains in the flea. Whereas the 9.5-kb pPst and the 70-kb pYV virulence plasmid were not required for normal infection and blockage of *X. cheopis*, strains lacking the 100-kb pFra plasmid failed to block the fleas (Hinnebusch et al., 1998a). Complementation of the $pFra^-$ strains with the *ymt* gene alone fully restored normal ability to infect and block fleas. Specific Ymt^- *Y. pestis* mutants were used for further analysis (Hinnebusch et al., 2002b). Within hours of being taken up in a blood meal by a flea, Ymt$^-$ *Y. pestis* assumed an aberrant, spheroplast-like cell morphology, and then rapidly disappeared from the flea midgut within the first day after infection. Rarely, the Ymt$^-$ bacteria established an initial foothold in the proventriculus, which is part of the foregut and physically separated from the midgut by the stomodeal valve except during the few minutes per week that the flea is actively feeding. Secluded in the proventriculus, the mutants could grow normally and eventually cause blockage. Because the proventriculus is rarely the primary site of infection, but is usually seeded secondarily from a prior midgut infection, the Ymt$^-$ mutant infected < 5% and blocked < 0.5% of fleas, compared to the normal infection and blockage rates of 50-60% and 25-45%, respectively (Hinnebusch et al., 2002b). Remarkably, introduction of the *ymt* gene into *Y. pseudotuberculosis* and *E. coli* significantly enhanced their ability to colonize the flea midgut also. Thus, Ymt may have a similar substrate and mechanism of action in both *Yersinia* and *E. coli*. Members of the PLD family of enzymes to which Ymt belongs are found in many different cell types and can have many different functions. Serendipitously, the PLD activity of Ymt enhances survival of Gram-negative bacteria in the flea midgut. Acquisition of this single gene by *Y. pestis* would have been a crucial step in the evolution of the flea-borne route of transmission.

MODELS FOR THE PROTECTIVE MECHANISM OF Ymt

How might an intracellular PLD protect *Y. pestis* in the flea midgut? The first option to consider is that Ymt might be secreted or released from lysed bacteria in the flea, and degrade an external cytotoxic agent in the midgut environment. Three types of experimental results argue against that: 1) Addition of exogenous Ymt protein to the infectious blood meal did not enhance the survival of Ymt$^-$ *Y. pestis* in the flea gut. 2) Coinfection of fleas with an equal mixture of Ymt$^+$ and Ymt$^-$ *Y. pestis* did not result in a coequal infection pattern. If active PLD were secreted, one might expect that enzyme from Ymt$^+$ bacteria would also protect Ymt$^-$ bacteria in the flea gut, resulting in infections consisting of an equal mixture of both strains. Instead, in these experiments the Ymt$^-$ mutant again survived primarily in the proventriculus. In the midgut, it persisted only in small clusters that were embedded within larger aggregates of wild-type bacteria. 3) In digestive tracts dissected from fleas infected with *Y. pestis* that synthesized a Ymt-GFP fusion protein, fluorescence localized only to the cytoplasm and was never detected extracellularly (Hinnebusch et al., 2002b).

An intracellular PLD conceivably could protect *Y. pestis* in the flea gut either by modifying an endogenous membrane component to make the bacteria impervious to the cytotoxic agent (prophylaxis model), or by neutralizing the agent, directly or indirectly, after it interacts with the bacteria (antidote model). In the prophylaxis model, the outer membrane of Ymt$^-$ *Y. pestis* would be differentially affected in the flea gut. Loss of outer membrane integrity could lead to the observed spheroplasty, because lysozymes are commonly secreted by insect midgut epithelium (Terra and Ferreira, 1994). Evidence for this model was sought by analyzing the outer membrane composition of Ymt$^+$ and Ymt$^-$ *Y. pestis*. Quantitative comparisons of membrane phospholipids and phosphodiester-linked substitutions of lipid A revealed no differences. The mutant was also no more susceptible than the parent *Y. pestis* to polymixin B, SDS and EDTA, cationic detergents, and other agents that target the Gram-negative outer membrane. Attempts to mimic the flea gut environment by culturing the bacteria in triturated flea gut contents; in whole or sonicated mouse blood containing proteases, lipase, and lysozyme; or under osmotic and oncotic pressure, oxidative stress, or low pH likewise failed to reveal any difference between mutant and wild type strains (B. J. Hinnebusch, unpublished). In sum, no phenotypic difference between Ymt$^-$ and Ymt$^+$ *Y. pestis* has been detected outside of the flea gut.

Alternatively, according to the antidote model, the toxic agent in the flea gut would interact with both Ymt$^-$ and Ymt$^+$ *Y. pestis*, but its effects then be neutralized by Ymt. In many bacteria, exposure to harmful environments induces autolytic pathways that result in self-digestion of the bacterial cell wall by endogenous peptidoglycan hydrolases (Lewis, 2000). The environmental stimuli, signal transduction mechanisms, and gene expression pathways leading to this programmed cell death are incompletely understood in bacteria. If an agent in the flea gut stimulates *Y. pestis* autolysis and leads to the observed rapid spheroplast formation, intracellular Ymt activity may block or redirect a step in the autolytic pathway. Such a role would be analogous to that of mammalian PLD, which is an intracellular effector in multiple signal transduction cascades (Gomez-Cambronero and Keire, 1998).

Whichever model is correct, the agent in the flea gut that is harmful to Ymt$^-$ *Y. pestis* appears to derive from a digestive product of blood plasma. Elimination from the flea gut during the first 24 hours after infection occurred if either filtered mouse plasma or whole blood was the source of the infectious meal fed to the fleas. However, if fleas were infected by feeding on an artificial plasma substrate consisting of PBS, pH 7.4, containing 7% bovine serum albumin, 6 mM glucose, 12 mM sodium bicarbonate, 10 mM $MgCl_2$, 2.5 mM $CaCl_2$, and 1 mM citric acid, the Ymt$^-$ *Y. pestis* survived as well as the Ymt$^+$ parent strain during the first 24 hours after infection. Identical results were obtained when washed mouse red blood cells were added to the artificial plasma substrate (Hinnebusch et al., 2002b). The fact that the artificial plasma meals were digested by the fleas further suggests that flea digestive enzymes, or the digestive milieu *per se*, do not directly harm the mutant. The mutant also survived as well as wild type *Y. pestis* after injection into the flea hemocoel. These results implicate a blood plasma digestive product as the cytotoxic agent, but the native substrate of the *Y. pestis* PLD and its protective mechanism remain to be discovered.

THE *hms* GENES AND THE BIOFILM MODEL OF PROVENTRICULAR BLOCKAGE

One of the first temperature-dependent phenotypes described for *Y. pestis* was the formation of densely pigmented colonies when incubated at 28°C or less on media containing hemin or the structurally analogous dye Congo red (Jackson and Burrows, 1956a; Surgalla and Beesley, 1969). The phenotype is not due to the production of a pigment by *Y. pestis*, but rather to the avid adsorption of the exogenous hemin or Congo red to the outer membrane (Perry et al., 1993). Despite the fact that the phenotype was not expressed at 37°C, pigmentation correlated with virulence. Spontaneous nonpigmented mutants had greatly reduced virulence for mice after peripheral routes of infection unless iron salts were injected simultaneously (Jackson and Burrows, 1956b). Further study of nonpigmented *Y. pestis* strains showed that they did not grow in iron-chelated media in vitro, failed to synthesize several iron-regulated proteins, and did not interact with pesticin, the bacteriocin encoded on the pPst plasmid (Brubaker, 1969; Sikkema and Brubaker, 1987, 1989).

The reason for the wide range of physiological effects associated with loss of pigmentation gradually emerged from a series of molecular genetics analyses. Robert Perry and colleagues used transposon mutagenesis to identify a 9.1-kb chromosomal locus required for the pigmentation phenotype (Lillard et al., 1997). Nucleotide sequence of this region, termed the hemin storage (*hms*) locus, revealed a 4-gene operon, *hmsHFRS*. A fifth, unlinked gene, *hmsT*, was later found also to be essential for pigmentation (Hare and McDonough, 1999; Jones et al., 1999). Concurrently, investigations by several groups into the iron-regulated proteins synthesized by pigmented *Y. pestis* culminated in the characterization of the *Yersinia* high-pathogenicity island (HPI), which encodes a siderophore-based iron acquisition system (see chapters 13 and 14). A key unifying discovery was reported by Fetherston et al. (1992) showing that most nonpigmented *Y. pestis* mutants resulted from spontaneous deletion of a 102-kb segment of the *Y. pestis* chromosome that was termed the pigmentation (Pgm) locus. The 102-kb Pgm locus contains not only the *hms* genes, but also the *Yersinia* HPI. It is flanked by IS*100* elements, and homologous recombination between these extensive direct repeat sequences likely accounts for the high spontaneous deletion rate (10^{-5} to 10^{-3}) of the entire 102-kb segment (Hare et al., 1999; Fetherston and Perry, 1994). Thus, elimination of this large locus by a single deletion event results not only in the nonpigmented (Hms^{-}) phenotype, but also in decreased virulence due to concomitant loss of the HPI. Not all nonpigmented mutants result from the loss of the entire 102-kb segment, however. The existence of certain nonpigmented *hmsHFSR*-negative, HPI-positive *Y. pestis* strains indicated that the *hmsHFRS* locus could be autonomously deleted and that pigmentation and iron acquisition phenotypes are clearly separable (Iteman et al., 1993; Buchrieser et al., 1998).

The incidental linkage of the *hmsHFRS* locus and the HPI within the same deletion-prone segment accounts for the long-held consideration of pigmentation as a virulence determinant, reinforced by referring to the entire 102-kb segment as the Pgm locus. However, the connection between pigmentation *per se* and virulence turns out to be merely "guilt by association" with the HPI. In retrospect, the temperature-dependence of the pigmentation phenotype provided an important clue as to its true biological role. As noted previously, pigmentation develops only at temperatures less than about 28°C, a temperature that matches the flea environment. It is not detected at 37°C, the mammalian body temperature. In fact, nonpigmented *Y. pestis* strains containing specific loss-of-function mutation of the *hms* genes are fully virulent, at least in mice, and the hypothesis that hemin storage is important nutritionally has also been disproven (Hinnebusch et al., 1996; Lillard et al., 1999). In contrast, nonpigmented *Y. pestis* strains lacking a functional *hmsHFRS* locus, or the entire 102-kb Pgm locus, were completely unable to produce proventricular blockage in *X. cheopis* fleas, although they survived in and established a chronic infection of the midgut at the same rate as the isogenic pigmented *Y. pestis*. The ability of Pgm^{-} *Y. pestis* to infect and block the proventriculus could be completely restored by reintroducing the *hmsHFRS* genes alone, indicating that the HPI and other genes in the 102-kb Pgm locus are not required in the flea (Hinnebusch et al., 1996). Earlier, working with genetically undefined strains, Bibikova (1977) correlated the pigmentation phenotype with the ability to cause proventricular blockage in *X. cheopis*; and Kutyrev et al. (1992) reported that a nonpigmented but pesticin sensitive and virulent *Y. pestis* strain failed to survive in the vole flea *Nosopsyllus laeviceps*. Whether physiological differences between the two flea species account for the ability of nonpigmented *Y. pestis* to colonize *X. cheopis* but not *N. laeviceps* is unknown.

ROLE OF THE *hms* GENES: PRODUCTION OF A *Y. PESTIS* BIOFILM REQUIRED FOR PROVENTRICULAR BLOCKAGE

Ironically, given its close phylogenetic relationship with enteric pathogens, *Y. pestis* does not penetrate or even adhere to the flea midgut epithelium, but remains confined to the lumen of the digestive tract. Because *Y. pestis* is not invasive in the flea, it is at constant risk of being eliminated by peristalsis and excretion in the feces. In fact, approximately half of *X. cheopis* fleas spontaneously rid themselves of infection in this way even if they feed on highly septicemic blood (Pollitzer, 1954; Hinnebusch et al., 1996). Success or failure in stable colonization of the flea gut depends on the ability of the bacteria to produce aggregates that are too large to be excreted (Figure 1B). Both Hms^+ and Hms^- *Y. pestis* are able to do this, and so achieve comparable infection rates in *X. cheopis*. However, transmission to a new host further requires that *Y. pestis*, which is nonmotile, move against the direction of blood flow when the flea feeds. As described above, this is accomplished by infecting the proventriculus, interfering with its valvular action in such a way as to generate backflow of blood into the bite site. The *hms* genes are required for proventricular infection, and recent evidence suggests that they synthesize an extracellular matrix required for biofilm formation.

HmsH and HmsF were characterized as surface-exposed outer membrane proteins (HmsF also contains a lipid attachment site typical of a lipoprotein), and HmsR, HmsS, and HmsT contain transmembrane domains and appear to be to be inner membrane proteins (Parkhill et al., 2001; Pendrak and Perry, 1993; Perry et al., 2004), but the first predictive clue as to the function of the *hms* genes came from database searches showing that they are similar to glycosyl transferase and polysaccharide deacetylase genes in other bacteria that are required to produce extracellular polysaccharides (Figure 3). Notably, similarity was detected between *hmsR* and *hmsF* and two genes in the *ica* (intercellular adhesion) operon of *Staphylococcus epidermidis* that is required to synthesize a linear β-1,6-linked glucosaminoglycan called the polysaccharide intercellular adhesin (PIA) (Heilmann et al., 1996; Lillard et al., 1999). PIA is an extracellular polysaccharide that leads to bacterial cell-cell aggregation and is required for the formation of staphylococcal biofilms. Interestingly, PIA as well as the extrapolysaccharide associated with several other bacterial biofilms binds Congo red (Heilmann and Götz, 1998; Weiner et al., 1999). The *ica* operon consists of four genes (*icaADBC*). HmsR has 39% identity and 58% amino acid sequence similarity to IcaA, an N-acetylglucosamine transferase that functions to polymerize UDP-N-acetylglusosamine units; and HmsF has 23% identity and 41% similarity to IcaB, a poly (β-1, 6) N-acetylglocosamine deacetylase that removes N-acetyl groups from the extracellular PIA polymer (Götz, 2002). HmsS and HmsH are not similar to *ica* gene products or other proteins of known function in the databases.

The *hmsT* gene is located 1.76 Mb away from the *hmsHFRS* operon on the *Y. pestis* chromosome but is also required for the pigmentation phenotype. The function of HmsT is unknown, although it contains putative transmembrane domains and a GGDEF domain (Jones et al., 1999). The GGDEF domain is widespread in bacteria and is homologous to the adenylyl cyclase catalytic domain. Interestingly, the GGDEF domain is present in several proteins that regulate the biosynthesis of extracellular cellulose. In *Acetobacter xylinum*, for example, the regulation is mediated by a novel effector molecule that consists of two covalently linked cGMP moieties, whose formation may be GGDEF-dependent (Ross et al., 1987). The GGDEF domain is also commonly found in the response regulators of bacterial two-component signal transduction systems, such as WspR of

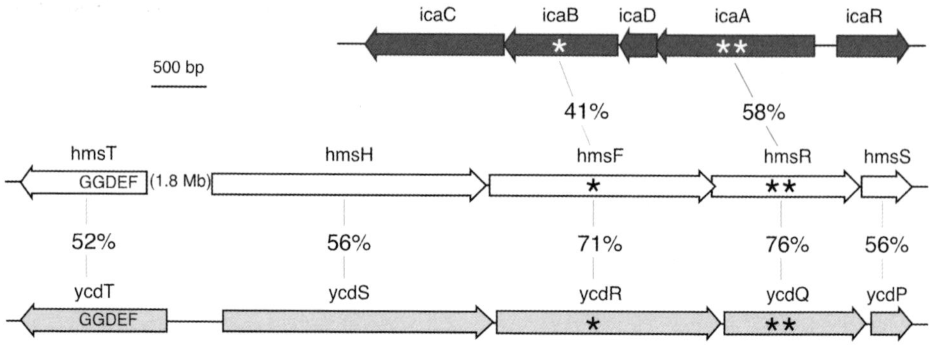

Figure 3. Comparison of the *ica, hms*, and *ycd* operons of *S. epidermidis*, *Y. pestis*, and *E. coli*, respectively. Numbers indicate the percent amino acid similarity of the predicted products of *Y. pestis hms* genes with *ica* and *ycd* gene products. Single asterisks indicate polysaccharide deacetylase domains, double asterisks indicate glycosyl transferase domains, and GGDEF indicates diguanylate cyclase domains.

Pseudomonas aeruginosa and MbaA of *Vibrio cholerae*, both of which regulate genes important for biofilm production (D'Argenio et al., 2002; Bomchil et al., 2003). Jones et al. (1999) detected four discrete domains shared by HmsT and a group of homologous proteins that included the response regulator PleD of *Caulobacter crescentus*. Unlike these known regulators, however, HmsT does not contain other signature elements of response regulators such as EAL or a helix-turn-helix motif, and there is no adjacent sensor component gene associated with *hmsT* as would be expected for a two-component regulatory system. A potential binding site for the iron uptake repressor Fur was detected in the *hmsT* promoter region (Jones et al., 1999), but upregulation of *hmsT* in iron-limiting conditions has not yet been demonstrated. Transcription of the *hmsT* and the *hmsHFRS* operons is not affected by growth temperature; however, protein levels of HmsT, HmsH, and HmsR are much lower in *Y. pestis* grown at 37° than at 26°C, which likely accounts for the temperature-dependence of the pigmentation phenotype (Perry et al., 2004).

The *E. coli* operon *ycdSRQP* is homologous to *Y. pestis hmsHFRS*, and the adjacent *ycdT* is an *hmsT* homolog (Lillard et al., 1997; Jones et al., 1999). The function of the *ycd* genes in *E. coli*, which is nonpigmented, is unknown, although *ycdQ* and *ycdP* restore pigmentation to *Y. pestis hmsR* and *hmsS* mutants, respectively (Jones et al., 1999). Homologs of *hmsH*, *F*, and *R* have also been found in the genome sequences of *Chromobacterium violaceum*, *Ralstonia solanacearum*, *Xanthomonas* spp., *Bordetella* spp., and *Pseudomonas fluorescens*. These have been labeled hemin storage genes based on their similarity to the *Yersinia* genes, but a phenotype for the *hms* homologs in *E. coli* or any of the other bacteria has not been described.

The amino acid sequence comparisons suggest that the *hms* genes encode products that synthesize the extracellular matrix of a biofilm. A bacterial biofilm is a complex, compact community of cells enclosed in an extracellular matrix, often attached to a surface (Costerton et al., 1995). Biofilms can form in spite of high shear forces and rapid currents, and are produced in vivo, particularly on implanted medical devices, by many bacterial pathogens (Costerton et al., 1999). Previous investigations have shown that the dense aggregates of *Y. pestis* that develop in the flea midgut and block the proventriculus are surrounded by an extracellular matrix, fitting the operational definition of a biofilm

(Hinnebusch et al., 1998a; 2002a; Jarrett et al., 2004). The ability of *Y. pestis* to produce an extracellular matrix in the flea, along with the ability to block the proventriculus, depends on the *hms* genes. The role of the individual *hms* genes in this in vivo phenotypes have not been systematically studied, but mutation of *hmsR* or *hmsT* eliminates or greatly reduces the ability of *Y. pestis* to block fleas (Hinnebusch et al., 1996; and unpublished data). The *hms* genes are also required for the ability of *Y. pestis* to produce an adherent biofilm on the surface of a glass flowcell, and to synthesize an extracellular material observed by scanning electron microscopy (Jarrett et al., 2004). Like pigmentation, the in vitro biofilm and extracellular material are only produced at low temperatures and not at 37°C. Darby et al. (2002) and Joshua et al. (2003) have also shown that *Y. pestis* and *Y. pseudotuberculosis* produce biofilm-like growth on agar plates that accumulates on the external mouthparts of *Caenorhabditis elegans* nematodes placed on them, and that this phenotype is *hms*-dependent.

Taken together, the genetic, in vitro, and in vivo observations strongly suggest that *Y. pestis* forms an *hms*-dependent biofilm to infect the hydrophobic, acellular surface of the flea's proventricular spines, and in this way overcomes the rhythmic, pulsating action of the proventricular valve and the inward flow of blood during feeding that would otherwise counteract transmission by washing the bacteria backwards into the midgut. Given the homology between the *Y. pestis hms* genes and the staphylococcal *ica* genes, it seems likely that the function of the *hms* gene products is to synthesize an extracellular polysaccharide required for biofilm development. The composition of the extracellular matrix that surrounds the *Y. pestis* biofilm in the flea is unknown, but appears to contain flea midgut-derived lipid components as well as *hms*-dependent components (Jarrett et al., 2004). The *hms* genes do not appear to be required in the mammal; thus, their primary biological function is to enable flea-borne transmission. Transmission of *Leishmania* parasites also depends on a foregut-blocking phenomenon in the sandfly vector (Stierhof et al., 1999), but *Y. pestis* is unique among bacteria characterized to date in using a biofilm mechanism to enable arthropod-borne transmission. In retrospect, the first hint that the *hms* genes pertained to biofilm formation was the observation in the original paper by Jackson and Burrows (1956a) that cells in pigmented colonies resist resuspension and remain bound together in densely packed masses. Surgalla (1960) also observed that another aspect of the pigmentation phenotype is the production of a substance in liquid cultures at room temperature that promotes autoaggregation and pellicle formation on the sides of the culture vessel– typical of biofilm formation.

THE *Y. PESTIS* PLASMINOGEN ACTIVATOR AND DISSEMINATION FOLLOWING FLEA-BORNE TRANSMISSION

Like murine toxin and the Hms pigmentation phenotype, the biological functions attributed to the *Y. pestis* plasminogen activator (Pla) have undergone revision. The *pla* gene is on the 9.5-kb *Y. pestis* plasmid referred to as pPCP1, pPst or pPla (Sodeinde and Goguen, 1988). It encodes a surface protease whose role in virulence is the subject of chapter 17 in this volume. Pla is considered to be an essential factor for the flea-borne route of transmission because it greatly enhances dissemination following subcutaneous injection, which is assumed to mimic transmission by fleas (Sodeinde et al., 1992). The requirement for Pla for dissemination from peripheral infection sites may not be universally true for all *Y. pestis* strains or for all animals, however (Samoilova et al., 1996; Welkos et al., 1997).

A prominent role for Pla in proventricular blockage of the flea has been proposed previously. The extracellular matrix that embeds the blocking masses of *Y. pestis* in the flea has often been assumed to be a fibrin clot derived from the flea's blood meal. It was also known that proventricular blockage does not develop normally in fleas kept at elevated temperatures, which helps explain striking epidemiological observations that flea-borne bubonic plague epidemics terminate abruptly with the onset of hot, dry weather (Cavanaugh and Marshall, 1972). Cavanaugh (1971) hypothesized that Pla activity could explain both phenomena. Pla synthesis is not temperature-dependent, but its plasminogen activator ability that leads to fibrinolysis is much greater at 37°C than at temperatures below 28°C (McDonough and Falkow, 1989). In fact, Pla has an opposite procoagulant activity at low temperatures, although this fibrin clot-forming ability is weak and is detected only in rabbit plasma and not in mouse, rat, guinea pig, squirrel, or human plasma (Jawetz and Meyer, 1944; Beesley et al., 1967). Nevertheless, it was hypothesized that this low-temperature activity of Pla formed what was presumed to be the fibrin matrix of the blocking mass of *Y. pestis* in the flea. The clot-dissolving plasminogen activator function was invoked to explain why blockage does not develop in fleas at higher temperatures. McDonough et al. (1993) later reported that Pla+ *Y. pestis* caused greater mortality in fleas than an isogenic Pla- mutant, and attributed this to an increased blockage rate. Blockage was not directly monitored in that study, however, and the mortality occurred only four days after infection, well before blockage would be expected to occur.

When the Cavanaugh hypothesis was put to the test, it was found that Pla is not required for normal proventricular blockage to develop in the flea (Hinnebusch et al., 1998a). Hms^+ *Y. pestis* strains lacking pPst were able to infect and block *X. cheopis* fleas as well as the wild-type parent strain. Both Pla^+ and Pla^- strains failed to block fleas kept at 30°C, even though midgut colonization rates were little affected by temperature; in other words, the identical in vivo phenotype as seen for Hms^- *Y. pestis*. Thus, the inability of *Y. pestis* to block fleas kept at 30°C can be fully explained by temperature dependence of the Hms phenotype. Furthermore, the presumption that flea-blocking masses of *Y. pestis* are embedded in a fibrin matrix is inconsistent with the fact that the matrix is not degraded by proteases or the fibrinolytic enzyme plasmin. Therefore, the *hms*-dependent biofilm model of proventricular blockage better fits the available data than the Pla-based fibrin clot model.

INSECT PATHOGEN-RELATED GENES IN *Y. PESTIS* AND *Y. PSEUDOTUBERCULOSIS*

Like most bacterial genomes, the *Y. pestis* genome contains several loci that appear to have been introduced by lateral transfer from unrelated organisms. Among these are several homologs of known insecticidal toxin complex (Tc) genes of bacterial pathogens of insects, and a homolog of a baculovirus enhancin protease gene required for insect pathogenesis, which conceivably could influence *Y. pestis* interaction with the flea (Parkhill et al., 2001). In beginning efforts to assess the role of these genes, fleas were infected with *Y. pestis* strains containing specific mutations in the baculovirus enhancin homolog and *tcaA*, a homolog of one of the Tc genes. Both mutants infected and blocked *X. cheopis* fleas normally, indicating that these two genes are not important for interaction with the flea (B. J. Hinnebusch and R. D. Perry, unpublished). Many of the insect pathogen-related genes are also present in *Y. pseudotuberculosis*; therefore, their acquisition appears to predate the divergence of *Y. pestis*. When outside the host in soil and water, *Y. pseudotuberculosis* would be expected to come into contact with and even be ingested by insects and other invertebrates, and the insecticidal toxins may help the bacteria survive those encounters. Because *Y. pestis* transmission depends on chronic infection of the flea gut, overt toxicity would be counterproductive. Thus, it seems likely that insect toxicity would be lost or moderated in *Y. pestis*.

Y. PESTIS AT THE HOST-VECTOR INTERFACE

Successful transmission of an arthropod-borne agent and subsequent infection depends on a complex co-evolved interaction between pathogen, vector, and host that has not been well-characterized for any arthropod-borne disease. Plague is initiated during the brief encounter between an infectious flea and a vertebrate host. For practical reasons, intradermal or subcutaneous inoculation by needle and syringe of in vitro-grown *Y. pestis* is routinely used for pathogenesis studies in animal models in lieu of flea-borne transmission. While this is a reasonable challenge method which may be adequate for most purposes, certain aspects of the flea-bacteria-host transmission interface are unique, and have unknown effects on the host-pathogen interaction and the initiation of disease.

FEEDING MECHANISM OF FLEAS AND THE MICROENVIRONMENT OF THE TRANSMISSION SITE

Two basic strategies have been described for the manner in which blood-feeding arthropods acquire a blood meal (Lavoipierre, 1965). Vessel feeders, such as triatomine bugs, penetrate the skin and cannulate a superficial blood vessel with their mouthparts before they begin to feed. In contrast, pool feeders, such as ticks and tsetse flies, lacerate blood vessels with their mouthparts as they probe, and feed from the resulting extravascular hemorrhage. Fleas, like mosquitoes, were originally classified as vessel feeders, but this may be an oversimplification. The flea mouthparts include a pair of thin serrated laciniae that act as cutting blades to perforate the dermis. Alternating, rapid contractions and thrusts of the left and right laciniae pierce the skin, and this pneumatic drill-like cutting motion continues as the mouthparts move vertically and laterally in the dermal tissue during probing (Wenk, 1980), an activity that can cause hemorrhage. When blood is located, the laciniae and epipharynx come together to form a feeding channel, and feeding ensues. Whether the tip of the mouthparts is inserted into a vessel or in an extravascular pool of blood has obvious implications for plague pathogenesis. If intravascular feeding occurs, *Y. pestis* might be regurgitated directly into the blood stream (i.v. transmission). If extravascular feeding occurs, intradermal transmission is the appropriate model (the flea mouthparts are not long enough to penetrate into the subcutaneous tissue). The most careful observations of flea feeding (Deoras and Prasad, 1967; Lavoipierre and Hamachi, 1961) suggest that fleas can suck extravascular blood that leaks from a capillary, but prefer to feed directly from a blood vessel. Whether blocked fleas show the same discretion, however, is unknown. The usual progression of bubonic plague, in which *Y. pestis* can produce a primary lesion at the fleabite site and disseminates first to the local draining lymph node, seems to better fit an intradermal transmission model.

Flea saliva is also secreted into the bite site. The saliva of all blood-feeding arthropod vectors contains anticoagulants, and may contain other factors that influence the outcome of transmission. For example, a component of sandfly saliva greatly enhances the infectivity of *Leishmania* (Titus and Ribeiro, 1988). Flea saliva is known to contain the anticoagulant apyrase, an enzyme which acts to inhibit platelet and neutrophil aggregation (Ribeiro et al., 1990), but this is the only component that has been identified to date.

THE TRANSMISSION PHENOTYPE OF *Y. PESTIS*

The phenotype of *Y. pestis* as it exits the flea and enters the mammal is clearly different from in vitro growth phenotypes. As described in previous sections, *Y. pestis* growth in the flea resembles a biofilm and is associated with an extracellular membrane. The infectious units transmitted by the flea may consist not only of individual *Y. pestis*, but small clumps of bacteria derived from the periphery of the proventriculus-blocking mass. If pieces of the biofilm are regurgitated by fleas, the bacteria within them may be protected from the initial encounter with the host innate immune response, because bacteria embedded in a biofilm have been shown to be more resistant to uptake or killing by phagocytes (Donlan and Costerton, 2002). Because known antiphagocytic factors such as the F1 capsule and the Type III secretion system are not produced by *Y. pestis* at the low temperature of the flea gut (Straley and Perry, 1995; Perry and Fetherston, 1997), the extracellular matrix associated with growth in the flea may provide initial protection until the known antiphagocytic virulence factors are synthesized. Secondarily, regurgitated aggregates that are larger than the diameter of the intradermal blood vessels would preclude direct intravenous transmission.

In nature, *Y. pestis* in a particular phenotype is transmitted along with flea saliva into an intradermal microenvironment. Details of the flea-bacteria-host interface during and after transmission have not been characterized, and cannot be satisfactorily mimicked by transmission using a needle and syringe. Consequently, aspects of host-parasite interactions specific to the unique context of the fleabite site are unknown and merit future investigation.

EVOLUTION OF ARTHROPOD-BORNE TRANSMISSION

Y. pestis provides a fascinating case study of how a bacterial pathogen can evolve a vector-borne route of transmission. Given the short evolutionary timeframe in which it occurred, the change from an enteric, food- and water-borne pathogen to systemic, insect-borne pathogen was too abrupt to result from the slow evolutionary process of random mutation of individual genes leading to natural selection. Instead, more rapid evolutionary processes were responsible, such as horizontal gene transfer and the fine-tuning of existing genetic pathways to perform new functions.

Carniel (2003) has proposed a sequential evolutionary scenario for the switch to vector-borne transmission in the yersiniae. Because the *ymt* (*pldA*) gene enhances survival in the flea digestive tract, a likely first step was acquisition of the 100-kb pFra plasmid by horizontal transfer to generate a *Y. pseudotuberculosis* (pFra) or pre-pestis 1 clone. *Y. pseudotuberculosis* can cause septicemia in rodents, so it probably was taken up by fleas periodically. This would have been a dead end, until the hypothetical pre-pestis 1 clone acquired the PLD activity encoded on the pFra replicon, which allowed it to survive in and colonize the flea midgut.

A second necessary step in the evolution of flea-borne transmission took advantage of a pre-existing biofilm-forming capacity, which involves the *hms* genes. *Y. pestis* makes an *hms*-mediated biofilm to produce an obstructing infection in the proventricular valve, which is required for efficient transmission. The *hms* genes are present in *Y. pseudotuberculosis*, but, curiously, most isolates that have been tested do not exhibit the pigmentation phenotype (Brubaker, 1991). Some *Y. pseudotuberculosis* strains do show the Congo red-binding pigmentation phenotype in vitro, but all *Y. pseudotuberculosis* that have been tested, whether pigmented or not, are unable to block the proventriculus of *X. cheopis* (B. J. Hinnebusch, unpublished). Thus, a separate, as yet undiscovered genetic change may have occurred in pre-pestis 1 to extend its biofilm-forming capacity to include the flea gut environment. The presumptive change likely affected the outer membrane in such a way as to enhance aggregate formation on the hydrophobic proventricular spines in the context of the flea digestive tract milieu.

A third important step in the evolution of flea-borne transmission occurred when the progenitor clone acquired the small plasmid containing the *pla* gene, which is thought to enable *Y. pestis* to disseminate from the fleabite site after transmission. The clone containing both of the new *Y. pestis*-specific plasmids has been referred to as pre-pestis 2 by Carniel (2003). Given the ecology of the *Y. pseudotuberculosis* ancestor, horizontal transfer of pFra and pPst could have occurred in a mammal, a flea, or the environment. Of course, it is likewise impossible to know with certainty the order in which the plasmids were transferred, or the plasmid donors, although molecular biology analyses may provide some clues. For example, the 100-kb pFra shares major sequence identity with a *Salmonella* Typhi plasmid, suggesting that the *Y. pseudotuberculosis* ancestor acquired what became pFra from a *Salmonella* donor (Prentice et al., 2001). This horizontal transfer to generate the *Y. pseudotuberculosis* (pFra) clone may have occurred in the digestive tract of a rodent, since both bacteria are enteric pathogens. On the other hand, plasmid transfer by conjugation occurs readily in mixed bacterial biofilms, both in the environment and in the flea gut (Hinnebusch et al., 2002a).

COEVOLUTION OF FLEA-BORNE TRANSMISSION AND INCREASED VIRULENCE IN *Y. PESTIS*

The evolutionary path that led to flea-borne transmission also led to *Y. pestis* becoming one of the most virulent and feared pathogens of human history. It is probably no accident that increased virulence coevolved with vector-borne transmission. In fact, reliance on the flea for transmission imposed new selective pressures that would have strongly favored this. Some consideration of the dynamics of the *Y. pestis*-flea relationship serve to reinforce this point (Figure 4). First, flea-borne transmission is actually quite inefficient, which may reflect the fact that *Y. pestis* has only recently adapted to its insect host. The number of *Y. pestis* needed to infect 50% of susceptible mammals (the ID_{50}, often referred to as the minimum infectious dose) is the same as the 50% lethal dose (LD_{50})– less than 10 (Perry and Fetherston, 1997). In contrast, the ID_{50} of *Y. pestis* for *X. cheopis* is about 10^4 bacteria (Hinnebusch et al., 1996; Pollitzer, 1954). Fleas take small blood meals (0.1-0.3 μl), so *Y. pestis* must achieve levels approaching 10^8 per milliliter in the peripheral blood in order to have a 50% chance of infecting its vector. Bacteremias of 10^9 per milliliter are routinely present in moribund white laboratory mice (Douglas and Wheeler, 1943). The concept of a very high threshold level of bacteremia, below which infection of feeding fleas does not occur or is rare, is supported by the observations of several investigators (Douglas and Wheeler, 1943; Pollitzer, 1954; Kartman and Quan, 1964). Thus, *Y. pestis* does not infect the flea very efficiently in the first place, and this would have been strong selective pressure favoring more invasive, and consequently, more virulent strains able to produce the severe bacteremia that typifies plague.

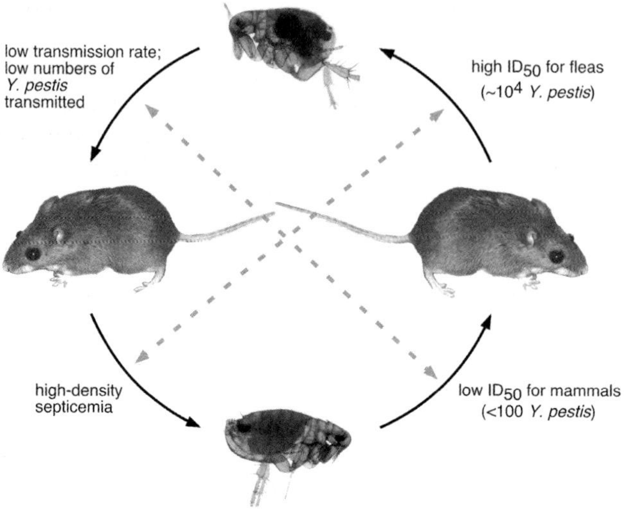

Figure 4. Dynamics of flea-borne transmission of plague. The high ID_{50} of *Y. pestis* for fleas is compensated for by the ability to produce a high-density septicemia in the rodent. The inefficient transmission by blocked fleas is compensated for by a low ID_{50} (LD_{50}) from a peripheral inoculation site.

A second weak link in the *Y. pestis* life cycle is that, even after successful infection of the vector, subsequent transmission is not very efficient. Not all infected fleas develop transmissible proventricular infections– for *X. cheopis*, the rate is only about 50%, and this rate can be much lower for other flea species (Wheeler and Douglas, 1945; Burroughs, 1947; Pollitzer, 1954). Furthermore, it is well established that the bite of a blocked flea does not always result in disease. Past studies report that individual blocked *X. cheopis* fleas that feed on a susceptible host transmit plague only about 50% of the time (Burroughs, 1947). Very few studies have addressed the number of *Y. pestis* transmitted by a single infectious fleabite. Burroughs (1947) triturated and plated mouse-skin biopsies taken from the site where a blocked flea had been allowed to attempt to feed. Only one of thirty samples was positive, which contained 88 *Y. pestis* colony-forming units (CFU). This number was multiplied by a factor to compensate for the low plating efficiency from other skin samples into which known numbers of *Y. pestis* had been injected, to calculate an estimate of 11,000 to 24,000 bacteria transmitted by the flea. However, this estimate was derived indirectly from a single positive sample, which seems less than satisfactory. We recently reexamined this topic by allowing individual blocked *X. cheopis* fleas to feed on a mouse, collecting a skin biopsy of the bite site after the flea had attempted to feed, and evaluating the number of *Y. pestis* deposited in the skin by quantitative PCR or by CFU count. Only half of the skin samples were positive, and the median number of *Y. pestis* transmitted was less than 100. This rather erratic transmission, frequently involving only a small number of bacteria, would have favored the selection of more invasive clones (such as the ancestral $pPla^+$ clone), able to successfully disseminate from the fleabite site even from a small initial dose.

Because reliance on the flea vector to complete the *Y. pestis* life cycle is contingent upon the ability to produce a high-density bacteremia and to invade from a small infectious dose in the skin, the evolution of vector-borne transmission and of increased virulence would have been mutually reinforcing. Parasite evolution is often assumed to evolve towards commensalism, or a state of peaceful coexistence with the host. Actually, the degree of virulence that is evolved by a successful parasite is functionally coupled with transmissibility (Ewald, 1983). That is, pathogens will tend to evolve to a level of virulence that optimizes their chance of successful transmission. Paul Ewald (1983) has argued that parasites transmitted by blood-feeding arthropods pay little cost for harming their hosts, and may actually benefit by being virulent. The extensive reproduction and hematogenous spread associated with severe disease increases the probability that a blood-feeding vector would acquire an infectious dose; and host immobilization and morbidity would not hinder, and may even enhance, the ability of a vector to find the host and feed to repletion.

The risk for a pathogen as virulent as *Y. pestis* is killing the host too quickly for transmission to occur. *Y. pestis* may have gotten away with the risky strategy of being rapidly fatal because most mammals maintain a more or less permanent flea population in their fur and in their burrows. It is not unusual for a single rodent to carry 5 to 80 fleas, with more than 100 fleas present in the nest or burrow, and it is simply a fact of life for most rodents to experience daily fleabites throughout their lives (Traub, 1972). Thus, *Y. pestis* does not have to rely on a chance encounter with a flying vector like a mosquito. Even if there are only a few hours between the development of high-density septicemia and death, there is good probability that one or more resident fleas will take an infectious blood meal during that short window of time. Because fleas do not readily leave their host mammal, killing the host may actually be important to *Y. pestis* transmission. Brubaker (2000) has made the point that death of the host compels the resident infected fleas to seek a new, healthy host– a necessary step to complete the transmission cycle.

Arthropod-transmitted pathogens may be indifferent to or even benefit from producing morbidity and mortality in their mammalian hosts, but infections in the arthropod vector are rarely symptomatic. As a rule it would be counterproductive to impair the health of the vector they rely on for transmission. For example, malaria parasites do not cause morbidity in the mosquitoes that transmit them to new hosts. *Y. pestis* breaks this rule– it causes an overwhelming, fatal infection in the flea as well as the mammal. A single blocked flea contains greater than 10^6 *Y. pestis* on average, and proventricular blockage leads to dehydration and death by starvation within a few days (Hinnebusch et al., 1998b; Kartman and Prince, 1956; Pollitzer, 1954)). During this time, however, a blocked flea will make persistent, frequent attempts to feed. A normal flea probes the skin and feeds to repletion within a few minutes. A blocked flea, in contrast, will try to take a blood meal in one location for a period of several minutes, then withdraw, move to another location, and try again. As it gradually starves, this process is repeated many times. Finally, the dehydrated blocked flea often remains with its mouthparts embedded in the skin for an hour or more, expending its last bit of energy in a futile attempt to satisfy its appetite. This anomalous feeding behavior provides multiple opportunities for transmission to occur before the flea dies, somewhat analogous to the manner in which the altered behavior of a rabid dog enhances the transmission of rabies virus.

CONCLUSIONS

Y. pestis exhibits a distinct life stage in its flea vector. The barriers to infection and the environmental conditions it faces in the flea digestive tract are quite different from those encountered in the mammalian host. Consequently, the development of a transmissible infection requires a distinct subset of genes. None of the known mammalian virulence factors that have been tested (the F1 capsule, *pla*, the type III secretion system encoded on the pYV virulence plasmid, and the HPI) are required in *X. cheopis* (Hinnebusch et al., 1996; 1998). Conversely, the two genetic loci (*ymt* and *hms*) that have been shown to be required for the flea-specific phenotype are not required for virulence in the mammal. Thus, a distinction can be made between *Y. pestis* genes required for pathogenesis in the mammal (virulence factors) and the genes required to produce a transmissible infection in the flea (transmission factors).

When a flea takes up *Y. pestis* in a blood meal, the bacteria experience a drop in temperature from 37°C to the ambient temperature of the flea. This temperature shift appears to be an important environmental signal for the bacteria to regulate gene expression appropriate to the invertebrate or vertebrate host. The expression of many *Y. pestis* virulence factors is upregulated at 37°C compared to room temperature (Straley and Perry, 1995; Perry and Fetherston, 1997), whereas the upregulation of the Hms phenotype and the *ymt* gene at room temperature compared to 37°C was a predictive initial clue that their biological role might occur in the flea. With the advent of DNA microarray and proteomics technologies, the global effect of the temperature shift from mammal to flea on *Y. pestis* gene expression can be analyzed, which may identify new candidate transmission factors.

In this chapter, I have focused on recent work using *X. cheopis* as the animal model and genetically defined strains of *Y. pestis* KIM (biovar Medievalis) and 195/P (biovar Orientalis). Earlier work, particularly by Russian investigators, led to often contradictory conclusions, with some studies suggesting a role in flea infection or blockage for the F1 capsule and Pla virulence factors, and others finding no role (Kutyrev et al., 1992; McDonough et al., 1993; Anisimov, 1999). Such reported differences in host-parasite relationships have yet to be resolved, but if nothing else serve as a reminder of the ecological complexity of plague, which can involve over 200 species of mammals and their fleas (Pollitzer, 1954). Factors that could explain contradictory data include differences in: 1) the digestive tract physiology and proventricular anatomy of the various flea species investigated; 2) the biochemical composition of the blood of different rodent species; 3) *Y. pestis* strains used; and 4) the temperature at which the fleas were maintained (Table 1).

Three discrete genetic steps that led to the recent evolutionary transition of *Y. pestis* from an enteric to a flea-borne route of transmission can now be identified. Two of them involved horizontal transfer of the *ymt*- and *pla*-harboring plasmids that are unique to *Y. pestis*. The third step involved adapting the pre-existing *hms* chromosomal genes to a new function– biofouling of the proventriculus to interfere with its normal valvular operation. Several unanswered questions remain. The biochemical mechanisms of action of the PLD and the Hms proteins in the flea have yet to be fully characterized. Additional transmission factors probably remain to be discovered before a complete recounting of the adaptation to the flea vector can be told. The *Y. pestis* genome also contains many pseudogenes, and the consequence, if any, of this large-scale gene loss on the interaction with the flea remains to be explored. Careful comparison of *Y. pseudotuberculosis* and *Y. pestis* genomes should provide important insights into these and other questions. Identifying and characterizing the molecular mechanisms that ensued from the specific genetic changes responsible for flea-borne transmission will ultimately provide an instructive case study in the evolution of bacterial pathogenesis.

ACKNOWLEDGEMENTS

I thank Clayton Jarrett, Roberto Rebeil, and Florent Sebbane for their contributions to the work described in this review and for discussions about it. Robert Perry and Andrey Anisimov kindly provided preprints of in press manuscripts. Research in my laboratory is supported in part by a New Scholars in Global Infectious Diseases award from the Ellison Medical Foundation.

REFERENCES

Achtman, M., Zurth, K., Morelli, G., Torrea, G., Guiyoule, A., and Carniel, E. 1999. *Yersinia pestis*, the cause of plague, is a recently emerged clone of *Yersinia pseudotuberculosis*. Proc. Natl. Acad. Sci. USA. 96: 14043-14048.

Ajl, S.F., Reedal, J.S., Durrum, E.L., and Warren, J. 1955. Studies on plague. I. Purification and properties of the toxin of *Pasteurella pestis*. J. Bacteriol. 70: 158-169.

Anisimov, A.P. 1999. [Factors providing the blocking activity of *Yersinia pestis*]. Mol. Gen. Mikrobiol. Virusol. (Russ.) 4: 11-15.

Anisimov, A.P., Lindler, L.E., and Pier, G.B. 2004. Intraspecific diversity of *Yersinia pestis*. Clin. Microbiol. Rev. in press.

Azad, A.F., Radulovic, S., Higgins, J.A., Noden, B.H., and Troyer, J.M. 1997. Flea-borne rickettsioses: ecologic considerations. Emerg. Infect. Dis. 3: 319-327.

Bacot, A.W. 1915. Further notes on the mechanism of the transmission of plague by fleas. J. Hygiene Plague Suppl. 4: 14: 774-776.

Bacot, A.W., and Martin, C.J. 1914. Observations on the mechanism of the transmission of plague by fleas. J. Hygiene Plague Suppl. 3. 13: 423-439.

Barbour, A.G., and Hayes, S.F. 1986. Biology of *Borrelia* species. Microbiol. Rev. 50: 381-400.

Beard, C.B., Butler, J.F., and Hall, D.W. 1990. Prevalence and biology of endosymbionts of fleas (Siphonaptera: Pulicidae) from dogs and cats in Alachua County, Florida. J. Med. Entomol. 27: 1050-1061.

Beaucournu, J.C. 1999. [Diversity of flea vectors as a function of plague foci]. Bull. Soc. Pathol. Exot. 5: 419-421.

Beesley, E.D., Brubaker, R.R., Janssen, W.A., and Surgalla, M.J. 1967. Pesticins. III. Expression of coagulase and mechanism of fibrinolysis. J. Bacteriol. 94: 19-26.

Bibikova, V.A. 1977. Contemporary views on the interrelationships between fleas and the pathogens of human and animal diseases. Ann. Rev. Entomol. 22: 23-32.

Bomchil, N., Watnick, P., and Kolter, R. 2003. Identification and characterization of a *Vibrio cholerae* gene, *mbaA*, involved in maintenance of biofilm architecture. J. Bacteriol. 185: 1384-1390.

Brown, S.D., and Montie, T.C. 1977. Beta-adrenergic blocking activity of *Yersinia pestis* murine toxin. Infect. Immun. 18: 85-93.

Brubaker, R.R. 1969. Mutation rate to nonpigmentation in *Pasteurella pestis*. J. Bacteriol. 98: 1404-1406.

Brubaker, R.R. 1991. Factors promoting acute and chronic diseases caused by yersiniae. Clin. Microbiol. Rev. 4: 309-324.

Brubaker, R.R. 2000. *Yersinia pestis* and bubonic plague. In: The Prokaryotes, an evolving electronic resource for the microbiological community. M. Dworkin, S. Falkow, E. Rosenberg, K.-H. Schleifer, and E. Stackelbrandt, ed. Springer Verlag, New York. Online.

Buchrieser, C., Prentice, M., and Carniel, E. 1998. The 102-kilobase unstable region of *Yersinia pestis* comprises a high-pathogenicity island linked to a pigmentation segment which undergoes internal rearrangement. J. Bacteriol. 180: 2321-2329.

Burroughs, A.L. 1947. Sylvatic plague studies. The vector efficiency of nine species of fleas compared with *Xenopsylla cheopis*. J. Hygiene. 45: 371-396.

Carniel, E. 2003. Evolution of pathogenic *Yersinia*, some lights in the dark. In: The Genus *Yersinia*: Entering the Functional Genomic Era. M. Skurnik, J.A. Bengoechea, and K. Granfors, ed. Plenum, New York. p. 3-11.

Cavanaugh, D.C. 1971. Specific effect of temperature upon transmission of the plague bacillus by the oriental rat flea, *Xenopsylla cheopis*. Am. J. Trop. Med. Hyg. 20: 264-273.

Cavanaugh, D.C., and Marshall, J.D., Jr. 1972. The influence of climate on the seasonal prevalence of plague in the Republic of Vietnam. J. Wildl. Dis. 8: 85-94.

Chapple, P.J., and Lewis, N.D. 1965. Myxomatosis and the rabbit flea. *Nature*. 207: 388-389.

Cherepanov, P.A., Mikhailova, T.G., Karimova, G.A., Zakharova, N.M., Ershov, I.V., and Volkovoi, K.I. 1991. [Cloning and detailed mapping of the fra-ymt region of the *Yersinia pestis* pFra plasmid]. Mol. Gen. Mikrobiol. Virusol. 12: 19-26.

Chomel, B.B., Kasten, R.W., Floyd-Hawkins, K., Chi, B., Yamamoto, K., Roberts-Wilson, J., Gurfield, A.N., Abbott, R.C., Pedersen, N.C., and Koehler, J.E. 1996. Experimental transmission of *Bartonella henselae* by the cat flea. J. Clin. Microbiol. 34: 1952-1956.

Costerton, J.W., Lewandowski, Z., Caldwell, D.E., Korber, D.R., and Lappin-Scott, H.M. 1995. Microbial biofilms. Ann. Rev. Microbiol. 49: 711-745.

Costerton, J.W., Stewart, P.S., and Greenberg, E.P. 1999. Bacterial biofilms: a common cause of persistent infections. Science. 284: 1318-1322.

D'Argenio, D.A., Calfee, M.W., Rainey, P.B., and Pesci, E.C. 2002. Autolysis and autoaggregation in *Pseudomonas aeruginosa* colony morphology mutants. J. Bacteriol. 184: 6481-6489.

Darby, C., Hsu, J.W., Ghori, N., and Falkow, S. 2002. *Caenorhabditis elegans*: plague bacteria biofilm blocks food intake. Nature. 417: 243-244.

Deoras, P.J., and Prasad, R.S. 1967. Feeding mechanism of Indian fleas *X. cheopis* (Roths) and *X. astia* (Roths). Indian J. Med. Res. 55: 1041-1050.

Dimopoulus, G., Richman, A., Müller, H.-M., and Kafatos, F.C. 1997. Molecular immune responses of the mosquito *Anopheles gambiae* to bacteria and malaria parasites. Proc. Natl. Acad. Sci. USA. 94: 11508-11513.

Donlan, R.M., and Costerton, J.W. 2002. Biofilms: survival mechanisms of clinically relevant microorganisms. Clin. Microbiol. Rev. 15: 167-193.

Douglas, J.R., and Wheeler, C.M. 1943. Sylvatic plague studies. II. The fate of *Pasteurella pestis* in the flea. J. Inf. Dis. 72: 18-30.

Drozdov, I.G., Anisimov, A.P., Samoilova, S.V., Yezhov, I.N., Yeremin, S.A., Karlyshev, A.V., Krasilnikova, V.M., and Kravchenko, V.I. 1995. Virulent non-capsulate *Yersinia pestis* variants constructed by insertion mutagenesis. J. Med. Microbiol. 42: 264-268.

Du, Y., Galyov, E., and Forsberg, A. 1995. Genetic analysis of virulence determinants unique to *Yersinia pestis*. Contrib. Microbiol. Immunol. 13: 321-324.

Ewald, P.W. 1983. Host-parasite relations, vectors, and the evolution of disease severity. Ann. Rev. Ecol. Syst. 14: 465-485.

Fetherston, J.D., and Perry, R.D. 1994. The pigmentation locus of *Yersinia pestis* KIM6+ is flanked by an insertion sequence and includes the structural genes for pesticin sensitivity and HMWP2. Mol. Microbiol. 13: 697-708.

Fetherston, J.D., Schuetze, P., and Perry, R.D. 1992. Loss of the pigmentation phenotype in *Yersinia pestis* is due to the spontaneous deletion of 102 kb of chromosomal DNA which is flanked by a repetitive element. Mol. Microbiol. 6: 2693-2704.

Gomez-Cambronero, J., and Keire, P. 1998. Phospholipase D: a novel major player in signal transduction. Cell Signal. 10: 387-397.

Götz, F. 2002. *Staphylococcus* and biofilms. Mol. Microbiol. 43: 1367-1378.

Hare, J.M., and McDonough, K.A. 1999. High-frequency RecA-dependent and -independent mechanisms of Congo red binding mutations in *Yersinia pestis*. J. Bacteriol. 181: 4896-4904.

Heilmann, C., and Götz, F. 1998. Further characterization of *Staphylococcus epidermidis* transposon mutants deficient in primary attachment or intercellular adhesion. Zentralbl. Bakteriol. 287: 69-83.

Heilmann, C., Schweitzer, O., Gerke, C., Vanittanakom, N., Mack, D., and Götz, F. 1996. Molecular basis of intercellular adhesion in the biofilm-forming *Staphylococcus epidermidis*. Mol. Microbiol. 20: 1083-1091.

Hinnebusch, B.J., Fischer, E.R., and Schwan, T.G. 1998a. Evaluation of the role of the *Yersinia pestis* plasminogen activator and other plasmid-encoded factors in temperature-dependent blockage of the flea. J. Inf. Dis. 178: 1406-1415.

Hinnebusch, B.J., Gage, K.L., and Schwan, T.G. 1998b. Estimation of vector infectivity rates for plague by means of a standard curve-based competitive polymerase chain reaction method to quantify *Yersinia pestis* in fleas. Am. J. Trop. Med. Hyg. 58: 562-569.

Hinnebusch, B.J., Perry, R.D., and Schwan, T.G. 1996. Role of the *Yersinia pestis* hemin storage (*hms*) locus in the transmission of plague by fleas. Science. 273: 367-370.

Hinnebusch, B.J., Rosso, M.-L., Schwan, T.G., and Carniel, E. 2002a. High-frequency conjugative transfer of antibiotic resistance genes to *Yersinia pestis* in the flea midgut. Mol. Microbiol. 46: 349-354.

Hinnebusch, B.J., Rudolph, A.E., Cherepanov, P., Dixon, J.E., Schwan, T.G., and Forsberg, Å. 2002b Role of *Yersinia* murine toxin in survival of *Yersinia pestis* in the midgut of the flea vector. Science. 296: 733-735.

Hinnebusch, J., Cherepanov, P., Du, Y., Rudolph, A., Dixon, J.D., Schwan, T., and Forsberg, A. 2000. Murine toxin of *Yersinia pestis* shows phospholipase D activity but is not required for virulence in mice. Int. J. Med. Microbiol. 290: 483-487.

Hirst, L.F. 1923. On the transmission of plague by fleas of the genus *Xenopsylla*. Indian J. Med. Res. 10: 789-820.

Hopla, C.E. 1974. The ecology of tularemia. Adv. Vet. Sci. Comp. Med. 18: 25-52.

Iteman, I., Guiyoule, A., de Almeida, A.M., Guilvout, I., Baranton, G., and Carniel, E. 1993. Relationship between loss of pigmentation and deletion of the chromosomal iron-regulated *irp2* gene in *Yersinia pestis*: evidence for separate but related events. Infect. Immun. 61: 2717-2722.

Jackson, S., and Burrows, T.W. 1956a. The pigmentation of *Pasteurella pestis* on a defined medium containing haemin. Br. J. Exp. Pathol. 37: 570-576.

Jackson, S., and Burrows, T.W. 1956b. The virulence-enhancing effect of iron on nonpigmented mutants of virulent strains of *Pasteurella pestis*. Brit. J. Exptl. Pathol. 37: 577-583.

Jarrett, C.O., Deak, E., Isherwood, K.E., Oyston, P.C., Fischer, E. R., Whitney, A.R., Kobayashi, S.D., DeLeo, F.R., and Hinnebusch, B.J. 2004. Transmission of *Yersinia pestis* from an infectious biofilm in the flea vector. J. Inf. Dis. in press.

Jawetz, E., and Meyer, K.F. 1944. Studies on plague immunity in experimental animals. II. Some factors of the immunity mechanism in bubonic plague. J. Immunol. 49: 15-29.

Jones, H.A., Lillard, J.W., Jr., and Perry, R.D. 1999. HmsT, a protein essential for expression of the haemin storage (Hms+) phenotype of *Yersinia pestis*. Microbiology. 145: 2117-2128.

Joshua, G.W.P., Karlyshev, A.V., Smith, M.P., Isherwood, K.E., Titball, R.W., and Wren, B.W. 2003. A *Caenorhabditis elegans* model of *Yersinia* infection: biofilm formation on a biotic surface. Microbiology. 149: 3221-3229.

Kartman, L. 1957. The concept of vector efficiency in experimental studies of plague. Exp. Parasitol. 6: 599-609.

Kartman, L., and Prince, F.M. 1956. Studies on *Pasteurella pestis* in fleas. V. The experimental plague-vector efficiency of wild rodent fleas compared with *Xenopsylla cheopis*, together with observations on the influence of temperature. Am. J. Trop. Med. Hyg. 5: 1058-1070.

Kartman, L., and Quan, S.F. 1964. Notes on the fate of avirulent *Pasteurella pestis* in fleas. Trans. R. Soc. Trop. Med. Hyg. 58: 363-365.

Kutyrev, V.V., Filippov, A.A., Oparina, O.S., and Protsenko, O.A. 1992. Analysis of *Yersinia pestis* chromosomal determinants Pgm^+ and Pst^s associated with virulence. Microb. Pathog. 12: 177-186.

Lavoipierre, M.M.J. 1965. Feeding mechanism of blood-sucking arthropods. Nature. 208: 302-303.

Lavoipierre, M.M.J., and Hamachi, M. 1961. An apparatus for observations on the feeding mechanism of the flea. Nature. 192: 998-999.

Lehane, M.J., Wu, D., and Lehane, S.M. 1997. Midgut-specific immune molecules are produced by the blood-sucking insect *Stomoxys calcitrans*. Proc. Natl. Acad. Sci. USA. 94: 1502-11507.

Lewis, K. 2000. Programmed death in bacteria. Microbiol. Mol. Biol. Rev. 64: 503-514.

Lewis, R.E. 1998. Résumé of the Siphonaptera (Insecta) of the world. J. Med. Entomol. 35: 377-389.

Lillard, J.W., Bearden, S.W., Fetherston, J.D., and Perry, R.D. 1999. The haemin storage (Hms+) phenotype of *Yersinia pestis* is not essential for the pathogenesis of bubonic plague in mammals. Microbiology. 145: 197-209.

Lillard, J.W., Fetherston, J.D., Pedersen, L., Pendrak, M.L., and Perry, R.D. 1997. Sequence and genetic analysis of the hemin storage (*hms*) system of *Yersinia pestis*. Gene. 193: 13-21.

McDonough, K.A., Barnes, A.M., Quan, T.J., Montenieri, J., and Falkow, S. 1993. Mutation in the *pla* gene of *Yersinia pestis* alters the course of the plague bacillus-flea (Siphonaptera: Ceratophyllidae) interaction. J. Med. Entomol. 30: 772-780.

McDonough, K.A., and Falkow, S. 1989. A *Yersinia pestis*-specific DNA fragment encodes temperature-dependent coagulase and fibrinolysin-associated phenotypes. Mol. Microbiol. 3: 767-775.

Montie, T.C., and Ajl, S.J. 1970. Nature and synthesis of murine toxins of *Pasteurella pestis*. In: Microbial Toxins, vol. 3. T.C. Montie, S. Kadis, and S.J. Ajl, ed. Academic Press, New York. p. 1-37.

Munshi, D.M. 1960. Micro-anatomy of the proventriculus of the common rat flea *Xenopsylla cheopis* (Rothschild). J. Parasitol. 46: 362-372.

Parkhill, J., Wren, B.W., Thomson, N.R., Titball, R.W., Holden, M.T., Prentice, M.B., Sebaihia, M., James, K.D., Churcher, C., Mungall, K.L., Baker, S., Basham, D., Bentley, S.D., Brooks, K., Cerdeno-Tarraga, A.M., Chillingworth, T., Cronin, A., Davies, R.M., Davis, P., Dougan, G., Feltwell, T., Hamlin, N., Holroyd, S., Jagels, K., Karlyshev, A.V., Leather, S., Moule, S., Oyston, P.C., Quail, M., Rutherford, K., Simmonds, M., Skelton, J., Stevens, K., Whitehead, S. and Barrell, B.G. 2001. Genome sequence of *Yersinia pestis*, the causative agent of plague. Nature. 413: 523-527.

Paskewitz, S.M. 1997. Transmission factors for insect-vectored microorganisms. Trends Microbiol. 5: 171-173.

Pendrak, M.L., and Perry, R.D. 1993. Proteins essential for the expression of the Hms+ phenotype of *Yersinia pestis*. Mol. Microbiol. 8: 857-864.

Perry, R.D., and Fetherston, J.D. 1997. *Yersinia pestis*-etiologic agent of plague. Clin. Microbiol. Rev. 10: 35-66.

Perry, R.D., Lucier, T.S., Sikkema, D.J., and Brubaker, R.R. 1993. Storage reservoirs of hemin and inorganic iron in *Yersinia pestis*. Infect. Immun. 61: 32-39.

Perry, R.D., Bobrov, A.G., Kirillina, O., Jones, H.A., Pedersen, L., Abney, J. and Fetherston, J.D. 2004. Temperature regulation of the Hemin storage (Hms+) phenotype of *Yersinia pestis* is posttranscriptional. J. Bacteriol. 186: in press.

Pollitzer, R. 1954. Plague. World Health Organization, Geneva.

Ponting, C.P., and Kerr, I.D. 1996. A novel family of phospholipase D homologues that includes phospholipid synthases and putative endonucleases: identification of duplicated repeats and potential active site residues. Protein Sci. 5: 914-922.

Prentice, M.B., James, K.D., Parkhill, J., Baker, S.G., Stevens, K., Simmonds, M.N., Mungall, K.L., Churcher, C., Oyston, P.C.F., Titball, R.W., Wren, B.W., Wain, J., Pickard, D., Hien, T.T., Farrar, J.J., and Dougan, G. 2001. *Yersinia pestis* pFra shows biovar-specific differences and recent common ancestry with a *Salmonella enterica* serovar Typhi plasmid. J. Bacteriol. 183: 2586-2594.

Ribeiro, J.M.C., Vaughan, J. A., and Azad, A. F. 1990. Characterization of the salivary apyrase activity of three rodent flea species. Comp. Biochem. Physiol. 95B: 215-219.

Ross, P., Weinhouse, H., Aloni, Y., Michaeli, D., Weinberger-Ohana, P., Mayer, R., Braun, S., de Vroom, E., van der Marel, G.A., van Boom, J.H., and Benziman, M. 1987. Regulation of synthesis in *Acetobacter xylinum* by cyclic diguanylic acid. Nature. 325: 279-281.

Rudolph, A.E., Stuckey, J.A., Zhao, Y., Matthews, H.R., Patton, W.A., Moss, J., and Dixon, J.E. 1999. Expression, characterization, and mutagenesis of the *Yersinia pestis* murine toxin, a phospholipase D superfamily member. J. Biol. Chem. 274: 11824-11831.

Sacks, D., and Kamhawi, S. 2001. Molecular aspects of parasite-vector and vector-host interactions in Leishmaniasis. Ann. Rev. Microbiol. 55: 453-483.

Samoilova, S.V., Samoilova, L.V., Yezhov, I.N., Drozdov, I.G., and Anisimov, A.P. 1996. Virulence of pPst+ and pPst- strains of *Yersinia pestis* for guinea-pigs. J. Med. Microbiol. 45: 440-444.

Savalev, V.N., Kozlov, M.P., Nadeina, V.P., and Reitblat, A.G. 1978. [A study of the microflora of fleas of rodents]. Med. Parazitol. Parazit. Bol. 47: 73-74.

Schär, M., and Meyer, K.F. 1956. Studies on immunization against plague. XV. The pathophysiologic action of the toxin of *Pasteurella pestis* in experimental animals. Schweiz. Z. Path. Bakt. 19: 51-70.

Sikkema, D.J., and Brubaker, R.R. 1987. Resistance to pesticin, storage of iron, and invasion of HeLa cells by yersiniae. Infect. Immun. 55: 572-578.

Sikkema, D.J., and Brubaker, R.R. 1989. Outer membrane peptides of *Yersinia pestis* mediating siderophore-independent assimilation of iron. Biol. Metals. 2: 174-184.

Sodeinde, O.A., and Goguen, J.D. 1988. Genetic analysis of the 9.5-kilobase virulence plasmid of *Yersinia pestis*. Infect. Immun. 56: 2743-2748.

Sodeinde, O.A., Subrahmanyam, Y.V., Stark, K., Quan, T., Bao, Y., and Goguen, J.D. 1992. A surface protease and the invasive character of plague. Science. 258: 1004-1007.

Stierhof, Y.-D., Bates, P.A., Jacobson, R.L., Rogers, M.E., Schlein, Y., Handman, E., and Ilg, T. 1999. Filamentous proteophosphoglycan secreted by *Leishmania* promastigotes forms gel-like three-dimensional networks that obstruct the digestive tract of infected sandfly vectors. Eur. J. Cell Biol. 78: 675-689.

Straley, S.C., and Perry, R.D. 1995. Environmental modulation of gene expression and pathogenesis in *Yersinia*. Trends Microbiol. 3: 310-317.

Stuckey, J.A., and Dixon, J.E. 1999. Crystal structure of a phospholipase D family member. Nat. Struct. Biol. 6: 278-284.

Surgalla, M.J. 1960. Properties of virulent and avirulent strains of *Pasteurella pestis*. Ann. N. Y. Acad. Sci. 88: 1136-1145.

Surgalla, M.J., and Beesley, E.D. 1969. Congo red agar plating medium for detecting pigmentation in *Pasteurella pestis*. Appl. Microbiol. 18: 834-837.

Terra, W.R., and Ferreira, C. 1994. Insect digestive enzymes: properties, compartmentalization and function. Comp. Biochem. Physiol. 109B: 1-62.

Titus, R.G., and Ribeiro, J.M.C. 1988. Salivary gland lysates from the sand fly, *Lutzomyia longipalpis*, enhance *Leishmania* infectivity. Science. 239: 1306-1308.

Traub, R. 1972. Notes on fleas and the ecology of plague. J. Med. Entomol. 9: 603.

Vaughan, J.A., and Azad, A.F. 1993. Patterns of erythrocyte digestion by bloodsucking insects: constraints on vector competence. J. Med. Entomol. 30: 214-216.

Walker, R.V. 1967. Plague toxins- a critical review. Curr. Top. Microbiol. Immunol. 41: 23-42.

Welkos, S.L., Friedlander, A.M., and Davis, K.J. 1997. Studies on the role of plasminogen activator in systemic infection by virulent *Yersinia pestis* strain C092. Microb. Pathogen. 23: 211-223.

Weiner, R., Seagren, E., Arnosti, C. and Quintero, E. 1999. Bacterial survival in biofilms: probes for exopolysaccharide and its hydrolysis, and measurements of intra- and interphase mass fluxes. Meth. Enzymol. 310: 403-418.

Wenk, P. 1980. How bloodsucking insects perforate the skin of their hosts. In: Fleas. R. Traub and H. Starcke, eds. A.A. Balkema, Rotterdam.

Wheeler, C.M., and Douglas, J.R. 1945. Sylvatic plague studies. V. The determination of vector efficiency. J. Inf. Dis. 77: 1-12.

Wigglesworth, V.B. 1972. The Principles of Insect Physiology. Chapman and Hall, London.

Chapter 5

N-Acylhomoserine Lactone-Mediated Quorum Sensing In *Yersinia*

Steven Atkinson, R. Elizabeth Sockett, Miguel Cámara and Paul Williams

ABSTRACT

Bacterial cell-to-cell communication ('quorum sensing') is mediated by structurally diverse, small diffusible signal molecules which regulate gene expression as a function of cell population density. Many different Gram-negative animal, plant and fish pathogens employ *N*-acylhomoserine lactones (AHLs) as quorum sensing signal molecules which control a variety of physiological processes including bioluminescence, swarming, antibiotic biosynthesis, plasmid conjugal transfer, biofilm development and virulence. AHL-dependent quorum sensing is highly conserved in both pathogenic and non-pathogenic members of the genus *Yersinia*. *Yersinia pseudotuberculosis* for example, produces at least six different AHLs and possesses two homologues of the LuxI family of AHL synthases and two members of the LuxR family of AHL-dependent response regulators. In all *Yersinia* species so far examined, the genes coding for LuxR and LuxI homologues are characteristically arranged convergently and overlapping. In *Y. pseudotuberculosis* AHL-dependent quorum sensing is involved in the control of cell aggregation and swimming motility, the latter via the flagellar regulatory cascade. This is also the case for swimming and also swarming motility in *Yersinia enterocolitica*. However the role of AHL-dependent quorum sensing in *Yersinia pestis* remains to be determined.

INTRODUCTION

Until relatively recently, cell-to-cell communication was rarely considered to constitute a major mechanism for facilitating bacterial adaptation to an environmental challenge. However, it is now clear that diverse bacterial genera communicate using specific, extracellular signal molecules, which facilitate the coordination of gene expression in a multi-cellular fashion. Signalling can be linked to specific environmental or physiological conditions and is employed by bacteria to monitor their cell population density. The term quorum sensing is commonly used to describe the phenomenon whereby the accumulation of a diffusible, low molecular weight signal molecule (sometimes called an 'autoinducer') enables individual bacterial cells to sense when the minimum number or quorum of bacteria has been achieved for a concerted response to be initiated (Williams et al., 2000; Swift et al., 2001; Cámara et al., 2002). The term 'autoinducer', implies a positive feedback or autoregulatory mechanism of action. However this is frequently not the case and therefore the term can be misleading and will be avoided here (Cámara et al., 2002). The accumulation of a diffusible signal molecule also indicates the presence of a diffusion barrier, which ensures that more molecules are produced than lost from the micro-habitat

(Winzer et al., 2002b; Redfield, 2002). This could be regarded as a type of 'compartment sensing', where signal molecule accumulation is both the measure for the degree of compartmentalisation and the means to distribute this information among the entire population. Similarly, diffusion of quorum sensing signal molecules between spatially separated bacterial sub-populations may convey information about their physiological state, their numbers, and the individual environmental conditions encountered.

At the molecular level, quorum sensing requires a synthase together with a signal transduction system for producing and responding to the signal molecule respectively. The latter usually involves a response regulator and/or sensor kinase protein (Swift et al., 2001; Cámara et al., 2002). While quorum sensing systems are ubiquitous in both Gram negative and Gram positive bacteria, there is considerable chemical diversity in the nature of the signal molecules involved which range from post-translationally modified peptides to quinolones, lactones and furanones (Cámara et al., 2002). In addition, siderophores, which previously were considered only in the context of iron transport, may also function in the producer organism as signal molecules capable of controlling genes unrelated to iron acquisition (Lamont et al., 2002). While there is, as yet, no clear cut evidence for a molecularly conserved quorum sensing system throughout the bacterial kingdom, the LuxS protein and the furanone generated from the ribosyl moiety of *S*-ribosylhomocysteine (termed AI-2 for autoinducer-2) have been suggested to fulfil such a role (Bassler, 2002). However, many bacteria (e.g. the pseudomonads) do not possess *luxS* and hence do not produce AI-2 (Winzer et al., 2002a). Furthermore, as LuxS is a key metabolic enzyme in the activated methyl cycle responsible for recycling *S*-adenosylmethionine (SAM), (Winzer et al., 2002b; Winzer et al., 2002a) phenotypes associated with mutation of *luxS* are often not a consequence of a defect in cell-to-cell communication but the result of the failure to recycle SAM metabolites (Winzer et al., 2002b; Winzer et al., 2002a). To date, only in *Vibrio harveyi* is there any direct experimental data to support the function of LuxS as a quorum sensing signal molecule synthase. Thus although *Yersinia* spp. possess a *luxS* gene and produce AI-2 (unpublished data) this does not constitute evidence for the presence of a quorum sensing system based on AI-2. In this chapter, the nature and contribution of *N*-acylhomoserine lactone (AHL)-mediated quorum sensing to the lifestyle of *Yersinia* spp. will be explored.

N-ACYLHOMOSERINE LACTONE-DEPENDENT QUORUM SENSING

The most intensively investigated family of quorum sensing signal molecules in Gram negative bacteria are the AHLs. These are produced by many different bacterial genera including *Aeromonas, Agrobacterium, Brucella, Burkholderia, Chromobacterium, Erwinia, Enterobacter, Pseudomonas, Rhizobium, Serratia, Vibrio* and *Yersinia* although not in all strains and species of the same genus (Swift et al., 2001). AHLs consist of a homoserine lactone ring covalently linked *via* an amide bond to an acyl side chain (Figure 1). To date, naturally occurring AHLs with chain lengths between 4 and 18 carbons have been identified which may be saturated or unsaturated and with or without a hydroxy-, oxo- or no substituent on the carbon at the 3 position of the *N*-linked acyl chain (Figure 1). AHLs are usually synthesised by enzymes belonging to the LuxI family, which employ the appropriately charged acyl acyl-carrier protein (acyl-ACP) as the major acyl chain donor while *S*-adenosyl methionine (SAM) provides the homoserine lactone moiety (Jiang et al., 1998; Watson et al., 2002; Moré et al., 1996). Recent data derived from the crystal structure of EsaI from *Pantoea stewartii* indicates that LuxI proteins

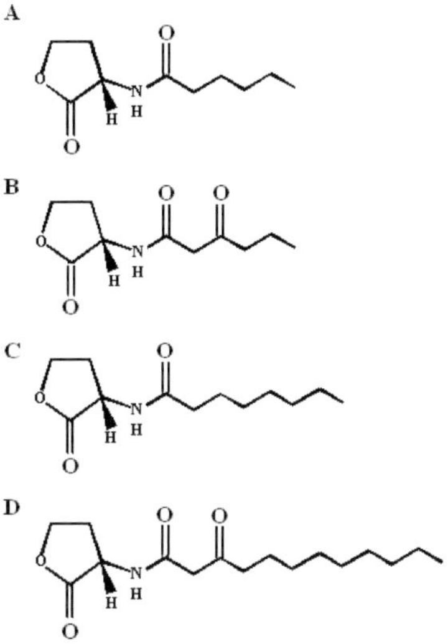

Figure. 1. Structures of AHLs produced by *Y. pseudotuberculosis*.
(A). *N*-hexanoyl homoserine lactone.
(B). *N*-(3-oxohexanoyl)homoserine lactone.
(C). *N*-octanoyl homoserine lactone.
(D). *N*-(3-oxododecanoyl)homoserine lactone.

share structural similarities with eukaryotic *N*-acetyltransferases which employ similar fatty acid precursors as substrates (Watson et al., 2002). Once synthesised, AHLs either move across the bacterial cell envelope by simple diffusion in the case of AHLs with a short acyl side chain, or are pumped out of the cell if the AHLs have a long e.g. C12 acyl side chain (Pearson et al., 1999).

Apart from the AHL synthase, the second regulatory component of an AHL-dependent quorum sensing circuit is usually a member of the LuxR family of transcriptional regulators (Swift et al., 2001; Cámara et al., 2002; Fuqua et al., 1996). AHLs bind to and activate LuxR proteins such that the LuxR/AHL complex is responsible for either activating or repressing target structural genes (Swift et al., 2001; Cámara et al., 2002; Zhang et al., 2002). The *V. fischeri* LuxR protein for example, consists of an N-terminal, cytoplasmic membrane-associated regulatory domain that binds AHLs and a cytoplasmic C-terminal domain that binds DNA. LuxR is considered to activate transcription as a dimer and a number of studies have identified regions within the protein required for multimerisation (Choi et al., 1992a), AHL binding (Shadel et al., 1990; Slock et al., 1990; Hanzelka et al., 1995), DNA binding, transcriptional activation (Choi et al., 1991; Choi et al., 1992b) and autorepression (Choi et al., 1991). The *V. fischeri* LuxR protein appears unable to associate with DNA until a conformational change has been induced *via* an interaction between the N-terminal AHL binding-domain and the cognate AHL molecule(s) (Choi et al., 1992b). However, among the growing family of LuxR homologues, it is now clear that there are variations between their respective mechanisms of transcriptional activation (Luo et al., 1999). In addition, certain LuxR homologues function as repressors rather than activators (e.g. EsaR) (von Bodman et al., 1998). To date the structure of only one LuxR homologue has been solved, that of TraR from *Agrobacterium tumefaciens* (Zhang et al., 2002). These data revealed that TraR forms a dimer and confirmed that the amino-terminal domain contains the AHL-binding region. However, the cognate AHL, *N*-3-(oxooctanoyl)homoserine lactone (3-oxo-C8-HSL) is completely buried within the protein and it appears that the AHL associates with nascent, actively folding TraR during protein synthesis. However this 'co-translational' mechanism may prove to be unique to TraR and does not appear to be representative of all LuxR homologues.

The DNA sequence of the LuxR binding site has been determined and is called the *lux* box. This is a 20 nucleotide inverted repeat, which, in *V. fischeri* is situated within the intergenic region between *luxR* and *luxI*, and is required for the primary regulation of *lux* gene expression. This DNA element has features that are conserved with the binding sites of other LuxR-type proteins and share the consensus RNSTGYAXGATNXTRCASRT where N = A, T, C or G; R = G or A; S = C or G; Y = C or T and X = N or a gap in the sequence (Stevens et al., 1997 Stevens et al., 1999). In *Pseudomonas aeruginosa* for example, *lux* box-like elements have been identified in the promoter regions of many different target structural genes (Whiteley et al., 2001) and purified LuxR proteins such as TraR (from *A. tumefaciens*) and ExpR (from *Erwinia chrysanthemi*) have been shown to bind *in vitro* to *lux* box-type sequences (Zhu et al., 1999; Nasser et al., 1998; Reverchon et al., 1998).

AHL-controlled multicellular behaviour in Gram negative bacteria includes a variety of physiological processes such as bioluminescence, swarming, swimming and sliding motility, antibiotic biosynthesis, biofilm maturation, plasmid conjugal transfer and the production of virulence determinants in animal, fish and plant pathogens (for reviews see Swift S. et al., 1999b; Swift et al., 2001; Williams P., 2002).

THE SEARCH FOR AHL-MEDIATED QUORUM SENSING IN *YERSINIA*

Because of their relative hydrophobicities, AHLs can readily be concentrated from culture supernatants by partitioning into organic solvents such as dichloromethane or ethyl acetate. However, since they lack good chromophores, relatively high concentrations are required for detection *via* their ultra violet spectroscopic properties following separation by HPLC. As a consequence, a number of sensitive bioassays have been developed which facilitate detection of AHL production by bacterial colonies on plates, in microtitre plate assays, on thin layer chromatograms (TLC) and in fractions collected after HPLC-based separation (Bainton et al., 1992; Throup et al., 1995; McClean et al., 1997; Shaw et al., 1997; Winson et al., 1998). For example, Throup et al., (1995) constructed a recombinant AHL reporter plasmid which coupled *luxR* and the *luxI* promoter region from *Vibrio fischeri* to the *luxCDABE* structural operon of *Photorhabdus luminescens*. When introduced into *E. coli*, this *lux*-based AHL reporter (termed pSB401) is dark but responds to the presence of exogenous AHLs by emitting light. The advantage of this AHL reporter compared with earlier versions employing *luxAB* alone (Bainton et al., 1992; Swift et al., 1993) is that it does not require the addition of a long chain fatty aldehyde (which is required by the luciferase) since the inclusion of the *luxCDE* genes provides for *in situ* synthesis of the aldehyde.

Using this AHL bioassay, cell free supernatants from several species of *Yersinia* including isolates of *Yersinia enterocolitica* belonging to serotypes O:3, O:8, O:9, O:10K, O:1(2a, 3), a non-typeable strain along with *Yersinia pseudotuberculosis* serotype III pIB1, *Yersinia frederiksenii*, *Yersinia kristensenii,* and *Yersinia intermedia* were screened. All of these strains induced bioluminescence in the reporter offering preliminary evidence that *Yersinia* spp. were AHL producers (Throup et al., 1995). Subsequently, *Yersinia pestis* (Swift et al., 1999a) and *Yersinia ruckeri* (unpublished data) were also confirmed as AHL-producers using the same strategy.

AHL-MEDIATED QUORUM SENSING IN *Y. ENTEROCOLITICA*

Although *E. coli* [pSB401] incorporates LuxR from *V. fischeri*, the cognate AHL for which is *N*-(3-oxohexanoyl)homoserine lactone (3-oxo-C6-HSL), it responds, with differing sensitivities, to a variety of AHLs (Winson et al., 1998). While a positive bioassay result confirms the presence of AHLs, it does not indicate either which AHL or indeed how many different AHLs may be present in the sample. To characterise unequivocally the AHLs present in spent *Y. enterocolitica* culture supernatants, they were first concentrated by solvent extraction and subjected to preparative HPLC. The fractions found to be positive in the AHL bioassay were subjected to high-resolution tandem mass spectrometry (MS-MS) which revealed the presence of two AHLs in an approximately 50:50 ratio. These AHLs were identified as 3-oxo-C6-HSL and *N*-hexanoyl homoserine lactone (C6-HSL) respectively (Figure 1) (Throup et al., 1995) and are easily separated and visualised by thin layer chromatography (TLC) as purple spots on a white background using the *Chromobacterium violaceum* CV026 biosensor in an agar overlay (See Figure 2 for an example of this TLC overlay originally described by McClean et al., 1997).

Since AHL synthesis depends on the presence of an AHL synthase, a plasmid-based *in trans* complementation strategy in which a *Y. enterocolitica* chromosomal gene library was transformed into *E. coli* [pSB401] was devised (Swift et al., 1993; Throup et al., 1995). This approach yielded a highly bioluminescent clone indicative of an insert carrying a gene whose product was capable of inducing the AHL reporter fusion. This clone was subsequently shown to be producing both 3-oxo-C6-HSL and C6-HSL and to contain two convergently transcribed and overlapping open reading frames (Figure 3). The first of these genes was termed *yenI*, the product of which exhibited homology to the LuxI protein family while the product of the second, termed *yenR* belongs to the LuxR protein family. TLC in conjunction with the CV026 biosensor showed that inactivation of *yenI* by deletion mutagenesis abolished AHL production in *Y. enterocolitica* confirming the function of *yenI* as an AHL synthase.

In *V. fischeri luxI* expression is governed by a positive feedback loop as a consequence of its induction by LuxR activated by 3-oxo-C6-HSL (Meighen, 1994). However, in *Y. enterocolitica*, *yenI* is not subject to such autoinduction since inactivation of *yenI* did not abolish production of *yenI* mRNA (Throup et al., 1995). This observation was confirmed by examining the expression of *yenI'::luxAB* transcriptional fusions as chromosomal

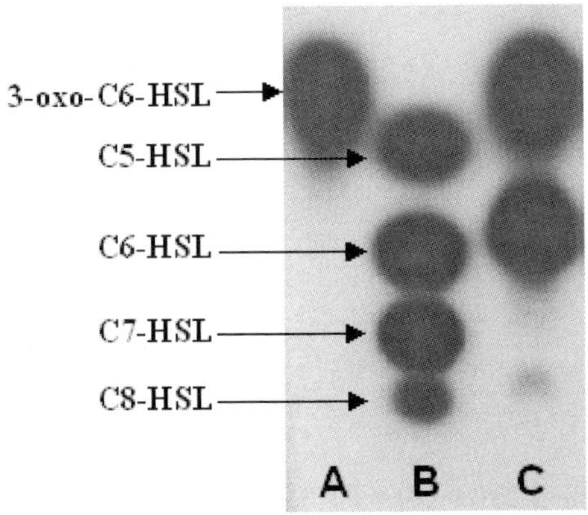

Figure 2. TLC chromatogram of the AHLs present in cell-free supernatants of *Y. pseudotuberculosis* grown with shaking in LB at 22°C and detected using the *C. violaceum* CV026 AHL biosensor. Lane A, 3-oxo-C6-HSL; 1.4 x 10^{-8} mol; lane B from top, AHL standards, *N*-pentanoylhomoserine lactone (C5-HSL) (5.4 x 10^{-8} mol), C6-HSL (5 x 10^{-9} mol), C7-HSL (4.7 x 10^{-7} mol) and C8-HSL (2.2 x 10^{-8} mol); lane C, *Y. pseudotuberculosis* supernatant extract from cultures grown at 22° C showing the presence of 3-oxo-C6-HSL, C6-HSL, and C8-HSL. It should be noted that CV026 does not respond to long chain AHLs and the size of the spots are not equivalent to the concentration of AHL present.

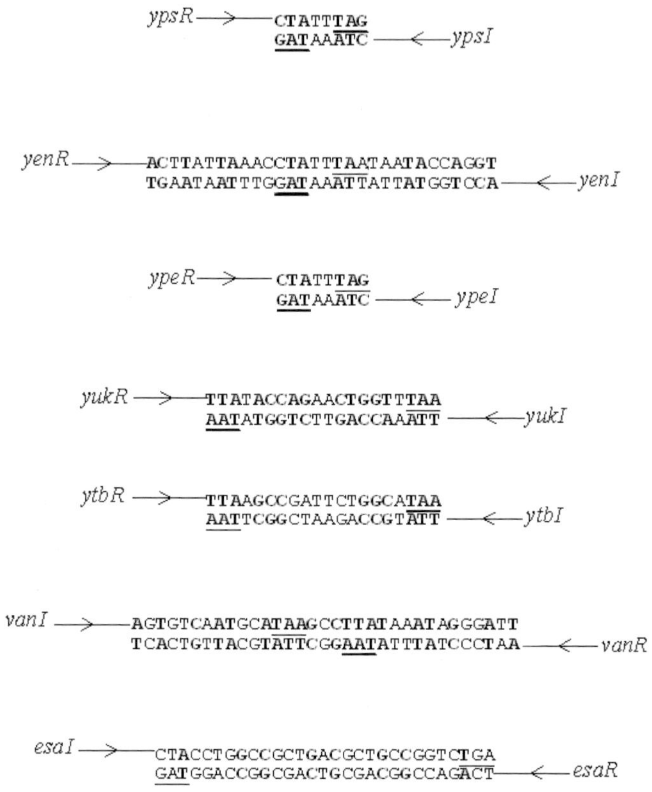

Figure 3. Organisation of genes coding for LuxR and LuxI homologues in *Yersinia* spp. Hyphenated inverted repeat sequences identified within convergently transcribed *luxR/I* homologues. Stop codons are underlined. Bold bases indicate inverted repeats. Key: *ypsR/I* and *ytbR/I*, *Yersinia pseudotuberculosis; yenR/I*, *Yersinia enterocolitica*; *ypeR/I*, *Yersinia pestis*; *yukR/I*, *Yersinia ruckeri*; *vanR/I*, *Vibrio anguillarum*; *esaR/I*, *Pantoea* (*Erwinia*) *stewartii.*

insertions in both the parent and *yenI* mutant respectively. In addition, the promoter region of *yenI* contains no matches for the *lux* box consensus sequence. Thus the expression of *yenI* is not part of an autoinducible system analogous to that of *V. fischeri* but is independent of growth phase.

A preliminary analysis of potential *Y. enterocolitica* target genes which may be controlled by AHL-dependent quorum sensing was undertaken by comparing 2D SDS-PAGE profiles of the *Y. enterocolitica* strains 90/54 (serotype O:9) and 10460 (biotype 3, serotype O:1,2a,3; NCTC 10460) and their corresponding *yenI* mutants. While no obvious differences between the 10460 parent and mutant were observed, the 90/54 *yenI* mutant profile lacked a number of proteins that were present in the parent strain. However, as yet, none of these proteins has been identified. The major differences between 10460 and 90/54 relate to virulence since the former lacks the pYV virulence plasmid and this implies a possible role for AHL-mediated quorum sensing in controlling plasmid-borne functions. Although we were unable to demonstrate a role for AHL-mediated quorum sensing in regulating type III Yop secretion in *Y. enterocolitica* (Throup et al.,

1995) the *yenI* mutant of 90/54, in contrast to the 10460 *yenI* mutant was unable to swim or swarm (unpublished data). Swarming motility is a flagellum-dependent behaviour that facilitates bacterial migration over solid surfaces and is distinct from swimming motility, which occurs in fluid environments. Such multicellular behaviour has been implicated in the formation of biofilms and in bacterial pathogenesis (Fraser et al., 1999). In the *Y. enterocolitica yenI* mutant, swimming and swarming could both be restored by complementation with a plasmid-borne copy of *yenI*. Although Y. *enterocolitica* strains 10460 and 90/54 are not isogenic, these data do provide preliminary evidence linking quorum sensing with the virulence plasmid even though the loss of motility and swarming are both likely to involve changes in expression of the chromosomally encoded flagellar genes. In bacteria such as *Serratia liquefaciens* (Lindum et al., 1998), *Burkholderia cepacia* (Huber et al., 2001) and *Pseudomonas aeruginosa* (Kohler et al., 2000) the loss of swarming motility in quorum sensing mutants is associated with the loss of biosurfactant production. For mutants defective in AHL production, swarming can be restored by the provision of either the appropriate AHL signal molecule(s) or a biosurfactant. In addition, although such mutants do not swarm they remain capable of swimming motility. In the *Y. enterocolitica* 90/54 *yenI* mutant, swarming could not be restored by providing an exogenous surfactant (unpublished data) and since the mutant also failed to swim, this suggests that AHL-mediated quorum sensing in this organism is directly linked to flagellin expression/function given that flagellar are required for both swimming and swarming motility. When analysed by SDS-PAGE the 90/54 *yenI* mutant fails to produce flagellin when compared to the parent (unpublished data). Furthermore, given that motility is required to initiate host cell invasion by *Y. enterocolitica* (Young et al., 2000), it is possible that AHL-mediated quorum sensing may also turn out to play a role in this context.

AHL-MEDIATED QUORUM SENSING IN *Y. PSEUDOTUBERCULOSIS*

TLC analysis of *Y. pseudotuberculosis* spent culture supernatants grown at (22°C) extracted with dichloromethane using both *C. violaceum* CV026 (Figure 2) and *E. coli* [pSB401] revealed the presence of at least three AHLs, the chemical identities of which were subsequently confirmed using HPLC and MS as 3-oxo-C6-HSL, C6-HSL and *N*-octanoyl-L-homoserine lactone (C8-HSL) (Figure 2). However, when grown to stationary phase at 37°C in LB medium, *Y. pseudotuberculosis* appears to produce extremely low levels of these AHLs. Since temperature plays an important role in the regulation of virulence genes in *Yersinia* (Kapperud et al., 1985; Cornelis et al., 1987; Pepe et al., 1993; Pierson et al., 1993; Cornelis et al., 1997; Bleves et al., 2000), AHL production was examined in stationary phase cultures grown for 24 h at 28°C and 22°C (Yates et al., 2002). By comparing the AHL profiles of cultures grown at the three different temperatures, AHL levels were found to be markedly dependent on growth temperature. The higher levels appeared to be greater after growth at 22°C, substantially reduced at 28°C and almost undetectable after growth at 37°C. Initially this suggested that either the expression or function of the corresponding AHL synthase(s) or substrate availability might be temperature dependent, implying that quorum sensing was temperature dependent. However, it subsequently became clear that there was a simple physico-chemical explanation. This is because the AHL ring is highly susceptible to pH-dependent lactonolysis. Ring opened compounds are unable to activate LuxR-type proteins and therefore do not activate AHL biosensors. By simply acidifying spent culture supernatants to below pH 2.0, lactonolysis can be reversed and the AHLs recovered from cultures grown at 37°C (Yates et al., 2002). In LB, which is unbuffered, the growth of *Y. pseudotuberculosis* rapidly renders the pH alkaline as a consequence of the metabolism of amino acids and the release of ammonia. The rate of pH-dependent lactonolysis is further enhanced by raising the incubation temperature. Thus, these parameters must be taken into consideration for any study which seeks to quantitate AHL levels in culture supernatants. Furthermore, the AHL profile is also dependent on the structure of the AHLs present. In the context of lactonolysis, the longer the acyl side chain of an AHL then the more stable it will be towards increases in pH and temperature such as those encountered *in vivo* in mammalian hosts. Indeed the data obtained by Yates et al. (2002) revealed that to be functional under physiological conditions in mammalian tissue fluids, AHLs require an acyl side chain of at least 4 carbons in length. This explains why homoserine lactone itself cannot function as a quorum sensing signal molecule as it rapidly undergoes lactonolysis at pH's above 2 (Yates et al., 2002). This in turn suggests that the production of a range of AHL signal molecules with different acyl side chain lengths provides a pathogen with greater versatility in adapting to changing environments.

Y. PSEUDOTUBERCULOSIS PRODUCES MULTIPLE SHORT AND LONG CHAIN AHLS

Using the same *trans*-complementation strategy described by Throup et al. (1995) i.e. the introduction of a genomic library into the AHL biosensor *E. coli* [pSB401], Atkinson et al (1999) cloned a *luxI* homologue from *Y. pseudotuberculosis* by screening for bioluminescent clones. This approach facilitated the identification of a LuxI homologue, YpsI and a LuxR homologue, YpsR. Deletion insertion mutations were introduced into both *ypsI* and *ypsR* and an analysis of the AHL profiles of both mutants revealed that after growth at 37°C, the mutants produced the same three AHLs (3-oxo-C6-HSL, C6-HSL and C8-HSL) as the wild type. However, the *ypsI* mutant produced no 3-oxo-C6-HSL at 28°C while the *ypsR* mutant produced reduced levels of 3-oxo-C6-HSL and C6-HSL and no C8-HSL could be detected (Atkinson et al., 1999). These data suggested that *Y. pseudotuberculosis* possessed an additional AHL synthase, the expression of which was likely therefore to be controlled by YpsR.

To locate this second putative AHL synthase, the same complementation strategy used above was employed but using genomic DNA prepared from the *Y. pseudotuberculosis ypsI* mutant (Atkinson et al., 1999). This approach yielded a clone containing two genes termed *ytbI* and *ytbR* which, when translated, exhibited homology to YpsI and YpsR and were therefore additional members of the LuxR and LuxI protein families. When expressed in *E. coli*, YtbI is responsible for the production of C6-HSL and C8-HSL and a molecule we tentatively identified as, *N*-heptanoylhomoserine lactone (C7-HSL) while YpsI generates 3-oxo-C6-HSL and C6-HSL (Atkinson et al., 1999 and unpublished data). These data suggest that there is a difference in the AHLs made *via* YtbI and YpsI in an *E. coli* genetic background and those made in *Y. pseudotuberculosis* since in the latter homologous background, YtbI can clearly produce all three major AHLs in the absence of YpsI (Atkinson et al., 1999). This finding was unexpected since the AHL profile of recombinant LuxI proteins from other bacteria expressed in *E. coli* has previously corresponded closely with the AHL profile in the original organism (Atkinson et al., 1999). Notwithstanding our observations with respect to pH-dependent lactonolysis, these data indicate that YtbI is able to vary the AHL profile produced in *Y. pseudotuberculosis* in a temperature-dependent manner and in the absence of YpsI. AHL production in *Y. pseudotuberculosis* is abolished when both *ypsI* and *ytbI* are mutated but not after deletion of both *ypsR* and *ytbR* indicating that, as in *Y. enterocolitica,* AHL production is not dependent on the presence of a functional YpsR or YtbR protein (Buckley, 2002).

Much of the analysis of AHL production in *Y. pseudotuberculosis* was undertaken using AHL biosensors which preferentially respond to short chain (C4-C8) AHLs (Figure 2). Recently, we have re-investigated AHL production in *Y. pseudotuberculosis* using an AHL biosensor (pSB1075) based on the LuxR homologue, LasR from *Pseudomonas aeruginosa*, the cognate AHL for which is the long chain *N*-(3-oxododecanoyl)homoserine lactone (3-oxo-C12-HSL). In pSB1075, we have incorporated *lasR* and the *lasI* promoter fused to *luxCDABE* (Winson et al., 1998). When *E. coli* [pSB1075] was used as an overlay in conjunction with a TLC system optimised for the chromatography of long chain AHLs employing reverse phase RP-2 TLC plates and 45% v/v methanol in water as the mobile phase (Yates et al., 2002), dichloromethane extracts of the *Y. pseudotuberculosis* wild type were found to contain three long chain AHLs. These were tentatively identified as *N*-(3-oxodecanoyl)homoserine lactone (3-oxo-C10-HSL), *N*-(3-oxododecanoyl)homoserine lactone (3-oxo-C12-HSL) and *N*-(3-oxotetradecanoyl) homoserine lactone (3-oxo-C14-HSL). None of these AHLs were made by the *ypsI/ytbI* double mutant nor by the *ytbI* mutant suggesting that YtbI was the synthase responsible for their synthesis. This was confirmed by demonstrating that *E. coli* expressing *ytbI* but not *ypsI* also produces the long chain AHLs (Buckley, 2002).

AHL-MEDIATED QUORUM SENSING CONTROLS MOTILITY AND CELL AGGREGATION IN *Y. PSEUDOTUBERCULOSIS*

When grown in LB at 28°C or 37°C, *Y. pseudotuberculosis ypsR* mutant cells aggregate, a phenomenon which does not occur at 22°C or in either the parent or *ypsI* mutant at any temperature (Atkinson et al., 1999) (Figure 4). This phenotype implied the involvement of surface components which were likely to be regulated *via* AHL-dependent quorum sensing. SDS-PAGE of cell surface extracts prepared from parent, *ypsI* and *ypsR* mutants revealed that the expression of numerous proteins was affected. In particular, a prominent protein of approximately 42 kDa was substantially up-regulated in the *ypsR* mutant (Atkinson et al., 1999). N-terminal sequence analysis indicated that this protein exhibited significant homology to the major flagellin structural proteins FleA, FleB and FleC of *Y. enterocolitica* (Kapatral et al., 1995). These data implicated *ypsR* in the control of flagellar mediated motility. On swimming motility plate assays, both the *ypsR* and *ypsI* mutants, but not the parent strain, were motile after 24h when grown at 22°C. None of the strains was motile after 24 h or 48 h at 37°C by phase contrast microscopy or on swarming motility plate assays. By following the induction of motility in liquid medium over the growth curve, it is clear that at 22°C, motility is induced after 6 h in both *ypsI* and *ypsR* mutants but delayed until 26 h in the parent (Atkinson et al., 1999). Thus cell population density appears to be one important environmental cue for activating swimming motility in *Y. pseudotuberculosis* (Atkinson et al., 1999). Furthermore, although the *ypsR/I* locus is involved in the repression of motility, mutation of this locus does not overcome the temperature dependence of swimming motility.

To determine whether the *ytbR/I* locus is also involved in the control of flagellar-mediated motility, both *ytbI* and *ytbR* genes were inactivated. We speculated that if the

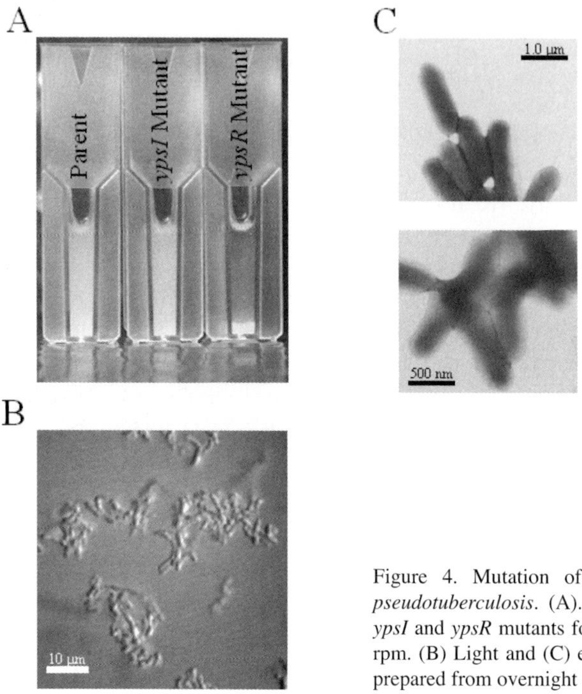

Figure 4. Mutation of *ypsR* causes cell aggregation in *Y. pseudotuberculosis*. (A). 24 h Static LB cultures of the parent *ypsI* and *ypsR* mutants following growth for 24 h shaking at 200 rpm. (B) Light and (C) electron micrographs of the *ypsR* mutant prepared from overnight cultures grown with shaking at 22°C.

ytbR/I system was not involved in the regulation of motility, the mutants would display the hypermotile phenotype characteristic of the *ypsR/I* mutants. However, in common with the parent strain, both *ytbI* and *ytbR* mutants were non motile even after 48 h at 22°C. While this initially implied that the *ytbR/I* system was not involved in controlling motility, the corresponding double I or double R mutants i.e. the *ypsI ytbI* and *ypsR ytbR* mutants were also non-motile. These data indicate that the *ytbR/I* locus may be an activator of motility while the *ypsR/I* locus functions to repress motility. The two quorum sensing circuits also appear to be linked since YpsR appears to be involved in the control of *ytbI* expression since *ypsR* mutants fail to produce the YtbI-generated AHL, C8-HSL during growth at 28°C.

Since *ypsI/ytbI* and *ypsR/ytbR* double mutants are non-motile while the *ypsR* and *ypsI* single mutants exhibit early motility in batch culture when compared to the wild type, it is likely that the quorum sensing circuitry controls the expression of key regulatory genes involved in the control of swimming motility. In enteric bacteria, flagellar gene transcription is arranged as a hierarchy with genes in three regulatory levels or classes which are responsible for producing and maintaining functional flagella and the chemotaxis apparatus. (MacNab, 1996). FlhD and FlhC activate the class II structural genes encoding the basal body and hook proteins, along with an activator FliA, (σ^F) and a repressor FliM (anti-σ^F) of class III genes. Class III genes encode the proteins which constitute the mature flagellar filament, chemotactic sensors and motor and are transcribed from FliA-dependent promoters in a temperature dependent manner (MacNab, 1996). Although the *flhDC* master regulatory system was once considered to be exclusively involved in regulating the production of flagella for swimming motility (Fraser et al., 1999) it is now known to play a more global role in regulating diverse physiological processes (Pearson et al., 2000; Tan et al., 1999). In *Y. enterocolitica* (Young et al., 1999) as well as *S. liquefaciens* (Givskov et al., 1997) and *P. mirabilis* (Fraser et al., 1999), *flhDC* is required for flagellin production, swimming motility and swarming motility.

An analysis of the *Y. pseudotuberculosis flhDC* and *fliA* promoter/operator region revealed the presence of putative *lux*-box like elements. By employing reporter gene fusions in single and double *ypsR/I* and *ytbR/I* mutants we have established that, for example *flhDC* expression is repressed by YpsR and YtbR at 37°C. In addition, YtbI appears to be involved as an activator of *flhDC* expression at temperatures below 28°C (unpublished data).

In a *Y. enterocolitica flhDC* mutant, Bleves et al., (2002) noted an increase in Yop expression and a loss of the temperature dependency usually associated with Yop expression at 37°C. Since the *Y. pseudotuberculosis* quorum sensing circuitry appears to control *flhDC*, this implies the existence of a regulatory link between type III secretion and motility. However such a link has to be demonstrated experimentally.

THE BIOLOGICAL SIGNIFICANCE OF LONG CHAIN AHLS PRODUCED BY *Y. PSEUDOTUBERCULOSIS*

Although only C6-HSL, 3-oxo-C6-HSL and C8-HSL were initially identified and fully chemically characterised (Atkinson et al., 1999), it is also clear, as indicated above that *Y. pseudotuberculosis* produces three long chain AHLs, 3-oxo-C10-HSL, 3-oxo-C12-HSL and 3-oxo-C14-HSL. These AHLs which are also produced *via* recombinant YtbI expressed in *E. coli* are not produced by recombinant YpsI. They are present in spent supernatants prepared from *ypsI* mutants but absent from spent culture supernatants prepared from *ytbI* mutants or *ypsI ytbI* double mutants, the latter of which make no detectable AHLs (Buckley, 2002). While the significance of this finding in the context of quorum sensing dependent gene regulation in *Y. pseudotuberculosis* is not yet clear, it is worth noting that 3-oxo-C12-HSL (which is also produced *via* LasI in *P. aeruginosa*, Passador et al., 1993; Pearson et al., 1994) possesses potent, dose dependent, pro-inflammatory, immune modulatory and vasorelaxant properties (Telford et al., 1998; Lawrence et al., 1999; Smith et al., 2001; Chhabra et al., 2003). For example 3-oxo-C12-HSL modulates both T and B cell functions, blocks lipopolysaccharide-stimulated production of tumour necrosis factor alpha (TNF-α) by peritoneal macrophages and mediates switching of the T-helper-cell response from the antibacterial Th-1 response (characterised by IL-12 and gamma interferon production) to a Th-2 response. In addition, 3-oxo-C12-HSL also exerts a pharmacological effect on the cardio-vascular system suggesting that host cardiovascular function may be modulated, or influenced, by bacterial quorum sensing molecules. In isolated porcine coronary arteries, 3-oxo-C12-HSL caused a concentration dependent relaxation effect on thromboxane mimetic-induced contractions (Lawrence et al., 1999) and also induces a marked bradycardia in live conscious rats.(Gardiner et al., 2001). Structure activity studies have indicated that the immune modulatory properties of AHLs are optimal in compounds with 11 to 13 acyl chain carbons and a 3-oxo substituent (Chhabra et al., 2003). Short chain (4 to 8 acyl carbons) AHLs have little activity in this context. Thus, 3-oxo-C12-HSL and to a lesser extent 3-oxo-C10-HSL and 3-oxo-C14-HSL may contribute to virulence directly through manipulation of eukaryotic cells and tissues to maximize the provision of nutrients *via* the bloodstream while down regulating host defence mechanisms

QUORUM SENSING IN *Y. PESTIS* AND *Y. RUCKERI*

Given the close genetic relationship between *Y. pseudotuberculosis* and *Y. pestis*, it is perhaps not surprising to find two pairs of LuxRI homologues, the genes for which have been termed *ypeRI* and *yepRI* in the latter (Isherwood, 2001; Swift et al., 1999a). In addition, preliminary studies on *Y. ruckeri*, which is responsible for enteric red mouth disease in salmonid fish, have also revealed the presence of a *luxRI* pair, termed *yukRI* (unpublished data). Each of these *Yersinia* LuxR and LuxI homologues share significant homology with each other and each genetic locus is organized similarly in that the two genes are convergently transcribed and overlapping by either 8 or 20 bp in the respective host species (Figure 3). Neither the genetic nor physiological consequences of such a gene arrangement are clear since *luxR* and *luxI* homologues in different organisms can be transcribed in tandem, divergently or convergently (Salmond et al., 1995; Swift et al., 1999a)

Using AHL biosensors such as *C. violaceum* CV026, initial cross streak analysis of *Y. pestis* isolates taken from a range of environmental and geographic sources revealed that AHL production is conserved in this *Yersinia* species. Subsequent analysis indicated that *Y. pestis*, in common with *Y. pseudotuberculosis* produces C6-HSL, 3-oxo-C6-HSL and C8-HSL (Swift et al., 1999a). Furthermore, YpeI (which is more closely related to YpsI and YenI than YtbI), directs the synthesis of C6-HSL and 3-oxo-C6-HSL suggesting that the second *Y. pestis* LuxI homologue, YepI is responsible for C8-HSL synthesis. Sequence analysis indicates that YepI is more closely related to YtbI than YpsI which is consistent with the nature of the respective AHLs produced. Whether YpeI in *Y. pestis* is also responsible for producing additional AHLs including compounds with long (C10-C14) acyl side chains is not yet known. For *Y. ruckeri*, preliminary TLC analysis of the recombinant YukI expressed in *E. coli* suggests that this fish pathogen also produces C6-HSL, 3-oxo-C6-HSL and C8-HSL (unpublished data). Indeed although analysis of the translated sequence for a given LuxI homologue provides little information on the nature of the corresponding AHL(s) produced, the presence of a threonine at position 143 in the carboxy terminal region relative to the *V. fischeri* LuxI protein (Watson et al., 2002) is characteristic of synthases which direct the synthesis of 3-oxo-substituted AHLs. This observation appears to hold for all six *Yersinia* LuxI homologues all of which are capable of 3-oxo-C6-HSL production.

As yet, the target structural genes controlled *via* the *yukR/I* locus in *Y. ruckeri* have not been identified and no mutants have been constructed. However in *Y. pestis*, both *ypeI* and *ypeR* have both been mutated. Despite the observation that quorum sensing is a key regulator of motility in both *Y. pseudotuberculosis* and *Y. enterocolitica*, *Y. pestis* does not exhibit any motility under the conditions tested at all temperatures even though the genome contains one full set of flagella genes plus a second partial set (Parkhill et al., 2001; Isherwood, 2001). Despite the similarity between *ypsI* and *ypeI*, *Y. pestis ypeI* mutants are also non-motile (Isherwood, 2001). Analysis of *Y. pestis* virulence gene expression in parent, *ypeI* and *ypeR* mutants grown at 28°C and 37°C did not reveal any differences with respect to V-antigen, pH 6 antigen, the coagulase/fibrinolytic protein Pla or lipopolysaccharide profile (Swift et al., 1999a). However in a mouse infection model an increase in time to death was observed for mice challenged with the mutant when compared to the parent. For example at a challenge dose of approximately 10^4 cfu the mean time to death increased from 5.4 days for mice challenged with *Y. pestis* GB to 6.7 days for mice challenged with the *ypeR* mutant implying a possible role for AHL-dependent quorum sensing in the pathogenesis of *Y. pestis* infections (Swift et al., 1999a).

CONCLUSIONS

AHL-dependent quorum sensing is clearly highly conserved within the genus *Yersinia*. Together with temperature, bacterial cell population density is at least one additional environmental parameter which the organism must sense during adaptation to life in different ecological niches. In *Y. pseudotuberculosis* and *Y. enterocolitica*, motility depends on the integration of both parameters. In *Y. pestis*, the identity of the target structural genes regulated *via* AHL-dependent quorum sensing is not yet apparent and may reflect the recent gene/promoter loss or rearrangement as a consequence of its recent divergence from the *Y. pseudotuberculosis* infection route to the insect vector route. For all three species of human pathogenic yersiniae, genome wide microarray analysis should rapidly facilitate mapping of the quorum sensing regulons and the subsequent definition of their contribution to survival and virulence.

ACKNOWLEDGEMENTS

Work in the authors' laboratories has been supported by grants and studentships from the Wellcome Trust, Biotechnology and Biological Sciences Research Council, U.K. and the Medical Research Council, U.K. which are gratefully acknowledged.

REFERENCES

Atkinson, S., Throup, J.P., Stewart, G.S.A.B., and Williams, P. 1999. A hierarchical quorum-sensing system in *Yersinia pseudotuberculosis* is involved in the regulation of motility and clumping. Mol. Microbiol. 33: 1267-1277.

Bainton, N.J., Bycroft, B.W., Chhabra, S.R., Stead, P., Gledhill, L., Hill, P.J., Rees, C.E.D., Winson, M.K., Salmond, G.P.C., Stewart, G.S.A.B., and Williams, P. 1992. A general role for the Lux autoinducer in bacterial-cell signaling-control of antibiotic biosynthesis in *Erwinia*. Gene. 116: 87-91.

Bassler, B.L. 2002. Small talk: Cell-to-cell communication in bacteria. Cell. 109: 421-424.

Bleves, S. and Cornelis, G.R. 2000. How to survive in the host: the *Yersinia* lesson. Microbes Infect. 2: 1451-1460.

Bleves, S., Marenne, M.N., Detry, G., and Cornelis, G.R. 2002. Up-regulation of the *Yersinia enterocolitica yop* regulon by deletion of the flagellum master operon *flhDC*. J. Bacteriol. 184: 3214-3223.

Buckley, C.M.F.G. Quorum Sensing in *Yersinia pseudotuberculosis*. 2002. Nottingham University. Thesis.

Cámara, M., Williams, P., and Hardman, A. 2002. Controlling infection by tuning in and turning down the volume of bacterial small-talk. Lancet Infect. Dis. 2: 667-676.

Chhabra, S.R., Harty, C., Hooi, D.S.W., Daykin, M., Williams, P., Telford, G., Pritchard, D.I., and Bycroft, B.W. 2003. Synthetic analogues of the bacterial signal (quorum sensing) molecule *N*-(3-oxododecanoyl)-L-homoserine lactone as immune modulators. J. Med. Chem. 46: 97-104.

Choi, S.H. and Greenberg, E.P. 1991. The C-terminal region of the *Vibrio fischeri* LuxR protein contains an inducer-independent *lux* gene activating domain. Proc. Nat. Acad. Sci. USA. 88: 11115-11119.

Choi, S.H. and Greenberg, E.P. 1992a. Genetic evidence for multimerization of LuxR, the transcriptional regulator of *Vibrio fischeri* luminescence. Mol. Marine Biol. Biotechnol. 1: 408-413.

Choi, S.H. and Greenberg, E.P. 1992b. Genetic dissection of DNA-binding and luminescence gene activation by the *Vibrio fischeri* LuxR protein. J. Bacteriol. 174: 4064-4069.

Cornelis, G., Vanooteghem, J.-C., and Sluiters, C. 1987. Transcription of the *yop* regulon from *Y. enterocolitica* requires trans acting pYV and chromosomal genes. Microb. Pathogen. 2: 367-379.

Cornelis, G.R. and Wolf-Watz, H. 1997. The *Yersinia* Yop virulon: A bacterial system for subverting eukaryotic cells. Mol. Microbiol. 23: 861-867.

Fraser, G.M. and Hughes, C. 1999. Swarming motility. Curr. Opin. Microbiol. 2: 630-635.

Fuqua, C., Winans, S.C., and Greenberg, E.P. 1996. Census and consensus in bacterial ecosystems: The LuxR-LuxI family of quorum-sensing transcriptional regulators. Ann. Rev. Microb. 50: 727-751.

Gardiner, S.M., Chhabra, S.R., Harty, C., Williams, P., Pritchard, D.I., Bycroft, B.W., and Bennett, T. 2001. Haemodynamic effects of the bacterial quorum sensing signal molecule, *N*-(3-oxododecanoyl)-L-homoserine lactone, in conscious, normal and endotoxaemic rats. Brit. J. Pharmacol. 133: 1047-1054.

Givskov, M., Eberl, L., and Molin, S. 1997. Control of exoenzyme production, motility and cell differentiation in *Serratia liquefaciens*. FEMS Microbiol. Lett. 148: 115-122.

Hanzelka, B.L. and Greenberg, E.P. 1995. Evidence that the N-terminal region of the *Vibrio fischeri* luxR protein constitutes an autoinducer-binding domain. J. Bacteriol. 177: 815-817.

Huber, B., Riedel, K., Hentzer, M., Heydorn, A., Gotschlich, A., Givskov, M., Molin, S., and Eberl, L. 2001. The *cep* quorum-sensing system of *Burkholderia cepacia* H111 controls biofilm formation and swarming motility. Microbiology. 147: 2517-2528.

Isherwood, K. E. 2001. Quorum sensing in *Yersinia pestis*. Nottingham University. Thesis.

Jiang, Y., Cámara, M., Chhabra, S.R., Hardie, K.R., Bycroft, B.W., Lazdunski, A., Salmond, G.P.C., Stewart, G.S.A.B., and Williams, P. 1998. In vitro biosynthesis of the *Pseudomonas aeruginosa* quorum-sensing signal molecule *N*-butanoyl-L-homoserine lactone. Mol. Microbiol. 28: 193-203.

Kapatral, V. and Minnich, S.A. 1995. Co-ordinate, temperature-sensitive regulation of the three *Yersinia enterocolitica* flagellin genes. Mol. Microbiol. 17: 49-56.

Kapperud, G., Namork, E., and Skarpeid, H.J. 1985. Temperature-inducible surface fibrillae associated with the virulence plasmid of *Yersinia enterocolitica* and *Yersinia pseudotuberculosis*. Infect. Immunity. 47: 561-566.

Kohler, T., Curty, L.K., Barja, F., Van Delden, C., and Pechere, J.C. 2000. Swarming of *Pseudomonas aeruginosa* is dependent on cell-to-cell signaling and requires flagella and pili. J. Bacteriol. 182: 5990-5996.

Lamont, I.L., Beare, P.A., Ochsner, U., Vasil, A.I., and Vasil, M.L. 2002. Siderophore-mediated signaling regulates virulence factor production in *Pseudomonas aeruginosa*. Proc. Nat. Acad. Sci. USA. 99: 7072-7077.

Lawrence, R.N., Dunn, W.R., Bycroft, B., Cámara, M., Chhabra, S.R., Williams, P., and Wilson, V.G. 1999. The *Pseudomonas aeruginosa* quorum-sensing signal molecule, *N*-(3-oxododecanoyl)-L-homoserine lactone, inhibits porcine arterial smooth muscle contraction. Brit. J. Pharmacol. 128: 845-848.

Lindum, P.W., Anthoni, U., Christophersen, C., Eberl, L., Molin, S., and Givskov, M. 1998. N-acyl-L-homoserine lactone autoinducers control production of an extracellular lipopeptide biosurfactant required for swarming motility of *Serratia liquefaciens* MG1. J. Bacteriol. 180: 6384-6388.

Luo, Z.Q. and Farrand, S.K. 1999. Signal-dependent DNA binding and functional domains of the quorum-sensing activator TraR as identified by repressor activity. Proc. Nat. Acad. Sci. USA. 96: 9009-9014.

MacNab, R. M. 1996. In *Escherichia coli* and *Salmonella typhimurium*. Eds. Neidhardt, F.C., Curtiss, R III., Ingraham, J.L., Lin, E.C.C., Low, K.B., et al American Society for Microbiology, Washington DC. 123-145.

McClean, K.H., Winson, M.K., Fish, L., Taylor, A., Chhabra, S.R., Cámara, M., Daykin, M., Lamb, J.H., Swift, S., Bycroft, B.W., Stewart, G.S.A.B., and Williams, P. 1997. Quorum sensing and *Chromobacterium violaceum*: exploitation of violacein production and inhibition for the detection of *N*-acylhomoserine lactones. Microbiology. 143: 3703-3711.

Meighen, E.A. 1994. Genetics of bacterial bioluminescence. Ann. Rev. Genet. 28: 117-139.

Moré, M.I., Finger, L.D., Stryker, J.L., Fuqua, C., Eberhard, A., and Winans, S.C. 1996. Enzymatic synthesis of a quorum-sensing autoinducer through use of defined substrates. Science. 272: 1655-1658.

Nasser, W., Bouillant, M.L., Salmond, G., and Reverchon, S. 1998. Characterization of the *Erwinia chrysanthemi expI-expR* locus directing the synthesis of two *N*-acyl-homoserine lactone signal molecules. Mol. Microbiol. 29: 1391-1405.

Parkhill, J., Wren, B.W., Thomson, N.R., Titball, R.W., Holden, M.T.G., Prentice, M.B., Sebaihia, M., James, K.D., Churcher, C., Mungall, K.L., Baker, S., Basham, D., Bentley, S.D., Brooks, K., Cerdeno-Tarraga, A.M., Chillingworth, T., Cronin, A., Davies, R.M., Davis, P., Dougan, G., Feltwell, T., Hamlin, N., Holroyd, S., Jagels, K., Karlyshev, A.V., Leather, S., Moule, S., Oyston, P.C.F., Quail, M., Rutherford, K., Simmonds, M., Skelton, J., Stevens, K., Whitehead, S., and Barrell, B.G. 2001. Genome sequence of *Yersinia pestis*, the causative agent of plague. Nature. 413: 523-527.

Passador, L., Cook, J.M., Gambello, M.J., Rust, L., and Iglewski, B.H. 1993. Expression of *Pseudomonas aeruginosa* virulence genes requires cell-to-cell communication. Science. 260: 1127-1130.

Pearson, J.P., Feldman, M., Iglewski, B.H., and Prince, A. 2000. *Pseudomonas aeruginosa* cell-to-cell signaling is required for virulence in a model of acute pulmonary infection. Infect. Immunity. 68: 4331-4334.

Pearson, J.P., Gray, K.M., Passador, L., Tucker, K.D., Eberhard, A., Iglewski, B.H., and Greenberg, E.P. 1994. Structure of the autoinducer required for expression of *Pseudomonas aeruginosa* virulence genes. Proc. Nat. Acad. Sci. USA. 91: 197-201.

Pearson, J.P., Van Delden, C., and Iglewski, B.H. 1999. Active efflux and diffusion are involved in transport of *Pseudomonas aeruginosa* cell-to-cell signals. J. Bacteriol. 181: 1203-1210.

Pepe, J.C. and Miller, V.L. 1993. *Yersinia enterocolitica* invasin: A primary role in the initiation of infection. Proc. Nat. Acad. Sci. USA. 90: 6473-6477.

Pierson, D.E. and Falkow, S. 1993. The *ail* gene of *Yersinia enterocolitica* has a role in the ability of the organism to survive serum killing. Infect. Immunity. 61: 1846-1852.

Redfield, R.J. 2002. Is quorum sensing a side effect of diffusion sensing? Trends in Microbiology. 10: 365-370.

Reverchon, S., Bouillant, M.L., Salmond, G., and Nasser, W. 1998. Integration of the quorum-sensing system in the regulatory networks controlling virulence factor synthesis in *Erwinia chrysanthemi*. Mol. Microbiol. 29: 1407-1418.

Salmond, G.P.C., Bycroft, B.W., Stewart, G.S.A.B., and Williams, P. 1995. The bacterial enigma - cracking the code of cell-cell communication. Mol. Microbiol. 16: 615-624.

Shadel, G.S., Young, R., and Baldwin, T.O. 1990. Use of regulated cell-lysis in a lethal genetic selection in *Escherichia coli* - identification of the autoinducer-binding region of the LuxR protein from *Vibrio fischeri* ATCC-7744. J. Bacteriol. 172: 3980-3987.

Shaw, P.D., Ping, G., Daly, S.L., Cha, C., Cronan, J.E., Rinehart, K.L., and Farrand, S.K. 1997. Detecting and characterizing *N*-acyl-homoserine lactone signal molecules by thin-layer chromatography. Proc. Nat. Acad. Sci. USA. 94: 6036-6041.

Slock, J., Vanriet, D., Kolibachuk, D., and Greenberg, E.P. 1990. Critical regions of the *Vibrio fischer*i LuxR protein defined by mutational analysis. J. Bacteriol. 172: 3974-3979.

Smith, R.S., Fedyk, E.R., Springer, T.A., Mukaida, N., Iglewski, B.H., and Phipps, R.P. 2001. IL-8 production in human lung fibroblasts and epithelial cells activated by the *Pseudomonas* autoinducer *N*-3-oxododecanoyl homoserine lactone is transcriptionally regulated by NF-kappa B and activator protein-2. J. Immunol. 167: 366-374.

Stevens, A. M. and Greenberg E.P. 1999. Cell-Cell Signalling in Bacteria. Eds. Dunny G.M.; Winans S.C. American Society for Microbiology, Washington DC. 231-242.

Stevens, A.M. and Greenberg, E.P. 1997. Quorum sensing in *Vibrio fischeri*: Essential elements for activation of the luminescence genes. J. Bacteriol. 179: 557-562.

Swift, S., Downie, J.A., Whitehead, N.A., Barnard, A.M.L., Salmond, G.P.C., and Williams, P. 2001. Quorum sensing as a population-density-dependent determinant of bacterial physiology. Advan. Microb. Physiol. 45: 199-270.

Swift, S., Isherwood, K. E., Atkinson, S., Oyston, P., and Stewart, G. S. A. B. 1999a. Quorum sensing in *Aeromonas* and *Yersinia*. In Microbial Signalling and Communication. Eds. England, R., Hobbs, G., Bainton, N.J., and Roberts, D.M. Cambridge University Press, Cambridge, UK. 85-104

Swift S., Williams P., and Stewart, G. S. A. B. 1999b. *N*-acylhomoserine lactones and quorum sensing in proteobacteria. In Cell-Cell Signaling in Bacteria. Eds. G. M. Dunny and S. C. Winans. ASM Press, Washington DC. 291-314.

Swift, S., Winson, M.K., Chan, P.F., Bainton, N.J., Birdsall, M., Reeves, P.J., Rees, C.E.D., Chhabra, S.R., Hill, P.J., Throup, J.P., Bycroft, B.W., Salmond, G.P.C., Williams, P., and Stewart, G.S.A.B. 1993. A novel strategy for the isolation of *luxI* homologs - evidence for the widespread distribution of a *luxR luxI* superfamily in enteric bacteria. Mol. Microbiol. 10: 511-520.

Tan, M.W., Rahme, L.G., Sternberg, J.A., Tompkins, R.G., and Ausubel, F.M. 1999. *Pseudomonas aeruginosa* killing of *Caenorhabditis elegans* used to identify *P. aeruginosa* virulence factors. Proc. Nat. Acad. Sci. USA. 96: 2408-2413.

Telford, G., Wheeler, D., Williams, P., Tomkins, P.T., Appleby, P., Sewell, H., Stewart, G.S.A.B., Bycroft, B.W., and Pritchard, D.I. 1998. The *Pseudomonas aeruginosa* quorum-sensing signal molecule *N*-(3-oxododecanoyl)-L-homoserine lactone has immunomodulatory activity. Infect. Immunity. 66: 36-42.

Throup, J.P., Cámara, M., Bainton, N.J., Briggs, G.S., Chhabra, S.R., Bycroft, B.W., Williams, P., and Stewart, G.S.A.B. 1995. Characterisation of the *yenI/yenR* locus from *Yersinia enterocolitica* mediating the synthesis of two *N*-acyl homoserine lactone signal molecules. Mol. Microbiol. 17: 345-356.

von Bodman, S.B., Majerczak, D.R., and Coplin, D.L. 1998. A negative regulator mediates quorum-sensing control of exopolysaccharide production in *Pantoea stewartii* subsp. *stewartii*. Proc. Nat. Acad. Sci. USA. 95: 7687-7692.

Watson, W.T., Minogue, T.D., Val, D.L., von Bodman, S.B., and Churchill, M.E.A. 2002. Structural basis and specificity of acyl-homoserine lactone signal production in bacterial quorum sensing. Mol. Cell. 9: 685-694.

Whiteley, M. and Greenberg, E.P. 2001. Promoter specificity elements in *Pseudomonas aeruginosa* quorum-sensing-controlled genes. J. Bacteriol. 183: 5529-5534.

Williams P. 2002. Quorum sensing: an emerging target for antimicrobial chemotherapy? Expert Opin. Ther. Targets. 6: 257-274.

Williams, P., Cámara, M., Hardman, A., Swift, S., Milton, D., Hope, V.J., Winzer, K., Middleton, B., Pritchard, D.I., and Bycroft, B.W. 2000. Quorum sensing and the population-dependent control of virulence. Phil. Trans. Roy. Soc. London. B. 355: 667-680.

Winson, M.K., Swift, S., Hill, P.J., Sims, C.M., Griesmayr, G., Bycroft, B.W., Williams, P., and Stewart, G.S.A.B. 1998. Engineering the *luxCDABE* genes from *Photorhabdus luminescens* to provide a bioluminescent reporter for constitutive and promoter probe plasmids and mini-Tn*5* constructs. FEMS Microbiol. Lett. 163: 193-202.

Winzer, K., Hardie, K.R., Burgess, N., Doherty, N., Kirke, D., Holden, M.T.G., Linforth, R., Cornell, K.A., Taylor, A.J., Hill, P.J., and Williams, P. 2002a. LuxS: its role in central metabolism and the in vitro synthesis of 4-hydroxy-5-methyl-3(2H)-furanone. Microbiology. 148: 909-922.

Winzer, K., Hardie, K.R., and Williams, P. 2002b. Bacterial cell-to-cell communication: Sorry, can't talk now - gone to lunch! Curr. Opin. Microb. 5: 216-222.

Yates, E.A., Philipp, B., Buckley, C., Atkinson, S., Chhabra, S.R., Sockett, R.E., Goldner, M., Dessaux, Y., Cámara, M., Smith, H., and Williams, P. 2002. *N*-acylhomoserine lactones undergo lactonolysis in a pH-, temperature-, and acyl chain length-dependent manner during growth of *Yersinia pseudotuberculosis* and *Pseudomonas aeruginosa*. Infect. Immun. 70: 5635-5646.

Young, G.M., Badger, J.L., and Miller, V.L. 2000. Motility is required to initiate host cell invasion by *Yersinia enterocolitica*. Infect. Immun. 68: 4323-4326.

Young, G.M., Smith, M.J., Minnich, S.A., and Miller, V.L. 1999. The *Yersinia enterocolitica* motility master regulatory operon, *flhDC*, is required for flagellin production, swimming motility, and swarming motility. J. Bacteriol. 181: 2823-2833.

Zhang, R.G., Pappas, T., Brace, J.L., Miller, P.C., Oulmassov, T., Molyneauz, J.M., Anderson, J.C., Bashkin, J.K., Winans, S.C., and Joachimiak, A. 2002. Structure of a bacterial quorum-sensing transcription factor complexed with pheromone and DNA. Nature. 417: 971-974.

Zhu, J. and Winans, S.C. 1999. Autoinducer binding by the quorum-sensing regulator TraR increases affinity for target promoters in vitro and decreases TraR turnover rates in whole cells. Proc. Nat. Acad. Sci. USA. 96: 4832-4837.

Chapter 6

The Invasin Protein of Enteropathogenic *Yersinia* Species: Integrin Binding and Role in Gastrointestinal Diseases

Ka-Wing Wong, Penelope Barnes and Ralph R. Isberg

ABSTRACT

The invasin of enteropathogenic *Yersinia* is an outer membrane protein that promotes the attachment of bacteria to mammalian cells via binding to multiple members of the integrin superfamily of cell adhesion molecules. The protein is an example of a group of proteins of high sequence similarity involved in mediating attachment to host cell surfaces that are found on the surface of many Gram-negative organisms. Attachment of the bacteria via invasin can result in promoting internalization of the bacterium into a phagosome, in an event that is antagonized by the *Yersinia* Yop proteins. After oral inoculation of model host animals with enteropathogenic *Yersinia*, this uptake process appears to occur after attachment to M cells, which are located over Peyer's patches and interface with the lumen of the small intestine. Efficient colonization of local lymph nodes, therefore, requires the presence of the invasin protein. Studies on the molecular mechanism of invasin-promoted uptake indicate that high affinity binding is a critical determinant of uptake, and clustering of integrin receptors results in activation of Rho family members, which is a necessary prelude to the formation of a phagosome encompassing the bacterium.

INTRODUCTION

During the course of systemic disease promoted by enteropathogenic *Yersinia*, the bacteria appear to be largely extracellular after they translocate out of an intestinal locale and enter deep tissue sites (Heesemann et al., 1993). This is consistent with the fact that many of the known translocated protein substrates of the *Yersinia* Type III secretion system (TTSS) are highly effective at antagonizing phagocytosis by virtue of inactivating key cytoskeletal signaling cascade components (Cornelis, 2000). In this regard, the enteropathogens are probably similar to *Yersinia pestis* during growth within lymphoid tissue. *Yersinia enterocolitica* and *Yersinia pseudotuberculosis*, however, are not transmitted by the flea, and disease by these organisms does not initiate after an insect bite. Instead, disease is initiated after oral ingestion of contaminated foodstuffs, which is followed shortly afterward by the entry of the bacteria into M cells overlying the gut-associated lymphoid tissue (GALT) (Kraehenbuhl and Neutra, 2000). This is the only stage in the disease process on which there is general agreement that enteropathogenic *Yersinia* species are found in an intracellular locale. Therefore, the enteropathogens must encode proteins that allow initiation of disease after oral ingestion of the microorganisms, and have strategies that allow their dissemination to other hosts after colonization is initiated within the gastrointestinal tract.

One such protein that appears uniquely associated with *Yersinia* enteric disease is the bacterial outer membrane protein invasin (Isberg et al., 1987). Invasin was identified using a genetic screen in which molecular clones from *Y. pseudotuberculosis* were identified that had the ability to convert *Escherichia coli* into an organism capable of entering cultured mammalian cells (Isberg and Falkow, 1985). The gene for invasin is intact in enteropathogenic *Yersinia* species, but is disrupted and nonfunctional in *Y. pestis* (Simonet et al., 1996), arguing that expression of this protein is somehow related to the ability to establish gastrointestinal disease or facilitate dissemination between hosts via the oral route. In this chapter we review studies on the mechanism of action of the invasin protein, the response of its receptors to engagement by invasin, and the role that invasin-dependent and invasin-independent pathways play in establishing disease in animal models. Finally we discuss an updated view on the intracellular signaling pathway induced by invasin during the entry of bacteria into a host cell.

INVASIN AND RECEPTOR ENGAGEMENT

Enteropathogenic *Yersinia* synthesize at least four proteins that promote bacterial adhesion to host cells. These include the 986 residue invasin protein, the plasmid-encoded YadA protein (El Tahir and Skurnik, 2001), the pH 6.0 antigen pilus (Yang and Isberg, 1997), and, in *Y. enterocolitica*, the Ail protein (Wachtel and Miller, 1995). Of these, invasin is by far the most efficient at promoting bacterial uptake, and evidence has been presented that after oral inoculation of mice, the protein is required for entry into M cells overlying Peyer's patches in the small intestine (see below; Marra and Isberg, 1997; Pepe and Miller, 1993b). From these results, it appears that at least one of the roles of invasin during disease is to promote uptake into host cells, as is suggested by the behavior of invasin during bacterial interaction with cultured cells. Invasin is sufficient to promote uptake in cell lines, and no other bacterial products are necessary for this activity, because invasin-coated latex beads can be internalized efficiently by cultured cells, as can *Staphylococcus aureus* coated by hybrid proteins containing the invasin cell binding domain (Dersch and Isberg, 2000; Rankin et al., 1992).

The host receptors that are recognized by invasin and are necessary for uptake promoted by this protein were identified biochemically as multiple members of the integrin superfamily of cell adhesion molecules (Isberg and Leong, 1990). Integrins are α/β heterodimers that bind to a variety of substrates in a divalent metal ion-dependent manner (Hynes, 2002). The identified natural substrates of integrins have diverse roles in cell-cell and cell-matrix communication in animal cells. Substrate binding to integrins can trigger intracellular signaling events that lead to a variety of responses, such as cell survival, proliferation, motility, and reorganization of cytoskeleton. How such signals are initiated from ligand binding is poorly understood, although it is believed that clustering of the receptor into polyvalent arrays is important (Hogg et al., 2003). In addition, intracellular signals can also modulate the affinity of integrin for substrates (Hynes, 2002). For instance, $\alpha_4\beta_1$ integrin, which is an invasin receptor expressed on T cells, requires activating signals to allow it to assume a high affinity conformation that allows binding to natural substrates (Ennis et al., 1993).

The binding of invasin to integrin receptors is of extremely high affinity. Invasin has a 100 fold higher affinity than does fibronectin for the $\alpha_5\beta_1$ integrin receptor, although fibronectin is the natural substrate for the receptor (Tran Van Nhieu and Isberg, 1993). Similar results were observed when the binding of laminin to the $\alpha_3\beta_1$ receptor was analyzed (Eble et al., 1998). High affinity binding to integrins is delivered by the C-terminal 497 residues (Inv497) of invasin (Rankin et al., 1992), as coating of latex bead or the Gram-positive *S. aureus,* with Inv497, allows high efficiency uptake (Dersch and Isberg, 1999; Rankin et al., 1992). The N-terminal of invasin likely functions to properly localize the protein in the outer membrane, anchor the protein once it is established within this locale, and facilitate presentation of the C-terminal adhesion site.

The crystal structure of the C terminal Inv497 fragment demonstrates that the C-terminal cell adhesion domain of invasin is arrayed like a string of beads, with five contiguous domains situated side-by-side in an extended structure 170 Å in length (Figure 1; Hamburger et al., 1999). The amino terminal four domains (D1 to D4) are each members of the immunoglobulin fold (Ig) family, although there is no clear sequence similarity to other domains that assume such a folded structure. The C-terminal D5 domain resembles a C-type lectin-like domain, similar to receptors found on the surface of NK cells (Weis et al., 1998). Although invasin does not show any sequence similarity to any other integrin substrates, the overall two domain D4-D5 structure is reminiscent of the integrin-binding region of fibronectin, which also consists of two domains. In the case of fibronectin, however, the cell adhesion region consists of a pair of Ig folds, called the fibronectin type III repeats 9 and 10 (Fn-III 9- 10) (Aota et al., 1994; Bowditch et al., 1994b; Danen et al., 1995), that have a considerably smaller interdomain interface than observed with invasin. Although the surfaces presented to the receptor by each of these substrates appear to be different, critical residues for receptor binding found on the two proteins must interact in a similar fashion with the receptor. This is because invasin and fibronectin appear to recognize the same site on the integrin (Tran Van Nhieu and Isberg, 1991). Invasin is a competitive inhibitor of fibronectin, and peptides containing the RGD integrin-recognition sequence from fibronectin block the binding of both substrates to the $\alpha_5\beta_1$ integrin, although invasin does not contain an RGD sequence (Tran Van Nhieu and Isberg, 1991). There must be some explanation other than the site of binding that is the reason for the difference in affinity of these two substrates. It is possible that the nature of interdomain contacts may dictate the relative affinity of fibronectin and invasin.

The loop in FN-III-10 that contains the RGD integrin recognition sequence of fibronectin extends ~12 Å from the highly flexible interdomain interface between FN-III 9 and 10. An aspartate in invasin, D911 in located in D5, is also the most critical residue for integrin binding and bacterial entry (Leong et al., 1995), and is similarly positioned in a small loop ~12 Å away from the interdomain region of D4-D5, where there are extensive contacts that form a rigid superdomain containing D4 and D5. The fact that the interdomain region of FN-III 9-10 is highly flexible may contribute to the important affinity difference between invasin and fibronectin (TranVanNhieu and Isberg, 1993). It is also possible that additional surface contacts within D4-D5 that are not present in fibronectin may contribute to the high-affinity integrin binding. It has been proposed that the presence of aromatic amino acids in invasin that are not found on the surface of fibronectin could be the key residues that facilitate high affinity binding. Evidence consistent with this interpretation has been presented for the *Listeria monocytogenes* internalin protein, which also binds a host cell receptor, promoting bacterial uptake (Machner et al., 2003).

As a model for receptor recognition, fibronectin has residues involved in integrin binding that are noncontiguous with the RGD sequence in the so-called synergy region on FN-III 9, and this relative array of critical residues appears to be found similarly in invasin (Bowditch et al., 1994a). The synergy region and the RGD loop are on the same surface

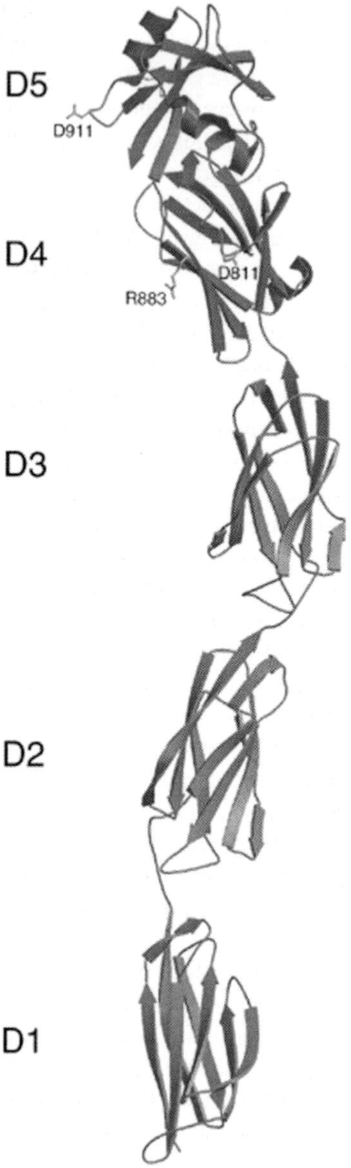

Figure 1. Crystal structure of Inv497, the carboxyl terminal of invasin. The region of invasin presented to the mammalian cell consists of 497 amino acid carboxyl terminal fragment that is exposed on the surface of the bacterium. The cell binding domain consists of a D4-D5 superdomain that binds integrin receptors. Domains D1-D4 are all members of the IgSF fold family, whereas as D5 is a C-type lectin-like domain. See text for details. Taken from (Hamburger et al., 1999).

of fibronectin and are separated by ~32 Å. Crystallographic and mutagenesis data indicate invasin also contains a synergy-like region centered around the D811 residue, which is located 32 Å away from D911 and on the same face of invasin as Inv497 (Saltman et al., 1996). Although it appears that invasin and fibronectin employ similar recognition strategies, there are clearly elements involved in binding that differ for the two substrates. First, there is the issue of very different relative affinities of the substrates. Secondly, the spectrum of receptors recognized by the two substrates differs significantly. For instance, the $\alpha_V\beta_3$ integrin recognizes fibronectin but does not bind invasin, while $\alpha_6\beta_1$

integrin binds to invasin but not fibronectin. Therefore, the integrin binding specificity is determined by unidentified regions of invasin. Identification of such regions could be pursued by searching for gain-of-function mutants that bind to the $\alpha_V\beta_3$ integrin or loss-of-function mutations that would affect some but not all invasin-binding integrins.

INTEGRIN DETERMINANTS THAT ALLOW RECOGNITION OF INVASIN

Several structural determinants of invasin recognition by the α and β chain integrins have been identified. Integrins require divalent cations for recognition of most substrates, including invasin. Previous studies implicated sequences within integrins that are responsible for divalent cation coordination called metal ion-dependent adhesion sites (MIDAS) (Lee et al., 1995). Mutations within these motifs were isolated in the α_3 and β_1 chains that selectively blocked invasin binding, or which attenuated both invasin and fibronectin binding (Krukonis and Isberg, 2000). Comparing the sites of these changes to the corresponding residues on the recently-solved structure of the related integrin $\alpha_V\beta_3$ in complex with RGD peptide reveals that all except one of these mutations affects residues that are predicted to participate directly in coordinating divalent cations on either the α or β integrin chains (Xiong et al., 2002). Mutations in the MIDAS sequence in the β_1 chain (D154A, S156A, and S158A) and in the α_3 chain (N283, D346E, D408E) all blocked both invasin and natural substrate binding (Krukonis et al., 1998; Krukonis and Isberg, 2000). In contrast, selected residues in sites other than the MIDAS region affected invasin recognition more drastically than natural substrate recognition. For instance the α3 Y218A mutation resulted in a partial defect in invasin binding but did not affect binding of epiligrin, a natural ligand of $\alpha_3\beta_1$ integrins (Krukonis et al., 1998). The residue in α_V that corresponds to α3-Y218A appears to make contact with the Arg of the RGD peptide. A second phenotype was uncovered in this study, as a β_1 chain double mutation (KDD160RDV) was found to have wild-type binding to fibronectin but was severely defective for invasin binding (Krukonis and Isberg, 2000). The corresponding residues in the β_3 chain are involved in coordinating a second Mn^{2+} ion adjacent to the MIDAS that had been unexpectedly found in the β_3 chain crystal structure. This second Mn^{2+}-binding site has been called the "ligand-associated metal binding site" (LIMBS) because the novel binding site does not exist in the absence of the RGD ligand. Single mutations in the predicted β_1 chain LIMBS residues K160A and D162A in β1 also led to selective phenotypes, in that fibronectin binding was enhanced without affecting binding to invasin, whereas the D161I mutation enhanced both invasin and fibronectin adhesion (Krukonis and Isberg, 2000). Although it is impossible to state with assurance that invasin binding generates a LIMBS site on the β1 chain, it does appear that sites predicted to form a LIMBS modulate the relative affinity of the integrin for invasin compared to natural substrates. Ideally, the availability of a crystal structure of a β1 integrin in complex with invasin will not only help to further define the structural basis of invasin-integrin interaction and but allow insight into how the LIMBS region modulates substrate recognition.

Recent crystallographic work and electron microscopy studies of purified integrins have indicated that a conformational switch occurs upon substrate binding in the presence of divalent cations. This appears to result in straightening of a bent conformation that the receptor assumes in the absence of substrate engagement (Hynes, 2002; Takagi et al., 2002). It is unclear whether monomeric D4-D5 of invasin can induce this unbending of integrins, but multivalent engagement of integrins resulting in such large scale conformational changes could conceivably induce a robust signal to be sent from the surface of the mammalian cell in response to bacterial binding.

INTEGRIN SIGNALING STIMULATED BY INVASIN

High efficiency bacterial uptake promoted by invasin requires not only the high-affinity binding of invasin to integrins but also self-association of invasin, presumably leading to clustering of the integrin receptor (Dersch and Isberg, 1999). Cultured cells do not efficiently internalize latex beads that are coated by the size-fractionated monomeric D4-D5 fragment of invasin. In contrast, crosslinking of this monomeric D4-D5 fragment by antibody restores the efficient uptake observed when beads are coated by the D1-D5 C-terminal fragment of invasin (Dersch and Isberg, 1999). The region within Inv497 that is responsible for self-association is believed to be the D2 domain (Figure 1). Interestingly, the invasin ortholog from *Y. enterocolitica* naturally lacks the D2 domain (Dersch and Isberg, 2000). It had been noted previously that the *Y. enterocolitica* invasin derivative was less efficient at promoting uptake than the *Y. pseudotuberculosis* protein (Pepe and Miller, 1990) and the absence of D2 contributes to this deficiency. Interestingly, the *Y. enterocolitica* protein is expressed at higher levels than the *Y. pseudotuberculosis* protein, and this appears to compensate for the lack of D2 in this protein (Dersch and Isberg, 2000).

There are many examples of receptor binding events that require oligomerization to generate intracellular signaling. Heptameric Protective Antigen (PA), which is the receptor binding domain of anthrax toxin, clusters the anthrax toxin receptor and triggers its endocytic uptake (Abrami et al., 2003). In addition, oligomeric, but not monomeric, class II major histocompatility complex-peptide engages T-cell antigen receptors to trigger $CD4^+$ T-cell activation (Cochran et al., 2000). Similarly for invasin, it is likely that D2-dependent multimerization is used to increase the local density of integrin receptors. Several lines of evidence indicate that increased receptor density enhances the efficiency of bacterial uptake. First, overexpression of the integrin α_5 chain in host cells enhances the uptake of latex bead coated with low-affinity ligands such as fibronectin (TranVanNhieu and Isberg, 1993). Secondly, high concentration of invasin lacking the D2 self-association domain coated on latex beads or expressed on the surface of enteropathogenic *Yersinia* can compensate for the lower uptake efficiency promoted by this derivative (Dersch and Isberg, 1999; Dersch and Isberg, 2000). Finally, reduced availability of integrin receptors for invasin, as a result of plating cells on coverslips coated with anti-integrin, interferes with the invasin-promoted uptake (TranVanNhieu and Isberg, 1993). It is not known whether multimerization of invasin has an additional function other than clustering. It may be fruitful to determine if the large-scale conformational change that results in unbending of integrins is enhanced by invasin self-association or by crosslinking invasin monomers with antibody, and if high-affinity binding of invasin oligomers can induce a conformational state of integrins that has not been previously observed during the binding to other substrates.

The mechanism of phagocytic uptake that results from integrin clustering is poorly understood, particularly in regards to the critical signaling events that occur in the first few seconds after integrin engagement. Both the α and β integrin chains have short cytoplasmic tails that presumably modulate these early events. The α chain cytoplasmic domain is not required for invasin-mediated internalization, whereas mutations within β1 chain cytoplasmic domain define regions that control the efficiency of bacterial uptake (Gustavsson et al., 2002). Loss-of-function mutations that abolish the binding of cytoskeletal proteins to the cytoplasmic tail often increase the efficiency of invasin-promoted uptake (Tran Van Nhieu et al., 1996). This rather surprising result can be explained by proposing that integrin receptors in tight association with cytoskeletal components have reduced mobility in the plane of plasma membrane, limiting their ability to move to bacterial adhesion sites. Alternatively, by lowering the affinity for cytoskeletal components, the receptors may interact with cytoskeletal remodeling proteins at a high turnover rate that would otherwise be impossible if integrin receptors were already occupied. This scenario is plausible because extensive reorganization of cytoskeleton structure is likely to take place in a dynamic and flexible environment rather than on a rigid immobilized surface. Given that invasin would likely trigger a large-scale conformational change in the receptor, it is possible that such a change could be relayed through plasma membrane to the cytoplasmic domain, down-modulating the affinity for cytoskeletal proteins. In this case, a flexible local environment is available for cytoskeletal remodeling to initiate reorganization of the actin cytoskeleton. This model also predicts that cytoskeletal proteins associated with integrin receptors prior to contact of the host cell with bacterium would dissociate from the integrin receptor and generate the signaling complex that is ultimately responsible for invasin-mediated internalization.

IgSF DOMAINS AS BUILDING BLOCK FOR OTHER INVASIN PARALOGS

Three paralogs of invasin exist in *Y. pestis,* while one is present in *Y. enterocolitica* (Deng et al., 2002; Dersch and Isberg, 2000; Parkhill et al., 2001). All paralogs have the N-terminal conserved region (~500 amino acids) for outer membrane localization that is predicted to have a β-barrel structure found in a large number of proteins involved in interaction of Gram negative bacteria with host cells (Touze et al., 2004). The best known members of this family are the intimins, encoded by enteropathogenic *E. coli* strains (Nougayrede et al., 2003). The C-terminal region of each paralog is composed of a series of immunoglobulin superfamily (IgSF) domains of varying numbers. Invasin of *Y. enterocolitica* lacks a D2 IgSF domain (part of a subfamily of IgSF domains that assume the I1 fold set; (Harpaz and Chothia, 1994)) that mediates self-association (Figure 2). A *Y. pestis* paralog that is almost identical to the *Y. pseudotuberculosis* invasin is inactivated by an insertion sequence in the middle of the open reading frame, which is otherwise almost identical to invasin from *Y. pseudotuberculosis* (Simonet et al., 1996). The second *Y. pestis* paralog is missing the D5 C type lectin-like domain, and also is likely to have no activity, as all active members of this family of proteins appear to have an intact D5-like domain (Figure 2). The third paralog, which is quite large, 3013 amino acids, contains twenty-five IgSF tandem domains, and appears to have all the hallmarks of a protein involved in mediating interaction with host cells. The minimal integrin-binding domain D4-D5, together with D3, is on the C-terminus. Between D3-D5 and the outer membrane localization region are eight self-association D2 domains scattered along fourteen D1 domains (also known as members of the IgSF fold set I2; Harpaz and Chothia, 1994). The multiple D2 domains suggests that this very long protein can self-associate and can therefore cluster receptor. A recent study indicated that *Y. pestis* KIM strain can be internalized at high efficiency into HeLa cells in an actin-dependent manner (Cowan et al.,

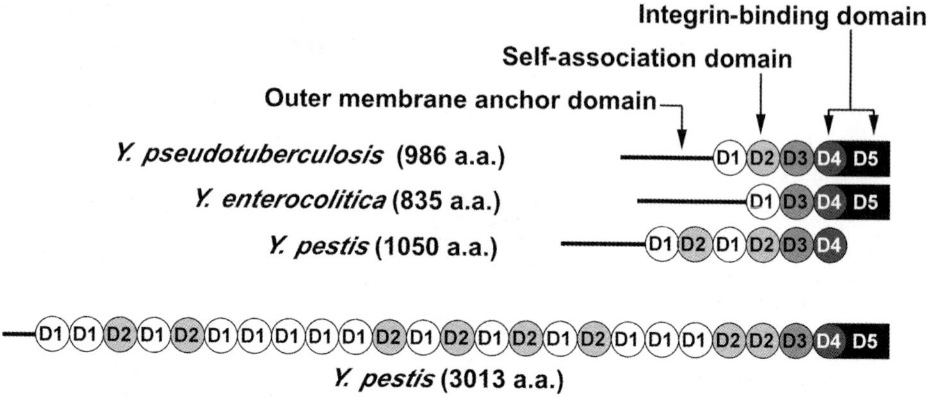

Figure 2. Paralogs of the *Y. pseudotuberculosis* invasin protein. The domain structures are described in Figure 1 and in the text. The large *Y. pestis* paralog is predicted to have 25 domains arrayed end-to-end.

2000). Whether this long invasin paralog contributes to internalization into tissue culture cells or plays a role in establishing systemic infection in mammalian hosts is unknown. Perhaps this very large structure mimics a pilus, facilitating access of the D4-D5 cell adhesion region to mammalian receptors.

HOW DOES INVASIN-BOUND INTEGRIN TRIGGER INTRACELLULAR SIGNAL TRANSDUCTION TO DIRECT ACTIN CYTOSKELETON REARRANGEMENT?

Invasin-promoted uptake is highly sensitive to cytochalasin D, a potent inhibitor of actin polymerization, and tyrosine kinase poisons, indicating that reorganization of the actin cytoskeleton is required for uptake (Rosenshine et al., 1992). Many factors have been implicated in coupling rearrangement of actin cytoskeleton to integrin engagement, and some of them are known to be p130Cas (Weidow et al., 2000), phosphatidylinositol 3-kinase, and members of the Rho small GTPase family (Keely et al., 1997). Although all of these have been demonstrated to have a role in invasin-mediated internalization of bacteria, it is still unclear if they participate directly or simply modulate the uptake competency of the host cell.

ROLE OF FAK

Focal adhesion kinase (FAK) is a tyrosine kinase that participates in events that link integrins with the cytoskeleton (Guan, 1997). A portion of FAK is able to directly bind the integrin β_1 cytoplasmic domain, and the protein appears to modulate cellular migration events, perhaps by aiding in the turnover of focal adhesions, which are large actin-rich structures that result from integrin adhesion to extracellular matrix proteins (Burridge and Chrzanowska, 1996). FAK also appears to be involved in invasin-promoted internalization, as a dominant-negative inhibitor of FAK called FAK-related nonkinase (FRNK) or a FAK point mutant lacking an autophosphorylation site necessary for binding of Src family kinases significantly block invasin-promoted uptake (Alrutz et al., 2001). One interpretation of these data is that FAK directly mediates the events resulting from integrin clustering that lead to the recruitment of cytoskeletal remodeling molecules to the phagocytic cup. Some data are consistent with this proposition. During cell migration on integrin substrates, FAK is regulated by adhesion (Sieg et al., 2000), resulting in the formation of a heterotetrameric signaling complex that includes FAK, Src, p130Cas and Dock180 (Hsia et al., 2003). Dock180 is a guanine exchange factor (GEF) that works specifically on the small GTPase Rac1, which is known to mediate the formation of actin-rich lamellipodia, and is involved in invasin-promoted uptake (to be discussed below). Indeed, increased Rac1 activation and accumulation of FAK at lamellipodia have been detected under FAK-overexpressing condition (Hsia et al., 2003). Therefore, FAK may provide a molecular link between integrin signaling and Rac1-dependent actin cytoskeleton rearrangement. However, FAK may also have an indirect role during invasin-promoted uptake. FAK-dependent tyrosine phosphorylation may disrupt focal adhesions at the bacterial binding site, preventing immobilization of the cytoskeleton and facilitating rearrangement of actin necessary for bacterial entry. Alternatively, FAK may enhance the accessibility of integrin receptors by invasin-coated bacteria through disrupting focal adhesion sites that may be quite distant from the site of bacterial binding (Isberg et al., 2000).

THE ROLE OF Rac1

Most aspects of cytoskeletal rearrangements are regulated by members of the Rho GTPase family, which include RhoABC, Rac1 and Cdc42 (Etienne-Manneville and Hall, 2002). These small GTPases cycle between an active GTP-bound state and an inactive GDP-bound state. Guanine nucleotide exchange factors (GEFs) convert the GDP-bound state into the GTP-bound state, which can then directly bind downstream effectors that actively modulate actin remodeling (Karnoub et al., 2001). The inefficient GTPase activity of Rho family members is stimulated by GTPase-activating proteins (GAPs). When activated after GEF binding, Rac1 and Cdc42 mediate the formation of lamellipodia (ruffle) and filopodia (microspikes), respectively (Etienne-Manneville and Hall, 2002). The localization and/or activity of Rac1 and Cdc42 can be regulated by integrins during cell spreading and cell motility, although it is unclear what proteins link Rac1/Cdc42 activation to integrin engagement (Del Pozo et al., 2002; Keely et al., 1997). In the case of invasin-promoted uptake, an activated form of Rac1 stimulates uptake, whereas a dominant negative form of Rac1 blocks entry (Alrutz et al., 2001). Interestingly, a dominant negative form of Cdc42 does not block invasin-stimulated uptake, indicating that this process is a distinctly Rac1-promoted event (Alrutz et al., 2001). Consistent with this interpretation, work by Bliska and coworkers (Black and Bliska, 2000) showed that a constitutively active RhoA protein depressed bacterial uptake. Interestingly, they showed that uptake can take place under conditions in which multiple Rho family members are inactivated if the cell overexpresses a constitutively active form of Rac1. Constitutively active Cdc42 or RhoA, however, could not overcome inactivation of Rho family members, indicating that activation of Rac1 is necessary to allow uptake (Black and Bliska, 2000). Using human macrophages, a somewhat different result was obtained, although the cell type used and the assay for uptake were different than in other work (Wiedemann et al., 2001). In this latter study, the combined contribution of several Rho family members in the entry process was posited (Wiedemann et al., 2001).

Entry is also linked to transient Rac1 activation and efficient accumulation of Rac1 around the phagocytic cup in a fashion that resembles membrane ruffles (Alrutz et al.,

2001). Rac1 activation appears to occur specifically at the site of bacterial binding. Using fluorescence resonance energy transfer (FRET) to detect protein-protein interaction in vivo (Miyawaki, 2003), it can be shown that the Rac1 molecules accumulated around nascent phagosomes are in an active, GTP-bound state (Wong and Isberg, unpublished data).

THE REGULATION OF Rac1 TARGETING AND ACTIVATION DURING INVASIN-MEDIATED INTERNALIZATION

The fact that dominant inhibitory derivatives of Rho family members only interfere with specific actin rearrangements can be explained by their ability to recognize only a subset of GEFs. Dominant negative small GTPases are usually able to form a highly stable complex with cognate GEFs, and as a result prevent the GEF from activating the endogenous small GTPase in the cell (Feig, 1999). It is very likely that a Rac1-specific GEF plays a direct role in connecting integrin signaling to actin cytoskeleton reorganization during invasin-promoted bacterial entry, although it is unclear which GEF plays this role. In support of the GEF model, a novel dominant negative mutant of Cdc42 that recognizes Rac1-GEFs, but not Cdc42-GEFs (Gao et al., 2001; Karnoub et al., 2001), is able to block invasin-promoted uptake (Wong and Isberg, unpublished data). There are at least four known proteins that contain a Rac1-GEF domain: Dock180, Trio, Asef, and Sos1 family members (Bellanger et al., 1998; Brugnera et al., 2002; Kawasaki et al., 2000). Only Dock180, however, has an established role in both integrin-dependent events as well as promoting actin rearrangements (Brugnera et al., 2002; Hsia et al., 2003). Dock180 is a particularly good candidate for linking invasin-mediated clustering to Rac1 activation because its localization and activity are regulated by FAK, which has been implicated in invasin-promoted uptake (Alrutz and Isberg, 1998).

The nature of the biochemical signals that target and activate Rac1 at membrane sites having engaged integrin receptors is unknown. Cytosolic Rac1 is associated with RhoGDI, which inhibits GDP/GTP exchange reactions as well as GTP hydrolysis *in vitro* (Hancock and Hall, 1993; Sasaki et al., 1993). Three events are necessary for the activation of Rac1 at plasma membrane: dissociation of RhoGDI from Rac1, membrane translocation of Rac1, and nucleotide exchange activity on Rac1. There must be some way to coordinate these events after invasin binds the integrin receptor. Earlier work indicates that membrane translocation of Rac1 is accomplished by a membrane-localized nucleotide exchange activity (Bokoch et al., 1994). GTPγS and a membrane-associated protein appear to be necessary for the dissociation of RhoGDI from Rac1(Bokoch et al., 1994). These data suggest a model in which Rac1-GDP is delivered by RhoGDI to plasma membrane, and then subjected to a subsequent nucleotide exchange reaction. It is possible, but not proven, that RhoGDI dissociation occurs before nucleotide exchange activities. During intracellular trafficking events involving the Rab GTPase family, a GDI-displacement factor releases RabGDI so that the GTPase can undergo nucleotide exchange (Dirac-Svejstrup et al., 1997). Perhaps such a GDI-displacement factor is involved in integrin-promoted uptake. Alternatively, a membrane-associated GEF may promote nucleotide exchange by Rac1, causing dissociation of Rac1 from RhoGDI as a result of the lower affinity of Rac1-GTP for RhoGDI relative to that observed for Rac1-GDP (Sasaki et al., 1993). A third model is also possible, based on the finding that a significant fraction of constitutively active Rac1 mutant inside a cell is associated with RhoGDI, instead of downstream effectors (Del Pozo et al., 2002). It is possible that nucleotide exchange occurs prior to RhoGDI binding, and an active RhoGDI-Rac1-GTP complex is triggered and ready to promote actin rearrangements once RhoGDI delivers Rac1-GTP to the plasma membrane in response to integrin engagement (Del Pozo et al., 2002). Further experiments using different forms of Rac1 mutants coupled to detailed biochemical studies will help to determine which model is the best description for what is happening inside a cell when challenged with invasin-coated bacteria.

CONSEQUENCES OF Rac1 ACTIVATION

Activated Rac1 can direct actin polymerization via multiple pathways (Etienne-Manneville and Hall, 2002). The limiting factor for actin polymerization is the generation of free barbed ends of actin filaments, where free actin monomer can be incorporated to elongate filaments. Three ways exist to create free barbed ends: release of capping proteins from protected barbed ends, severing of actin filaments, and de novo filament formation through Arp2/3-mediated nucleation from actin monomers or via branching from existing filaments (Blanchoin et al., 2000). Dissociation of the capping protein gelsolin from barbed ends can be stimulated by the lipid phosphoinositol-4,5-phosphate (PIP_2). Formation of this lipid at cell edge is a Rac1-regulated process, which appears to be the result of recruitment of phosphatidylinositol-4-phosphate 5-kinase (PIP5K) to the site of Rac1 localization (Janmey and Stossel, 1987; Tolias et al., 2000). In fact, type Iα PIP5K is recruited to bacterial adhesion sites to form PIP_2 (Wong and Isberg, 2003). The production of this lipid appears to play a significant role in uptake, as a membrane-targeted PIP_2-specific phosphatase has an inhibitory effect on invasin-promoted bacterial uptake (Wong and Isberg, 2003).

A second pathway for initiation of actin nucleation promoted by Rac1 activation occurs via enhancing the activity of the heteroheptameric protein Arp2/3. Activation of Rac1 has been proposed to lead to Arp2/3 activation via interaction with IRSp53 which in turn stimulates the activity of WAVE2, a direct activator for Arp2/3 (Miki et al., 2000). Other models exist for how Rac1-GTP activates Arp2/3, which better explain the fact that purified WAVE, in the absence of Rac1, can stimulate Arp2/3-dependent actin nucleation. This property suggests that within cells, WAVE proteins are negatively regulated, and Rac1 acts to inactivate the negative regulators. Results supporting this latter model indicate that in mammalian cells, the WAVE1 protein is found in a heterotetrameric complex that is inactive and cannot activate Arp2/3. Rac1-GTP can dissociate the negative regulatory subunits from WAVE1, allowing WAVE1 to stimulate Arp2/3-dependent actin nucleation (Eden et al., 2002). Several of the players in this pathway appear to be involved in invasin-promoted uptake. Endogenous WAVE1 as well as Arp2/3 accumulate around bacterial entry sites (Alrutz et al., 2001) (Wong and Isberg, unpublished data). PIP_2, on the other hand, probably participates in uptake at some other level than Arp2/3 activation, because PIP2 has no effect on the activity of the WAVE1 complex (Eden et al., 2002). Instead, PIP_2 may have a role in recruiting Arp2/3 to integrin-dependent cytoskeleton assembly structures. PIP_2 together with an Arp2/3-activating WAVE fragment (VCA domain) stimulates the binding of the Arp2/3 complex to vinculin, a component of integrin-associated cytoskeletal structures (DeMali et al., 2002). Recruitment of the Arp2/3 complex to vinculin is enhanced by activated Rac1 and blocked by dominant negative Rac1 (DeMali et al., 2002). Future work should be focused on defining the biochemical activities of Rac1 and PIP_2 that spatially and temporally control invasin-stimulated actin polymerization.

In addition to regulating cytoskeletal rearrangements, activation of Rho family members leads to activation of MAP kinase pathways in the cell (Minden et al., 1995). The results of such activation include the transcriptional induction of a number of products that potentially modulate the course of an infectious disease within the host. Among these products include cytokines that influence the extent and severity of the inflammatory response. Purified invasin and bacteria expressing surface-exposed invasin from *Y. enterocolitica* stimulate the production of interleukin-8 (IL-8) by host cells (Schulte et al., 2000a). This induction appears to occur in response to engagement of integrin receptors followed by activation of Rac1 (Grassl et al., 2003). Dominant inhibitory forms of Rac1 block the induction of IL-8, and expression of the cytokine is in response to translocation of the transcriptional activator NF-κB into the nucleus (Grassl et al., 2003). Several MAP kinases appear to regulate this response.

REGULATION OF INVASIN EXPRESSION

The expression of invasin is under tight control by environmental factors such as temperature, growth phase, pH and osmolarity. In standard laboratory media, invasin expression in *Y. pseudotuberculosis* is maximal at 28°C but is repressed at 37°C (Krukonis et al., 1998). This has led to the argument that invasin is expressed maximally prior to contact with a mammalian host, as might be expected for a protein that only promotes events that occur immediately after ingestion of bacteria by the host. Once the microorganism is growing within the mammalian host, invasin expression is postulated to be shut down, and the protein would not be involved in later events after translocation across the intestine (Isberg, 1990). This model probably underestimates the complexity of invasin regulation. Manipulation of media conditions can allow robust expression of the *Y. enterocolitica* invasin at 37°C, and bacteria isolated from mouse tissue appear to express high levels of the protein, in spite of the results using broth-grown bacteria (Pepe et al., 1994).

Invasin expression is controlled by what appears to be a global regulator that is responsible for temperature-dependent invasin expression in broth-grown bacteria. This regulator is an 18.6 kDa protein called RovA, which is a member of a family of bacterial proteins that regulate virulence factors (Nagel et al., 2001; Revell and Miller, 2000). RovA is positively autoregulated and binds directly to the *inv* promoter region. The expression of *rovA* is regulated post-translationally and has a pattern of expression that is similar to the *inv* gene, at least in the case of broth grown bacteria at neutral pH (Nagel et al., 2001). Thus, RovA is a direct positive regulator of *inv* expression. It is clear that the protein regulates expression of several other proteins that contribute to establishing an enteric disease. In *Y. enterocolitica*, *rovA* mutants are defective for establishing systemic disease in the mouse model when the bacteria are inoculated via the oral route (Dube et al., 2003; Revell and Miller, 2000), and this mutant also induces a strikingly attenuated inflammatory response in local intestinal lymph nodes compared to what is observed with wild type strains (Dube et al., 2001). RovA, however, appears to play a secondary role in regards to growth in deep tissue sites. Intraperitoneal inoculation with the *rovA* mutant results in a systemic disease that is difficult to distinguish from infections with wild type organisms (Dube et al., 2003).

ROLE OF INVASIN DURING ENTERIC DISEASES

Initial work with *Y. pseudotuberculosis inv* mutants indicated that invasin is likely to be dispensable for systemic disease and, in fact, may antagonize the disease process. *Y. pseudotuberculosis inv* mutants were shown to cause lethal disease in mice at infectious doses that were similar to that observed with wild type strains, after either oral or intraperitoneal inoculations (Rosqvist et al., 1988). As *Y. pestis* does not have an intact copy of invasin (see above), the lack of effect of the *inv* mutant after intraperitoneal inoculation of either *Y. enterocoliica* or *Y. pseudotuberculosis* is probably not surprising (Pepe and Miller, 1993a; Rosqvist et al., 1988). It was also hypothesized that a double mutant, lacking invasin and the YadA adhesin was even more virulent than the wild type host, allowing *Y. pseudotuberculosis* to mimic the hypervirulent *Y. pestis* species (Rosqvist et al., 1988). It is not clear, however, if this result was due to other rearrangements suffered in the strain background harboring the *inv* mutant (Han and Miller, 1997).

Analysis of *Y. enterocolitica inv* mutants in mouse oral inoculation models indicates that simple lethal dose experiments do not yield sufficient insight into the behavior of strains lacking invasin (Pepe and Miller, 1993b). During oral inoculation of mice, there appears to be an ordered temporal relationship regarding colonization of the bacteria in various tissues, in which entry of bacteria into the Peyer's patches is followed by introduction into the mesenteric lymph nodes, which in turn precedes growth of bacteria in the liver and spleen (Heesemann et al., 1993). This suggests an ordered path in which bacteria translocate from the lumen of the intestine into Peyer's patches, and the bacteria present in this site eventually make their way into deep tissue sites. The behavior of *Y. enterocolitica inv* mutants turns this model on its head (Pepe and Miller, 1993b). In spite of the fact that the *inv* mutants promote levels of lethal systemic disease that appear similar to that observed with wild type strains, the rest of the infection process appears very different from that promoted by the wild type. Colonization of the Peyer's patches and mesenteric lymph nodes by *inv* mutants is delayed by several days compared to wild type strains, and the mutant rarely appears in these localized lymph nodes prior to detection in deep tissue sites (Pepe and Miller, 1993b). This implies that local disease, and in particular *Yersinia* mesenteric lymphadenitis, requires the invasin protein to establish colonization. These results also indicate that the ordered pathway model is likely to be incorrect, as systemic disease occurs with *inv* mutants in the absence of prior colonization of regional lymph nodes. Therefore it seems highly likely that during animal infection models, two disease paths are taking place simultaneously. One, which is invasin-dependent, results in entry into Peyer's patches and mesenteric lymph nodes, while the other pathway, which occurs in the absence of invasin, involves translocation across the intestine via another site, resulting in seeding of the liver and spleen and eventual death of the animal.

In *Y. pseudotuberculosis*, defective colonization of *inv* mutants in the Peyer's patches appears to be the result of an inability of mutants to traverse the intestinal epithelium and enter into cells overlying the Peyer's patches. In most mammalian species, the layer of cells that interfaces with the lumen of the intestine are called M cells, which are specialized cells that appear adapted to sampling small amounts of intestinal contents and facilitating presentation of translocated antigens to immune effector cells (Neutra et al., 2001). When ligated intestinal loop experiments with *Y. pseudotuberculosis* were performed in mice, bacterial binding appeared to be directed toward M cells located over the Peyer's patches (Marra and Isberg, 1997). On the other hand, *Y. pseudotuberculosis inv* mutants were highly defective for entry into M cells, indicating that the lack of colonization of the mutants in localized lymph nodes is a result of lack of entry into M cells. The tropism of wild type enteropathogenic *Yersinia* for M cells can be easily explained, as most intestinal epithelial cells do not present β_1 chain integrins on their apical face to the lumen of the intestine (Clark et al., 1998). In contrast, M cells appear to have luxurious amounts of invasin receptors on their apical surfaces, effectively acting as sinks for bacterial interaction. An interesting cell culture system reproduces this phenomenon with the related *Y. enterocolitica* (Schulte et al., 2000b). Using a polarized epithelial cell model, in which a subpopulation of cells differentiate into M cells, the bacteria are found to be tropic for this subpopulation, and translocate through the monolayer in an invasin-dependent fashion (Schulte et al., 2000b).

Recent results indicate that invasin may play a role in persistent colonization of bacteria within some site in the intestine. Using a mutagenesis strategy that identifies random mutants that cannot efficiently colonize the intestine has allowed the isolation of strains lacking invasin. Such mutants poorly compete with wild type strains to establish replication sites within some unknown intestinal site. Furthermore, in contrast to previous results (Mecsas et al., 2001), the growth of these mutants in deep tissue sites seems impaired, most likely because they are delayed in movement across the epithelium into deep tissue sites relative to wild type strains.

CONCLUSIONS

Work on the enteropathogenic *Yersinia* invasin protein has contributed much to our understanding of how bacteria promote uptake into normally nonphagocytic cells. It has provided a simple system to probe bacterial uptake *in vitro*, and the analysis of mutants defective for this protein has given surprising insights into the events associated with systemic disease after oral inoculation of an enteric pathogen. The challenges for the future involve uncovering the events that link integrin engagement to activation of Rho family members prior to uptake, and determining why persistent colonization of the bacteria in the intestine requires the presence of the invasin protein. Emerging techniques in image analysis and the use of clonal analysis of bacterial infections in animal models will greatly facilitate studies on the role played by this protein in both the cellular and organismal interactions of enteropathogenic *Yersinia* species.

REFERENCES

Abrami, L., Liu, S., Cosson, P., Leppla, S. H., and van der Goot, F. G. 2003. Anthrax toxin triggers endocytosis of its receptor via a lipid raft-mediated clathrin-dependent process. J. Cell Biol. 160: 321-328.

Alrutz, M. A., and Isberg, R. R. 1998. Involvement of focal adhesion kinase in invasin-mediated uptake. Proc. Natl. Acad. Sci. USA. 95: 13658-13663.

Alrutz, M. A., Srivastava, A., Wong, K. W., D'Souza-Schorey, C., Tang, M., Ch'Ng, L. E., Snapper, S. B., and Isberg, R. R. 2001. Efficient uptake of *Yersinia pseudotuberculosis* via integrin receptors involves a Rac1-Arp 2/3 pathway that bypasses N-WASP function. Mol. Microbiol. 42: 689-703.

Aota, S., Nomizu, M., and Yamada, K. M. 1994. The short amino acid sequence Pro-His-Ser-Arg-Asn in human fibronectin enhances cell-adhesive function. J. Biol. Chem. 269: 24756-24761.

Bellanger, J. M., Lazaro, J. B., Diriong, S., Fernandez, A., Lamb, N., and Debant, A. 1998. The two guanine nucleotide exchange factor domains of Trio link the Rac1 and the RhoA pathways in vivo. Oncogene. 16: 147-152.

Black, D. S., and Bliska, J. B. 2000. The RhoGAP activity of the *Yersinia pseudotuberculosis* cytotoxin YopE is required for antiphagocytic function and virulence. Mol. Microbiol. 37: 515-527.

Blanchoin, L., Amann, K. J., Higgs, H. N., Marchand, J. B., Kaiser, D. A., and Pollard, T. D. 2000. Direct observation of dendritic actin filament networks nucleated by Arp2/3 complex and WASP/Scar proteins. Nature. 404: 1007-1011.

Bokoch, G. M., Bohl, B. P., and Chuang, T. H. 1994. Guanine nucleotide exchange regulates membrane translocation of Rac/Rho GTP-binding proteins. J. Biol. Chem. 269: 31674-31679.

Bowditch, R. D., Hariharan, M., Tominna, E. F., Smith, J. W., Yamada, K. M., Getzoff, E. D., and Ginsberg, M. H. 1994a. Identification of a novel integrin binding site in fibronectin. Differential utilization by β_3 integrins. J. Biol. Chem. 269: 10856-10863.

Bowditch, R. D., Hariharan, M., Tominna, E. F., Smith, J. W., Yamada, K. M., Getzoff, E. D., and Ginsberg, M. H. 1994b. Identification of a novel integrin binding site in fibronectin. Differential utilization by beta 3 integrins. J. Biol. Chem. 269: 10856-10863.

Brugnera, E., Haney, L., Grimsley, C., Lu, M., Walk, S. F., Tosello-Trampont, A. C., Macara, I. G., Madhani, H., Fink, G. R., and Ravichandran, K. S. 2002. Unconventional Rac-GEF activity is mediated through the Dock180-ELMO complex. Nat. Cell Biol. 4: 574-582.

Burridge, K., and Chrzanowska, W. M. 1996. Focal adhesions, contractility, and signaling. Ann. Rev. Cell Develop. Biol. 12: 463-518.

Clark, M. A., Hirst, B. H., and Jepson, M. A. 1998. M-cell surface beta1 integrin expression and invasin-mediated targeting of *Yersinia pseudotuberculosis* to mouse Peyer's patch M cells. Infect. Immun. 66: 1237-1243.

Cochran, J. R., Cameron, T. O., and Stern, L. J. 2000. The relationship of MHC-peptide binding and T cell activation probed using chemically defined MHC class II oligomers. Immunity. 12: 241-250.

Cornelis, G. R. 2000. Molecular and cell biology aspects of plague. Proc. Natl. Acad. Sci. USA. 97: 8778-8783.

Cowan, C., Jones, H. A., Kaya, Y. H., Perry, R. D., and Straley, S. C. 2000. Invasion of epithelial cells by *Yersinia pestis*: evidence for a *Y. pestis*-specific invasin. Infect. Immun. 68: 4523-4530.

Danen, E. H., Aota, S., van, K. A., Yamada, K. M., Ruiter, D. J., and van, M. G. 1995. Requirement for the synergy site for cell adhesion to fibronectin depends on the activation state of integrin alpha 5 beta 1. J. Biol. Chem. 270: 21612-21618.

Del Pozo, M. A., Kiosses, W. B., Alderson, N. B., Meller, N., Hahn, K. M., and Schwartz, M. A. 2002. Integrins regulate GTP-Rac localized effector interactions through dissociation of Rho-GDI. Nat. Cell Biol. 4: 232-239.

DeMali, K. A., Barlow, C. A., and Burridge, K. 2002. Recruitment of the Arp2/3 complex to vinculin: coupling membrane protrusion to matrix adhesion. J. Cell Biol. 159: 881-891.

Deng, W., Burland, V., Plunkett, G., 3rd, Boutin, A., Mayhew, G.F., Liss, P., Perna, N.T., Rose, D.J., Mau, B., Zhou, S., Schwartz, D.C., Fetherston, J.D., Lindler, L.E., Brubaker, R.R., Plano, G.V., Straley, S.C., McDonough, K.A., Nilles, M.L., Matson, J.S., Blattner, F.R. and Perry, R.D. 2002. Genome sequence of *Yersinia pestis* KIM. J. Bacteriol. 184: 4601-11.

Dersch, P., and Isberg, R. R. 1999. A region of the *Yersinia pseudotuberculosis* invasin protein enhances integrin-mediated uptake into mammalian cells and promotes self-association. EMBO J. 18: 1199-1213.

Dersch, P., and Isberg, R. R. 2000. An immunoglobulin superfamily-like domain unique to the *Yersinia pseudotuberculosis* invasin protein is required for stimulation of bacterial uptake via integrin receptors. Infect. Immun. 68: 2930-2938.

Dirac-Svejstrup, A. B., Sumizawa, T., and Pfeffer, S. R. 1997. Identification of a GDI displacement factor that releases endosomal Rab GTPases from Rab-GDI. EMBO J. 16: 465-472.

Dube, P. H., Handley, S. A., Revell, P. A., and Miller, V. L. 2003. The *rovA* mutant of *Yersinia enterocolitica* displays differential of virulence depending on the route of infection. Infect. Immun. 71: 3512-3520.

Dube, P. H., Revell, P. A., Chaplin, D. D., Lorenz, R. G., and Miller, V. L. 2001. A role for IL-1 alpha in inducing pathologic inflammation during bacterial infection. Proc. Natl. Acad. Sci. USA. 98: 10880-10885.

Eble, J. A., Wucherpfennig, K. W., Gauthier, L., Dersch, P., Krukonis, E., Isberg, R. R., and Hemler, M. E. 1998. Recombinant soluble human alpha 3 beta 1 integrin: purification, processing, regulation, and specific binding to laminin-5 and invasin in a mutually exclusive manner. Biochemistry. 37: 10945-10955.

Eden, S., Rohatgi, R., Podtelejnikov, A. V., Mann, M., and Kirschner, M. W. 2002. Mechanism of regulation of WAVE1-induced actin nucleation by Rac1 and Nck. Nature. 418: 790-793.

El Tahir, Y., and Skurnik, M. 2001. YadA, the multifaceted *Yersinia* adhesin. Int. J. Med Microbiol. 291: 209-218.

Ennis, E., Isberg, R. R., and Shimizu, Y. 1993. Very late antigen 4-dependent adhesion and costimulation of resting human T cells by the bacterial beta 1 integrin ligand invasin. J. Exp. Med. 177: 207-212.

Etienne-Manneville, S., and Hall, A. 2002. Rho GTPases in cell biology. Nature. 420: 629-635.

Feig, L. A. 1999. Tools of the trade: use of dominant-inhibitory mutants of Ras-family GTPases. Nat. Cell Biol. 1: E25-27.

Gao, Y., Xing, J., Streuli, M., Leto, T. L., and Zheng, Y. 2001. Trp(56) of rac1 specifies interaction with a subset of guanine nucleotide exchange factors. J. Biol. Chem. 276: 47530-47541.

Grassl, G. A., Kracht, M., Wiedemann, A., Hoffmann, E., Aepfelbacher, M., Von Eichel-Streiber, C., Bohn, E., and Autenrieth, I. B. 2003. Activation of NF-kappaB and IL-8 by *Yersinia enterocolitica* invasin protein is conferred by engagement of Rac1 and MAP kinase cascades. Cell Microbiol. 5: 957-971.

Guan, J. L. 1997. Role of focal adhesion kinase in integrin signaling. Int. J. Biochem. Cell Biol. 29: 1085-1096.

Gustavsson, A., Armulik, A., Brakebusch, C., Fassler, R., Johansson, S., and Fallman, M. 2002. Role of the β1-integrin cytoplasmic tail in mediating invasin-promoted internalization of *Yersinia*. J. Cell Sci. 115: 2669-2678.

Hamburger, Z. A., Brown, M. S., Isberg, R. R., and Bjorkman, P. J. 1999. Crystal structure of invasin: a bacterial integrin-binding protein. Science. 286: 291-295.

Han, Y. W., and Miller, V. L. 1997. Reevaluation of the virulence phenotype of the *inv yadA* double mutants of *Yersinia pseudotuberculosis*. Infect. Immun. 65: 327-330.

Hancock, J. F., and Hall, A. 1993. A novel role for RhoGDI as an inhibitor of GAP proteins. EMBO J. 12: 1915-1921.

Harpaz, Y., and Chothia, C. 1994. Many of the immunoglobulin superfamily domains in cell adhesion molecules and surface receptors belong to a new structural set which is close to that containing variable domains. J. Mol. Biol. 238: 528-539.

Heesemann, J., Gaede, K., and Autenrieth, I. B. 1993. Experimental Yersinia enterocolitica infection in rodents: a model for human yersiniosis. APMIS. 101: 417-429.

Hogg, N., Laschinger, M., Giles, K., and McDowall, A. 2003. T-cell integrins: more than just sticking points. J. Cell Sci. 116: 4695-4705.

Hsia, D. A., Mitra, S. K., Hauck, C. R., Streblow, D. N., Nelson, J. A., Ilic, D., Huang, S., Li, E., Nemerow, G. R., Leng, J., et al. 2003. Differential regulation of cell motility and invasion by FAK. J. Cell Biol. 160: 753-767.

Hynes, R. O. 2002. Integrins: bidirectional, allosteric signaling machines. Cell. 110: 673-687.

Isberg, R. R. 1990. Pathways for the penetration of enteroinvasive *Yersinia* into mammalian cells. Mol. Biol. Med. 7: 73-82.

Isberg, R. R., and Falkow, S. 1985. A single genetic locus encoded by *Yersinia pseudotuberculosis* permits invasion of cultured animal cells by *Escherichia coli* K-12. Nature. 317: 262-264.

Isberg, R. R., Hamburger, Z., and Dersch, P. 2000. Signaling and invasin-promoted uptake via integrin receptors. Microbes Infect. 2: 793-801.

Isberg, R. R., and Leong, J. M. 1990. Multiple β_1 chain integrins are receptors for invasin, a protein that promotes bacterial penetration into mammalian cells. Cell. 60: 861-871.

Isberg, R. R., Voorhis, D. L., and Falkow, S. 1987. Identification of invasin: a protein that allows enteric bacteria to penetrate cultured mammalian cells. Cell. 50: 769-778.

Janmey, P. A., and Stossel, T. P. 1987. Modulation of gelsolin function by phosphatidylinositol 4,5-bisphosphate. Nature. 325: 362-364.

Karnoub, A. E., Worthylake, D. K., Rossman, K. L., Pruitt, W. M., Campbell, S. L., Sondek, J., and Der, C. J. 2001. Molecular basis for Rac1 recognition by guanine nucleotide exchange factors. Nat. Struct. Biol. 8: 1037-1041.

Kawasaki, Y., Senda, T., Ishidate, T., Koyama, R., Morishita, T., Iwayama, Y., Higuchi, O., and Akiyama, T. 2000. Asef, a link between the tumor suppressor APC and G-protein signaling. Science. 289: 1194-1197.

Keely, P. J., Westwick, J. K., Whitehead, I. P., Der, C. J., and Parise, L. V. 1997. Cdc42 and Rac1 induce integrin-mediated cell motility and invasiveness through PI(3)K. Nature. 390: 632-636.

Kraehenbuhl, J. P., and Neutra, M. R. 2000. Epithelial M cells: differentiation and function. Annu Rev. Cell Dev. Biol. 16: 301-332.

Krukonis, E. S., Dersch, P., Eble, J. A., and Isberg, R. R. 1998. Differential effects of integrin alpha chain mutations on invasin and natural ligand interaction. J. Biol. Chem. 273: 31837-31843.

Krukonis, E. S., and Isberg, R. R. 2000. Integrin beta1-chain residues involved in substrate recognition and specificity of binding to invasin. Cell Microbiol. 2: 219-230.

Lee, J. O., Rieu, P., Arnaout, M. A., and Liddington, R. 1995. Crystal structure of the A domain from the alpha subunit of integrin CR3 (CD11b/CD18). Cell. 80: 631-638.

Leong, J. M., Morrissey, P. E., Marra, A., and Isberg, R. R. 1995. An aspartate residue of the *Yersinia pseudotuberculosis* invasin protein that is critical for integrin binding. EMBO J. 14: 422-431.

Machner, M. P., Frese, S., Schubert, W. D., Orian-Rousseau, V., Gherardi, E., Wehland, J., Niemann, H. H., and Heinz, D. W. 2003. Aromatic amino acids at the surface of InlB are essential for host cell invasion by *Listeria monocytogenes*. Mol. Microbiol. 48: 1525-1536.

Marra, A., and Isberg, R. R. 1997. Invasin-dependent and invasin-independent pathways for translocation of *Yersinia pseudotuberculosis* across the Peyer's patch intestinal epithelium. Infect. Immun. 65: 3412-3421.

Mecsas, J., Bilis, I., and Falkow, S. 2001. Identification of attenuated *Yersinia pseudotuberculosis* strains and characterization of an orogastric infection in BALB/c mice on day 5 postinfection by signature-tagged mutagenesis. Infect. Immun. 69: 2779-2787.

Miki, H., Yamaguchi, H., Suetsugu, S., and Takenawa, T. 2000. IRSp53 is an essential intermediate between Rac and WAVE in the regulation of membrane ruffling. Nature. 408: 732-735.

Minden, A., Lin, A., Claret, F. X., Abo, A., and Karin, M. 1995. Selective activation of the JNK signaling cascade and c-Jun transcriptional activity by the small GTPases Rac and Cdc42Hs. Cell. 81: 1147-1157.

Miyawaki, A. 2003. Visualization of the spatial and temporal dynamics of intracellular signaling. Dev. Cell. 4: 295-305.

Nagel, G., Lahrz, A., and Dersch, P. 2001. Environmental control of invasin expression in *Yersinia pseudotuberculosis* is mediated by regulation of RovA, a transcriptional activator of the SlyA/Hor family. Mol. Microbiol. 41: 1249-1269.

Neutra, M. R., Mantis, N. J., and Kraehenbuhl, J. P. 2001. Collaboration of epithelial cells with organized mucosal lymphoid tissues. Nat. Immunol. 2: 1004-1009.

Nougayrede, J. P., Fernandes, P. J., and Donnenberg, M. S. 2003. Adhesion of enteropathogenic *Escherichia coli* to host cells. Cell. Microbiol. 5: 359-372.

Parkhill, J., Wren, B. W., Thomson, N. R., Titball, R. W., Holden, M. T., Prentice, M. B., Sebaihia, M., James, K. D., Churcher, C., Mungall, K. L., et al. 2001. Genome sequence of *Yersinia pestis*, the causative agent of plague. Nature. 413: 523-527.

Pepe, J. C., Badger, J. L., and Miller, V. L. 1994. Growth phase and low pH affect the thermal regulation of the *Yersinia enterocolitica inv* gene. Mol. Microbiol. 11: 123-135.

Pepe, J. C., and Miller, V. L. 1990. The *Yersinia enterocolitica inv* gene product is an outer membrane protein that shares epitopes with *Yersinia pseudotuberculosis* invasin. J. Bacteriol. 172: 3780-3789.

Pepe, J. C., and Miller, V. L. 1993a. The biological role of invasin during a *Yersinia enterocolitica* infection. Infect. Agents Dis. 2: 236-241.

Pepe, J. C., and Miller, V. L. 1993b. *Yersinia enterocolitica* invasin: a primary role in the initiation of infection. Proc. Natl. Acad. Sci. USA. 90: 6473-6477.

Rankin, S., Isberg, R. R., and Leong, J. M. 1992. The integrin-binding domain of invasin is sufficient to allow bacterial entry into mammalian cells. Infect. Immun. 60: 3909-3912.

Revell, P. A., and Miller, V. L. 2000. A chromosomally encoded regulator is required for expression of the *Yersinia enterocolitica inv* gene and for virulence. Mol. Microbiol. 35: 677-685.

Rosenshine, I., Duronio, V., and Finlay, B. B. 1992. Tyrosine protein kinase inhibitors block invasin-promoted bacterial uptake by epithelial cells. Infect. Immun. 60: 2211-2217.

Rosqvist, R., Skurnik, M., and Wolf, W. H. 1988. Increased virulence of *Yersinia pseudotuberculosis* by two independent mutations. Nature. 334: 522-524.

Saltman, L. H., Lu, Y., Zaharias, E. M., and Isberg, R. R. 1996. A region of the *Yersinia pseudotuberculosis* invasin protein that contributes to high affinity binding to integrin receptors. J. Biol. Chem. 271: 23438-23444.

Sasaki, T., Kato, M., and Takai, Y. 1993. Consequences of weak interaction of rho GDI with the GTP-bound forms of rho p21 and rac p21. J. Biol. Chem. 268: 23959-23963.

Schulte, R., Grassl, G. A., Preger, S., Fessele, S., Jacobi, C. A., Schaller, M., Nelson, P. J., and Autenrieth, I. B. 2000a. *Yersinia enterocolitica* invasin protein triggers IL-8 production in epithelial cells via activation of Rel p65-p65 homodimers. FASEB J. 14: 1471-1484.

Schulte, R., Kerneis, S., Klinke, S., Bartels, H., Preger, S., Kraehenbuhl, J. P., Pringault, E., and Autenrieth, I. B. 2000b. Translocation of *Yersinia entrocolitica* across reconstituted intestinal epithelial monolayers is triggered by *Yersinia* invasin binding to beta1 integrins apically expressed on M-like cells. Cell. Microbiol. 2: 173-185.

Sieg, D. J., Hauck, C. R., Ilic, D., Klingbeil, C. K., Schaefer, E., Damsky, C. H., and Schlaepfer, D. D. 2000. FAK integrates growth-factor and integrin signals to promote cell migration. Nat. Cell Biol. 2: 249-256.

Simonet, M., Riot, B., Fortineau, N., and Berche, P. 1996. Invasin production by *Yersinia pestis* is abolished by insertion of an IS*200*-like element within the *inv* gene. Infect. Immun. 64: 375-379.

Takagi, J., Petre, B. M., Walz, T., and Springer, T. A. 2002. Global conformational rearrangements in integrin extracellular domains in outside-in and inside-out signaling. Cell. 110: 599-511.

Tolias, K. F., Hartwig, J. H., Ishihara, H., Shibasaki, Y., Cantley, L. C., and Carpenter, C. L. 2000. Type Ialpha phosphatidylinositol-4-phosphate 5-kinase mediates Rac-dependent actin assembly. Curr. Biol. 10: 153-156.

Touze, T., Hayward, R. D., Eswaran, J., Leong, J. M., and Koronakis, V. 2004. Self-association of EPEC intimin mediated by the beta-barrel-containing anchor domain: a role in clustering of the Tir receptor. Mol. Microbiol. 51: 73-87.

Tran Van Nhieu, G., and Isberg, R. R. 1991. The *Yersinia pseudotuberculosis* invasin protein and human fibronectin bind to mutually exclusive sites on the alpha 5 beta 1 integrin receptor. J. Biol. Chem. 266: 24367-24375.

Tran Van Nhieu, G., and Isberg, R. R. 1993. Bacterial internalization mediated by beta 1 chain integrins is determined by ligand affinity and receptor density. EMBO J. 12: 1887-1895.

Tran Van Nhieu, G., Krukonis, E. S., Reszka, A. A., Horwitz, A. F., and Isberg, R. R. 1996. Mutations in the cytoplasmic domain of the integrin beta1 chain indicate a role for endocytosis factors in bacterial internalization. J. Biol. Chem. 271: 7665-7672.

Wachtel, M. R., and Miller, V. L. 1995. In vitro and in vivo characterization of an *ail* mutant of *Yersinia enterocolitica*. Infect. Immun. 63: 2541-2548.

Weidow, C. L., Black, D. S., Bliska, J. B., and Bouton, A. H. 2000. CAS/Crk signalling mediates uptake of *Yersinia* into human epithelial cells. Cell. Microbiol. 2: 549-560.

Weis, W. I., Taylor, M. E., and Drickamer, K. 1998. The C-type lectin superfamily in the immune system. Immunol. Rev. 163: 19-34.

Wiedemann, A., Linder, S., Grassl, G., Albert, M., Autenrieth, I., and Aepfelbacher, M. 2001. *Yersinia enterocolitica* invasin triggers phagocytosis via beta1 integrins, CDC42Hs and WASp in macrophages. Cell. Microbiol. 3: 693-702.

Wong, K. W., and Isberg, R. R. 2003. Arf6 and phosphoinositol-4-phosphate-5-kinase activities permit bypass of the Rac1 requirement for beta1 integrin-mediated bacterial uptake. J. Exp. Med. 198: 603-614.

Xiong, J. P., Stehle, T., Zhang, R., Joachimiak, A., Frech, M., Goodman, S. L., and Arnaout, M. A. 2002. Crystal structure of the extracellular segment of integrin alpha Vbeta3 in complex with an Arg-Gly-Asp ligand. Science. 296: 151-155.

Yang, Y., and Isberg, R. R. 1997. Transcriptional regulation of the *Yersinia pseudotuberculosis* pH6 antigen adhesin by two envelope-associated components. Mol. Microbiol. 24: 499-510.

Chapter 7

Transcriptional Regulation in *Yersinia*: An Update

Michaël Marceau

ABSTRACT

In response to the ever-present need to adapt to environmental stress, bacteria have evolved complex (and often overlapping) regulatory networks that respond to various changes in growth conditions, including entry into the host. The expression of most bacterial virulence factors is regulated; thus the question of how bacteria orchestrate this process has become a recurrent research theme for every bacterial pathogen, and the three pathogenic *Yersinia* species are no exception. The earliest studies of regulation in these species were prompted by the characterization of plasmid-encoded virulence determinants, and those conducted since have continued to focus on the principal aspects of virulence in these pathogens. Most *Yersinia* virulence factors are thermally regulated, and are active at either 28°C (the optimal growth temperature) or 37°C (the host temperature). However, regulation by this omnipresent thermal stimulus occurs through a wide variety of mechanisms, which generally act in conjunction with (or are modulated by) additional controls for other environmental cues such as pH, ion concentration, nutrient availability, osmolarity, oxygen tension and DNA damage. *Yersinia*'s recent entry into the genome sequencing era has given scientists the opportunity to study these regulators on a genome-wide basis. This has prompted the first attempts to establish links between the presence or absence of regulatory elements and the three pathogenic species' respective lifestyles and degrees of virulence.

INTRODUCTION

Compared to cells of multicellular organisms, microorganisms face a significant additional challenge: they encounter a wide array of sudden, intense and sometimes even life-threatening environmental changes, and must therefore rapidly modify their structure and metabolism accordingly. Although other mechanisms exist, these changes in bacterial physiology mainly occur by regulating the production of the appropriate structural proteins and enzymes. Adaptation of gene expression in response to such situations appears to be essential for bacterial survival, and thus the regulators involved in these processes should be treated as being as important as the effectors themselves. In bacterial pathogens like those of the *Yersinia* genus, most outside-to-inside stress-induced responses lead to changes in the expression of virulence factors. In fact, most of the known *Yersinia* virulence genes are regulated, and elements controlling their expression are thus also virulence factors.

All regulatory systems have a common purpose: to create an interface between the perception of one or several stimuli and to activate or repress expression of their cognate effectors. As we will see by reviewing what is known about *Yersinia*, the means used to regulate the production of a given bacterial factor range from very simple mechanisms (where the DNA-binding properties of the transcriptional regulator are directly altered by the stimulus) to extremely complex systems which sometimes require lengthy signal transduction cascades and/or simultaneous contribution of multiple regulators that may act at different levels, indeed all the way from initiation of gene transcription to protein turn-over. As in most bacteria, *Yersinia* regulatory networks are generally organized hierarchically, with a global regulatory system involving a master regulator such as a sigma factor or a histone-like protein; and lower-level, downstream-acting secondary regulators that control only a subset of a regulon's genes in response to more specific stress situations. Regulatory circuits may interfere with each other, leading to the discovery of unexpectedly dense, overlapping regulons.

The present chapter is divided into two parts. The first aims to provide a comprehensive overview of the better-known regulatory systems in *Yersinia* (summarized in Figure 1); many of these systems regulate virulence factors which are described in further detail in the other chapters of this book. The second part of this chapter is dedicated to what we can (and cannot) learn from genomic analysis. The genomic sequences of two *Y. pestis* strains have recently been released into public databases, and at least one *Y. pseudotuberculosis* sequence will be available in the very near future. This prompts opportunities to search for potentially new genus- or species-specific regulators. What will *Yersinia* genome-wide analysis tell us about regulation?

PART ONE: THE PRE-GENOMIC ERA: WHAT WE KNEW ABOUT *YERSINIA* REGULATION

TRANSCRIPTIONAL REGULATION OF THE pYV PLASMID-BORNE ANTIHOST GENES

The 70-kb pYV (Yersinia virulence) plasmid is found in all human pathogenic *Yersinia* strains and governs the synthesis of two major virulence factors: the first consists of the various Yop (Yersinia outer membrane protein) effector proteins (YopE, YopH, YopM, YopO/YpkA, YopP/YopJ, YopT) along with the type III secretion system (TTSS) subunits required for their delivery into the eukaryotic cytosol. Yops play a major role during the course of *Yersinia* host infection by contributing to phagocyte resistance, triggering macrophage apoptosis and provoking disorders in cytokine release patterns (see Chapter 16 for a review of the genetics and functions of the Yops). The second virulence determinant encoded by the pYV plasmid is YadA, which is principally known as a major adhesin involved in bacterial adherence to various eukaryotic extracellular matrix elements. YadA is also involved in resistance to the host's non-specific defences: it protects *Yersinia* from certain antimicrobial peptide classes synthesized by polymorphonuclear leukocytes (PMNs), and also interrupts formation of the Molecular Attack Complex (MAC). It is noteworthy that the *yadA* gene is inactivated in *Y. pestis* but not in the two other pathogenic species. A recent review of YadA has been published by El Tahir and Skurnik (2001).

The genes governing the synthesis of the pYV-encoded virulence factors belong to the same stimulon (the *yop* stimulon), which means they are all upregulated by the same environmental stimulus, an increase in temperature to 37°C upon entry into the mammalian host (Bölin et al., 1988). Some members of the *yop* stimulon are also

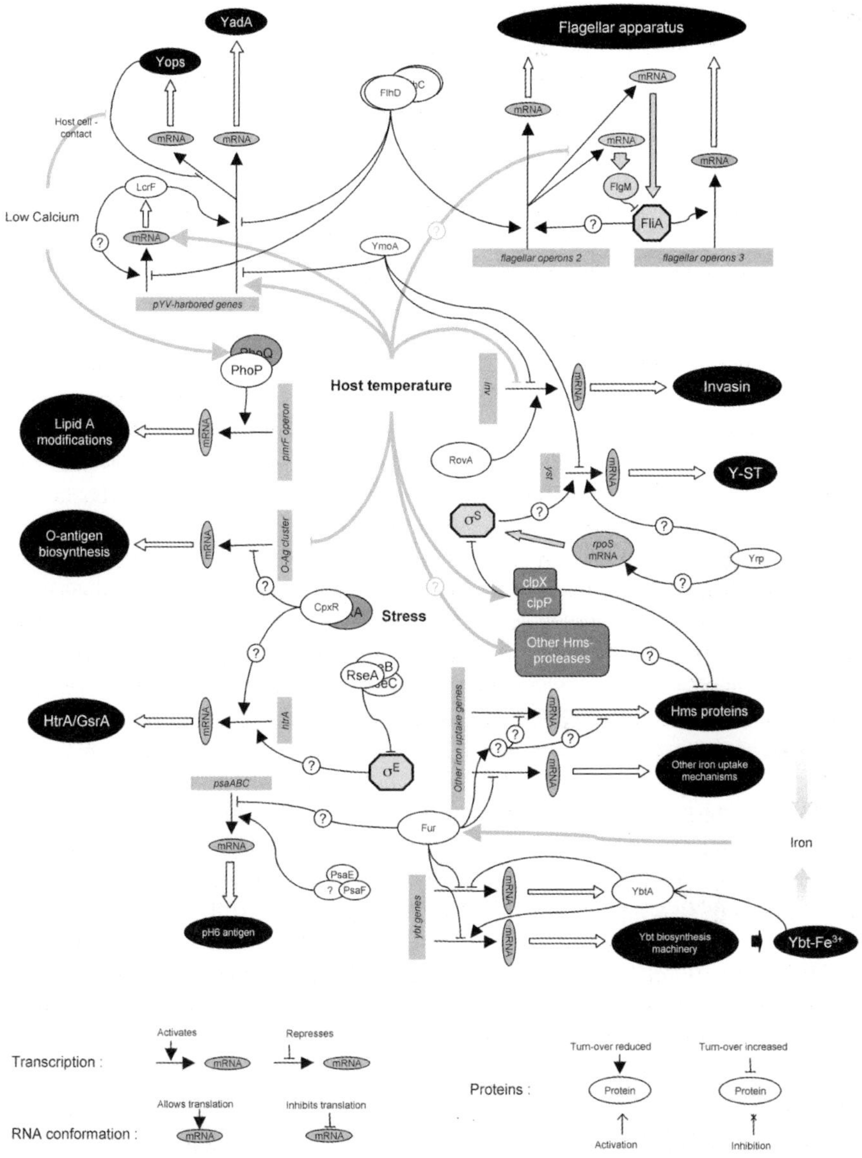

Figure 1. Overview of regulation networks discussed in this chapter. Question marks have been added for regulation processes either yet uncharacterized or predicted to exist based on similarity with known systems in other enterobacteria.

controlled by an additional stimulus: Ca^{2+} availability. This gene subset is often referred to as the LCRS (Low Calcium Response Stimulon) (Goguen et al., 1984; Straley et al., 1993). Transcriptional control of the Yop-encoding genes is a complex process, and provides a good example of how non-related stimuli may be taken into account hierarchically by the regulatory system.

TEMPERATURE AND THE OVERALL REGULATION OF pYV-HARBOURED VIRULENCE GENES

The 8.1 kDa YmoA protein was first characterized by Cornelis et al. (1991). Transposon mutagenesis revealed *Y. entrocolitica ymoA* mutants which displayed unusually high *yop* and *yadA* transcription levels at 28°C, suggesting that this protein behaves as a global repressor. The YmoA protein is highly similar (82% identity) to the haemolysin expression modulating protein Hha from *Escherichia coli*, and (at least partial) restoration of haemolysin synthesis in an *hha* mutant by complementation *in trans* with *ymoA* showed that these two molecules share similar functions (Mikulskis and Cornelis, 1994; Balsalobre et al., 1996). In light of their amino acid composition, Hha and YmoA were predicted to be histone-like proteins that modulate gene expression through control of DNA topology. Like Hha, YmoA may interact with H-NS (another chromatin-associated protein known to play a role in the thermal modulation of virulence factor expression) and thus form a nucleoid-protein complex responsible for thermoregulation (Nieto et al., 2002). In addition to its effect on pYV-borne genes, the YmoA histone is also involved in the silencing of *yst*, which encodes a thermostable toxin (see below).

However, studies carried out by Lambert de Rouvroit et al. (1992) with *yopH::cat* fusions showed that transcription of the *yop* genes is enhanced by temperature upshifts even after inactivation of *ymoA*, suggesting that YmoA modulates rather than regulates transcription of the *yadA* and *yop* genes, and that other means of thermoregulation might exist. Novobiocin is a compound that affects DNA superhelicity by interacting with gyrases. The fact that Yop genes were still expressed by wild type *Y. enterocolitica* at 30°C in the presence of sub-inhibitory levels of this drug (i.e., the *ymoA* mutant phenotype) argues in favour of mechanisms based on DNA conformation (Rohde et al., 1994). Consistent with this hypothesis is the demonstration that thermo-induced Yop expression coincided with variations of pYV DNA supercoiling (Rohde et al., 1994). Five years later, the same authors showed that these dramatic topological changes could be initiated by the melting of local DNA bends as a consequence of a temperature shift from 30°C to 37°C (Rohde et al., 1999).

However, in contrast to the *ysc* genes involved in the synthesis of the Yop secretion apparatus itself, transcription of most *yop*s, as well as other pYV-borne genes such as *ylpA yadA, sycE* (Skurnik and Toivanen, 1992; Wattiau and Cornelis, 1993) and the *virC* operon (Michiels et al., 1991), has been shown to require an additional regulator: LcrF (Yother et al., 1986; Lambert de Rouvroit et al., 1992). The 30.9-kDa *Y. enterocolitica* LcrF/VirF (Cornelis et al., 1989) transcriptional activator belongs to the AraC family: it contains two adjacent C-terminal Helix-Turn-Helix (H-T-H) DNA binding motifs. The *lcrF* gene is pYV-borne and, as with the other Yop and Ysc encoding genes, its transcription is induced at 37°C following changes in DNA superhelicity (Cornelis et al., 1989). Additionally, translation of LcrF can be enhanced, at least in *Y. pestis*, by the melting of heat-unstable mRNA secondary structures at the Ribosome Binding Site (RBS) (Hoe and Goguen, 1993). The VirF-binding region deduced from footprinting assays is 13-bp long, with the following consensus sequence: TTTTaGYcTtTat (where nucleotides conserved in 60% or more of the sequences are in uppercase letters, and Y indicates C or T) (Wattiau and Cornelis, 1994). Yop promoters are expressed constitutively by *ymoA virF* double mutants. Accordingly, *yop* expression has been found to be VirF-independent in *ymoA* mutants (Lambert de Rouvroit et al., 1992). These observations suggest that the role of VirF may be to counteract the negative effects of YmoA. However, the fact that *yop* expression is still enhanced by temperature upshifts (Lambert de Rouvroit et al., 1992) suggests the involvement of other thermoregulation mechanisms.

FROM LOW CALCIUM LEVELS TO *yop* TRANSCRIPTIONAL REGULATION

During infection, the *Yersinia* pYV-encoded type III secretion machinery is activated by a succession of environmental signals. The initial cue is a temperature upshift to 37°C (the host-entry signal), which promotes transcription of all Yop and Ysc proteins. Other possible signals encountered within the host may be required for the complete assembly of the secretion apparatus (Lee et al., 2001). The final signal that triggers the translocation of the Yops *in vivo* is contact with the host cell (Rosqvist et al., 1994; Sory and Cornelis, 1994; Pettersson et al., 1996). Biosynthesis and activation of the Yop machinery can be induced in *vitro* by incubating bacteria at 37°C in calcium-poor media (i.e., a calcium concentration lower than 80 μM). This experimentally induced response is often referred to as the Low Calcium Response (LCR). However, Ca^{2+} concentrations are dramatically lower inside the target cell cytoplasm than outside. Thus, as suggested by Lee et al. (2001), low calcium may also be the real triggering stimulus sensed by *Yersinia in vivo*. In the presence of millimolar calcium concentrations, the type III secretion apparatus remains blocked, and Yops are not secreted (Forsberg et al., 1991; Yother and Goguen, 1985). Furthermore, transcription of genes governing Yop synthesis is repressed (Cornelis et al., 1987; Forsberg and Wolf-Watz, 1988; Straley et al., 1993). Despite its name, LcrF alone cannot account for this downregulation, since some genes of its regulon, including YadA, are not elements of the LCR Stimulon (Skurnik and Toivanen, 1992). This clearly indicates that temperature and calcium concentrations regulate Yop expression via two independent systems.

How then does calcium regulate the LCR stimulon? Genes encoding the Yop effectors are only actively transcribed when their respective products are not present in high amounts in the bacterial cytoplasm. This is typically the case in calcium deprivation, when Yops are expelled from the bacterium via the Ysc secretion apparatus. Calcium-dependent blockage of Yop secretion depends on a set of at least three TTSS subunits: LcrG, the channel gatekeeper (Skrzypek and Straley, 1993), TyeA, required for polarized delivery of Yop effectors (Iriarte et al., 1998) and YopN/LcrE, the cell contact sensor (Forsberg et al., 1991). Mutations inactivating any of these proteins will result in the massive leakage of the Yop effectors from the bacterium and the subsequent derepression of the *yop* transcription, regardless of the calcium concentration (Boland et al., 1996; Cheng and Schneewind, 2000; Forsberg et al., 1991; Iriarte et al., 1998; Skryzpek and Straley, 1993; Yother and Goguen, 1985).

How does the amount of Yops regulate yop transcription? In *Y. pseudotuberculosis*, another subunit, LcrQ, is co-injected into the target cell along with the Yop effectors (Cambronne et al., 2000). In *Y. enterocolitica,* two LcrQ counterparts, called YscM1 and YscM2, have been identified (Stainier et al., 1997). LcrQ is a negative regulator of the LCR stimulon. Its accumulation in the bacterial cytoplasm, resulting from blockage of the secretion apparatus, is thought to be the first step in the LCR stimulon downregulation cascade (Pettersson et al., 1996).

Regulation of *yop* transcription by this kind of negative feedback system allows advance synthesis and cytoplasmic storage of ready-to-use Yop molecules prior to cell contact. In the absence of such a system, neosynthesized effectors (i.e. produced upon contact with macrophages) would never be available on time to prevent phagocytosis.

Yops and flagella are both temperature-regulated in *Yersinia*. However, the former are produced at 37°C but not the latter, and the opposite situation is observed at 28°C. Interestingly, LcrD of the Yop secretion machinery shares some structural similarity with elements involved in the flagellar apparatus assembly. Identification *in silico* of a putative binding site upstream of *lcrD* suggested that σ^{28} (an alternative sigma factor also called FliA and which controls flagellum assembly, see below), may be involved in these temperature-induced, physiological modifications. Studies performed by Iriarte et al. (1995c) showed that FliA as such is not involved in Yop regulation. However, these results do not rule out the possible contribution of an as yet unknown, FliA-like, global regulator. In accordance with these observations and the regulation of other flagellar regulatory system, FlhDC may well play such a role, as discussed below.

Yst BIOSYNTHESIS: THE EXTREME COMPLEXITY OF ITS REGULATION BY A SIGMA FACTOR

Y. enterocolitica synthesises and secretes an enterotoxin (Pai and Mors, 1978; Delor et al., 1990) which affects the digestive tract of its mammalian hosts by causing an overproduction of cyclic GMP within the intestinal epithelial cells (Robins-Browne et al., 1979). In light of its significant similarity to the heat stable *E. coli* ST-I toxin, this chromosome-encoded molecule was originally called Y-ST but is currently referred to as Y-STa or Y-STb, depending on the subtype (Ramamurthy et al., 1997). Transcription of *yst*, the Y-ST-encoding gene, is growth phase-regulated and is influenced by environmental cues such as pH, osmolarity and temperature (Mikulskis et al., 1994). However, with the discovery of the sigma factor RpoS (also known as σ^S, σ^{38} or katF) as one of the *yst* regulators (Iriarte et al., 1995b), it has become increasingly evident that control of *yst* expression may be one of the most complex regulatory systems ever encountered in bacteria. RpoS regulation has been the subject of intense study, especially in *E. coli* and pseudomonads. The idea that *yst* can be expressed in a σ^S-dependent manner arose from the observation that Y-ST synthesis was initially considered as stationary phase-specific and that its promoter region contains strong, σ^S-recognized consensus motifs (Iriarte et al., 1995b). However, direct regulation of *yst* transcription by this sigma factor has not been experimentally demonstrated.

Although first described as a stationary phase-specific regulator, it is now recognized that RpoS's influence extends beyond the stationary phase-related response. In light of what is currently known, RpoS should rather be considered as a global stress-response regulator required for tolerance of a variety of potentially lethal conditions, such as hyperosmolarity, temperature shocks, oxidative and UV stresses, most of the time in conjunction with other regulatory systems. The extreme complexity of RpoS-based control also stems from the fact that its own expression is finely controlled at each possible regulation level: transcription, translation, and turnover of *rpoS* mRNA and RpoS protein. (For a recent comprehensive review see Hengge-Aronis, 2002).

rpoS transcription may increase approximately 5- to 10-fold during the stationary phase (Lange and Hengge-Aronis, 1991). In *E. coli*, *rpoS* transcriptional upregulation has been shown to result (at least partially) from interferences with previously characterized regulatory networks. These include the catabolite repression system, in which the CRP-cAMP complex negatively regulates *rpoS* transcription (Lange and Hengge-Aronis, 1991; 1994), and the BarA two-component sensor-kinase- phosphotransferase system. In *E. coli*, the latter acts positively (Mukhopadhyay et al., 2000) but probá independently of its only known cognate response regulator, UvrY (also referred to a YecB). Accumulation of ppGpp (Lange et al., 1995) and polyphosphate (Shiba et al., 1997) may also trigger *rpoS* transcription, although the mechanisms by which these compounds induce transcription are poorly understood. Comparison between the *E. coli* and *Pseudomonas* models (reviewed in Venturi, 2003) strongly suggests the existence of core *rpoS* transcription regulators. However, some other regulators controling *rpoS* transcription appear to be optional with regards to the bacterial species, possibly as the result of the adaptation to divergent lifestyles. *rpoS* transcriptional regulation has not so far been studied in the *Yersinia* genus.

Translation of *E. coli rpoS* mRNA is both osmolarity- and temperature-dependent. The most likely site of *rpoS* translation regulation is a very long mRNA leader sequence (approximately 560 bp between the transcript start and the AUG codon) suspected of forming defined secondary structures (Cunning et al., 1998). The switch of this transcript from an inactive to an active state may occur through stabilization of this region in a translationally competent conformation that provides access to the ribosome binding sites. As demonstrated by a series of studies performed in *E. coli*, this *cis*-regulation results from complex interplays between several *trans*-acting elements, some of which act positively, like the Hfq (HF-1) RNA-binding protein, the nucleoid HU protein and DsrA, a low temperature-induced small RNA (Brown and Elliott, 1997; Lease et al., 1998). In contrast, other elements have been shown to exert an opposite effect. For example, the *oxyS* gene transcript may inhibit *rpoS* translation by binding HF-1 and sequestering this molecule from *rpoS* leader RNA (Zhang et al., 1998). The H-NS histone-like protein has also been reported to antagonize *rpoS* mRNA translation (Yamashino et al., 1995).

Recent studies conducted in *Y. enterocolitica* are consistent with this model: Yrp, the recently characterized HF-1 (Hfq) counterpart in *Yersinia* (also referred to as Ymr in certain databases) has been shown to control *yst* expression at the transcriptional level (Nakao et al., 1995). In line with these results, it was hypothesized that Yrp may exert its effect though the control of DNA topology (Nakao et al., 1995). With regard to the *E. coli* model, it is also possible that this control may occur through regulation of *rpoS*.

Reducing protein turnover by preventing proteolysis is a very efficient means of promoting regulator accumulation. According to studies performed in *E. coli*, RpoS has a very short half-life (less than 2 min) during the exponential phase (Lange et al., 1994), due to rapid degradation by the cytoplasmic ATP-dependent ClpXP protease complex; this is not the case during the stationary phase, where the half-life can achieve values of over 30 min (Schweder et al., 1996). The ClpXP-catalysed degradation of RpoS depends on a two-component system response regulator, RssB, also called MviA in *Salmonella* (Bearson et al., 1996) and referred to or identified as Hnr by the *Y. pestis* CO92 and KIM genome annotation groups. Unlike most two-component systems, this regulator does not appear to directly regulate gene expression, but promotes σ^S degradation by directly and specifically interacting with a domain of the RpoS protein (Muffler et al., 1996; Pratt and Silhavy, 1996). Accordingly, RssB's regulation of proteolysis does not extend to any other ClpXP substrate. The RssB cognate sensor has not been characterized yet in any bacterial species, suggesting that RssB may be regulated by another two-component system sensor kinase.

In *Y. enterocolitica*, RpoS is not essential for virulence in the murine model (Badger and Miller, 1995), in contrast to what has been observed in *Salmonella* (Nickerson and lowever, it has been shown that RpoS is required (only at the 37°C host adaptation to at least some of the environmental stresses mentioned nd Miller, 1995). In agreement with these findings, growth phase *st* gene was found to be thermo-dependent, and once again YmoA (the ...stone-like regulator of the yop regulon, see above) seems to play a critical role in this process (Mikulskis et al., 1994). It is noteworthy that RpoS does not regulate *yop* transcription (Iriarte et al., 1995b). Y-ST host tissue-specific expression (most likely in the ileum) may thus result principally from the complex interplay between the RpoS and YmoA global regulators. To date, the role of RpoS in the virulence of *Y. pestis* and *Y. pseudotuberculosis* has not been analyzed.

THE YERSINIA pH6 ANTIGEN: REGULATED BY A ToxRST-LIKE SYSTEM?

In order to produce *E. coli* Pap-like fimbrial adhesins termed Psa, *Y. pseudotuberculosis* and *Y. pestis* must be grown at 34°C or higher, consistent with their contribution to virulence (Lindler et al., 1990; Yang et al., 1996). These fimbrial adhesins are termed "Psa" (pH Six Antigen) due to the fact that, in addition to the temperature requirement, maximal expression of these appendages is obtained at pH 6. Psa biosynthesis requires two neighbouring gene clusters. The first (*psaABC*) encodes the structural subunit (A), along with its chaperone (B) and membrane usher (C); whereas the second (*psaEF*) is required for the transcriptional regulation of *psaA* (and possibly other genes) (Lindler et al.,1990; Yang et al., 1996; Yang and Isberg, 1997). Two similar genetic clusters displaying identical organisation and functions have been identified in *Y. enterocolitica* and designated *myfABC* and *myfEF* (Iriarte and Cornelis, 1995a). The 24-kDa PsaE and 18.5-kDa PsaF proteins are respectively 52% and 54% identical to their *Y. enterocolitica* counterparts, MyfE and MyfF (Iriarte and Cornelis, 1995a). Both elements are constitutively expressed and are essential for *psaA/myfA* transcription in the three pathogenic *Yersinia species*: no *psaA/myfA* mRNA could be detected in cultures following mutation of either of these two elements, even during growth in highly permissive conditions, unless *psaA* was under control of a constitutive promoter (Yang and Isberg, 1997). Surprisingly, these positive regulatory elements do not exhibit any obvious DNA-binding motifs. However, based on topological predictions and the results of fusions with *phoA*, it has been hypothesized that PsaE/MyfE and PsaF/MyfF may be functionally similar to ToxR and ToxS respectively, i.e. two of the three elements required for transcriptional activation of the *V. cholereae tcp* (toxin co-regulated pili) operon, and that their regulation may be similar (Yang and Isberg, 1997). The ways in which temperature and pH may influence this regulatory system are still unknown. In addition, it has been recently been reported that the *psaEF* operon is possibly regulated at the transcription level by Fur (Panina et al., 2001a; see below).

IRON HOMEOSTASIS SYSTEMS: GLOBAL AND SPECIFIC REGULATION

Iron, an essential cofactor for many enzymatic processes, plays a vital role in most living species. Iron sources and availability vary from one environment to another: this metallic ion exists as insoluble ferric (Fe^{3+}) iron hydroxides in aerobic conditions, soluble ferrous iron (Fe^{2+}) in anaerobic environments or complexed with iron-binding molecules (siderophores) within the host (reviewed in Weinberg, 1978). Thus, bacterial pathogens display a wide arsenal of uptake systems adapted to these various iron sources, some of which are encoded by pathogenicity islands (see Chapters 13 and 14). Although numerous, these mechanisms are not as redundant as they first appear, and some experimental evidence encourages the belief that to ensure optimal iron uptake, bacteria preferentially activate (via specific regulatory systems) the most appropriate mechanism for their environment. In contrast, iron overload can be deleterious for the bacterial cell, leading to the accumulation of strongly oxidizing hydroxyl radicals that damage DNA and provoke cell death (Halliwell and Gutteridge, 1984). Hence, all iron uptake mechanisms are ultimately repressed by iron. This global downregulation involves the Fur repressor, a regulator with an unexpectedly wide potential sphere of influence.

Fur: MORE THAN JUST AN IRON UPTAKE REGULATOR

Fur (for Ferric Uptake Regulator) plays a central role (Staggs and Perry, 1991; 1992; Staggs et al., 1994) in directly or indirectly regulating the expression of most of the genes involved in iron metabolism (for general reviews, see Crosa, 1997; Crosa and and Walsh 2002; Escolar et al., 1999), although separate Fur-independent iron regulatory system may exist in *Yersinia*. The 17-kDa Fur protein is a Fe^{2+}-dependent transcriptional repressor. When cytoplasmic Fe^{2+} is in excess, two overlapping dimers of Fur binds to the operator sequence ("Fur boxes" or "iron boxes") of iron-repressible (*irp*) genes, including *fur* itself. In contrast, in the absence of this micronutrient, the Fe(II)-free Fur aporepressor is released from DNA, leading to gene derepression. Fur exhibits two major differences from classical substrate-binding repressors, and thus it has been speculated that Fur may be more than just a transcriptional regulator. Firstly, the amount of Fur is more than a hundred times higher than typical repressor levels: in *E. coli*, around 5,000 copies of Fur per cell may be achieved in normal growing conditions, and up to 10,000 following oxidative stress (Zheng et al., 1999). Secondly, the 19-bp Fur box is unusually long, i.e. 7 bp more than the box recognized by classical regulators containing helix-turn-helix motifs (Harrison et al., 1990). The 19-bp minimal consensus sequence (5'-GATAATGATAATCATTATC-3') has been shown to consist of a 5'-GATAAT-3' hexamer tandem repeat followed by a third hexamer in the opposite orientation (F-F-x-R configuration) (De Lorenzo et al., 1987). Some Fur boxes may contain additional hexamers, but these motifs make minor contributions to the Fur-DNA interaction (Escolar et al., 1998; 1999). Each dimer binds a 13 nucleotide-spanning region on opposite faces of the helix, with fewer phosphate contacts than observed for classical regulators (Baichoo and Helmann, 2002).

Based on experimental and/or computational evidence, all the iron scavenging systems in *Yersinia* studied to date have been shown to be Fur-controlled: the Hmu/Hem (Thompson et al., 1999) and *Serratia*-like Has haemophore-dependent heme acquisition machineries (Rossi et al., 2001), the siderophore-dependent yersiniabactin (Ybt) inorganic iron transport system, the YfeABCD iron and manganese uptake system (Bearden et al., 1998; Bearden and Perry, 1999), and the *yfuABC* operon-encoded iron transporters (Saken et al., 2000). All these mechanisms are described in detail in Chapter 13.

In the last few years, genome-wide computational analyses and the use of biochemical and genetic techniques have revealed a number of potentially Fur-controlled genes, showing that the initial size of the Fur regulon was probably underestimated. In other bacteria, a systematic search for Fur-controlled genes has revealed that this molecule may also regulate physiological functions that go beyond iron uptake. In *E. coli*, for example, Fur has been shown to directly control the expression of SodA and SodB, two oxidative stress-combative superoxide dismutases, and *fur* has also been shown to be controlled by *oxyS* in this species, demonstrating that Fur also contributes to protection against oxidative damage and mutagenesis (Zhang et al., 1998). Fur may also exert (mainly indirect) negative or positive control of a broad range of cellular processes, such as acid shock and redox-stress responses, chemotaxis, metabolic pathways (e.g. glycolysis and TCA cycle) and the production of toxins and virulence factors (McHugh et al., 2003). In *Y. pestis*, Fur is suspected of directly controlling expression of the pH6 antigen at the transcriptional level (Panina et al., 2001). Given this recent information, it is tempting to consider Fur as a global regulator, rather than just an iron uptake-specific transcription factor, that controls other aspects of bacterial metabolism besides extracellular iron availability.

HOW TO FAVOUR USE OF THE MOST APPROPRIATE IRON UPTAKE SYSTEM: THE YbtA LESSON

Upon iron starvation, the three pathogenic *Yersinia* species release a high-affinity iron-binding compound called yersiniabactin (Ybt), which captures ferric iron from the environment. The resulting iron-siderophore complex is then transported back in to the bacterial cytosol in a TonB-dependent manner via a specific surface receptor, termed FyuA in *Y.enterocolitica* and Psn in the two other pathogenic species. *psn/fyuA* and the other genes required for yersiniabactin biosynthesis and secretion are located on a 36 to 43-kb chromosomal region within the unstable *pgm* locus. This region is known as the High-Pathogenicity Island (HPI) and is required for full virulence in highly pathogenic isolates of *Y. pestis*, *Y. pseudotuberculosis* and *Y. enterocolitica*. It has been shown to be essential during the early stages of infection in the mouse model. Like most other iron uptake systems in *Yersinia*, it is Fur-controlled (see Carniel, 2001; and Chapters 13 and 14).

As with other siderophores, like pyochelin and pyoverdin produced by pseudomonads (Crosa, 1997), yersiniabactin plays a regulatory role by enhancing its own synthesis along with that of its Psn/FyuA receptor. This activation occurs via binding to an HPI-encoded 36-kDa transcriptional activator, YbtA (Fetherston et al., 1996), a member of the AraC/XylS family which contain two adjacent C-terminal Helix-Turn-Helix (H-T-H) DNA binding motifs. Like AraC and PchR (the latter being the YbtA homologue for the uptake and synthesis of pyochelin in *P. aeruginosa*), YbtA inhibits transcription of its own gene. However, unlike these two regulators, YbtA does not seem to act as a repressor in the absence of its cognate ligand. In addition to *ybtA* itself, YbtA-binding DNA inverted repeats have been identified immediately upstream from the –35 box of *psn/fyuA*, the *ybtPQXS* operon and the Ybt biosynthesis gene *irp2*. (Fetherston et al., 1996; Bearden et al., 1997; Fetherston et al., 1999). YbtA is thought to bind to these sequences as a dimer and to positively regulate their transcription. Conversely, it has also been shown that the *psn/fyuA*, *ybtA*, *irp2* and *ybtPQXS* operons are repressed by Fur (Carniel et al., 1992; Gehring et al., 1998; Fetherston et al., 1996; Staggs et al., 1994; Panina et al., 2001).

Although its free form is thought to exhibit residual activity, YbtA must bind yersiniabactin to be fully active (Fetherston et al., 1996). In the bacterial cytosol, the only yersiniabactin source comes from the Ybt-Fe^{3+} recovered from the surrounding medium. Thus, the Ybt iron-uptake system will be maximally active when Fe^{3+} is available in the surrounding medium and there is no cytoplasmic iron overload, but will be less active in the presence of iron sources other than Fe^{3+}. In other words, by this kind of positive feedback, the yersiniabactin system will be most efficient when best suited to the iron source.

YERSINIA IRON UPTAKE SYSTEMS: ARE THEY DIFFERENTIALLY EXPRESSED?

In pseudomonads, production of exogenous siderophore receptors is selectively upregulated by cognate ligands in the environment, mostly via processes requiring extracellular sigma factors (Poole et al., 2003). It has been suggested that the *Y. pestis* Yfe and Ybt systems function during different stages of the infectious process in bubonic plague (Bearden and Perry, 1999). Thus, the various iron uptake systems available in one given *Yersinia* cell may be differentially regulated according to the iron source available in the environment, with preferential expression of the most suitable systems at the expense of the others. In an attempt to verify this hypothesis, Jacobi et al. (2001) used translational fusions with reporter genes to monitor the expression of *yfuA* and *hemR* (encoding the Fe^{3+}-Yersiniabactin and haem receptors respectively) in *Y. enterocolitica* during its course of infection in the murine model. Expression of these two genes was found to fluctuate from one organ to another, with the highest expression levels in the peritoneal cavity and the lowest in the intestinal lumen and liver. However, because identical variations were observed for both *yfuA* and *hemR,* these genes may be coordinately regulated. Nevertheless, the anticipated environment-driven, differential expression of *Yersinia* iron-uptake systems still awaits direct experimental confirmation.

HAEMIN STORAGE: ITS UNUSUAL REGULATION BY TEMPERATURE AND IRON

The haemin absorption system, also known as the haemin storage system or Hms (Carniel et al., 1989; Fetherston et al., 1992), is essential for the flea-mediated transmission of *Y. pestis* but does not contribute to the pathogenesis of bubonic plague in mammals (Lillard et al., 1999; and Chapter 4). At least two haemin storage loci (*hmsHFRS* and *hmsT*) enable *Yersinia pestis* to accumulate haemin, ferric iron and Congo Red in a process that is only active at temperatures lower than 34°C (Perry et al., 1990; Lillard et al., 1999). Interestingly, as recently shown by Perry et al. (2004), thermal control of the Hms+ phenotype is unlikely to depend on transcriptional regulation for at least two reasons. Firstly, levels of mRNA transcribed from both loci do not change significantly with growth temperature. Secondly, levels of the HmsH, F, R and S proteins seem to result from independent regulation processes, whereas HmsH, and HmsR protein levels appear to decrease upon temperature upshifts, levels of HmsF and HmsS do not change, even though their genes belong to the same polycistronic unit. This second observation also suggests that translational control is unlikely. Temperature-regulated expression of the Hms^{+} phenotype may rather depend on differential lability of the Hms proteins to proteases (Perry et al., 2004). Except in the case of HmsT, which is degraded by the Lon, ClpXP, and/or ClpAP complexes at 37°C, candidate proteases have not been identified yet.

Earlier studies had shown that the Fur repressor was an essential actor in the regulation of the Hms phenotype, since *fur* mutants were found to be constitutively Hms+ (Hms^{c}, i.e. formation of red colonies even at 37°C on CR agar) (Staggs et al., 1994). The detection of a putative upstream Fur binding site for *hmsT* but not for *hmsHFRS* strongly suggested that the Hms phenotype could be regulated by Fur through the expression of HmsT (Jones et al., 1999). Given the recent finding that *hmsT* transcription is not regulated by temperature (Perry et al., 2004), the mechanism for Fur control of the Hms+ phenotype still needs to be elucidated.

FLAGELLAR BIOSYNTHESIS: A HIGHLY HIERARCHICAL REGULATORY SYSTEM

Y. enterocolitica and *Y. pseudotuberculosis* are flagellated and motile, whereas *Y. pestis* is not. Type III flagella are critical (at least in *Y. enterocolitica*) for swarming motility, migration and adherence to host cells (Young et al., 2000; and Chapter 12). However, in *Y. enterocolitica*, flagella are synthesized at 30°C or below but not at the host temperature, strongly suggesting that their contribution may not last beyond the early stages of infection. Biosynthesis of the bacterial flagellum requires more than 40 genes, including those necessary for flagellar rotation and those encoding the chemosensory apparatus. *Yersinia* flagellar operons (also called motility operons) display significant homologies and similar arrangements to those of *E. coli* and *Salmonella*, and even though there is often no direct experimental evidence, it is commonly presumed that flagellar synthesis in *Yersinia* is identical in most aspects to the paradigm established from studies in these two species. Flagellar operons (sometimes referred to as flagellar regulons) fall into three classes which are expressed in a hierarchical manner: expression of the class 2 operons required for basal rod and hook assembly depends on products encoded by the unique class 1 operon, *flhDC* (also known as the master operon). In turn, transcription of the class 3 genes required for i) biogenesis of the filament and motor torque generator, and ii) motility and chemotaxis regulation, depends on the complete and correct assembly of the class 2 subunits: mutants lacking these components do not express the subunits needed for the later stages. In addition to assembly monitoring, expression of motility genes may be further regulated in response to environmental signals. The key transcriptional regulators involved in this complex processes are the products of the master flhD/flhC regulatory operon and the FliA/FlgM sigma/anti-sigma factors (Chilcott and Hughes, 2000; Shapiro, 1995, for review).

The *flhDC* operon (the sole class I operon) is at the top of the regulatory cascade and is thus required for expression of all the class 2 and class 3 genes in the flagellar regulon as well as its own transcription (see below). Mutations within *flhDC* completely abolish swimming and swarming motilities. FlhD and FlhC form a heterotetrameric (C2D2) complex in which FlhC may act as an allosteric activator of FlhD, the DNA-binding subunit (Campos and Matsumura, 2001). Very recent reports strongly suggest that this regulatory complex is not only a motility-specific activator (as initially thought) but probably also a global regulator. In *E. coli*, the heterotetrameric complex has been shown to regulate flagellum-unrelated physiological functions, such as membrane transport, respiration, sugar metabolism and other enzymatic processes (Pruss et al., 2001; 2003). In pathogens like *Proteus mirabilis* and the insect pathogen *Xenorhabdus nematophilus*, amongst others, it may control various virulence-associated phenotypes including invasion and production of proteases, haemolysins and phospholipases (Fraser et al., 2002; Givaudan et al., 2000).

This operon has also been designated *flhDC* in *Yersinia*, because of its high degree of identity to the master operons of other Gram-negative bacteria and because of its similar contributions to regulation of flagellar biosynthesis (Young et al., 1999a). As in some other enterobacterial pathogens, the operon has been shown to modulate the expression of other virulence factors, either (i) associated with flagella synthesis, like YplA, a phospholipase that requires the type III flagellum to be secreted (Young et al., 1999b; Young and Young, 2002) and is considered to form part of the flagellar regulon; or (ii) unrelated to flagella synthesis, like the Yops and their secretion apparatus (Bleves et al., 2002). Recent results demonstrate that transcription of the flagellar master operon is also environmentally controlled. In *Salmonella and E. coli*, it has been shown that expression of the *flhDC* operon is controlled by a wide array of regulatory systems, including the catabolite repression cascade via cAMP-CRP, the histone-like H-NS protein, and at least two signal transduction systems (also referred to as two-component systems): OmpR-EnvZ (Shin and Park, 1995) and RscC-RscB-RscA-YojN (Francez-Charlot et al., 2003). Additionally, flagellar biosynthesis has been seen to depend on cell density via two pairs of of LuxR/I-type quorum sensing homologues (Atkinson et al., 1999; and Chapter 5).

FliA AND FlgM

FliA, also known as Sigma 28 (σ^{28}), is required for the master regulon-dependent expression of most class 3 operons, although at least two of them (*flg*KL and *fli*DS) may also be directly activated by FlhD-FlhC (Bartlett et al., 1988). FliA-dependent operons are only expressed upon complete and correct assembly of the class 2 gene products (Hughes et al., 1993). How then can the bacterial cell sense completion of the hook-basal body intermediate structure? The exact mechanism of this phenomenon remained obscure until FlgM, the FliA-cognate anti-sigma factor, was characterized (Ohnishi et al., 1992). As already shown in several enterobacteria, *flgM* mutants exhibit high transcription levels of the class 3 operons. Conversely, cytosolic accumulation of FlgM leads to class 3 gene silencing (Hughes et al., 1993). Thus, completion of the flagellar apparatus requires low FlgM levels. FlgM is normally secreted from the bacterial cell upon assembly of functional hooks and basal rods. If these latter structures are not functional for any reason, the anti-σ^{28} factor accumulates within the cytoplasm, blocking the later steps of flagellar synthesis by inactivating FliA (Gillen and Hughes, 1991; Hughes et al., 1993). *fliA* and *flgM* have been recently characterized in *Y. enterocolitica* and shown by functional complementation to exhibit properties similar to counterparts previously studied in other enterobacterial species (Kapatral et al., 1996). Interestingly, neither *fliA* nor *flgM* is transcriptionally active at 37°C and, consequently, most class 3 operons remain silent (Kapatral and Minnich, 1995; Kapatral et al., 1996), arguing for the presence of flagella only during the very first steps of host infection. However, unlike the FlhC-FlhD complex, it is currently believed that FliA has no impact on the expression of virulence factors other than those associated with flagella, including the flagella themselves and YplA (Schmiel et al., 2000), and is not involved in the temperature-sensitive regulation of the pYV-harboured genes (Iriarte et al., 1995c). Despite high structural similarity among enterobacterial flagella, slight differences exist: The fact that flagella are expressed at 37° in *Salmonella* but not in *Yersinia* is probably the best example. In agreement with this observation is the finding that a *Y. enterocolitica flgM* mutant is fully virulent but *Salmonella flgM* mutants display attenuated virulence, probably due to the abnormal expression level of FliC flagellin (Schmitt et al., 1994; 1996).

CHEMOTAXIS REGULATION

Swimming behaviour of bacteria such as *E. coli* and *Salmonella* depends on the direction of flagellar rotation: the flagellar apparatus fluctuates between clockwise rotation (causing jerky movements known as tumbling motility) and counter-clockwise rotation (associated with straight swimming). The signal transduction system that mediates bacterial chemotaxis allows cells to modify the frequency of transition between these two states as a function of the environmental conditions (for review, see Macnab, 1996). In most cases, the absence of environmental input signals is sensed at the periplasmic level by specific, inner-membrane anchored receptors and is then transmitted to the chemotaxis regulation apparatus by the activation (via methylation) of membrane-spanning molecules called MCPs (for Methyl-accepting Chemotaxis Proteins). MCPs are often associated with several receptors, and may thus function as specific intermediates for several environmental cues. In some cases, they may be directly regulated by the environmental stimulus itself. In *E. coli*, the cytoplasmic chemotaxis regulation apparatus *per se* is composed of 6 subunits: CheA, CheB, CheR CheW, CheY and CheZ. Each MCP may assemble with two of these subunits (CheA and CheW), and activation occurs through conformational changes in this ternary complex (Gegner et al., 1992; Ninfa et al., 1991). CheA functions as a histidine-kinase which, once activated, phosphorylates CheY. In turn, this latter subunit binds to FliM, which belongs to the flagellar motor/switch complex (consisting of the three proteins FliG, FliM, and FliN), forcing a rotation change from counter-clockwise to clockwise and thus causing tumbling motility. Conversely, the sensing of an attractant stimulus will abrogate signal transduction and facilitate swimming behaviour. The high switch frequency is facilitated by controlling the methylation of MCPs by CheR (positive) and CheB (negative) and the dephosphorylation of CheY by CheZ (Hess et al., 1988).

A complete chemotaxis system, including the six Che signal transduction proteins, 8 MCP and the three flagellar motor/switch complex subunits was found in the three pathogenic *Yersinia* species (Deng et al., 2002; Hinchliffe et al., 2003; and results obtained from *Y. enterocolitica* genome BLAST searches - http://www.sanger.ac.uk). All six Che subunit-encoding genes share 70 to 90% identity with their respective *E. coli* and *Salmonella* counterparts. In constrast, FliG, FliM, and FliN amino-acid sequences were found to be much less conserved. Whether and how this evolution of the flagellar motor/switch complex might impact *Yersinia* chemotaxis remains to be assessed.

F1 ANTIGEN REGULATION

Expression of the *Y. pestis* antiphagocytic capsule, also known as F1 antigen (for review, see Perry and Fetherston, 1997; and Chapter 18), requires four genes harboured by the *Y. pestis*-specific 110kb pFra virulence plasmid. The first three are clustered in the *caf* operon and encode the molecular chaperone (Caf1M), the outer membrane anchor (Caf1A) and the F1 structural subunit (Caf1). The fourth gene, encoding the Caf1R AraC family regulatory protein, is located just upstream of, but in opposite orientation, to the *caf* operon, suggesting that both the regulator-encoding gene and the *caf* operon may have a common operating region. Expression of the *caf* operon is thermally sensitive, as evidenced by an increase in Caf1A levels following a temperature shift from 28°C to 37°C (Karlyshev et al., 1992). Caf1R is apparently required for this process. However, whether induction by temperature occurs through Caf1R has not yet been investigated. Recently, it has been proposed that Caf1A may be associated with a *Y. pestis*-specific galactolipid (Feodorova and Devdariani, 2001). This moiety can also be extracted from pFra-less strains, suggesting that its biosynthesis depends on as yet unidentified, chromosome-harboured genes. Like CafA1, higher amounts of this galactolipid are recovered from bacteria grown at 37°C than at 28°C, raising the possibility that synthesis of these two capsular compounds may be similarly regulated.

THE PLEIOTROPIC PhoP-PhoQ REGULATORY SYSTEM

Calcium (Ca^{2+}) and magnesium (Mg^{2+}) are essential for stabilizing the negatively-charged lipopolysaccharide (LPS) in the outer membrane. Hence, reduced availability of these two cations may be considered as a source of stress for the bacterial cell. The maintenance of LPS integrity under these conditions requires the PhoP-PhoQ two-component system (Groisman et al., 1997; Guo et al., 1997). PhoP-PhoQ responds not only to external Mg^{2+} and Ca^{2+} ion concentrations but also to Mn^{2+} (García Véscovi et al., 1996; Soncini et al., 1996). As in most two-component systems, activation of PhoP requires autophosphorylation of PhoQ upon Ca^{2+}/Mg^{2+} deprivation and subsequent phosphoryl transfer to PhoP. The phospho-PhoP response regulator switches from an inactive to active state, and its binding to DNA then promotes both transcription of PhoP-activated genes (*pags*) - including the *phoPQ* operon itself (Soncini et al., 1995) - and repression of PhoP-repressed genes (*prgs*). Conversely, when bound to Mg^{2+} and/or Ca^{2+}, PhoQ is able to inactivate phospho-PhoP by dephosphorylation. As with almost all two-component systems, the means by which the sensor recognizes the response regulator is still poorly understood.

In *Salmonella enterica* serovar Typhimurium, PhoP-PhoQ has been shown to control the transcription of a wide array of unlinked genes that contribute to various modifications of cellular physiology. It became rapidly apparent that some of these were essential for survival inside the host, since *Salmonella phoP* mutants were found to show highly attenuated virulence in mice (Miller et al., 1989; Galán and Curtiss, 1989). In this species, the most intensively studied PhoP-PhoQ controlled phenotype is the ability to modify the lipid A moiety of LPS by performing at least two substitutions: one with palmitate (catalysed by a palmitoyl transferase termed PagP (Bishop et al., 2000) and a second with 4-amino-arabinose, through the upregulation of the 7-ORF *pmrHFIGKLM* operon, also referred to as *pmrF* (Gunn et al., 1998). This latter modification contributes to a decrease in the net negative charge of LPS, thus promoting stabilization of the outer membrane during calcium and magnesium starvation (Groisman et al., 1997). Both modifications have been shown to promote *Salmonella*'s resistance to a broad range of cationic antimicrobial peptides (essential components of the innate immune response) and facilitate the pathogen's survival within acidified macrophage phagosomes (Guo et al., 1998; Baker et al., 1999).

Operons encoding *Salmonella* PhoP and PhoQ orthologues have been identified in the three pathogenic *Yersinia* species. Recent studies have shown that the *Y. pestis* PhoP-PhoQ system is involved in infection of mice challenged by subcutaneous injection: although a *Y. pestis phoP* mutant was still virulent, its LD50 increased by 75 fold. PhoP was also shown to contribute to intra-macrophage survival, though to a lesser extent than in *Salmonella*, since *Yersiniae* mostly remain extracellular during infection (Oyston et al., 2000). Accordingly, tolerance of the three pathogenic *Yersinia* species to antimicrobial peptides was found to depend (at least partially) on elements of the PhoP-PhoQ regulons, such as the *pmrF* operon (Marceau et al., unpublished). In *Y. pestis*, PhoP-PhoQ may also control the production of an alternative lipo-oligosaccharide (LOS) form containing terminal galactose instead of heptose (Hitchen et al. 2002).

Over the past decade, the central role of the PhoP-PhoQ two-component system in virulence regulation has become increasingly clear in many bacterial pathogens. In *Salmonella*, more than forty genes have been shown to fall into the PhoP regulon (Miller and Mekalanos, 1990), and the first results from 2D protein gel analyses strongly suggest the existence of a regulon (including PhoP-activated and -repressed genes) of at least equal size in *Y. pestis* (Oyston et al., 2000). A subset of this regulon may be further regulated by temperature.

RpoE: REGULATION OF VIRULENCE BY OUTER MEMBRANE STRESS-RESPONSE SYSTEMS

Extracytoplasmic function (ECF) sigma factors (a subgroup of the σ^{70} family) regulate a range of physiological processes, including envelope homeostasis, folding, assembly and degradation of Outer Membrane Proteins (OMPs), in response to envelope-damaging environmental stresses (for general reviews of ECF sigma factors and their regulation, see Helmann, 2002; Raivio and Silhavy, 2001).

In contrast to the general heat shock transcription factor σ^{32} which directly senses the misfolding of cytoplasmic proteins, the activity of the ECF sigma factor depends on one or several other signal transducing proteins called anti-sigma factors. The term anti-sigma arose from the fact that, in non-inducing conditions, these molecules complex with their cognate sigma factor and maintain it in an inactive form (for review, see Helmann, 1999). Of these envelope stress responsive systems, σ^E (also referred to as RpoE) and its cognate anti-sigma factors are, by far, the most intensively studied. In *E. coli*, the activity of σ^E is tightly controlled by two negative regulators encoded by the *rpoErseABC* operon (Missiakas and Raina., 1997). The first of these (RseA, a regulator of σ^E), spans the inner membrane and is referred to as the anti-sigma factor itself, in light of its cytoplasmic N terminal σ^E-binding domain. RseA is probably necessary and sufficient for downregulating the activity of σ^E. The second regulator (RseB) also exerts a negative effect on σ^E, possibly upon sensing misfolded OMPs. However, its periplasmic location and its affinity for the C-terminal periplasmic domain of RseA suggest that this molecule is more likely to stabilize the σ^E-RseA complex than regulate σ^E by distinct means. RseC is encoded by the last gene of the *rpoErseABC* operon and regulates σ^E activity in a positive manner. The respective roles and modes of action of RcsB and RcsC remain to be clarified.

In a wide range of living organisms, the σ^E-dependent stress response includes the synthesis of a periplasmic chaperone/heat shock serine protease called DegP and also known as HtrA, for high temperature requirement A. Inactivation of *htrA* leads to a decrease in tolerance to high temperatures (i.e. exceeding 39°C) and osmotic and oxidative stresses. Additionally, pathogens such as *Brucella*, *Salmonella*, and *Legionella*, also need this enzyme for survival within macrophage phagosomes (for review, see Pallen and Wren, 1997; Pedersen et al., 2001). The 49.5-kDa GsrA (global stress requirement) protein is the *Yersinia* counterpart of HtrA (Wren et al., 1995; Yamamoto et al, 1996). *Y. enterocolitica gsrA* mutants display similar virulence phenotypes as those observed for *Brucella* and *Salmonella* (Li et al., 1996; Elzer et al., 1996; Chatfield et al., 1992). Contrasting with this result, a *Y. pestis htrA* mutant was attenuated and exhibited increased sensitivity to oxidative stress, but to a much lesser extent than seen for mutants of the three above-mentioned species (Williams et al., 2000). HtrA/GsrA is most likely present in *Y. pseudotuberculosis*, but how it contributes to these phenotypes has not been reported so far.

Transcription levels of *gsrA* and *rpoE* were found to be significantly increased following pathogen uptake by macrophages, suggesting that GsrA is induced (probably by σ^E as judged by the presence of a specific binding motif) in response to stresses encountered within phagosomes (Yamamoto et al., 1996; 1997). The discovery of the three anti-sigma orthologues in *Yersinia* suggest that GsrA and HtrA homologues found in other bacterial species may have identical regulation processes.

In *E. coli*, the HtrA-mediated protective response also depends on an overlapping regulation mechanism involving the CpxA-CpxR two-component system (Missiakas and Raina, 1997). CpxA-CpxR has been described for its ability (among others) to monitor and, if necessay, downregulate the synthesis of P-pili, as a consequence of an excess of misfolded Pap subunits (Jones et al., 1997; Hung et al., 2001). The CpxA sensor and its cognate regulator have been identified in the three pathogenic *Yersinia* species. However, whether they contribute identically to this physiological process has not been demonstrated.

O-ANTIGEN SYNTHESIS: ALSO REGULATED BY A STRESS RESPONSE SYSTEM

O-antigen, the outermost part of the LPS, is required for the full virulence of *Y. pseudotuberculosis* and *Y. enterocolitica*, in contrast to *Y. pestis*, which is unable to produce this moiety. The structure and genetic determinism of *Yersinia* LPS are reviewed in detail in Chapter 11. In *Yersinia*, expression of the O-Antigen is repressed at 37°C (Al-hendy et al., 1991; Bengoechea et al., 2002). Although this fact has been known for over ten years, the first steps towards characterization of the mechanisms behind *Yersinia* O-antigen biosynthesis thermoregulation cascade were only initiated very recently. Studies performed with *Y. enterocolitica* serotype O:8 indicate that control of O-antigen synthesis involves finely balanced regulatory mechanisms at both the transcriptional and post-transcriptional levels, and is also associated with stress-response systems. (Bengoechea et al., 2002). Promoter cloning experiments reveal that a promoter, termed Pwb1, drives the transcription of most of the O-antigen biosynthetic genes; whereas a second promoter, named Pwb2, controls the last two genes of the O-antigen gene cluster, *gne* and *wzz*. The activity of both promoters is temperature regulated and is repressed at 37°C. A molecular analysis of the loci involved in O-antigen regulation demonstrated that Wzz avalability and the activity of the thermo-inducible RosAB system (Bengoechea et al., 2000) are key elements in the regulatory network. RosA and RosB are an inner membrane efflux pump and a K+ antiporter respectively, i.e. functions which argue against direct regulation based on interaction with DNA. This inconsistency could be explained by a mechanism whereby RosA and RosB indirectly repress the transcription of *gne* and *wzz* by modifying H+ and K+ intracellular levels, similarly to what has been observed for transcription of their own genes.

Even though overexpression of *wzz* causes an increase in O-antigen expression at 37°C, this upregulation does not result from the activation of P*wb1* and P*wb2*. This observation initially prompted researchers to suppose that Wzz played its regulatory role at a post-transcriptional level, probably by interfering with the O-antigen assembly and export machinery. However, it has since been shown that Wzz downregulates the transcription of of *rosA* and *rosB*. In other words, the less Wzz is produced, the more transcription of its gene is repressed (Bengoechea et al., 2002). This regulatory circuit may thus explain the almost complete abolishment of O-antigen synthesis at 37°C.

How Wzz downregulates *rosA* and *rosB* transcription is still unknown. However, recent data suggest that this could occur via the envelope stress-combative CpxA-CpxR two component system, with CpxA possibly sensing outer-membrane disorders resulting from an excess of Wzz. The detection of potential binding motifs in the P*wb1*, P*wb2* and *rosAB* promoter region may suggest direct transcriptional regulation by CpxR, the CpxA cognate response regulator (Bengoechea et al., 2002).

REGULATION OF INVASIN EXPRESSION

Invasin, produced only by the two enteropathogenic species, is one of the most studied *Yersinia* virulence factors, along with the Yops, and is reviewed in Chapter 6. However, in contrast to the Yops, little was known about the regulation of the invasin gene (*inv*) until recently. Invasin expression is thermoregulated (Isberg et al., 1988) but differs from the other host temperature-induced *Yersinia* adhesins (like YadA, Ail or the newly characterized YAPI encoded type IV pilus) in being poorly expressed at 37°C, pH8. A *Y. enterocolitica inv* mutant displays delayed Peyer's patch-colonization, but no change in LD_{50} for mice (Pepe and Miller, 1993). It was therefore initially proposed that this adhesin contributed only to an acceleration of the early stages of host infection. However, two pieces of evidence suggested that regulation of *inv* transcription during infection may well depend on stimuli other than temperature (Pepe et al., 1994). Firstly, invasin can still be detected in murine Peyer's patches two days after oral challenge with *Y. enterocolitica.* Secondly, its expression can be modified *in vitro* in response to several environmental cues other than temperature that may be encountered in the host, such as mildly acidic pH, nutrient availability, growth phase and oxidizing and osmotic stresses. In view of these environmental cues, it was initially thought that the sigma factor RpoS might be a key regulator of invasin expression. However, experimental evidence has ruled out this preliminary hypothesis (Badger et al., 1995).

The transcriptional regulator RovA (regulator of virulence) was characterized in *Y. enterocolitica* and *Y. pseudotuberculosis* by two different groups using opposite approaches (Revell et al., 2000; Nagel et al., 2001). It was shown to be essential for production of high invasin levels *in vitro* in both species. RovA belongs to the MarR family (which mostly contains non-specific, antibiotic resistance regulators), and is comparable in size (with around 75% identity at the amino acid level) to the *Salmonella typhimurium* pleiotropic transcriptional regulator SlyA. Potential RovA/SlyA orthologues have also been identified in a wide range of bacterial species. Based on recent structural studies performed with SlyA, RovA may contain a winged-helix DNA-binding domain (i.e. two helixes separated by a glycine-rich hinge region) and may function as a dimer (Wu et al., 2003). In *Y. pseudotuberculosis*, it has been proposed that RovA binds with unequal affinities to two similar palindromic motifs within the *inv* promoter region and that it positively regulates transcription of this gene in response to low temperature, mild acidic pH, and growth in stationary phase. This environment-dependent activation is thought to occur mainly through post-transcriptional control of RovA biosynthesis (Nagel et al., 2001). It can be enhanced by auto-activation, since RovA also promotes the transcription of its own gene. In contrast to the results obtained in the invasin studies, the *Y. enterocolitica rovA* mutant displays a 70 to 500-fold increase in the LD_{50} mice (depending on the mouse lineage) when compared to its wild type counterpart (Revell et al., 2000; Dube et al., 2003). RovA may thus be a pleiotropic regulator and, control one or several as yet unknown virulence factors in addition to invasin. Interestingly, such a dramatic virulence decrease was not observed when mice where challenged by routes bypassing the Peyer's patches (Dube et al., 2003). This strongly suggests that these yet unkonwn members of the RovA regulon may be required to improve survival whithin these tissues.

Other studies suggest that other *inv* regulators exist: Tn*5* insertions in *sspA* or *uvrC* caused a significant decrease in invasin expression, whereas inactivation of these two genes had the opposite effect on flagellin transcription, strongly suggesting the existence of mechanisms regulating both *inv* expression and flagella biosynthesis (Badger et al., 1998). Recently, expression of *slyA* has been shown to be PhoP-controlled in *Salmonella typhimurium* (Norte et al., 2003).

PART TWO: THE POST-GENOMIC ERA: WHAT MIGHT WE LEARN FROM *YERSINIA* GENOME SEQUENCES?

The field of molecular biology has changed dramatically over the last ten years. In particular, advances in DNA sequencing have provided data with ever-increasing accuracy and speed. Genome sequence analyses of the two *Yersinia pestis* strains CO92 and KIM (biovar Orientalis and Medievalis respectively) have recently been completed (Parkhill et al, 2001; Deng et al., 2002), and the release of the *Y. pseudotuberculosis* and *Y. enterocolitica* sequences is imminent. Within the next decade, genome-wide analyses and derived experimental techniques will undoubtedly provide important clues to the two following fundamental and recurrent questions: how do pathogenic Yersiniae cause disease, and how did *Y. pestis* diverge so rapidly from *Y. pseudotuberculosis* (switching from an environmental enteropathogenic lifestyle to a host-dependent, septicaemic lifestyle in less than 20,000 years)? Whole genome sequences and associated annotation are prerequisites for the assessment of genome-wide transcript profiling, and also offer unprecedented opportunities for opening up new fields of investigation in gene regulation. This will enable research strategies to move from conventional regulator hunting (i.e. starting from the effector genes and trying to characterize their regulators) towards systematic searching for the cognate target regulons on the basis of each identified transcriptional regulator. There is no doubt that comparison of the huge amount of new data being generated by the use of these new technologies with our current physiological knowledge will lead to a wealth of discoveries - providing new insights into how pathogenic *Yersinia* regulate the expression of their virulence gene arsenal and how these mechanisms may differ from those in other pathogenic bacteria.

At this point in time, what can we already learn in terms of regulation from the currently available *Yersinia* genome sequences, from comparison with the genomes of other enterobacteria and from genomic divergences within the pathogenic *Yersinia* themselves?

YERSINIA REGULATORS: AN OVERVIEW

Genome-wide screening in *Y. pestis* has revealed the presence of approximately 250 transcriptional regulators (including sigma and anti-sigma factors, two component systems and histone-like molecules), which is probably slightly less than the estimated number of regulators in *E. coli* K12 (reviewed by Perez-Rueda and Collado-Vides, 2000). Not surprisingly, less than half of them (the 79 listed in Table 1) have an assignable function, based on either experimental evidence and/or high similarity (i.e. >50 % identity over the length of the whole molecule: the high stringency of this criteria being justified by the fact that some non-orthologue regulators exhibit around 30% baseline identity due to the presence of highly conserved domains). The remaining regulatory elements (except for obvious phage-related transcriptional regulators YPO0878, YPO1904, YPO2785, YPO2823, YPO3485, YPO3612 and YPO4031) are listed in Table 2 and fall into two categories: firstly, those which have been previously identified in other bacterial species but have an as yet uncharacterised function (interestingly, several of these display the highest similarities with regulators found in *Photorhabdus luminescens*, an insect pathogen); and secondly, those identified as transcriptional regulators because they contain canonical DNA binding motifs but are not highly similar to regulators currently found in protein databases. Among these latter molecules, some have no known counterparts other than in *Yersinia*, and will therefore be of great interest for deciphering potentially new aspects of *Yersinia*-specific physiology.

Table 1. CO92 transcriptional regulators with assigned functions.

CO92 ID	Name	Regulated function	Possible regulon found in *Yersinia*	Function in Yersinia ?	Relevant DB entry
YPO0002	AsnC	Amino acid metabolism	*asnA, GidA*	By sim.	P03809
YPO0046	Ttk	Resistance to antibiotics and detergents?	??	By sim.	P06969
YPO0072	Ada	Bifunctional : regulatory / DNA repair	??	By sim.	P06134
YPO0108	CytR	Catabolizing enzymes	*deoCA*BD, udp, cdd, nupC, nupG*…*	By sim.	P06964
YPO0114	MetJ	Met and Sam Biosynthesis	*met* regulon	By sim.	P08338
YPO0120	GlpR	Glycerol-3-phosphate metabolism	*glpEGR*	By sim.	P09392
YPO0123	MalT	Maltose regulon.	Maltose regulon	By sim.	P06993
YPO0175	Crp	Catabolic repression	Pleitropic regulation	By sim.	P06170
YPO0236	ZntR	Zn(II)-responsive regulator	zntA	By sim.	P36676
YPO0314	LexA	SOS system	SOS regulon	By sim.	P03033
YPO0315	Zur?	Zinc-uptake	*znuACB*	By sim.	P32692
YPO0332	RhaS	L-rhamnose metabolism	*rha* genes	By sim.	P27029
YPO0333	RhaR	L-rhamnose metabolism	*rha* genes	By sim.	P09378
YPO0373	Yrp	*Yersinia* multiple regulator	*yst*, in *Y. enterocolitica*	Nakao et al., 1995	P25521
YPO0444	NadR	Transcriptional regulator NadR	*nadA, nadB, pncB*	By sim.	P24518
YPO0453	TrpR	Trp operon repressor	*trp* operon	By sim.	P03032
YPO0471	NhaR	Na+/H+ antiporter system	*nha*A	By sim.	P10087
YPO0535	LeuO	LysR-family transcriptional regulator LeuO	*leuABCD*	By sim.	P46924
YPO0543	FruR	Putative fructose repressor	*fruAKB*	By sim.	P21168
YPO0576	ExuR	Hexuronate utilization repressor	*exuT, uxaCA, uxuR, uxuA* and *uxuB*	By sim.	P42608
YPO0795	GalR	Galactose operon repressor	gal regulon (*mgl*)	By sim.	P03024
YPO0797	LysR	Diaminopimelate decarboxylase.	*lysA*	By sim.	P03030
YPO0985	YspR	Quorum-sensing regulator	Flagellar regulon	Atkinson et al., 1999	O87971

YPO1029	GcvA	Glycine cleavage (activator)	*gcv* operon	By sim.	P32064
YPO1167	BetI	Choline-glycine betaine pathway	*BetABT*	By sim.	P17446
YPO1279	ExuR	Sugar interconversion regulator	*uxuR, uxuA and uxuB*	By sim.	P39161
YPO1301	PsaE	Psa type pili regulatory protein	*psaABC*	Yang et al., 1997.	P31524
YPO1308	RscR	Possible role in virulence	See Nelson et al. 2001	Young and Miller, 1997	AAK81923
YPO1322	DeoR	Nucleotide and deoxynucleotide catabolism	deoCA*BD, udp, cdd, nupC, nupG*...	By sim.	P06217
YPO1375	Lrp	Mediates a global response to leucine.	*ilvIH* operon and others	By sim.	P19494
YPO1642	CscR	Sucrose utilization	*csc* operon (not found)	By sim.	P40715
(YPO1662)	FlhD	Flagellum biosynthesis	Flagellar regulon	Young et al., 1999a	P11164
YPO1663	FlhC	Flagellum biosynthesis	Flagellar regulon	Young et al., 1999a	P11165
YPO1714	KdgR	Pectinolysis and pectinase secretion.	*kdgK, kdgT*	By sim.	P37728
YPO1760	HpcR	Homoprotocatechuate degradation	*hpaBC* and *hpaGEDFHI*	By sim.	Q07095
YPO1857	WrbA	Possible Trp repressor binding protein	*trp* operon	By sim.	P30849
YPO1912	YbtA	Yersiniabactin synthesis	Ybt biosynthesis genes	Fetherston et al., 1996.	T17438
YPO1973	HutC	Histidine utilization	*hut* operon	By sim.	P22773
YPO2065	HexR	Hex regulon repressor.	*zwf, eda, glp*...	By sim.	P46118
YPO2144	FadR	Fatty acid metabolism.	*fadA, fadB, fadD, fadL* and *fadE*	By sim.	P09371
YPO2175	Hns	Pleiotropic	Pleitropic regulation	Bertin et al., 2001	P08936
YPO2219	CysB	Biosynthesis of L-cysteine	*cys* regulon	By sim.	P06613
YPO2258	AraC	Arabinose operon regulatory protein	Arabinose operon	By sim.	P07642
YPO2268	Mlc	Glucose uptake or glycolysis	?	By sim.	P50456
YPO2300	Fnr	Fumarate and nitrate reduction	Global regulation of over 100 genes	By sim.	P03019
YPO2344	TyrR	Aromatic amino acid biosynthesis	8 operons in *E.coli*	By sim.	P07604
YPO2352	PspF	Phage shock protein F	*pspA, B, C* and *E*	Darwin and Miller, 2001	P37344
YPO2374	RovA	Inv and virulence regulation	*inv* + yet uncharacterized	Nagel et al., 2001	P55740
YPO2387	PurR	Purine metabolism	*pur* genes	By sim.	P15039
YPO2445	YfeE	Inorganic iron transport	*yfeABCD*	Bearden et al., 1999	Q56956
YPO2457	YpeR	Quorum-sensing	Flagellar regulon	Atkinson et al., 1999	O87971

YPO2556	PecT	Pectinase gene expression	Pectate lyase genes and others	By sim.	P52662
YPO2625	NagC	Uptake and degradation of GlcN and GlcNac	*nagE, A, B*	By sim.	P15301
YPO2634	Fur	Iron uptake	Global regulation	Staggs et al., 1991	P06975
YPO2681	ChbR	Possible diacetylchitobiose repressor	previously annotated as *cel A, B, C*	By sim.	P17410
YPO3063	GcvR	Glycine cleavage (repressor)	*gcv* operon	By sim.	P23483
YPO3085	CueR	Copper efflux regulator	*ybaR*?	By sim.	P23483
YPO3131	AcrR	Multidrug efflux pump	*acrAB* operon repressor	By sim.	P34000
YPO3138	YmoA	Pleiotropic	Pleitropic regulation	Cornelis et al., 1991	P27720
YPO3143	GlnK	Nitrogen assimilation	*glnA*	By sim.	P38504
YPO3266	EmrR	Drug resistance	*emr* operon	By sim.	P24201
YPO3346	ArsR	Arsenical resistance	*ArsB*?	By sim. - see * (table legend)	P15905
YPO3396	SfsA	Sugar fermentation stimulation	*mal* genes	By sim.	P18273
YPO3420	PdhR	Pyruvate dehydrogenase complex	*aceEF* and *lpdA*	By sim.	P06957
YPO3456	PhnF	Carbon-phosphorus bond cleavage	*phn* operon	By sim.	P16684
YPO3517	ArgR	Arginine biosynthesis	*carAB* operon	By sim.	P15282
YPO3561	SspA	Survival during starvation	Invasin / motility - global response	Badger et al., 1998	P05838
YPO3695	Rnk	Nucleoside diphosphate kinase activity	*ndk*??	By sim.	P40679
YPO3698	TreR	Trehalose utilization	*tre*BC operon	By sim.	P36673
(YPO3723)	IclR	Glyoxylate bypass	*aceBAK* operon	By sim. - inactive	P16528
YPO3759	BirA	Biotin synthesis	*bio A, bioBFCD...*	By sim.	P06709
YPO3770	RfaH	K antigen and lipopolysaccharide	Hemolysin?	By sim.	P26614
YPO3789	MetR	Methionine biosynthesis	*met A, E, H*	By sim.	P19797
YPO3889	IlvY	Isoleucine-valine biosynthetic pathway	*ilvGMEDA*	By sim.	P05827
YPO3904	HfdR	Control of the flagellar master operon	*flhDC*	By sim.	Q8ZAA7
YPO3915	OxyR	Oxydative stress	Catalases, glutathione-reductases	By sim.	P71318
YPO3955	GntR	Gluconate utilization	*gntRK, edd, eda*	By sim.	P46860
YPO4034	XylR	Xylose transport and metabolism	*xylAB and xylFG*	By sim.	P37390
YPO4066	MtlR	Mannitol utilization	*mtlD* and manitol operon	By sim.	P36563
Includes transcriptional regulators (including histone-like proteins) with putatively assignable functions identified in *Y. pestis* CO92. Unless mentioned (written in bold characters and with relevant reference to experimental work), the function and target regulons have been deduced by similarity (by sim.). Putative pseudogenes are in parentheses. * A pYV plasmid-harbored arsenic resistance operon (with an ArsR-like transcriptional repressor) was characterized by Neyt et al. (1997) in low-virulence *Y. enterocolitica* strains.					

Table 2. Putative CO92 transcriptional regulators with unknown functions.						
CO92 ID	Size	Family	% id	Overlap (aa)	Closest species	Specific comments
YPO3913	211	TetR	96	206	*E. coli* K12	YijC
YPO3545	297	LysR	89	294	*E. coli* K12	YhaJ
YPO2807	297	LysR	84	297	*E.coli* CFT073	Possible Xanthosine operon or exotoxin regulation
YPO3146	153	AsnC	83	153	*P. luminescens*	Possible LRP-like transcriptional regulator
(YPO0414)	306	SorC	81	306	*P. luminescens*	Regulation of sugar utilization, (sorbose?) - Inactive in CO92
YPO3683	303	LysR	80	301	*E. coli* K12	
YPO2283	305	LacI	79	305	*P. luminescens*	
YPO2568	344	LacI	78	344	*E.coli* CFT073	
YPO1929	294	LysR	78	292	*S. enteritidis*	Possible regulator of pathogenicity island genes
YPO2685	175	-	72	175	*E. coli* K12	Involved DNA replication, possible transcription factor as well
YPO2497	313	LysR	74	299	*E. coli* K12	
YPO2388	310	LysR	70	304	*E. coli* K12	YdbH
YPO3211	304	ROK	69	299	*E. coli* K12	YajF - doubtfull : alternatively, possible sugar kinase
YPO3348	319	DeoR	68	316	*P. multocida*	
YPO3017	292	RpiR	69	284	*S. typhimurium* LT2	
YPO0010	229	GntR	67	229	*E. coli* K12	
YPO3651	224	GntR	68	219	*B. fungorum*	
YPO0341	191	TetR	65	191	*P. luminescens*	
YPO0669	303	LysR	66	293	*R. solanacearum*	
YPO2150	301	LysR	64	300	*B. parapertussis*	
YPO1938	256	DeoR	64	252	*E. coli* (plasmid)	
YPO2169	286	LysR	62	286	*P. luminescens*	
YPO3223	133	-	62	133	*S. typhimurium* LT2	Refered to as Crl - regulon (curli) not found on chromosome
YPO0799	302	LysR	63	291	*P. syringae*	
YPO2926	279	RpiR	60	279	*E. coli* K12	
YPO2880	345	XRE	59	345	*S. typhimurium*	
YPO3310	314	DeoR	59	314	*P. syringae*	
YPO0679	297	AraC	61	285	*E. coli* O157:H7	
YPO0841	408	-	61	382	*P. multocida*	Possible arylsulfatase regulator
YPO2979	292	LysR	58	281	*S. typhimurium*	
(YPO2267)	304	LysR	59	287	*E. coli* K12	Possible als operon regulator - Inactive in CO92
YPO1960	473	GntR	56	467	*P. luminescens*	Possible pyridoxal-phosphate dependent enzyme.
YPO0846	360	LacI	54	359	*E.coli* CFT073	
YPO1651	149	AsnC	56	142	*S. meliloti*	
YPO2324	318	DeoR	54	313	*B. fungorum*	
YPO0758	331	LacI	53	331	*V. parahaemolyticus*	
YPO0883	132	XRE	56	124	*P. luminescens*	Possibly phage-related
YPO0401	291	AraC	53	287	*E. coli* K12	

YPO1237	270	DeoR	55	256	*P. luminescens*	Possibly involved in sugar metabolism
YPO2845	501	GntR	54	472	*B. parapertussis*	
YPO2378	199	TetR	51	196	*E. coli* O157:H7	
(YPO2449)	194	LuxR	50	194	*P. luminescens*	Inactive in CO92
YPO2537	330	LacI	50	330	*S. typhimurium* LT2	
YPO1503	289	LysR	51	282	*E. coli* K12	
(YPO3840)	221	TetR	54	200	*P. luminescens*	Inactive in CO92
(YPO1728)	338	LacI	49	336	*E. coli* K12	Raffinose utilization? - Inactive in CO92
YPO1934	320	LysR	52	299	*B. parapertussis*	
(YPO1671)	338	LacI	49	333	*E. coli* K12	Inactive in CO92
YPO0165	328	LacI	48	328	*V. vulnificus*	
YPO3978	375	-	46	375	*V. vulnificus*	Possible sugar diacid utilization regulator
YPO0084	411	LysR	47	394	*P. putida*	
YPO0831	258	DeoR	46	252	*P. luminescens*	Possible transcriptional regulator of aga operon
YPO1253	246	RpiR	46	240	*B. halodurans*	
YPO2762	261	AraC	48	243	*V. parahaemolyticus*	
YPO0849	357	LacI	43	353	*E. coli* K12	LacI?
YPO0631	318	LysR	43	307	*B. pertussis*	
YPO1810	325	DeoR	41	303	*S. meliloti*	
YPO1737	128	AraC	50	97	*P. vulgaris*	Possible regulator of blood coagulation
YPO1169	297	LysR	39	287	*S. typhimurium*	
YPO1890	265	GntR	41	233	*C. crescentus*	
YPO0611	328	LacI	38	308	*L. innocua*	
YPO3259	277	RpiR	35	277	*B. fungorum*	
YPO3327	269	DeoR	35	264	*B. halodurans*	
YPO0260	259	AraC	35	246	*C. violaceum*	LcrF-like N-term domain - PI harbored
YPO2458	308	LysR	34	299	*C. crescentus*	
YPO2498	334	LacI	35	295	*M. morganii*	
YPO2243	297	AraC	33	265	*P. luminescens*	
YPO2036	384	RpiR	31	364	*A. tumefaciens*	
YPO0276	327	LysR	31	305	*R. solanacearum*	
YPO1837	291	AraC	29	258	*S. typhimurium*	
YPO2478	346	LacI	27	314	*E. coli* K12	
YPO3228	303	LysR	28	252	*P. luminescens*	*Yersinia*-specific?
YPO0720	88	-	30	68	*V. parahaemolyticus*	Possible FlgM-like anti-Sigma factor
YPO3682	288	LysR	26	232	*P. luminescens*	
YPO3619	292	AraC	23	263	*P. aeruginosa* PA01	
YPO0804	219	-	41	94	*V. vulnificus*	*Yersinia*-specific?
YPO0736	348	-	26	189	*L. anguillarum*	*Yersinia*-specific? Possible Response Regulator (doubtfull)
YPO2593	205	LuxR	51	52	*E. coli* K12	*Yersinia*-specific?
YPO2955	200	LuxR	43	55	*S. coelicolor*	*Yersinia*-specific?
YPO2337	279	MerR	35	67	*C. tetani*	*Yersinia*-specific?

SIGMA AND ANTI-SIGMA FACTORS

The bacterial DNA-dependent, RNA polymerase contains five core enzyme structural subunits that associate with sigma factors to provide transcription specificity. In *E. coli*, seven such molecular species have been identified and extensively studied: σ^D (sigma 70), the four alternative sigma factors (σ^N, σ^S, σ^H and σ^F, also referred to as FliA), the extracellular function regulating (ECF) σ^E, and FecI (a fur-repressed regulator of ferric citrate (Fec) transport system (Angerer et al., 1995). All these factors except FecI have been identified in the two *Y. pestis* genomes. However, there are no extra sigma factors compared to *E. coli*, although two copies of FliA have been identified, consistent with the existence of two distinct flagellar apparatuses in *Y. pestis*. In addition to the FliA and σ^E cognate anti-sigma factors (FlgM and RseA respectively), eight sigma factor modulating proteins have been identified in *Y. pestis*: one for σ^D, another possibly modulating at least one of the two FliAs, two for σ^E (discussed above) and three possible σ^N modulators (one of which is a pseudogene in *Y. pestis*). One last element (YPO3571/y0142) has also been considered as a sigma factor regulator by the two *Y. pestis* annotation teams, in light of its weak homology with RsbV, an agonist of the σ^B alternative sigma factor that controls the general stress response in Gram-positive bacteria, but its role and potential cognate sigma factor remain unknown.

TWO-COMPONENT SYSTEMS: TRYING TO REASSEMBLE THE PUZZLE

In bacteria, the most rapid and efficient means of transcriptional adaptation to extracellular signals occur through sophisticated and powerful systems based on phospho-transfers between conserved transmitter domains of (generally transmembrane) molecules that sense the input signal and the receiver regions of cytoplasmic, regulatory elements which exhibit DNA-binding properties in most cases. These sensor kinases and response regulators are generally arranged in cognate pairs, referred to as two-component systems (TCSs). In *E. coli* K12, 62 TCS-subunits have been identified by genome-wide scanning. 32 are response regulators and 23 are canonical sensor kinases with one histidine kinase (HK) domain; the remaining seven are also sensory kinases but ones that display more complex structures, i.e., containing additional phosphotransfer (HPt) or response regulator (RR) modules, or both (Mizuno, 1997). At least 26 sensors (including 7 hybrid molecular species) and 29 molecular species harboring response-regulator modules have been identified in *Y. pestis* CO92, resulting in 24 possible complete TCSs. More than half of these (16 sensors and 20 response regulators) are highly similar (i.e. >60% identity at the amino acid level, for the same reasons as mentioned above: non-orthologue TCS subunits can exhibit around 40% identity) to elements of previously well-documented systems in other bacteria; they are thus expected to have identical functions in *Yersinia*. On this basis and building on the fact that two partner subunit-encoding genes are often closely linked, 13 of these TCSs can be reconstructed *in silico* with a fair degree of confidence (Table 3a). Four other systems (UhpB-UhpA, RcsC-RcsB/A, PmrB-PmrA and ZraS-ZraR) may be assembled according to the same criteria, although in these cases either the sensor or the regulator may exhibit more marked differences (<60% identity) from their putative counterparts in protein databases (Table 3b). In some cases, functional assignation may be further facilitated by the detection of additional peptide signatures at consistent positions. The presence of a conserved ExxxE motif (reported as binding Fe^{3+}) in PmrB (Wosten et al., 2000) perfectly illustrates this point. ZraS and ZraR (previously referred to as HydH and HydG) were found only in *Y. pseudotuberculosis*[(1)] and in partially sequenced strains of the *Y. pestis* biovar Antiqua (Radnedge et al., 2002). It is noteworthy that *Yersinia* ZraS differs from its *E. coli* counterpart by an insertion of over one hundred amino acids containing a putative PAS domain, which is found in many TCS sensor subunits (for review, see Taylor and Zhulin, 1999). Two other systems (listed in Table 3e) may be deduced from further computational analyses. The first consists of the NarX sensor and the NarP regulator. In other enterobacteria, these elements are part of distinct TCSs (NarX-NarL and NarQ-NarP), both of which reportedly contribute to regulation of nitrate/nitrite metabolism (Stewart, 1994). The fact that NarQ and NarL are absent from the *Yersinia* genomes and that NarX can cross-activate NarP in *E. coli* strongly suggests that these apparent orphans may work together as cognate partners in *Yersinia*. The second system is constituted by the YehT and YehU orthologues. In *E. coli*, genes encoding the sensor kinase and response regulator reside next to each other, and probably belong to the same monocistronic unit. Although *yehT* and *yehU* are found at separate locations on the *Yersinia* chromosomes, these two regulatory elements may also be cognate partners by analogy with the ArcB-ArcA, NarQ-NarP, RscC-RscB-RscA-YojN and BarA-UvrY two component systems in *E. coli*. The functions of all currently identified *Yersinia* TCSs and the possible cross regulations are represented in Figure 2. Readers should bear in mind that with the exception of PhoP-PhoQ, all the depicted relationships are theoretical, and, even though highly probable, have not been verified by experimental evidence. The three last putative TCSs (and all the orphan subunits, except for RssB) are probably the most interesting regulatory TCS elements, since they have no obvious counterparts in *E. coli*. Although this must be checked experimentally, the first TCS may be a potential virulence regulation system in *Yersina*. It is encoded by two *ssrA* (*spiR*) and *ssrB*-like tandem genes that have already been reported to belong to a *Salmonella*-like pathogenicity island (PI) encoding a putative type III secretion system (Deiwick et al., 1999; Garmendia et al., 2003). The two other systems, YPO2997-2998 and YPO3008-YPO3009 (Table 3d), have no counterparts in the currently available databases. Their function(s) and role(s) in virulence and regulation are thus certainly worth investigating. Interestingly, as for ZraR-ZraS, the latter of these two TCS could not be found in the *Y. enterocolitica* strain 8081 genome using BLAST searches [(1)].

Of the five orphan subunits, only YPO0712 has a predictable function. This regulator is similar (51% identity across 70% of the protein) to the *P. aeruginosa* FleR flagellar regulator. The immediate vicinity of genes encoding a flagellar apparatus pleads for a similar contribution in *Yersinia*. However, that fact that YPO0712 appears to be truncated at its N-terminus (i.e. the region supposedly encoding the response regulator domain) strongly suggests that it is inactive.

Lastly, four TCS subunits with similarity to UhpB, YojN, BaeS, and EvgS have also been predicted to be inactive in CO92 (Table 3, shown in parentheses) due to frameshifts or IS insertions in their coding sequences (Parkhill et al., 2001). With the exception of EvgS, these elements were also predicted to be inactive in *Y. pestis* KIM also (Deng et al., 2002). In contrast, it has been assumed that all four proteins are functional proteins in the fully sequenced *Y. pseudotuberculosis* strain [(1)]. How this may impact the physiological evolution of *Y. pestis* will be discussed below.

[(1)] Determined by BLAST searches run on the *Y. pseudotuberculosis* and *Y. enterocolitica* genome sequence data available at the websites: http://bbrp.llnl.gov/bbrp/bin/y.pseudotuberculosis_blast and http://www.sanger.ac.uk/cgi-bin/blast/submitblast/y_enterocolitica.

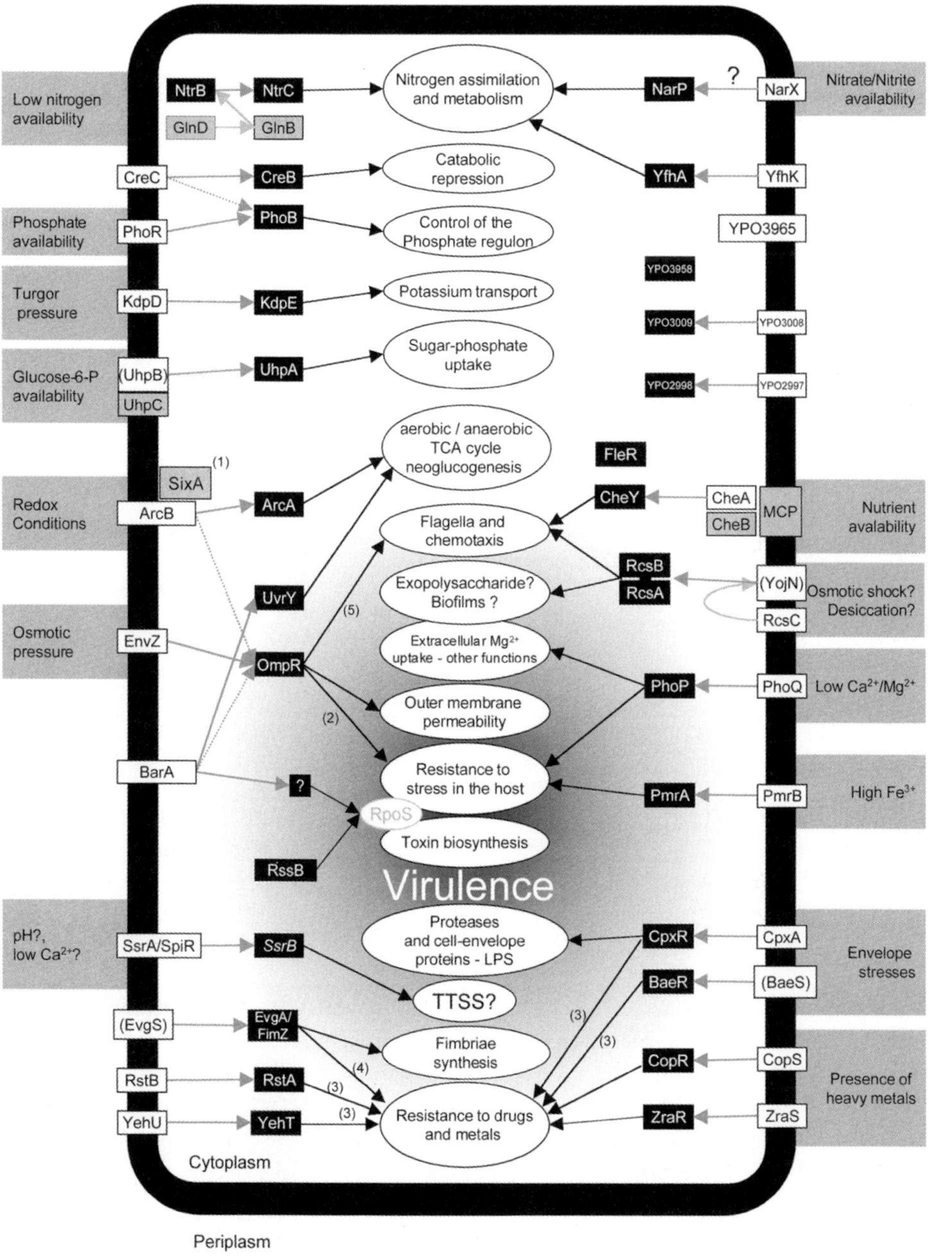

Figure 2. Putative two-component systems of *Y. pestis, Y. pseudotuberculosis,* and *Y. enterocolitica.* Sensors and regulators are represented in white and black boxes respectively along with their their possible functions and input signals (when known) deduced by similarity to regulation networks studied in other enterobacteria. Elements which are parts of the signaling pathways but are not TCS subunits *per se* are represented in gray. Sensors with names in parentheses are thought to be inactive in CO92. For clarity, only the major direct regulations (plain arrows) or cross-regulations (dotted arrows) between these systems are represented. For the same reason, transcriptional regulation of two-component systems by other two-component systems are not included. Additional references: Ogino et al., 1998; Brzostek et al. 2003; Hirakawa et al. 2003; Masuda and Church, 2002; Shin and Park, 1995.

Table 3. Two component systems predicted in *Yersinia*, according to the currently available literature.

locus	CO92	gene	modules	function	DB entry	Relevant references [1]
3a : > 60% aminoacid identity and similar genetic organization						
1	YPO0022	NtrC	RR	Nitrogen assimilation (Sigma54 dependent)	P06713	Ninfa et al., 1995
	YPO0023	NtrB	HK		P06712	
2	YPO0073	CpxA	HK	Protein misfolding	P08336	DiGiuseppe et al., 2003
	YPO0074	CpxR	RR		P16244	
3	YPO0136	OmpR	RR	Osmotic regulation	P03025	Forst et al., 1994
	YPO0137	EnvZ	HK		P24242	
4a	YPO0458	ArcA	RR	Aerobic respiration control	P03026	Luchi et al., 1993
4b	YPO3555	ArcB	HK, Hpt		P22763	
5	YPO0896	CreB	RR	Catabolic regulation.	P08368	Wanner, 1996
	YPO0895	CreC	HK		P08401	
6	YPO1633	PhoQ	HK	Ca2+/Mg2+ metabolism Virulence	P23837	Oyston et al., 2000
	YPO1634	PhoP	RR		P23836	
7a	YPO1666	CheA	Hpt, HK	Chemotaxis regulation	P07363	Djordjevic et al., 1998
	YPO1667	CheW	HK		P07365	
7b	YPO1680	CheY	RR		P06143	
8a	YPO3381	BarA	HK, RR, Hpt	Hypdrogen peroxyde sensitivity	P26607	Pernestig et al., 2001
8b	YPO1865	UvrY	RR		P07027	
9	YPO2308	RstA	RR	Possibly involved in stress response.	P52108	Hirakawa et al., 2003
	YPO2309	RstB	HK		P18392	
10	YPO2688	KdpE	RR	Low turgor pressure-dependent potassium transport	P21866	Walderhaug et al. , 1992
	YPO2689	KdpD	HK		P21865	
11	(YPO2851)	BaeS	HK	Resistance to extracellular stresses	P30847	Raffa et al., 2002
	YPO2853	BaeR	RR		P30846	
12	YPO2914	YfhA	RR	Sigma-54 dependent regulation	P21712	-
	YPO2916	YfhK	HK		P52101	
13	YPO3204	PhoR	HK	Control of Phosphate regulon	P08400	Tommassen et al., 1982
	YPO3205	PhoB	RR		P08402	
3b : <60% aminoacid identity for one of the TCS subunits but similar genetic organization						
14a	YPO1217	RcsC	HK, RR	Control of exopolysaccharide biosynthesis ?	P14376	Stout et al., 1990
14b	YPO1218	RcsB	RR		P14374	
	(YPO1219)	YojN	HK, HPt		P39838	
14c	YPO2449	RcsA	RR		P14374	
15	(YPO4008)	UhpB	SK	Sugar phosphate transport	P09835	Island et al., 1992
	YPO4012	UhpA	RR		P10940	
16	YPO3507	PmrA	RR	Possibly involved in Fe3+ induced regulation	P30843	Roland et al., 1993
	YPO3508	PmrB	SK		P30844	
17	see text	ZraR	RR	Response to Zinc and Lead.	P14375	Leonhartsberger et al., 2001
		ZraS	SK		P14377	

3c : <60% aminoacid identity for both TCS subunits but similar genetic organization						
18	YPO0255	SsrB	RR	Possible pathogenicity island regulating system	AE0700	Garmendia et al., 2003
	YPO0256	SsrA	HK, RR		AD0700	
19a	(YPO1923) (2)	EvgS	HK, RR, Hpt	Possible regulator of virulence in response to diverse environmental signals	P30855	Masuda et al., 2002
19b	YPO1925	EvgA/ FimZ	RR		P21502	
20	YPO2000	CopS?	HK	Copper resistance	P76339	Mills et al., 1993
	YPO2001	CopR	RR		P77380	
3d : unknown						
21	YPO2997	-	HK	Unknown function	-	-
	YPO2998	-	RR		-	
22	YPO3008	-	SK	Unknown function.	-	-
	YPO3009	-	RR		-	
3e : Possible TCSs reconstructed in silico						
23a	YPO1959	NarX	HK	Nitrate-nitrite metabolism	P10956	Rabin et al., 1993
23b	YPO3041	NarP	RR		P31802	
24a	YPO3943	YehU	HK	Unknown function.	P33357	Hirakawa et al., 2003
24b	YPO3287	YehT	RR		P33356	
3f : Orphans						
25	YPO3965	CvgSY?	HK, RR	Unknown function	-	-
26	YPO3958	-	RR	Unknown function	-	-
27	YPO2173	RssB	RR	RpoS regulator (MviA, Hnr)	P37055	Bearson et al., 1996
28	YPO0712	FleR	RR	Sigma-54 dependent flagellar regulatory protein (partial)	P17899	Ritchings et al., 1995
29	YPO1576	-	RR	Unknown function	-	-

Pseudogenes in CO92 are shown between parentheses.
(1) not necessarily princeps publication
(2) Pseudogene in CO92, but not in KIM.

DID EVOLUTION OF THE REGULATORS CONTRIBUTE TO EMERGENCE OF *Y. PESTIS*?

Y. pestis and *Y. pseudotuberculosis* are genetically very close but are responsible for very different diseases in terms of severity. Compared to its *Y. pseudotuberculosis* ancestor, *Y. pestis* displays a restricted ability to grow outside the host but, on the other hand, has become highly pathogenic. According to recent analyses, this change in lifestyle took place very recently on the evolutionary time scale - 20,000 years ago at most (Achtman, et al., 1999). The switch from a bimodal (environment + host) to a host-restricted lifestyle is the easiest to explain because it is consistent with the ongoing reductive evolution of the *Y. pestis* genome: this latter phenomenon probably constitutes the most salient information that has emerged from both CO92 and KIM genome analyses. One can reasonably expect the inactivation of certain regulators to greatly accelerate this evolutionary process, since the silencing of regulons is likely to be as dramatic as complete gene-block deletions. On the other hand, inactivation of transcriptional repressors leading to gene overexpression may account (at least partially) for an increase in pathogenicity. However, the issue of how regulator inactivation may have directed the evolution of *Y. pestis* physiology will never be clear until the regulons have been comprehensive characterised.

Around 150 genes are thought to be pseudogenes, due mainly to IS insertions, frameshift or nonsense mutations, in *Y. pestis* but not in *Y. pseudotuberculosis*. Fourteen encode transcriptional regulators, and inactivation of at least seven of these could have effects on bacterial phenotype. Of these seven regulators, three are parts of TCS sensors with high similarity to *E. coli* BaeS, UphB and YojN, respectively. In view of the situation in *E. coli*, the *Yersinia* BaeR-BaeS-like TCS may act as an envelope stress adaptive system, distinct from those controlled by Sigma E and the CpxR-CpxA TCS: in other words, it is probably an as yet uncharacterised stress response system. In *E. coli*, YojN is a phosphotransfer intermediate associated with RcsC-RcsB, a TCS required for colanic acid biosynthesis assumed to play a role in the remodeling of the bacterial surface (Ferrieres and Clarke, 2003). Again, the inactivation of YojN observed in the CO92 and KIM strains may argue for the evolution of *Y. pestis* from an environmental lifestyle towards a strictly parasitic role. Evidence for changes in the regulation of certain metabolic pathways is also consistent with this evolutionary scenario. Firstly, inactivation in *Y. pestis* of the third TCS mutated subunit (UphB) probably causes downregulation of genes involved in the uptake and metabolism of hexose phosphates. Secondly, frameshifts in *sorC* and *rafR* homologues (YP00414 and YPO1728) may possibly lead to deficiencies in utilization of sorbose and raffinose. Thirdly (and potentially of greater impact) is the inactivation of YPO3583 (an YhbH-like σ^{54} modulator), which probably results in dramatically altered regulation of the σ^{N}-dependent genes in the two species (for review on σ^{N}, see Reitzer and Schneider, 2001).

How, then, might evolution of the regulator pool have contributed to the increase of *Y. pestis* pathogenicity? The recurrent questions concerning the determinism of and reasons for the very rapid evolution of a contrasting virulence phenotype in *Y. pestis* can only be partially explained by the acquisition of the pPCP1 (9.6 Kb) and pMT1 (102 Kb) plasmid-harboured, virulence determinant-encoding genes. Acquisition of chromosomal fragments of exogenous origin (predicted to have been a frequent occurrence) can hardly account for this dramatic evolution either, since it is thought that all the known fragments were acquired prior to *Y. pestis* speciation (Hinchliffe et al., 2003). An additional, simple explanation may be provided by the selective deregulation of genes common to both species. This will be discussed below through two examples. Conversely, gene silencing through the inactivation of regulators may also be of great importance, particularly in the case of physiological functions that are necessary for the *Y. pseudotuberculosis* lifecycle but not for that of *Y. pestis* - in a similar manner to what has been demonstrated previously with other pathogens (Parish et al., 2003). At least two repressor inactivations may have promoted virulence in *Y. pestis*. The first is the frameshift mutation of IclR, the *aceBAK* operon repressor: inactivation in both CO92 and KIM strains (but not in *Y. pseudotuberculosis* 32953) is consistent with the constitutive glyoxylate bypass previously reported in *Y. pestis* (reviewed in Perry and Fetherson, 1997). It is has now become apparent that this biochemical pathway is closely associated with survival in the host in an increasing range of bacterial and fungal pathogens (Lorenz and Fink, 2001). Thus, it is possible that overexpression of the glyoxylate bypass enzymes in *Y. pestis* may enhance virulence. The second case is that of FlhD, encoded by the flagellar master operon (*flhDC*). As mentioned in the first part of this chapter, transcription of this polycistronic unit is essential for initiating transcription of most flagellar subunits and their subsequent assembly into a motile organelle. In both the sequenced *Y. pestis* strains, FlhD was found to be inactivated by a nonsense mutation leading to the production of a truncated protein

(60% of its wild-type length). Inactivation of this regulator in *Y. enterocolitica* led to partial derepression of Yop expression (Bleves et al., 2002). Besides loss of flagella, expression of these molecules at higher levels may confer a selective advantage to *Y. pestis* during its course of infection. By analogy with other bacterial pathogens, the FlhDC complex may co-ordinately regulate the expression of other as yet unknown virulence factors (Pruss et al. 2001; 2003). Hence, it is worth investigating the impact of FlhD inactivation on *Y. pestis*.

CONCLUSION: THE LIMITATIONS AND PITFALLS OF *IN SILICO* ANALYSIS

Attempts to explain phenotypic differences between *Y. pestis* and *Y. pseudotuberculosis* by the evolution of their regulators, based on comparing a limited amount of genomic sequences, must never be considered as more than a starting point for experimental investigations. This is true for several reasons, the most evident of which is that strain-to-strain differences may exist within a given species, as illustrated below by two examples. For example, the ZraS-ZraR TCS, although found in *Y. pestis* strain 32953 and at least some biovar Antiqua strains, was apparently lost in both CO92 and KIM strains. Similarly, EvgS (one of the four TCS sensor subunits) is active in 32953 but not in CO92, and is also predicted to be functional in the sequenced KIM strain. A second major reason is that predicting function in terms of similarity is often very speculative. By anticipating the role of a given regulator based on what is known about its role in other bacterial species, one assumes that the element controls a similar regulon and that its inactivation has the same physiological consequences, which in fact depend on the presence or integrity of other potential co-regulatory systems (i.e. independent, overlapping or cross-regulatory systems). One striking example is the role of YPO3223, designated as *crl* by the two *Y. pestis* annotation teams in light of the 62.1% identity of its product with the *E. coli* curlin genes transcriptional activatory protein. Unexpectedly, none of the genomes of the three pathogenic *Yersinia* species harbour genes (csgA and B-like) coding for such appendages [(1)]. One may then suppose that either these genes have been eliminated or that YPO3223 may serve other purposes. Similarly, the presence of the Rsc-YojN TCS suggests the synthesis of exopolysaccharide in *Yersinia*. However, nothing is known about this antigen in *Yersinia*, and some genes thought to be essential for its biosynthesis have not been detected by *Y. pestis* genome annotation or by BLAST searches of the genomes of the two other pathogenic species. In other, even more complex cases, target regulons have been identified but are seen to differ among the three species, despite the fact that the regulators share > 90% identity at the peptide level. The comparative analysis of the respective roles of the CtrA response regulator in *Caulobacter crescentus* and *Brucella abortus* reported by Bellefontaine et al. (2002) perfectly illustrates the possibility of regulation network plasticities when comparing one bacterial species to another.

Computer-assisted identification of operating regions in sequenced bacterial genomes may be useful in refining prediction of co-regulated genes and thus the regulated physiological function. Although this is extremely difficult most of the time due to the low degree of sequence conservation in prokaryotic operating regions, some apparently

[(1)] Determined by BLAST searches run on the *Y. pseudotuberculosis* and *Y. enterocolitica* genome sequence data available at the websites: http://bbrp.llnl.gov/bbrp/bin/y.pseudotuberculosis_blast and http://www.sanger.ac.uk/cgi-bin/blast/submitblast/y_enterocolitica.

successful attempts (based on the simultaneous analysis of several gamma proteobacterial genomes) have been reported already for regulons of highly structurally conserved regulators, including Fur. In *Y. pestis*, the presence of upstream Fur-boxes has lead to the identification of several previously characterized Fur-controlled operons (including, very surprisingly, the *psaEF* operon that is discussed above), as well as potentially new operons (Panina et al, 2001a). Similar studies have also been performed in *Y. pestis* with other regulators, such those of aromatic amino acid, ribose, arabinose, and xylose metabolism (Laikova et al., 2001; Panina et al., 2001b).

ACKOWLEDGEMENTS

I would like to thank M. L. Rosso for the critical reading of this chapter, J. A. Bengoechea for the careful proofreading of the O-antigen regulation section, and R. D. Perry for the preprint of his publication and for helpful discussions.

REFERENCES

Achtman, M., Zurth, K., Morelli, G., Torrea, G., Guiyoule, A., and Carniel, E. 1999. *Yersinia pestis*, the cause of plague, is a recently emerged clone of *Yersinia pseudotuberculosis*. Proc. Natl. Acad. Sci. USA. 96: 14043-14048.

Al-Hendy, A., Toivanen, P., and Skurnik, M. 1991. The effect of growth temperature on the biosynthesis of *Yersinia enterocolitica* O:3 lipopolysaccharide: temperature regulates the transcription of the rfb but not of the rfa region. Microb. Pathog. 10: 81-86.

Angerer, A., Enz, S., Ochs, M., and Braun, V. 1995. Transcriptional regulation of ferric citrate transport in *Escherichia coli* K-12. Fecl belongs to a new subfamily of sigma 70-type factors that respond to extracytoplasmic stimuli. Mol. Microbiol. 18: 163-174.

Atkinson, S., Throup, J.P., Stewart, G.S., and Williams, P. 1999. A hierarchical quorum-sensing system in *Yersinia pseudotuberculosis* is involved in the regulation of motility and clumping. Mol. Microbiol. 33: 1267-1277.

Badger, J.L., and Miller, V.L. 1995. Role of RpoS in survival of *Yersinia enterocolitica* to a variety of environmental stresses. J. Bacteriol. 177: 5370-5373.

Badger, J.L., and Miller, V.L. 1998. Expression of invasin and motility are coordinately regulated in *Yersinia enterocolitica*. J. Bacteriol. 180: 793-800.

Baichoo, N., and Helmann, J.D. 2002. Recognition of DNA by Fur: a reinterpretation of the Fur box consensus sequence. J. Bacteriol. 184: 5826-5832.

Baker, S.J., Gunn, J.S., and Morona, R. 1999. The *Salmonella typhi* melittin resistance gene *pqaB* affects intracellular growth in PMA-differentiated U937 cells, polymyxin B resistance and lipopolysaccharide. Microbiology. 145: 367-378.

Balsalobre, C., Juarez, A., Madrid, C., Mourino, M., Prenafeta, A., and Munoa, F.J. 1996. Complementation of the hha mutation in *Escherichia coli* by the *ymoA* gene from *Yersinia enterocolitica*: dependence on the gene dosage. Microbiology. 142: 1841-1846.

Bartlett, D.H., Frantz, B.B., and Matsumura, P. 1988. Flagellar transcriptional activators FlbB and FlaI: gene sequences and 5' consensus sequences of operons under FlbB and FlaI control. J. Bacteriol. 170: 1575-1581.

Bearden, S.W., Fetherston, J.D., and Perry, R.D. 1997. Genetic organization of the Yersiniabactin biosynthetic region and construction of avirulent mutants in *Yersinia pestis*. Infect. Immun. 65: 1659-1668.

Bearden, S.W., Staggs, T.M., and Perry, R.D. 1998. An ABC transporter system of *Yersinia pestis* allows utilization of chelated iron by *Escherichia coli* SAB11. J. Bacteriol. 180: 1135-1147.

Bearden, S.W., and Perry R.D. 1999. The Yfe system of *Yersinia pestis* transports iron and manganese and is required for full virulence of plague. Mol. Microbiol. 32: 403-414.

Bearson, S.M., Benjamin, W.H. Jr, Swords, W.E., and Foster, J.W. 1996. Acid shock induction of RpoS is mediated by the mouse virulence gene mviA of *Salmonella typhimurium*. J. Bacteriol. 178: 2572-2579.

Bellefontaine, A.F., Pierreux, C.E., Mertens, P., Vandenhaute, J., Letesson, J.J., and De Bolle, X. 2002. Plasticity of a transcriptional regulation network among alpha-proteobacteria is supported by the identification of CtrA targets in Brucella abortus. Mol. Microbiol. 43: 945-960.

Bengoechea, J.A., and Skurnik, M. 2000. Temperature-regulated efflux pump/potassium antiporter system mediates resistance to cationic antimicrobial peptides in *Yersinia*. Mol. Microbiol. 37: 67-80.

Bengoechea, J.A., Zhang, L., Toivanen, P., and Skurnik, M. 2002. Regulatory network of lipopolysaccharide O-antigen biosynthesis in *Yersinia enterocolitica* includes cell envelope-dependent signals. Mol. Microbiol. 44: 1045-1062.

Bertin, P., Hommais, F., Krin, E., Soutourina, O., Tendeng, C., Derzelle, S. and Danchin, A. 2001. H-NS and H-NS-like proteins in Gram-negative bacteria and their multiple role in the regulation of bacterial metabolism. Biochimie. 83: 235-241.

Bishop, R.E., Gibbons, H.S., Guina, T., Trent, M.S., Miller, S.I., and Raetz., C.R. 2000. Transfer of palmitate from phospholipids to lipid A in outer membranes of Gram-negative bacteria. EMBO J. 19: 5071-5080.

Bleves, S., Marenne, M.N., Detry, G., and Cornelis, G.R. 2002. Up-regulation of the *Yersinia enterocolitica* yop regulon by deletion of the flagellum master operon flhDC. J. Bacteriol. 184: 3214-3223.

Boland, A., Sory, M.P., Iriarte, M., Kerbourch, C., Wattiau, P., and Cornelis, G.R. 1996. Status of YopM and YopN in the *Yersinia* Yop virulon: YopM of *Y.enterocolitica* is internalized inside the cytosol of PU5-1.8 macrophages by the YopB, D, N delivery apparatus. EMBO J. 15: 5191-5201.

Bölin, I., Forsberg, A., Norlander, L., Skurnik, M.and Wolf-Watz, H. 1988. Identification and mapping of the temperature-inducible, plasmid-encoded proteins of *Yersinia* spp. Infect. Immun. 56: 343-348.

Brown, L., and Elliott, T. 1997. Mutations that increase expression of the rpoS gene and decrease its dependence on hfq function in *Salmonella typhimurium*. J. Bacteriol. 179: 656-662.

Brzostek, K., Raczkowska, A., and Zasada, A. 2003. The osmotic regulator OmpR is involved in the response of *Yersinia enterocolitica* O:9 to environmental stresses and survival within macrophages. FEMS Microbiol. Lett. 228: 265-271.

Cambronne, E.D., Cheng, L.W., and Schneewind, O. 2000. LcrQ/YscM1, regulators of the *Yersinia* yop virulon, are injected into host cells by a chaperone dependent mechanism. Mol. Microbiol. 37: 263-273

Campos, A., and Matsumura, P. 2001. Extensive alanine scanning reveals protein-protein and protein-DNA interaction surfaces in the global regulator FlhD from *Escherichia coli*. Mol. Microbiol. 39: 581-594.

Carniel, E., Mercereau-Puijalon, O., and Bonnefoy, S. 1989. The gene coding for the 190,000-dalton iron-regulated protein of *Yersinia* species is present only in the highly pathogenic strains. Infect. Immun. 57: 1211-1217.

Carniel, E., Guiyoule, A., Guilvout, I., and Mercereau-Puijalon, O. 1992. Molecular cloning, iron-regulation and mutagenesis of the *irp2* gene encoding HMWP2, a protein specific for the highly pathogenic *Yersinia*. Mol. Microbiol. 6: 379-388.

Carniel, E. 2001. The *Yersinia* high-pathogenicity island: an iron-uptake island. Microbes Infect. 3: 561-569.

Chatfield, S.N., Strahan, K., Pickard, D., Charles, I.G., Hormaeche, C.E., and Dougan, G. 1992. Evaluation of *Salmonella typhimurium* strains harbouring defined mutations in *htrA* and *aroA* in the murine salmonellosis model. Microb. Pathog. 12: 145-151.

Cheng, L.W., and Schneewind, O. 2000. *Yersinia enterocolitica* TyeA, an intracellular regulator of the type III machinery, is required for specific targeting of YopE, YopH, YopM, and YopN into the cytosol of eukaryotic cells. J. Bacteriol. 182: 3183-3190.

Chilcott, G.S., and Hughes, K.T. 2000. Coupling of flagellar gene expression to flagellar assembly in *Salmonella enterica* serovar *typhimurium* and *Escherichia coli*. Microbiol. Mol. Biol. Rev. 64: 694-708.

Cornelis, G., Vanootegem, J.C., and Sluiters, C. 1987. Transcription of the yop regulon from Y. *enterocolitica* requires trans acting pYV and chromosomal genes. Microb. Pathog. 2: 367-379.

Cornelis, G.R., Sluiters, C., Lambert de Rouvroit, C.L., and Michiels, T. 1989. Homology between virF, the transcriptional activator of the *Yersinia* virulence regulon, and AraC, the *Escherichia coli* arabinose operon regulator. J. Bacteriol. 171: 254-262.

Cornelis, G.R., Sluiters, C., Delor, I., Geib, D., Kaniga, K., Lambert de Rouvroit, C., Sory, M.P., Vanooteghem, J.C. and Michiels, T. 1991. *ymoA*, a *Yersinia enterocolitica* chromosomal gene modulating the expression of virulence functions. Mol. Microbiol. 5: 1023-1034.

Crosa, J.H. 1997. Signal transduction and transcriptional and posttranscriptional control of iron-regulated genes in bacteria. Microbiol. Mol. Biol. Rev. 61: 319-336.

Crosa, J.H., and Walsh, C.T. 2002. Genetics and assembly line enzymology of siderophore biosynthesis in bacteria. Microbiol. Mol. Biol. Rev. 66: 223-249.

Cunning, C., Brown, L., and Elliott, T. 1998. Promoter substitution and deletion analysis of upstream region required for rpoS translational regulation. J. Bacteriol. 180: 4564-4570.

Darwin A.J., and Miller, V.L. 2001. The *psp* locus of *Yersinia enterocolitica* is required for virulence and for growth in vitro when the Ysc type III secretion system is produced. Mol. Microbiol. 39: 429-444.

Deiwick, J., Nikolaus, T., Erdogan, S., and M. Hensel. 1999. Environmental regulation of *Salmonella* pathogenicity island 2 gene expression. Mol. Microbiol. 31: 1759-1774.

de Lorenzo, V., Wee, S., Herrero, M., and Neilands, J.B. 1987. Operator sequences of the aerobactin operon of plasmid ColV-K30 binding the ferric uptake regulation (fur) repressor. J. Bacteriol. 169: 2624-2630.

Delor, I., Kaeckenbeeck, A., Wauters, G., and Cornelis, G.R. 1990. Nucleotide sequence of yst, the *Yersinia enterocolitica* gene encoding the heat-stable enterotoxin, and prevalence of the gene among pathogenic and nonpathogenic Yersiniae. Infect. Immun. 58: 2983-2988.

Deng, W., Burland, V., Plunkett, G. 3rd, Boutin, A., Mayhew, G.F., Liss, P., Perna, N.T., Rose, D.J., Mau, B., Zhou, S., Schwartz, D.C., Fetherston, J.D., Lindler, L.E., Brubaker, R.R., Plano, G.V., Straley, S.C., McDonough, K.A., Nilles, M.L., Matson, J.S., Blattner, F.R., and Perry, R.D. 2002. Genome sequence of *Yersinia pestis* KIM. J. Bacteriol. 184: 4601-4611.

DiGiuseppe, P.A., and Silhavy, T.J. 2003. Signal detection and target gene induction by the CpxRA two-component system. J. Bacteriol. 185: 2432-2440.

Djordjevic, S., and Stock, A.M. 1998. Structural analysis of bacterial chemotaxis proteins: components of a dynamic signaling system. J. Struct. Biol. 124: 189-200.

Dube, P.H., Handley, S.A., Revell, P.A., and Miller, V.L. 2003. The rovA mutant of *Yersinia enterocolitica* displays differential degrees of virulence depending on the route of infection Infect. Immun. 71: 3512-3520.

El Tahir, Y.,and Skurnik, M. 2001. YadA, the multifaceted *Yersinia* adhesin. Int. J. Med. Microbiol. 291: 209-218.

Escolar, L., Perez-Martin, J., and de Lorenzo, V. 1998. Binding of the Fur (ferric uptake regulator) repressor of *Escherichia coli* to arrays of the GATAAT sequence. J. Mol. Biol. 283: 537-547.

Escolar, L., Perez-Martin, J., and de Lorenzo, V. 1999. Opening the iron box: transcriptional metalloregulation by the Fur protein. J. Bacteriol. 181: 6223-9.

Elzer, P.H., Phillips, R.W., Robertson, G.T., and Roop, R.M. 2nd. 1996. The HtrA stress response protease contributes to resistance of *Brucella abortus* to killing by murine phagocytes. Infect. Immun. 64: 4838-4841.

Feodorova, V.A., and Devdariani, Z.L. 2001. New genes involved in *Yersinia pestis* fraction I biosynthesis. J. Med. Microbiol. 50: 969-978.

Ferrieres, L., and Clarke, D.J. 2003. The RcsC sensor kinase is required for normal biofilm formation in *Escherichia coli* K-12 and controls the expression of a regulon in response to growth on a solid surface. Mol. Microbiol. 50: 1665-1682.

Fetherston, J.D., Schuetze, P., and Perry, R.D. 1992. Loss of the pigmentation phenotype in *Yersinia pestis* is due to the spontaneous deletion of 102 kb of chromosomal DNA which is flanked by a repetitive element. Mol. Microbiol. 6: 2693-2704.

Fetherston, J.D, Bearden, S.W, and Perry, R.D. 1996. YbtA, an AraC-type regulator of the *Yersinia pestis* pesticin/*Yersinia*bactin receptor. Mol. Microbiol. 22: 315-325.

Fetherston, J.D., Bertolino, V.J., and Perry, R.D. 1999. YbtP and YbtQ: two ABC transporters required for iron uptake in *Yersinia pestis*. Mol. Microbiol. 32: 289-299.

Forsberg, A., and Wolf-Watz, H. 1988. The virulence protein Yop5 of *Yersinia pseudotuberculosis* is regulated at transcriptional level by plasmid-plB1-encoded trans-acting elements controlled by temperature and calcium. Mol. Microbiol. 2: 121-133.

Forsberg, A., Viitanen, A.-M., Skurnik, M., and Wolf-Watz, H. 1991. The surface-located YopN protein is involved in calcium signal transduction in *Yersinia pseudotuberculosis*. Mol. Microbiol. 5: 977-986.

Forst, S.A., and Roberts, D.L. 1994. Signal transduction by the EnvZ-OmpR phosphotransfer system in bacteria. Res. Microbiol. 145: 363-373.

Francez-Charlot, A., Laugel, B., Van Gemert, A., Dubarry, N., Wiorowski, F., Castanie-Cornet, M.P., Gutierrez, C., and Cam, K. 2003. RcsCDB His-Asp phosphorelay system negatively regulates the flhDC operon in *Escherichia coli*. Mol. Microbiol. 49: 823-832.

Francis, M.S., Lloyd, S.A., and Wolf-Watz, H. 2001. The type III secretion chaperone LcrH co-operates with YopD to establish a negative, regulatory loop for control of Yop synthesis in *Yersinia pseudotuberculosis*. Mol. Microbiol. 42: 1075-1093.

Fraser, G.M., Claret, L., Furness, R., Gupta, S., and Hughes, C. 2002. Swarming-coupled expression of the *Proteus mirabilis* hpmBA haemolysin operon. Microbiology. 148: 2191-2201.

Galán, J.E., and Curtiss, R., III. 1989. Virulence and vaccine potential of *phoP* mutants of *Salmonella typhimurium*. Microb. Pathog. 6: 433-443.

García Véscovi, E., Soncini, F.C., and Groisman, E.A. 1996. Mg2+ as an extracellular signal: environmental regulation of *Salmonella* virulence. Cell. 84: 165-174.

Garmendia, J., Beuzon, C.R., Ruiz-Albert; J., and Holden, D.W. 2003. The roles of SsrA-SsrB and OmpR-EnvZ in the regulation of genes encoding the *Salmonella typhimurium* SPI-2 type III secretion system. Microbiology. 149: 2385-2396.

Gegner, J.A., Graham, D.R., Roth, A.F., and Dahlquist, F.W. 1992. Assembly of an MCP receptor, CheW, and kinase CheA complex in the bacterial chemotaxis signal transduction pathway. Cell. 70: 975-82.

Gehring, A.M., Demoll, E., Fetherston, J.D., Mori, I., Mayhew, G.F., Blattner, F.R., Walsh, C.T., and Perry, R.D. 1998. Iron acquisition in plague - modular logic in enzymatic biogenesis of *Yersinia*bactin by *Yersinia pestis*. Chem. Biol. 5: 573–586.

Gillen, K.L., and Hughes, K.T. 1991. Molecular characterization of *flgM*, a gene encoding a negative regulator of flagellin synthesis in *Salmonella typhimurium*. J. Bacteriol. 173: 6453-6459.

Givaudan, A, and Lanois, A. 2000. flhDC, the flagellar master operon of *Xenorhabdus nematophilus*: requirement for motility, lipolysis, extracellular hemolysis, and full virulence in insects. J. Bacteriol. 182: 107-115.
Goguen, J.D., Yother, J., and Straley, S. C. 1984. Genetic analysis of the low calcium response in *Yersinia pestis* mu d1(Ap lac) insertion mutants. J. Bacteriol. 160: 842-848.
Groisman, E. A., Kayser, J., and Soncini, F.C. 1997. Regulation of polymyxin resistance and adaptation to low-Mg2+ environments. J. Bacteriol. 179: 7040-7045.
Gunn, J.S., Lim, K.B., Krueger, J., Kim, K., Guo, L., Hackett, M., and Miller S. I. 1998. PmrA-PmrB-regulated genes necessary for 4-aminoarabinose lipid A modification and polymyxin resistance. Mol. Microbiol. 27: 1171-1182.
Guo, L., Lim, K.B., Poduje, C.M., Daniel, M., Gunn, J.S., Hackett, M., and Miller, S.I. 1998. Lipid A acylation and bacterial resistance against vertebrate antimicrobial peptides. Cell. 95: 189-198.
Harrison, S.C., and Aggarwal, A.K. 1990. DNA recognition by proteins with the helix-turn-helix motif. Annu. Rev. Biochem. 59: 933-969.
Halliwell, B., and Gutteridge, J.M. 1984. Role of iron in oxygen radical reactions. Methods. Enzymol. 105: 47-56.
Helmann, J.D. 1999. Anti-sigma factors. Curr. Opin. Microbiol. 2: 135-141.
Helmann, J.D. 2002. The extracytoplasmic function (ECF) sigma factors. Adv. Microb. Physiol. 46: 47-110.
Hengge-Aronis, R. 2002. Signal transduction and regulatory mechanisms involved in control of the sigma(S) (RpoS) subunit of RNA polymerase. Microbiol. Mol. Biol. Rev. 66: 373-395.
Hess, J.F., Oosawa, K., Kaplan, N., and Simon, M.I. 1988. Phosphorylation of three proteins in the signaling pathway of bacterial chemotaxis. Cell. 53: 79-87.
Hirakawa, H., Nishino, K., Hirata, T., Yamaguchi, A. 2003. Comprehensive studies of drug resistance mediated by overexpression of response regulators of two-component signal transduction systems in *Escherichia coli*. J. Bacteriol. 185: 1851-1856.
Hinchliffe, S.J., Isherwood, K.E., Stabler, R.A., Prentice, M.B., Rakin, A., Nichols, R.A., Oyston, P.C., Hinds, J., Titball, R.W., and Wren, B.W. 2003. Application of DNA microarrays to study the evolutionary genomics of *Yersinia pestis* and *Yersinia pseudotuberculosis*. Genome Res. 13: 2018-2029.
Hitchen, P.G., Prior, J.L., Oyston, P.C., Panico, M., Wren, B.W., Titball, R.W., Morris, H.R., and Dell, A. 2002. Structural characterization of lipo-oligosaccharide (LOS) from *Yersinia pestis*: regulation of LOS structure by the PhoPQ system. Mol. Microbiol. 44: 1637-1650.
Hoe, N.P., and Goguen, J.D. 1993. Temperature sensing in *Yersinia pestis*: translation of the LcrF activator protein is thermally regulated. J. Bacteriol. 175: 7901-7909.
Hughes, K.T., Gillen, K.L., Semon, M.J., and Karlinsey, J.E. 1993. Sensing structural intermediates in bacterial flagellar assembly by export of a negative regulator. Science. 262: 1277-1280.
Hung, D.L., Raivio, T.L., Jones, C.H., Silhavy, T.J., and Hultgren, S.J. 2001. Cpx signaling pathway monitors biogenesis and affects assembly and expression of P pili. EMBO J. 20: 1508-1518.
Iriarte, M., Cornelis, G.R. 1995a. MyfF, an element of the network regulating the synthesis of fibrillae in *Yersinia enterocolitica*. J. Bacteriol. 177: 738-744.
Iriarte, M., Stainier, I., and Cornelis, G.R. 1995b. The rpoS gene from *Yersinia enterocolitica* and its influence on expression of virulence factors. Infect. Immun. 63: 1840-1847.
Iriarte, M., Stainier, I., Mikulskis; A.V., and Cornelis, G.R. 1995c. The fliA gene encoding sigma 28 in *Yersinia enterocolitica*. J. Bacteriol. 177: 2299-2304.
Iriarte, M., Sory, M.P., Boland, A., Boyd, A.P., Mills, S.D., Lambermont, I., and Cornelis, G.R. 1998. TyeA a protein involved in control of Yop release and in translocation of *Yersinia* Yop effectors. EMBO J. 17: 1907–1918.
Isberg, R,R., Swain; A., and Falkow, S. 1988. Analysis of expression and thermoregulation of the *Yersinia pseudotuberculosis* inv gene with hybrid proteins. Infect. Immun. 56:2133-2138.
Island, M.D., Wei, B.Y., and Kadner, R.J. 1992. Structure and function of the uhp genes for the sugar phosphate transport system in *Escherichia coli* and *Salmonella typhimurium*. J. Bacteriol. 174: 2754-2762.
Jacobi, C.A., Gregor, S., Rakin, A., and Heesemann, J. 2001. Expression analysis of the *Yersinia*bactin receptor gene *fyuA* and the heme receptor *hemR* of *Yersinia enterocolitica* in vitro and in vivo using the reporter genes for green fluorescent protein and luciferase. Infect. Immun. 69: 7772-7782.
Jones, C.H., Danese, P.N., Pinkner, J.S., Silhavy, T.J., and Hultgren, S.J. 1997. The chaperone-assisted membrane release and folding pathway is sensed by two signal transduction systems. EMBO J. 16: 6394-6406.
Jones, H.A., Lillard, J.W. Jr, and Perry, R.D. 1999. HmsT, a protein essential for expression of the haemin storage (Hms+) phenotype of *Yersinia pestis*. Microbiology. 145: 2117-2128.
Kapatral, V., and Minnich, S.A. 1995. Co-ordinate, temperature-sensitive regulation of the three *Yersinia enterocolitica* flagellin genes. Mol. Microbiol. 17: 49-56.

Kapatral, V., Olson, J.W., Pepe, J.C., Miller, V.L., and Minnich, S.A. 1996. Temperature-dependent regulation of *Yersinia enterocolitica* Class III flagellar genes. Mol. Microbiol. 19: 1061-1071.

Karlyshev, A.V., Galyov; E.E., Abramov; V.M., and Zav'yalov; V.P. 1992. Caf1R gene and its role in the regulation of capsule formation of Y. *pestis*. FEBS. Lett. 305: 37-40.

Laikova, O.N., Mironov, A.A., and Gelfand, M.S. 2001. Computational analysis of the transcriptional regulation of pentose utilization systems in the gamma subdivision of Proteobacteria. FEMS Microbiol. Lett. 205: 315-322.

Lambert de Rouvroit, C., Sluiters, C., and Cornelis, G.R. 1992. Role of the transcriptional activator, VirF, and temperature in the expression of the pYV plasmid genes of *Yersinia enterocolitica*. Mol. Microbiol. 6: 395-409.

Lange, R., and Hengge-Aronis, R. 1991. Identification of a central regulator of stationary-phase gene expression in *Escherichia coli*. Mol. Microbiol. 5: 49-59.

Lange, R., and Hengge-Aronis, R. 1994. The cellular concentration of the σ^S subunit of RNA-polymerase in *Escherichia coli* is controlled at the levels of transcription, translation and protein stability. Genes Dev. 8: 1600-1612.

Lange, R., Fischer, D., and Hengge-Aronis, R. 1995. Identification of transcriptional start sites and the role of ppGpp in the expression of rpoS, the structural gene for the sigma S subunit of RNA polymerase in *Escherichia coli*. J. Bacteriol. 177: 4676-4680.

Lease, R.A., Cusick, M.E., and Belfort, M. 1998. Riboregulation in *Escherichia coli*: DsrA RNA acts by RNA: RNA interaction at multiple loci. Proc. Natl. Acad. Sci. USA. 95: 12456-12461.

Lee, V. T., Mazmanian, S. K., and Schneewind, O. 2001. A program of *Yersinia enterocolitica* type III secretion reactions is triggered by specific host signals. J. Bacteriol. 183: 4970-4978.

Leonhartsberger, S., Huber, A., Lottspeich, F., and Bock, A. 2001. The *hydH/G* genes from *Escherichia coli* code for a zinc and lead responsive two-component regulatory system. J. Mol. Biol. 307: 93-105.

Li, S.R., Dorrell, N., Everest, P.H., Dougan, G., and Wren, B.W. 1996. Construction and characterization of a *Yersinia enterocolitica* O:8 high-temperature requirement (*htrA*) isogenic mutant. Infect. Immun. 64: 2088-2094.

Lillard, J.W. Jr, Bearden; S.W., Fetherston, J.D., and Perry, R.D. 1999. The haemin storage (Hms+) phenotype of *Yersinia pestis* is not essential for the pathogenesis of bubonic plague in mammals. Microbiology. 145: 197-209.

Lindler, L.E., Klempner, M.S., and Straley, S.C. 1990. *Yersinia pestis* pH 6 antigen: genetic, biochemical and virulence characterization of a protein involved in the pathogenesis of bubonic plague. Infect. Immun. 58: 2569–2577.

Lorenz, M.C., and Fink, G.R. 2001. The glyoxylate cycle is required for fungal virulence. Nature. 412: 83-86.

Luchi, S., and Lin, E.C.C. 1993. Adaptation of *Escherichia coli* to redox environments by gene expression. Mol. Microbiol. 9: 715-727.

Macnab, R.M. 1996. Flagella and motility. In: Neidhardt, F.C., Curtis, R., III, Ingraham, J.L., Lin, E.C.C., Low, K.B., Magasanik, B., Reznikoff, W.S., Riley, M., Schaechter, M., and Umbarger, H.E., Editors. *Escherichia coli* and *Salmonella*: Cellular and Molecular Biology (2nd edit. ed.). American Society for Microbiology. Washington, DC. 123–145.

Masuda, N., and Church, G.M. 2002. *Escherichia coli* gene expression responsive to levels of the response regulator EvgA. J. Bacteriol. 184: 6225-6234.

McHugh, J.P., Rodriguez-Quinones, F., Abdul-Tehrani, H., Svistunenko, D.A., Poole, R.K., Cooper, C.E., and Andrews, S.C. Global iron-dependent gene regulation in *Escherichia coli*. A new mechanism for iron homeostasis. J. Biol. Chem.. 278: 29478-29486.

Michiels, T., Vanooteghem, J.C., Lambert de Rouvroit, C.L., China, B., Gustin, A., Boudry, P.and Cornelis, G. R.. 1991. Analysis of virC, an operon involved in the secretion of Yop proteins by *Yersinia enterocolitica*. J. Bacteriol. 173: 4994-5009.

Mikulskis, A.V., and Cornelis, G.R. 1994. A new class of proteins regulating gene expression in enterobacteria. Mol Microbiol. 11: 77-86.

Miller, S.I., Kukral, A.M., and Mekalanos, J.J. 1989. A two-component regulatory system (phoP-phoQ) controls *Salmonella typhimurium* virulence. Proc. Natl. Acad. Sci. USA. 86: 5054-5058.

Miller, S. I., and J. J. Mekalanos. 1990. Constitutive expression of the phoP regulon attenuates *Salmonella* virulence and survival within macrophages. J. Bacteriol. 172: 2485-2490.

Mills, S.D., Jasalavich, C.A., and Cooksey, D.A. 1993. A two-component regulatory system required for copper-inducible expression of the copper resistance operon of *Pseudomonas syringae*. J. Bacteriol. 175: 1656-1664.

Missiakas, D., and Raina, S. 1997. Signal transduction pathways in response to protein misfolding in the extracytoplasmic compartments of E. coli: role of two new phosphoprotein phosphatases PrpA and PrpB. EMBO J. 16: 1670-1685.

Mizuno, T. 1997. Compilation of all genes encoding two-component phosphotransfer signal transducers in the genome of *Escherichia coli*. DNA Res. 28: 161-168.

Muffler, A., Fischer, D., Altuvia, S. Storz, G. and Hengge-Aronis, R. 1996. The response regulator RssB controls stability of the {sigma}S subunit of RNA polymerase in *Escherichia coli*. EMBO J. 15: 1333-1339.

Mukhopadhyay, S., Audia, J.P., Roy, R. N., and Schellhorn, H.E. 2000. Transcriptional induction of the conserved alternative sigma factor RpoS in *Escherichia coli* is dependent on BarA, a probable two-component regulator. Mol. Microbiol. 37: 371-381.

Nagel, G., Lahrz, A., and Dersch, P. 2001. Environmental control of invasin expression in *Yersinia pseudotuberculosis* is mediated by regulation of RovA, a transcriptional activator of the /Hor family. Mol. Microbiol. 41: 1249-1269.

Nakao, H, Watanabe, H, Nakayama, S, and Takeda, T. 1995. *yst* gene expression in *Yersinia enterocolitica* is positively regulated by a chromosomal region that is highly homologous to *Escherichia coli* host factor 1 gene (*hfq*). Mol. Microbiol. 18: 859-865.

Neyt, C., Iriarte, M., Thi, V.H., and Cornelis, G.R. 1997. Virulence and arsenic resistance in *Yersinia*e. J. Bacteriol. 179: 612-619.

Nickerson, C.A., and Curtiss, R., III. 1997. Role of sigma factor RpoS in initial stages of *Salmonella typhimurium* infection. Infect. Immun. 65: 1814-1823.

Nieto, J.M., Madrid, C., Miquelay, E., Parra, J.L., Rodriguez, S., and Juarez, A. 2002. Evidence for direct protein-protein interaction between members of the enterobacterial Hha/YmoA and H-NS families of proteins. J. Bacteriol. 184: 629-635.

Ninfa, E.G., Stock, A., Mowbray, S., and Stock, J. 1991. Reconstitution of the bacterial chemotaxis signal transduction system from purified components. J. Biol. Chem. 266: 9764-9770.

Ninfa, A.J., Atkinson, M.R., Kamberov, E. S., Feng, J., and Ninfa, E.G. 1995. Control of nitrogen assimilation by the NRI-NRII two-component system of enteric bacteria, p.67–88. In J.A. Hoch and T.J. Silhavy (ed.), Two-component signal transduction. American Society for Microbiology, Washington, D.C.

Norte, VA, Stapleton, MR, and Green, J. 2003. PhoP-responsive expression of the *Salmonella* enterica serovar *typhimurium* slyA gene. J. Bacteriol. 185: 3508-3514.

Ogino, T., Matsubara, M., Kato, N., Nakamura, Y., and Mizuno, 1998. An *Escherichia coli* protein that exhibits phosphohistidine phosphatase activity towards the HPt domain of the ArcB sensor involved in the multistep His-Asp phosphorelay. Mol. Microbiol. 27: 573-585.

Ohnishi, K., Kutsukake, K., Suzuki, H., and Lino, T. 1992. A novel transcriptional regulation mechanism in the flagellar regulon of *Salmonella typhimurium*: an anti-sigma factor inhibits the activity of the flagellum-specific sigma factor, sigma F. Mol. Microbiol. 6: 3149-3157.

Oyston, P.C., Dorrell, N., Williams, K., Li, S.R., Green, M., Titball, R.W., and Wren, B.W. 2000. The response regulator PhoP is important for survival under conditions of macrophage-induced stress and virulence in *Yersinia pestis*. Infect. Immun. 68: 3419-3425.

Pai, C.H., and Mors, V. 1978. Production of enterotoxin by *Yersinia enterocolitica*. Infect. Immun. Mar;19: 908-911.

Pallen, M.J., and Wren, B.W. 1997. The HtrA family of serine proteases. Mol. Microbiol. 26: 209-221.

Panina, E.M., Mironov, A.A., and Gelfand, M.S. 2001a. Comparative analysis of FUR regulons in gamma-proteobacteria. Nucleic Acids Res. 29: 5195-5206.

Panina, E.M, Vitreschak, A.G, Mironov, A.A, and Gelfand, M.S. 2001b. Regulation of aromatic amino acid biosynthesis in gamma-proteobacteria. J. Mol. Microbiol. Biotechnol. 3: 529-543.

Parish, T., Smith, D.A., Kendall, S., Casali, N., Bancroft, G.J., and Stoker, N.G. 2003. Deletion of two-component regulatory systems increases the virulence of *Mycobacterium tuberculosis*. Infect. Immun. 71: 1134-1140.

Parkhill, J., Wren, B.W., Thomson, N.R., Titball, R.W., Holden, M.T., Prentice, M.B., Sebaihia, M., James, K.D., Churcher, C., Mungall, K.L., Baker, S., Basham, D., Bentley, S.D., Brooks, K., Cerdeno-Tarraga, A.M., Chillingworth, T., Cronin, A., Davies, R.M., Davis, P., Dougan, G., Feltwell, T., Hamlin, N., Holroyd, S., Jagels, K., Karlyshev, A.V., Leather, S., Moule, S., Oyston, P.C., Quail, M., Rutherford, K., Simmonds, M., Skelton, J., Stevens, K., Whitehead, S., and Barrell, B.G. 2001. Genome sequence of *Yersinia pestis*, the causative agent of plague. Nature. 413: 523-527.

Pedersen, L.L., Radulic, M., Doric, M., and Abu Kwaik, Y. 2001. HtrA homologue of *Legionella pneumophila*: an indispensable element for intracellular infection of mammalian but not protozoan cells. Infect. Immun. 69: 2569-2579.

Pepe; J.C., and Miller, V.L. 1993. *Yersinia enterocolitica* invasin: a primary role in the initiation of infection. Proc. Natl. Acad. Sci. USA. 90: 6473-6477.

Pepe, J.C., Badger, J.L., and Miller, V.L. 1994. Growth phase and low pH affect the thermal regulation of the *Yersinia enterocolitica* inv gene. Mol. Microbiol. 11: 123-135.

Perez-Rueda, E., and Collado-Vides, J. 2000. The repertoire of DNA-binding transcriptional regulators in *Escherichia coli* K-12. Nucleic Acids Res. 28: 1838-1847.

Pernestig, A.K., Melefors, O., and Georgellis, D. 2001. Identification of UvrY as the cognate response regulator for the BarA sensor kinase in *Escherichia coli*. J. Biol. Chem. 276: 225-231.

Perry, R.D., Pendrak, M.L., and Schuetze, P. 1990. Identification and cloning of a hemin storage locus involved in the pigmentation phenotype of *Yersinia pestis*. J. Bacteriol. 172:5929-5937.

Perry, R.D., and Fetherston, J.D. 1997. *Yersinia pestis*, the etiologic agent of plague. Clin. Microbiol. Rev. 10: 35-66.

Perry, R.D., Bobrov, A.G., Kirillina, O., Jones, H.A., Pedersen, L., Abney, J., and Fetherston, J.D. 2004. Temperature regulation of the Hemin storage (Hms^+) phenotype of *Yersinia pestis* is posttranscriptional. J. Bacteriol. 186: in press.

Pettersson, J., Nordfelth, R., Dubinina, E., Bergman, T., Gustafsson, M., Magnusson, K.E., and Wolf-Watz, H. 1996. Modulation of virulence factor expression by pathogen target cell contact. Science. 273: 1231-1233.

Poole, K., and McKay, G.A. 2003. Iron acquisition and its control in *Pseudomonas aeruginosa*: many roads lead to Rome. Front. Biosci. 8: 661-686.

Pratt, L.A., and Silhavy, T.J. 1996. The response regulator, SprE, controls the stability of RpoS. Proc. Natl. Acad. Sci. USA. 93: 2488-2492.

Pruss, B.M., Liu, X., Hendrickson, W., and Matsumura, P. 2001. FlhD/FlhC-regulated promoters analyzed by gene array and lacZ gene fusions. FEMS. Microbiol. Lett. 197: 91-97.

Pruss, B.M., Campbell. J.W., Van Dyk, T.K., Zhu, C., Kogan, Y., and Matsumura, P. 2003. FlhD/FlhC is a regulator of anaerobic respiration and the Entner-Doudoroff pathway through induction of the methyl-accepting chemotaxis protein Aer. J. Bacteriol. 185: 534-43.

Rabin, R.S., and Stewart, V. 1993. Dual response regulators (NarL and NarP) interact with dual sensors (NarX and NarQ) to control nitrate- and nitrite-regulated gene expression in *Escherichia coli* K-12. J. Bacteriol. 175: 3259-3268.

Radnedge, L., Agron, P.G., Worsham, P.L., and Andersen, G.L. 2002. Genome plasticity in *Yersinia pestis*. Microbiology. 148: 1687-1698.

Raffa, R.G., and Raivio, T.L. 2002. A third envelope stress signal transduction pathway in *Escherichia coli*. Mol. Microbiol. 45: 1599-1611.

Raivio, T.L., and Silhavy, T.J. 2001. Periplasmic stress and ECF sigma factors. Annu. Rev. Microbiol. 55: 591-624.

Ramamurthy, T., Yoshino, K., Huang, X., Balakrish Nair, G., Carniel, E., Maruyama, T., Fukushima, H., and Takeda, T. 1997. The novel heat-stable enterotoxin subtype gene (*ystB*) of *Yersinia enterocolitica*: nucleotide sequence and distribution of the *yst* genes. Microb. Pathog. 23: 189-200.

Reitzer, L., and Schneider, B.L. 2001. Metabolic context and possible physiological themes of sigma(54)-dependent genes in *Escherichia coli*. Microbiol. Mol. Biol. Rev. 65: 422-444.

Revell, P.A., and Miller, V.L. 2000. A chromosomally encoded regulator is required for expression of the *Yersinia enterocolitica inv* gene and for virulence. Mol. Microbiol. 35: 677-685.

Ritchings, B.W., Almira, E.C., Lory, S., and Ramphal, R. 1995. Cloning and phenotypic characterization of *fleS* and *fleR*, new response regulators of *Pseudomonas aeruginosa* which regulate motility and adhesion to mucin. Infect. Immun. 63: 4868-4876.

Robins-Browne, R.M., Still, C.S., Miliotis, M.D., and Koornhof, H.J. 1979. Mechanism of action of *Yersinia enterocolitica* enterotoxin. Infect. Immun. 25: 680-684.

Rohde, J.R., Fox, J.M., and Minnich, S.A. 1994. Thermoregulation in *Yersinia enterocolitica* is coincident with changes in DNA supercoiling. Mol. Microbiol. 12: 187-199.

Rohde, J.R., Luan, X.S., Rohde, H., Fox, J.M., and Minnich, S.A. 1999. The *Yersinia enterocolitica* pYV virulence plasmid contains multiple intrinsic DNA bends which melt at 37 degrees C. J. Bacteriol. 181: 4198-4204.

Roland, K.L., Martin, L.E., Esther, C.R., and Spitznagel, J.K. 1993. Spontaneous *pmrA* mutants of *Salmonella typhimurium* LT2 define a new two-component regulatory system with a possible role in virulence. J. Bacteriol. 175: 4154-4164.

Rosqvist, R., Magnusson, K.-E., and Wolf-Watz, H. 1994. Target cell contact triggers expression and polarized transfer of *Yersinia* YopE cytotoxin into mammalian cells. EMBO J. 13: 964-972.

Rossi, M.S., Fetherston, J.D., Letoffe, S., Carniel, E., Perry, R.D., Ghigo, J.M. 2001. Identification and characterization of the hemophore-dependent heme acquisition system of *Yersinia pestis*. Infect. Immun. 69: 6707-6717.

Saken, E., Rakin; A., and Heesemann, J. 2000. Molecular characterization of a novel siderophore-independent iron transport system in *Yersinia*. Int J. Med. Microbiol. 290: 51-60.

Schmiel, D.H., Young, G.M., and Miller, V.L. 2000. The *Yersinia enterocolitica* phospholipase gene *yplA* is part of the flagellar regulon. J. Bacteriol. 182:2314-2320.

Schmitt, C.K., Darnell, S.C., Tesh, V.L., Stocker, B.A., and O'Brien, A.D. 1994. Mutation of *flgM* attenuates virulence of *Salmonella typhimurium*, and mutation of *fliA* represses the attenuated phenotype. J. Bacteriol. 176: 368-377.

Schmitt, C.K., Darnell, S.C., and O'Brien, A.D. 1996. The attenuated phenotype of a *Salmonella typhimurium flgM* mutant is related to expression of FliC flagellin J. Bacteriol. 178: 2911-2915.

Schweder, T., Lee, K.-H., Lomovskaya, O., and Matin, A. 1996. Regulation of *Escherichia coli* starvation sigma factor ({sigma}S) by ClpXP protease. J. Bacteriol. 178: 470-476.

Shapiro, L. 1995. The bacterial flagellum: from genetic network to complex architecture. Cell. 80: 525–527.

Shiba, T., Tsutsumi, K., Yano, H., Ihara, Y., Kameda, A., Tanaka, K., Takahashi, H., Munekata, M., Rao, N. N., and Kornberg, A.. 1997. Inorganic polyphosphate and the induction of rpoS expression. Proc. Natl. Acac. Sci. USA. 94: 11210-11215.

Shin, S., and Park, C. 1995. Modulation of flagellar expression in *Escherichia coli* by acetyl phosphate and the osmoregulator OmpR. J. Bacteriol. 177: 4696-4702.

Skrzypek, E., and Straley, S.C. 1993. LcrG, a secreted protein involved in negative regulation of the low-calcium response in *Yersinia pestis*. J. Bacteriol. 175: 3520–3528.

Skurnik, M., and Toivanen, P. 1992. LcrF is the temperature-regulated activator of the *yadA* gene of *Yersinia enterocolitica* and *Yersinia pseudotuberculosis*. J. Bacteriol. 174: 2047-2051.

Soncini, F.C., García Véscovi, E., and Groisman, E.A. 1995. Transcriptional autoregulation of the *Salmonella typhimurium* phoPQ operon. J. Bacteriol. 177: 4364-4371.

Soncini, F.C., García Véscovi, E., Solomon, F., and Groisman, E.A. 1996. Molecular basis of the magnesium deprivation response in *Salmonella typhimurium*: identification of PhoP-regulated genes. J. Bacteriol. 178: 5092-5099.

Sory, M.P., and Cornelis, G.R. 1994. Translocation of a hybrid YopE-adenylate cyclase from *Yersinia enterocolitica* into HeLa cells. Mol. Microbiol. 14: 583-594.

Staggs, T.M., and Perry, R.D. 1991. Identification and cloning of a fur regulatory gene in *Yersinia pestis*. J. Bacteriol. 173: 417-425.

Staggs, T.M., and Perry, R.D. 1992. Fur regulation in *Yersinia* species. Mol. Microbiol. 6: 2507-2516.

Staggs, T.M., Fetherston, J.D., and Perry, R.D. 1994. Pleiotropic effects of a *Yersinia pestis fur* mutation. J. Bacteriol. 176: 7614-7624.

Stainier, I., Iriarte, M., and Cornelis, G.R. 1997. YscM1 and YscM2, two *Yersinia enterocolitica* proteins causing downregulation of yop transcription. Mol. Microbiol. 26: 833-843.

Stewart, V. 1994. Dual interacting two-component regulatory systems mediate nitrate- and nitrite-regulated gene expression in *Escherichia coli*. Res. Microbiol. 145: 450-454.

Stout, V, and Gottesman, S. 1990. RcsB and RcsC: a two-component regulator of capsule synthesis in *Escherichia coli*. J. Bacteriol. 172: 659-669.

Straley, S.C., Plano, G.V., Skrzypek, E., Haddix, P.L., and Fields, K.A. 1993. Regulation by Ca2+ in the *Yersinia* low-Ca2+ response. Mol. Microbiol. 8: 1005-1010.

Taylor, B.L., and Zhulin, I.B. 1999. PAS domains: internal sensors of oxygen, redox potential, and light. Microbiol. Mol. Biol. Rev. 63: 479-506.

Thompson, J.M., Jones, H.A, and Perry R.D. 1999. Molecular characterization of the hemin uptake locus (*hmu*) from *Yersinia pestis* and analysis of hmu mutants for hemin and hemoprotein utilization. Infect. Immun. 67: 3879-3892.

Tommassen, J., de Geus, P., Lugtenberg, B., Hackett J., and Reeves, P. 1982. Regulation of the pho regulon of *Escherichia coli* K-12. Cloning of the regulatory genes *phoB* and *phoR* and identification of their gene products. J. Mol. Biol. 157: 265-274.

Venturi V. 2003. Control of rpoS transcription in *Escherichia coli* and *Pseudomonas*: why so different? Mol. Microbiol. 49: 1-9.

Walderhaug, M.O., Polarek, J.W., Voelkner, P., Daniel, J.M., Hesse, J.E., Altendorf, K., and Epstein, W. 1992. KdpD and KdpE, proteins that control expression of the kdpABC operon, are members of the two-component sensor-effector class of regulators. J. Bacteriol. 174: 2152-2159.

Wanner, B.L. 1996. Phosphorus assimilation and control of the phosphate regulon. In *Escherichia coli* and *Salmonella typhimurium* : Cellular and Molecular biology. Neidhardt, F.C., Curtiss, R, III, Ingraham, J.L., Lin, E.C.C., Low, K.B., Jr, and Magasanik, B., et al (eds). Washington DC: American Society for Microbiology Press.

Wattiau, P., and Cornelis, G.R. 1993. SycE, a chaperone-like protein of *Yersinia enterocolitica* involved in the secretion of YopE. Mol. Microbiol. 8: 123-131.

Wattiau, P., and Cornelis, G.R. 1994. Identification of DNA sequences recognized by VirF, the transcriptional activator of the *Yersinia* yop regulon. J. Bacteriol. 176: 3878-3884.

Weinberg, E.D. 1978. Iron and infection. Microbiol. Rev. 42: 45-66.

Williams, A.W., and Straley, S.C. 1998. YopD of *Yersinia pestis* plays a role in negative regulation of the low-calcium response in addition to its role in translocation of Yops. J. Bacteriol. 180: 350-358.

Williams, K., Oyston, P.C., Dorrell, N., Li, S., Titball, R.W., and Wren, B.W. 2000. Investigation into the role of the serine protease HtrA in *Yersinia pestis* pathogenesis. FEMS. Microbiol. Lett. 186: 281-286.

Wosten, M.M., Kox, L.F., Chamnongpol, F.S., Soncini, F.C., and Groisman, E.A. 2000. A signal transduction system that responds to extracellular iron. Cell. 103: 113-125.

Wren, B.W., Olsen, A.L., Stabler, R., and Li, S.R. 1995. A PCR-based strategy for the construction of a defined *Yersinia enterocolitica* O:8 htrA mutant. Contrib. Microbiol. Immunol. 13: 290-293.

Wu, R.Y., Zhang, R.G., Zagnitko, O., Dementieva, I., Maltzev, N., Watson, J.D., Laskowski, R., Gornicki, P., and Joachimiak, A. 2003. Crystal structure of *Enterococcus faecalis* SlyA-like transcriptional factor. J. Biol. Chem. 278: 20240-20244.

Yamamoto, T., Hanawa, T., Ogata, S., and Kamiya, S. 1996. Identification and characterization of the *Yersinia enterocolitica gsrA* gene, which protectively responds to intracellular stress induced by macrophage phagocytosis and to extracellular environmental stress. Infect. Immun. 64: 2980-2987.

Yamamoto, T., Hanawa, T., Ogata, S., and Kamiya, S. 1997. The *Yersinia enterocolitica* GsrA stress protein, involved in intracellular survival, is induced by macrophage phagocytosis. Infect. Immun. 65: 2190-2196.

Yamashino, T., Ueguchi, C., and Mizuno, T. 1995. Quantitative control of the stationary phase-specific sigma factor, {sigma}S, in *Escherichia coli*: involvement of the nucleoid protein H-NS. EMBO J. 14: 594-602.

Yang, Y., Merriam, J.J., Mueller, J.P., and Isberg, R.R. 1996. The psa locus is responsible for thermoinducible binding of *Yersinia pseudotuberculosis* to cultured cells. Infect. Immun. 64: 2483-2489.

Yang, Y., and Isberg, R.R. 1997. Transcriptional regulation of the *Yersinia pseudotuberculosis* pH6 antigen adhesin by two envelope-associated components. Mol. Microbiol. 24: 499-510.

Yother, J., and Goguen, J.D. 1985. Isolation and characterization of Ca2+-blind mutants of *Yersinia pestis*. J. Bacteriol. 164: 704-711.

Yother, J., Chamness, T.W., and Goguen, J.D. 1986. Temperature-controlled plasmid regulon associated with low calcium response in *Yersinia pestis*. J. Bacteriol. 165: 443-447.

Young, G.M., Smith, M.J., Minnich, S.A, and Miller, V.L. 1999a. The *Yersinia enterocolitica* motility master regulatory operon, flhDC, is required for flagellin production, swimming motility, and swarming motility. J. Bacteriol. 181:2823-2833.

Young, G.M., Schmiel, D.H., and Miller, V.L. 1999b. A new pathway for the secretion of virulence factors by bacteria: the flagellar export apparatus functions as a protein-secretion system. Proc. Natl. Acad. Sci. USA. 96: 6456-6461.

Young, G.M., Badger, J.L., Miller, V.L. 2000. Motility is required to initiate host cell invasion by *Yersinia enterocolitica*. Infect. Immun. 68: 4323-4326.

Young, B.M., and Young, G.M. 2002. YplA is exported by the Ysc, Ysa, and flagellar type III secretion systems of *Yersinia enterocolitica*. J. Bacteriol. 184: 1324-1334.

Zhang, A., Altuvia, S., Tiwari, A. Argaman, L., Hengge-Aronis, R. and Storz, G. 1998. The OxyS regulatory RNA represses rpoS translation and binds the Hfq (HF-I) protein. EMBO J. 17: 6061-6068.

Zheng, M., Doan, B., Schneider, T.D., and Storz, G. 1999. OxyR and SoxRS regulation of fur. J. Bacteriol. 81: 4639-4643.

Chapter 8

Identification of *Yersinia* Genes Expressed During Host Infection

Andrew J. Darwin

ABSTRACT

One of the obvious goals during the study of bacterial pathogenesis is to identify the bacterial genes required for growth within the host. Historically, this has presented a significant technological challenge. However, with this goal in mind, the *in vivo* expression technology (IVET) and signature-tagged mutagenesis (STM) techniques were developed during the 1990s. Both techniques have been used to identify virulence genes in *Y. enterocolitica* and *Y. pseudotuberculosis*, using variations of their mouse models of infection. In this chapter, each of these studies is described individually, including the pertinent details of how each was done, and a brief discussion of the genes identified. In addition, the results of these IVET and STM screens are compared, and the striking lack of overlap between the genes identified is discussed. Most of these studies were only recently published, which means that there have been few follow-up studies on some of the novel virulence genes identified. However, the *Y. enterocolitica hreP, rscR* and *psp* genes have become the subject of further publications, which are also summarized here. Finally, I briefly describe the use of the genome-wide (but not *in vivo*) technology, subtractive hybridization, to identify *Yersinia* virulence genes.

INTRODUCTION

This chapter is devoted to the use of "genome-wide" approaches to identify *Yersinia* virulence genes, with an emphasis on those technologies that can be directly applied during host infection. A number of these *in vivo* technologies are available. However, it is not the purpose of this chapter to go into the technical details of each available technique. For a comprehensive review of the available *in vivo* strategies, I refer the reader elsewhere (Handfield and Levesque, 1999).

In *Yersinia* spp, two such *in vivo* technologies have been used on multiple occasions, and these studies will be covered in detail. The first was *in vivo* expression technology (IVET), which was developed in John Mekalanos' laboratory (Mahan et al., 1993). The second was signature-tagged transposon mutagenesis (STM), developed in David Holden's laboratory (Hensel et al., 1995). In a few cases, some of the genes identified in these screens have become the subjects of subsequent studies, which will also be summarized in this chapter.

I have listed most of the genes found in a separate table for each screen. In several cases, the authors of the original studies reported that analysis of some of the DNA sequence information revealed no similarity to the databases. I have decided to omit these from the tables, simply because it is not informative in the context of this chapter. Of course, this does not mean that I think these potentially novel genes are not interesting. When the genomes of all three pathogenic *Yersinia* species are complete and annotated, I think that it would be a very useful endeavor to re-publish the gene lists with the relevant genome annotation information. This includes all of the genes identified in each screen, whether they were shown to have homologues in other species or not. An update of this nature would be a useful resource for the community, as it would unequivocally identify the genes found in each screen. This will require the cooperation of the authors of each study, either to supply their sequence data, or to do the annotation and supply the results. I am confident that this cooperation would be forthcoming from all of us.

IVET SCREENS

IVET identifies bacterial genes that are expressed during an animal infection, but not during selected laboratory growth conditions. The hypothesis is that genes that meet these criteria are likely to be required for virulence, but unlikely to be so-called housekeeping genes. However, IVET does not directly reveal whether or not the genes identified are required for virulence. For this, a null mutant must be constructed and its phenotype determined. This has been done for only a small minority of the genes identified by IVET, in a number of different bacterial species. In the case of *Yersinia* spp., IVET has been used to study *Y. enterocolitica* only, in both intestinal and systemic mouse models of infection (Young and Miller, 1997; Gort and Miller, 2000).

IVET IDENTIFICATION OF GENES EXPRESSED EARLY IN INFECTION

The first *Y. enterocolitica* IVET screen studied a derivative of strain 8081 (biotype 1B, serotype O8). The screen identified genes expressed in murine Peyer's patch tissue throughout at least the first 46 hours after an orogastric infection (Young and Miller, 1997). A library of strains with random chromosomal *cat* operon fusions (encoding chloramphenicol acetyltransferase) was constructed. The source of genomic DNA for this library was a strain lacking the virulence plasmid (pYV). Whilst this was understandably done to avoid the reidentification of known pYV virulence genes, in future studies it might also be interesting to know which pYV genes would meet the subsequent selection criteria.

The *cat* operon fusion library was used to infect BALB/c mice. Chloramphenicol was administered to the animals throughout the first 46 hours of infection. This enriched for those strains with *cat* fusions expressed during infection. Strains that survived two rounds of this enrichment were recovered from the Peyer's patches and further characterized to identify *cat* fusions that were not expressed on rich or minimal agar plates in the laboratory at 26°C. This class of strains was designated to contain *cat* fusions to host responsive elements (*hre*). 61 different *hre* allelic groups were identified, and the DNA sequence of 48 was determined in the original report (Table 1; Young and Miller, 1997).

Following the original publication, the DNA sequence of the remaining 13 was determined, and the information has been incorporated into Table 1 (G.M. Young and V. L. Miller, personal communication). In addition, some of the fusions that originally revealed no homology to the databases were analyzed further (Heusipp et al., 2003). One is a fusion to a homologue of the *E. coli nadB* gene (Table 1). Another is a fusion to a region with similarity to the *E. coli rpoE* promoter. Further sequence analysis confirmed that

Table 1. IVET identification of *Y. enterocolitica* genes (*hre*) expressed in Peyer's patches and early in infection.

Encoded protein/homologue [a]	Predicted function/property	Role in virulence [b]
Stress response		
70% Gsh, *E. coli*	Glutathione synthesis	ND
91% YdhD, *E. coli*	Glutaredoxin	ND
56% MtpS, *Providencia stuartii*	DNA methylase	ND
69% MutL, *E. coli*	Methylation-dependent DNA repair	ND
79% HflX, *E. coli*	GTP-binding protein	ND
67% RecB, *E. coli*	Exodeoxyribonuclease V	ND
61% AcrR, *E. coli*	Stress response regulator	ND
Similarity to *E. coli rpoE* promoter	Extracytoplasmic stress response sigma factor	Essential gene
100% ClpX, *Y. enterocolitica*	ATP-binding subunit of Clp protease	ND
Iron aquisition		
100% Irp2, *Y. enterocolitica*	HMWP2, Iron acquisition	
100% Irp3, *Y. enterocolitica*	Unknown function (probable iron acquisition role)	Yes
85% FoxA, *Y. enterocolitica*	Siderophore receptor	ND
100% FyuA, *Y. enterocolitica*	Yersiniabactin receptor	Yes
85% YfuB, *Y. enterocolitica*	Iron transport	ND
59% HemD, *E. coli*	Uroporphrynigen III synthase	ND
Cell envelope maintenance		
74% MdoG, *E. coli*	Membrane-derived oligosaccharide synthesis	ND
79% MdoH, *E. coli*	Membrane-derived oligosaccharide synthesis	Yes
66% LpxA, *Y. enterocolitica*	Acyl-transferase	Yes
Miscellaneous		
99% Tnp, *E. coli*	Transposase	ND
65% AceB, *E. coli*	Malate synthase	ND
59% CpdP, *Vibrio fisheri*	3'-5' cAMP phosphodiesterase	ND
31% HoxQ, *E. coli*	Nickel transport/Hydrogenase activity	ND
88% Tgt, *E. coli*	tRNA-guanine transglycosylase	ND
100% RscR, *Y. enterocolitica*	Transcriptional regulator	Yes
56% KpyI, *E. coli*	Pyruvate kinase	ND
100% HreP, *Y. enterocolitica*	Protease	Yes
80% NadB, *E. coli*	quinolinate synthetase, B protein	ND
92% QueA, *E. coli*	synthesis of queuine in tRNA	ND
54% MioC, *E. coli*	Unknown function	ND
81% YrbA, *E. coli*	Unknown function	ND
46% Rub, *Desulfovibrio vulgaris*	Unknown function	ND
32% Orf, *E. coli* (Accession U73857)	Unknown function	ND

[a] The closest homologous protein is shown, along with the percent amino acid identity. Some fusion sequences revealed no similarity to the databases, and they are omitted from this table.
b. Indicates whether a null mutation affects *Y. enterocolitica* virulence (as measured by altered LD_{50} or kinetics of infection).
ND = not determined.

this is a fusion to the *Y. enterocolitica rpoE* promoter. Interestingly, attempts to construct a *Y. enterocolitica rpoE* null mutant have been unsuccessful (Heusipp et al., 2003). This indicates that *rpoE* is an essential gene in *Y. enterocolitica*, as is the case in *E. coli* (De Las Penas et al., 1997).

The authors of this IVET study divided the *hre* loci into functional groups (stress response, iron acquisition, cell envelope maintenance and miscellaneous functions; Table 1). Some of the *hre* loci were known or suspected to be required for normal virulence. These include the iron acquisition proteins HMWP2 and FyuA, which are part of the high pathogenicity island (see chapter 14 in this book and Carniel, 2001). However, the majority of the *hre* loci had not previously been implicated in virulence. Therefore, as part of their study, Young and Miller went on to show that null mutations in four of these novel genes (*mdoH, lpxA, hreP* and *rscR*) affected the course of a mouse infection. In addition, the *rscR* and *hreP* genes became the subjects of more extensive studies, which are described below.

MdoH HOMOLOGUE

One of the genes studied further, originally designated as *hre-13*, is a homologue of *E. coli mdoH.* This gene is predicted to encode a protein involved in the synthesis of cyclic β-glucans (membrane derived oligosaccharides). These molecules are essential for virulence of the plant pathogen *Pseudomonas syringae* pv. *syringae* (Mukhopadhyay et al., 1988), and also affect bacterial survival in environments of low osmotic strength (Kennedy, 1996). The *hre-13* null mutant was slightly delayed in the ability to colonize the Peyer's patches and mesenteric lymph nodes early in infection (Young and Miller, 1997). However, the LD_{50} dose after oral infection was unchanged. This phenotype is consistent with a role early in the infectious process, as would be expected for genes identified in this screen.

LpxA HOMOLOGUE

Another gene that was studied further is also predicted to be involved in cell envelope biosynthesis. The gene, originally designated *hre-14*, encodes a homologue of the acyltransferase LpxA, which is required for lipid A synthesis (Galloway and Raetz, 1990; Vuorio et al., 1991; Kelley et al., 1993; Vuorio et al., 1994). The *hre-14* null mutant was delayed in its ability to colonize the Peyer's patches and mesenteric lymph nodes, and this phenotype appeared to be more severe than that of the *mdoH* homologue mutant (Young and Miller, 1997). Furthermore, the *hre-14* null mutant had a slight increase in oral LD_{50} dose (a little over 5 times that of the wild type) and the average day of death increased by approximately 25%. These phenotypes are consistent with a role throughout the course of infection.

HreP

A *hreP* null mutant (originally designated *hre-22*) has a 33-fold increase in LD_{50} following oral infection of BALB/c mice (Young and Miller, 1997). The predicted HreP protein is most similar to the cyanobacterial calcium-stimulated protease PrcA of *Anabena variabilis.* In addition, it was noted that HreP had significant similarity to several subtilisin-like proteases, which belong to the family of eukaryotic subtilisin/kexin-like proprotein convertases (Heusipp et al., 2001). These proteases are initially synthesized as a single protein, consisting of the N-terminal proprotein and the C-terminal mature protease. Consistent with this, it was shown that a HreP-6×His fusion protein undergoes a single autocatalytic cleavage event (Heusipp et al., 2001). The amino acid sequence of the cleavage site matched the consensus for the subtilisin/kexin-like proprotein convertases. However, the ability of HreP to cleave other proteins could not be demonstrated. The authors offered some possible explanations for this, including lack of knowledge about the optimal conditions for enzyme activity, or substrate specificity, and inhibitory effects of the proprotein, which could not be purified away from the mature protease. The identification of the normal substrate for HreP holds the promise of revealing insight into its physiological function. However, this will not be trivial, especially if the substrate is a host protein. Unfortunately, conditions for *hreP* gene expression from its native promoter in the laboratory have not yet been determined. This has prevented some important studies, such as examining the subcellular location of this protease.

It is also interesting to note that *hreP* is located in a cluster of flagellar biosynthesis and chemotaxis genes (Heusipp et al., 2001). These flanking genes are organized differently in enteric species. This suggests that *Y. enterocolitica* may have acquired *hreP* by horizontal transfer. The gene is also closely linked to *inv*, encoding the Invasin protein required for entry into host cells.

RscR

A polar *rscR* null mutant (originally called *hre-20*) was shown to have interesting *in vivo* phenotypes. There was a modest five-fold increase in LD_{50} following oral infection of BALB/c mice. However, when the bacterial load was monitored over time, the mutant was consistently found to be present in increased numbers in the liver and spleen when compared to wild type (Young and Miller, 1997). This kinetic phenotype was reproduced with a non-polar *rscR* in frame deletion mutant (Nelson et al., 2001). Therefore, the gene designation *rscR* was chosen to indicate the effect of having the wild type gene intact (reduced splenic colonization).

RscR is predicted to be a member of the LysR family of transcriptional regulators, most closely related to the uncharacterized hypothetical YeiE protein of *E. coli* K-12 (68% identity). It was hypothesized that the *rscR* null mutant phenotype was probably due to altered expression of one or more RscR-regulated genes, which prompted a transposon-based screen to identify them (Nelson et al., 2001). This led to the identification of the *rscBAC* locus, which is homologous to the *hmwABC* operon of *Haemophilus influenzae* (although the gene order is different). In the transposon screen, various *rscB::lacZ* insertion mutants were isolated, and analysis of them indicated that *rscB::lacZ* expression was induced 5 to 40-fold when RscR was overproduced.

The similarity to the *H. influenzae* operon suggested that RscA is an extracellular adhesin, which may rely on RscB and RscC for its processing and secretion. Therefore, an *rscA* null mutant was constructed and analyzed (Nelson et al., 2001). The *rscA* null mutant had a similar kinetic phenotype in mice to that of the *rscR* null mutant (increased dissemination to the spleen, but normal colonization of the Peyer's patches and mesenteric lymph nodes). Although the authors conceded that their evidence was indirect, they came to the reasonable conclusion that RscR regulates the *rscBAC* locus during host infection.

It is intriguing that *rscR* or *rscA* null mutations cause increased systemic spread, but do not decrease the LD_{50} (in fact, it may be slightly increased; Young and Miller, 1997). Further study will be required to determine the role of the putative RscA adhesin in the pathogenesis of a *Y. enterocolitica* infection.

IVET IDENTIFICATION OF GENES EXPRESSED DURING SYSTEMIC INFECTION

The second *Y. enterocolitica* IVET screen studied the same strain as the first (a derivative of strain 8081, which is biotype 1B, serotype O8). This screen identified genes expressed in murine spleen throughout at least the first 24 hours after an intraperitoneal infection (Gort and Miller, 2000). The same chromosomal *cat* operon fusion library from the first IVET screen was used. Once again, chloramphenicol was administered to the animals to enrich for those strains with *cat* fusions expressed during infection. Strains that survived two rounds of this enrichment were recovered from the spleen and further characterized to identify host responsive elements, exactly as described for the first IVET screen. In this case the host responsive elements identified were designated as *sif* (systemic infection factor). 31 different *sif* allelic groups were identified, and the DNA from at least one representative of each group was determined (Table 2). Subsequently, *in vivo* expression of some of the *sif* genes was confirmed by chloramphenicol-mediated enrichment of the Φ(*sif-cat*) fusion strain when it was used to infect a mouse in competition with the wild type strain (Gort and Miller, 2000).

The *sif* genes apparently encode proteins that play roles in general physiology, transcriptional regulation, and various other functions. Some of the genes were already known to play a role in virulence (*fyuA* and *manB*), or strongly suspected to do so (the *rffG* homologue). However, once again the majority of the genes identified had not previously been shown to play a role in virulence. The authors constructed a null mutation in one of these genes, originally designated as *sif15* (see below; Gort and Miller, 2000).

There was little overlap between the genes identified in the two *Y. enterocolitica* IVET screens (compare Tables 1 and 2). In fact, the only gene identified in both screens

Table 2. IVET identification of *Y. enterocolitica* genes (*sif*) expressed in spleen during systemic infection.

Encoded protein/homologue [a]	Predicted function/property	Role in virulence [b]
37% SitC, *Staphylococcus epidermidis*	ABC transporter component	ND
80% LepA, *E. coli*	Membrane-bound GTPase	ND
76% RffG, *Erwinia carotovora*	LPS biosynthesis	ND
84% FrdA, *Proteus vulgaris*	Fumarate reductase	ND
80% MetL, *E. coli*	Aspartokinase/homoserine reductase	ND
91% YohI, *E. coli*	Putative transcriptional regulator	ND
82% BioH, *E. coli*	Biotin synthesis	ND
100% FyuA, *Y. enterocolitica*	Yersiniabactin receptor	Yes
100% ManB, *Y. enterocolitica*	O-antigen biosynthesis	Yes
78% YicD, *E. coli*	Unknown function	ND
70% YifJ, *Bacillus subtilis*	Pyruvate-flavodoxin oxidoreductase	ND
68% RhlB, *E. coli*	RNA helicase	ND
69% Orf *Salmonella*		
(20% HP0694 *Helicobacter pylori*)	Putative outer membrane protein	Yes
32% GacA *P. syringae*	Transcriptional regulator	ND

[a] The closest homologous protein is shown, along with the percent amino acid identity. Some fusion sequences revealed no similarity to the databases, and they are omitted from this table.
[b] Indicates whether a null mutation affects *Y. enterocolitica* virulence (as measured by altered LD_{50} or kinetics of infection).
ND = not determined.

was *fyuA*, which is part of the high pathogenicity island. The lack of overlap may indicate that different genes are expressed during early (intestinal) and late (systemic) stages of infection. It strongly suggests that there is much to be gained by doing IVET screens at different stages of infection, in different tissues, and following different modes of infection. The subject of overlap between the various IVET and STM screens will be discussed in more detail later.

Sif15

The predicted Sif15 protein is 69% identical to a putative outer membrane protein in various *Salmonella enterica* serovars, and 20% identical to HP0694 of *Helicobacter pylori*. Following a mixed intraperitoneal infection with the wild type, a *sif15* null mutant had a competitive defect in the spleen (competitive index of 0.08; see Gort and Miller, 2000). However, following orogastric infection the mutant had a significantly less severe competitive defect in the Peyer's patches (competitive index of 0.27; Gort and Miller, 2000). The authors concluded that Sif15 plays an important role during systemic infection, but is less important during colonization of the Peyer's patch tissue. A role for Sif15 during systemic infection might be expected to increase the LD_{50}, but this has not yet been tested. In the laboratory, *sif15* expression was shown to be higher at 37°C than at 26°C, consistent with a role in the host.

STM SCREENS

STM solves the ethical, financial and labor-related concerns associated with screening large numbers of null mutants for decreased virulence in an animal model of infection (Hensel et al., 1995). This is because STM allows relatively large groups of transposon-insertion mutants (e.g. 96 different mutants) to be screened in a single animal. STM will not identify essential genes because the mutants must be able to grow in the laboratory. Furthermore, mutants that can be complemented by wild type bacteria may not be identified in the mixed infections. The major advantage of STM is that it is a direct screen for decreased virulence of null mutants. It can identify mutants with either severe or subtle virulence defects, as demonstrated by the screens done with *Yersinia* spp. STM has been used successfully in both *Y. enterocolitica* (Darwin and Miller, 1999) and *Y. pseudotuberculosis* (Karlyshev et al., 2001; Mecsas et al., 2001). Together, these three studies also covered the three commonly used routes of infection in animal studies (oral, intraperitoneal and intravenous).

Y. ENTEROCOLITICA STM SCREEN (INTRAPERITONEAL INFECTION)

The first *Yersinia* STM screen was done with *Y. enterocolitica*, using the same strain 8081 derivative as the IVET screens described above (Darwin and Miller, 1999). In this study, attempts to use an oral route of infection were unsuccessful. This was apparently due to the so-called "bottleneck" problem, which has been reported by a number of investigators using STM to study enteric pathogens (Mecsas, 2002). In the case of *Y. enterocolitica*, it appears that as few as 30 bacteria seed the Peyer's patches following an oral infection. Therefore, it was not possible to use a pool containing 96 different mutants in oral infections, even at a very high dose (Darwin and Miller, 1999). The effect of lowering the complexity of the pool (i.e. the number of different mutants) was not studied. However, an intraperitoneal (i.p.) route of infection, with pools of 96 mutants, was successful when a high dose was used (approximately 10^4 times greater than the LD_{50} dose for a *Y. enterocolitica* i.p. infection).

The *Y. enterocolitica* signature-tagged transposon insertion mutants were initially isolated on minimal agar so that auxotrophs would not be present. These mutants were assembled into pools of 96 and used to infect mice by i.p. injection. Surviving bacteria were isolated from the spleen after 48 hours. Putative attenuated mutants identified from this first screen were then reassembled into new pools and screened a second time. After this double-screening procedure, attenuation was confirmed for 81% of the strains identified. This was done by detecting a defect in the ability of the mutant to compete with the wild type, when mice were infected with an equal mixture of the two strains (i.p. infection, bacteria recovered from the spleen). A total of 2015 random transposon insertion mutants were screened, and 55 attenuated mutants were identified. Subsequent DNA sequence analysis indicated that 27 different virulence loci had been identified (Table 3). Of these, nine were encoded on the virulence plasmid, and 18 on the chromosome.

Most of the virulence plasmid genes that were identified encode components of the Ysc type III secretion apparatus, or are involved in the regulation of its production. Surprisingly, the screen identified only one of the *yop* genes, which encode the effector proteins secreted by the Ysc type III system (*yopP*, which is known as *yopJ* in *Y. pestis* and *Y. pseudotuberculosis*). The *yopP* null mutant had only a relatively subtle virulence defect (Table 3). YopP/J has been studied extensively using *in vitro* models, where it has been shown to play a role in inducing host cell apoptosis (reviewed by Orth, 2002). The reason(s) why more *yop* genes were not identified in the screen is unknown. One intriguing possibility is that some *yop* mutations can be complemented by a corresponding *yop*$^+$ strain during the mixed infections. However, this has not been tested experimentally. A less interesting, but perhaps more likely possibility is that the transposon did not insert randomly in the virulence plasmid. The failure to identify *yop* mutants is a common theme for all three of the *Yersinia* STM studies (see below).

Nine of the 18 chromosomal virulence loci are in a single locus that is involved in biosynthesis of the O-antigen component of lipopolysaccharide (Table 3). All of these mutants had significant virulence defects as measured by the competition assay following intraperitoneal infection. This is consistent with the observation that a spontaneous *Y. enterocolitica* O-antigen mutant has a significantly increased LD_{50} dose when administered by the oral route of infection (Zhang et al., 1997). Furthermore, the O-antigen is known to be important for a number of enteric pathogens, and O-antigen mutants had previously been isolated in STM screens of *S. typhimurium* and *Vibrio cholerae* (Hensel et al., 1995; Chiang and Mekalanos, 1998). *Y. pseudotuberculosis* O-antigen mutants were also isolated in a subsequent STM screen (see below).

The remaining nine chromosomal loci are predicted to be involved in a variety of functions (Table 3). These include the biosynthesis of cell envelope components (*yifH* and *nlpD*), phosphate (*pstC*) and iron (*irp1*) acquisition, and stress response (*dnaJ*). Two independent mutants had transposon insertions in a homologue of the *E. coli yibP* gene. This gene has no known function, but may be co-transcribed with an upstream gene encoding a putative 2,3-diphosphoglycerate-independent phosphoglyceromutase. These two genes are conserved in the same order in *Pseudomonas syringae* pv. *tomato*, in which a transposon insertion in the upstream gene causes attenuation in a tomato plant infection (Morris et al., 1995). It is interesting that mutations in homologous loci, which might

Table 3. *Y. enterocolitica* virulence genes identified by STM after intraperitoneal infection and recovery of the bacteria from the spleen.

Encoded protein/homologue [a]		Function or property	Competitive index [b]
Virulence plasmid			
yscU	(100%, Ye)	Ysc-Yop Type III secretion	0.00025
lcrV	(100%, Ye)	Ysc-Yop Type III secretion	0.000079
yscR	(100%, Ye)	Ysc-Yop Type III secretion	0.0043
yscC	(100%, Ye)	Ysc-Yop Type III secretion	0.00071
yscL	(100%, Ye)	Ysc-Yop Type III secretion	0.00053
virF	(100%, Ye)	Yop regulon transcriptional activator	0.00011
virG	(100%, Ye)	Ysc-Yop Type III secretion	ND
yopP	(100%, Ye)	Yop effector protein	0.17
sycT-yopM	(100%, Ye)	Intergenic region	0.000031
O-antigen			
ddhA	(100%, Ye)	Glucose-1-phosphate cytidyltransferase	0.00013
ddhB	(100%, Ye)	CDP-glucose 4,6-dehydratase	< 0.00011
wbcC	(100%, Ye)	Abequosyltransferase	0.0065
wbcF	(100%, Ye)	Unknown	< 0.081
wbcH	(100%, Ye)	Galactoside 2-L-fucosyltransferase	< 0.000056
wbcI	(100%, Ye)	Galactosyltransferase	0.000078
manC	(100%, Ye)	GDP-mannose pyrophosphorylase	< 0.00038
manB	(100%, Ye)	Phosphomannomutase	0.00011
galE	(100%, Ye)	UDP-glucose 4-epimerase	< 0.011
Miscellaneous chromosomal mutants			
yifH	(67%, Ec)	Enterobacterial common antigen synthesis	0.07
dnaJ	(79%, St)	Heat shock response	0.0052
pstC	(72%, Ec)	Inorganic phosphate importer	0.005
topA	(93%, Ec)	DNA topoisomerase I	0.0068
nlpD	(72%, Ec)	Outer membrane lipoprotein	0.10
pspC	(62%, Ec)	Regulation of phage shock protein operon	0.000041
irp1	(100%, Ye)	Siderophore synthesis	0.017
yibP	(40-55%, Ec)	Unknown	0.18
yspC	(100% Ye)	Ysa type III secretion system effector	0.36

[a] Amino acid identity (over the region sequenced) compared to the most similar homologue (Ye, *Yersinia enterocolitica*; Ec, *Escherichia coli*; St, *Salmonella typhimurium*).
[b] Mice were infected i.p. with an input ratio of approximately 1:1 (mutant : wild type bacteria). Survivors were recovered from the spleen after 48 hours, and the output ratio of mutant to wild type was determined. Competitive index is the output ratio divided by input ratio. A competitive index of less than one indicates that the mutant is less virulent than the wild type. The lower the competitive index, the more severe is the virulence defect. A number beginning with "<" indicates that no mutant bacteria were recovered in one or more of the test animals. ND = not determined. In some cases multiple mutations in the same gene were isolated. In this case, the mutant with the lowest competitive index is shown.

be involved in carbon metabolism, cause attenuation in both animal and plant models of infection, without affecting growth *in vitro*. It is also interesting to note that the *E. coli* YibP protein is approximately 40% identical to the C-terminal domain of NlpD, also identified in the STM screen.

When the *Y. enterocolitica* STM screen was originally published, one of the chromosomal loci was found to have no significant homology to database entries (Darwin and Miller, 1999). However, with the completion of the *Y. enterocolitica* genome sequence, I have re-examined the site of this transposon insertion. It is now clear that the transposon is inserted into the *Y. enterocolitica yspC* gene, which encodes one of the secreted effectors of the recently discovered Ysa type III secretion system (Foultier et al., 2003). The *yspC* null mutant had the most subtle virulence defect of all of the mutants isolated in the screen. However, this is consistent with the subtle effect of a *ysaV* null mutation on virulence following either oral or i.p. infections (Haller et al., 2000). Together, all of these observations support a role for the Ysa type III secretion system in the pathogenesis of a *Y. enterocolitica* infection.

One of the chromosomal virulence genes identified is homologous to the *pspC* gene of *E. coli*. The *Y. enterocolitica pspC* mutation caused a level of attenuation that was equivalent to that of a virulence plasmid-cured strain. Essentially, the *pspC* mutant is completely avirulent as measured by the i.p. infection competition assay. The *Y. enterocolitica psp* locus became the subject of a subsequent study (see below), and continues to be an area of interest in my laboratory.

THE Psp SYSTEM

The *E. coli* K-12 phage shock protein (*psp*) locus was discovered because infection with a filamentous phage caused massive production of the PspA protein (Brissette et al., 1990). Subsequent work demonstrated that PspA protein synthesis is induced by the mislocalization of a secretin protein, which is an outer membrane pore-forming protein used by the phage to secrete progeny from the infected cell. Homologous secretins are also essential components of a number of bacterial systems, including type II and type III secretion systems, and type IV pilus biosynthesis. The overproduction of a number of these secretin proteins has been shown to induce *E. coli* PspA synthesis. The role of the *E. coli* Psp system is unknown, but it has been hypothesized to be a stress response system that is activated by an unknown signal (reviewed by Model et al., 1997).

The *Y. enterocolitica pspC* null mutant was avirulent in mice, but grew normally in LB broth at 26°C (Darwin and Miller, 1999). However, subsequent analysis indicated that the mutant had a growth defect under conditions that induce production of the virulence-plasmid encoded Ysc type III secretion system. Specifically, mislocalization of the YscC secretin protein caused a complete growth arrest of the *pspC* mutant (Darwin and Miller, 2001). Therefore, it was concluded that the attenuation of the *pspC* mutant was due, at least in part, to stress resulting from mislocalization of the YscC secretin.

This conclusion assumes that a certain proportion of YscC protein does not become correctly localized in the outer membrane during host infection. This has not been directly tested, and it would not be easy to do so. However, it is somewhat reminiscent of the situation involving the assembly of the P pilus of uropathogenic *E. coli*. In this case, it is proposed that some P pilus subunits become mislocalized. This is sensed by the Cpx extracytoplasmic stress response pathway, which itself can control P pilus biosynthesis (Hung et al., 2001). However, whilst the Psp extracytoplasmic stress response system somehow senses mislocalized YscC protein, there is no evidence that the *psp* system directly controls biosynthesis of the Ysc type III secretion system itself. In fact, the *Y. enterocolitica pspC* null mutant still assembles a functional Ysc system, at least *in vitro* (Darwin and Miller, 2001).

Analysis of the *Y. enterocolitica* Psp system is ongoing in my laboratory, both with respect to its role in virulence, and its role in the physiology of the bacterial cell. We are attempting to understand the nature of the inducing signal(s), whether other virulence factors besides YscC might also be inducers, the molecular details of the signal transduction mechanisms, and to identify genes controlled by the Psp response.

Y. PSEUDOTUBERCULOSIS STM SCREEN (ORAL INFECTION)

The results from two *Y. pseudotuberculosis* STM screens were reported in 2001, both of which studied the same strain (strain YPIII pIB1; Karlyshev et al., 2001; Mecsas et al., 2001). The first of these was able to overcome the "bottleneck" problem reported above, which allowed the oral route of infection to be used (this was achieved by reducing the complexity of the mutant pool from 96 to 48 different mutants). In an informative set of preliminary experiments, it was concluded that oral infection with a pool of 48 mutants would allow the identification of those mutants unable to survive in the cecum, and possibly the mesenteric lymph nodes. However, subsequent dissemination to the spleen was too inefficient to allow the identification of attenuated mutants in this tissue (Mecsas et al., 2001).

Another series of preliminary experiments indicated that a strain unable to secrete the Yops was not complemented by the wild type strain in mixed infections. However, this does not rule out the possibility that a defect in only one specific Yop protein might be complemented by the wild type strain. Of course, addressing this question for each of the secreted Yops would be a significant undertaking, beyond the scope of their study. However, it's an interesting question for the future.

The authors of this study also reported that, following oral infection, the *Y. pseudotuberculosis* Yop-secretion mutant grew less well than the wild type in all tissues, and was most significantly attenuated in the cecum, Peyer's patches and spleen. However, the Yop secretion mutant was occasionally detected in the mesenteric lymph nodes, where the fold enrichment of the wild type over the mutant was occasionally less than 20-fold (Mecsas et al., 2001).

Their preliminary experiments led the authors to conclude that the cecum may be the best selective environment to identify attenuated mutants following orogastric infection. Therefore, a library of 960 *Y. pseudotuberculosis* signature-tagged transposon insertion mutants was generated and screened for reduced virulence following oral infection of BALB/c mice. After a double-screening procedure, similar to that described above for *Y. enterocolitica*, 19 mutants were found that were not detected in the cecum and/or the Peyer's patches and mesenteric lymph nodes. However, only 13 of these 19 mutants had a virulence defect when they were tested individually in separate oral mouse infections. In particular, it was found that mutants that were absent from the Peyer's patches and mesenteric lymph nodes, but present in the cecum in the original screening procedure, were as virulent as wild type when tested individually. Therefore, the authors concluded that their screen was only able to reliably identify mutants defective for survival in the cecum.

DNA sequence analysis of the transposon-insertion sites in the attenuated mutants revealed that 13 different virulence loci had been identified (Table 4). Of these, six were on the virulence plasmid, and seven on the chromosome. Four of the virulence plasmid mutations affected structural components of the Ysc type III secretion system (*yscH, yscU, yscB* and *yscL*), and these mutants were unable to secrete any of the Yops (Mecsas et al.,

2001). One of the virulence plasmid insertions was in the *lcrV* gene, and this mutant secreted all of the Yops except for LcrV, YopB and YopD. An *lcrR* mutant was also isolated, but it behaved like the wild type strain *in vitro*, secreting all Yops under low-calcium conditions, but not under high-calcium conditions. The isolation of these virulence plasmid mutants, and the preliminary experiments with the Yop-secretion mutant, indicate that the Ysc type III secretion system is important for survival in the cecum. More specifically, it suggests that the secretion of one or more secreted Yop proteins plays a role in the cecum, which is an intriguing observation. However, no individual *yop* insertion mutants were isolated in the screen. This is similar to the *Y. enterocolitica* STM screen, in which only one *yop* gene mutant was identified amongst many mutations affecting structural components of the Ysc system, or its regulation (Darwin and Miller, 1999).

Three of the chromosomal mutations affected biosynthesis of the O-antigen component of lipopolysaccharide (Table 4). Surprisingly, these mutants were defective in the ability to invade epithelial cells *in vitro* (Mecsas et al., 2001). In fact, the invasion defect of the O-antigen mutants was similar to that of an invasin-defective mutant (*inv*), which was also isolated in the screen. However, it is clear that this invasion defect is not the only deficiency of the O-antigen mutants, because they were severely attenuated in an i.p. infection, whereas the *inv* mutant was fully virulent (Mecsas et al., 2001). Furthermore, O-antigen mutants were also isolated in the *Y. enterocolitica* STM screen, which used an i.p. route of infection (Darwin and Miller, 1999).

The remaining four chromosomal mutations affected *inv* (described above), and three other genes that had not previously been implicated in virulence (*ksgA, sufI* and *cls*). None

Table 4. *Y. pseudotuberculosis* virulence genes identified by STM after oral infection and recovery of the bacteria from the cecum.

Encoded protein/ homologue [a]	Function or property	Cecum virulence defect [b]
Virulence plasmid		
yscH	Ysc-Yop Type III secretion	ND
yscU	Ysc-Yop Type III secretion	ND
yscB	Ysc-Yop Type III secretion	ND
yscL	Ysc-Yop Type III secretion	ND
lcrR	Ysc-Yop Type III secretion	- 2.01
lcrV	Ysc-Yop Type III secretion	ND
O-antigen		
ddhC	Glucose-1-phosphate cytidyltransferase	ND
wzx	O-antigen flippase	- 2.71
gmd	GDP-D-mannose dehydratase	- 2.97
Miscellaneous chromosomal mutants		
inv	Invasin, entry into host cells	- 1.63
sufI	Unknown function (peptidoglycan biosynthesis?)	ND
cls	Cardiolipin synthesis	- 3.42
ksgA	Ribosomal protein, kasugamycin resistance	- 0.59

[a] For homologous genes, the level of amino acid identity, or the organism with the homologue, was not reported.
[b] Mice were infected with an equal mixture of mutant and wild type strains. 5 days later, the numbers of wild type and mutant bacteria in the cecum was determined. The virulence defect is expressed as: log CFU of mutant – log CFU of wild type.

of these latter three mutants were defective in Yop synthesis or delivery, or in the ability to invade cultured epithelial cells *in vitro*. KsgA is a ribosomal protein that is targeted by the drug kasugamycin. The *ksgA* mutant had a growth defect under some conditions, which might explain its virulence defect (although, mice shed the *ksgA* mutant 40 days postinfection, suggesting that it survives and replicates in the cecum for long periods). The authors also discussed the possibility that the virulence phenotype of the *ksgA* mutant could be due to polar effects on downstream genes, which are predicted to be involved in metal ion transport.

The *sufI* and *cls* mutations are both likely to affect the cell envelope. The *E. coli sufI* gene was originally identified as a multicopy suppressor of a temperature-sensitive *ftsI* (PBP-3) mutation. Although the role of *sfiI* has not been studied, it seems likely that it plays a role in peptidoglycan biosynthesis. The *cls* gene is predicted to encode cardiolipin synthase, which makes one of the primary phospholipids in the cell envelope.

Finally, the authors of this study made two additional observations worthy of note. First, 12 of the 13 mutants identified as attenuated in the cecum were also defective for growth in the spleen following an i.p. infection. This indicates that these mutants would have been isolated regardless of the route of infection used in the screen. However, there was one exception (the *inv* mutant), which suggests some benefit to doing STM screens using different routes of infection with the same pathogen-animal system. Second, in their preliminary experiments the authors reported that they occasionally found signature-tagged strains that were present in the spleen, but absent from the mesenteric lymph nodes of the same mouse. It is possible that these strains passed through the mesenteric lymph nodes but did not establish a persistent infection there, or that they were able to infect the spleen without passing through the mesenteric lymph nodes at all. These observations point to the usefulness of signature-tagged strains in studying the dynamics of an animal infection, in addition to their use in identifying attenuated mutants.

Y. PSEUDOTUBERCULOSIS STM SCREEN (INTRAVENOUS INFECTION)

The authors of the second *Y. pseudotuberculosis* STM screen made some modifications to the technique (Karlyshev et al., 2001). First, each transposon had two different signature tags, which had been pre-selected on the basis of uniform hybridization efficiency, and lack of cross hybridization. Second, a high-density oligonucleotide array was used for the detection of signature tags, which allowed quantification of the hybridization signals. An intravenous route of infection was used (tail vein injection). A dose 30 to 300 times higher than the LD_{50} was used to avoid any potential bottleneck problem.

A library of 603 signature-tagged transposon insertion mutants was generated and used to infect mice in pools of 30 or 60 different mutants. Surviving bacteria were recovered from the spleen after three days. From this, 31 putative attenuated mutants were identified, which were in 30 different loci. This is a higher percentage of attenuated mutants than occurred in the other *Yersinia* STM screens, as well as some of the STM screens in other species. The authors suggested that this might have been the result of their modifications to the STM technique, which allowed quantification of hybridization signals and increased overall reliability due to the double tags. However, it should also be noted that a lower stringency of criteria was used to define attenuation in their study when compared to the other *Yersinia* STM screens. Of the 31 different mutants, the competitive

Table 5. *Y. pseudotuberculosis* virulence genes identified by STM after intravenous infection and recovery of the bacteria from the spleen.

Encoded protein/ homologue [a]	Function or property	Competitive index [b]
LPS biosynthesis		
YPO0054, 97%	Glycosyltransferase	0.03*
YPO1382, 97%	LpsA, glycosyltransferase	0.08*
YPO2174, 98%	UDP-glucose-6-dehydrogensae	ND
YPO3099, 96%	ManC, mannose-1-P guanylyltransferase	0.43*
YPO3100, 98%	Fcl, fucose synthetase	0.13
YPO3104, 90%	O-antigen polymerase	0.29
YPO3114, 98%	DdhB, CDP-D-glucose-dehydratase	ND
YPO3116, 95%	AscD, ascarylose biosynthesis	0.04
S. enterica Wzx, 80%	O-antigen flippase	0.003
Miscellaneous chromosomal mutants		
YPO0702, 99%	Putative lipoprotein	ND
YPO1108, 98%	Citrate synthase	0.48*
YPO1174, 96%	Putative adhesin	0.53
YPO1186, 98%	Amino acid transport	0.055
YPO1987, 95%	Unknown function	0.021
YPO1994, 97%	Unknown function	0.084
YPO2287, 96%	Amino acid transport	0.0036
YPO2440, 98%	Iron transport	0.25
YPO2532, 100%	Unknown function	ND
YPO2712, 98%	RseA, negative regulation of *rpoE*	ND
YPO3004, 97%	Prodipeptidase	0.27
YPO3144, 97%	MdlB, mutidrug resistance protein	0.21
YPO3572, 98%	Putative transcriptional regulator	0.44
YPO3657-8, 97%	Unknown function	ND
(intergenic)		
YPO3834, 99%	PldA, phospholipase A	0.017
YPO3965, 96%	VirA, His kinase	0.41
Xyella fastidiosa orf, 70%	Phage-related transcriptional activator	ND
Phage HP1 *orf*, 59%	Unknown	0.89

[a] Amino acid identity (over the region sequenced) compared to the most similar homologue. In most cases, this was a *Y. pestis* CO92 orthologue (YPO). Some transposon insertions were in regions with no sequence similarity to the databases, and they are omitted from this table.
[b] Mice were infected i.v. with an input ratio of approximately 1:1 (mutant : wild type bacteria). Survivors were recovered from the spleen after three days, and the output ratio of mutant to wild type was determined. Competitive index is the output ratio divided by input ratio. A competitive index of less than one indicates that the mutant is less virulent than the wild type. The lower the competitive index, the more severe is the virulence defect. ND = not determined. An asterisk indicates mutants where the *in vitro* competitive index (not shown) was equal or less than this mouse competitive index. In cases where more than one mutation in the same gene was isolated, the mutant with the lowest competitive index is shown.

index of 20 was determined individually. Of these, the authors noted that only 14 had a competitive index of less than 0.3 (one mutant CI was 0.89, and it is omitted from Table 5). Of these 14 mutants, two had an *in vitro* competitive defect that was the same or even more severe than the competitive defect in mice. Therefore, if the criteria used to define attenuation were the same as in the other *Yersinia* studies (see above), then the percentage of attenuated mutants isolated in this screen would probably be quite similar.

Twenty seven of the putative virulence loci that were identified in the screen are listed in Table 5. However, as mentioned above, it is not clear if all of these meet the criteria used in the other *Yersinia* STM screens. The virulence defect of each individual mutant would have to be determined in order to clarify this.

Many of the genes identified are predicted to be involved in lipopolysaccharide biosynthesis, which is consistent with the other *Yersinia* STM screens. However, unlike the other STM screens, none of the mutations were in the virulence plasmid. As the authors pointed out, this is probably because the transposon mutant library was pre-screened on Congo red magnesium oxalate plates to confirm the presence of the virulence plasmid. Disruption of structural or regulatory components of the Ysc type III secretion system would probably cause an abnormal phenotype on these plates, resulting in their elimination from the mutant library. However, insertions in *yop* genes would be unlikely to cause an abnormal phenotype on magnesium oxalate plates (for example, a *yopP* null mutant has a wild type phenotype; see Darwin and Miller, 1999). Therefore, the pre-screening does not explain the fact that individual *yop* mutants were not identified in this STM screen. However, this result is consistent with the rarity of individual *yop* mutants in the other *Yersinia* STM screens.

The other genes identified in the screen are predicted to encode a variety of different functions. These include amino acid and iron transport proteins, transcriptional regulators and a putative adhesin. One of the transposon insertions disrupted the *rseA* gene, which encodes a negative regulator of the RpoE sigma factor. The *Y. enterocolitica rpoE* promoter was identified in an IVET screen (Table 1). Together, these observations suggest that modulation of the *Yersinia* RpoE extracytoplasmic stress response is important during host infection.

A phospholipase gene was also identified in this screen. Phospholipases had been identified as virulence determinants in other bacterial pathogens, but not in *Y. pseudotuberculosis*. Therefore, the authors went on to characterize the phospholipase mutant in more detail, as discussed below.

During the characterization of the transposon insertion sites, the authors noted whether or not there was an orthologue of each gene in *Y. pestis* (Table 5; Karlyshev et al., 2001). Most of the genes identified did have a *Y. pestis* orthologue. However, there were some exceptions. The authors suggested that these exceptions might contribute to the different tropisms of the closely related *Y. pseudotuberculosis* and *Y. pestis* pathogens. In this regard, once the genome sequence of *Y. pseudotuberculosis* is complete, it will be very interesting to determine the full extent of the genetic differences between these two pathogens, which are predicted to have had a relatively recent common ancestor (Achtman et al., 1999).

PldA

One of the attenuated mutants had a transposon insertion in a gene encoding a 282 amino acid protein with homology to a family of bacterial outer membrane phospholipases A. The mutant had a mouse competitive index of 0.017, indicating significant attenuation, and an *in vitro* competitive index of 0.46. Furthermore, following i.v. injection the median lethal dose of the mutant was approximately 200-fold higher than that of the wild type strain (Karlyshev et al., 2001). Note that the median lethal dose is the expected median dose required to produce morbidity or death. This is different to the LD_{50}, which measures the dose expected to kill 50% of infected animals.

The *pldA* mutant was shown to have significantly reduced phospholipase activity (172 units) when compared to the wild type (449 units). The remaining activity was postulated to be due to other phospholipases. Indeed, a non-homologous phospholipase A (YplA) has been characterized in *Y. enterocolitica*, and implicated in its virulence (Schmiel et al., 1998). PCR analysis indicated that a homologous *yplA* gene is also present in *Y. pseudotuberculosis* (Karlyshev et al., 2001), possibly explaining the residual activity of the *pldA* mutant. In a previous study the *yplA* gene was not detected in *Y. pseudotuberculosis* by southern hybridization analysis with a *Y. enterocolitica yplA* probe (Schmiel et al., 1998). However, this may simply have been due to the level of stringency used in the experiment.

Phospholipases C are known to be important virulence factors for a number of bacterial pathogens (Titball, 1993). However, the role of phospholipases A in virulence is less well studied, with YplA of *Y. enterocolitica* being an exception. The authors postulated that PldA might mimic the effects of mammalian phospholipases A, which release arachidonic acid from host cell membrane phospholipids. The released arachidonic acid can serve as a substrate for the generation of a variety of inflammatory mediators.

It was also postulated that PldA might play a role in the invasion of host cells, although this was not tested. The authors based this hypothesis on two observations. First, PldA of *Helicobacter pylori* is essential for the colonization of gastric mucosa (Dorrell et al., 1999). Second, activation of host phospholipase A2 in cultured epithelial cells was required for invasion by *S. enterica* serovar *Typhimurium* (Pace et al., 1993). Although this involves a host phospholipase A, it is possible that the surface-bound PldA of *Y. pseudotuberculosis* could contribute to this process. This is an interesting hypothesis that will have to be tested in future experiments.

IVET AND STM: COMPARISON AND CONCLUSIONS

The IVET and STM techniques have now been used to study a number of different pathogens. Therefore, it seems appropriate to consider the question of how the two compare. For this chapter, I will confine this comparison to the case of *Y. enterocolitica* where both have been used. The most striking observation is the almost complete lack of overlap between the genes identified (compare Tables 1 - 2 with Table 3). This is true, even when only comparing the IVET and STM studies in which the i.p. route of infection was used, and bacteria were harvested from the spleen after two days (Tables 2 and 3). However, there is one important caveat to this observation. Both of the *Y. enterocolitica* IVET screens used operon fusion libraries generated from the genomic DNA of a virulence plasmid-cured strain. This was done so that novel virulence genes might be identified, rather than the known virulence plasmid genes. If the virulence plasmid DNA had been included in the library, it seems likely that some of the genes would have been identified in the IVET screens, which would increase the amount of overlap with the STM screen.

Aside from the special case of virulence plasmid genes, why was there so little overlap in the genes identified by the two techniques? One reason is that the techniques ask different questions. The IVET screens were designed to identify genes expressed in the animal, but not in the laboratory. However, some of the genes identified by STM (e.g. the O-antigen biosynthesis genes) are expected to be expressed significantly in the laboratory. Another difference is that IVET identifies promoter fusions, whereas STM identifies null mutants. Therefore, STM cannot identify essential genes. IVET can identify this class of genes, provided that their expression in the laboratory is quite low and less than during infection of the host. An example is provided by the *rpoE* gene, identified by IVET, and later shown to be an essential *Y. enterocolitica* gene (Heusipp et al., 2003). Therefore, it is possible that some of the other genes identified by IVET are also essential, which would explain their absence amongst the genes found by STM. However, it is clear that not all of the IVET genes are essential, and in the cases where null mutants were shown to reduce bacterial load in the host, one would expect these genes to be identified by STM. Of course, there are some technical considerations that may also explain the lack of overlap. For example, the *Y. enterocolitica* STM and IVET screens were not comprehensive. Mutants that could not grow on minimal media were excluded from the STM screen. The transposon insertion library (STM) and operon fusion libraries (IVET) may not have been completely random, and were almost certainly not comprehensive. It is possible for some of the IVET operon fusion integrants to be polar within operons that might be required for virulence, preventing the survival of these strains in the animal.

Regardless of the reasons behind the lack of overlap, both IVET and STM have been successful in identifying novel *Yersinia* virulence genes. Many of the genes identified are involved in metabolic functions, and this is beginning to give us a picture of the physiology of *Yersinia* cells during host infection. Whilst these may not be so-called "classical virulence factors", their identification and future characterization is of great importance and interest. After all, the most successful antibacterial agents to date are antibiotics, which do not specifically target virulence factors.

Some of the genes identified are known or thought to encode proteins involved in stress responses (e.g. *acrA*, *clpX*, *dnaJ*, *pspC rpoE*). This apparently points to the importance of the ability of the infecting organism to respond to the environmental changes encountered in the host, such as changes in temperature, pH and osmolarity. The RpoE and PspC proteins are apparently involved in separate extracytoplasmic stress responses, which might be triggered by the production of virulence proteins located in the cell envelope, or that must pass through the cell envelope during their secretion. Indeed, induction of the Psp response is triggered by an essential component of the Ysc type III secretion system (Darwin and Miller, 2001), and by other secretin proteins that play a role in virulence (M. E. Maxson and A. J. Darwin, unpublished data). It would be interesting to know whether specific virulence proteins also induce the RpoE response of *Yersinia* species.

The results of the various *Yersinia* IVET and STM screens also demonstrate the importance of using different methods, and different infection models, to identify virulence genes. With the availability of annotated genome sequence information, the characterization of loci identified by these techniques can proceed much more rapidly than previously. Therefore, I do not think we have seen the last of the application of these techniques to the study of *Yersinia* pathogenesis.

A BRIEF WORD ABOUT SUBTRACTIVE HYBRIDIZATION

It seems appropriate here to briefly discuss another genome-wide approach that has recently been used to identify *Yersinia* virulence genes, even though it is not an *in vivo* technology. This method, known as suppressive subtractive hybridization, was used to identify a novel chromosomal locus that was unique to the a *Y. enterocolitica* serotype O: 8 biotype 1B strain (Iwobi et al., 2003).

Total genomic DNA of a non-pathogenic *Y. enterocolitica* biotype 1A strain was subtracted from the genome of a highly pathogenic *Y. enterocolitica* serotype O:8 biotype 1B strain. The success of the technique was validated by PCR analysis of the resulting clones, which led to the identification of the *ail* (attachment and invasion locus) and *inv* (invasin) genes, and also several genes from the high pathogenicity island. Therefore, 200 subtracted clones were analyzed, and they generally fell into three categories: sequences similar to known genes from *Yersinia* or other species; sequences similar to phages and mobile genetic elements; sequences with no similarity to the databases. The authors focused on one of the sequences that was homologous to the *epsE* gene of *V. cholerae*, which encodes part of a type II secretion system.

Further analysis indicated that the *epsE* homologue is part of a cluster of genes (named the *yts1* cluster), which is homologous to loci from various other species that encode type II secretion systems. The system shared the most similarity with the Eps system of *V. cholerae*, which exports a protease, a chitinase, and the most important *V. cholerae* virulence factor, cholera toxin. Southern hybridization analysis of various *Yersinia* strains confirmed that the Yts1D secretion system is only present in the highly pathogenic *Y. enterocolitica* strains, and is also not found in any other *Yersinia* species. Interestingly, analysis of the *Y. enterocolitica* strain 8081 genome sequence revealed the presence of another locus encoding a type II secretion system (Yts2). Unlike *yts1*, the *yts2* locus was found to be present in all *Y. enterocolitica* strains tested. However, like *yts1*, the *yts2* locus was not found in *Y. pestis* or *Y. pseudotuberculosis*.

A *yts1E* null mutant was shown to be significantly attenuated in a mouse model of oral infection. 48 hours after infection the bacterial CFU recovered from the liver and spleen were approximately 100-fold lower for the mutant than for the wild type strain. However, the mutant showed only a minor reduction in the CFU recovered from the small intestine and Peyer's patches. Therefore, the Yts1 secretion system appears to be important for later stages of the infection. This could be either the process of dissemination, or the ability to survive in tissues such as the liver and spleen. The authors of this study postulated that the Yts1 secretion system must be responsible for the secretion of one or more virulence factors.

FUTURE PROSPECTS

All future studies intended to identify *Yersinia* virulence genes should be greatly impacted by the availability of genome sequence information. At the time this chapter was written, the genome sequences of two *Y. pestis* strains have been published (Parkhill et al., 2001; Deng et al., 2002), the complete genome sequence of the highly pathogenic *Y. enterocolitica* strain 8081 is being annotated (http://www.sanger.ac.uk/Projects/Y_enterocolitica/), and the genome sequence of a *Y. pseudotuberculosis* strain is in progress (http://bbrp.llnl.gov/bbrp/html/microbe.html). In terms of locating virulence genes, there will be obvious advances facilitated by the availability of this information. First, it will greatly aid any future IVET and STM studies. Only a few base pairs of DNA sequence information will be required to determine the genomic context of an IVET operon fusion, or STM transposon insertion. Not only will this save time and money, but it also increases the ability to interpret limited data. An example of this is demonstrated by one of the mutants from the *Y. enterocolitica* STM screen. Initial sequence analysis revealed no homology to database entries (Darwin and Miller, 1999). However, the analysis of a much larger contig from the genome revealed that the transposon insertion was in the *yspC* gene (see above).

Another advance is that genome sequences provide the ability to use other methods to identify *Yersinia* virulence genes. For example, microarray studies to identify all members of a virulence regulon, or genes that are specifically expressed under conditions found in the host. Bioinformatics will also become important by allowing the identification of candidate virulence genes based on their homology to known virulence factors from other organisms. In the future, it will also be interesting to determine the genome sequences of multiple strains of the same species that differ in their levels of virulence. An example would be to compare the genome sequences of non-pathogenic, low-level pathogenic, and highly pathogenic *Y. enterocolitica* strains.

The screens described in this chapter have already told us a great deal about the *Yersinia* genes that must be expressed during host infection. Genomics holds the promise of extending these conclusions further. Perhaps it is now reasonable to set a somewhat ambitious long-term goal. This would be to have a comprehensive understanding of all genes that must be expressed during infection in an animal model, for all three of the pathogenic *Yersinia* species. We are already on the way to achieving this goal.

ACKNOWLEDGEMENTS

Work in my laboratory is supported by National Institutes of Health grant AI52148, and by a grant from the Speaker's Fund for Biomedical Research: Toward the Science of Patient Care, awarded by the City of New York. I thank Virginia Miller and Glenn Young for their permission to cite unpublished data.

REFERENCES

Achtman, M., Zurth, Z., Morelli, G., Torrea, G., Guiyoule, A., and Carniel, E. 1999. *Yersinia pestis*, the cause of plague, is a recently emerged clone of *Yersinia pseudotuberculosis*. Proc. Natl. Acad. Sci. USA 96: 14043-14048.

Brissette, J.L., Russel, M., Weiner, L., and Model, P. 1990. Phage shock protein, a stress protein of *Escherichia coli*. Proc. Natl. Acad. Sci. USA 87: 862-866.

Carniel, E. 2001. The *Yersinia* high-pathogenicity island: an iron-uptake island. Microbes. Infect. 3: 561-569.

Chiang, S., and Mekalanos, J. 1998. Use of signature-tagged transposon mutagenesis to identify *Vibrio cholerae* genes critical for colonization. Mol. Microbiol. 27: 797-805.

Darwin, A.J., and Miller, V.L. 1999. Identification of *Yersinia enterocolitica* genes affecting survival in an animal host using signature-tagged transposon mutagenesis. Mol. Microbiol. 32: 51-62.

Darwin, A.J., and Miller, V.L. 2001. The *psp* locus of *Yersinia enterocolitica* is required for virulence and for growth *in vitro* when the Ysc type III secretion system is produced. Mol. Microbiol. 39: 429-444.

De Las Penas, A., Connolly, L., and Gross, C.A. 1997. σ^E is an essential sigma factor in *Escherichia coli*. J. Bacteriol 179: 6862-6864.

Deng, W., Burland, V., Plunkett, G., 3rd, Boutin, A., Mayhew, G.F., Liss, P., Perna, N.T., Rose, D.J., Mau, B., Zhou, S., Schwartz, D.C., Fetherston, J.D., Lindler, L.E., Brubaker, R.R., Plano, G.V., Straley, S.C., McDonough, K.A., Nilles, M.L., Matson, J.S., Blattner, F.R., and Perry, R.D. 2002. Genome sequence of Yersinia pestis KIM. J Bacteriol. 184: 4601-4611.

Dorrell, N., Martino, M.C., Stabler, A., Ward, S.J., Zhang, Z.W., McColm, A.A., Farthing, M.J., and Wren, B.W. 1999. Characterization of *Helicobacter pylori* PldA, a phospholipase with a a role in colonization of the gastric mucosa. Gastroenterology 117: 1098-1104.

Foultier, B., Troisfontaines, P., Vertommen, D., Marenne, M.N., Rider, M., Parsot, C., and Cornelis, G.R. 2003. Identification of substrates from the *Yersinia enterocolitica* 1B Ysa type III secretion system. Infect. Immun. 71: 242-253.

Galloway, S.M., and Raetz, C.R. 1990. A mutant of *Escherichia coli* defective in the first step of endotoxin biosynthesis. J. Biol. Chem. 265: 6394-6640.

Gort, A.S., and Miller, V.L. 2000. Identification and characterization of *Yersinia enterocolitica* genes induced during systemic infection. Infect. Immun. 68: 6633-6642.

Haller, J.C., Carlson, S., Pederson, K.J., and Pierson, D.E. 2000. A chromosomally encoded type III secretion pathway in *Yersinia enterocolitica* is important in virulence. Mol. Microbiol. 36: 1436-1446.

Handfield, M., and Levesque, R.C. 1999. Strategies for isolation of in vivo expressed genes from bacteria. FEMS Microbiol. Rev. 23: 69-91.

Hensel, M., Shea, J.E., Gleeson, C., Jones, M.D., Dalton, E., and Holden, D.W. 1995. Simultaneous identification of bacterial virulence genes by negative selection. Science 269: 400-403.

Heusipp, G., Schmidt, M.A., and Miller, V.L. 2003. Identification of *rpoE* and *nadB* as host responsive elements of *Yersinia enterocolitica*. FEMS Microbiol. Lett. 226: 291-298.

Heusipp, G., Young, G.M., and Miller, V.L. 2001. HreP, an *in vivo*-expressed protease of *Yersinia enterocolitica*, is a new member of the family of subtilisin/kexin-like proteases. J. Bacteriol. 183: 3556-3563.

Hung, D.L., Raivio, T.L., Jones, C.H., Silhavy, T.J., and Hultgren, S.J. 2001. Cpx signaling pathway monitors biogenesis and affects assembly and expression of P pili. EMBO J. 20: 1508-1518.

Iwobi, A., Heesemann, J., Garcia, E., Igwe, E., Noelting, C., and Rakin, A. 2003. Novel virulence-associated type II secretion system unique to high-pathogenicity *Yersinia enterocolitca*. Infect. Immun. 71: 1872-1879.

Karlyshev, A.V., Oyston, P.C.F., K., W., Clark, G.C., Titball, R.W., Winzeler, E.A., and Wren, B.W. 2001. Application of high-density array-based signature-tagged mutagenesis to discover novel *Yersinia* virulence-associated genes. Infect. Immun. 69: 7810-7819.

Kelley, T.M., Stachula, S.A., Raetz, C.R.H., and Anderson, M.S. 1993. The *firA* gene of *Escherichia coli* encodes UDP-3-(R-3-hydroxymyristol)-glucosamine N-acyltransferase. J. Biol. Chem. 268: 19866-19874.

Kennedy, E. 1996. Membrane-derived oligosaccharides (periplasmic beta-D-glucans) of *Escherichia coli*. In: *Escherichia coli* and *Salmonella*: cellular and molecular biology. F. C. Neidhardt, R. Curtis III, J. L. Ingraham, E. C. C. Lin, K. Brooks Low, B. Magasanik, W. S. Reznikoff, M. Riley, M. Schaechter and H. E. Umbarger, eds. ASM Press, Washington, D.C. p. 1064-1071.

Mahan, M.J., Slauch, J.M., and Mekalanos, J.J. 1993. Selection of bacterial virulence genes that are specifically induced in host tissues. Science 259: 686-688.

Mecsas, J. 2002. Use of signature-tagged mutagenesis in pathogenesis studies. Curr. Opin. Microbiol. 5: 33-37.

Mecsas, J., Bilis, I., and Falkow, S. 2001. Identification of attenuated *Yersinia pseudotuberculosis* strains and characterization of an orogastric infection in BALB/c mice on day 5 postinfection by signature-tagged mutagenesis. Infect. Immun. 67: 2779-2787.

Model, P., Jovanovic, G., and Dworkin, J. 1997. The *Escherichia coli* phage-shock-protein (*psp*) operon. Mol. Microbiol. 24: 255-261.

Morris, V.L., Jackson, D.P., Grattan, M., Ainsworth, T., and Cuppels, D.A. 1995. Isolation and sequence analysis of the *Pseudomonas syringae* pv. tomato gene encoding a 2,3-diphosphoglycerate-independent phosphoglyceromutase. J. Bacteriol. 177: 1727-1733.

Mukhopadhyay, P., Williams, J., and Mills, D. 1988. Molecular analysis of a pathogenicity locus in *Pseudomonas syringae* pv. syringae. J. Bacteriol. 170: 5479-5488.

Nelson, K.M., Young, G.M., and Miller, V.L. 2001. Identification of a locus involved in systemic dissemination of *Yersinia enterocolitica*. Infect. Immun. 69: 6201-6208.

Orth, K. 2002. Function of the *Yersinia* effector YopJ. Curr. Opin. Microbiol. 5: 38-43.

Pace, J., Hayman, M.J., and Galan, J.E. 1993. Signal transduction and invasion of epithelial cells by *S. typhimurium*. Cell 72: 505-514.

Parkhill, J., Wren, B.W., Thomson, N.R., Titball, R.W., Holden, M.T., Prentice, M.B., Sebaihia, M., James, K.D., Churcher, C., Mungall, K.L., Baker, S., Basham, D., Bentley, S.D., Brooks, K., Cerdeno-Tarraga, A.M., Chillingworth, T., Cronin, A., Davies, R.M., Davis, P., Dougan, G., Feltwell, T., Hamlin, N., Holroyd, S., Jagels, K., Karlyshev, A.V., Leather, S., Moule, S., Oyston, P.C., Quail, M., Rutherford, K., Simmonds, M., Skelton, J., Stevens, K., Whitehead, S., and Barrell, B.G. 2001. Genome sequence of *Yersinia pestis*, the causative agent of plague. Nature 413: 523-527.

Schmiel, D.H., Wagar, E., Karamanou, L., Weeks, D., and Miller, V.L. 1998. Phospholipase A of *Yersinia enterocolitica* contributes to pathogenesis in a mouse model. Infect. Immun. 66: 3941-3951.

Titball, R.W. 1993. Bacterial phospholipases C. Microbiol. Rev. 57: 347-366.

Vuorio, R., Harkonen, T., Tolvannen, M., and Vaara, M. 1994. The novel hexapeptide motif found in the acetyltransferases LpxA and LpxD of lipid A biosynthesis is conserved in various bacteria. FEBS Letters 337: 289-292.

Vuorio, R., Hirvas, L., and Vaara, M. 1991. The Ssc protein of enteric bacteria has significant homology to the acyltransferases LpxA of lipid A biosynthesis, and to three acetyltransferases. FEBS Letters 292: 90-94.

Young, G.M., and Miller, V.L. 1997. Identification of novel chromosomal loci affecting *Yersinia enterocolitica* pathogenesis. Mol. Microbiol. 25: 319-328.

Zhang, L., Radziejewshka-Lebrecht, J., Krajewska-Pietrasik, D., Toivanen, P., and Skurnik, M. 1997. Molecular and chemical characterization of lipopolysaccharide O-antigen and its role in the virulence of *Yersinia enterocolitica* serotype O:8. Mol. Microbiol. 23: 63-76.

Chapter 9

Immune Responses to *Yersinia*

Erwin Bohn and Ingo B. Autenrieth

ABSTRACT

Most of the data available on immunity to enteric *Yersinia* come from experimental studies in rodents, particularly murine studies with *Yersinia enterocolitica*. *Yersinia* virulence factors (e.g., YadA, Yops) function to evade innate defense mechanisms such as complement lysis or phagocytosis, which allows a certain degree of bacterial replication and dissemination in host tissue. Enteric *Yersinia* transmigrate the intestinal epithelial barrier via M cells into the Peyer's patches. Infection of the Peyer's patches stimulates the recruitment of CD11b+ cells such as granulocytes and macrophages. These events and the secretion of cytokines such as IL-12, IFN-γ, IL-18 and TNF-α are crucial for clearance of the *Yersinia* infection. Besides these innate immune responses, adaptive immune defense mechanisms provided by T cells are necessary for clearance of *Yersinia* infections. Thus, because the virulence factors of enteric *Yersinia* partially subvert innate immune responses, extracellularly located enteric *Yersinia* need to be controlled by adaptive immune defense mechanisms that are typically required for resolution of intracellularly located pathogens. Production of immunglobulins by B cells directed against *Yersinia* virulence factors such as YadA, LcrV and, exclusively for *Y. pestis,* F1 provide a protective immunity upon a secondary challenge with *Yersinia*.

HOST DEFENSE MECHANISMS AGAINST EXTRACELLULAR AND INTRACELLULAR PATHOGENS

One important tool to study immune defense against pathogens in vivo is the availability of appropriate animal models. Because *Yersinia enterocolitica, Yersinia pseudotuberculosis,* and *Yersinia pestis* are pathogenic for man as well as for rodents, there are considerable data available on host responses against *Yersinia spp.*, particularly *Y. enterocolitica* and *Y. pseudotuberculosis,* from both clinical studies in humans and from experimental animal infection models. Most features of yersiniosis, such as enteritis, lymphadenopathy and reactive arthritis can be reproduced in various experimental animal model systems including mice, rats and rabbits (Heesemann et al., 1993). Therefore, it is conceivable that the features observed in these animal models are relevant to the pathogenesis of human yersiniosis, while keeping in mind that one cannot simply extrapolate data from animal models to human yersiniosis.

Immunity against the two enteric *Yersinia* species, Y. *enterocolitica* and Y. *pseudotuberculosis,* is based on innate and acquired host defense mechanisms. Most of the available data comes from murine studies with *Y. enterocolitica*; a few studies have also been done in rabbits and rats. Therefore, most of the data reviewed herein may hold for infections with *Y. enterocolitica* rather than for those caused by *Y. pseudotuberculosis*.

This is important to note, as *Y. enterocolitica* and *Y. pseudotuberculosis* may differ in the immune responses they induce. *Y. enterocolitica* appears to be located exclusively extracellularly in host tissues (Autenrieth and Firsching, 1996), but this is less clear and not yet completely established for *Y. pseudotuberculosis*. In fact, some of the data, in particular histopathology studies, even imply that *Y. pseudotuberculosis* might be located both extra- and intracellularly in host tissues. This conclusion is based on the fact that granulomatous lesions are typically induced by intracelluarly located pathogens whereas extracellularly located pathogens induce predominantly abscesses (Hahn and Kaufmann, 1981) and that *Y. pseudotuberculosis* is significantly more invasive in animal models than *Y. enterocolitica* (Cover, 1995). While *Y. enterocolitica* induces abcesses and few granuloma-like lesions in infected tissue, *Y. pseudotuberculosis* induces primarily granuloma-like lesions and atypical abcesses (Bradford et al., 1974; Hill, 1980; Hahn and Kaufmann, 1981; authors' unpublished observations). This is an important basic consideration, because the immune response to microorganisms varies depending on their location in host tissues (Autenrieth et al., 1999; Allen and Maizels, 1997). Compared to *Y. enterocolitica* and *Y. pseudotuberculosis*, relatively few data are available on cellular immune responses against *Y. pestis,* the causative agent of plague. In fact, the primary focus in this research field is the development of vaccines against plague, as reviewed below.

Previously, immunity to extracellular pathogens was thought to be due to polymorphonuclear leukocytes (PMN), the complement system, and antibodies; whereas intracellular pathogens evaded these host defense systems and were controlled by T cells and activated macrophages. It is now known that several extracellularly located pathogens such as *Yersinia* have evolved efficient mechanisms to evade host innate defense mechanisms, including phagocytosis by PMNs, macrophages and the complement system. In the Yersiniae these defense mechanisms are mediated by factors such as outer membrane proteins (e.g. YadA) and secreted antihost effector proteins (Yops). Consequently, mechanisms involving helper (CD4+) and cytotoxic (CD8+) T cell responses, particularly interferon-γ –producing CD4 T helper 1 (TH1) cells, probably in cooperation with activated macrophages, are required to control *Yersinia* infections (Autenrieth et al., 1992; Bohn and Autenrieth, 1996; Bohn et al., 1998b; Autenrieth et al., 1996b). Such host responses were considered to be classical host defense mechanisms against obligate or facultative intracellular pathogens such as *Leishmania*, *Toxoplasma*, *Listeria*, *Mycobacteria*, and *Salmonella* (Allen and Maizels, 1997). However, in addition to *Yersinia*, other extracellular pathogens such as *Bacteroides spp*. and *Candida albicans* subvert some of the innate defense mechanisms and are therefore also controlled by TH cell dependent host responses (Katz et al., 1990; Romani, 2000). Moreover, recent work has provided further evidence that the classical differentiation of host defense components with respect to their relevance for the control of intra- and extracelluar pathogens appears to be oversimplified. For example, antibodies might also play a role in neutralizing intracellular pathogens and mediate protection against the intracellular pathogen *Listeria monocytogenes* and viruses (Edelson and Unanue, 2001). Intra-epithelial cell neutralization of several viruses such as Sendai virus, influenza virus, rotavirus, and measles virus by IgA monoclonal antibodies (MAbs) has been demonstrated (Burns et al., 1996; Mazanec et al., 1995; Mazanec et al., 1992; Yan et al., 2002).

HOST DEFENSE AGAINST ENTERIC *YERSINIA* … INTESTINAL MUCOSA

MECHANISMS OF INFECTION AND INVASION OF THE INTESTINAL …

Enteric *Yersinia spp.* are enteroinvasive pathogens. Although they cannot i… intestinal epithelial cells from the apical side, and therefore cannot simpl… intestinal epithelial barrier (Schulte and Autenrieth, 1998a; Coconnier et al., 1… …y are taken up by M cells. M cells are specialized epithelial cells which occur wi… …n the follicle-associated epithelium of the Peyer's patches (Grutzkau et al., 1990; Autenrieth et al., 1996). M cells are specialized to take up particles from the intestinal lumen and transport these particles across the intestinal epithelium barrier. By this means, immune responses in the mucosa-associated lymphoid tissue are initiated, because the Peyer's patches are important components of the afferent limb of this system (Autenrieth and Schmidt, 2000).

For *Salmonella typhimurium,* which requires the invasion genes of the Salmonella pathogenicity island 1 (SPI-1) to penetrate M cells, alternative uptake mechanisms were suggested. Vazsquez-Torrez et al. (1999) reported that SPI-1 deficient *Salmonella* strains can also be transported directly from the gastrointestinal tract to the bloodstream by CD18-expressing phagocytes. More recent work demonstrated that dendritic cells may actively participate in microbial transport across mucosal surfaces. By opening the tight junctions between epithelial cells and sending dendrites through them, dendritic cells can directly sample lumenal bacteria via their dendrites (Rescigno et al., 2001). Whether this alternative uptake mechanism described for *Salmolnella* is involved in invasion, uptake, and dissemination of *Yersinia* is unknown. Moreover, it is not clear whether enterocytes, which are typical epithelial cells, contribute to mucosal anti-bacterial immune responses or not.

Recently, it was demonstrated in mice and with human intestinal epithelial M-like cell lines that M cells, in contrast to enterocytes, expose β1 integrins at their apical ("lumenal") surface, which may bind to the *Yersinia* invasin protein (Schulte et al., 2000; Clark et al., 1998). This interaction leads to the rapid transport of *Yersinia* across the epithelium (Figure 1). This strategy of entry is common in enteric pathogens including *Salmonella* and *Shigella spp.* (Siebers and Finlay, 1996; Jepson and Clark, 1998). However, the invasion strategy of pathogens via M cells is a paradox, because antigen sampling by M cells is an important event for initiating mucosal immune responses. In fact, the interaction of pathogens with M cells stimulates the afferent mucosal immune system which is represented by the Peyer's patches. Thus, entry of enteric *Yersinia spp.* by this route is an ambiguous process: on the one hand, *Yersinia* may use M cells as a structure of the mucosal immune system for invasion of host tissue, and most likely for subsequent dissemination. On the other hand, the mucosal immune system is stimulated upon entry of *Yersinia* at the very site where efficient immune responses are usually initiated. Consequently, host immune responses to *Yersinia* are triggered upon M cell invasion, which in most cases subsequently eliminate the invading pathogen. Interestingly, efficient colonization of Peyer's patches also occurs with *Yersinia* invasin protein-deficient mutants but with delayed kinetics, which suggests that there must be alternative uptake mechanims (Pepe and Miller, 1993). Eitel and Dersch (2002) recently demonstrated that under enviromental conditions in which the *inv* gene is not expressed, YadA of *Y. pseudotuberculosis* can

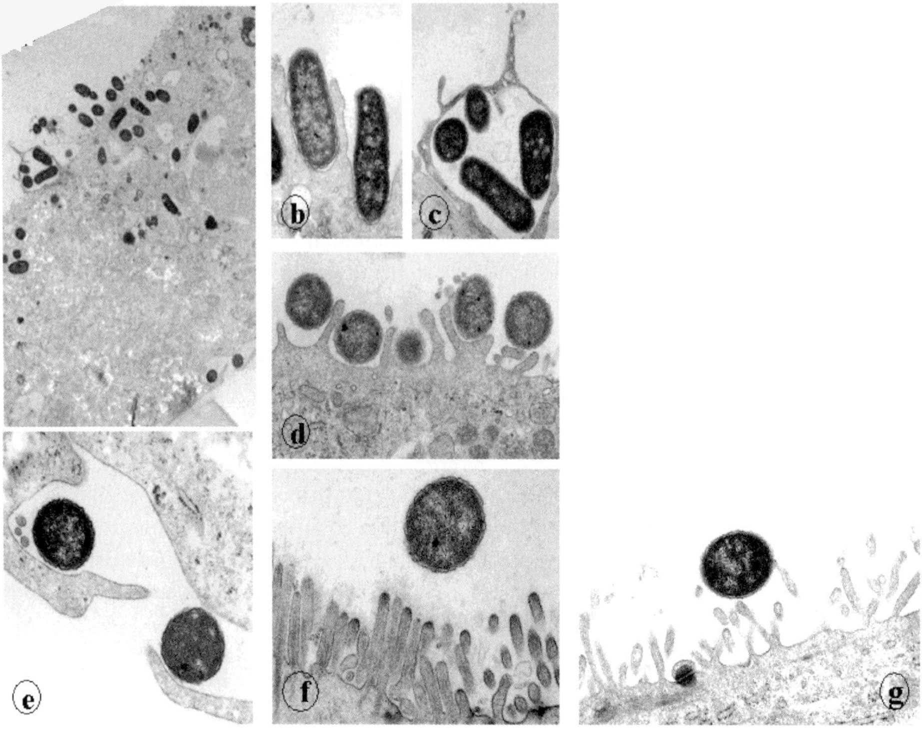

Figure 1. Transmission electron microscopy of Caco-2 cell monolayers cocultured with lymphocytes and infected with *Yersinia enterocolitica*. A. Transcytosis of *inv*-expressing yersiniae from apical to the basal compartment. After internalization, yersiniae are located within vacuoles and released at the basal membrane. Uptake by a zipper-like (B) and macropinocytosis-like (C) mechanism. D. Pedestal-like structures in cells exposed to yersiniae. E. Release of yersiniae at the basal side of Caco-2 cells. F. Enterocyte-like cells with a brush border are not in close contact with yersiniae. G. YadA-expressing *Y. enterocolitica* without close adherence to a M-like cell. Reprinted from Schulte et al., 2000.

mediate uptake of *Yersinia* by epithelial cells in a β1-integrin-dependent manner. Therefore, one might speculate that YadA could substitute for invasin in entry of M cells.

HOST RESPONSE TO INTESTINAL TRACT INVASION

One day after M cell invasion by *Y. enterocolitica*, microabcesses consisting of PMNs and extracellularly located *Yersinia* can be detected in Peyer's patch tissue (Hanski et al., 1989; Autenrieth et al., 1996b) (Figure 2). In *Y. pseudotuberculosis*, granulomatous lesions can be observed, and the location (intra- versus extracellular) of *Y. pseudotuberculosis* is less clear. Depending on the pathogenicity of the *Yersinia* strain, the size of the bacterial inoculum and host factors (e.g., genetic background, immune status) the inflammatory responses induced in Peyer's patches may lead to disruption of the epithelial barrier (Figure 3).

This reaction can be considered to be a net-effect resulting from two contrary events: *Yersinia spp.* obviously trigger the recruitment of inflammatory cells such as PMN, which are normally absent or present in very low numbers in Peyer's patches. This could be a

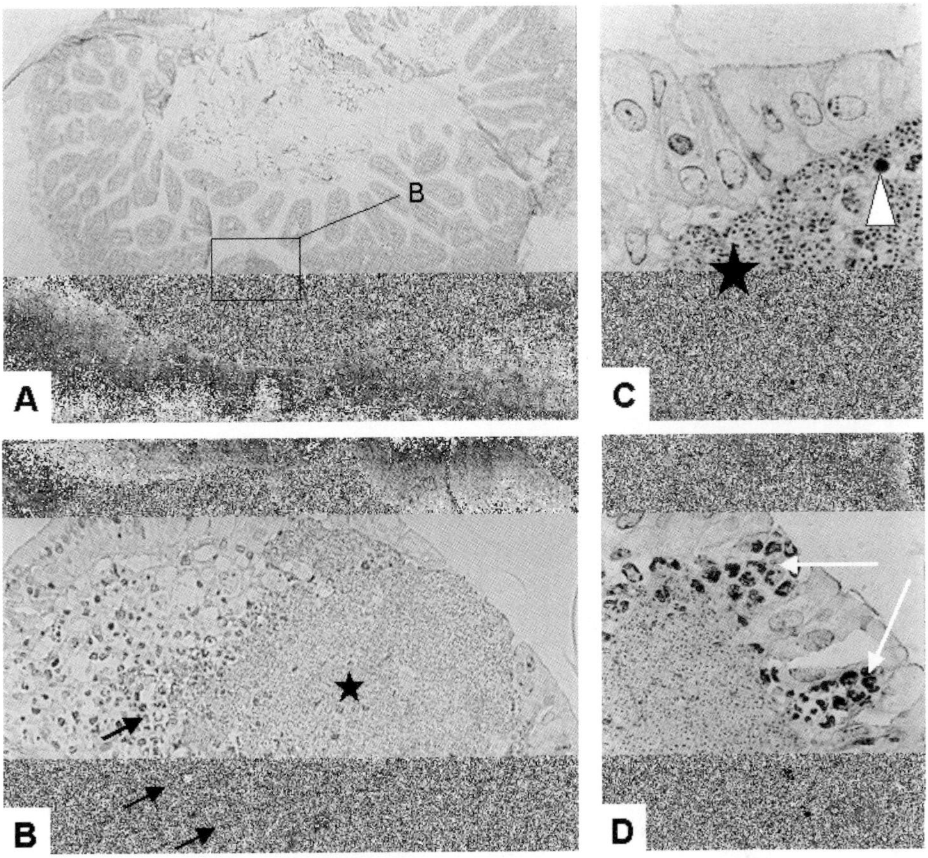

Figure 2. Microabcess in a Peyer's patch of a C57BL/6 mouse one day after oral *Yersinia enterocolitica* infection A. and B. Yersiniae (asterisk, dark staining) are extracellularly located beneath the dome, polymorphonuclear leukocytes (PMN; arrows) are recruited to the infection site. C. Yersiniae in close contact to the basal side of enterocytes. Only very few apoptotic PMNs can be observed (arrow head) D. PMN disrupt and migrate through the mucosal barrier (arrows).

result of the function of e.g., *Yersinia* invasin protein. Recent work demonstrated that invasin, by binding to host cell β1 integrins, can trigger activation of proteins of the Rel family, e.g., nuclear factor kappa B (NF-κB) in epithelial cells and macrophages (Figures 4 and 5), which leads to expression of various genes such as interleukin (IL)-8, granulocyte macrophage-colony stimulating factor (GM-CSF), and monocyte chemoattractant protein (MCP)-1 (Kampik et al., 2000; Schulte et al., 2000; Schulte et al., 1998b) (Figure 6). These cytokines, particularly the chemokine IL-8, are proinflammatory cytokines and may recruit PMN. Whether signaling via toll-like receptor (TLR)-4 mediated by LPS or via TLR-2 mediated by LcrV of *Yersinia spp.* is important for this proinflammatory host response is unclear. Besides Inv and LPS, other factors such as YadA (Müller et al., manuscript in preparation) and YopB (Viboud et al., 2003) have the ability to induce a proinflammatory response. Like Inv, YadA binding to β1 integrins triggers IL-8 production in epithelial cells (authors' unpublished results). On the other hand, translocation of Yop effectors such as YopP/J via the type III secretion apparatus may induce apoptosis and in

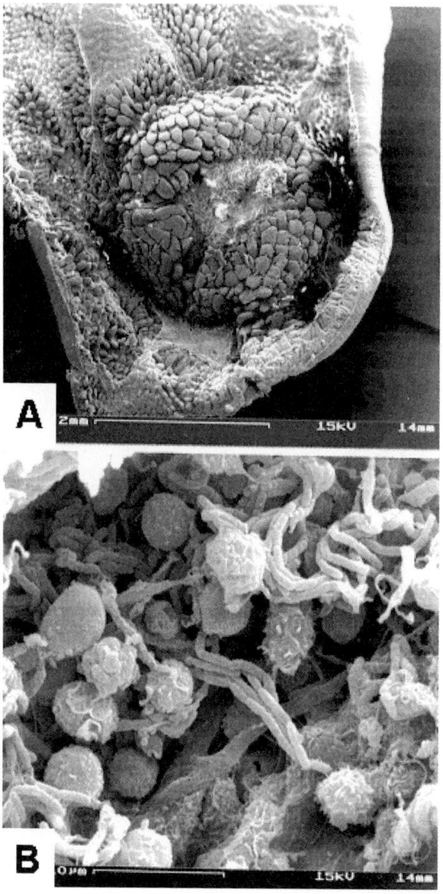

Figure 3. Scanning electron microscopy of Peyer's patch 7 days after *Y. enterocolitica* infection. A. Peyer's patch with enlarged adjacent vili destroyed dome. B. Chain-forming yersiniae and phagocytes on the surface of the destroyed follicle-associated epithelium. Reprinted from Autenrieth and Firsching, 1996.

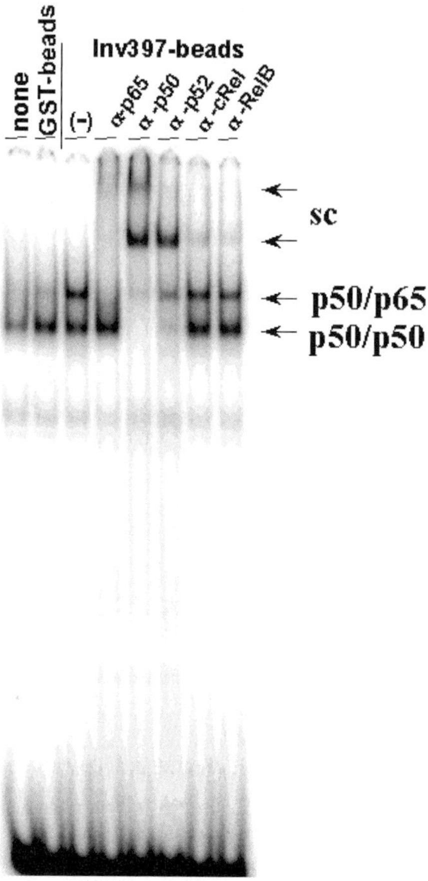

Figure 4. Electromobility shift assay with supershift analysis. Determination of NF-κB activation in nuclear extracts from HeLa epithelial cells exposed to beads coated with invasin protein (Inv397). Nuclear extracts were incubated with a NF-κB consensus probe and anti-Rel protein antibodies. Binding activity due to nuclear translocation of NF-κB is indicated by arrows. Reprinted from Schulte et al., 2000.

concert with other Yops may suppress expression of proinflammatory genes by inhibiting NF-κB activation in macrophages (Monack et al., 1997; Palmer et al., 1998; Schesser et al., 1998; Ruckdeschel et al., 1998) (Figure 7).

Recent work in an experimental mouse infection model provides more direct evidence that YopP/J may account for such an apoptotic effect in mesenteric lymph nodes and spleen (Monack et al., 1998). In general, plasmid-encoded virulence factors of *Y. enterocolitica* profoundly repress gene expression in macrophages as well as epithelial cells, at least in vitro (Sauvonnet et al., 2002; Bohn et al., 2004).

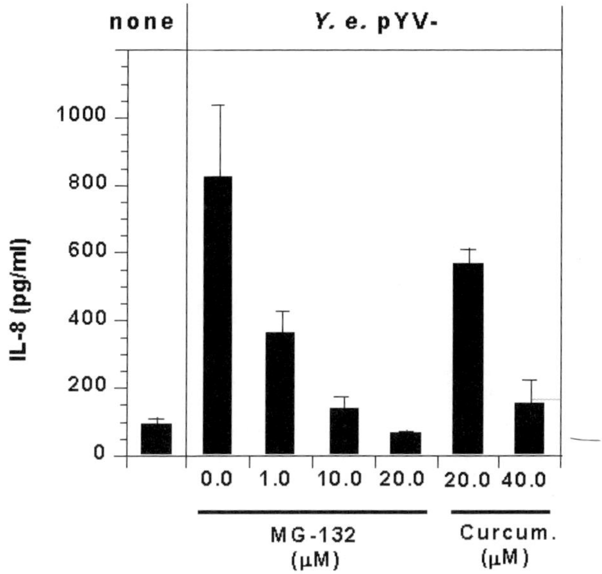

Figure 5. Inhibition of *Yersinia*-induced IL-8 production in HeLa epithelial cells by NF-κB inhibitors MG-132 and curcumin. Reprinted from Schulte et al., 2000.

Induction of apoptosis may also be important for a possible role of IL-1 in *Yersinia* infections. IL-1 production can be triggered by Inv-mediated NF-κB activation (Kampik et al., 2000). However, in contrast to IL-8, GM-CSF, or MCP-1, which are all Inv-induced and secreted by epithelial cells, IL-1 is not released by the host cells. In *Shigella* infection, however, it was shown that IL-1 may be passively released by macrophages upon induction of apoptosis by *Shigella* pathogenicity factors (Sansonetti et al., 1995). Thus,

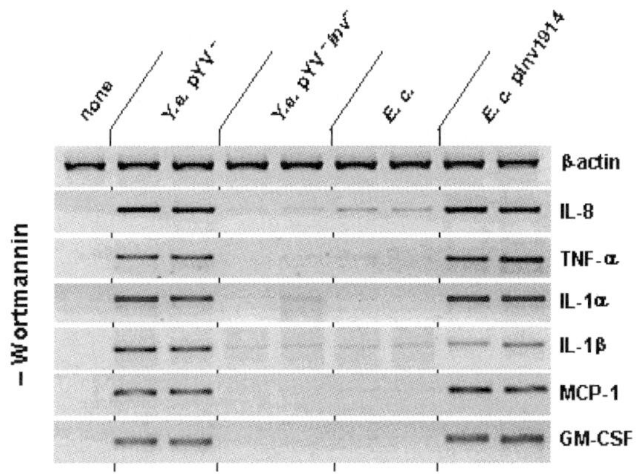

Figure 6. Cytokine (IL-8, TNF-α, IL-1α, IL-1β, MCP-1, GM-CSF) mRNA production by HeLa cells after infection with invasin-expressing *Y. enterocolitica* (Y.e. pYV-), invasin-deficient *Yersinia* mutant (Y.e. pYV-, inv-), *Escherichia coli* (E.c.), and *E. coli* (E.c. pInv1914) expressing the *Y. enterocolitica* invasin gene. Reprinted from Kampik et al., 2000.

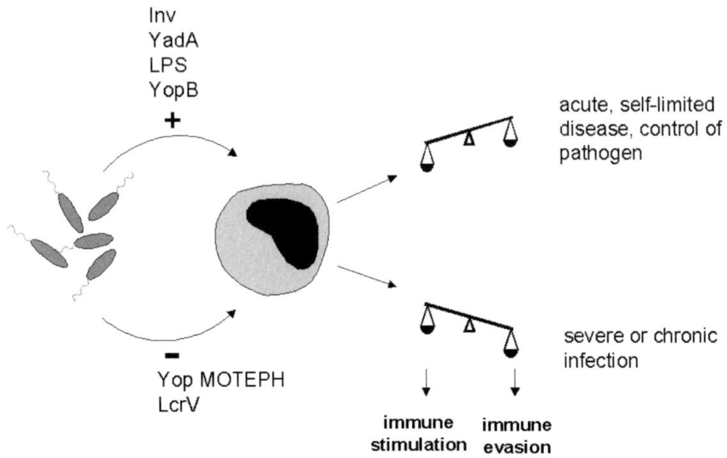

Figure 7. Consequences of pro- (Inv, YadA, LPS, YopB) and anti-inflammatory (Yop MOTEPH and LcrV) immunomodulation by *Yersinia* pathogenicity factors.

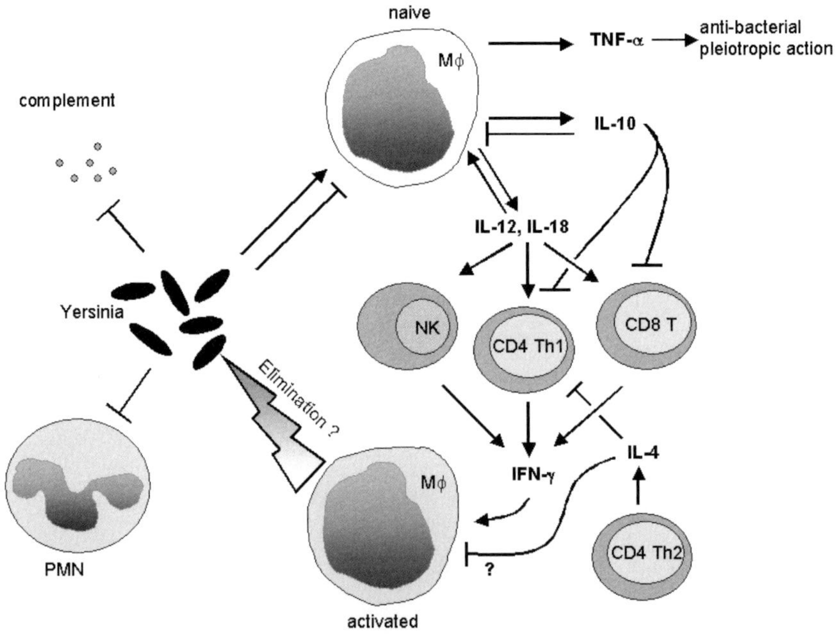

Figure 8. Simplified scheme of protective immune responses against *Yersinia spp.* according to the currently published literature on *Yersinia* immunology. *Yersinia spp.* evade innate immune mechanisms mediated by PMN or complement by e.g., YadA, SodA, LcrV, Yops at the early phase of infection. Nevertheless, data suggest that activation of macrophages occurs. Macrophages produce the proinflammatory cytokines TNF-α, IL-12 and IL-18, and eventually IL-10. The possible anti-inflammatory cytokine IL-10 may play an ambiguous role in *Yersinia* infections. Subsequently (day 3 to 7 post infection) NK cells as well as both CD4 and CD8 T cells are a main source of INF-γ which contributes to resolution of the infection, e.g. by activation of macrophages. Although it is not yet demonstrated, T-cell-activated macrophages are likely the most important effector mechanism that clear the infection. (→ stimulation, ⊣ inhibition).

pathogen-triggered apoptosis is a second process which leads to host cell disruption and IL-1 release, which might augment the host inflammatory response. It is conceivable that such a scenario might also occur in *Yersinia* infection. Furthermore, there is some evidence from in vivo studies that IL-1α induction mediated by chromosomally encoded genes of *Yersinia* which are under the control of the transcriptional regulator RovA contributes to pathological inflammation in Peyer's patches during *Yersinia* infection (Dube et al., 2001).

Taken together, present data suggest that M cells and epithelial cells are among the first cells encountered by enteric yersiniae. Epithelial cells can produce inflammatory cytokines by which they signal the presence of pathogenic microorganisms to the host immune system. Therefore, epithelial cells have been called "watchdogs for the natural immune system" (Eckmann et al., 1995). At present, it is not known whether M cells are also capable of producing cytokines upon interaction with microorganisms. As M cells expose β1 integrins at the luminal surface which are engaged by enteric Yersiniae, it is likely that M cell interaction with *Yersinia* results in cytokine production in addition to bacterial translocation. Alternatively, *Yersinia* might trigger cytokine responses by binding to β1 integrins exposed at the basal surface of enterocytes. Macrophages and lymphocytes are located in the "pocket" adjacent to the basal side of M cells, where *Yersinia* may also elicit cytokine production causing proinflammatory cell reactions.

INNATE IMMUNE RESPONSE

After invasion via M cells, *Yersinia* colonize Peyer's patches. Several virulence factors have been tested for their ability to promote bacterial colonization of Peyer's patches. Among the *Yersinia* mutants tested, YadA-deficient mutants were significantly attenuated in their ability to colonize Peyer's patches (di Genaro et al., 2003; Roggenkamp et al., 1995). Furthermore, mutation of *rovA*, a transcriptional regulator of several *Yersinia* genes such as *inv,* partially attenuates colonization of Peyer's patches by *Yersinia* (Revell and Miller, 2000). Likewise, *yopE* (Holmström et al., 1995; Iriarte and Cornelis, 1998) and *yopP* (Monack et al., 1998) mutants are less able to colonize Peyer's patches.

Histological and immunohistological analyses suggest that innate host defense mechanisms including PMN and macrophages are involved in control of *Yersinia* in Peyer's patches. In one experimental approach, the administration of neutralizing antibodies against adhesion molecules or cytokines into mice prior to *Yersinia* infection showed that some of these components are essential for control of *Yersinia.* Blocking the adhesion molecules VLA-4 (an α4β1 integrin expressed on T cells, B cells, monocytes, and involved in, e.g., cell trafficking to sites of inflammation by interaction with VCAM-1 on endothelial cells), Mac-1 (the complement receptor 3 CD11b, an αmβ2 integrin expressed on phagocytes and involved in binding and phagocytosis of pathogens and cell trafficking by binding to ICAM-1 expressed on endothelial cells); or the cytokines TNF-α, IFN-γ and IL-12 revealed that all these components play a protective role in the Peyer's patches (Autenrieth et al., 1996b; Bohn and Autenrieth, 1996). Thus, blocking of VLA-4 or Mac-1 reduced phagocytosis of yersiniae and resulted in increased bacterial numbers in the tissue. Administration of anti-cytokine antibodies prevented clearance of *Y. enterocolitica* from Peyer's patches and led to increased dissemination of bacteria to lymph nodes, spleen, liver, lung and death. However, it is not yet established which cell types in the Peyer's patches produce these cytokines in the early phase of the infection.

A role for complement in yersiniosis has been demonstrated by in vitro experiments. In particular, YadA is involved in mediating resistance to lysis by complement and defensins (Balligand et al., 1985; Pilz et al., 1992; Visser et al., 1996). Because YadA is essential for survival of *Yersinia* in host tissue and therefore is expressed in vivo (Autenrieth and Firsching, 1996a), it is not clear whether the complement system, which obviously is subverted by *Yersinia* virulence factors, is actually involved in host defense. However, work in an experimental rat infection model suggests that complement components are important for the development of reactive arthritis triggered by *Yersinia spp.* (Gaede et al., 1995). *Yersinia*-triggered reactive arthritis usually occurs after resolution of the infection and is considered to be the result of an immunopathological host reaction. Interestingly, in the Lewis rat model of reactive arthritis, decomplementation by the administration of cobra venom factor suppressed *Yersinia*-induced reactive arthritis (Gaede et al., 1995). Whether this is due to reduction of the inflammatory reaction mediated by immune complexes composed of killed or degraded yersiniae that precipitate in synovial tissue remains to be elucidated.

MUCOSAL IMMUNITY

Secretory IgA (sIgA) antibodies are secreted onto mucosal surfaces and may prevent invasion of pathogens. In fact, in a hybridoma tumor model, the transfer of hybridoma cells secreting monoclonal anti-*Salmonella* LPS antibodies prevented invasive salmonellosis in mice (Michetti et al., 1992). In patients with yersiniosis, IgA antibodies directed against various antigens including the Yop proteins can usually be detected in serum. In fact, anti *Yersinia* Yop IgA production is of considerable diagnostic value. The immunobiological role of sIgA antibodies in host defense against enteric *Yersinia spp.*, however, is not yet established. Oral inoculation of recombinant *Y. enterocolitica* expressing a YopH-CtxB hybrid protein induced sIgA antibodies against the cholera toxin B (CtxB) subunit in the intestinal and respiratory tract, and conferred protection in mice against a cholera toxin challenge (Van Damme et al., 1992). Moreover, preliminary work has shown that sIgA is produced upon *Y. enterocolitica* O:8 wild type infection in mice (Bielfeldt et al., unpublished). However, it is not clear whether *Yersinia*-specific sIgA antibodies prevent invasion of the intestine by yersiniae or otherwise contribute to the control or modulation of infection by yersiniae. It is also conceivable that the occurrence of a certain level of IgA antibodies in *Yersinia* infections reflects an inappropriate T cell response such as the development of strong Th2 responses, resulting in chronic yersiniosis and possibly reactive arthritis rather than in control of the pathogen.

After replication in Peyer's patches, enteric *Yersinia spp.* disseminate via the lymphatics and possibly via the blood stream to the mesenteric lymph nodes, spleen, liver, lungs and peripheral lymph nodes. The mechanism for the tropism of *Yersinia* for these organs is unclear, but may reflect specific adherence to or invasion of certain types of host cells or tissues. Preliminary data suggest that phagocytes expressing Mac-1 may be involved in the dissemination process of yersiniae (Autenrieh et al., 1996). In fact, administration of anti-Mac-1 antibodies resulted in increased bacterial counts in Peyer's patches but decreased bacterial counts in the spleen. Likewise, in *Salmonella* infection, $CD18^+$ host cells, possibly dendritic cells (DC), have been demonstrated to be involved in intestinal translocation and dissemination as well as in induction of immune responses (Vazquez-Torres et al., 1999). Recent work demonstrated that *Y. enterocolitica* can be internalized by dendritic cells and affect their maturation, but might not induce necrosis or apoptosis in these cells (Schoppet et al., 2000). In fact, upon interaction with DC, *Yersinia* modulates their function by upregulating CD83 and CD86 and transiently downregulating MHC class II molecule expression, resulting in decreased ability of the DCs to promote T cell proliferation (Schoppet et al., 2000). Preliminary data from our laboratory indicate that secreted proteins of *Y. enterocolitica* may interfere with the processing and presentation of antigens to MHC class II molecules. As CD4 T cell responses are crucial for resolution of *Yersinia* infection, modulation of DC functions such as antigen presentation might be a strategy to evade acquired immune responses by *Y. enterocolitica*. The role of DC in yersiniosis should be further defined, as these cells are important sensors of infection and central in the development of immune responses against pathogens (Reis e Sousa, 2001).

ADAPTIVE IMMUNITY TO *YERSINIA SPP.*

CYTOKINE INDUCTION AND CYTOKINE NETWORK

The adaptive immune response plays a crucial role in the clearance of a *Yersinia* infection. It has been clearly demonstrated that *Yersinia* infection leads to strong T cell responses, including activation and proliferation of CD4 and CD8 T cells, and that these T cells are involved in control of *Yersinia* (Autenrieth et al., 1992a; Autenrieth et al., 1993; Bohn et al., 1996; 1998b; Noll and Autenrieth, 1996; Falgarone et al., 1999a; Falgarone et al., 1999b). In fact, mice deficient for T cells are unable to control the pathogen, and therefore develop chronic progressive and fatal infection (Autenrieth et al., 1993). Adoptive transfer of *Yersinia*-specific CD4+ or CD8+ T cells consistently mediates resistance to a normally lethal challenge of *Yersinia* (Autenrieth et al., 1992a). As the protective CD4 or CD8 T cells produce cytokines such as IFN-γ and IL-2, but not IL-4 or IL-10, it can be concluded that TH1 or IFN-γ producing cytotoxic T cells are protective in yersiniosis (Figure 8).

TH1 cells produce predominantly IFN-γ, and it is established that IFN-γ can activate macrophages which in turn might be able to kill a pathogen. T cell-activated macrophages are likely to be the most important effector cells in yersiniosis, although no data are available that directly demonstrates this. Interestingly, *Yersinia* may induce apoptosis in macrophages via translocation of YopP/J (Monack et al., 1997; Palmer et al., 1998; Schesser et al., 1998; Ruckdeschel et al., 1998). Therefore, we can speculate that enteric *Yersinia spp.* may not only evade innate immunity including phagocytosis by PMN and macrophages, but also, at least to some extent, may evade acquired immunity in which macrophages are important mediators of effector functions and antigen presentation. By translocating YopH into the cytosol of T or B cells, *Yersinia* may also impair calcium flux in these cells and reduce their ability to be activated through their antigen receptors (Yao et al., 1999). Thereby *Yersinia* may impair the development of acquired immunity. Nevertheless, as enteric yersiniosis is usually a self-limited disease, these immune evasion strategies of *Yersinia* are only partially successsful, and in the end *Yersinia* are obviously not able to efficiently evade acquired immunity including phagocytosis and killing by T cell-activated macrophages.

The role of cytokines in yersiniosis has been extensively studied in mouse strains that are relatively susceptible (e.g., BALB/c) or resistant (e.g., C57BL/6) to *Yersinia* infection. One reason for this differential susceptibility of mice is their different ability to mount TH1 responses and produce IFNγ upon *Yersinia* infection (Autenrieth et al., 1994; Bohn et al., 1994). *Yersinia*-resistant C57BL/6 mice can be rendered *Yersinia*-susceptible by neutralizing IFN-γ in vivo with monoclonal antibodies (Autenrieth et al., 1994).

Conversely, *Yersinia*-susceptible BALB/c mice can be rendered resistant by treatment with IFN-γ. In keeping with these results, it was found that administration of neutralizing anti-IL-4 antibodies also rendered BALB/c mice resistant to *Yersinia* infection (Autenrieth et al., 1994). From these data it can be concluded that IFN-γ is a central protective cytokine in *Yersinia* infection (Figure 8).

More recent work with cytokine- or cytokine receptor-decificent mice clearly demonstrated that the cytokines TNF-α, IL-12, IL-18 and IFN-γ, and interferon consensus sequence binding protein, a transcription factor, are all essential for control of *Yersinia* infection (Bohn et al., 1996, 1998a, 1998b; Hein et al., 2000). Although recent work suggested that in certain experimental conditions involving i.v. inoculation of very high numbers of *Y. enterocolitica*, tumor necrosis factor receptor (TNFR) p55 chain-mediated mechanisms might be eventually detrimental to the host (Zhao et al., 2000), TNFR-dependent mechanisms in general are crucial for overcoming yersiniosis (Autenrieth and Heesemann, 1992b; Autenrieth and Firsching, 1996a; Bohn et al., 1998a). Oral infection of cytokine deficient mice reveals that certain virulence factors enable yersiniae to compete with distinct cytokine-dependent host defense mechanisms. Thus, a YadA deficient *Y. enterocolitica* mutant is not able to colonize intestinal tissues of C57BL/6 mice and is attenuated for systemic infection (Di Genaro et al., 2003) However, in TNFRp55-deficient mice YadA-deficient *Yersinia* are still able to induce a systemic infection (Di Genaro et al., 2003). These data demonstrate that in the absence of TNFR-mediated protection an immune evasion mechanism distinct from those which require YadA is sufficient to overcome the immune defense. In contrast, *Y. enterocolitica* lacking the chaperone SycH and therefore no longer able to translocate YopH into host cells can still colonize the Peyer's patches, but are attenuated in systemic infection even in the absence of IL-12, IL-18 or TNFR (Di Genaro et al., 2002). Thus, in this experimental model the lack of SycH reduced virulence of yersiniae in such a way that IL-12, IL-18 or TNFR were dispensable. In general, these examples demonstrate that the complexity of immune defense mechanisms required against a pathogen is dependent on the armament of the pathogen. This principle is evident even when comparing infections caused by closely related bacteria such as *Y. enterocolitica* O:8, O:9, O:3 or *Y. pseudotuberculosis*. Amino acid sequence differences in virulence factors, e.g. YopP, Inv, and the presence or absence of certain genes influences the net effect of the armament of a strain and may therefore have an impact on the quality of immune defense needed to protect against a certain strain.

IL-10, which is secreted by macrophages, B cells and several subtypes of T cells such as TH2 or regulatory T cells, has an antiprotective effect in yersiniosis. Thus, infection of IL-10 deficient mice with *Y. enterocolitica* resulted in a lower bacterial load in spleen and liver as well as an increased survival rate compared to wild type mice (Sing et al., 2002a). More recent data provided evidence that the *Yersinia* virulence factor LcrV seems to be actively involved in upregulation of IL-10 secretion in a TLR-2- and CD14-dependent fashion, which in turn suppresses TNF-α secretion in vitro (Sing et al., 2002b). Unfortunately, this work did not address whether LcrV-signaling through TLR-2 may cause a decrease of IFN-γ levels in *Yersinia* infections. Nevertheless, these observations are in line with previous findings that LcrV injection in vivo suppresses TNF-α and IFN-γ secretion upon infection with LcrV deficient *Y. pestis* (Nakajima et al., 1995), and indicate that suppression of these cytokines might be a general immune evasion mechanism by all pathogenic *Yersinia* species. Together, the data imply that LcrV apparently represents a strategy to suppress T cell-mediated immunity.

Table 1. Protective immune components against *Yersinia* ssp.

Protective components	Antigen	Pathogen	Reference
CD4 T cells	Hsp60	*Y. enterocolitica*	Noll et al., 1996a,b
CD8 T cells	unknown	*Y. enterocolitica*	Autenrieth et al., 1992
IgG antibodies	YadA	*Y. enterocolitica*	Vogel et al., 1993
IgG antibodies	LcrV	*Y. enterocolitica* *Y. pseudotuberculosis* *Y. pestis*	Anderson et al., 1996, Nakajima et al., 1995 Motin et al., 1994, Roggenkamp et al., 1997
IgG antibodies	F1 antigen	*Y. pestis*	Anderson et al., 1997
IgG antibodies	YopD	*Y. pestis*	Andrews et al., 1999

T CELL-MEDIATED IMMUNE RESPONSES

Epitope mapping of murine protective *Yersinia*-reactive T cell clones showed that the heat shock protein HSP60 of *Y. enterocolitica* is an immunodominant and protective antigen for CD4 T cells and is presented by MHC class II molecules (Noll et al., 1994; Noll and Autenrieth, 1996). Table 1 lists the protective T cell and antibody components against infection and the corresponding *Yersinia* antigens. Interestingly, the epitopes recognized by the T cells do not share significant homology to the corresponding mammalian HSP60 sequences. Recently, studies in humans confirmed these observations and demonstrated that *Yersinia* HSP60 is also an immunodominant antigen for CD4 and CD8 T cells in reactive arthritis in humans (Mertz et al., 2000). A 12-mer core epitope (aa 322-333) of the aa 319-342 peptide was recognized by T cell clones from one patient in the context of different HLA-DR alleles, suggesting that these findings might be generally relevant to human infection. Interestingly, although IFN-γ was the most abundant cytokine secreted by these human T cell clones, some clones also secreted considerable amounts of IL-10. As IL-10 is an inhibitory cytokine, *Yersinia*-HSP60-triggered IL-10 secretion might promote bacterial persistence in chronic infections. As in mice, none of the human T cell clones reactive with *Yersinia* HSP60 cross-reacted with human HSP60. These results argue against the hypothesis that T cell responses against shared epitopes of HSPs are involved in the pathogenesis of reactive arthritis after *Yersinia* infection. Whether *Yersinia* heat shock proteins might be released and stimulate innate immune responses by engaging Toll-like receptors (Wagner, 2001) is unclear. In summary, we can conclude that CD4 T cells are important and essential for mediating the cellular host mechanisms required for control of enteric *Yersinia spp*. The T cell responses against *Y. pestis* have not yet been studied in detail. The fact that administration of TNF-α and IFN-γ reduced bacterial replication in vivo and protected against a lethal challenge with *Y. pestis* (Nakajima et al., 1993) suggest that T cells and macrophages play a central protective role in plague.

CD8 T cell responses have also been observed in murine, rat and human yersiniosis. The adoptive transfer of *Y. enterocolitica*-reactive CD8 T cell clones can mediate immunity against a lethal challenge with *Y. enterocolitica* (Autenrieth et al., 1992a). However, the actual extent to which CD8 T cells are involved in control of *Yersinia* infection remains to be further defined. In vivo depletion studies suggest only a minor role of CD8 T cells as compared with CD4 T cells. However, in susceptible hosts such as BALB/c mice, IL-12-activated CD8 T cells appear to be essential for pathogen control (Bohn et al., 1998). In a *Yersinia* rat infection model, Lewis rats mounted a *Yersinia*-specific RT-1-restricted CTL response (Falgarone et al., 1999a). In this study, the *Yersinia* invasin protein was required to mediate binding to blast target cells. In turn, YopE appeared to be translocated and presented via MHC class I molecules to CD8 T cells. Both YopD and YopB were necessary for this process. The fact that Yop proteins are translocated into the cytosol of host cells suggests a possible pathway of MHC class I antigen presentation for antigens of an extracellular bacterial pathogen. On the other hand, however, it is surprising that host cells which are affected by several Yop proteins may still be capable of antigen processing and presenting and of T cell stimulation. This is an issue that requires further molecular studies.

Although acquired immune mechanisms usually succeed in controlling *Yersinia* infections, enteric *Yersinia spp.* might interfere with this system and at least partially evade acquired immune responses. For example, by translocating YopH into the cytosol of T or B cells, *Yersinia* may impair calcium fluxes and their ability to be activated through antigen receptors (Yao et al., 1999). This may subsequently inhibit cytokine production by T cells and the surface expression of B7.2, an important costimulatory molecule of B cells. In consequence, these inhibitory processes could significantly disturb cell-mediated immunity against the pathogen. Since *Yersinia* may bind to integrins of B and T cells via invasin protein (Ennis et al., 1993), it is conceivable that lymphocytes, and by extension the acquired immune response, are actually a target of the pathogenicity armament of yersiniae in vivo. Interestingly, it was recently reported that *Yersinia* can translocate heterologous proteins such MAGE-A1 or *L. monocytogenes* p60, fused to amino acids 1-130 of YopE, into antigen presenting cells including dendritic cells, which results in the presentation of antigenic peptides via MHC class I molecules (Chaux et al., 1999; Duffour et al., 1999; Ruessman et al., 2000). These findings suggest that attenuated *Yersinia* strains might be suitable as a live carrier to induce MHC class I-restricted CTL responses against heterologous antigens.

HUMORAL IMMUNE RESPONSES

In contrast to T cell responses, protective antibodies recognize the outer membrane protein YadA of *Y. enterocolitica* (Vogel et al., 1993) (Table 1). Anti-LPS antibodies do not appear to mediate significant protection. This might be explained by the fact that invasin or YadA, the two major outer membrane proteins of *Yersinia* might mask LPS structures. Studies in mice have consistently demonstrated that immunization with avirulent or virulent *Yersinia* mutant strains can confer immunity to challenge with highly pathogenic yersiniae, and that antibodies against YadA are crucial in this process. Immunity conferred by YadA-specific antibodies appears to be serotype-specific, as antibodies against YadA from *Y. enterocolitica* serotype O:8 strains which express the virulence plasmid including YadA from serotype O:8 do not mediate protection against challenge with a *Y. enterocolitica* O:8 strain in which the plasmid of the O:8 strain was replaced by a virulence plasmid of an O:9 strain expressing YadA of serotype O:9 (Vogel et al., 1993). Immunization of mice with a *S. typhimurium* strain expressing the outer membrane invasin protein from *Yersinia* induced anti-invasin antibodies in serum and intestinal secretions, giving rise to inhibition of intestinal translocation, but failed to prevent *Yersinia* dissemination from the gut lymphoid tissue (Simonet et al., 1994). In *Y. pestis* infections, antibodies against LcrV, F1 antigen and YopD have been demonstrated to be protective (Table 1). Together, these data suggest that different antigens are possibly involved in protective cellular and humoral immune responses.

THE SUPERANTIGEN OF *Y. PSEUDOTUBERCULOSIS*

Superantigens are microbial compounds that bind to conserved parts of MHC class II molecules and the Vβ region of T cell receptors, thereby triggering the activation of certain subtypes of T cells and antigen-presenting cells. A mitogenic activity was found in *Y. pseudotuberculosis,* culture supernatants, which exhibited superantigenic activity (Abe et al., 1993). The superantigen-encoding gene, *ypm*, was subsequently deleted in *Yersinia* strains and it was observed that these *Y. pseudotuberculosis ypm* mutants were not affected in their ability to colonize the spleen, liver or lungs (Carnoy et al., 2000). However, the virulence (time to death) of a *ypmA* mutant strain was significantly reduced, suggesting that YpmA contributes to the virulence of *Y. pseudotuberculosis* in systemic, but not intestinal, infections. Unfortunately, it is not yet clear whether the effect of YpmA on acquired immunity is responsible for the increased virulence. In contrast to this observation, in vitro studies provided evidence that YpmA may reduce active ion transport and permeability in human T84 epithelial cell monolayers (Donnelly et al., 1999). The significance of *Yersinia* superantigen for modulation of immunity is therefore not yet clear. For more information on the *Y. pseudotuberculosis* superantigen, see Chapter 10.

IMMUNOLOGICAL MECHANISMS INVOLVED IN *YERSINIA*-TRIGGERED REACTIVE ARTHRITIS

Reactive arthritis is a "sterile" joint inflammation that can follow infection with various microbial pathogens including enteric *Yersinia spp*. Although it is established that patients with HLA B27 develop reactive arthritis at a 20-fold higher frequency than patients with other HLA alleles, the pathogenesis of reactive arthritis is still unknown. It is unclear how HLA B27 is linked to reactive arthritis. Several possible mechanisms, including molecular mimicry or whether HLA B27 is related to susceptibility for *Yersinia* infections have been discussed. Interestingly, HLA B27 transgenic mice exhibit higher susceptibility for *Yersinia* infections than control mice (Nickerson et al., 1990).

More recently, it was reported that U937 cell transfectants expressing various HLA B27 constructs or other HLA alleles were unaltered in their ability to internalize *Salmonella* or *Yersinia* (Virtala et al., 1997; Ortiz-Alvarez et al., 1998), but were impaired in their ability to eliminate *Salmonella enteritidis* (Virtala et al., 1997), suggesting that HLA B27 monocytes might contribute to persistence of microorganisms that trigger reactive arthritis. Unfortunately, the mechanisms that account for this impaired intracellular killing have not been elucidated. In contrast to that study, no effect of endogenous HLA B27 on the survival of *Yersinia* or *Salmonella* was found in primary fibroblast cells (Huppertz and Heesemann, 1996a; 1996b).

Persistence of the pathogens is considered to be an important factor in reactive arthritis, and may depend on the genetic background of the host, the type of immune response of the host, as well as on the arthritogenic potential of a given *Yersinia* strain. Thus, both a certain type of immunodeficiency that allows the persistence of the pathogen in host tissue, particularly at mucosal sites, and the generation of a strong immunopathological inflammatory response are crucial in this process.

Immune complexes have been demonstrated in the synovial fluid of patients with reactive arthritis (Lahesmaa-Rantala et al., 1987). Whether microbial antigens reach the joints as free antigens or as part of immune complexes is unknown. Previous studies failed to detect *Yersinia* DNA or viable yersiniae in the joints of these patients. A recent study, however, demonstrated the presence of *Y. enterocolitica* 16S rRNA in the synovial fluid of a patient (Gaston et al., 1999). These observations are in keeping with results from an experimental rat model of *Yersinia*-triggered reactive arthritis, which indicated that complement as well as the pathogenicity (arthritogenicity) of *Yersinia spp.* were important cofactors of reactive arthritis. Interestingly, some rat strains (e.g., Lewis rats) develop *Yersinia*-triggered reactive arthrtitis while others (e.g., Fischer rat) do not (Gaede et al., 1992). Preliminary data from these models indicate that during the very early phase of the infection, *Yersinia* bacteria can be detected and even cultured from the joints of both rat strains, suggesting that during the early phase of infection the bacteria actually reach the joints (C. Kaiser, thesis, University of Würzburg, 1997). In both rat strains, the bacteria initially present in the joints were rapidly eliminated. However, in Lewis rats the bacterial numbers in the synovial tissue were higher than in Fischer rats. After two weeks, Lewis rats developed reactive arthritis; however, no viable bacteria could be grown from the joints at this time point.

As HLA B27 is a MHC class I molecule, it is conceivable that MHC class I-restricted immune responses may play a role in reactive arthritis. Patients with *Yersinia*-triggered reactive arthritis were found to recognize several antigens of *Yersinia* at the humoral or cellular level: the 19-kDa β-urease subunit protein of *Y. enterocolitica* O:3 is recognized by IgG and IgA antibodies as well as by synovial fluid mononuclear cells, and correlates with a synovial cellular immune response (Appel et al., 1999). Although these studies did not elucidate the actual pathogenic role of a particular antigen, the antigens identified might be useful in screening systems to identify cases of *Yersinia*-triggered reactive arthritis. Others found a common intra-articular oligoclonal T cell expansion in patients with reactive arthritis, providing further evidence for a molecular mimicry phenomenon at the T cell level in which arthritogenic microbial peptides may be presented by HLA B27 molecules (Dulphy et al., 1999). On the other hand, recent work showed that enterobacterial infection may downregulate expression of MHC class I molecules, particularly in patients with HLA B27 (Kirveskari et al., 1999). Taken together, the pathomechanism of reactive arthritis and the role of HLA B27 molecules in this process remains obscure. Moreover, the causative relationship between other autoimmune diseases such as Graves disease or Hashimoto's thyroiditis and *Yersinia* infections is still questionable and remains to be further elucidated (Arscott et al., 1992; Chatzipanagiotou et al., 2001).

VACCINATION AGAINST *YERSINIA PESTIS*

The worldwide incidence of plague from 1967-1993 averaged about 1600 cases per year and has increased during the last decade (Perry and Fetherston, 1997). Although this number of cases may seem to be relatively low, *Y. pestis* has the potential to spread in an epidemic manner. Moreover, unfortunately, bioterrorism has to be taken as an incalculable risk of modern ages. With this background, the availibility of a safe and efficacious vaccine against plague is highly desirable.

The previously licensed USP vaccine against plague consisted of killed whole cells of virulent *Y. pestis* (Meyer, 1970). Efficacy of this vaccine against bubonic, but not against pneumonic, plague was suggested by epidemiological studies and directly demonstrated in experimental animals (Cavanaugh et al., 1974; Meyer et al., 1974). A live attenuated *Y. pestis* vaccine strain (EV76) has also been used as a human vaccine. In mice this vaccine showed greater protection than the USP vaccine, but also caused severe adverse reactions (Russell et al., 1995), which prevented the live vaccine from gaining worldwide acceptance (Meyer, 1970).

Yersinia pestis fraction 1 capsular antigen (F1) is a plasmid-encoded, proteinaceous capsule synthesized in large quantities by the pathogen and reported to confer antiphagocytic properties on *Y. pestis* by interfering with complement mediated opsonization (Williams et al., 1972). F1 is thought to be the major immunodominant antigen in the USP plague vaccine, which is in line with the finding that highly purified F1 used to vaccinate mice resulted in very high titers and led to protection against either subcutaenous or aerosol challenge with *Y. pestis* that was superior to the USP vaccine against plague (Andrews et al., 1996). Similarly successful passive immunization of mice against plague with monoclonal against F1 was reported (Anderson et al., 1997). However, in rare cases F1$^-$ variants were isolated from some mice which survived challenge. Isogenic F1$^-$ strains of *Y. pestis* are virulent for mice (Drozdov et al., 1995; Worsham et al., 1995), guinea pigs (Drozdov et al., 1995) and green monkeys (Davis et al., 1996). Therefore, a multivalent vaccine composed of multiple immunogens should reduce the possibility of selecting variants that could produce a chronic infection and should protect against F1$^-$ strains.

One additional important immunogenic peptide is LcrV, a key component of *Yersinia* type III secretion apparatus located on the virulence plasmid of all three human pathogenic *Yersinia* species. Several studies showed that immunization with recombinant V antigen protects mice against pneumonic or bubonic plague caused by F1 capsule positive and negative strains of *Y. pestis* (Anderson et al., 1996; Nakajima et al., 1995). Additionally, successful passive immunization of mice against *Y. pestis* using antibodies against V-antigen was demonstrated (Motin et al., 1994). V-antigen, although highly similar in different *Yersinia* species and serotypes, can be subdivided in two major groups: (I) *Y. enterocolitica* serotype O:8; and (II) *Y. enterocolitica* serotype O:3, O:9, *Y. pestis* and *Y. pseudotuberculosis;* which differ in a hypervariable region in amino acid residues 225 to 232 (Roggenkamp et al., 1997). This difference may explain why passive immunization with LcrV of *Y. enterocolitica* O:8 (subtype I) does not protect against yersiniae of subtype II and vice versa, but is crossprotective within the same subgroup (Roggenkamp et al., 1997). Therefore, one can speculate that the failure of North American *Y. enterocolitica* strains represented by serogroup O:8 to protect against *Yersinia* spp. of LcrV subtype II may be a reason for the presence of plague foci in the Americas (Motin et al., 1994; Roggenkamp et al., 1997).

Besides LcrV and F1, so far only vaccination with recombinant YopD protein has been shown to at least partially protect mice against challenge with *Y. pestis* (Andrews et al., 1999). The most promising approach for the development of a new vaccine against plague combines recombinant F1 and LcrV antigens in a subunit vaccine and in which both components of the vaccine may be additive in effect (Williamson et al., 2000; Jones et al., 2000; Williamson, 2001). A similar vaccine consisting of a recombinant fusion protein of F1 and LcrV has also shown promise (Heath et al., 1998). Delivery of vaccines by a mucosal route such as nasal, oral or inhalational delivery would have important advantages, including induction of IgA antibodies as well as easier administration. For this purpose, subunit vaccines encapsulated within poly-L-lactide microspheres and subunit vaccines expressed by attenuated *Salmonella* strains, which are suitable for mucosal delivery, are in development (Titball and Williamson, 2001).

REFERENCES

Abe J., Takeda T., Watanabe Y., Nakao, H., Kobayashi, N., Leung, D.Y., and Kohsaka, T. 1993. Evidence for superantigen production by *Yersinia pseudotuberculosis*. J. Immunol. 151: 4183-4188.

Allen, J.E., and Maizels, R.M. 1997. Th1-Th2, reliable paradigm or dangerous dogma? Immunol. Today. 18: 387-392.

Anderson, G.W. Jr., Leary, S.E., Williamson, E.D., Titball, R.W., Welkos, S.L., Worsham, P.L., and Friedlander, A.M. 1996. Recombinant V antigen protects mice against pneumonic and bubonic plague caused by F1-capsule-positive and -negative strains of *Yersinia pestis*. Infect. Immun. 64: 4580-5.

Anderson, G.W. Jr., Worsham, P.L., Bolt, C.R., Andrews, G.P., Welkos, S.L., Friedlander, A.M., and Burans, J.P. 1997. Protection of mice from fatal bubonic and pneumonic plague by passive immunization with monoclonal antibodies against the F1 protein of *Yersinia pestis*. Am. J. Trop. Med. Hyg. 56: 471-3.

Andrews, G.P., Heath, D.G., Anderson, G.W. Jr., Welkos, S.L.,and Friedlander, A.M. 1996. Fraction 1 capsular antigen (F1) purification from *Yersinia pestis* CO92 and from an *Escherichia coli* recombinant strain and efficacy against lethal plague challenge. Infect. Immun. 64: 2180-7.

Andrews, G.P., Strachan, S.T., Benner, G.E., Sample, A.K., Anderson, G.W. J.r, Adamovicz, J.J., Welkos, S.L., Pullen, J.K., and Friedlander, A.M. 1999. Protective efficacy of recombinant *Yersinia* outer proteins against bubonic plague caused by encapsulated and nonencapsulated *Yersinia pestis*. Infect. Immun. 67: 1533-7.

Appel, H., Mertz, A., Distler, A., Sieper, J., and Braun, J. 1999. The 19 kDa protein of *Yersinia enterocolitica* O:3 is recognized on the cellular and humoral level by patients with *Yersinia* induced reactive arthritis. J. Rheumatol. 26: 1964-71.

Arscott, P., Rosen, E.D., Koenig, R.J., Kaplan, M.M., Ellis, T., Thompson, N., and Baker, J.R. Jr. 1992. Immunoreactivity to *Yersinia enterocolitica* antigens in patients with autoimmune thyroid disease. J. Clin. Endocrinol. 75: 295 – 300.

Autenrieth, I.B., Tingle, A., Reske-Kunz, A., and Heesemann, J. 1992a. T lymyphocytes mediate protection against *Yersinia enterocolitica* in mice: characterization of murine T cell clones specific for *Y. enterocolitica*. Infect. Immun. 3: 1140-1149.

Autenrieth, I.B., and Heesemann, J. 1992b. In vivo neutralization of tumor necrosis factor-alpha and interferon-gamma abrogates resistance to *Yersinia enterocolitica* in mice. Med. Microbiol. Immunol. 181: 333-338.

Autenrieth, I.B., Vogel, U., Preger, S., Heymer, B., and Heesemann, J. 1993. Experimental *Yersinia enterocolitica* infection in euthymic and T-cell-deficient athymic nude C57BL/6 mice: comparison of time course, histomorphology and immune response. Infect. Immun. 61: 2585-2595.

Autenrieth, I.B., Beer, M., Bohn, E., Kaufmann, S.H.E., and Heesemann, J. 1994. Immune responses to *Yersinia enterocolitica* in susceptible BALB/c and resistant C57BL/6 mice: an essential role for gamma interferon. Infect. Immun. 62: 2590-2599.

Autenrieth, I.B., and Firsching, R. 1996a. Penetration of M cells and destruction of Peyer's patches by *Yersinia enterocolitica*: an ultrastructural and histological study. J. Med. Microbiol. 44: 285-294.

Autenrieth, I.B., Kempf, V., Sprinz, T., Preger, S., and Schnell, A. 1996b. Defense mechanisms in Peyer's patches and mesenteric lymph nodes against *Yersinia enterocolitica* involve integrins and cytokines. Infect. Immun. 64: 1357-1368.

Autenrieth, I.B., Hein, J., and Schulte, R. 1999. Defense mechanisms against extracellular pathogens. In: Embryonic Encyclopedia of Life Sciences. Nature Publishing Group, London, www.els.net.

Autenrieth, I.B., and Schmidt, M.A. 2000. Bacterial interplay at intestinal mucosal surfaces: implications for vaccine development. Trends Microbiol. 8: 457-464.

Balligand, G., Laroche, Y., and Cornelis, G.R. 1985. Genetic analysis of virulence plasmid from a serogroup 9 *Yersinia enterocolitica* strain: role of outer membrane protein P1 in resistance to human serum and autoagglutination. Infect. Immun. 48: 782-786.

Bohn, E., Heesemann, J., Ehlers, S., and Autenrieth, I.B. 1994. Early gamma interferon mRNA expression is associated with resistance of mice against *Yersinia enterocoltica*. Infect. Immun. 62: 3027-3032.

Bohn, E., and Autenrieth, I.B. 1996. IL-12 is essential for resistance against *Yersinia enterocolitica* by triggering IFN-γ production in NK cells and CD4+ T cells. J. Immunol. 156: 1458-1468.

Bohn, E., Schmitt, E., Noll, A., Schulte, R., and Autenrieth, I.B. 1998a. Ambiguous role of IL-12 in *Yersinia enterocolitica* infection in susceptible and resistant mouse strains. Infect. Immun. 66: 2213-2220.

Bohn, E., Sing, A., Zumbihl, R., Okamura, H., Kurimoto, M., and Autenrieth, I.B. 1998b. Interleukin-18 (Interferon-γ inducing factor) regulates early cytokine production in, and promotes resolution of, bacterial infection in mice. J. Immunol. 160: 299-307.

Bohn, E. , Müller, S., Lauber, J., Geffers, R., Speer, N., Spieth, C., Krejci, J., Manncke, B., Buer, J., Zell, A., and Autenrieth, I.B. 2004. Gene expression patterns of epithelial cell by pathogenicity factors of *Yersinia enterocolitica*. Cell. Microbiol. 6: 129-141.

Bradford, W.D., Noce, P.S., Gutman, L.T. 1974. Pathologic features of enteric infection with *Yersinia enterocolitica*. Arch. Pathol. 98: 17-22.

Burns, J.W., Siadat-Pajouh, M., Krishnaney, A.A, and Greenberg HB. 1996. Protective effect of rotavirus VP6-specific IgA monoclonal antibodies that lack neutralizing activity. Science 272: 104-107.

Carnoy, C., Mullet, C., Müller-Alouf, H., Leteurtre, E., and Simonet, M. 2000. Superantigen YPMa exacerbates the virulence of *Yersinia pseudotuberculosis* in mice. Infect. Immun. 68: 2553-2559.

Cavanaugh, D.C., Elisberg, B.L., Llewellyn, C.H., Marshall, J.D, Jr., Rust, J.H. Jr., Williams, J.E., and Meyer, K.F. 1974. Plague immunization. V. Indirect evidence for the efficacy of plague vaccine. J. Infect. Dis. 129 Suppl: S37-40.

Chatzipanagiotou, S., Legakis, J.N., Boufidou, F., Petroyiami, V., and Nicolaou, C. 2001. Prevalence of *Yersinia* plasmid - encoded outer protein (Yop) class-specific antibodies in patients with Hashimoto's thyroiditis. Clin. Microbiol. Infect. Dis. 7: 138 - 143

Chaux, P., Luiten, R., Demotte, N., Vantomme, V., Stroobant, V., Traversari, C., Russo, V., Schultz E, Cornelis GR, Boon T. 1999. Identification of five MAGE - A1 epitopes recognized by cytolytic T lymphocytes obtained by in vitro stimulation with dendritic cells transduced with MAGE - A1. J. Immunol. 163: 2928 – 2936.

Clark, M.A., Barry, H.H., and Jepson M. A. 1998. M-cell surface β1 integrin expression and invasin-mediated targeting of *Yersinia pseudotuberculosis* to mouse Peyer's Patch M cells. Infect. Immun. 66: 1237-1243.

Coconnier MH, Bernet Cammard M.F., and Servin A. 1994. How intestinal epithelial cell differentiation inhibits the cell entry of *Yersinia pseudotuberculosis* in colon carcinoma Caco-2 cell line in culture. Differentiation 58: 87-94.

Cover, T.L. 1995. *Yersinia enterocolitica* and *Yersinia pseudotuberculosis*. In: Infections of the gastrointestinal tract. Ed.: Blaser, J.M., Smith, P.D., Ravdin, J.I., Greenberg, H.B., Guerrant, R.L. Raven Press, Ltd., New York.

Davis, K.J., Fritz, D.L., Pitt, M.L., Welkos, S.L., Worsham, P.L., and Friedlander, A.M. 1996.
Pathology of experimental pneumonic plague produced by fraction 1-positive and fraction 1-negative *Yersinia pestis* in African green monkeys (*Cercopithecus aethiops*). Arch. Pathol. Lab. Med. 120: 156-63.

Di Genaro, M.S., Waidmann, M., Kramer, U., Hitziger, N., Bohn, E., and Autenrieth, I.B. 2003. Attenuated *Yersinia enterocolitica* Mutant Strains Exhibit Differential Virulence in Cytokine-Deficient Mice: Implications for the Development of Novel Live Carrier Vaccines. Infect. Immun. 71: 1804-12.

Donnelly, G.A.E., Lu, J., Takeda, T., and McKay, D.M. 1999. Colonic epithelial physiology is altered in response to the bacterial superantigen *Yersinia pseudotuberculosis* mitogen. J. Infect. Dis. 180 : 1590-1596.

Drozdov, I.G. , Anisimov, A.P., Samoilova, S.V., Yezhov, I.N., Yeremin, S.A., Karlyshev, A.V., Krasilnikova, V.M., and Kravchenko, V.I. 1995. Virulent non-capsulate Yersinia pestis variants constructed by insertion mutagenesis. J. Med. Microbiol. 42: 264-8.

Dube, P.H., Revell, P.A., Chaplin, D.D., Lorenz, R.G., and Miller VL. 2001. A role for IL-1 alpha in inducing pathologic inflammation during bacterial infection. Proc. Natl. Acad. Sci. USA. 98: 10880-5.

Duffour, M-T., Lurquin, C., Cornelis, G., Boon, T., and Bruggen, Pi van der. A. 1999. MAGE - A4 peptide presented by HLA-A2 is recognized by cytolytic T lymphocytes. J. Immunol. 29: 3329 – 3337.

Dulphy, N., Peyrat, M.A., Tieng, V., Douay, C., Rabian, C., Tamouza, R., Laoussadi, S., Berenbaum F., Chabot A., Bonneville, M., Charron, D., and Toubert, A. 1999. Common intra-articular T cell expansions in patients with reactive arthritis: identical beta-chain junctional sequences and cytotoxicity toward HLA-B27. J. Immunol. 162: 3830-3839.

Eckmann, L., Kagnoff, M.F., and Fierer, J. 1995. Intestinal epithelial cells as watchdogs for the natural immune system. Trends. Microbiol. 3: 118-120.

Edelson, B.T., and Unanue, E.R. 2001. Intracellular antibody neutralizes *Listeria* growth. Immunity 14: 503-512.

Eitel, J., and Dersch, P. 2002. The YadA protein of *Yersinia pseudotuberculosis* mediates high-efficiency uptake into human cells under environmental conditions in which invasin is repressed. Infect. Immun. 70: 4880-91.

Ennis, E., Isberg, R., and Shimizu, Y. 1993. Very late antigen-4dependent adhesion and costimulation of resting human T cells by the bacterial beta 1 integrin ligand invasin. J. Exp. Med. 177: 207-212.

Falgarone, G., Blanchard, H. S., Virecoulon, F., Simonet, M., and Breban, M. 1999a Coordinate involvement of invasin and Yop proteins in a *Yersinia pseudotuberculosis*-specific class I-restricted cytotoxic T cell-mediated response. J. Immunol. 162: 2875 - 2883.

Falgarone G., Blanchard HS., Riot B., Simonet M., and Breban M. 1999b. Cytotoxic T-cell-mediated response against *Yersinia pseudotuberculosis* in HLA-B27 transgenic rat. Infect. Immun. 67: 3773-3779.

Gaede, K., Mack, D., and Heesemann, J. 1992. Experimental *Yersinia enterocolitica* infection in rats: analysis of the immune response to plasmid encoded antigens of arthritis-susceptible Lewis rats and arthritis-resistant Fisher rats. Med. Microbiol. Immunol. 181: 165-172.

Gaede, K., Baumeister, E., and Heesemann, J. 1995. Decomplementation by Cobra venom factor suppresses *Yersinia*-induced arthritis in rats. Infect. Immun. 63: 3697-3701.

Gaston, J.S., Cox, C., and Granfors, K. 1999. Clinical and experimental evidence for persistent *Yersinia* infection in reactive arthritis. Arthritis Rheum. 42: 2239-42.

Grutzkau, A., Hanski, C., Hahn, O., and Riecken, E.O. 1990. Involvement of M cells in the bacterial invasion of Peyer's patches: a common mechanism shared by *Yersinia enterocolitica* and other enteroinvasive bacteria Gut 31: 1011-1015.

Hahn, H., Kaufmann, S.H.E. 1981. The role of cell-mediated immunity in bacterial infections. Rev. Infect. Dis. 3: 1221-1250.

Hanski, C., Kutschka, U., Schmoranzer, H.P, Naumann, M., Stallmach, A., Hahn, H., Menge, H., and Riecken, E.O. 1989. Immunohistochemical and electron microcopic study of interaction of *Yersiniaenterocolitica* serotype O8 with intestinal mucosa during experimental enteritis. Infect. Immun. 57: 673-678.

Heath, D.G., Anderson G.W., Mauro J.M., Welkos, S.L., Andrews, G.P., Adamovicz, J., Friedlander, A.M. 1998. Protection against experimental bubonic and pneumonic plague by a recombinant capsular F1-V antigen fusion protein vaccine. Vaccine 16: 1131-1137.

Heesemann, J., Gaede, K., and Autenrieth, I.B. 1993. Experimental *Yersinia enterocolitica* infection in rodents: a model for human yersiniosis. APMIS. 101: 417-429.

Hein, J., Kempf, V.A.J., Diebold, J., Bücheler, N., Preger, S., Horak, I., Sing, A., and Autenrieth, I.B. 2000. Interferon consensus sequence binding protein confers resistance against *Yersinia enterocolitica*. Infect. Immun. 68: 1408-1417.

Hill., K. 1980. Histologische Befunde und Differentialdiagnose der *Yersinia pseudotuberculosis*-Ileitis. Pathologe. 1: 95-99.

Holmstrom, A., Rosqvist, R., Wolf-Watz, H., and Forsberg A. 1995. Virulence plasmid-encoded YopK is essential for *Yersinia pseudotuberculosis* to cause systemic infection in mice. Infect. Immun. 63: 2269-76.

Huppertz, H.I., and Heesemann, J. 1996a. The influence of HLA B27 and interferon-gamma on the invasion and persistence of yersinia in primary human fibroblasts. Med. Microbiol. Immunol. 185: 163-70.

Huppertz, H.I., and Heesemann, J. 1996b. Experimental *Yersinia* infection of human synovial cells: persistence of live bacteria and generation of bacterial antigen deposits including "ghosts," nucleic acid-free bacterial rods. Infect. Immun. 64: 1484-7.

Iriarte, M., and Cornelis, G.R. 1998. YopT, a new *Yersinia* Yop effector protein, affects the cytoskeleton of host cells. Mol. Microbiol. 29: 915-29.

Jepson, M.A., and Clark, M.A. 1998. Studying M cells and their role in infection. Trends Microbiol. 6: 359-365.

Jones, S.M., Day, F., Stagg, A.J., and Williamson, E.D. 2000. Protection conferred by a fully recombinant sub-unit vaccine against *Yersinia pestis* in male and female mice of four inbred strains. Vaccine 19: 358-66.

Kampik, D., Schulte, R, and Autenrieth, I.B. 2000. *Yersinia enterocolitica* invasin protein triggers differential production of IL-1, IL-8, MCP-1, GM-CSF, and TNF-α in epithelial cells: implications for understanding the early cytokine network in *Yersinia* infection. Infect. Immun. 68: 2484-2492.

Katz, J., Michalek, S.M., Beagly, K.W., Eldridge, J.H. 1990. Characterization of rat T helper cell clones specific for *Bacteroides gingivalis*. Infect. Immun. 58: 2785-2791.

Kirveskari, J., He, Q., Leirisalo-Repo, M., Mäki-Ikola, O., Wuorela, M., Putto-Laurila, A., and Gransfors K. 1999. Enterobacterial infection modulates major histocompatibility complex class I expression on mononuclear cells. Immunology 97: 420-428.

Lahesmaa-Rantala, R., Granfors, K., Isomaki, H., and Toivanen, A. 1987. *Yersinia* specific immune complexes in the synovial fluid of patients with yersinia triggered reactive arthritis. Ann. Rheum. Dis. 46: 510-4.

Mazanec, M.B., Coudret C.L., and Fletcher, D.R. 1995. Intracellular neutralization of influenza virus by immunoglobulin A anti-hemagglutinin monoclonal antibodies. J. Virol. 69: 1339-1343.

Mazanec, M.B., Katzel, C.S., Lamm, M.E., Fletcher, D., and Nedrud, J.G. 1992. Intracellular neutralization of virus by immunoglobulin A antibodies. Proc. Natl. Acad. Sci. USA. 89: 6901-6905

Mertz, A., Wu, P., Sturniolo, T., Stoll, D., Rudwaleit, M., Lauster, R., Braun, J., Sieper, J. 2000. Multispecific CD4+ T cell response to a single 12-mer epitope of the immunodominant heat-shock protein 60 of *Yersinia enterocolitica* in *Yesinia*-triggered reactive arthritis: overlap with the B27-restricted CD8 epitope, functional properties, and epitope presentation by multiple DR. J. Immunol. 164: 1529 – 1537.

Meyer, K.F. 1970. Effectiveness of live or killed plague vaccines in man. Bull. World Health Organ. 42: 653-66.

Meyer, K.F., Cavanaugh, D.C., Bartelloni, P.J., and Marshall, J.D. Jr. 1974. Plague immunization. I. Past and present trends. J. Infect. Dis. 129 Suppl: S13-8.

Michetti, P., Mahan, M.J., Slauch, J.M., Mekalanos, J.J., and Neutra, M.R. 1992. Monoclonal secretory immunoglobulin A protects mice against oral challenge with the invasive pathogen Salmonella typhimurium. Infect. Immun. 60: 1786-1792.

Monack D.M., Mecsas, N., Ghori, N., and Falkow S. 1997. *Yersinia* signals macrophages to undergo apoptosis and YopJ is necessary for this cell death. Proc. Natl. Acad. Sci. USA. 94: 12638-12643.

Monack, D.M., Mecsas, J., Bouley, D., and Falkow S. 1998. *Yersinia*-induced apoptosis in vivo aids in the establishment of a systemic infection of mice. J. Exp. Med. 188: 2127-2137.

Motin, V.L., Nakajima, R., Smirnov, G.B., and Brubaker, R.R. 1994. Passive immunity to yersiniae mediated by anti-recombinant V antigen and protein A-V antigen fusion peptide. Infect. Immun. 62: 4192-201.

Nakajima, R., and Brubaker, R.R. 1993. Association between virulence of Yersinia pestis and suppression of gamma interferon and tumor necrosis factor alpha. Infect. Immun. 61: 23-31.

Nakajima, R., Motin, V.L., and Brubaker RR. 1995. Suppression of cytokines in mice by protein A-V antigen fusion peptide and restoration of synthesis by active immunization. Infect. Immun. 63: 3021-9.

Nickerson, C.L., Luthra, H.S., Savarirayan, S., and David, C.S. 1990. Susceptibility of HLA-B27 transgenic mice to *Yersinia enterocolitica* infection. Hum. Immunol. 28: 382-396.

Noll, A., Roggenkamp, A., Heesemann, J., and Autenrieth, I.B. 1994. Protective role for heat shock protein-reactive $\alpha\beta$ T cells in murine yersiniosis. Infect. Immun. 62: 2784-2791.

Noll A, and Autenrieth, I.B. 1996. *Yersinia*-HSP60-reactive T cells are efficiently stimulated by peptides of 12 and 13 amino acid residues in a MHC class II (I-Ab) restricted manner. Clin. Exp. Immunol. 105: 231-237.

Noll, A., and Autenrieth, I.B. 1996. Immunity against *Yersinia enterocolitica* by vaccination with *Yersinia*-HSP60-ISCOM or *Yersinia*-HSP60 plus interleukin-12. Infect. Immun. 64: 2955-2961.

Ortiz-Alvarez, O., Yu, D.T., Petty, R.E., and Finlay, B.B. 1998. HLA-B27 does not affect invasion of arthritogenic bacteria into human cells. J. Rheumatol. 25: 1765-1771

Palmer, L.E., Hobbie, S., Galan, H., and Bliska, J.B. 1998. YopJ of *Yersinia pseudotuberculosis* is required for the inhibition of macrophage TNF-alpha production and downregulation of the MAP kinases p38 and JNK. Mol. Microbiol. 27: 953-965

Pepe, J.C., and Miller, V.L. 1993. *Yersinia enterocolitica* invasin: a primary role in the initiation of infection. Proc. Natl. Acad. Sci. USA. 90: 6473-6477.

Perry, R.D., and Fetherston, J.D. 1997. *Yersinia pestis*--etiologic agent of plague. Clin. Microbiol. Rev. 10: 35-66.

Pilz, D., Vocke, T., Heesemann, J., and Brade, V. 1992. Mechanism of YadA-mediated serum resistance of *Yersinia enterocolitica* serotype O3. Infect. Immun. 60: 189-195

Rescigno, M., Urbano, M., Valzasina, B., Francolini, M., Rotta, G., Bonasio, R., Granucci, F., Kraehenbuhl, J.P., and Ricciardi-Castagnoli, P. 2001. Dendritic cells express tight junction proteins and penetrate gut epithelial monolayers to sample bacteria. Nat. Immunol. 2: 361-367

Revell, P.A., and Miller, V.L. 2000. A chromosomally encoded regulator is required for expression of the *Yersinia enterocolitica* inv gene and for virulence. Mol. Microbiol. 35: 677-85.

Reis e Sousa, C. 2001. Dendritic cells as sensors of infection. Immunity. 14: 495-498

Roggenkamp, A., Neuberger, H.R., Flugel, A., Schmoll, T., and Heesemann, J. 1995. Substitution of two histidine residues in YadA protein of *Yersinia enterocolitica* abrogates collagen binding, cell adherence and mouse virulence. Mol. Microbiol. 166: 1207-1219.

Roggenkamp, A., Geiger, A.M., Leitritz, L., Kessler, A., and Heesemann, J. 1997. Passive immunity to infection with *Yersinia* spp. mediated by anti-recombinant V antigen is dependent on polymorphism of V antigen. Infect. Immun. 65: 446-451.

Romani, L. 2000. Innate and adaptive immunity in *Candida albicans* infections and saprophytism. J. Leukoc. Biol. 68: 175-179.

Ruckdeschel, K., Harb, S., Roggenkamp, A., Hornef, M., Zumbihl, R., Kohler, S., Heesemann, J., and Rouot, B. 1998. *Yersinia enterocolitica* impairs activation of transcription factor NF-kappaB: involvement in the induction of programmed cell death and in the suppression of the macrophage tumor necrosis factor alpha production. J. Exp. Med. 187: 1069-1079.

Ruessmann, H., Weissmueller, A., Geginat, G., Igwe, E., Roggenkamp, A., Bubert, A., Goebel, W., Hof, H., and Heesemann, J. 2000. *Yersinia enterocolitica*-mediated translocation of defined fusion proteins to the cytosol of mammalian cells results in peptide-specific MHC class-I-restricted antigen presentation. Eur J. Immunol. 30: 1375-1384.

Russell, P., Eley, S.M., Hibbs, S.E., Manchee, R.J., Stagg, A.J., and Titball, R.W. 1995. A comparison of Plague vaccine, USP and EV76 vaccine induced protection against *Yersinia pestis* in a murine model. Vaccine 13: 1551-1556.

Sansonetti, P.J., Arondel, J., Cavaillon, J.M., and Huerre, M. 1995. Role of interleukin-1 in the pathogenesis of experimental shigellosis. J. Clin. Invest. 96: 884-892

Sauvonnet, N., Pradet-Balade, B., Garcia-Sanz, J.A., and Cornelis, G.R. 2002. Regulation of mRNA expression in macrophages after *Yersinia enterocolitica* infection. Role of different Yop effectors. J. Biol. Chem. 277: 25133-25142.

Schesser, K., Splik, A., Dukuzumuremyi, J., Neurath, M., Petterson, S., and Wolf-Watz, H. 1998. The yopJ locus is required for *Yersinia*-mediated inhibition of NF-κB activation and cytokines expression, YopJ contains a eukaryotic SH2-like domian that is essential for its repressive activity. Mol. Microbiol. 28: 1067-1079.

Schoppet, M., Bubert, A., and Huppertz H-I. 2000. Dendritic cell function is perturbed by *Yersinia enterocolitica* infection in vitro. Clin. Exp. Immunol. 122: 316-323.

Schulte, R., and Autenrieth, I.B. 1998a. *Yersinia enterocolitica*-induced interleukin-8 secretion by human intestinal epithelial cells depends on cell differentiation. Infect. Immun. 66: 1216-1224.

Schulte, R., Zumbihl, R., Kampik, D., Fauconnier, A., and Autenrieth, I.B. 1998b. Wortmannin blocks *Yersinia* invasin-triggered internalization, but not IL-8 secretion by epithelial cells. Med. Microbiol. Immunol. 187: 53-60.

Schulte, R., Grassl, G., Preger, S., Fessele, S., Jacobi, C.A., Schaller, M., Nelson, P.J., Autenrieth, I.B. 2000a. *Yersinia enterocolitica* invasin protein triggers IL-8 production via activation of Rel p65-p65 homodimers. FASEB J. 14: 1471-1484.

Schulte, R., Kerneis, S., Klinke, S., Bartels, H., Preger, S., Kraehenbuhl, J.P., Pringault, E., and Autenrieth, I.B. 2000b. Translocation of *Yersinia enterocolitica* across reconstituted intestinal epithelial monolayer is triggered by *Yersinia* invasin binding to β1 integrins apically expressed on M-like cells. Cell. Microbiol. 2: 173-186.

Siebers, A., and Finlay, B.B. 1996. M cells and the pathogenesis of mucosal and systemic infections. Trends Microbiol. 4: 22-29.

Simonet, M., Fortineau, N., Beretti, J.L., and Berche, P. 1994. Immunization with live *aroA* recombinant *Salmonella typhimurium* producing invasin inhibits intestinal translocation of *Yersinia pseudotuberculosis*. Infect. Immun. 62: 863-867

Sing, A., Roggenkamp, A., Geiger, A.M., and Heesemann, J. 2002a. *Yersinia enterocolitica* evasion of the host innate immune response by V antigen-induced IL-10 production of macrophages is abrogated in IL-10-deficient mice. J. Immunol. 168: 1315-1321.

Sing, A., Rost, D., Tvardovskaia, N., Roggenkamp, A., Wiedemann, A., Kirschning, C.J., Aepfelbacher, M., and Heesemann, J. 2002b. *Yersinia* V-antigen exploits toll-like receptor 2 and CD14 for interleukin 10-mediated immunosuppression. J. Exp. Med. 196: 1017-1024.

Titball, R.W., and Williamson, E.D. 2001. Vaccination against bubonic and pneumonic plague. Vaccine. 19: 4175-4184.

Van Damme, M., Sory, M.P., Biot, T., Vaerman, J.P., and Cornelis, G.R. 1992. Oral immunization against cholera toxin with a live *Yersinia enterocolitica* carrier in mice. Gastroenterology 103: 520-531.

Vazquez-Torres A., Jones-Carson J., Bäumler A., Falkow S., Valdivia R., Brown, W., Le, M., Berggren, R., Parks, W.T., and Fang, F.C. 1999. Extraintestinal dissemination of *Salmonella* by CD18-expressing phagocytes. Nature 401: 804-808.

Viboud, G.I., So, S.S., Ryndak, M.B., and Bliska, J.B. 2003. Proinflammatory signalling stimulated by the type III translocation factor YopB is counteracted by multiple effectors in epithelial cells infected with *Yersinia pseudotuberculosis*. Mol. Microbiol. 47: 1305-1315.

Virtala, M., Kirveskari, J., and Granfors, K. 1997. HLA-B27 modulates the survival of Salmonella enteritidis in transfected L cells, possibly by impaired nitric oxide production. Infect. Immun. 65: 4236-4242.

Visser, L.G., Hiemstra, P.S., van den Barselaar, M.T., Ballieux, P.A., and van Furth, R. 1996. Role of YadA in resistance to killing of *Yersinia enterocolitica* by antimicrobial polypeptides of human granulocytes. Infect. Immun. 64: 1653-1658

Vogel, U., Autenrieth, I.B., Berner, R., and Heesemann, J. 1993. Role of plasmid-encoded antigens of *Yersinia enterocolitica* in humoral immunity against secondary *Y. enterocolitica* infection in mice. Microb. Pathog. 14: 1-14

Wagner H. 2001. Toll meets bacterial CpG-DNA. Immunity 14: 499-502

Williams, R.C. Jr., Gewurz, H., and Quie, PG. 1972. Effects of fraction I from *Yersinia pestis* on phagocytosis in vitro. J. Infect. Dis. 126 : 235-241.

Williamson, E.D. Plague vaccine research and development. 2001. J. Appl. Microbiol. 91: 606-608.

Williamson, E.D, Eley, S.M., Stagg, A.J., Green, M., Russell, P., and Titball, R.W. 2000. A single dose sub-unit vaccine protects against pneumonic plague. Vaccine 19: 566-571.

Worsham, P.L., Stein, M.P., and Welkos, S.L. 1995. Construction of defined F1 negative mutants of virulent *Yersinia pestis*. Contrib. Microbiol. Immunol. 13: 325-328.

Yan, H., Lamm, M.E., Bjorling, E., and Huang, Y.T. 2002. Multiple functions of immunoglobulin A in mucosal defense against viruses: an in vitro measles virus model. J. Virol. 76: 10972-10979.

Yao, T., Mecsas, J., Healy, J., Falkow, S., and Chien, Y. 1999. Suppression of T and B lymphocyte activation by a *Yersinia pseudotuberculosis* virulence factor, YopH. J. Exp. Med. 190: 1343-1350.

Zhao, Y.X., Lajoie, G., Zhang, H., Chiu, B., Payne, U., and Inman, R.D. 2000. Tumor necrosis factor receptor p55-deficient mice respond to actute *Yersinia enterocolitica* infection with less apoptosis and more effective host resistance. Infect. Immun. 68: 1243-1251.

Chapter 10

Superantigens of *Yersinia pseudotuberculosis*

Jun Abe

ABSTRACT

Bacterial superantigens have been implicated in the pathogenesis of several human diseases. Among them, toxic shock syndrome is a prototypic acute intoxication caused by the pyrogenic exotoxin family of superantigens. *Yersinia pseudotuberculosis* produces a superantigen, YPM, which is a 14.5-kDa protein able to stimulate human T cells bearing Vβ3, 9, 13.1, and 13.2 segments on the T cell receptor (TCR). There are three variant proteins of YPM (YPM-a, -b, -c) and genetic studies suggest that these variant genes came from an ancestral *ypm* gene transferred from another species into the genome of *Y. pseudotuberculosis*. The *ypm* genes are not present in all strains, and *ypm* positive strains are much more frequently isolated in Far East Asia, including Japan, than in Europe. In Japan, *Y. pseudotuberculosis* infection is frequently associated with systemic symptoms such as erythematous skin rash, cervical lymphadenopathy, and the late onset of interstitial nephritis and coronary aneurysms. The high incidence of the YPM-producing strains in Japan, together with the higher anti-YPM antibody titers in patients with the systemic manifestations, implicatesYPM in the pathogenesis of these symptoms.

INTRODUCTION

Immune cells recognize invading microorganisms through various receptors in their cell surface membranes. Toll-like receptor family members, which are present on macrophages and dendritic cells as well as on lymphocytes and neutrophiles, recognize lipopolysaccharide (LPS) of the Gram-negative bacteria, peptidoglycan of the Gram-positive bacteria, bacterial DNA containing unmethylated CpG dinucleotides, and double-stranded viral RNA by their specific molecular patterns, which do not exist in mammalian cells (Janeway et al., 2000; Akira et al., 2001). The immune reaction provoked by these receptors is called innate immunity because the response requires no memory cells against a specific antigen. On the other hand, antigen receptors that are present on T cells (TCR) and B cells (BCR) recognize specific peptide antigens (epitopes) derived from the microorganisms in conjunction with their own major histocompatibility complex (MHC) molecules. The response launched by these receptors is called acquired immunity.

Superantigens are highly potent immune stimulatory proteins produced by some bacteria and retroviruses (White et al., 1989; Janeway et al., 1989). Although, like typical peptide antigens, they stimulate T cells through the TCR, the characteristics of the stimulatory activities of superantigens are more like LPS and the other innate immune stimuli in that they stimulate a much wider range of cells without the need for antigen memory. Bacteria produce most of the superantigens that have pathogenic roles in human diseases. For example, *Staphylococcus aureus* produces a variety of enterotoxins such as staphylococcal enterotoxin (SE) A to E (Marrack et al., 1990) and toxic shock syndrome toxin-1 (TSST-1) (Schlievert et al., 1981; Bergdoll et al., 1981), and *Streptococcus pyogenes* produces several pyrogenic exotoxins such as SPEA and SPEC (Watson, 1959; Bohach et al., 1990). So far, *Y. pseudotuberculosis* is the only known Gram-negative bacterium that produces a potent superantigenic toxin, designated *Yersinia pseudotuberculosis*-derived mitogen (YPM) (Ito et al., 1995; Miyoshi-Akiyama et al., 1995).

The diseases caused by these three superantigen-producing bacteria share some clinical symptoms, such as high-grade fever, conjunctival and pharyngeal infection, and an erythematous skin rash followed by desquamation in the early convalescent phase (Todd et al., 1978; Sato et al., 1983; Cone et al., 1987). The pathophysiological mechanism of these symptoms may be partly explained by the action of superantigens. Superantigens stimulate a large proportion of T cells to produce TNF-α (Fast et al., 1989), IFN-γ (Jupin et al., 1988), and IL-2 (Uchiyama et al., 1986). These cytokines then stimulate macrophages, natural killer cells, vascular endothelial cells and fibroblasts to produce a variety of inflammatory cytokines and chemokines such as IL-1 (Ikejima et al., 1984), IL-6 (Miethke et al., 1993a), TNF-α (Miethke et al., 1993a), IL-12 (Leung et al., 1995), IL-8 (Konig et al., 1994), MIP-1α (Tessier et al., 1998), MIP-2α (Tessier et al., 1998), MCP-1 (Newman et al., 1996), and IP-10 (Neumann et al., 1998). These mediators of inflammation work together to cause high-grade fever, production of acute reactive proteins, and an increase in vascular permeability of peripheral blood vessels.

In this chapter, I summarize the biological activity, genetic and epidemiologic features of the superantigens of *Y. pseudotuberculosis*, and discuss the possible pathogenic roles that the superantigens may have in the disease caused by this microorganism.

GENETIC AND BIOCHEMICAL CHARACTERIZATION OF YPM

HISTORICAL ASPECTS OF SUPERANTIGENS AND THEIR BIOLOGICAL CHARACTERISTICS

The term "superantigen" was first described in 1989 (Kappler et al., 1989; White et al., 1989). Since then, the word has been used in innumerable papers and laboratory manuals. Prior to the discovery of superantigens in the mid-1980s, the genes for the TCR heterodimer were identified and their genetic organization was clarified (Saito et al., 1984; Chien et al., 1984.). In parallel with these findings, many variable segment genes for both the α- and β-chains of the TCR were cloned from T cell hybridomas and malignant T cell lines, followed by the production of monoclonal antibodies specific to these variable receptor components (Toyonaga et al., 1987; Wilson et al., 1988). In 1989, Kappler and colleagues, in experiments using these antibodies, found that the staphylococcal enterotoxin B (SEB) expands T cells that have Vβ3, Vβ 8.1, Vβ 8.2 and Vβ 8.3 segments in their TCR. They called the toxin a "superantigen" because it stimulated a large proportion of T cells in a TCR-Vβ dependent manner regardless of their antigen specificity (White et al., 1989). In the same paper, they reported that the murine minor lymphocyte stimulating (Mls) antigens had the same superantigenic activity. The Mls antigens were, at the time, unidentified genetic traits which acted in the mixed leukocyte reaction (MLR) to stimulate lymphocytes of MHC-identical strains of mice (Festenstein, 1976). In later studies, three independent research groups (Frankel et al., 1991; Dyson et al., 1991; Woodland et al., 1991) discovered that the Mls antigen genes were the murine mammary tumor provirus (MMTV) genes.

Bacterial superantigens are powerful mitogens of human T cells and are produced by *S. aureus*, *S. pyogenes* and *Y. pseudotuberculosis*. T cells respond to superantigens through their TCR as in the ordinary acquired immune response. However, the immune responses against superantigens more closely resemble the innate immune responses against conserved microbial products such as LPS and CpG oligomers. First of all, superantigens bind directly to MHC class II molecules on the antigen presenting cells (APC) (Carlsson et al., 1988; Fleischer et al., 1988; Jardetzky et al., 1994). Therefore, T cell activation occurs as rapidly as in innate immunity because there is no need for antigen processing by the APC. Secondly, superantigen binding to TCR is dependent mainly on the Vβ element and is hardly affected by other variable elements (Dβ, Jβ, Vα, Jα) of TCR (Pullen et al., 1990; Choi et al., 1990). Each superantigen has specificity for a set of Vβ families and can interact with all T cells expressing those Vβ elements independent of antigen specificity (Kappler et al., 1989; Choi et al., 1989). In the case of human peripheral blood T cells, unlike conventional antigens that interact with about one in every 10^2 - 10^6 T cells (Tan et al., 1999; Zippelius et al., 2002), superantigens are capable of interacting with as many as 5 to 20% of T cells. Thirdly, the interaction between T cells and APC is not MHC-restricted. For this reason, superantigens are able to stimulate CD8(+) T cells as well as CD4(+) T cells, in addition to the T cells from donors with different MHC class II allotypes (Dellabona et al., 1990; Scholl et al., 1990).

Thus far, most of the bacterial superantigens that have been identified are produced by either *S. aureus* or *S. pyogenes*. They share certain characteristics in their tertiary amino acid structure; and exhibit common biological properties such as pyrogenicity, erythrogenicity, and enhancement of susceptibility to LPS-induced shock (Bohach et al., 1990). They are generally called pyrogenic toxin superantigens (PTSA) and are classified into four major groups based on their amino acid sequence similarity (Proft et al., 2001). Some of the PTSA (e.g. SED) are encoded on mobile genetic elements such as a plasmid (Bayles et al., 1989), and others (e.g. SEA, SPEA, and SPEC) are encoded by bacteriophages (Betley et al., 1984; Johnson et al., 1986; Goshorn et al., 1989). Like these movable superantigen genes, chromosomally encoded superantigen genes are not equally distributed among strains either. For example, the TSST-1 gene is flanked by a 17-nucleotide direct repeat and is carried by a kind of mobile pathogenicity island which is readily transferred to a *recA* recipient (Lindsay et al., 1998). In addition, it is worth noting that these toxins have biological activities other than superantigenicity. For example, SEB has distinct domains responsible for superantigenicity and for its emetic effect (Spero et al., 1978), and the ability of exfoliative toxin to cause staphylococcal scalded skin syndrome (SSSS) depends on the serine protease activity of the toxin on a desmosomal cadherin which mediates cell-cell adhesion in the skin (Hanakawa et al., 2002). Recently, an increasing number of superantigens have been identified from both *S. aureus* and *S. pyogenes* by computer-based searches of the genome databases of the both species (Proft et al., 2001; Jarraud et al., 2001). The biologic activities and the effects on host immune system of these novel superantigens should be clarified in the near future, and the knowledge will give us additional insights into the pathophysiology of the infection caused by these bacteria.

CLONING AND BIOLOGICAL FUNCTIONS OF YPM

Superantigenic activity in *Y. pseudotuberculosis* was first reported by two research groups, including ours, in 1993 (Abe et al., 1993; Uchiyama et al., 1993). This was followed by the purification of the protein from culture supernatants of the productive strain and the determination of partial amino acid sequences of the protein (Yoshino et al., 1994). Figure 1 illustrates the result of semi-quantitative RT-PCR with 26 different Vβ-specific primer pairs, demonstrating a marked increase in Vβ3, Vβ9, Vβ13.1 and Vβ13.2 expression after stimulation with the purified YPM as compared to stimulation with anti-CD3 antibody (Abe et al., 1993). The same groups cloned the gene and completed the sequencing of the protein in 1995, and designated it YPM (Ito et al., 1995; Miyoshi-Akiyama et al., 1995).

The *ypm* gene has an open reading frame of 456 nucleotides encoding a protein of 151 amino acids. The protein has a 20 amino acid signal peptide in its N-terminal portion and is synthesized as a precursor protein. This precursor is subsequently cleaved at the Ala-Thr bond to produce a 131 amino acid mature secretory protein (Figure 2). The calculated molecular weight of the mature YPM is 14,529 Da and the pI value is 4.95. The molecular weight of YPM is substantially smaller than the molecular weight of 22,000 to 28,000 Da of the known members of the pyrogenic superantigen family (Marrack et al., 1990). According to Ito et al., (1995) homology search with the nucleotide sequence of the *ypm* gene using the EMBL-GDB database did not reveal any significantly similar gene, and no homologous protein was found in a survey of the NBRF-PDB (PIR) data bank with the YPM amino acid sequence. Alignment of the amino acid sequence of mature YPM with that of other known superantigens did not reveal any particular homologous sequences except for a 4 amino acid stretch between YPM and toxic shock syndrome toxin-1 (TSST-1) (Ito et al., 1995).

In subsequent experiments, recombinant YPM (rYPM) was purified and tested for its superantigenic property on human peripheral blood mononuclear cells (PBMC). The rYPM had strong mitogenic activity at concentrations as low as 1 pg/ml. A marked increase in the percentages of Vβ3, Vβ9, Vβ13.1 and Vβ13.2 cDNA was observed after stimulation with rYPM, as compared with stimulation with MBP-α-fragment control protein. In addition, rYPM stimulated highly purified human peripheral blood T cells in the presence of mouse fibroblasts expressing HLA-DR4, DR1, DQw6 or DPw9, but not with the fibroblasts alone. This mitogenic activity was not dependent on the processing of

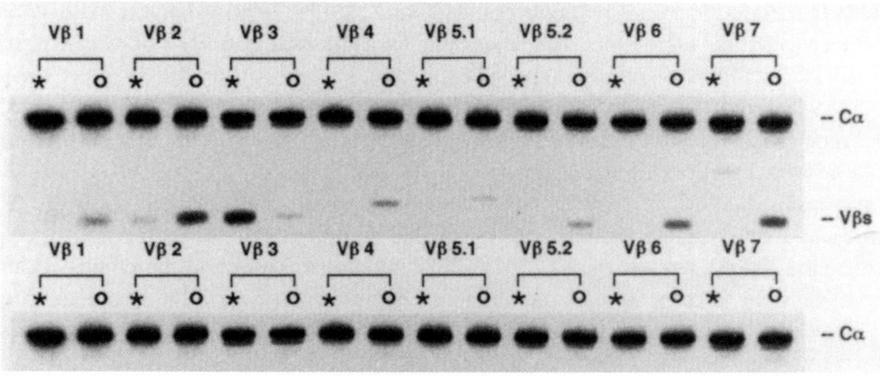

Figure 1. Autoradiograms of TCR transcripts amplified by RT-PCR. RNA was extracted from PBMC stimulated either with YPM or anti-CD3. RT-PCR was performed using a Vβ-specific oligomer (corresponding to one of the 22 Vβ families) and an oligomer from the downstream β chain constant region (Cβ primer) as one pair, and two Cα primers as the other pair.

```
GATGAACTGGTCCTGTTTTATCTGTTGGCTGCGCTTCAACTTTTGCTGACTTACC -142

TACGCTAATGCAACTGAGCCATTATTTCCACAACCAATCCCCCGAGGATGAGTTT -87

TATAAAATTTGATGTTAATCACAAAAAAAACAATAAAGATAGTGTAAATAATACA -32
               -35              -1 1      -10
AATGAGAGTGATTATATTTATAGGTTGAGTT ATG AAA AAT AAA CTT TTG  18
                      SD         M   K   N   K   L   L  -15

TCA TTA CTA ACA TTT ACA CTT TTC TCT GGA GTA GCG TTG GCG  60
 S   L   L   T   F   T   L   F   S   G   V   A   L   A  -1
                                                     -1
ACT GAT TAT GAT AAT ACA CTA AAT TCA ATC CCC TCT CTT CGG  102
↑T   D   Y   D   N   T   L   N   S   I   P   S   L   R  14
 1
ATA CCC AAT ATC GCA ACA TAT ACT GGT ACT ATC CAA GGA AAA  144
 I   P   N   I   A   T   Y   T   G   T   I   Q   G   K  28

GGA GAA GTA TGT ATT ATA GGA AAT AAA GAG GGC AAG ACG AGA  186
 G   E   V   C   I   I   G   N   K   E   G   K   T   R  42

GGA GGG GAA TTA TAT GCT GTA TTA CAT TCT ACC AAT GTA AAT  228
 G   G   E   L   Y   A   V   L   H   S   T   N   V   N  56

GCA GAT ATG ACG TTA ATT TTA CTA CGC AAT GTA GGA GGC AAT  270
 A   D   M   T   L   I   L   L   R   N   V   G   G   N  70

GGA TGG GGA GAG ATA AAG AGA AAC GAT ATT GAC AAA CCT CTT  312
 G   W   G   E   I   K   R   N   D   I   D   K   P   L  84

AAG TAT GAG GAT TAT TAT ACT TCA GGG CTT AGT TGG ATT TGG  354
 K   Y   E   D   Y   Y   T   S   G   L   S   W   I   W  98

AAA ATT AAA AAC AAT AGC TCT GAA ACA TCT AAT TAT TCA TTA  396
 K   I   K   N   N   S   S   E   T   S   N   Y   S   L  112

GAT GCT ACT GTA CAT GAT GAC AAG GAA GAT AGT GAC GTA TTG  438
 D   A   T   V   H   D   D   K   E   D   S   D   V   L  126

ACG AAA TGT CCT GTG TGA AAAATATTTAATGGCTCCTTTACTTAGTATA  487
 T   K   C   P   V   *                                  131

AGTTGTATCCCTCCTTTATTTATTTGAGTAGCACTTTATATTTTGAAGGGAGCCT  542

ATTGTCTGAGCCCCCCCGATCTGATTAGGTTCCATCACAGCATATAGCGGTGTTC  597

CGGTTGCTGGCAACTGACGATATTCAGTTCTTTCATTATCTTACTGGCTCGCCAA  652
```

Figure 2. Nucleotide sequence of *ypmA* and deduced amino acid sequence of YPMa. The deduced amino acids are given below the triplets using single letter codes. An asterisk indicates the stop codon. Numbering of nucleotide is in reference to the ATG start codon. The most likely cleavage site for prokaryote signal peptidase is denoted with a vertical arrow and the amino-terminal Thr in the mature protein is referred to the first amino acid. The pre-determined sequence of N-terminal 23 amino acids is underlined. The possible promoter (-10 and -35) and Shine-Dalgarno (SD) sequences are indicated by thick underlines.

the protein by APC, because the paraformaldehyde-fixed APCs, HLA-transfected mouse fibroblasts, and EBV-transformed human B cells were able to support T cell proliferation by the rYPM. Furthermore, antibodies against the known staphylococcal superantigens SEA, SEB, SEC2, SED, SEE, and TSST1 did not neutralize the mitogenic activity of rYPM. Based on these results, it was concluded that YPM of *Y. pseudotuberculosis* is a novel superantigen (Ito et al., 1995).

Miyoshi-Akiyama et al. (1997) reported the superantigenic activity of the purified YPM on mouse T cells. The mitogenicity of YPM for mouse splenocytes was less effective than for human PBMC, and 10 ng/ml or higher concentrations of YPM were needed for the proliferation of whole splenocytes. YPM stimulated mouse T cells bearing Vβ7, Vβ8.1 Vβ8.2, or Vβ8.3 segments in their TCR, and this stimulation required MHC class II positive APC such as I-A^b or I-A^k transfected mouse fibroblasts.

ANALYSIS OF FUNCTIONAL REGIONS OF YPM BY DNA MUTAGENESIS

The X-ray crystallographic analyses of SEA (Schad et al., 1995), SEB (Swaminathan et al., 1995), SEC2 (Papageorgiou et al., 1995), SED (Sundstrom et al., 1996), TSST-1 (Acharya et al., 1994), SPEA (Papageorgiou et al., 1999), SPEC (Roussel et al., 1997), and the computer-modeled protein structure analyses of SEE (Lamphear et al., 1996) revealed that all the pyrogenic superantigens possess a conserved protein folding pattern characterized by a N-terminal β-barrel globular domain and a C-terminal globular domain. The hydrophobic residues on the β1-β2 loop region in the N-terminal domain are shared by all the staphylococcal enterotoxins and TSST-1 and work as a generic binding site for the α-chain of MHC class II molecules (Jardetzky et al., 1994; Kim et al., 1994). In addition, SEA (Hudson et al., 1995), SED (Sundstrom et al., 1996), SEE (Fraser et al., 1992) and SPEC (Li et al., 1997) were found to have a different binding site in the C-terminal globular domain of the β-chain of MHC class II molecules. This site contains a zinc-binding motif, and the binding between the three residues in the C-terminal region of superantigen and the His81in the β-chain of MHC class II is mediated by the chelating effect of zinc ion (Schad et al., 1995). The binding affinity of this zinc-dependent β-chain to SEA has been shown to be more than 100 times higher than that of the N-terminal α-chain binding (Dowd et al., 1996). Thus, it is speculated that the role of MHC β-chain binding to SEA might be to recruit and stabilize the MHC α-chain binding so as to activate T cells at a concentration far below the dissociation constant of the α-chain binding alone (Hudson et al., 1995). With regard to the binding of SEC2 to TCR β-chain, it has been shown that the responsive elements reside in a cleft between the two globular domains of SEC2, and that the TCR β-chain interacts with these elements through the complementarity determining region (CDR) 2 (Fields et al., 1996).

In the case of YPM, X-ray crystallographic analysis has not succeeded so far. The distinct amino acid sequence of YPM compared to the other pyrogenic superantigens makes it difficult to investigate the protein structure by computer-modeling. In order to cope with these difficulties, Ito et al. (1999) carried out a DNA-based mutagenesis analysis of YPM and provided some important information on the structure/function relationship of YPM. First, they introduced random mutations along the whole nucleotide sequence of *ypm* and made an expression library of the mutated *ypm* genes. After screening by Western hybridization using anti-FLAG monoclonal antibodies, 230 clones were found to express recombinant proteins. They classified the clones by the mitogenic activity of the expressed proteins to human PBMC. Among the 230 recombinants, 29 clones expressed a mutated protein with mitogenic activity less than 25% of the wild-type YPM. By sequencing, 7 out of the 29 low-mitogenic clones were found to have mutations with some relation to Cys residues of wild-type YPM. They also proved that the two Cys residues (Cys32 and Cys129) in the wild-type YPM are connected by a disulfide bond, and that disruption of this bond almost completely abolished the mitogenic activity of the protein.

The other finding from the sequencing of the low-mitogenic mutants was that 13 out of the 29 clones had an amino acid exchange within the segment between residues 35 to 60 of the wild-type YPM. By a hydrophilicity search, this segment contained two hydrophilic peaks which were considered to be important in maintaining superantigenicity of YPM. They then tried to define the specific amino acid mutations responsible for binding to MHC class II or TCR Vβ. For this purpose, they utilized the co-culture system of mouse fibroblasts expressing HLA-DR1 molecules and purified human T cells. Pre-treating fibroblasts with an excess amount of mutant YPM should inhibit the mitogenic activity of wild-type YPM if the mutation did not affect the HLA-DR1 binding, but the mitogenicity should be unchanged if the mutation impaired the HLA-DR1 binding. In fact, the mutated proteins were divided into two groups after this competition analysis. However, the distribution of the mutated amino acids was unexpectedly dispersed. The clones with a point mutation in either Ser52 or in Leu112, or with a deletion of the N-terminal 11 residues were unable to block wild-type YPM and were considered to be defective in the binding to HLA-DR. The clones with a point mutation in either of Gly35, Asp88, or His117 inhibited the wild-type YPM and they were considered to be defective in the TCR-Vβ binding. Thus, the results did not clearly define the domains needed for either the MHC binding or the TCR Vβ binding. We will have to wait for the crystallographic data to further understand the structure/function relationship of YPM and to appreciate its unique character among the prototypic superantigens.

CHARACTERIZATION OF GENETIC STRUCTURE SURROUNDING YPM GENES

Since the cloning of *ypm* gene in 1995, two variant types of YPM protein have been reported. One variant gene was cloned in 1997 and designated as *ypmB* (Ramamurthy et al., 1997). YPMb consists of 150 amino acid residues (one residue less than the original YPM, now called YPMa). The amino acid sequence homology between YPMa and YPMb is 83% and YPMb displays the same extent of mitogenicity on human PBMC as YPMa. The purified YPMb stimulates human T cells bearing Vβ3, 9, 13.1 and 13.2, the same Vβ specificity as YPMa.(Ramamurthy et al., 1997) Another variant gene was reported in 1999 and designated as *ypmC* (Carnoy et al., 1999). YPMc differs from YPMa by only a single amino acid substitution that changes the His51 of YPMa to Tyr in YPMc.

The genetic location of these *ypm* genes has been a matter of interest because it was suspected that that the *ypm* genes might be located on a mobile genetic element, like some staphylococcal pyrogenic superantigens. One reason for this speculation is that not all *Y. pseudotuberculosis* strains have these *ypm* genes. Yoshino et al. (1995) reported that the *ypmA* gene was present in 59 out of 76 (78%) clinical strains, but in as few as 33 out of 128 (26%) strains from animal and environmental sources. They also reported that none of 225 strains of *Y. enterocolitica* had the *ypm* gene. The other reason to suspect a mobile element is based on the difference in the GC content between the *ypm* genes and the overall DNA from a strain of this species. Although the reported GC content of total DNA from a strain of *Y. pseudotuberculosis* is 46.5% (Bercovier et al., 1984), the GC content of the *ypmA* gene is 35.3%, suggesting that the gene might have been transferred from an unrelated organism on a mobile gene element (Ramamurthy et al., 1997).

Carnoy et al. investigated this subject and provided a definitive answer. They found that the *ypm* genes are located in the same position of the chromosome in all the superantigenic strains and are not integrated into an obvious mobile genetic element such as a plasmid, phage, transposon, or pathogenicity island (Carnoy et al., 2002). They cloned and sequenced the 17.1-kb DNA segment encompassing the *ypmA* gene from a *Y. pseudotuberculosis* strain bearing this gene. They also sequenced the corresponding DNA segments from two strains bearing either *ypmB* or *ypmC*, and a strain without any of the *ypm* genes. The DNA segment from a *ypmA*(+) strain contained 11 new open reading frames (ORFs) besides the *ypmA* gene. The locus of the *ypmB*(+) strain had a similar genetic organization as the *ypmA*(+) strain except for the lack of a newly found insertion sequence encompassing ORF5, which was present in the *ypmA*(+) strain. The gene organization of the *ypmC*(+) strain was much different. A 4,380-bp deletion involving ORF1, 2 and 3, and a 2,100-bp deletion including a part of ORF6 and ORF7 were detected in the genome of the *ypmC*(+) strain. However, the most important finding was that the variant *ypm* alleles were always located between the ORF3 and ORF4 in all three *ypm*(+) strains, and there was no bacteriophage gene, transposon or pathogenicity island in the DNA flanking the *ypm* genes. PCR analysis of the intergenic regions within this DNA segment from 30 other *ypm*(+) strains confirmed the consistency of these three types of gene organization, and further suggested that *ypm* is not mobile in *Y. pseudotuberculosis*. In regard to the *ypm*(-) strains, the sequencing of the corresponding DNA segment revealed that the 918-bp fragment between ORF3 and ORF4 encoding the *ypm* gene was absent from this DNA segment. Moreover, the intergenic sequences between these two ORFs were identical in 9 other *ypm*(-) strains, suggesting that the absence of the 918-bp *ypm*-containing fragment is due to the non-incorporation of a *ypm* gene rather than to a deletion of the gene.

How then was the *ypm* gene acquired? In this regard, Carnoy et al. (2002) found that, upstream of the *ypm* gene, there is a 26-bp sequence homologous to recombination sites of *Escherichia coli* (Clerget, 1991), *Xanthomonas camp*estris (Dai et al., 1988), and several other microorganisms. They designated this site as *yrs* for *Yersinia* recombination site, and determined that this site is preserved among all *Y. pseudotuberculosis* strains examined as well as a strain of *Y. pestis*. The presence of *yrs* seems to be suggestive of a phage or a plasmid-mediated transfer of an ancestor gene of *ypm* from another microorganism. To explain the facts that the GC content of both the *ypm* gene and ORF4 are lower than the other DNA segments in this area, and that ORF4 is present in the *ypm*(-) strain as well as in the three types of *ypm*(+) strains, these two genes might have been individually transferred from the other species on separate occasions; first ORF4, and then *ypm*.

EPIDEMIOLOGY OF YPM-PRODUCING *Y. PSEUDOTUBERCULOSIS*

It is known that the *ypm* genes are not present in all *Y. pseudotuberculosis* strains. The initial study of Yoshino et al. (1995) revealed that the distribution of the *ypm*(+) strains correlates strongly with both the geographical origin and the environmental source of isolation (Yoshino et al., 1995). They examined 76 clinical strains and 128 strains isolated from animal and environmental sources. Among the clinical strains, 96.6% of the 58 Far East Asian strains were *ypmA* positive, whereas only 16.7% of the 18 European strains were positive for *ypmA*. The same tendency was seen among the non-human borne strains, i.e., 59.8% of the 117 Far Eastern strains were positive for *ypmA* genes but none of the 11 European strains possessed the *ypmA* gene. Thus, the frequency of *ypmA* (+) strains seems to be higher in Far East Asia than in Europe, and to be higher in clinical isolates than in strains isolated from animals or the environment. Uesiba et al. (1998) also reported a higher frequency of the *ypmA* genes among clinical strains than in non-human borne isolates in Japan.

Recently, a more extensive study of 2,235 *Y. pseudotuberculosis* strains was reported by Fukushima et al. (2001). They investigated the distribution of the *ypmA*, *ypmB* and *ypmC* genes as well as the distribution of the high-pathogenicity islands (HPI) and the truncated right half of the high-pathogenicity islands (R-HPI) among these isolates. Their results are summarized in Table 1. In this table, new data regarding the relationships between serotypes and the *ypm* genotypes are added to the original data (courtesy of H. Fukushima). Comparing the *ypm* genotypes of the clinical isolates from Europe and Far East Asia, it is quite remarkable that the *ypmA*(+), HPI(-) strains predominate in Far East Asia, constituting 97.4% of the total clinical isolates. On the other hand, although the number of strains analyzed was much lower, the same *ypmA*(+), HPI(-) genotype was rare in Europe where *ypm*(-), HPI(+)strains are most prevalent (62.5%). These results may be explained by the unequal distribution of the different *Y. pseudotuberculosis* serotypes between the two geographic areas. In fact, in Far East Asia, serotypes 4a and 4b account for 59.4% of the total clinical isolates, and all strains of these two serotypes are *ypmA*(+). In contrast, in Europe serotype 1a is the most prevalent and all strains belonging to this serotype are *ypmA*(-), HPI(+). However, even within a single serotype, such as 1b, a higher frequency of *ypmA* is apparent in Far East Asia than in Europe (100% vs. 0.0%).

The other interesting findings in their studies were, firstly, that the *ypmB*(+) genotype was not detected in the clinical isolates, but was present in 5.3% and 11.7%, respectively, of animal and environmental isolates; and secondly, that the *ypmC*(+) strains were also mostly from the non-human sources and not from human hosts. The reason for these discrepancies is currently unknown. Fukushima et al. (2002) also investigated the ability of the isolates to ferment melibiose, and found that all 93 *ypmB*(+) strains except one and all 235 *ypmC*(+) strains were defective in melibiose fermentation. In contrast, almost all of the *ypmA*(+) strains were capable of melibiose fermentation (Fukushima et al., 2001). They suggested that the ability to ferment melibiose might be correlated to the high pathogenicity of the *ypmA*(+) strains. Thus, it would be of interest to determine if the host specificity or pathogenicity of *ypmB*(+) and *ypmC*(+) strains would be changed by transforming them with the genes required for melibiose fermentation. Alternatively, there might be strong linkage disequilibrium between *ypmA* and the gene(s) needed for melibiose fermentation.

The higher pathogenicity of the *ypmA*(+) strains has drawn attention to the disparity in the distribution of the *ypmA* gene, and that this disparity might be related to the differences in clinical manifestations observed in Europe and Far East Asia. The main clinical symptoms of *Y. pseudotuberculosis* infection in Europe are fever and gastroenteric symptoms (Carniel et al., 1990). In contrast, the clinical features in Japan (Sato et al., 1983, Inoue et al., 1984) and Far Eastern Russia (Somov et al., 1973) are usually more diverse and severe. The major differences in clinical symptoms between the two areas are skin rash and desquamation, which are not seen in European patients but which are common in Far East Asia. Moreover, approximately 10% of infected children in Japan fulfill the

Table 1. Prevalence of the *ypm* genes and the high pathogenicity islands among *Y. pseudotuberculosis* strains of different serotypes.

Serotype	Group 1	Group 2	Group 3	Group 4	Group 5	Group 6
	ypmA, HPI-	*ypmA, HPI+*	*ypmB, HPI-*	*ypmC, R-HPI+*	*ypm-, HPI-*	*ypm-, HPI+*
FAR EAST ASIA						
HUMANS						
1a	0	0	0	0	0	0
1b	61	1	0	0	0	0
1c	0	0	0	0	0	0
2a	9	0	0	0	1	0
2b	41	0	0	0	0	0
2c	10	0	0	0	0	0
3	19	1	0	1	2	1
4a	8	0	0	0	1	0
4b	253	0	0	0	0	0
5a	60	0	0	0	0	0
5b	133	0	0	0	7	0
others	39	2	0	0	0	0
total	633	4	0	1	11	1
(%)	97.4	0.6	0.0	0.2	1.7	0.2
ANIMAL						
total	651	5	48	74	112	16
(%)	71.9	0.6	5.3	8.2	12.4	1.8
ENVIRONMENT						
total	302	0	45	0	37	0
(%)	78.6	0.0	11.7	0.0	9.6	0.0
EUROPE						
HUMANS						
1a	0	0	0	0	0	10
1b	0	0	0	0	2	5
1c	0	0	0	0	0	0
2a	0	0	0	0	0	0
2b	0	0	0	0	2	0
2c	0	0	0	0	0	0
3	0	0	0	4	0	0
4a	0	0	0	0	0	0
4b	0	0	0	0	0	0
5a	0	0	0	0	1	0
5b	0	0	0	0	0	0
others	0	0	0	0	0	0
total	0	0	0	4	5	15
(%)	0.0	0.0	0.0	16.7	20.8	62.5
ANIMAL						
total	3	0	0	156	45	67
(%)	1.1	0.0	0.0	57.6	16.6	24.7
ENVIRONMENT						
total	0	0	0	0	0	0
(%)	0.0	0.0	0.0	0.0	0.0	0.0

diagnostic criteria of Kawasaki syndrome, including coronary vasculitis (Baba et al., 1991). Acute renal failure, mostly tubulointerstitial nephritis, is also seen in the Japanese patients (Takeda et al., 1991). Thus, it would be of interest and of clinical importance to know whether the difference in clinical manifestations of *Y. pseudotuberculosis* infection between Far East Asia and Europe is related to the different distribution patterns of the *ypm* genes.

CLINICAL IMPLICATIONS OF YPM

SYMPTOMS OF *Y. PSEUDOTUBERCULOSIS* INFECTION

Infection with *Y. pseudotuberculosis* can be accompanied by multiple systemic symptoms as well as gastrointestinal tract symptoms. According to the study by Takeda et al. (1995) in Japan, among 399 pediatric patients with *Y. pseudotuberculosis* infection between 1980 and 1993, high grade fever (98%) was the most frequent symptom followed by skin rash (77%), diarrhea (62%), abdominal pain (50%), strawberry tongue (40%), conjunctival *infection* (21%) and cervical lymphadenopathy (10%). In most cases, patients recovered after 1-2 weeks of illness with the appearance of typical desquamation of skin in the tips of the fingers and toes (77%). However, in some patients, extra-gastrointestinal complications, such as acute renal failure (10%), coronary aneurysms (2.2%) and juvenile idiopathic arthritis (0.7%) developed following the acute phase of the disease (Konishi et al., 1997). In Europe, on the other hand, the clinical manifestations of *Y. pseudotuberculosis* infection are mainly gastrointestinal, although systemic disease characterized by septicemia and high mortality has been reported rarely in patients with underlying disorders such as hemochromatosis, diabetes or hepatic cirrhosis (Van Noyen et al., 1995; Ljungberg et al., 1995). Tertti et al. (1989) reported an outbreak involving 34 children with *Y. pseudotuberculosis* infection in Finland, among whom fever (53%), abdominal pain (38%), diarrhea (15%) and nausea and vomiting (15%) were the main symptoms. The involvement of extra-gastrointestinal sites such as skin rash, mucosal *infection* and late onset renal insufficiency was not reported. As mentioned in the previous section, these differences in the clinical manifestations between the two areas, i.e., Europe and Far East Asia, may be related to the disparity in the distribution of highly pathogenic strains of *Y. pseudotuberculosis*. In this regard, it will be of interest to investigate whether and how the co-existence of the ability for melibiose fermentation and the *ypmA* gene, both of which are most frequently found among the Far Eastern strains (Fukushima et al., 2001), is actually related to the systemic manifestations in the acute phase of the disease. It is highly likely that YPM contributes to the manifestation of systemic illnesses by activating a large proportion of T cells, inducing excessive amounts of inflammatory cytokines in patients. We will briefly view the roles of superantigens in other diseases such as toxic shock syndrome (TSS) and the newly described neonatal TSS-like exanthematous disease (NTED) in the next section.

DISEASES ASSOCIATED WITH OTHER SUPERANTIGEN-PRODUCING BACTERIA

Staphylococcal TSS is a prototype of superantigen-related diseases characterized by high-grade fever, hypotension, skin rash and desquamation and multi-organ dysfunction involving more than three organ systems (Wiesenthal et al., 1985). From these foci of infection, *S. aureus* that produces TSST-1, or sometimes, other pyrogenic superantigens, can be isolated (Schlievert et al., 1981; Bergdoll et al., 1981). TSS is associated with marked macrophage and T cell activation, resulting in the elevation of inflammatory cytokines (Miethke et al., 1993). In addition, TSST-1 responsive Vβ2 bearing T cells are expanded in the peripheral circulation of human TSS patients (Choi et al., 1990). The experimental mouse model has been established using either the constant subcutaneous infusion of TSST-1 (Parsonnet et al., 1987) or pre-sensitization of mice with D-galactosamine (Miethke et al., 1992). In these animal models, TNF-α and TNF-β secreted from the TSST-1-stimulated T cells played an essential role in inducing shock, and cyclosporin A was able to prevent the onset of shock. The importance of T cells and T cell-derived cytokines in TSS was further proven by adoptive transfer experiments in which *scid* mice reconstituted with syngeneic T cells became susceptible to TSST-1-mediated shock (Miethke et al., 1993b).

The neonatal TSS-like exanthematous disease (NTED) was first described in the English literature in 1998 (Takahashi et al., 1998). NTED occurs mainly in newborn infants and the clinical symptoms simulate TSS in high-grade fever, infection of nasopharyngeal mucosa and skin rash, but without hypotension and multi-organ dysfunction. Thrombocytopenia is one of the major symptoms in the acute phase. TSST-1-producing *S. aureus*, frequently methicillin-resistant *S. aureus* (MRSA), is isolated from the nasopharynx and umbilicus. In acute-phase patients, expansion of the TSST-1-responsive Vβ2-bearing T cells was observed in PBMC (Takahashi et al., 2000), and the concentrations of inflammatory cytokines such as IL-1β and IFN-γ in sera were elevated (Okada et al., 2000). However, the percentages of Vβ2-bearing T cells in patients rapidly decreased below normal levels by a clonal deletion mechanism, and most of the symptoms resolved spontaneously within a week after the onset of the disease. The precise pathogenesis of NTED is still unknown and the reasons why the affected neonates do not develop hypotension and why the inflammation in this disease is self-limited are currently under investigation.

The findings concerning these superantigen-related diseases reveal several important conditions that should be considered in demonstrating a pathogenic role for a superantigen in a particular disease. Firstly, isolation of bacteria that are capable of producing a superantigen is required. Secondly, expansion or deletion of T cells that bear the corresponding Vβ elements should be demonstrated in patients. Thirdly, the disease is associated with T cell and macrophage activation. Animal experiments that reproduce similar illnesses by in vivo administration of a superantigen would be an additional requirement. With these considerations in mind, I will discuss the clinical role of YPM in *Y. pseudotuberculosis* infection in the following sections.

INVOLVEMENT OF YPM IN *Y. PSEUDOTUBERCULOSIS* INFECTION

A study of the role of superantigen in clinical *Y. pseudotuberculosis* infection was reported by us in 1997 (Abe et al., 1997). First, we attempted to detect antibodies against YPM in patient sera by ELISA and Western hybridization. By solid phase ELISA using rYPM, IgG anti-YPM antibodies were detected in 20 out of 33 (61%) patients. In most cases, a high titer was observed after the second week of the illness. In healthy control children and in intravenous immunoglobulin (IVIG) preparations, we were unable to detect anti-YPM antibodies, although the presence of IgG antibodies to staphylococcal superantigens has been reported in IVIG preparations (Takei et al., 1993). The difference may be explained by a very low frequency of the *Yersinia* infection compared with the high prevalence of *S. aureus* in Japan (Fukami, 1986). More importantly, the patients with systemic

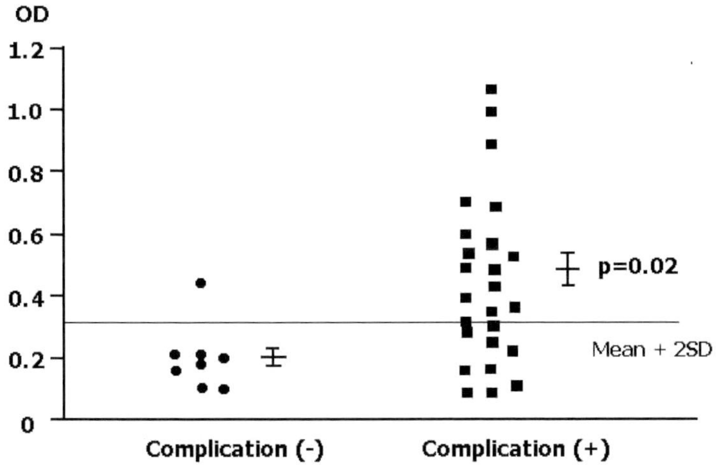

Figure 3. IgG anti-YPM antibodies in patients with *Y. pseudotuberculosis* infection measured by ELISA. Patients were classified by the presence (filled squares) or the absence (filled circles) of systemic complications (renal dysfunction, acute abdomen, arthritis, lymphadenopathy, and coronary aneurysms). Anti-YPM titer was compared between the two groups. Each bar indicates the mean ± 2SE. * p=0.009 by Student's *t*-test.

symptoms such as transient renal failure, arthritis, skin rash, and coronary aneurysms had significantly higher titers of anti-YPM than the patients with gastrointestinal symptoms alone (Figure 3). Although the higher anti-YPM titer does not necessarily mean a higher production of YPM in the host, it implies that the host's immune response against YPM had been more vigorous in the patients with systemic illnesses. The results of Western hybridizations confirmed the production of anti-YPM in the convalescent phase patients (Figure 4).

Next, we examined the TCR-Vβ repertoire of T cells from patients with acute *Y. pseudotuberculosis* infection. PBMC stimulated with antiCD3 antibody and IL-2

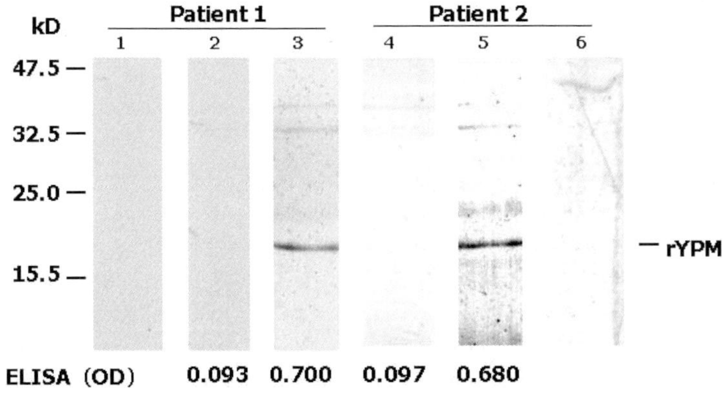

Figure 4. Western blot analysis of patient sera. rYPM (lanes 2-5), and maltose binding protein (MBP) control (lanes 1 and 6), were incubated with sera obtained from Patient 1 on day 6 (lanes 1 and 2) or day 21 (lane 3); and from Patient 2 on day 7 (lane 4) or day 16 (lanes 5 and 6). Anti-YPM ELISA results for each sample are also indicated. Molecular weight markers are noted on the left and the size of the rYPM is indicated on the right.

were analyzed by immunofluorescence and flow cytometry. A significant increase of Vβ3-bearing T cells was seen in the patients compared with control children. Other Vβ elements examined, such as Vβ2, Vβ8.1-2 and Vβ13.6, showed little difference between the two groups. By using semi-quantitative RT-PCR, in 7 out of 14 patients the percentages of Vβ3 positive T cells exceeded 2 SD above the mean value obtained for healthy children. However, the percentages of Vβ9, 13.1, and 13.2 bearing T cells were not increased in the patients, although these T cell repertoires were also stimulated and expanded by rYPM in the in vitro experiments (Figure 1). More importantly, the expansion of Vβ3-bearing T cells was restricted to the acute phase of the illness. When 7 paired samples from the acute and convalescent phase were examined by RT-PCR, all 3 patients who had expansion of Vβ3 T cells in the acute phase showed a significant decrease in the percentage of Vβ3 T cells during the convalescent phase of their illness (Figure 5). Currently, it is not known why the expansion of T cells in patients is confined to Vβ3-bearing lines and not to the other YPM-responsive Vβ repertoires such as Vβ9, 13.1, and 13.2. Recently, Chen et al. reported the varied response patterns to a superantigen by different TCR Vβ-expressing T cells in mice. They administered SEA or YPM to mice using subcutaneously implanted mini-osmotic pumps. Ten days after implantation, Vβ3(+)CD4(+) T cells were expanded more vigorously than Vβ11(+)CD4(+) T cells in mice that received SEA, and Vβ7(+)CD4(+) T cells were more enriched than Vβ8(+)CD4(+) T cells in mice that received YPM (Chen et al., 2002). Although the precise mechanism behind the disparity in the Vβ-specific responses in these prolonged exposure models is not known, the authors speculated that the different binding affinity of each Vβ-bearing TCR to a superantigen-MHC complex might affect the responsiveness and the life span of the target T cells. In our patients, it may also be possible that T cells bearing Vβ3, which has higher amino acid sequence homology to mouse Vβ7 than to mouse Vβ8.1-2 TCR element, are more likely to be stimulated by a limited amount of YPM than the Vβ9, Vβ13.1 or Vβ13.2 bearing T cells, and are preferentially expanded.

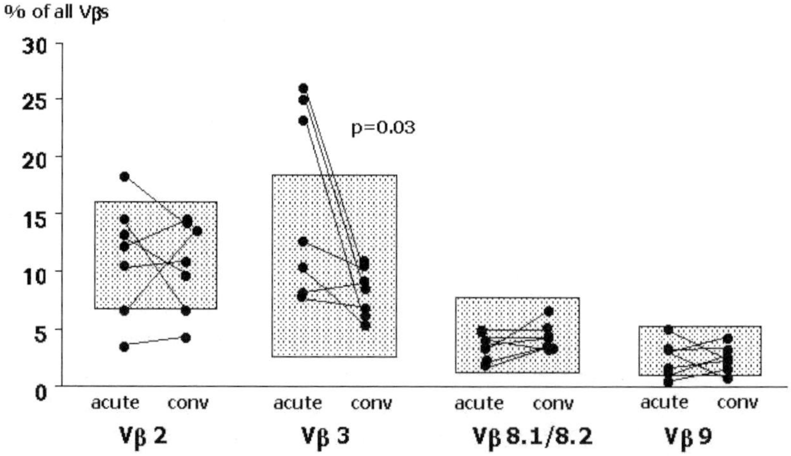

Figure 5. TCR Vß expression in T cells from acute (within 20 days of onset) and convalescent (later than 120 days) patients analyzed by RT-PCR. Each line indicates the paired samples. The shaded areas represent the control mean ± 2SD for the percentages of each Vß positive T cells determined for a group of healthy children. * p=0.03 by Student's paired *t*-test.

In the context of the pathogenicity of superantigens, it has been stressed that the activation of T cells by superantigens favors the overproduction of proinflammatory cytokines such as TNF-α, TNF-β, IL-1, IFN-γ and IL-6, which contribute to endothelial cell and organ damage (Miethke et al., 1993, Kotb et al., 1995). In this regard, it is of interest to study how and to what degree YPM influences the cytokine production by PBMC in patients with *Y. pseudotuberculosis* infection. In preliminary experiments, we measured the serum levels of proinflammatory cytokines such as IFN-γ, IL-6 and IL-12p40 by ELISA, and compared the results to the anti-YPM titers obtained from these patients (Abe et al., 2000). Although inflammatory cytokine levels in the acute sera were not significantly correlated to the anti-YPM titer of the same patients' sera in the convalescent phase, anti-YPM titer as well as IFN-γ, IL-6 and IL-12p40 levels in patients with acute renal failure were significantly elevated compared with patients without this complication. Concentrations of IL-12p40 in sera were also elevated in patients with erythematous skin rash compared with patients without skin rash or other complications. In addition, we found that human PBMC stimulated in vitro with rYPM secreted IL12p40 into the culture supernatant and that cutaneous lymphocyte antigen (CLA), a skin homing receptor, was expressed on T cells. These findings suggest that YPM may cause the activation of Th1 type T cells through the enhanced production of IL-12 from PBMC, which may be involved in the onset of the acute renal failure and skin rash complications of *Y. pseudotuberculosis* infection. So far, the functions and the kinetics of the YPM-responding T cells during the course of the infection are not fully understood. It would be of interest and of clinical importance to clarify whether and how these YPM-responsive T cells and APC contribute to the development of the systemic symptoms in the human infection by this microorganism.

PATHOGENIC ROLE OF YPM IN EXPERIMENTAL ANIMALS

The function of YPM has been studied in experimental animals by two modes. One is to administer YPM directly into the animals by injection or by osmotic pumps, and the other is to infect the animals by live bacteria that produce YPM. Uchiyama's group used the former in 1997 (Miyoshi-Akiyama et al., 1997) and in 2002 (Chen et al., 2002). They found that subcutaneously injected YPM (purified protein from a bacterial lysate) was able to induce toxic shock and lethality in mice pretreated with D-galactosamine. This toxicity was dependent on T cells and their production of the inflammatory cytokines, because monoclonal antibodies against TNF-α, or IFN-γ, as well as CD4, blocked the lethal effect of YPM in mice. Later (Chen et al., 2002), they utilized the osmotic pump system and implanted the pump subcutaneously to ensure the continuous exposure of mice to YPM for a longer period than by the simple injection method. The mice receiving YPM by this procedure did not develop lethal shock but showed a prolonged, polyclonal expansion of T cells in the spleen. However, as mentioned in the previous section, there was a discrepancy in the TCR Vβ associated with the T cells expanded in the spleen, i.e., Vβ7(+)CD4(+) T cells were preferentially expanded compared to Vβ8(+)CD4(+) T cells and CD8(+) T cells with either of the Vβ segments. More importantly, CD4(+) T cells from mice that received SEA by osmotic pump responded to a secondary in vitro stimulation with SEA by proliferation and by enhanced production of IL-4 and IFN-γ. Because the conditions used in these experiments seem to be more relevant to the situation seen in human infection, their results may shed light into the questions why and how, in human *Y. pseudotuberculosis* infection, the YPM-responding T cells were resistant to apoptosis and were able to produce inflammatory cytokines for days during the acute phase of illnesses. In this respect, it will be of interest to examine whether some of the systemic symptoms of human *Y. pseudotuberculosis* infection, such as interstitial nephritis and coronary artery aneurysm, could be reproduced by continuous exposure of mice to YPM.

A study of mice infected with YPM-producing bacteria has also been reported (Carnoy et al., 2000). In this model, the virulence of a YPM-producing wild-type strain and an isogenic YPM-deficient mutant strain was compared by infecting mice intravenously with the strains. The mice that received the YPM-deficient mutant strain survived longer than the mice that received the wild-type strain, although a similar degree of colonization of spleen, liver, and lungs was observed in the two groups of mice. The precise mechanism for the higher mortality in the mice infected with the wild-type strain is now under investigation, but these findings may further argue for a pathogenic role of YPM in *Y. pseudotuberculosis* infection.

CONCLUDING REMARKS

Since the finding of superantigen production by a strain of *Y. pseudotuberculosis* in 1993, much progress has been made in our understanding of the genetics, the epidemiology and the clinical roles of the superantigen, YPM. First of all, the recent advances in the genome sequencing projects of the *Yersinia* species revealed that the *ypm* gene is unique to *Y. pseudotuberculosis,* and that the ancestral *ypm* gene might have been transferred from another (superantigen-producing?) microorganism by a phage integration mechanism. Sequence analysis of the three variant strains (*ypmA*, *ypmB* and *ypmC*) also revealed that these *ypm* genes are located at a constant site of the chromosome but that the locus surrounding the *ypm* gene is rather unstable, resulting in the loss of several ORFs in *ypmB*(+) and *ypmC*(+) strains.

Epidemiological studies have focused on the different distribution patterns of the YPM-producing strains. It is now evident that the frequency of *ypmA*(+) strains is higher in Far East Asia than in Europe, and is higher among clinical isolates than in strains isolated from animal and environmental sources, while *ypmB*(+) strains, and to a lesser extent *ypmC*(+) strains, are found mostly in animals and the environment. The observed discrepancy in host association could result if *ypmA*(+) strains are more likely to cause serious manifestations in humans, and therefore more likely to be isolated in the clinical setting. Alternatively, the ability to ferment melibiose may be an advantage in human infection. However, because YPMa and YPMb have a similar superantigenic capacity towards human PBMC, it is still unknown which factor may be responsible for the different virulence between the two variant *ypm*(+) strains. Functional analysis of the proteins encoded by the ORFs surrounding the *ypm* genes, and their affect on *ypm* expression, might give us the answer to this question in the near future.

With regard to the clinical role of YPM in the *Y. pseudotuberculosis* infection, evidence is accumulating that YPM is produced in vivo in patients during the acute phase of the disease. It stimulates T cells bearing the Vβ3 element of TCR and seems to support the production of Th1-type cytokines during the acute phase. Thus, YPM may be involved in the acute systemic manifestations that are similar to those seen in infections by *S. aureus* and *S. pyogenes*, both of which are well known producers of the superantigens. However, many unsolved questions remain in our understanding of the pathophysiologic roles of YPM. Which factors (cytokines, chemokines, other chemical mediators of inflammation) are important in the manifestation of the individual symptoms, such as gastroenteritis, erythematous skin rash, acute renal failure and coronary artery aneurysm? Which immune cells (T cells, macrophages and B cells) are stimulated by YPM to produces such factors? Does YPM influence the Th1-Th2 homeostasis in the local tissues by inducing such factors? Why are young children more susceptible to the infection and more likely to present systemic manifestations? Why don't the patients with this infection develop hypotension as seen in TSS patients? Does YPM stimulate the development of auto-reactive T cells? Future studies directed to these questions will help to understand the possible roles of YPM in the *Yersinia*-related diseases. In addition, X-ray crystallographic analysis and molecular structure/function studies of YPM, along with studies of the other prototypic superantigens, will provide an understanding of unique aspects of YPM, compared to the prototypic Gram-positive superantigens, in the interaction with APC and T cells. This information will be important in directing new investigations into the treatment of infections caused by *Y. pseudotuberculosis*.

ACKNOWLEDGEMENTS

Preparation of this manuscript was supported by a Grant for Child Health and Development from the Ministry of Health, Labour and Welfare and a grant from the Japan Human Sciences Foundation. Dr. Hiroshi Fukushima is acknowledged for providing data included in Table 1 in this chapter.

REFERENCES

Abe, J., Takeda, T., Watanabe, Y., Nakao, H., Kobayashi, N., Leung, D.Y.M., and Kohsaka, T. 1993. Evidence for superantigen production by *Yersinia pseudotuberculosis*. J. Immunol. 151: 4183-4188.

Abe, J., Onimaru, M., Matsumoto, S., Noma, S., Baba, K., Ito, Y., Kohsaka, T., Takeda, T. 1997. Clinical role for a superantigen in *Yersinia pseudotuberculosis* infection. J. Clin. Invest. 99: 1823-1830.

Abe, J., Tatewaki, M., Nogami, H., Takeda, T., Matsumoto, S., and Baba, K. 2000. Increased Th1 type cytokines in *Yersinia pseudotuberculosis* infection complicated with acute renal failure. FASEB J. 14: A1058

Acharya, K.R., Passalacqua, E.F., Jones, E.Y., Harlos, K., Stuart, D.I., Brehm, R.D., and Tranter, H.S. 1994. Structural basis of superantigen action inferred from crystal structure of toxic-shock syndrome toxin-1. Nature 367: 94-97.

Akira, S., Takeda, K., and Kaisho, T. 2001. Toll-like receptors: critical proteins linking innate and acquired immunity. Nat. Immunol. 2: 675-680.

Baba, K., Takeda, N., and Tanaka, M. 1991. Cases of *Yersinia pseudotuberculosis* infection having diagnostic criteria of Kawasaki disease. Contrib. Microbiol. Immunol. 12: 292-296.

Bayles, K.W., and Iandolo, J.J. 1989. Genetic and molecular analysis of the gene encoding staphylococcal enterotoxin D. J. Bacteriol. 171: 4799-4806.

Bercovier, H., and Mollaret, H.H. 1984. Genus *Yersinia*. In: Bergey's manual of systemic bacteriology, Vol. 1. N.R. Krieg, and J.G. Holt, ed. The Williams and Wilkins Co., Baltimore. p. 498-506.

Bergdoll, M.S., Crass, B.A., Reiser, R.F., Robbins, R.N., and Davis, J.P. 1981. A new staphylococcal enterotoxin, enterotoxin F, associated with toxic-shock-syndrome *Staphylococcus aureus* isolates. Lancet 1: 1017-1021.

Betley, M.J., and Mekalanos, J.J. 1985. Staphylococcal enterotoxin A is encoded by a phage. Science 229: 185-187.

Bohach, G.A., Fast, D.J., Nelson, R.D., and Schlievert, P.M. 1990. Staphylococcal and streptococcal pyrogenic toxins involved in toxic shock syndrome and related illnesses. Crit. Rev. Microbiol. 17: 251-272.

Carlsson, R., Fischer, H., and Sjogren, H.O. 1988. Binding of staphylococcal enterotoxin A to accessory cells is a requirement for its ability to activate human T cells. J. Immunol. 140: 2484-2488.

Carniel, E., Mollaret, H.H. 1990. Yersiniosis. Comp. Immunol. Microbiol. Infect. Dis. 13: 51-58

Carnoy, C., and Simonet, M.1999. *Yersinia pseudotuberculosis* superantigenic toxins, In: Bacterial protein toxins: a comprehensive sourcebook. J.E Alouf, and J.H. Freer, ed. Academic Press, London. p. 611-622.

Carnoy, C., Mullet, C., Muller-Alouf, H., Leteurtre, E., and Simonet, M. 2000. Superantigen YPMa exacerbates the virulence of *Yersinia pseudotuberculosis* in mice. Infect. Immun. 68: 2553-2559.

Carnoy, C., Floquet, S., Marceau, M., Sebbane, F., Haentjens-Herwegh, S., Devalckenaere, A., and Simonet, M. 2002. The superantigen gene *ypm* is located in an unstable chromosomal locus of *Yersinia pseudotuberculosis*. J. Bacteriol. 184: 4489-4499.

Chen, L., Koyanagi, M., Fukuda, K., Imanishi, K., Yagi, J., Kata, H., Miyoshi-Akiyama, T., Zhang, R., Miwa, K., and Uchiyama, T. 2002. Continuous exposure of mice to superantigenic toxins induces a high-level protracted expansion and an immunological memory in the toxin-reactive CD4+ T cells. J. Immunol. 168: 3817-3824.

Chien, Y.H., Gascoigne, N.R., Kavaler, J., Lee, N.E., and Davis, M.M. 1984. Somatic recombination in a murine T-cell receptor gene. Nature 309: 322-326.

Choi, Y., Kotzin, B., Herron, L., Callahan, J., and Marrack, P. 1989. Interaction of *S. aureus* toxin superantigens with human T cells. Proc. Natl. Acad. Sci. USA 86: 8941-8945.

Choi, Y.W., Herman, A., DiGiusto, D., Wade, T., Marrack, P., and Kappler, J. 1990. Residues of the variable region of the T-cell-receptor beta-chain that interact with *S. aureus* toxin superantigens. Nature 346: 471-473.

Choi, Y., Jafferty, J.A., Clements, J.R., Todd, J.K., Gelfand, E.W., Kappler, J., Marrack, P., and Kotzin, B.L. 1990. Selective expansion of T cells expressing Vβ2 in toxic shock syndrome. J. Exp. Med. 172: 981-984.

Clerget, M. 1991. Site-specific recombination promoted by a short DNA segment of plasmid R1 and by a homologous segment in the terminus region of the Escherichia coli chromosome. New Biol. 3: 780-788.

Cone, L.A., Woodard, D.R., Schlievert, P.M., Tomory, G.S. 1987. Clinical and bacteriologic observations of a toxic shock-like syndrome due to *Streptococcus pyogenes*. N. Engl. J. Med. 317: 146-149.

Dai, H., Chow, T.Y., Liao, H.J., Chen, Z.Y., and Chiang, K.S. 1988. Nucleotide sequences involved in the neolysogenic insertion of folamentous phage Cf16-v1 into the *Xanthomonas campestris* pv. *citri* chromosome. Virology 167: 613-620.

Dellabona, P., Peccoud, J., Kappler, J., Marrack, P., Benoist, C., and Mathis, D. 1990. Superantigens interact with MHC class II molecules outside of the antigen groove. Cell 62: 1115-1121.

Dowd, J.E., Karr, R.W., and Karp, D.R. 1996. Functional activity of staphylococcal enterotoxin A requires interactions with both the alpha and beta chains of HLA-DR. Mol. Immunol. 33: 1267-1274.

Dyson, P.J., Knight, A.M., Fairchild, S., Simpson, E., Tomonari, K. 1991. Genes encoding ligands for deletion of V beta 11 T cells cosegregate with mammary tumor virus genomes. Nature 349: 531-532.

Fast, D.J., Schlievert, P.M., and Nelson, R.D. 1989. Toxic shock syndrome-associated staphylococcal and streptococcal pyrogenic toxins are potent inducers of tumor necrosis factor production. Infect. Immun. 57: 291-294.

Festenstein, H. 1976. The Mls system. Transplant. Proc. 8: 339-42.

Fields, B.A., Malchiodi, E.L., Li, H., Ysern, X., Stauffacher, C.V., Schlievert, P.M., Karjalainen, K., and Mariuzza, R.A. 1996. Crystal structure of a T-cell receptor beta-chain complexed with a superantigen. Nature 384: 188-192.

Fleischer, B., and Schrezenmeier, H. 1988. T cell stimulation by staphylococcal enterotoxins. J. Exp. Med. 167: 1697-1707.

Frankel, W.N., Rudy, C., Coffin, J.M., and Huber, B.T. 1991. Linkage of Mls genes to endogenous mammary tumor viruses of inbred mice. Nature 349: 526-528.

Fraser, J.D., Urban, R.G., Strominger, J.L., and Robinson, H. 1992. Zinc regulates the function of two superantigens. Proc. Natl. Acad. Sci. USA. 89: 5507-5511.

Fukami, T. 1986. [Frequency of *Yersinia* species isolated from sporadic patients with diarrhea] (in Japanese). Media Circle 31: 314-318.

Fukushima, H., Matsuda, Y., Seki, R., Tsubokura, M., Takeda, N., Shubin, F.N., Paik, I.K., and Zheng, X.B. 2001. Geographical heterogeneity between Far Eastern and Western countries in prevalence of the virulence plasmid, the superantigen *Yersinia pseudotuberculosis*-derived mitogen, and the high-pathogenicity island among *Yersinia pseudotuberculosis* strains. J. Clin. Microbiol. 39: 3541-3547.

Goshorn, S.C., and Schlievert, P.M. 1989. Bacteriophage association of streptococcal pyrogenic exotoxin type C. J. Bacteriol. 171: 3068-3073.

Hanakawa, Y., Schechter, N.M., Lin, C., Garza, L., Li, H., Yamaguchi, T., Fudaba, Y., Nishifuji, K., Sugai, M., Amagai, M., and Stanley, J.R. 2002. Molecular mechanisms of blister formation in bullous impetigo and staphylococcal scalded skin syndrome. J. Clin. Invest. 110: 53-60.

Hudson, K.R., Tiedemann, R.E., Urban, R.G., Lowe, S.C., Strominger, J.L., and Fraser, J.D. 1995. Staphylococcal enterotoxin A has two cooperative binding sites on major histocompatibility complex class II. J. Exp. Med. 182: 711-720.

Ikejima, T., Dinarello, C.A., Gill, D.M., and Wolff, S.M. 1984. Induction of human interleukin-1 by a product of *Staphylococcus aureus* associated with toxic shock syndrome. J. Clin. Invest.73: 1312-1320.

Inoue, M., Nakashima, H., Ueba, O., Ishida, T., Date, H., Kobashi, S., Takagi, K., Nishu, T., and Tsubokura, M. 1984. Community outbreak of *Yersinia pseudotuberculosis*. Contrib. Microbiol. Immunol. 28: 883-891.

Ito, Y., Abe, J., Yoshino, K., Takeda, T., and Kohsaka, T. 1995. Sequence analysis of the gene for a novel superantigen produced by *Yersinia pseudotuberculosis* and expression of the recombinant protein. J. Immunol. 154: 5896-5906.

Ito, Y., Seprenyi, G., Abe, J., and Kohsaka, T. 1999. Analysis of functional regions of YPM, a superantigen derived from gram-negative bacteria. Eur. J. Biochem. 263: 326-337.

Jardetzky, T.S., Brown, J.H., Gorga, J.C., Stern, L.J., Urban, R.G., Chi, Y.I., Stauffacher, C., Strominger, J.L., and Wiley, D.C. 1994. Three-dimensional structure of a human class II histocompatibility molecule complexed with superantigen. Nature 368: 711-718.

Janeway, C.A.Jr., Yagi, J., Conrad, P.J., Katz, M.E., Jones, B., Vroegop, S., and Buxser, S. 1989. T-cell responses to Mls and to bacterial proteins that mimic its behavior. Immunol. Rev. 107: 61-88.

Janeway, C.A. Jr., and Medzhitov, R. 2000. Innate immune recognition. Annu. Rev. Immunol. 20: 197-216.

Jardetzky, T.S., Brown, J.H., Gorga, J.C., Stern, L.J., Urban, R.G., Chi, Y.I., Stauffacher, C., Strominger, J.L., and Wiley, D.C. 1994. Three-dimensional structure of a human class II histocompatibility molecule complexed with superantigen. Nature 368: 711-718.

Jarraud, S., Peyrat, M.A., Lim, A., Tristan, A., Bes, M., Mougel, C., Etienne, J., Vandenesch, F., Bonneville, M., and Lina, G. 2001. egc, a highly prevalent operon of enterotoxin gene, form a putative nursery of superantigens in *Staphylococcus aureus*. J. Immunol. 166: 669-677.

Johnson, L.P., Tomai, M.A., and Schlievert, P.M. 1986. Bacteriophage involvement in group A streptococcal pyrogenic exotoxin A production. J. Bacteriol. 166: 623-627.

Jupin, C., Anderson, S., Damais, C., Alouf, J.E., and Parant, M. 1988. Toxic shock syndrome toxin 1 as an inducer of human tumor necrosis factors and gamma interferon. J. Exp. Med. 167: 752-761.

Kappler, J., Kotzin, B., Herron, L., Gelfand, E.W., Bigler, R.D., Boylston, A., Carrel, S., Posnett, D.N., Choi, Y.W., and Marrack, P. 1989. Vβ-specific stimulation of human T cells by staphylococcal toxins. Science 244: 811-813.

Kim, J., Urban, R.G., Strominger, J.L., and Wiley, D.C. 1994. Toxic shock syndrome toxin-1 complexed with a class II major histocompatibility molecule HLA-DR1. Science 266: 1870-1874.

Konig, B., Koller, M., Prevost, G., Piemont, Y., Alouf, J.E., Schreiner, A., and Konig, W. 1994. Activation of human effector cells by different bacterial toxins (leukocidin, alveolysin, and erythrogenic toxin A): generation of interleukin-8. Infect. Immun. 62: 4831-4837.

Konishi, N., Baba, K., Abe, J., Maruko, T., Waki, K., Takeda, N., and Tanaka, M. 1997. A case of Kawasaki disease with coronary artery aneurysms documenting *Yersinia pseudotuberculosis* infection. Acta Paediatr. 86: 661-664.

Kotb, M. 1995. Bacterial pyrogenic exotoxins as superantigens. Clin. Microbiol. Rev. 8: 411-426.

Lamphear, J.G., Mollick, J.A., Reda, K.B., and Rich, R.R. 1996. Residues near the amino and carboxyl termini of staphylococcal enterotoxin E independently mediate TCR V beta-specific interactions. J. Immunol. 156: 2178-2185.

Leung, D.Y.M., Gately, M., Trumble, A., Ferguson-Darnell, B., Schlievert, P., and Picker, L.J. 1995. Bacterial superantigens induce T cell expression of the skin-selective homing receptor, the cutaneous lymphocyte-associated antigen, via stimulation of interleukin 12 production. J. Exp. Med. 181: 747-753.

Li, P.L., Tiedemann, R.E., Moffat, S.L., and Fraser, J.D. 1997. The superantigen streptococcal pyrogenic exotoxin C (SPE-C) exhibits a novel mode of action. J. Exp. Med. 186: 375-383.

Lindsay J.A., Ruzin A., Ross H.F., Kurepina N., Novick R.P. 1998. The gene for toxic shock toxin is carried by a family of mobile pathogenicity islands in *Staphylococcus aureus*. Mol. Microbiol. 29:527-43.

Ljungberg, P., Valtonen, M., Harjola, V.P., Kaukoranta-Tolvanen, S.S., and Vaara, M. 1995. Report of four cases of *Yersinia pseudotuberculosis* septicemia and a literature review. Eur. J. Clin. Microbiol. Infect. Dis. 14: 804-810.

Marrack, P., Kappler, J. 1990. The staphylococcal enterotoxins and their relatives. Science 248: 705-711.

Miethke, T., Wahl, C., Heeg, K., Echtenacher, B., Krammer, P.H., and Wagner, H. 1992. T cell-mediated lethal shock triggered in mice by the superantigen staphylococcal enterotoxin B: critical role of tumor necrosis factor. J. Exp. Med. 175: 91-98.

Miethke, T., Wahl, C., Regele, D., Gaus, H., Heeg, K., and Wagner, H. 1993a. Superantigen mediated shock: a cytokine release syndrome. Immunobiol. 189: 270-284.

Miethke, T., Duschek, K., Wahl, C., Heeg, K., and Wagner, H. 1993b. Pathogenesis of the toxic shock syndrome: T cell mediated lethal shock caused by the superantigen TSST-1. Eur. J. Immunol. 23: 1494-1500.

Miyoshi-Akiyama.T., Abe, A., Kato, H., Kawahara, K., Narimatsu, H., and Uchiyama, T. 1995. DNA sequencing of the gene encoding a bacterial superantigen, *Yersinia pseudotuberculosis*-derived mitogen (YPM), and characterization of the gene product, cloned YPM. J. Immunol. 154: 5228-5234.

Miyoshi-Akiyama, T., Fujimaki, W., Yan, X.J., Yagi, J., Imanishi, K., Kato, H., Tomonari, K., and Uchiyama, T. 1997. Identification of murine T cells reactive with the bacterial superantigen *Yersinia pseudotuberculosis*-derived mitogen (YPM) and factors involved in YPM-induced toxicity in mice. Microbiol. Immunol. 41: 345-352.

Neumann, B., Emmanuilidis, K., Stadler, M., and Holzmann, B. 1998. Distinct functions of interferon-gamma for chemokine expression in models of acute lung inflammation. Immunology 95: 512-21.

Newman, I., and Wilkinson, P.C. 1996. The bacterial superantigen staphylococcal enterotoxin B stimulates lymphocyte locomotor capacity during culture in vitro. Immunology 87: 428-433.

Okada, T., Makimoto, A., Kitamura, A., Furukawa, M., Miwa, T., and Sakai, R. 2000. [An emerging neonatal exanthematous disease induced by a toxin-producing MRSA] (in Japanese). Kansenshougakuzasshi 74: 573-578.

Papageorgiou, A.C., Acharya, K.R., Shapiro, R., Passalacqua, E.F., Brehm, R.D., and Tranter, H.S. 1995. Crystal structure of the superantigen enterotoxin C2 from *Staphylococcus aureus* reveals a zinc-binding site. Structure 3: 769-779.

Papageorgiou, A.C., Collins, C.M., Gutman, D.M., Kline, J.B., O'Brien, S.M., Tranter, H.S., and Acharya, K.R. 1999. Structural basis for the recognition of superantigen streptococcal pyrogenic exotoxin A (SpeA1) by MHC class II molecules and T-cell receptors. EMBO J. 18: 9-21.

Parsonnet, J., Gillis, Z.A., Richter, A.G., and Pier, G.B. 1987. A rabbit model of toxic shock syndrome that uses a constant, subcutaneous infusion of toxic shock syndrome toxin 1. Infect. Immun. 55: 1070-1076.

Proft, T., Arcus, V.L., Handley, V., Baker, E.N., and Fraser, J.D. 2001. Immunological and biochemical characterization of streptococcal pyrogenic exotoxins I and J (SPE-I and SPE-J) from *Streptococcus pyogenes*. J. Immunol. 166: 6711-6719.

Pullen, A.M., Wade, T., Marrack, P., and Kappler, J.W. 1990. Identification of the region of T cell receptor beta chain that interacts with the self-superantigen MIs-1a. Cell 61: 1365-1374.

Ramamurthy, T., Yoshino, K., Abe, J., Ikeda, N., and Takeda, T. 1997. Purification, characterization and cloning of a novel variant of the superantigen *Yersinia pseudotuberculosis*-derived mitogen. FEBS Lett. 413: 174-176.

Roussel, A., Anderson, B.F., Baker, H.M., Fraser, J.D., and Baker, E.N. 1997. Crystal structure of the streptococcal superantigen SPE-C: dimerization and zinc binding suggest a novel mode of interaction with MHC class II molecules. Nat. Struct. Biol. 4: 635-643.

Saito, H., Kranz, D.M., Takagaki, Y., Hayday, A.C., Eisen, H.N., and Tonegawa, S. 1984. Complete primary structure of a heterodimeric T-cell receptor deduced from cDNA sequences. Nature 309: 757-762.

Sato, K., Ouchi, K., and Taki, M. 1983. *Yersinia pseudotuberculosis* infection in children, resembling Izumi fever and Kawasaki syndrome. Pediatr. Infect. Dis. 2: 123-126.

Schad, E.M., Zaitseva, I., Zaitsev, V.N., Dohlsten, M., Kalland, T., Schlievert, P.M., Ohlendorf, D.H., and Svensson, L.A. 1995. Crystal structure of the superantigen staphylococcal enterotoxin type A. EMBO J. 14: 3292-3301 .

Schlievert, P.M., Shands, K.N., Dan, B.B., Schmid, G.P., and Nishimura, R.D. 1981. Identification and characterization of an exotoxin from *Staphylococcus aureus* associated with toxic-shock syndrome. J. Infect. Dis. 143: 509-516.

Scholl, P.R., Diez, A., Karr, R., Sekaly, R.P., Trowsdale, J., and Geha, R.S. 1990. Effect of isotypes and allelic polymorphism on the binding of staphylococcal exotoxins to MHC class II molecules. J. Immunol. 144: 226-230.

Somov, G.P., Martinevsky, I.L. 1973. New facts about *pseudotuberculosis* in the USSR. Contrib. Microbiol. Immunol. 2: 214-216.

Spero, L., and Morlock, B.A. 1978. Biological activities of the peptides of staphylococcal enterotoxin C generated by limited tryptic hydrolysis. J. Biol. Chem. 253: 8787-8791.

Sundstrom, M., Abrahmsen, L., Antonsson, P., Mehindate, K., Mourad, W., and Dohlsten, M. 1996. The crystal structure of staphylococcal enterotoxin type D reveals Zn^{2+}-mediated homodimerization. EMBO. J. 15: 6832-6840.

Swaminathan, S., Furey, W., Pletcher, J., and Sax, M. 1995. Residues defining V beta specificity in staphylococcal enterotoxins. Nat. Struct. Biol. 2: 680-686.

Takahashi, N., Nishida, H., Kato, H., Imanishi, K., Sakata, Y., and Uchiyama, T. 1998. Exanthematous disease induced by toxic shock syndrome toxin 1 in the early neonatal period. Lancet 351: 1614-1619.

Takahashi, N., Kato, H., Imanishi, K., Miwa, K., Yamanashi, S., Nishida, H., and Uchiyama, T. 2000. Immunopathophysiological aspects of an emerging neonatal infectious disease induced by a bacterial superantigen. J. Clin. Invest. 106: 1409-1415.

Takeda, N., Usami, I., Fujita, A., Baba, K., and Tanaka, M. 1991. Renal complications of Yersinia pseudotuberculosis infection in children. Contrib. Microbiol. Immunol. 12: 301-306.

Takeda, N., Konishi, N., and Nishida, Y. 1996. [Similarities of clinical symptoms between Kawasaki disease and infection by *Yersinia pseudotuberculosis*] (in Japanese). Shounikashinryou 59: 47-53.

Takei, S., Arora, Y.K., and Walker, S.M. 1993. Intravenous immunoglobulin contains specific antibodies inhibitory to activation of T cells by staphylococcal toxin superantigens. J. Clin. Invest. 91: 602-607.

Tan, L.C., Gudgeon, N., Annels, N.E., Hansasuta, P., O'Callaghan, C.A., Rowland-Jones, S., McMichael, A.J., Rickinson, A.B., and Callan, M.F. 1999. A re-evaluation of the frequency of CD8+ T cells specific for EBV in healthy virus carriers. J Immunol 162: 1827-1835.

Tertti, R., Vuento, R., Mikkola, P., Granfors, K., Makela, A.L., and Toivanen, A. 1989. Clinical manifestation of *Yersinia pseudotuberculosis* infection in children. Eur. J. Clin. Microbiol. Infect. Dis. 8: 587-591.

Tessier, P.A., Naccache, P.H., Diener, K.R., Gladue, R.P., Neote, K.S., Clark-Lewis, I., and McColl, S.R. 1998. Induction of acute inflammation in vivo by staphylococcal superantigens. II. Critical role for chemokines, ICAM-1, and TNF-alpha. J. Immunol. 161: 1204-1211.

Todd, J., Fishaut, M., Kapral, F., and Welch, T. 1978. Toxic-shock syndrome associated with phage-group-I *Staphylococci*. Lancet 2: 1116-1118.

Toyonaga, B., and Mak, T.W. 1987. Genes of the T-cell antigen receptor in normal and malignant T cells. Annu. Rev. Immunol. 5: 585-620.

Uchiyama, T., Kamagata, Y., Wakai, M., Yoshioka, M., Fujikawa, H., and Igarashi, H. 1986. Study of the biological activities of toxic shock syndrome toxin-1. I. Proliferative response and interleukin 2 production by T cells stimulated with the toxin. Microbiol. Immunol. 30: 469-483.

Uchiyama, T., Miyoshi-Akiyama, T., Kato, H., Fujimaki, W., Imanishi, K., and Yan, X.J. 1993. Superantigenic properties of a novel mitogenic substance produced by *Yersinia pseudotuberculosis* isolated from patients manifesting acute and systemic symptoms. J. Immunol. 151: 4407-4413.

Ueshiba, H., Kato, H., Miyoshi-Akiyama, T., Tsubokura, M., Nagano, T., Kaneko, S., and Uchiyama, T. 1998. Analysis of the superantigen-producing ability of *Yersinia pseudotuberculosis* strains of various serotypes isolated from patients with systemic or gastroenteric infections, wildlife animals and natural environments. Zentbl. Bakteriol. 288: 277-291.

Van Noyen, R., Selderslaghs, R., Bogaerts, A., Verhaegen, J., and Wauters, G. 1995. *Yersinia pseudotuberculosis* in stool from patients in a regional Belgian hospital. Contrib. Microbiol. Immunol. 13: 19-24.

Watson, D.W., 1959. Host-parasite factors in group A streptococcal infections: pyrogenic and other effects on immunologic distinct exotoxins related to scarlet fever toxins. J. Exp. Med. 111: 255-283.

White, J., Herman, A., Pullen, A.M., Kubo, R., Kappler, J.W., and Marrack, P. 1989. The V beta-specific superantigen staphylococcal enterotoxin B: stimulation of mature T cells and clonal deletion in neonatal mice. Cell 56: 27-35.

Wiesenthal, A.M., Ressman, M., Caston, S.A., and Todd, J.K. 1985. Toxic shock syndrome. I. Clonical exclusion of other syndromes by strict and screening definitions. Am. J. Epidemiol. 122: 847-856.

Wilson, R.K., Lai, E., Concannon, P., Barth, R.K., and Hood, L.E. 1988. Structure, organization and polymorphism of murine and human T-cell receptor alpha and beta chain gene families. Immunol. Rev. 101: 149-172.

Woodland, D.L., Happ, M.P., Gollob, K.J., and Palmer, E. 1991. An endogenous retrovirus mediating deletion of alpha beta T cells? Nature 349: 529-530.

Yoshino, K., Abe, J., Murata, H., Takao, T., Kohsaka, T., Shimonishi, Y., and Takeda, T. 1994. Purification and characterization of a novel superantigen produced by a clinical isolate of *Yersinia pseudotuberculosis*. FEBS Lett. 356: 141-144.

Yoshino, K., Ramamurthy, T., Nair, G.B., Fukushima, H., Ohtomo, Y., Takeda, N., Kaneko, S., and Takeda, T. 1995. Geographical heterogeneity between Far East and Europe in prevalence of *ypm* gene encoding the novel superantigen among *Yersinia pseudotuberculosis*. J. Clin. Microbiol. 33: 3356-3358.

Zippelius, A., Pittet, M.J., Batard, P., Rufer, N., de Smedt, M., Guillaume, P., Ellefsen, K., Valmori, D., Lienard, D., Plum, J., MacDonald, H.R., Speiser, D.E., Cerottini, J.C., and Romero, P. 2002. Thymic selection generates a large T cell pool recognizing a self-peptide in humans. J. Exp. Med. 195: 485-494.

Chapter 11

Lipopolysaccharides of *Yersinia*

Mikael Skurnik

ABSTRACT

All Gram-negative bacteria including the members of the genus *Yersinia* possess an outer membrane (OM). Most of the outer leaflet of the OM is occupied by lipopolysaccharide (LPS), which is composed of two biosynthetic entities: (i) lipid A–core oligosaccharide; and attached to it (ii) O-specific polysaccharide, also known as O-antigen. This biosynthetic division has relevance for the understanding of the biology and genetics of LPS; therefore, the biosynthesis of these two LPS entities will be first briefly discussed in this chapter. The O-antigens are highly variable and form the basis for serotyping of strains. *Y. pseudotuberculosis* strains have been assigned to 21 and *Y. enterocolitica* and *Y. enterocolitica*–like strains to more than 70 serotypes, while *Y. pestis* strains lack O-antigen and are therefore serologically homogeneous. Chemical structures of LPS and the genetic basis of their biosynthesis have been determined for a number of *Yersinia* strains representing different species and seotypes, and an overall picture of the relationship between genetics and structure is emerging.

LPS BIOSYNTHESIS

Purified LPS is traditionally visualized by silver staining after SDS-PAGE or DOC-PAGE[1] analysis (Figure 1), which reveals that bacteria express a heterogeneous population of LPS molecules with a M_r range of ~2000 to 50000. This is because some LPS molecules do not carry any O-specific polysaccharide chains at all (M_r ~2000), while those that do vary greatly in the length of the individual polysaccharide chains (M_r >2000 to 50000). (For recent reviews on LPS biosynthesis see (Raetz and Whitfield, 2002; Whitfield and Valvano, 1993).

POLYSACCHARIDE BIOSYNTHESIS PRECURSORS

Biosynthesis of oligo- or polysaccharides involves NDP[2]-activated sugars that function as sugar donors and corresponding glycosyltransferases that transfer the sugar moiety from NDP to the acceptor structure. Each sugar and its glycosidic linkage needs a specific glycosyltransferase.

LPS biosynthesis starts by the activation of a sugar-1-P[3], such as Glc-1-P or Fru-1-P, by reaction with NTP to form NDP-activated sugars. The activation reactions take place in the cytoplasm and are catalyzed by soluble enzymes. The various NDP-sugars are then modified in several enzymatic steps involving epimerases, reductases, hydratases

[1] SDS, sodium dodecyl sulphate; PAGE, polyacrylamide gel electrophoresis; DOC, deoxycholate.
[2] NTP, nucleoside triphosphate; NDP, nucleoside diphosphate.
[3] P, phosphate; Glc, D-glucose; Fru, D-fructose; Man, D-mannose; Rha, rhamnose; Gal, D-galactose.

etc. GTP is used to activate Man and D-Rha; CTP for Abe and Par; UTP for Gal and Glc; and dTTP for L-Rha; the reason for the NTP specificity is unclear. Some NDP-sugars are intermediates of general metabolism; however, most used in LPS biosynthesis are synthesized specifically for this purpose.

LIPID A–CORE BIOSYNTHESIS

Lipid A is assembled on the cytoplasmic face of the inner membrane (IM). In several enzymatic steps UDP-GlcNAc and fatty acids (supplied by acyl carrier proteins) are converted into lipid A, a polar lipid with a disaccharide backbone of β–GlcN–(1→6)-GlcN substituted with 4 to 7 N- and O-linked saturated fatty acid residues and 1 or 2 phosphate groups. The core oligosaccharide, which typically contains 10-15 sugar residues, is built on lipid A sequentially by specific glycosyltransferases. The completed lipid A-core molecules are translocated to the periplasmic face of the IM, where O-antigen ligase, WaaL, ligates the O-specific polysaccharide (see below) to some of them. From the IM the LPS molecules are finally translocated to the OM (Raetz and Whitfield, 2002; Whitfield and Valvano, 1993).

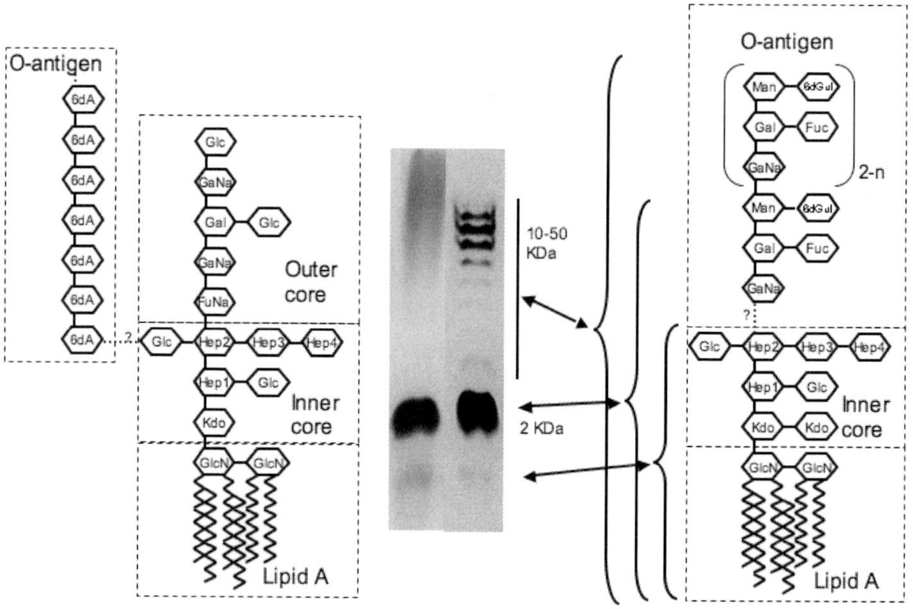

Figure 1. Examples of silver-stained LPS separated by DOC-PAGE showing the heterogeneity of the LPS molecule populations and the corresponding schematic representations of the structures. Examples of LPS with homo-polymeric (left, *Y. enterocolitica* O:3) and heteropolymeric (right, *Y. enterocolitica* O:8) O-antigens are shown. The locations of the different forms of LPS molecules in the gel are indicated for the *Y. enterocolitica* O:8 LPS. The approximate sizes of the LPS-molecules in the gel are indicated. Abbreviations: GlcN, glucosamine; Kdo, 2-keto-3-deoxyoctulosonic acid; Hep, L or D-*glycero*-α-D-*manno*-heptopyranose; FuNa, N-acetylfucosamine; GaNa, N-acetylgalactosamine, Gal, galactose; Glc, glucose; 6dA, 6-deoxyaltrose; 6d-Gul, 6-deoxygulose; Man, mannose; Fuc, fucose. The question marks and dashed lines indicate that the linkage has not been confirmed.

O-ANTIGEN BIOSYNTHESIS

The O-polysaccharide can be either a homo- or a heteropolymer (Figure 1), i.e., a polymer of a single sugar or a polymer of repeats of different sugars (O-units), respectively, and their biosynthesis differs in some respects. As shown in the figures, both types of O-polysaccharides have been found in the genus *Yersinia*.

THE HETEROPOLYMERIC PATHWAY

Most steps in O-antigen biosynthesis take place at the cytoplasmic face of the IM. In the heteropolymeric pathway, each identical repeat unit is first assembled on a carrier lipid, Und-P[4]. Remarkably, the initiation reaction (transfer of the first sugar-1-P to Und-P) is accomplished in many strains by WecA, a GlcNAc-1-P transferase that is encoded in the ECA[5] gene cluster. In a few strains, a dedicated transferase such as a Gal-1-P transferase is used for initiation. The O-unit assembly is then continued by dedicated glycosyltransferases that use NDP-activated sugar precursors. The 'flippase' protein Wzx (Liu et al., 1996) translocates the completed Und-PP-O-unit to the periplasmic side of IM. There, O-antigen polymerase, Wzy, polymerises the O-units into long chains still carried by Und-P. The chain length determinant protein Wzz regulates the chain length, and the completed chains are transferred from Und-P to the preformed lipid A-core structure by O-antigen ligase, WaaL.

THE HOMOPOLYMERIC PATHWAY

In the homopolymeric pathway, the O-antigen polymer is synthesized on Und-P and completely elongated to full length in the cytoplasmic side of IM by sequential transfer of sugar residues to the nonreducing end of the growing polysaccharide chain. The completed O-antigen is translocated to the periplasmic side by an ATP-driven transporter system, composed of Wzt and Wzm, which belong to the ABC[6] transporter family (Whitfield, 1995). O-antigen ligase, WaaL, transfers and also ligates the homopolymeric O-antigen onto lipid A-core.

GENETICS

LPS biosynthesis requires enzymes that are specific for each LPS component and linkage, therefore at least 50 genes, which are chromosomally located in most Gram-negative bacteria, are involved in LPS biogenesis. Lipid A biosynthesis genes are usually scattered around the genome, whereas those involved in core and O-antigen biosynthesis form distinct gene clusters. In a few bacteria, LPS genes are located on plasmids or within temperate bacteriophage genomes.

ECA AND LPS

ECA is a polysaccharide with a trisaccharide repeat unit expressed by all *Enterobacteriaceae* including *Yersinia* (Kuhn et al., 1988; Radziejewska-Lebrecht et al., 1998; Rick and Silver, 1996). ECA can occur in three different structural forms: (i) linked to L-glycerophosphatide, (ii) linked to LPS, or (iii) as a cyclic structure (Rick and Silver, 1996). When linked to LPS, ECA is ligated specifically to those lipid A–core molecules that do not carry O-antigen. ECA biosynthesis on Und-P takes place using

[4] Und-P, undecaprenol phosphate, a 55-carbon isoprenoid carrier lipid.
[5] ECA, enterobacterial common antigen.
[6] ABC, ATP binding cassette.

Table 1. Characteristics of *Y. pseudotuberculosis* serotypes (Bogdanovich et al., 2003a; Bogdanovich et al., 2003b; Reeves et al., 2003; Skurnik, 2003; Tsubokura and Aleksic, 1995).

O-Serotype	**O-factors** (serotype specific factor in bold)	Presence of virulence plasmid	*ddh*- genes	DDH
O:1a	2, **3**, 23	pYV	*ddhDABC*	Par
O:1b	2, **4**, 23	pYV	*ddhDABC*	Par*f*
O:1c	2, 3, 17, **24**	pYV	*ddhDABC*	(Par)?
O:2a	5, **6**, 16	pYV	*ddhDABC*	Abe
O:2b	5, **7**, 16, 17	pYV	*ddhDABC*	Abe
O:2c	5, 7, 11, **18**	pYV	*ddhDABC*	Abe
O:3	**8**, 15	pYV	*ddhDABC*	Par
O:4a	9, **11**	pYV	*ddhDABC*	Tyv
O:4b	9, **12**	pYV	*ddhDABC*	Tyv
O:5a	10, **14**	pYV	*ddhDABC*	Asc
O:5b	10, **15**	pYV	*ddhDA*	6d-Alt*f*
O:6	**13**, 19, 26	pYV	*ddhDABC*	Col
O:7	13, **19**			Col
O:8	11, **20**	pYV	*ddhDABC*	?
O:9	10, **25**			?
O:10	**26**	pYV		?
O:11	4, 15, **27**		*ddhDA*	(6d-Alt*f*)?
O:12	18, 27, **28**		*ddhDABC*	?
O:13	28, **29**		*ddhDABC*	?
O:14	13, **30**		*ddhDA*	(6d-Alt*f*)?
O:15			*ddhDABC*	(Par)?

reactions similar to those described above for heteropolymeric O-antigen, and ligation to LPS is WaaL-dependent (Rick and Silver, 1996). LPS-linked ECA is immunogenic and antibodies against ECA are produced during infection.

O-SEROTYPES OF *YERSINIA*

Serotyping as a means to classify *Salmonella* isolates began as early as 1926 (Kauffmann, 1967; White, 1926) and is known as the Kauffmann-White scheme. The primary antigens in the scheme are the cell surface lipopolysaccharides (O-antigens) and the flagellin proteins (H-antigens). The O-antigen epitopes are determined by the type, arrangement, and condition of sugar residues in the repeated O-units. O-serotyping was first applied to *Y. pseudotuberculosis* (5 serotypes) by Thal (1954) and to *Y. enterocolitica* (8 serotypes) by Winblad (1973).

At present, 21 serotypes of *Y. pseudotuberculosis* are recognized, whereas *Y. pestis* strains are serologically very homogeneous and cannot be serotyped because their LPS lacks O-antigen (Brubaker, 1991). *Y. pestis* is a recently evolved clone of *Y. pseudotuberculosis* O:1b (Achtman et al., 1999; Skurnik et al., 2000) carrying a cryptic O-antigen gene cluster that is almost identical to that of *Y. pseudotuberculosis* O:1b (Skurnik et al., 2000).

Y. PSEUDOTUBERCULOSIS SEROTYPES

The current *Y. pseudotuberculosis* typing scheme lists 15 major O-serotypes (Tsubokura and Aleksic, 1995) of which the first five contain 2 or 3 O-subtypes. The division into the serotypes is based on ca. 30 different O-factors (Table 1). Differences in the first seven serotypes are in part determined by the presence of different DDHs[7] in the O-antigen (Tsubokura et al., 1993; Tsubokura et al., 1984). Epidemiologically, there are some differences in prevalence of the serotypes isolated from humans. In Europe, most isolates belong to serotypes O:1 - O:3 (Aleksic et al., 1995), while in Japan serotypes O:4b, O:3, O:5a, and O:5b are most common (Tsubokura et al., 1989). Apparently the more recently recognized serotypes represent environmental *Y. pseudotuberculosis,* since not all of them carry the virulence plasmid (Table 1) (Nagano et al., 1997; Tsubokura and Aleksic, 1995).

SEROTYPES OF *Y. ENTEROCOLITICA* AND RELATED SPECIES

In *Y. enterocolitica* and related species there are more than 70 O-serotypes (Wauters et al., 1991). Human and animal pathogenic strains of *Y. enterocolitica* that carry the virulence plasmid, pYV[8], belong to certain serotypes (Table 2): O3 and O9 in Scandinavia and Europe, Canada, Japan, and South Africa; and O:8 in the United States. Less frequently encountered pathogenic serotypes are O:4,32, O:5,27, O:13a,18, and O:21. Certain *Y. enterocolitica* serotypes do not carry pYV, and these are collectively referred to as non-pathogenic or environmental serotypes. In some cases, however, infection with these strains may cause symptoms. The virulence of the pathogenic serotypes varies; for example, serotype O:8 strains are more virulent than O:3 or O:9 strains (Aulisio et al., 1983; Mäki et al., 1983). This is most evident in mouse lethality models. Serotype O:9 O-antigen cross-reacts with *Brucella abortus* O-antigen, causing diagnostic problems in serological confirmation of *B. abortus* infections (Ahvonen et al., 1969; Kittelberger et al., 1995a; Kittelberger et al., 1995b).

CHEMICAL STRUCTURES AND GENETICS OF *YERSINIA* LPS BIOSYNTHESIS

The chemical structures of a number of *Yersinia* LPS molecules or different parts of the molecules have been determined. In most of the cases the structure of the repeating O-unit has been resolved but in only a few cases are the lipid A and core structures known (Tables 3-6 and Figures 2-6). Structures for which genetic information is not yet available are listed in Table 6. It should be noted that only limited attention has been paid to the variability of the chemical structure of the LPS molecule with regard to different growth conditions, e.g., growth temperature, pH, ionic strength, growth phase, etc. This will become necessary in the future because there are already many indications that LPS structure is under complex regulatory control, and that structural variations may profoundly affect the biological activities of LPS (Bengoechea et al., 2003; Guo et al., 1998).

[7] DDH, 3,6-dideoxyhexose; Par, paratose; Abe, abequose; Tyv, tyvelose; Asc, ascarylose; Col, colitose; 6d-Alt*f*, 6-deoxy-L-altrofuranose.

[8] pYV, 70-75 kb virulence plasmid of *Yersinia*.

Table 2. Some characteristics of the serotypes of *Y. enterocolitica* and related species.* (Data compiled from Skurnik, 1985; Skurnik and Toivanen, 1991; Skurnik et al., 1995).

Serotype	Species	Presence of virulence plasmid	Presence of IVS in 23S rDNA	Presence of O:3 type LPS outer core
O:1	*Y. enterocolitica*	pYV	IVS	OC
O:1,2,3	*Y. enterocolitica*	pYV	n.t.	-
O:1,3	*Y. enterocolitica*	pYV	n.t.	n.t.
O:2	*Y. enterocolitica*	pYV	IVS	OC
O:3	*Y. enterocolitica*	pYV	IVS	OC
O:3	*Y. kristensenii*	-	n.t.	
O:4	*Y. enterocolitica*	pYV	n.t.	-
O:4,32	*Y. enterocolitica*	pYV	IVS	-
O:4,33	*Y. enterocolitica*	-	n.t.	n.t.
O:5	*Y. enterocolitica*			
O:5,27	*Y. enterocolitica*	pYV	IVS	OC
O:6,30	*Y. enterocolitica*	-	-	-
O:6,31	*Y. enterocolitica*	-	-	OC
O:7,8	*Y. enterocolitica*	-	-	-
O:8	*Y. enterocolitica*	pYV	IVS/-	-
O:9	*Y. enterocolitica*	pYV	IVS	OC
O:10	*Y. enterocolitica*	-	-	-
O:11,23	*Y. enterocolitica*	-	n.t.	n.t.
O:12,25	*Y. kristensenii*	-	-	-
O:13,7	*Y. enterocolitica*	pYV	n.t.	-
O:13,18	*Y. enterocolitica*	pYV	IVS	-
O:13a,13b	*Y. enterocolitica*	pYV	IVS	-
O:14	*Y. enterocolitica*	-	n.t.	-
O:16	*Y. kristensenii*	-	-	-
O:16	*Y. frederiksenii*	-	-	-
O:16,21	*Y. intermedia*	-	-	-
O:20	*Y. enterocolitica*	pYV	IVS	-
O:21	*Y. enterocolitica*	pYV	IVS	-
O:25	*Y. enterocolitica*	-	-	-
O:25,26,44	*Y. enterocolitica*	-	n.t.	OC
O:26,44	*Y. enterocolitica*	-	n.t.	-
O:28,50	*Y. enterocolitica*	-	-	-
O:34	*Y. enterocolitica*	pYV	IVS	-
O:35	*Y. frederiksenii*	-	n.t.	-
O:35,36	*Y. enterocolitica*	-	n.t.	-
O:35,52	*Y. enterocolitica*	-	-	-
O:41(27),42	*Y. enterocolitica*	-	-	-
O:41(27)43	*Y. enterocolitica*	-	n.t.	OC
O:41(27)K1	*Y. enterocolitica*	-	n.t.	-
O:41,43	*Y. enterocolitica*	-	n.t.	OC
O:48	*Y. frederiksenii*	-	-	-
O:50	*Y. enterocolitica*	-	-	OC
O:52,54	*Y. intermedia*	-	-	OC
O:58,16	*Y. frederiksenii*	-	n.t.	-
O:58,16	*Y. bercovieri*	-	IVS/-	-
O:59(20,36,7)	*Y. mollaretii*	-	-	-

*Abbreviations: pYV, ~70 Kb virulence plasmid of *Yersinia*; IVS, a ~100 bp intervening sequence in the 23S rRNA gene; n.t., not tested; IVS/-, IVS is not present in all rRNA operons; OC, outer core oligosaccharide.

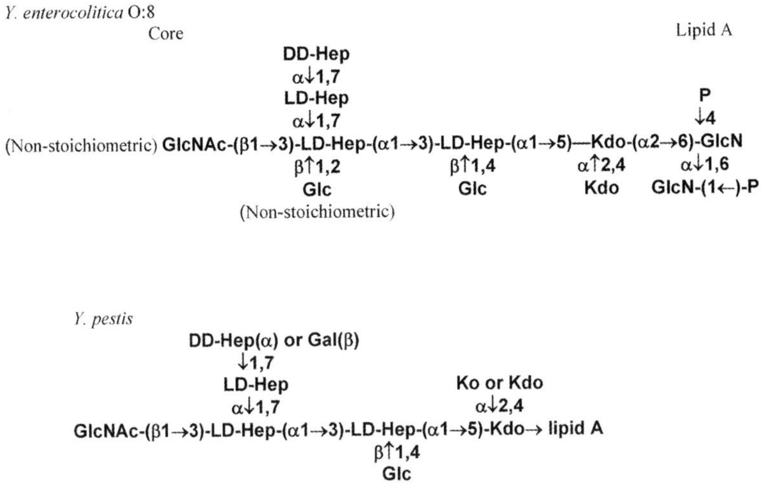

Figure 2. Structures of the lipid A disaccharide backbone and core regions of *Y. enterocolitica* serotype O:8 (Gronow et al., 2001) and *Y. pestis* (Hitchen et al., 2002; Vinogradov et al., 2002) LPSs. Abbreviations (see also Figure 1 legend): Ko, 2-keto-octulosonic acid; DD-Hep, D-*glycero*-α-D-*manno*-heptopyranose; LD-Hep, L-*glycero*-α-D-*manno*-heptopyranose; GlcNAc, N-acetylglucosamine: P, phosphate.

GENETICS OF LIPID A AND CORE BIOSYNTHESIS

No direct genetic experiments have been reported concerning the lipid A and inner core biosynthesis genes of *Yersinia*. The basic genetic setup most likely resembles that of *E. coli* and *Salmonella;* however, very likely there will be differences in the regulation of the biosynthetic activities under different environmental conditions. Elucidating the details of these questions will be substantially easier now that the genomic sequences of *Y. pestis, Y. pseudotuberculosis* O:1b and *Y. enterocolitica* O:8 are known.

THE *HEMH* AND *GSK* INTERGENIC LOCUS OF *YERSINIA*

In *Yersinia*, all the known gene clusters responsible for the heteropolymeric type of O-antigen are located between the *hemH* and *gsk* genes on the chromosome. Upstream of *hemH* is located the *adk* gene, and the *rosB-rosA* and *ushA*[9] genes are located downstream of the *gsk* gene. The *rosB-rosA* and *ushA* genes are not directly involved in O-antigen biosynthesis, but the *rosAB* operon is involved in regulation of O-antigen expression (Bengoechea et al., 2002b). In *Y. enterocolitica* serotypes O:3 and O:9, which both express homopolymeric O-antigens, the O-antigen gene clusters are not located between the *hemH* and *gsk* genes; this locus is instead occupied by the outer core gene cluster (Skurnik et al., 1995). The genomic locations of the homopolymeric O-antigen gene clusters of O:3 and O:9 are unknown.

Y. PSEUDOTUBERCULOSIS AND *Y. PESTIS* LPS

No genetic work has been reported on the lipid A and core biosynthesis in *Y. pseudotuberculosis*. The core structure of *Y. pseudotuberculosis* has been reported but

[9] *adk*, adenylate kinase gene; *hemH*, ferrochelatase gene; *gsk*, inosine-guanosine kinase gene; *ushA*, UDP-sugar hydrolase gene.

it is missing important details of the linkages and their configurations (Ovodov et al., 1992). The lipid A and core structure for *Y. pestis* was, however, determined recently by two groups, although some discrepancies were present between the reported structures (Figure 2) (Hitchen et al., 2002; Vinogradov et al., 2002).

In silver-stained SDS-PAGE, LPS of most *Y. pseudotuberculosis* serotypes shows the ladder-like staining pattern indicative of heteropolymeric O-antigen. However, detection of the ladder in a few serotypes, e.g. O:1b, is very difficult by silver-staining. This is due to absence of vicinal hydroxyl-groups in the chemical structure of the O-unit (a requirement for silver-staining) and/or expression of very low amounts of the O-antigen by the strain. *Y. pestis* does not express any O-antigen due to five pseudogenes in its O-antigen gene cluster (Skurnik et al., 2000).

Y. PSEUDOTUBERCULOSIS O-ANTIGENS

The O-antigen structures of a number of *Y. pseudotuberculosis* serotypes are known (Tables 3-5). The chromosomal location of the O-antigen gene cluster in *Y. pseudotuberculosis* and also of the cryptic O-antigen gene cluster in *Y. pestis* is between the *hemH* and *gsk* genes (Reeves et al., 2003; Skurnik et al., 2000). Based on sequencing, gene specific PCR analysis, and hybridization studies, it has been established that all except serotype O:7, O:9 and O:10 carry the *ddhDABC* genes or their homologues at the beginning of the gene cluster downstream of the *hemH* gene and the JUMPstart sequence (Table 1). The *ddhDABC*-encoded enzymes direct the biosynthesis of an intermediate and the completion of the specific DDH in each serogroup. One or two additional enzymes encoded by genes that determine the serogroup specificity are required, e.g., *abe* in serogroups O:2a, O:2b and O:2c; *prt* and *tyv* in serogroups O:4a and O:4b; *prt* in serogroups O:1a and O:1b and *ascEF* in serotype O:5a (Thorson and Liu, 1993; Thorson et al., 1993a; Thorson et al., 1993b; Thorson et al., 1994). PCR analysis also predicted a specific DDH for serotypes O:1c, O:11, O:14 and O:15, for which no structural data is available (Table 1). In addition to the DDH pathway genes, the O-antigen clusters contain the genes coding for the biosynthetic enzymes of the other sugar components, the respective glycosyltransferases, and the translocation and polymerization proteins Wzx, Wzy and Wzz (Tables 3-5). Interestingly, based on PCR analysis, all serotypes appear to use an almost identical Wzz protein as a chain length determinant (Skurnik et al., 2000).

Comparative analyses of a number of *Y. pseudotuberculosis* O-antigen gene clusters just downstream of the *ddh* genes revealed footprints of homologous recombinantion events and relics of the insertion sequence IS630. This was also observed in *Shigella sonnei,* which may reflect the evolution of the clusters (Hobbs and Reeves, 1995; Reeves et al., 2003). Another characteristic feature of the O-antigenic clusters of *Y. pseudotuberculosis* is the presence of numerous intergenic 6-8 nucleotide repeats (Skurnik et al., 2000).

Y. ENTEROCOLITICA SEROTYPE O:3 LPS

The lipid A structure of serotype O:3 is not known, but the inner and outer core structures as well as that of O-antigen are known (Figure 3 and 4) (Radziejewska-Lebrecht et al., 1994; Radziejewska-Lebrecht et al., 1998). Silver stained LPS of *Y. enterocolitica* O:3 is shown in Figure 1, along with a schematic structure of the wild type LPS.

Table 3. O-antigen structures and O-antigen gene cluster genes of *Y. pseudotuberculosis* serotypes with 6-deoxy-*manno*-heptopyranose (6d-*manno*-Hep) in the O-antigen. (Komandrova et al., 1984; Reeves et al., 2003; Samuelsson et al., 1974).

Serotype	O-antigen structure	Order of O-antigen cluster genes	Function of gene product(s)
O:1a	**Par** **α↓1,3** **6d-*manno*-Hep** **β↓1,4** **→3)-Gal-(α1→3)-GlcNAc-(β1→**	*ddhD-ddhA-ddhB-ddhC-prt-*	CDP-paratose biosynthesis
		wbyH-	pyranose-furanose mutase
		wzx-	O-unit flippase
		wbyM-	Paratosyltransferase
		wbyB-	6d-*manno*-heptosyltransferase
		wzy-	O-antigen polymerase
		wbbP-	Galactosyltransferase
		dmhB-dmhA-hddA-gmhA-hddC-gmhB-	GDP-6d-*manno*-heptose biosynthesis
		wzz	O-antigen chain length determinant
O:2a	**Abe** **(α)↓1,3** **6d-*manno*-Hep** **(β)↓1,4** **→3)-Gal-(α1→3)-GlcNAc-(β1→**	*ddhD-ddhA-ddhB-ddhC-abe-*	CDP-abequose biosynthesis
		tyv-*	*tyv*-homologous pseudogene
		wzx-	O-unit flippase
		wbyA-	Abequosyltransferases
		wbyB-	6d-*manno*-heptosyltransferase
		wzy-	O-antigen polymerase
		wbbP-	Glycosyltransferase
		dmhB-dmhA-hddA-gmhA-hddC-gmhB-	GDP-6d-*manno*-Hep biosynthesis
		wzz	O-antigen chain length determinant
O:4b	**Tyv** **(α)↓1,3** **6d-*manno*-Hep** **(β)↓1,4** **→3)-Gal-(α1→3)-GlcNAc-(β1→**	*ddhD-ddhA-ddhB-ddhC-prt-tyv-*	CDP-tyvelose biosynthesis
		wzx-	O-unit flippase
		wbyA-	Tyvelosyltransferase
		wbyB-	6d-*manno*-heptosyltransferase
		wzy-	O-antigen polymerase
		wbbP-	Glycosyltransferase
		dmhB-dmhA-hddA-gmhA-hddC-gmhB-	GDP-6d-*manno*-Hep biosynthesis
		wzz	O-antigen chain length determinant

Table 4. O-antigen structures and O-antigen gene cluster genes of *Y. pseudotuberculosis* serotypes with mannose and fucose in the O-antigen backbone. (Gorshkova et al., 1991; Gorshkova et al., 1980; Gorshkova et al., 1983a; Gorshkova et al., 1983c; Isakov et al., 1983; Korchagina et al., 1982; Reeves et al., 2003; Samuelsson et al., 1974; Skurnik et al., 2000; Tomshich et al., 1976).

Sero-type	O-antigen structure	Order of O-antigen cluster genes	Function of gene product(s)
O:1b	**Par*f*** **↓1,3** **→2)-Man-(1→4)-Man-(1→3) -Fuc-(1→3)-GlcNAc-(1→**	*ddhD-ddhA-ddhB-ddhC-prt-* *wbyH-* *wzx-* *wbyI-* *wbyJ-* *wzy-* *wbyK-* *gmd-fcl-* *manC-* *wbyL-* *manB-* *wzz*	CDP-paratose biosynthesis pyranose-furanose mutase O-unit flippase Paratosyltransferase Mannosyltransferase O-antigen polymerase Mannosyltransferase GDP-fucose biosynthesis GDP-mannose biosynthesis Fucosyltransferase GDP-mannose biosynthesis O-antigen chain length determinant
O:2b	**Abe** (β)**↓1,3** **→2)-Man-(α1→3)-Fuc-(α1→3)-GalNAc-(1→**	*ddhD-ddhA-ddhB-ddhC-abe-* *tyv*-* *wzx-* *wbyE-* *wzy-* *wbyK-* *gmd-fcl-* *manC-* *wbyQ-* *gne-* *manB-* *wzz*	CDP-abequose biosynthesis *tyv*-homologous pseudogene O-unit flippase Abequosyltransferase O-antigen polymerase Mannosyltransferase GDP-fucose biosynthesis GDP-mannose biosynthesis Fucosyltransferase UDP-N-acetylglucosamine-4-epimerase GDP-mannose biosynthesis O-antigen chain length determinant

O:2c	**Abe** **α↓1,6** **→3)-Man-(α1→2)-Man-(β1→2)-Man-(α1→3)-GalNAc-(α1→**	*ddhD-ddhA-ddhB-ddhC-abe-*	CDP-abequose biosynthesis
		tyv-*	*tyv*-homologous pseudogene
		wzx-	O-unit flippase
		wbyD-	Abequosyltransferase
		wbyN-	Mannosyltransferase
		wzy-	O-antigen polymerase
		wbyO-	Mannosyltransferase
		manC-	GDP-mannose biosynthesis
		gne-	UDP-N-acetylglucosamine-4-epimerase
		manB-	GDP-mannose biosynthesis
		wzz	O-antigen chain length determinant
O:3	**Par** **β↓1,4** **→3)-Fuc-(α1→3)-GalNAc-(α1→2)-Man-(α1→**	*ddhD-ddhA-ddhB-ddhC-prt-*	CDP-paratose biosynthesis
		wbyP-	Paratosyltransferase
		wzx-	O-unit flippase
		wzy-	O-antigen polymerase
		wbyK-	Mannosyltransferase
		gmd-fcl-	GDP-fucose biosynthesis
		manC-	GDP-mannose biosynthesis
		wbyQ-	Fucosyltransferase
		gne-	UDP-N-acetylglucosamine-4-epimerase
		manB-	GDP-mannose biosynthesis
		wzz	O-antigen chain length determinant
O:4a	**Tyv** **α↓1,6** **→3)-Man-(β1→3)-Man-(α1→2)-Man-(α1→3)-GalNAc-(α1→**	*ddhD-ddhA-ddhB-ddhC-prt-tyv-*	CDP-tyvelose biosynthesis
		wzx-	O-unit flippase
		wbyD-	Abequosyltransferase
		wbyN-	Mannosyltransferase
		wzy-	O-antigen polymerase
		wbyO-	Mannosyltransferase
		manC-	GDP-mannose biosynthesis
		gne-	UDP-N-acetylglucosamine-4-epimerase
		manB-	GDP-mannose biosynthesis
		wzz	O-antigen chain length determinant

O:5a	**Asc** **↓1,3** **→2)-Fuc-(1→3)-Man-(1→4)-Fuc-(1→3)-GalNAc-(1→**	*ddhD-ddhA-ddhB-ddhC-ascE-ascF-*	CDP-abequose biosynthesis
		wzx-	O-unit flippase
		wbyS-	Ascarosyltransferase
		wbyT-	Fucosyltransferase
		wzy-	O-antigen polymerase
		wbyU-	Mannosyltransferase
		gmd-fcl-	GDP-fucose biosynthesis
		manC-	GDP-mannose biosynthesis
		wbyQ-	Fucosyltransferase
		gne-	UDP-N-acetylglucosamine-4-epimerase
		manB-	GDP-mannose biosynthesis
		wzz	O-antigen chain length determinant
O:5b	**6d-Alt*f*** **↓1,3** **→2)-Fuc-(1→3)-Man-(1→4)-Fuc-(1→3)-GalNAc-(1→**	*ddhD-ddhA-orfX-wbyR-*	CDP-6-deoxyaltrose biosynthesis
		wbyH-	pyranose-furanose mutase
		wzx-	O-unit flippase
		wbyS-	Ascarosyltransferase
		wbyT-	Fucosyltransferase
		wzy-	O-antigen polymerase
		wbyU-	Mannosyltransferase
		gmd-fcl-	GDP-fucose biosynthesis
		manC-	GDP-mannose biosynthesis
		wbyQ-	Fucosyltransferase
		gne-	UDP-N-acetylglucosamine-4-epimerase
		manB-	GDP-mannose biosynthesis
		wzz	O-antigen chain length determinant

Table 5. O-antigen structures and O-antigen gene cluster genes of *Y. pseudotuberculosis* serotypes with colitose in the O-antigen. (Gorshkova et al., 1983b; Gorshkova et al., 1976; Kotandrova et al., 1989).

Sero-type	O-antigen structure	Order of O-antigen cluster genes	Function of gene product(s)
Y. pstb O:6	**→3)-GlcNAc-(β1→6)-GalNAc-(α1→3)-GalNAc-(β1→** **β↑1,3** (on the first GalNAc) **Col-(α1→2)-YerA**	Genes and order are not known except that *ddhD, ddhA, ddhB, ddhC* and *wzz* genes are present (Skurnik, 2003)	
Y. pstb O:7	**→6)-Glc-(β1→3)-GalNAc-(α1→3)-GalNAc-(β1→** **α↑1,2** (on Glc) **Col**; **α↑1,6** (on the second GalNAc) **Glc**	*wzx-*	O-unit flippase
		gmd-colB-colA-manC-manB-	GDP-colitose biosynthesis
		wbyV-	transferase
		wzy-	O-antigen polymerase
		wbyW-	transferase
		wzz-	O-antigen chain length determinant
		IS-	Insertion sequence
		gtrA-gtrB-wgtJ-	Glucosylation
		gne-	UDP-N-acetylglucosamine-4-epimerase
		wbyX-	transferase
		galE	UDP-glucose-4-epimerase

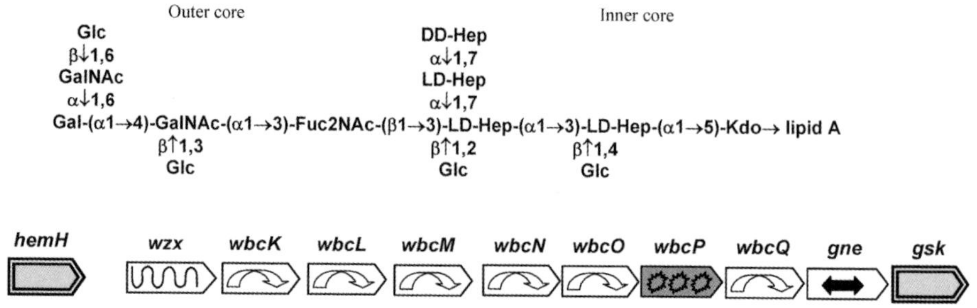

Figure 3. The inner and outer core structure of *Y. enterocolitica* O:3 and O:9 LPS (Müller-Loennies et al., 1999; Ovodov et al., 1992; Radziejewska-Lebrecht et al., 1998; Shashkov et al., 1995). The genetic organization of the outer core gene cluster of serotype O:3 is shown at bottom (Skurnik, 1999; Skurnik et al., 1995). In this figure and in figures 4-6, individual genes are drawn as arrows but not to scale. The flanking genes are drawn with double-borders and shading. The names of the genes are given above the genes. Symbols in the gene arrows indicate functions (explosion symbol indicates biosynthetic enzyme; curved arrow, a glycosyltransferase; and serpentine, a membrane-associated protein). The CDP-DDH pathway genes contain black explosion symbols while the GDP-mannose, GDP-fucose and GDP-perosamine pathway genes are indicated by shading and triple explosions. Abbreviations (see also Figures 1 and 2 legends): FucNAc or Fuc2NAc, N-acetylfucosamine or 2-acetamido-2,6-dideoxy-L-galactose; GalNAc, N-acetylgalactosamine.

Molecular genetic studies indicate that the outer core hexasaccharide of O:3 is an ancestral heteropolymeric O-unit (Skurnik et al., 1995). The structure of the *Y. enterocolitica* O:3 LPS, furthermore, is unique since the O-antigen, which is a homopolymer of 6d-Alt[10] (Hoffman et al., 1980), is not attached to the outermost tip of the outer core but instead to the inner core (Skurnik et al., 1995). Similarly, in serotype O:3 ECA-LPS, the ECA moiety is ligated to the inner core (Radziejewska-Lebrecht et al., 1998). In both cases the most likely residue for the linkage is the Glc β1,2 linked to LD-Hep of the inner core (Figure 3).

Y. ENTEROCOLITICA SEROTYPE O:3 OUTER CORE

Biosynthesis of the Outer Core

The biosynthesis of the O:3 outer core hexasaccharide (Figure 3) needs (i) enzymes for the biosynthesis of NDP-sugar precursors, i.e., Gne and WbcP for UDP-GalNAc and UDP-Fuc2Nac, respectively; (ii) six glycosyltransferases; and (iii) a flippase (Wzx). It is predicted that biosynthesis of the hexasaccharide unit follows the heteropolymeric O-antigen biosynthetic scheme and begins on the cytoplasmic face of the IM with the transfer of Fuc2NAc-1-P[11] from UDP-Fuc2NAc to the carrier lipid Und-P (Skurnik, 1999). The rest of the hexasaccharide is then sequentially completed on Und-P-P-Fuc2NAc, where additional sugar residues are transferred to the non-reducing end of the growing oligosaccharide. In the end, carbon 4 of the Fuc2NAc residue is decorated with a phosphate group (Radziejewska-Lebrecht et al., 1998). After completion, Wzx flips the Und-P-P-hexasaccharide unit to the periplasmic face of the IM, where WaaL ligates the hexasaccharide onto a lipid A-core structure.

[10] 6d-Alt, 6-deoxy-L-altropyranose.

[11] Fuc2NAc, N-acetylfucosamine or 2-acetamido-2,6-dideoxy-L-galactose; GalNAc, N-acetylgalactosamine; Glc, glucose; Gal, galactose; Und-P, undecaprenol phosphate, a 55-carbon isoprenoid carrier lipid.

→2)-6d-Alt-(β1→

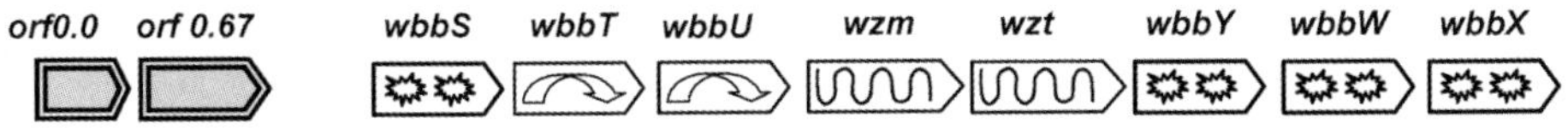

Figure 4. The homopolymeric O-antigen structure of *Y. enterocolitica* O:3 (Gorshkova et al., 1985; Hoffman et al., 1980) and the corresponding O-antigen gene cluster (Zhang et al., 1993). 6-d-Alt, 6-deoxyaltrose.

The O:3 Outer Core Gene Cluster

The outer core gene cluster (Figure 3) contains nine genes: *wzx, wbcK, wbcL, wbcM, wbcN, wbcO, wbcP, wbcQ* and *gne* (Skurnik et al., 1995). Upstream of *wzx* is a noncoding region of 400 bp, preceded by the *hemH* gene. It is predicted that WbcK and WbcL are Glc-transferases, WbcM is a Gal-transferase, WbcN and WbcQ are GalNAc-transferases, and WbcO is a Fuc2NAc-transferase. WbcP is apparently involved in UDP-Fuc2NAc biosynthesis, and Gne is UDP-N-acetylglucosamine-4-epimerase (EC 5.1.3.7) (Bengoechea et al., 2002a), thus providing UDP-GalNAc for outer core biosynthesis. Since UDP-Glc and UDP-Gal are products of general metabolism, the gene cluster appears to contain all necessary genes required for the outer core biosynthesis with the exception of the kinase needed to phosphorylate the C4-position of Fuc2NAc. The 39 bp regulatory JUMPstart sequence (Hobbs and Reeves, 1994) is present in the non-coding 400 bp region between *hemH* and *wzx* (Skurnik et al., 1995).

The O:3 Outer Core is the Receptor for Bacteriophage ϕR1-37 and is Present in Other *Y. enterocolitica serotypes*

The phage ϕR1-37 receptor is the outer core oligosaccharide, not only in *Y. enterocolitica* serotype O:3 strains, but also in serotypes O:1, O:2, O:5, O:5,27, O:6, O:6,31, O:9, O:21, O:25,26,44, O:41(27)43, O:41,43, and O:50, and in *Y. intermedia* O:52,54 strains (Table 2) (Skurnik et al., 1995). Phage resistant variants of some of these strains were analysed by DOC-PAGE, and all had lost the outer core band similar to serotype O:3 outer core mutants (Müller-Loennies et al., 1999; Skurnik et al., 1995) (E. Pinta and M. Skurnik, unpublished).

Y. ENTEROCOLITICA SEROTYPE O:3 O-ANTIGEN

The *Y. enterocolitica* O:3 O-antigen (Figure 4) is a homopolymer of 6d-Alt[12] which is linked to the inner core of LPS (Hoffman et al., 1980; Skurnik et al., 1995). Genetic evidence indicates that the reducing end sugar is GlcNAc. 6d-Alt is a carbon-3 epimer of L-Rha and their biosynthetic pathways start similarly, with dTDP-4-keto-6-deoxy-D-Glc as an intermediate. Biosynthesis of dTDP-L-Rha takes place in four steps, as is expected also for 6d-Alt (Skurnik et al., 1999; Zhang et al., 1993). However, details of dTDP-6d-Alt biosynthesis are not yet known and need to be elucidated in the future.

The O:3 O-Antigen Gene Cluster Contains 8 Genes

The O-antigen cluster is organized into two operons, both of which are transcribed from tandem promoters (Figure 4); one pair is located 1.2 kb upstream of the *wbbS* gene and the other pair between the *wbbU* and *wzm* genes (Zhang et al., 1993). O-antigen expression is under temperature-regulation, and optimal expression takes place below 30°C; however,

[12] 6d-Alt, 6-deoxy-L-altropyranose.

the temperature regulation is only functional in bacteria approaching or in the stationary phase of growth. In exponentially growing bacteria, O-antigen biosynthesis is turned on and unaffected by temperature (Lahtinen et al., 2003). The first operon contains three genes designated *wbbS, wbbT* and *wbbU,* and the second operon contains five: *wzm, wzt, wbbV, wbbW,* and *wbbX*. All eight genes in these two operons are essential for O-antigen synthesis (Al-Hendy et al., 1991; Al-Hendy et al., 1992; Zhang et al., 1993). There is a non-coding 1.5 kb region upstream of *wbbS* that contains relics of the *rmlB* gene and the regulatory JUMPstart sequence.

Biosynthesis of the O:3 O-Antigen

WbbS, WbbV and WbbW are similar to RmlC, RmlD, and RmlA of *Salmonella* (63%, 54% and 30% identical, respectively), the enzymes involved in dTDP-L-Rha biosynthesis, indicating that they function in dTDP-6d-Alt biosynthesis. The initiation of the O-antigen biosynthesis is WecA (undecaprenyl phosphate:GlcNAc-1-phosphate transferase) dependent (Zhang, 1996). Thus O-antigen biosynthesis utilizes an initiation step in which a single GlcNAc residue is transferred to Und-P. After this, a specific 6d-Alt-transferase transfers the first 6d-Alt residue to GlcNAc-Und-PP, upon which the homopolymer is then sequentially built up by a second 6d-Alt-transferase that uses $(6d\text{-}Alt)_n$-GlcNAc-Und-PP as an acceptor. Both WbbT and WbbU share conserved local motifs with a number of glycosyltransferases (Morona et al., 1995; Skurnik et al., 1995), and fulfill the need for two glycosyltransferases.

O-Antigen Transport

Wzm and Wzt make up an ATP-driven polysaccharide transporter system that transports the Und-PP-bound O-antigen to the periplasmic space (Whitfield, 1995; Zhang et al., 1993). Transposon insertions into the *wzm* and *wzt* genes result in accumulation of cytoplasmic O-antigen (Zhang et al., 1993).

Y. ENTEROCOLITICA SEROTYPE O:9 LPS

The overall structure of *Y. enterocolitica* O:9 LPS is very much similar to that of serotype O:3. The inner core structure of O:9 is identical to that of O:3 (Figure 3) (Müller-Loennies et al., 1999), as appears to be also the outer core. The O-antigen of *Y. enterocolitica* O:9 is a homopolymer of N-formylperosamine (Caroff et al., 1984) and its O-antigen gene cluster has been almost completely sequenced (Figure 5) (Lübeck et al., 2003).

Y. ENTEROCOLITICA SEROTYPE O:8 LPS

STRUCTURE OF *Y. ENTEROCOLITICA* O:8 LPS

Silver stained DOC-PAGE and a schematic structure of the LPS of *Y. enterocolitica* O:8 are shown in Figure 1. Details of the chemical structure are given in Figures 2 and 5. The inner core structure is very similar to that of O:3 and O:9. The O-unit is a branched pentasaccharide. The linkage between the O-antigen and the core is not known. Interestingly, the O-antigens of serotypes O:7,8 and O:19,8 are almost identical to that of O:8 (Table 6).

BIOSYNTHESIS OF SEROTYPE O:8 O-ANTIGEN

The *Y. enterocolitica* O:8 O-antigen gene cluster contains 18 genes (Figure 6) (Zhang et al., 1997). These include genes encoding enzymes for the biosynthesis of UDP-GalNAc

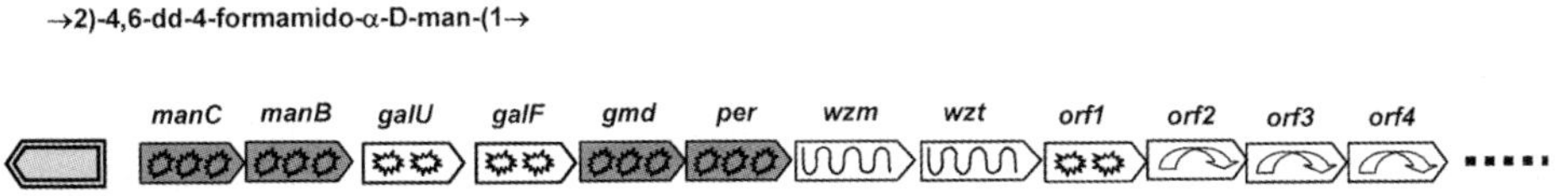

Figure 5. The O-antigen structure of *Y. enterocolitica* O:9 (Caroff et al., 1984) having a repeat unit of N-formyl-perosamine (perosamine = 4-amino-4,6-dideoxy-D-mannopyranose) and the corresponding O-antigen gene cluster (Lübeck et al., 2003).

(*gne)* (Bengoechea et al., 2002a), GDP-Man (*manB* and *manC*), GDP-Fuc (*gmd* and *fcl*) and CDP-6d-Gul (*ddhA, ddhB, wbcA* and *wbcB*). The *wbcC, wbcD, wbcG, wbcH,* and *wbcI* genes encode glycosyltransferases, the *wzx* gene encodes flippase, *wzy* encodes O-antigen polymerase, and *wzz* encodes the chain length determinant. One gene, *wbcF,* still remains unassigned (Zhang et al., 1997).

REGULATION OF LPS BIOSYNTHESIS

JUMPSTART AND *OPS* SEQUENCES

The gene clusters for *Y. enterocolitica* O:3 outer core and O-antigen, for *Y. enterocolitica* O:8 O-antigen, and for all *Y. pseudotuberculosis* O-antigens studied to date, carry a characteristic sequence called the JUMPstart sequence. The JUMPstart sequence has been identified in all surface-polysaccharide biosynthesis gene clusters and also in gene clusters for sex-pilus assembly and haemolysin production (Hobbs and Reeves, 1994; Nieto et al., 1996). The sequence is generally located downstream of functional promoters of the gene cluster. Nieto et al. (1996) reported that an 8 bp sequence within the JUMPstart sequence plays a role in gene regulation, and this element was named *ops* (operon polarity suppressor). *ops* was shown to function as a *cis*-acting element that together with RfaH, a transcriptional regulator, allows efficient transcription of long operons, especially of the distal genes (Bailey et al., 1996). It is postulated that *ops* provides specificity for RfaH for the subset of operons it regulates. The role of *ops* and RfaH in regulation of *Yersinia* LPS operons has not yet been studied but they are very likely to be involved.

TEMPERATURE AND OTHER ENVIRONMENTAL STIMULI

A general feature of *Yersinia* is temperature-dependency of LPS biosynthesis. Most striking is the almost complete down-regulation of O-antigen expression in *Y. pseudotuberculosis* and *Y. enterocolitica* grown at 37°C, another is the modification of the lipid A structure with respect to substitutions such as phosphate, aminoarabinose and acyl chains (Aussel et al., 2000; Lakshmi et al., 1989). Details of the regulatory mechanisms are not known.

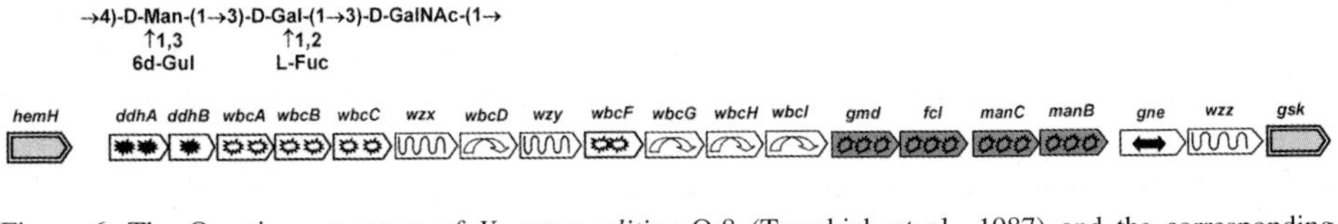

Figure 6. The O-antigen structure of *Y. enterocolitica* O:8 (Tomshich et al., 1987) and the corresponding O-antigen gene cluster (Zhang et al., 1997). For abbreviations see legends to Figures 1-3.

Table 6. Chemical structures of O-antigens of genus *Yersinia* for which no corresponding genetic data is available yet.

Species, serotype	Structure. Unless otherwise indicated the structure of the O-unit is given.	References
Y. aldovae (Fuc3NR[13])	**→6)-GlcNAc-(α1→4)-GalNAc-(β1→3)-GlcNAc-(β1→2)-Glc-(β1→2)-Fuc3NR-(α1→** **β↑1,3** **Glc**	(Zubkov et al., 1991)
Y. bercovieri O:10 (YerA[14])	**→3)-D-Rha-(α1→3)-D-Rha-(α1→** **α↑1,2** **YerA**	(Gorshkova et al., 1994)
Y. enterocolitica O:1,2a,3 (6d-Alt[15])	**→2)-6d-Alt-(β1→2)-6d-Alt-(β1→3)-6d-Alt-(β1→**	(Gorshkova et al., 1985; Hoffman et al., 1980)
Y. enterocolitica O:2a,2b,3	**→2)-6d-Alt-(β1→2)-6d-Alt-(β1→3)-6d-Alt-(β1→** **↑3 ↑3** **OAc OAc**	(Gorshkova et al., 1985; Hoffman et al., 1980)
Y. enterocolitica O:4,32 (YerB[16])	**(1'AcO)YerB** **α↓1,4** **→3)-GalNAc-(α1→3)-GalNAc-(β1→**	(Gorshkova et al., 1987b; Zubkov et al., 1989)
Y. enterocolitica O:5,27 (Xluf[17])	**→3)-L-Rha-(β1→3)-L-Rha-(α1→** **β↑2,2 β↑2,2** **Xlu*f* Xlu*f*** or **→3)-L-Rha-(α1 → 3)-L-Rha-(α1→3)-L-Rha -(β1→** **β↑2,2 β↑2,2** **Xlu*f* Xlu*f***	(Gorshkova et al., 1986; Perry and MacLean, 1987)

[13] Fuc3NR, 3,6-dideoxy-3-[(R)-3-hydroxybutyramido]-D-galactose.
[14] YerA, yersiniose A or 3,6-dideoxy-4-C-(L-glysero-1'-hydroxyethyl)-D-xylohexose.
[15] 6d-Alt, 6-deoxy-L-altropyranose.
[16] YerB, yersiniose B or 3,6-dd-4-C-(1-hydroxyethyl)-D-xylo-hexose.
[17] Xlu*f*, D-xylulofuranose.

Y. enterocolitica O:6,31 (6d-Gul[18])	**→2)-D-Gal-(α1→3)-6d-Gul-(β1→**	(Kalmykova et al., 1988)
Y. enterocolitica O:7,8 and O:19,8	**6d-Gul** **↓1,4** **→2)-D-Man-(1→3)-D-Gal-(1→3)-D-GalNAc-(1→** **↑1,3 ↑1,2** **6d-Gul L-Fuc**	(Ovodov et al., 1992; Tomshich et al., 1987)
Y. enterocolitica O:10 (L-Xlu[19])	**→3)-D-Rha-(α1→3)-D-Rha-(α1→** **α↑1,3** **L-Xlu**	(Gorshkova et al., 1995)
Y. enterocolitica O:11,23 and O:11,24[20]	**→3)-*L*-QuiNAc-(1α→4)-*D*-GalNAcA-(α1→3)-*L*-QuiNAc-(β1→3)-*D*-GlcNAc-(α1→**	(Marsden et al., 1994)
Y. enterocolitica O:28	**→3)-L-Rha*p*-(β1→3)-D-GlcpNAc-(α1→3)-L-Rha*p*-(α1→3)-L-Rha*p*-(α1→** **α↑1,2** **L-Rha*p*-(α1→4)-D-Gal*p*NAcA**	(Perry and MacLean, 2000)
Y. frederiksenii O:16,29	**→3)-D-Rha-(α1→3)-D-Rha-(α1→2)-D-Rha-(β1→** **β↑1,2** **YerA** **(33%)**	(Gorshkova et al., 1989)
Y. intermedia strain 680 (fructane[21])	**→1)-Fru*f*-(β2→1)-Fru-(α1→** **↑6 ↑6** **AcO AcO**	(Gorshkova et al., 1987a)

[18] 6d-Gul, 6-deoxy-D-gulose.
[19] L-Xlu, L-xylulose.
[20] QuiNAc, N-acetylquinosamine or 2-acetamido-2,6-dideoxy-L-glucose; GalNAcA, 2-acetamido-2-deoxy-D-galacturonic acid.
[21] Fru*f*, D-fructofuranose; Fru, D-fructopyranose.

Y. intermedia O:4,33	**YerB** **α↓1,4** **→3)-GalNAc-(α1→3)GalNAc-(β1→**	(Zubkov et al., 1988)
Y. kristensenii O:25,35	**→6)-Glc-(α1→4)-FucNAc-(β1→3)-GlcNAc-(1→2)-Gro-(β1-P→** **α↑1,3 α↑1,4** **GalNac Glc** **α↑1,6** **Glc**	(Gorshkova et al., 1993)
Y. kristensenii O:12,25 hexaosylglycerol phosphate structures[22]	**→6)-Glc-(α1→4)-FucNAc-(β1→3)-GlcNAc-(1→2)-Gro-(β1-P→** **α↑1,3 β↑1,4** **GalNac GlcNAc** **α↑1,6** **Glc**	(L'vov et al., 1992)
Y. kristensenii O:12,26	**→6)-Glc-(β1→6)-GalNac-(α1→3)-FucNAc-(α1→3)-GlcNAc-(β1→2)-Gro-(1-P→** **α↑1,2 α↑1,4** **Glc Glc**	(L'vov et al., 1990)
Y. rohdei	**→3)-L-Rha-(α1→3)-L-Rha-(β1→3)-L-Rha-(α1→**	(Zubkov et al., 1993)
Y.ruckerii 0:1[23]	**→3)-FucAm-(α1→3)-GlcNAc-(α1→8)-L-Sug-(α2→** **β↑1,4** **GlcNAc**	(Beynon et al., 1994)

[22] Gro, glycerol.
[23] FucAm, 2-acetamidino-2,6-dideoxy-L-galactose; L-Sug, 7-acetamido-3,5,7,9-tetradeoxy-5-(4-hydroxybutyramido)-D-glycero-L-galacto-nonulosonic acid.

REGULATION OF O:8 O-ANTIGEN BIOSYNTHESIS

The expression of *Y. enterocolitica* O:8 O-antigen is similarly regulated by temperature such that more abundant O-antigen is expressed on the cell surface at 25°C than at 37°C. The regulation mechanism involves repression of the O-antigen cluster transcription at 37°C (Bengoechea et al., 2002b; Zhang et al., 1995). The *rosAB* locus downstream of the O-antigen cluster, between the *gsk* and *ushA* genes, together with the *wzz* gene is involved in the temperature-regulated expression of O:8 O-antigen by an unknown mechanism that apparently involves sensing membrane stress (Bengoechea and Skurnik, 2000; Bengoechea et al., 2002b; Zhang et al., 1995). RosA is a efflux pump and RosB is a K^+/H^+ antiporter (Bengoechea and Skurnik, 2000).

YERSINIA LPS AND VIRULENCE

LPS MUTANTS

The role of LPS in the virulence of *Y. pseudotuberculosis* has been shown by signature-tagged-mutagenesis (STM) experiments (Karlyshev et al., 2001; Mecsas, 2002; Mecsas et al., 2001). A number of LPS mutants have been constructed in *Y. enterocolitica* strains and tested for virulence in animal experiments. These include both O-antigen negative (rough) and outer core negative mutants. The results indicate that LPS mutant strains, either missing O-antigen or outer core, are attenuated. In STM screens, mutations in the O-antigen gene cluster of *Y. enterocolitica* serotype O:8 were frequently detected, also indicating the critical role that O-antigen plays during infection (Darwin and Miller, 1999; Gort and Miller, 2000; Mcshan et al., 1997).

The LD_{50} values for rough *Y. enterocolitica* mutants were determined using intragastrically infected DBA/2 mice. All the mutants were less virulent than the wild type strain; the LD_{50} values for the rough mutants were about 50-100 times higher (Al-Hendy et al., 1992; Zhang et al., 1997), and the outer core mutants were nearly avirulent (Skurnik et al., 1999). The outer core provides resistance to cationic bactericidal peptides, and in coinfection experiments with wild type bacteria, the outer core mutant bacteria did not survive in deeper tissues as well as the wild type bacteria, although it could colonize the Peyer's patches for 2-5 days. Similar coinfection experiments with rough mutants indicated that O-antigen is required for the initial colonization of the Peyer's patches (Skurnik et al., 1999).

BIOLOGICAL ACTIVITIES

The endotoxic activity of *Yersinia* LPS has been addressed in some reports (Brubaker, 1972; Hartiala et al., 1989), but the role of *Yersinia* LPS in bacterial pathogenesis has received very little attention, so details of its biological role are far from known. LPS isolated from different *Yersinia* species show differences in acyl-chain fluidity that correlate with OM permeability to hydrophobic agents (Bengoechea et al., 2003; Bengoechea et al., 1998). Complete LPS also plays an important role in resistance against cationic antimicrobial peptides (Skurnik et al., 1999). There are indications that O-antigen may play a role in resistance to complement-mediated killing (Wachter and Brade, 1989), and LPS may even have roles in adhesion and invasion to host tissues and in the pathogenesis of reactive arthritis (Granfors et al., 1989; Granfors et al., 1998). LPS isolated from *Y. pseudotuberculosis* grown at 37°C and integrated into artificial membranes protected the membrane vesicles from complement-mediated lysis (Porat et al., 1995). Passive immunization has been used to show that antibodies against the O-antigen, but not those against the core, protect mice against intravenous challenge of virulent *Y. enterocolitica* (Skurnik et al., 1996).

REFERENCES

Achtman, M., Zurth, K., Morelli, G., Torrea, G., Guiyoule, A., and Carniel, E. 1999. *Yersinia pestis,* the cause of plague, is a recently emerged clone of *Yersinia pseudotuberculosis*. Proc. Natl. Acad. Sci. USA. 96: 14043-14048.

Ahvonen, P., Jansson, E., and Aho, K. 1969. Marked cross-agglutination between Brucellae and a subtype of *Yersinia enterocolitica*. Acta Path Microbiol Scand. 75: 291-295.

Aleksic, S., Bockemühl, J., and Wuthe, H.-H. 1995. Epidemiology of *Y. pseudotuberculosis* in Germany, 1983-1993. Contrib. Microb. Immunol. 13: 55-58.

Al-Hendy, A., Toivanen, P., and Skurnik, M. 1991. Expression cloning of *Yersinia enterocolitica* O:3 *rfb* gene cluster in *Escherichia coli* K12. Microb. Pathog. 10: 47-59.

Al-Hendy, A., Toivanen, P., and Skurnik, M. 1992. Lipopolysaccharide O side chain of *Yersinia enterocolitica* O:3 is an essential virulence factor in an orally infected murine model. Infect. Immun. 60: 870-875.

Aulisio, C.C.G., Hill, W.E., Stanfield, J.T., and Sellers, R.L.J. 1983. Evaluation of virulence factor testing and characteristics of pathogenicity in Yersinia enterocolitica. Infect Immun. 40: 330-335.

Aussel, L., Thérisod, H., Karibian, D., Perry, M.B., Bruneteau, M., and Caroff, M. 2000. Novel variation of lipid A structures in strains of different *Yersinia* species. FEBS Lett. 465: 87-92.

Bailey, M.J.A., Hughes, C., and Koronakis, V. 1996. Increased distal gene transcription by the elongation factor RfaH, a specialized homologue of NusG. Mol. Microbiol. 22: 729-737.

Bengoechea, J.A., Brandenburg, K., Arraiza, M.D., Seydel, U., Skurnik, M., and Moriyon, I. 2003. Pathogenic *Yersinia enterocolitica* strains increase the outer membrane permeability in response to environmental stimuli by modulating lipopolysaccharide fluidity and lipid A structure. Infect. Immun. 71: 2014-2021.

Bengoechea, J.A., Brandenburg, K., Seydel, U., Díaz, R., and Moriyón, I. 1998. *Yersinia pseudotuberculosis* and *Yersinia pestis* show increased outer membrane permeability to hydrophobic agents which correlates with lipopolysaccharide acyl-chain fluidity. Microbiology. 144: 1517-1526.

Bengoechea, J.A., Pinta, E., Salminen, T., Oertelt, C., Holst, O., Radziejewska-Lebrecht, J., Piotrowska-Seget, Z., Venho, R., and Skurnik, M. 2002a. Functional characterization of Gne (UDP-N-acetylglucosamine-4-epimerase), Wzz (chain length determinant), and Wzy (O-antigen polymerase) of *Yersinia enterocolitica* serotype O:8. J. Bacteriol. 184: 4277-4287.

Bengoechea, J.A., and Skurnik, M. 2000. Temperature-regulated efflux pump / potassium antiporter system mediates resistance to cationic antimicrobial peptides in *Yersinia*. Mol. Microbiol. 37: 67-80.

Bengoechea, J.A., Zhang, L., Toivanen, P., and Skurnik, M. 2002b. Regulatory network of lipopolysaccharide O-antigen biosynthesis in *Yersinia enterocolitica* includes cell envelope-dependent signals. Mol. Microbiol. 44: 1045-1062.

Beynon, L.M., Richards, J.C., and Perry, M.B. 1994. The structure of the lipopolysaccharide O antigen from *Yersinia ruckeri* serotype 01. Carbohydr. Res. 256: 303-317.

Bogdanovich, T., Carniel, E., Fukushima, H., and Skurnik, M. 2003a. Use of O-antigen gene cluster-specific PCRs for the identification and O-genotyping of *Yersinia pseudotuberculosis* and *Yersinia pestis*. J. Clin. Microbiol. 41: 5103-5112.

Bogdanovich, T.M., Carniel, E., Fukushima, H., and Skurnik, M. 2003b. Genetic (sero)typing of *Yersinia pseudotuberculosis*. In: The Genus *Yersinia*: entering the functional genomic era. M. Skurnik, K. Granfors and J.A. Bengoechea (Eds). Kluwer Academic/Plenum Publishers, New York. p. 337-340.

Brubaker, R.R. 1972. The genus *Yersinia*: Biochemistry and genetics of virulence. Curr. Top. Microbiol. Immunol. 57: 111-158.

Brubaker, R.R. 1991. Factors promoting acute and chronic diseases caused by *Yersiniae*. Clin. Microbiol. Rev. 4: 309-324.

Caroff, M., Bundle, D.R., and Perry, M.B. 1984. Structure of the O-chain of the phenol-phase soluble cellular lipopolysaccharide of Y*ersinia enterocolitica* serotype 0:9. Eur. J. Biochem. 139: 195-200.

Darwin, A.J., and Miller, V.L. 1999. Identification of *Yersinia enterocolitica* genes affecting survival in an animal host using signature-tagged transposon mutagenesis. Mol. Microbiol. 32: 51-62.

Gorshkova, R.P., Isakov, V.V., Kalmykova, E.N., and Ovodov, Y.S. 1995. Structural studies of O-specific polysaccharide chains of the lipopolysaccharide from *Yersinia enterocolitica* serovar 0:10. Carbohydr. Res. 268: 249-255.

Gorshkova, R.P., Isakov, V.V., Nazarenko, E.L., Ovodov, Y.S., Guryanova, S.V., and Dmitriev, B.A. 1993. Structure of the O-specific polysaccharide of the lipopolysaccharide from *Yersinia kristensenii* O:25.35. Carbohydr. Res. 241: 201-208.

Gorshkova, R.P., Isakov, V.V., Shevchenko, L.S., and Ovodov, Y.S. 1991. Structure of the O-specific polysaccharide chain of the *Yersinia pseudotuberculosis* lipopolysaccharide (serovar II C). Bioorganicheskaya Khimiya. 17: 252-257.

Gorshkova, R.P., Isakov, V.V., Zubkov, V.A., and Ovodov, Y.S. 1989. The structure of O-specific polysaccharide of *Yersinia frederiksenii* serotype O:16,29 lipopolysaccharide. Bioorganicheskaya Khimiya. 15: 1627-1633.

Gorshkova, R.P., Isakov, V.V., Zubkov, V.A., and Ovodov, Y.S. 1994. Structure of O-specific polysaccharide of lipopolysaccharide from *Yersinia bercovieri* 0:10. Bioorganicheskaya Khimiya. 20: 1231-1235.

Gorshkova, R.P., Kalmykova, E.N., Isakov, V.V., and Ovodov, Y.S. 1985. Structural studies on O-specific polysaccharides of lipopolysaccharides from *Yersinia enterocolitica* serovars O:1,2a,3, O:2a,2b,3 and O:3. Eur. J. Biochem. 150: 527-531.

Gorshkova, R.P., Kalmykova, E.N., Isakov, V.V., and Ovodov, Y.S. 1986. Structural studies on O-specific polysaccharides of lipopolysaccharides from *Yersinia enterocolitica* serovars O:5 and O:5,27. Eur. J. Biochem. 156: 391-397.

Gorshkova, R.P., Komandrova, N.A., Kalinovsky, A.I., and Ovodov, Y.S. 1980. Structural studies on the O-specific polysaccharide side-chains of *Yersinia pseudotuberculosis,* type III, lipopolysaccharides. Eur. J. Biochem. 107: 131-135.

Gorshkova, R.P., Korchahina, N.I., and Ovodov, Y.S. 1983a. Structural studies on the O-specific side-chain polysaccharide of lipopolysaccharide from the *Yersinia pseudotuberculosis* VA serovar. Eur. J. Biochem. 131: 345-347.

Gorshkova, R.P., Kovalchuk, S.V., Isakov, V.V., Frolova, G.M., and Ovodov, Y.S. 1987a. The structure of O-specific polysaccharide isolated from the lipopolysaccharide of *Yersinia intermedia* strain 180. Bioorganicheskaya Khimiya. 13: 818-824.

Gorshkova, R.P., Zubkov, V.A., Isakov, V.V., and Ovodov, Y.S. 1983b. Structure of O-specific polysaccharide from *Yersinia pseudotuberculosis* of serotype VI lipopolysaccharide. Bioorganicheskaya Khimiya. 9: 1068-1073.

Gorshkova, R.P., Zubkov, V.A., Isakov, V.V., and Ovodov, Y.S. 1983c. Study of a lipopolysaccharide from *Yersinia pseudotuberculosis* of the IVA serotype. Bioorganicheskaya Khimiya. 9: 1401-1407.

Gorshkova, R.P., Zubkov, V.A., Isakov, V.V., and Ovodov, Y.S. 1987b. A new branched-chain monosaccharide from the *Yersinia enterocolitica* serotype O:4,32 lipopolysaccharide. Bioorganicheskaya Khimiya. 13: 1146-1147.

Gorshkova, R.P., Zubkov, V.A., and Ovodov, Y.S. 1976. Chemical and immunochemical studies on lipopolysaccharide from *Yersinia pseudotuberculosis* type VI. Immunochemistry. 13: 581-583.

Gort, A.S., and Miller, V.L. 2000. Identification and characterization of *Yersinia enterocolitica* genes induced during systemic infection. Infect. Immun. 68: 6633-6642.

Granfors, K., Jalkanen, S., von Essen, R., Lahesmaa-Rantala, R., Isomäki, O., Pekkola-Heino, K., Merilahti-Palo, R., Saario, R., Isomäki, H., and Toivanen, A. 1989. *Yersinia* antigens in synovial-fluid cells from patients with reactive arthritis. N. Engl. J. Med. 320: 216-221.

Granfors, K., Merilahti-Palo, R., Luukkainen, R., Möttönen, T., Lahesmaa, R., Probst, P., Marker-Hermann, E., and Toivanen, P. 1998. Persistence of *Yersinia* antigens in peripheral blood cells from patients with *Yersinia enterocolitica* O:3 infection with or without reactive arthritis. Arthritis Rheum. 41: 855-862.

Gronow, S., Oertelt, C., Ervelä, E., Zamyatina, A., Kosma, P., Skurnik, M., and Holst, O. 2001. Characterization of the physiological substrate for lipopolysaccharide heptosyltransferases I and II. J. Endotoxin Res. 7: 263-270.

Guo, L., Lim, K.B., Poduje, C.M., Daniel, M., Gunn, J.S., Hackett, M., and Miller, S.I. 1998. Lipid A acylation and bacterial resistance against vertebrate antimicrobial peptides. Cell. 95: 189-98.

Hartiala, K., T., Granberg, I., Toivanen, A., and Viljanen, M. 1989. Inhibition of polymorphonuclear leucocyte functions in vivo by *Yersinia enterocolitica* lipopolysaccharide. Ann. Rheum. Dis. 48: 42-47.

Hitchen, P.G., Prior, J.L., Oyston, P.C., Panico, M., Wren, B.W., Titball, R.W., Morris, H.R., and Dell, A. 2002. Structural characterization of lipo-oligosaccharide (LOS) from *Yersinia pestis*: regulation of LOS structure by the PhoPQ system. Mol. Microbiol. 44: 1637-1650.

Hobbs, M., and Reeves, P. 1995. Genetic organization and evolution of *Yersinia pseudotuberculosis* 3,6-dideoxyhexose biosynthetic genes. Biochim. Biophys. Acta. 1245: 273-277.

Hobbs, M., and Reeves, P.R. 1994. The JUMPstart sequence: A 39 bp element common to several polysaccharide gene clusters. Mol. Microbiol. 12: 855-856.

Hoffman, J., Lindberg, B., and Brubaker, R.R. 1980. Structural studies of the O-specific side-chains of the lipopolysaccharide from *Yersinia enterocolitica* Ye 128. Carbohydr. Res. 78: 212-214.

Isakov, V.V., Komandrova, N.A., Gorshkova, R.P., and Ovodov, Y.S. 1983. 13C-NMR spectrum of O-specific polysaccharide from the lipopolysaccharide of *Yersinia pseudotuberculosis* of serotype III. Bioorganicheskaya Khimiya. 9: 1565-1567.

Kalmykova, E.N., Gorshkova, R.P., Isakov, V.V., and Ovodov, I.S. 1988. Structural studies of side-chains of the O-specific polysaccharide from lipopolysaccharide of *Yersinia enterocolitica* serovar O:6,31. Bioorganicheskaya Khimiya. 14: 652-657.

Karlyshev, A.V., Oyston, P.C., Williams, K., Clark, G.C., Titball, R.W., Winzeler, E.A., and Wren, B.W. 2001. Application of high-density array-based signature-tagged mutagenesis to discover novel *yersinia* virulence-associated genes. Infect. Immun. 69: 7810-7819.

Kauffmann, F., 1967. The bacteriology of Enterobacteriaceae. The Williams and Wilkins Co., Baltimore.

Kittelberger, R., Hilbink, F., Hansen, M.F., Penrose, M., Delisle, G.W., Letesson, J.J., Garinbastuji, B., Searson, J., Fossati, C.A., Cloeckaert, A., and Schurig, G. 1995a. Serological crossreactivity between Brucella abortus and Yersinia enterocolitica 0:9 .1. Immunoblot analysis of the antibody response to Brucella protein antigens in bovine brucellosis. Vet. Microbiol. 47: 257-270.

Kittelberger, R., Hilbink, F., Hansen, M.F., Ross, G.P., Joyce, M.A., Fenwick, S., Heesemann, J., Wolfwatz, H., and Nielsen, K. 1995b. Serological crossreactivity between Brucella abortus and Yersinia enterocolitica 0:9 .2. The use of Yersinia outer proteins for the specific detection of Yersinia enterocolitica infections in ruminants. Vet. Microbiol. 47: 271-280.

Komandrova, N.A., Gorshkova, R.P., Isakov, V.V., and Ovodov, Y.S. 1984. Structure of O-specific polysaccharide isolated from the *Yersinia pseudotuberculosis* serotype 1A lipopolysaccharide. Bioorganicheskaya Khimiya. 10: 232-237.

Korchagina, N.I., Gorshkova, R.P., and Ovodov, Y.S. 1982. Studies on O-specific polysaccharide from *Yersinia pseudotuberculosis* VB serovar. Bioorganicheskaya Khimiya. 8: 1666-1669.

Kotandrova, N.A., Gorshkova, R.P., Zubkov, V.A., and Ovodov, Y.S. 1989. The structure of the O-specific polysaccharide chain of the lipopolysaccharide of *Yersinia pseudotuberculosis* serovar VII. Bioorganicheskaya Khimiya. 15: 104-110.

Kuhn, H.-M., Meier-Dieter, U., and Mayer, H. 1988. ECA, the enterobacterial common antigen. FEMS Microbiol. Rev. 54: 195-222.

Lahtinen, P., Brzezinska, A., and Skurnik, M. 2003. Temperature and growth phase regulate the transcription of the O-antigen gene cluster of *Yersinia enterocolitica* O:3. In: The Genus *Yersinia*: entering the functional genomic era. M. Skurnik, K. Granfors and J.A. Bengoechea (Eds). Kluwer Academic/Plenum Publishers, New York. p. 289-292.

Lakshmi, S.K.B., Bhat, U.R., Wartenberg, K., Schlecht, S., and Mayer, H. 1989. Temperature-dependent incorporation of 4-amino-L-arabinose in lipid A of distinct Gram-negative bacteria. FEMS Microbiol. Lett. 60: 317-322.

Liu, D., Cole, R.A., and Reeves, P.R. 1996. An O-antigen processing function for Wzx (RfbX): a promising candidate for O-unit flippase. J. Bacteriol. 178: 2102-7.

L'vov, V.L., Gur'ianova, S.V., Rodionov, A.V., Dmitriev, B.A., Shashkov, A.S., Ignatenko, A.V., Gorshkova, R.P., and Ovodov, Y.S. 1990. The structure of a repetitive unit of the glycerolphosphate- containing O-specific polysaccharide chain from *Yersinia kristensenii* strain 103 (0:12,26) lipopolysaccharide. Bioorganicheskaya Khimiya. 16: 379-389.

L'vov, V.L., Guryanova, S.V., Rodionov, A.V., and Gorshkova, R.P. 1992. Structure of the repeating unit of the O-specific polysaccharide of the lipopolysaccharide of *Yersinia kristensenii* strain 490 (O:12,25). Carbohydr. Res. 228: 415-422.

Lübeck, P.S., Hoorfar, J., Ahrens, P., and Skurnik, M. 2003. Cloning and characterization of the *Yersinia enterocolitica* serotype O:9 lipopolysaccharide O-antigen gene cluster. In: The Genus *Yersinia*: entering the functional genomic era. M. Skurnik, K. Granfors and J.A. Bengoechea (Eds). Kluwer Academic/Plenum Publishers, New York. p. 207-209.

Marsden, B.J., Bundle, D.R., and Perry, M.B. 1994. Serological and structural relationships between *Escherichia coli* O:98 and *Yersinia enterocolitica* O:11,23 and O:11,24 lipopolysaccharide O-antigens. Biochem. Cell Biol. 72: 163-168.

Mcshan, W.M., Tang, Y.F., and Ferretti, J.J. 1997. Bacteriophage t12 of streptococcus pyogenes integrates into the gene encoding a serine tRNA. Mol. Microbiol. 23: 719-728.

Mecsas, J. 2002. Use of signature-tagged mutagenesis in pathogenesis studies. Curr. Opin. Microbiol. 5: 33-37.

Mecsas, J., Bilis, I., and Falkow, S. 2001. Identification of attenuated *Yersinia pseudotuberculosis* strains and characterization of an orogastric infection in BALB/c mice on day 5 postinfection by signature-tagged mutagenesis. Infect. Immun. 69: 2779-2787.

Morona, R., Macpherson, D.F., Vandenbosch, L., Carlin, N.I.A., and Manning, P.A. 1995. Lipopolysaccharide with an altered O-antigen produced in *Escherichia coli* K-12 harbouring mutated, cloned *Shigella flexneri rfb* genes. Mol. Microbiol. 18: 209-223.

Müller-Loennies, S., Rund, S., Ervelä, E., Skurnik, M., and Holst, O. 1999. The structure of the carbohydrate backbone of the core-lipid A region of the lipopolysaccharide from a clinical isolate of *Yersinia enterocolitica* O:9. Eur. J. Biochem. 261: 19-24.

Mäki, M., Vesikari, T., Rantala, I., Sundqvist, C., and Grönroos, P. 1983. Pathogenicity of 42-44 Mdal plasmid positive and negative Yersinia pseudotuberculosis I and Yersinia enterocolitica O:8 and O:9 studied in the guinea pig eye model (Sereny test). Acta path microbiol immunol scand Sect B. 91: 241-244.

Nagano, T., Kiyohara, T., Suzuki, K., Tsubokura, M., and Otsuki, K. 1997. Identification of pathogenic strains within serogroups of Yersinia pseudotuberculosis and the presence of non-pathogenic strains isolated from animals and the environment. J. Vet. Med. Sci. 59: 153-158.

Nieto, J.M., Bailey, M.J.A., Hughes, C., and Koronakis, V. 1996. Suppression of transcription polarity in the *Escherichia coli* haemolysin operon by a short upstream element shared by polysaccharide and DNA transfer determinants. Mol. Microbiol. 19: 705-713.

Ovodov, Y.S., Gorshkova, R.P., Tomshich, S.V., Komandrova, N.A., Zubkov, V.A., Kalmykova, E.N., and Isakov, V.V. 1992. Chemical and immunochemical studies on lipopolysaccharides of some *Yersinia* species - a review of some recent investigations. J. Carbohydr. Chem. 11: 21-35.

Perry, M.B., and MacLean, L.L. 1987. Structure of the lipopolysaccharide O-chain of *Yersinia enterocolitica* serotype O:5,27. Biochem. Cell Biol. 65: 1-7.

Perry, M.B., and MacLean, L.L. 2000. Structural identification of the lipopolysaccharide O-antigen produced by *Yersinia enterocolitica* serotype O:28. Eur. J. Biochem. 267: 2567-2572.

Porat, R., McCabe, W.R., and Brubaker, R.R. 1995. Lipopolysaccharide-associated resistance to killing of *Yersiniae* by complement. J. Endotoxin Res. 2: 91-97.

Radziejewska-Lebrecht, J., Shashkov, A.S., Stroobant, V., Wartenberg, K., Warth, C., and Mayer, H. 1994. The inner core region of *Yersinia enterocolitica* Ye75R (0:3) lipopolysaccharide. Eur. J. Biochem. 221: 343-351.

Radziejewska-Lebrecht, J., Skurnik, M., Shashkov, A.S., Brade, L., Rozalski, A., Bartodziejska, B., and Mayer, H. 1998. Immunochemical studies on R mutants of *Yersinia enterocolitica* O:3. Acta Biochimica Polonica. 45: 1011-1019.

Raetz, C.R., and Whitfield, C. 2002. Lipopolysaccharide endotoxins. Annu. Rev. Biochem. 71: 635-700.

Reeves, P.R., Pacinelli, E., and Wang, L. 2003. O antigen Gene Clusters of *Yersinia pseudotuberculosis*. In: The Genus *Yersinia*: entering the functional genomic era. M. Skurnik, K. Granfors and J.A. Bengoechea (Eds). Kluwer Academic/Plenum Publishers. p. 199-206.

Rick, P., D., and Silver, R., P. 1996. Enterobacterial common antigen and capsular polysaccharides. In: *Escherichia coli* and *Salmonella:* cellular and molecular biology. F.C. Neidhardt et al. (Eds). American Society for Microbiology, Washington, D.C. p. 104-118.

Samuelsson, K., Lindberg, B., and Brubaker, R.R. 1974. Structure of O-specific side chains of lipopolysaccharides from *Yersinia pseudotuberculosis*. J. Bacteriol. 117: 1010-1016.

Shashkov, A.S., Radziejewska-Lebrecht, J., Kochanowski, H., and Mayer, H. 1995. The chemical structure of the outer core region of the *Yersinia enterocolitica* O:3 lipopolysaccharide. In: Abstract B017, 8th European Carbohydrate Symposium. Sevilla, Spain,.

Skurnik, M., 1985. Studies on the virulence plasmids of *Yersinia* species. PhD Thesis, University of Oulu, Oulu, 61+39 pp.

Skurnik, M. 1999. Molecular genetics of *Yersinia* lipopolysaccharide. In: Genetics of Bacterial Polysaccharides. J. Goldberg (Ed.). CRC Press, Boca Raton, FL. p. 23-51.

Skurnik, M. 2003. Molecular genetics, biochemistry and biological role of *Yersinia* lipopolysaccharide. In: The Genus *Yersinia*: entering the functional genomic era. M. Skurnik, K. Granfors and J.A. Bengoechea (Eds). Kluwer Academic/Plenum Publishers, New York. p. 187-189.

Skurnik, M., Mikkola, P., Toivanen, P., and Tertti, R. 1996. Passive immunization with monoclonal antibodies specific for lipopolysaccharide (LPS) O-side chain protects mice against intravenous *Yersinia enterocolitica* serotype O:3 infection. APMIS. 104: 598-602.

Skurnik, M., Peippo, A., and Ervelä, E. 2000. Characterization of the O-antigen gene clusters of *Yersinia pseudotuberculosis* and the cryptic O-antigen gene cluster of *Yersinia pestis* shows that the plague bacillus is most closely related to and has evolved from *Y. pseudotuberculosis* serotype O:1b. Mol. Microbiol. 37: 316-330.

Skurnik, M., and Toivanen, P. 1991. Intervening sequences (IVSs) in the 23S ribosomal RNA genes of pathogenic *Yersinia enterocolitica* strains. The IVSs in *Y. enterocolitica* and *Salmonella typhimurium* have common origin. Mol. Microbiol. 5: 585-593.

Skurnik, M., Venho, R., Bengoechea, J.-A., and Moriyón, I. 1999. The lipopolysaccharide outer core of *Yersinia enterocolitica* serotype O:3 is required for virulence and plays a role in outer membrane integrity. Mol. Microbiol. 31: 1443-1462.

Skurnik, M., Venho, R., Toivanen, P., and Al-Hendy, A. 1995. A novel locus of *Yersinia enterocolitica* serotype O:3 involved in lipopolysaccharide outer core biosynthesis. Mol. Microbiol. 17: 575-594.

Thal, E., 1954. Untersuchungen ûber Past. pseudotuberculosis unter besonderer Berücksichtigung ihres immunologischen Verhaltens. Berlingske Tryckeriet.

Thorson, J.S., and Liu, H.W. 1993. Characterization of the First PMP-Dependent Iron Sulfur-Containing Enzyme Which Is Essential for the Biosynthesis of 3,6-Dideoxyhexoses. J. Am. Chem. Soc. 115: 7539-7540.

Thorson, J.S., Lo, S.F., and Liu, H.W. 1993a. Molecular basis of 3,6-dideoxyhexose biosynthesis - elucidation of CDP-ascarylose biosynthetic genes and their relationship to other 3,6-dideoxyhexose pathways. J. Am. Chem. Soc. 115: 5827-5828.

Thorson, J.S., Lo, S.F., Liu, H.W., and Hutchinson, C.R. 1993b. Biosynthesis of 3,6-Dideoxyhexoses - New Mechanistic Reflections upon 2,6-Dideoxy, 4,6-Dideoxy, and Amino Sugar Construction. J. Am. Chem. Soc. 115: 6993-6994.

Thorson, J.S., Lo, S.F., Ploux, O., He, X.M., and Liu, H.W. 1994. Studies of the biosynthesis of 3,6-dideoxyhexoses: Molecular cloning and characterization of the *asc* (Ascarylose) region from *Yersinia pseudotuberculosis* serogroup VA. J. Bacteriol. 176: 5483-5493.

Tomshich, S.V., Gorshkova, R.P., Elkin, Y.N., and Ovodov, Y.S. 1976. Lipopolysaccharide from *Yersinia pseudotuberculosis*, type 1B. A structural study of O-specific chains. Eur. J. Biochem. 65: 193-9.

Tomshich, S.V., Gorshkova, R.P., and Ovodov, Y.S. 1987. Structural investigation of the lipopolysaccharide of *Y. enterocolitica* serovar O:8. Khimia Prirodnykh Soedinenii: 657-664.

Tsubokura, M., and Aleksic, S. 1995. A simplified antigenic scheme for serotyping of *Yersinia pseudotuberculosis*: Phenotypic characterization of reference strains and preparation of O and H factor sera. Contrib. Microb. Immunol. 13: 99-105.

Tsubokura, M., Aleksic, S., Fukushima, H., Schulze, G., Someya, K., Sanekata, T., Otsuki, K., Nagano, T., Kuratani, Y., Inoue, M., Zheng, X., and Nakajima, H. 1993. Characterization of *Yersinia pseudotuberculosis* serogroups O9, O10 and O11 - subdivision of O1 serogroup into O1a, O1b, and O1c subgroups. Int. J. Med. Microb. 278: 500-509.

Tsubokura, M., Otsuki, K., Kawaoka, Y., Fukushima, H., Ikemura, K., and Kanazawa, Y. 1984. Addition of new serogroups and improvement of the antigenic designs of *Yersinia pseudotuberculosis*. Curr. Microbiol. 11: 89-92.

Tsubokura, M., Otsuki, K., Sato, K., Tanaka, M., Hongo, T., Fukushima, H., Maruyama, T., and Inoue, M. 1989. Special features of distribution of *Yersinia pseudotuberculosis* in Japan. J. Clin. Microbiol. 27: 790-791.

Wachter, E., and Brade, V. 1989. Influence of surface modulations by enzymes and monoclonal antibodies on alternative complement pathway activation by *Yersinia enterocolitica*. Infect. Immun. 57: 1984-1989.

Wauters, G., Aleksic, S., Charlier, J., and Schulze, G. 1991. Somatic and flagellar antigens of *Yersinia enterocolitica* and related species. Contrib. Microb. Immunol. 12: 239-243.

White, P.B. 1926. Further studies on the *Salmonella* group. Med. Res. Spec. Rep. Counc: (Great Britain) *Ser.* 103: 3-160.

Whitfield, C. 1995. Biosynthesis of lipopolysaccharide O antigens. Trends Microbiol. 3: 178-185.

Whitfield, C., and Valvano, M.A. 1993. Biosynthesis and expression of cell-surface polysaccharides in Gram-negative bacteria. In: Advances in Microbial Physiology. A.H. Rose (Ed.). Academic Press Limited, 24-28 Oval Road/London NW1 7DX/United Kingdom. p. 135-246.

Winblad, S. 1973. Studies of the O-serotypes of *Yersinia enterocolitica*. Contrib. Microb. Immunol. 2: 27-37.

Vinogradov, E.V., Lindner, B., Kocharova, N.A., Senchenkova, S.N., Shashkov, A.S., Knirel, Y.A., Holst, O., Gremyakova, T.A., Shaikhutdinova, R.Z., and Anisimov, A.P. 2002. The core structure of the lipopolysaccharide from the causative agent of plague, *Yersinia pestis*. Carbohydr. Res. 337: 775-777.

Zhang, L., 1996. Molecular genetics of the O-antigen of *Yersinia enterocolitica* lipopolysaccharide. PhD thesis., University of Turku, Turku, Finland, 118 pp.

Zhang, L., Al-Hendy, A., Toivanen, P., and Skurnik, M. 1993. Genetic organization and sequence of the *rfb* gene cluster of *Yersinia enterocolitica* serotype O:3: Similarities to the dTDP-L-rhamnose biosynthesis pathway of *Salmonella* and to the bacterial polysaccharide transport systems. Mol. Microbiol. 9: 309-321.

Zhang, L., Radziejewska-Lebrecht, J., Krajewska-Pietrasik, D., Toivanen, P., and Skurnik, M. 1997. Molecular and chemical characterization of the lipopolysaccharide O-antigen and its role in the virulence of *Yersinia enterocolitica* serotype O:8. Mol. Microbiol. 23: 63-76.

Zhang, L.J., Toivanen, P., and Skurnik, M. 1995. Genetic characterization of a novel locus of *Yersinia enterocolitica* serotype O:8 for down-regulation of the lipopolysaccharide O side chain at 37 degrees C.

In: Yersiniosis: Present and Future. G. Ravagnan and C. Chiesa (Eds), Contributions to Microbiology and Immunology. Karger, Basel, Switzerland. p. 310-313.

Zubkov, V.A., Gorshkova, R.P., Burtseva, Isakov, V.V., and Ovodov, Y.S. 1989. Structure of the O-specific polysaccharide of the *Yersinia enterocolitica* serovar O:4.32 lipopolysaccharide. Serologic relations of lipopolysaccharides of *Y. enterocolitica* O:4.32 and *Y. intermedia* O:4.33. Bioorganicheskaya Khimiya. 15: 187-191.

Zubkov, V.A., Gorshkova, R.P., Nazarenko, E.L., Shashkov, A.S., and Ovodov, Y.S. 1991. Structure of O-specific polysaccharide chain of the *Yersinia aldovae* lipopolysaccharide. Bioorganicheskaya Khimiya. 17: 831-838.

Zubkov, V.A., Gorshkova, R.P., and Ovodov, Y.S. 1988. Structural studies of the O-specific polysaccharide chain of the *Yersinia intermedia* 0:4.33 lipopolysaccharide. Bioorganicheskaya Khimiya. 14: 65-68.

Zubkov, V.A., Nazarenko, E.L., Gorshkova, R.P., and Ovodov, Y.S. 1993. Structure of the O-specific Polysaccharide of *Yersinia rohdei*. Bioorganicheskaya Khimiya. 19: 729-732.

Chapter 12

Flagella: Organelles for Motility and Protein Secretion

Glenn M. Young

ABSTRACT

Flagella represent an excellent trait to study the evolution of the genus *Yersinia* and have provided insight on the function of bacterial type III secretion systems. Flagella are organelles produced by enteropathogenic species of the genus *Yersinia* that provide a means for motility. This bacterial organelle also forms a type III secretion system for the export of non-flagellar proteins to the outside of the bacterium. Recent studies of flagellar functions have provided remarkable insight on how both flagellar-dependent motility and flagellar-dependent protein secretion contribute to the biology of the flagellated *Yersinia* species. The ability to migrate to a favorable environment certainly contributes to the survival of enteropathogenic *Yersinia* during free-living stages of their life cycle, but this attribute also can affect interactions with host organisms. Flagella promote the invasion of mammalian cells and, in some cases, appear to affect the pathogenic outcomes. Interactions with the host may also be influenced by the proteins delivered into the host by the flagellar type III secretion system.

PREVALENCE OF FLAGELLA AMONG THE GENUS *YERSINIA*

Flagella are dynamic organelles that provide bacteria with a means for motility and, as will be discussed further below, flagella also constitute a pathway for the export of proteins to the environment. From a historical perspective, the ability to produce flagella has been a key factor that distinguishes the gastrointestinal pathogens, *Y. enterocolitica* and *Y. pseudotuberculosis*, from the plague pathogen, *Y. pestis*. Essentially all pathogenic isolates of *Y. enterocolitica* and *Y. pseudotuberculosis* exhibit the ability to produce flagella whereas the production of flagella has not been observed for any isolate of *Y. pestis*. The lack of flagella production by *Y. pestis* may reflect the adaptation of this species to a specialized lifestyle in which the organism is directly spread from one animal host to another by flea-mediated transmission or aerosol exposure. The recently completed annotation of the *Y. pestis* genome provides clues that support this widely held inference (Parkhill et al., 2001; Deng et al., 2002). However, this does not mean that flagella are not important for *Yersinia*. Recent studies of *Y. enterocolitica* and *Y. pseudotuberculosis* suggest flagella play a role in virulence (Young et al., 1999; Young et al., 2000; Young and Young, 2002). At this time, we have the most complete picture of how flagella contribute to the lifestyle of *Y. enterocolitica*. Therefore, this chapter primarily focuses on this species with additional information about *Y. pseudotuberculosis* and *Y. pestis* included to provide a complete perspective.

ROLE OF FLAGELLA IN MOTILITY

The basic mechanism of flagellar mediated motility is the same as has been described for many other bacterial genera. Flagella protrude from the surface of the cell and rotate, which produces the thrust needed to migrate. Motility itself is not an unusual trait as 80% of bacterial species are known to have flagella. When combined with sensory perception, motility provides the bacterium with the ability to actively seek out favorable environments and escape unfavorable circumstances. The role of flagella in motility of *Yersinia* species has been primarily restricted to the studies of bacteria swimming in liquid environments (Rohde et al., 1994; Iriarte et al., 1995; Kapatral and Minnich, 1995; Kapatral et al., 1996; Fauconnier et al., 1997; Atkinson et al., 1999), but it is also clear that flagella provide a means for bacteria to translocate over surfaces as well (Young et al., 1998). The production of flagella and, therefore motility itself, occurs in the laboratory at temperatures below 30°C. This may indicate that motility has a predominant role in survival of *Yersinia* during free living stages of their life cycle, but it does not eliminate the possibility that flagella also contribute to pathogenic stages. Both *Y. enterocolitica* and *Y. pseudotuberculosis* retain flagella on their surface and motility for many hours after they are transferred from low temperature conditions to higher temperatures similar to those of a warm blooded animal ((Venediktov et al., 1988)and G. M. Young, unpublished observation). During this period of time there is plenty of opportunity for the bacterium to utilize motility to seek out favorable host niches in which to survive the assault of host innate immune responses.

THE STRUCTURE OF FLAGELLA

Flagella produced by *Y. enterocolitica* and *Y. pseudotuberculosis* are similar to what has been described for other members of the family of *Enterobacteriaceae*. Flagellar biogenesis in this group of bacteria has been best studied for *Salmonella enterica* and several studies of flagellar synthesis in *Y. enterocolitica* indicate this bacterium utilizes similar mechanisms to coordinate organelle assembly. Approximately 3 – 5 flagella are present on the surface of *Y. enterocolitica*, but the number can increase under some culture conditions (Young et al., 1998). Increased numbers of flagella are brought about by conditions known to promote swarming; a behavioral response that occurs for many bacteria resulting in the mass migration of cells over a solid surface. *Y. enterocolitica* displays this form of motility when cultivated on a nutrient-rich semi-solid agar medium (Young et al., 1998).

The flagellar organelle consists of the basal body, hook, filament and motor torque generating complex (Aizawa, 1996; Macnab, 1996) (Figure 1). The process of organelle biogenesis is highly ordered such that the basal body is initially assembled and traverses from the cytoplasm across the entire cell envelope to the outside of the cell (Figure 1, Stage 1). The hook (Figure 1, Stage 2) and filament (Figure 1, Stage 3) are sequentially assembled at the tip of the growing organelle to form its external portions. Also, at the late stages of flagellum biogenesis the motor torque-generating complex is assembled (Figure 1, Stage 3). During flagellum biogenesis many of the subunits destined for a location beyond the cytoplasm are transported by a flagellar-specific type III secretion (TTS) system (Macnab, 1999). The apparatus responsible for this type III secretion pathway is an integral part of the flagellar basal body (Fan et al., 1997; Hueck, 1998; Minamino and Macnab, 1999). Ten subunits of the basal body share a high degree of amino acid sequence similarity with subunits that form the secretion machinery of contact-

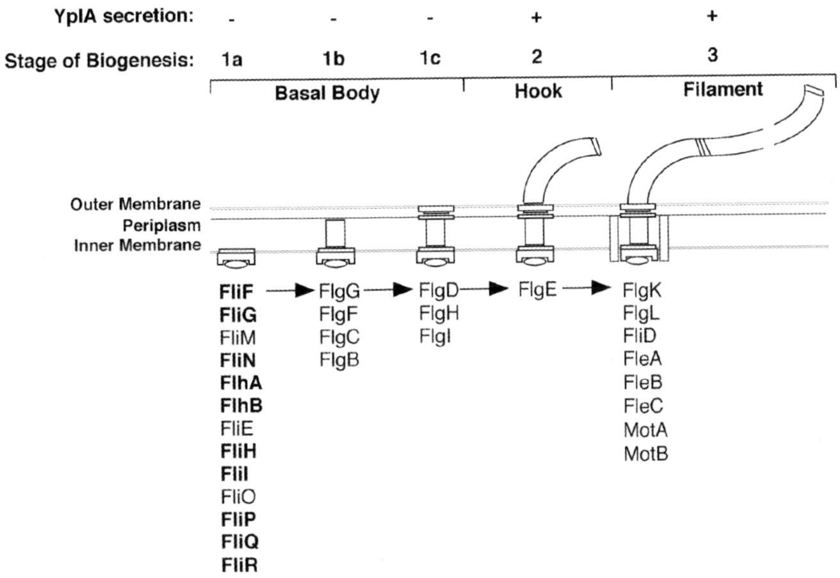

Figure 1. The morphogenesis pathway of the flagellum is described in three stages. Stage 1: The basal body spans the cell envelope. Its formation involves assembly of the TTS apparatus within the inner membrane (Stage 1a); the rod, which crosses the peptidoglycan in periplasm (Stage 1b); and the P and L rings within the outer membrane (Stage 1c). Transition to hook formation requires the hook scaffolding protein, FlgD, which is located at the distal end of the hook as it is assembled (Stage 1c). Stage 2: The hook is formed by the addition of FlgE to the basal body. Stage 3: The junction between the hook and the filament is formed by FlgK and FlgL. This is followed by the addition of FliD which is located at the tip of the growing filament. FliD acts as a scaffolding protein for the assembly of the flagellins, FleA, FleB and FleC. Also shown, formation of the filament coincides with morphogenesis of the flagellar motor torque-generating complex. Proteins that exhibit similarity with components of other type III secretion systems are highlighted in bold text. The ability of each of the indicated flagellar structures to export YplA is indicated at the top of the diagram.

dependent TTS systems (Figure 2)(Hueck, 1998). This similarity is quite apparent when flagellar proteins are compared to proteins that constitute the plasmid-encoded Ysc TTS apparatus produced by all pathogenic species of *Yersinia* and the chomosomally-encoded Ysa TTS apparatus produced by high-virulence strains of *Y. enterocolitica* (Haller et al., 2000; Snellings et al., 2001; Foultier et al., 2002). The identity between corresponding proteins ranges from 20-35% and their similarity reaches 50%. The relatedness of these proteins appears to reflect structural and functional conservation (Figure 2). It also suggests that the secretion machineries of flagella and contact-dependent TTS systems is evolutionarily related and has helped guide researchers in developing a detailed model of the machinery that constitutes contact-dependent TTS systems (Figure 2). Consistent with this notion, the superstructure of flagellar and contact-dependent TTS systems is conserved as microscopic procedures have revealed remarkable images of the *Salmonella* flagellar basal body (Aizawa et al., 1985), the *Salmonella* SPI-1 TTS system (Kubori et al., 1998) and the *Shigella flexneri mxi/spa* encoded TTS system (Blocker et al., 1999).

Eleven flagellar proteins are known to be transported by the flagellar TTS system: FliE (basal body); FlgB, FlgC, FlgF, FlgG (basal body rod); FlgE (hook); FlgD (hook scaffold protein); FlgK, FlgL (hook-filament junction); FleA, FleB and FleC (filament

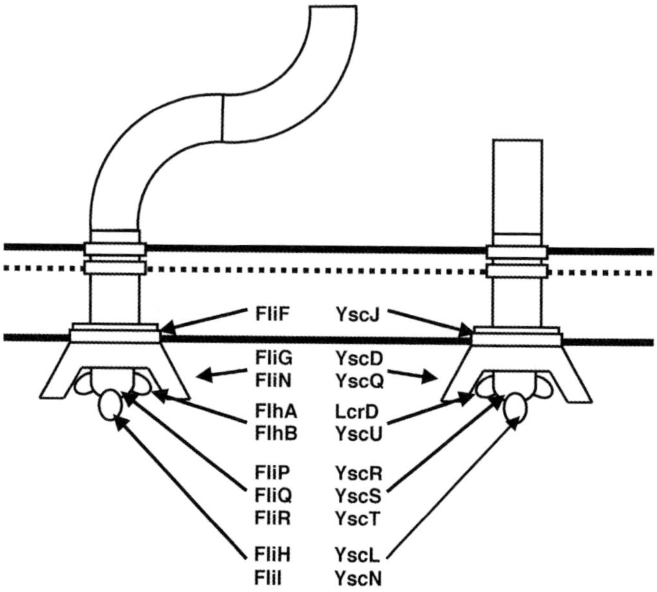

Figure 2. A comparison of components of the flagellar and Ysc TTS systems. All TTS systems span the bacterial cell envelope consisting of the inner membrane, the periplasm containing a peptidoglycan layer, and the outer membrane. The *Y. enterocolitica* flagellar TTS system (left) and a proposed structure for the YSC TTS system (right). The base of the TTS system is believed to be required for recognition and targeting of polypeptides for secretion. Protein subunits that are conserved among all TTS systems are located at the base as listed and their locations within the superstructure of the apparatus is indicated by the arrows.

proteins); and FliD (filament cap or scaffold protein) (Minamino and Macnab, 1999). Two other flagellar proteins transported by the flagellar TTS system are FlgM and FliK, which are secreted from the cell (Hughes et al., 1993; Macnab, 1996). FlgM and FliK are not structural components of the mature organelle, but function to coordinate flagellar gene expression and subunit assembly. Recently, it was realized that several other proteins, that are not structural subunits of the flagella, are also transported out of the cell by the flagellar TTS system in *Y. enterocolitica*. These results indicate that *Y. enterocolitica*, and potentially, other species of *Yersinia*, has evolved to use the flagellar TTS pathway to target proteins to sites outside of the cell where they may perform functions unrelated to motility per se.

GENETICS OF FLAGELLAR BIOGENESIS

The production of flagella by *Y. enterocolitica* and *Y. pseudotuberculosis* is controlled in response to a number of environmental cues such as temperature, salt concentration, and the availability of a utilizable carbohydrate (Rohde et al., 1994; Kapatral and Minnich, 1995; Kapatral et al., 1996; Young et al., 1998; Petersen and Young, 2002). There is also some indication that the production of flagella is modulated in response to cell density by a mechanism of quorum sensing involving the production and perception of extracellular signaling molecules of the homoserine lactone family (see chapter 5 and Atkinson et al., 1999). However, the mechanisms that control regulation of flagellar gene expression in response to these external stimuli have not been precisely determined. Presumably, most or all regulatory cues affecting flagellar gene expression are integrated through precise modulation of the *flhDC* master regulatory operon as discussed below.

While it remains to be formally established, the regulation of flagellar genes in *Y. enterocolitica* and *Y. pseudotuberculosis* is thought to be organized into a hierarchical transcriptional cascade. This regulatory cascade is based on the detailed knowledge that has already accumulated from studies of flagellar gene regulation in *S. enterica* and *Escherichia coli* (Macnab, 1996; Fernandez and Berenguer, 2000). Many flagellar genes of *Y. enterocolitica* have been cloned and characterized including *fleA, fleB, fleC, flhB, flhA, flhE, flhD, flhC, fliA* and *flgM* (Iriarte et al., 1995; Kapatral and Minnich, 1995; Kapatral et al., 1996; Fauconnier et al., 1997; Young et al., 1998). Each of these genes is homologous to their counter parts in *E. coli* and *S. enterica.* In addition, the recently completed sequence of the *Y. enterocolitica* (www.sanger.ac.uk/Projects/Y_enterocolitica) and *Y. pseudotuberculosis* (bbrp.llnl.gov/bbrp/html/microbe.html) genomes has revealed that these species maintain genes that are homologous to essentially all known flagellar genes of *S. enterica.* Furthermore, strains of *Y. enterocolitica* with mutations in many of these genetic loci have the same phenotypes as *E. coli* and *S. enterica* carrying similar types of mutations.

Expression of flagellar genes in *Y. enterocolitica* and *Y. pseudotuberculosis* is thought to be similar to that which has been documented for other members of the family *Enterobacteriaceae* (Hughes and Aldridge, 2002). Expression is coordinated with flagellar assembly and in response to environmental signals and consists of a hierarchical regulatory cascade of three major flagellar gene classes: I, II, and III. Class I genes, consisting of *flhD* and *flhC*, form a single operon which is expressed at the top of the hierarchy and is required for the expression of all other flagellar genes (Young et al., 1998). Mutations that inactivate these genes result in a transcriptional block in the flagellar cascade. Thus strains harboring a mutation in this locus do not produce a flagella (Figure 1, blocked before Stage 1a). The FlhD and FlhC proteins form a heteromultimeric transcriptional activator that is required for the expression of class II genes. Class II genes encode structural and accessory proteins required for assembly of the basal body and hook components of the flagellum as well as two genes (*flgM* and *fliA*) that encode the FlgM and σ^{28} regulatory proteins (Iriarte et al., 1995; Kapatral et al., 1996). Class III genes are transcribed from σ^{28}-dependent promoters and encode proteins required for maturation of the flagellum and chemosensory system. The flagellar type III secretion machinery exports some of these proteins, such as flagellin, to the outer surface of the cell. Mutations in *fliA* prevent the transcription of genes involved in assembly of extracellular portions of the flagellum (Figure 1, blocked after Stage 2). FlgM functions to limit σ^{28} activity until flagellar basal body assembly is complete and competent for the export of flagellin. A mutation in *flgM* results in unregulated expression of σ^{28}-dependent genes. These mutants proceed through all steps of flagellum biogenesis in an unregulated fashion and can lead to overproduction of flagellum subunits such as the filament proteins FleA, FleB and FleC (Young et al., 1999).

THE GENOMIC ORGANIZATION OF FLAGELLAR GENES

The genomic sequence has been determined for *Y. enterocolitica* (www.sanger.ac.uk/Projects/Y_enterocolitica/), *Y. pseudotuberculosis* (bbrp.llnl.gov/bbrp/html/microbe.html) and two different strains of *Y. pestis* (Parkhill et al., 2001; Deng et al., 2002). The availability of this genetic information has provided insight on how flagellar genes are organized and maintained in the chromosomes of pathogenic *Yersinia*. Bioinformatic analysis revealed that *Y. enterocolitica* contains one set of flagellar genes (Flagellar System 1) that are also present in *Y. pseudotuberculosis* and, except for the presence of inactivating mutations, in *Y. pestis* (Figure 3). These common genes display high homology to the flagellar genes of other members of the family of *Enterobacteriaceae*. From an evolutionary perspective, it appears that this flagellar system probably was carried by an ancestral bacterium of all three pathogenic species of *Yersinia*. In *Y. enterocolitica* the flagellar genes are grouped together in a single region of the chromosome, but it appears that chromosomal rearrangements have lead to dispersal of some flagellar genes in *Y. pseudotuberculosis* and *Y. pestis* (Figure 3). There are a few exceptions to the general clustering of flagellar genes in the chromosome of all the pathogenic species of *Yersinia*. Each species carries genes distributed throughout the chromosome that display strong homology to previously described chemotaxis genes and all carry a copy of *yplA* (a gene that will be described below). The presence of a frameshift mutation in *flhD* of both sequenced strains of *Y. pestis* provides a possible explanation for the lack of flagella for this species (Figure 3). Strain CO92 also carries two other mutations that likely affect flagellar genes and would

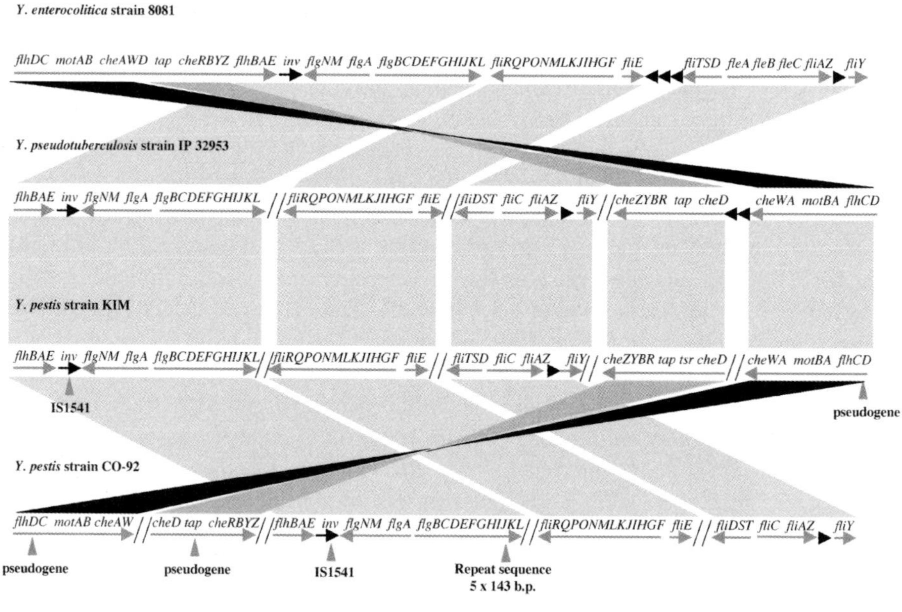

Figure 3. The genomic organization of the flagellar system 1 of pathogenic *Yersinia* species. All pathogenic *Yersinia* species have genes homologous to flagellar system 1. The diagram, not to scale, shows the orientation and organization of flagellar genes in the chromosome of each indicated *Yersinia* strain. Genes are listed relative to their position in the chromosome. Grey arrows indicate genes whose transcriptional orientation is the same, but are not necessarily genes belonging to the same operon. Black arrows indicate the location of genes and putative open reading frames that do not share homology to known flagellar genes. Double slashes indicate a location where groups of flagellar genes are separated by more than three putative open reading frames. Differences in the organization of flagellar genes for different strains of *Yersinia* are highlighted and likely represent major genetic rearrangements. The location of mutations within the flagellar gene clusters of *Y. pestis* strain KIM and CO-92 are indicated by the upturned arrowheads.

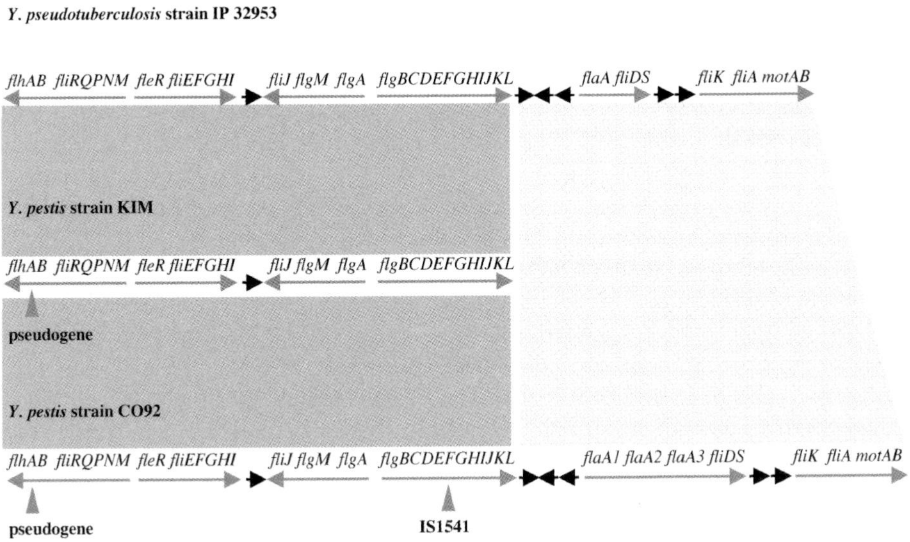

Figure 4. The genomic organization of the flagellar system 2 of pathogenic *Yersinia* species. Genes belonging to flagellar system 2 are found only in *Y. pseudotuberculosis* and *Y. pestis*. The diagram, not to scale, shows the orientation and organization of flagellar genes in the chromosome of each indicated *Yersinia* strain. Genes are listed relative to their position in the chromosome. Grey arrows indicate genes whose transcriptional orientation is the same, but are not necessarily genes belonging to the same operon. Black arrows indicate the location of genes and putative open reading frames that do not share homology to known flagellar genes. Double slashes indicate a location where groups of flagellar genes are separated by more than three putative open reading frames. Differences in the organization of flagellar genes for different strains of *Yersinia* are highlighted and likely represent major genetic rearrangements. The location of mutations within the flagellar gene clusters of *Y. pestis* strain KIM and CO-92 are indicated by the upturned arrowheads.

have consequences on the functions of flagella. One is an insertion mutation between *flgJ* and *flgL* and the other is a frameshift mutation in *tap* (Figure 3).

Interestingly, bioinformatic analysis has also revealed that *Y. pseudotuberculosis* has a second set of flagellar genes (Flagellar System 2) with homology to genes from *Pseudomonas* and *Vibrio* species (Figure 4). Many of these genes are represented in the *Y. pestis* genome as well (Parkhill et al., 2001; Deng et al., 2002). This second set of flagellar genes appears capable of encoding the components of flagella, but its functionality remains to be determined. Nonetheless, assuming that this second flagellar system is complete, the presence of two functional flagellar systems in *Y. pseudotuberculosis* has interesting implications for understanding the ecology of this species. In *Y. pestis* this system is not predicted to function since *flhA* is interrupted by a frameshift mutation in strains KIM10+ and CO92 (Figure 4). Furthermore, each of these strains of *Y. pestis* carries additional mutations which are predicted to have detrimental effects. An insertion element disrupts *flgF* of strain CO92 and an entire segment of flagellar genes appears to have been deleted from the chromosome of strain KIM10+ (Figure 4). The identification of two different flagellar gene clusters in *Y. pestis* has been noted previously, but it was not clear if this was a property exclusive to this species (Parkhill et al., 2001; Deng et al., 2002). With the availability of the genomic sequence of *Y. pseudotuberculosis*, it is now clear that acquisition of the second flagellar system was probably an ancestral event.

SECRETION OF NON-FLAGELLAR PROTEINS BY THE FLAGELLAR TYPE III SECRETION SYSTEM

The homology between components of the TTS apparatus of the flagellum and contact-dependent TTS systems combined with evidence that substrates of these apparatus are transported without modification suggest that these systems are functionally related and target proteins by a similar mechanism. This similarity has led to the hypothesis that these two types of apparatus are evolutionary related (Hueck, 1998). The flagellar TTS apparatus previously was thought to have a dedicated role in organelle biogenesis (Macnab, 1999). More recently, however, studies with *Y. enterocolitica* have demonstrated that the flagellar TTS system is also required for the transport of proteins to the extracellular environment that have no apparent role in motility (Young et al., 1999; Young and Young, 2002). Therefore the flagellum appears to have two functions; it provides a means for motility and functions as a bona fide TTS system. The secreted proteins or Fops (Flagellar Outer Proteins) define a new set of extracellular proteins. One Fop exported by the flagellar TTS system is the extracellular phospholipase YplA (Young et al., 1999). Identification of YplA also implicates the flagellum in having a role in virulence because YplA is known to be required for survival of *Y. enterocolitica* in experimentally infected mice (Schmiel et al., 1998). In addition, YplA has been shown to induce lecithin-dependent hemolysis of rabbit erythrocytes and HeLa cell cytotoxicity (Tsubokura et al., 1979).

The identification of Fops came about during a study of *flhDC*, the flagellar master regulatory operon (Young et al., 1998). An examination of proteins released by *Y. enterocolitica* into culture supernatants was conducted to confirm that a *flhDC* mutant was blocked for production of FleA, B and C flagellins. The results did show that flagellin production was affected, but also revealed that several other secreted proteins were not exported by the *flhDC* mutant (Young et al., 1998; Young et al., 1999). Complementation of the *flhDC* mutation with a plasmid encoded copy of *flhDC* restored the export of the flagellins and the other secreted proteins (Young et al., 1999). Similar results were seen when *flhDC* was placed under control of an exogenous P*tac* promoter such that expression became dependent on the presence of the inducer isopropyl β-D-thiogalactopyranoside. When the P*tac*-*flhDC* construct was expressed in the *flhDC* mutant, the amount of extracellular proteins detected in culture supernatants corresponded to levels of *flhDC* expression. These proteins were designated Fops (Flagellar outer proteins) because expression of the flagellar regulon was required for their appearance in the culture supernatant.

Further analysis revealed that other types of flagellar mutants were also affected in the export of Fops, implicating the flagellar TTS system as the apparatus responsible for their export. Fop export was affected by a mutation in *fliA*, which encodes σ^{28} and is necessary for flagellar class III gene transcription (Kapatral et al., 1996). The *fliA* mutant did not produce flagellins, as was previously shown (Kapatral et al., 1996), and it also did not secrete the Fops (Young et al., 1999). Complementation of the *fliA* mutation restored Fop and flagellin production. Examination of a strain with a mutation in *flgM*, which encodes a negative regulator of flagellar genes, revealed this regulatory defect caused overproduction of flagellin and increased secretion of the Fops (Young et al., 1999). Complementation of the *flgM* mutation reduced extracellular protein production to wild type levels. These results indicate expression of flagella and the secretion of Fops is coupled. Secretion of Fops was similarly affected by mutations in genes encoding components of the flagellar TTS system. Strains carrying mutations in *flhA* or *flhB* do not produce Fops (Young et al., 1999).

A survey of known secreted proteins also showed that the Fops were a novel set of extracellular proteins (Young et al., 1999). The Fops were demonstrated to be antigenically different from the FleA, FleB and FleC flagellins because they did not cross react with a flagellin specific monoclonal antibody. Further analysis indicated Fops are secreted by *Y. enterocolitica* and do not accumulate in culture supernatants as a result of general cell lysis, since culture supernatants were not contaminated by proteins that are normally cell associated. Examination of culture supernatants prepared from *Y. enterocolitica* strains revealed no contamination by the outer membrane protein invasin (Young et al., 1999). There was also no detectable contamination for strains engineered to express the periplasmic protein β-lactamase or the cytoplasmic protein chloramphenicol acetyl transferase. Another possibility was that the Fops were secreted by another protein secretion pathway that is coexpressed with the flagellar system. Such a situation occurs in *Salmonella enterica*, where it is known that flagellar mutations can limit the expression of the SPI-1 TTS system thereby affecting the profile of secreted proteins (Eichelberg and Galan, 2000; Lucas et al., 2000). However, this does not appear to be the case for *Y. enterocolitica* since Fop production was not affected in mutants defective for both the Ysc and Ysa TTS system (Young and Young, 2002). Taken together these results provided the first indication that secretion of the Fops requires the flagellar TTS secretion machinery.

SECRETION OF YplA REQUIRES THE FLAGELLAR TYPE III SECRETION SYSTEM

To more clearly determine if the flagellar TTS system is required for the export of Fops it was necessary to examine one Fop in more detail and YplA provided the opportunity (Young et al., 1999; Young and Young, 2002). However, the situation was complicated by the fact that *yplA* expression is coupled to regulation of the flagellar TTS system as a class III flagellar gene (Schmiel et al., 2000). To separate regulatory effects from secretion effects, *yplA* expression was placed under control of the P*tac* or P*cat* promoter (Young et al., 1999; Young and Young, 2002). This allowed transcription of *yplA* to occur independently of mechanisms that control flagellar gene expression. Furthermore, this provided the opportunity to directly examine how specific flagellar mutations affect YplA secretion. Two assays were used to evaluate YplA secretion by different flagellar mutants: (i) YplA activity was determined for different strains on phospholipase indicator media, and (ii) YplA export was determined by localization of the protein to whole cell or supernatant fractions of cultures by immunoblot analysis. The results from this analysis indicated that the flagellar TTS system was necessary for YplA export. Mutations that blocked production of the core components of the flagellar TTS system did not secrete YplA (Figure 1). This included strains carrying mutations in the flagellar master regulator operon, *flhDC,* or in genes encoding TTS apparatus components (*flgA*, *flgB*, *flgF*, *flhA*, *flhB* and *fliE*). Interestingly, genes that could be inactivated without affecting YplA secretion included *fliA*, *flgM*, *fleA* and *motA* (Figure 1). This indicated the extracellular portions of the flagellum and the torque generating motor complex are not necessarily required for YplA secretion. Collectively, these results make it unlikely that YplA is secreted by another protein secretion pathway, but definitive evidence for export of YplA by a type III mechanism awaited further analysis.

YplA IS SECRETED BY A TYPE III MECHANISM

The prima-facie evidence supporting the hypothesis that YplA is a protein that is exported by a type III mechanism came from studies to determine whether this protein could serve as a substrate for two other TTS systems in *Y. enterocolitica* (Young and Young, 2002). The Ysa and Ysc TTS systems are contact-dependent systems that have the capacity to deliver into targeted host cells effector proteins called Ysps and Yops, respectively. These two systems can also be induced to export proteins under specific laboratory conditions in the absence of host cells. When YplA was produced in wild type *Y. enterocolitica* under conditions that induce the Ysc TTS system, it was found that YplA was secreted along with the Yop proteins (Young and Young, 2002). Control experiments confirmed that mutations that affect Ysc TTS system function blocked YplA export. Likewise, YplA was also found to be secreted by the Ysa TTS system and its export was blocked by mutations that affect this protein secretion pathway (Young and Young, 2002). Further investigations have determined that recognition of YplA by the flagellar, Ysa and Ysc TTS systems involves a secretion signal that maps to the amino-terminal coding region of *yplA* (S. Warren and G. M. Young, unpublished observations). Taken together, these results not only revealed that YplA is exported by a type III mechanism, but also additionally indicate that each of the three different TTS systems present in high-pathogenic strains of *Y. enterocolitica* recognizes substrates in a similar manner. We can conclude from these studies that previous inferences suggesting there is evolutionary relatedness between flagellar and contact-dependent TTS systems are now supported by direct functional evidence.

ROLE OF FLAGELLA IN PATHOGENESIS

Flagella should be considered virulence factors or at the very least virulence effectors. Numerous studies have shown that eukaryotic hosts maintain a sensitive innate immune system capable of sensing bacterial flagellins though a ligand receptor called TLR5 (Smith and Ozinsky, 2002). This member of the Toll-like receptor family is required for the initiation of a robust host response to an invading pathogen. While it needs to be fully established that flagella are indeed produced by the enteropathogenic species of *Yersinia* in the host, there is no doubt that flagella are present on the surface of bacteria when they enter the host. Given that invasion of intestinal tissue by enteropathogenic *Yersinia* is detected within 90 minutes following oral administration of bacteria to BALB/c mice there is little time for flagella to be shed or lost due to bacterial outgrowth (Marra and Isberg, 1997). There is also some evidence to suggest flagella play an important role during the first stages of infection.

Recently, it was determined that expression of flagellar genes influences invasion of human epithelial cells by *Y. enterocolitica* (Young et al., 2000). Invasion primarily requires the bacterial outer membrane protein invasin which is encoded by the chromosomal *inv* locus (Isberg et al., 1987; Miller and Falkow, 1988). Invasin promotes host cell entry by binding host cell β_1 integrins (Isberg and Leong, 1990). The role of flagella in *Y. enterocolitica* invasion appears to be to ensure the bacterium migrates to and initiates host cell contact (Young et al., 2000). Non-motile strains of *Y. enterocolitica*, such as *flhDC* and *fliA* mutants are less invasive than motile strains by as much as 100-fold, but the reduction in invasion can be overcome by artificially bringing the bacteria into host cell contact by centrifugation. Thus flagella appear to be required to ensure the bacterium migrates to and initiates host cell contact. These results are consistent with the apparent need for motility by other species of enteropathogenic bacteria as well. For *S. enterica*, invasion of epithelial cells requires the function of the SPI-1 encoded TTS system and is enhanced by flagellar-mediated motility (Liu et al., 1988; Khoramian-Falsafi et al., 1990). In this case, however, bacterial invasion can not be fully restored to *flhDC* and *fliA* mutants by centrifugation because these regulatory genes are required for expression of the genes encoding the SPI-1 TTS system (Eichelberg and Galan, 2000; Lucas et al., 2000). These results indicate that *Y. enterocolitica* and *S. enterica* both appear to require expression of flagellar genes for epithelial cell invasion even though these bacteria promote cellular uptake by different mechanisms. In addition, these results indicate that there may be a strong natural selection for the maintenance of motility by enteropathogenic bacteria such as *Y. enterocolitica* and *Y. pseudotuberculosis*.

This hypothesis is supported by recent experimental evidence indicating that the flagellar TTS system might operate in the host environment, since the flagellar TTS system is required for export of the virulence factor YplA (Young et al., 1999). During infection, YplA contributes to the survival of *Y. enterocolitica* in host tissues and stimulation of the acute inflammation of gut-associated lymphoid tissue characteristic of yersiniosis (Schmiel et al., 1998). Using the mouse model of infection, a *Y. enterocolitica* YplA mutant exhibited reduced virulence and fewer pathological features compared to wild type *Y. enterocolitica* (Schmiel et al., 1998). The mutant did not survive well in host lymphatic tissues, Peyer's patches and mesenteric lymph nodes, where *Y. enterocolitica* normally manifests during the earliest stages of infection. The survival defect of the YplA mutant could be overcome by infecting mice with an exceptionally high dose of the bacterium. However, despite the restoration of *Y. enterocolitica* survival in the Peyer's patches and mesenteric lymph nodes, the pathology of the infection was dramatically altered (Schmiel et al., 1998). Normally *Y. enterocolitica* infection results in acute inflammation of the Peyer's patches and the mesenteric lymph nodes due to the infiltration of macrophages and polymophonuclear cells to the site of the infection (Carter, 1975; Autenrieth et al., 1996). As the disease progresses, the inflamed abscesses develop into necrotic lesions. In contrast, the Peyer's patches and mesenteric lymph nodes of mice infected with the YplA mutant often exhibited no apparent inflammation and necrosis (Schmiel et al., 1998). Other Fop proteins may also contribute to survival in the host or modulate the host inflammatory response. However, this hypothesis has not been fully evaluated. To keep this analysis in perspective, it should also be noted that a direct role for the flagellar TTS system in secreting YplA during infection of Peyer's patches or mesenteric lymph nodes has not been directly tested. The recent observation that YplA can, in addition to the flagellar TTS system, be secreted by the Ysa and Ysc TTS systems suggest the possibility that YplA is targeted by more than one TTS pathway (Young and Young, 2002).

Other evidence using the mouse model of infection also implicates flagella in having a role in pathogenesis. Preliminary experiments have shown that an insertion mutation in *flhB* of *Y. pseudotuberculosis* flagellar system 1 attenuated virulence of the bacterium when administered orally to BALB/c mice (H. Wolf-Watz, Umeå University, unpublished data). The experiment was completed by adding bacteria to the drinking water of mice at a concentration of 1 x 10^7, 1 x 10^8 and 1 x 10^9 cfu/ml. Mice were allowed to consume 3 – 5 ml of the contaminated water, which was then replaced with clean water and the animals were monitored for 12 days. The results of the experiment revealed that all of the mice exposed to the two highest doses of wild type bacteria died, but that all of the mice survived exposure to the *flhB* mutant. In contrast to these results, experiments with *Y. enterocolitica* failed to demonstrate a change in lethality of flagellar mutants toward

BALB/c mice (G. M. Young, unpublished data). However, the experiments with *Y. enterocolitica* differed from those performed with *Y. pseudotuberculosis* in terms of the method used to orally administer bacteria to animals. Mice infected with *Y. enterocolitica* were given bacteria by direct intragastric inoculation, a method that bypasses normal animal feeding behaviors and can reduce retention of bacteria in the stomach. At this time, it not possible to know whether there are differences in the role of flagella in pathogenesis caused by *Y. enterocolitica* and *Y. pseudotuberculosis*, but this is a tantalizing area of research that remains understudied.

CONCLUSIONS

Flagella are remarkable organelles used by *Yersinia* species for motility and for protein secretion. For the enteropathogenic species *Y. enterocolitica* and *Y. pseudotuberculosis* flagella may contribute to disease pathogenesis. It remains unlikely that *Y. pestis* produces full-length flagella for the function of motility, which may be an indication of how this pathogen has adapted to survival exclusively in association with its eukaryotic hosts. However, additional experimentation will be required to determine whether *Y. pestis* retains the ability to produce a remnant structure of the flagellum that can function in protein export. Continued efforts to understand the role of flagella in the biology of the *Yersinia* genus should provide insight as to how bacteria have evolved an array of strategies to survive encounters with eukaryotic hosts.

ACKNOWLEDGEMENTS

The author would like to thank Emilio Garcia, Julian Parkhill and Hans Wolf-Watz for providing access to unpublished data.

REFERENCES

Aizawa, S. 1996. Flagellar assembly in *Salmonella typhimurium*. Mol. Microbiol. 19: 1-5.

Aizawa, S.I., Dean, G.E., Jones, C.J., Macnab, R.M. and Yamaguchi, S. 1985. Purification and characterization of the flagellar hook-basal body complex of *Salmonella typhimurium*. J. Bacteriol. 161: 836-849.

Atkinson, S.A., Throup, J.P., Gordon, S.A.B. and Williams, P. 1999. A hierarchical quorum-sensing system in *Yersinia pseudotuberculosis* is involved in the regulation of motility and clumping. Mol. Microbiol. 33: 1267-1277.

Autenrieth, I.B., Kempf, V., Sprinz, T., Preger, S. and Schnell, A. 1996. Defense mechanisms in Peyer's patches and mesenteric lymph nodes against *Yersinia enterocolitica* involve integrins and cytokines. Infect. Immun. 64: 1357-1368.

Blocker, A., Gounon, P., Larquet, E., Niebuhr, K., Cabaux, V., Parsot, P. and Sansonetti, P. 1999. The tripartite type III secreton of *Shigella flexneri* inserts IpaB and IpaC into host membranes. J. Cell Biol. 147: 683-693.

Carter, P.B. 1975. Pathogenicity of *Yersinia enterocolitica* for mice. Infect. Immun. 11: 164-170.

Deng, W., Burland, V., Plunkett, G., 3rd, Boutin, A., Mayhew, G.F., Liss, P., Perna, N.T., Rose, D.J., Mau, B., Zhou, S., Schwartz, D.C., Fetherston, J.D., Lindler, L.E., Brubaker, R.R., Plano, G.V., Straley, S.C., McDonough, K.A., Nilles, M.L., Matson, J.S., Blattner, F.R. and Perry, R.D. 2002. Genome sequence of *Yersinia pestis* KIM. J Bacteriol. 184: 4601-4611.

Eichelberg, K. and Galan, J.E. 2000. The flagellar sigma factor FliA (sigma 28) regulates the expression of *Salmonella* genes associated with the centisome 63 type III secretion system. Infect. Immun. 68: 2735-2743.

Fan, F., Ohnishi, K., Francis, N.R. and Macnab, R.M. 1997. The FliP and FliR proteins of *Salmonella typhimurium*, putative components of the type III export apparatus, are located in the flagellar basal body. Mol Microbiol. 26: 1035-1046.

Fauconnier, A., Allaoui, A., Campos, A., Van Elsen, A., Cornelis, G. and Bollen, A. 1997. Flagellar *flhA*, *flhB*, and *flhE* genes, organized in an operon, cluster upstream of the *inv* locus in *Yersinia enterocolitica*. Microbiol. 143: 3461-3471.

Fernandez, L.A. and Berenguer, J. 2000. Secretion and assembly of regular surface structures in Gram-negative bacteria. FEMS Microbiol. Rev. 24: 21-44.

Foultier, B., Troisfontaines, P., Muller, S., Opperdoes, F.R. and Cornelis, G.R. 2002. Characterization of the ysa Pathogenicity Locus in the Chromosome of Yersinia enterocolitica and Phylogeny Analysis of Type III Secretion Systems. J Mol Evol. 55: 37-51.

Haller, J.C., Carlson, S., Pederson, K.J. and Pierson, D.E. 2000. A chromosomally encoded type III secretion pathway in Yersinia enterocolitica is important in virulence. Mol Microbiol. 36: 1436-1446.

Hueck, C.J. 1998. Type III secretion systems in bacterial pathogens of animals and plants. Microbiol. Mol. Biol. Rev. 62: 379-433.

Hughes, K.T. and Aldridge, P. 2002. Regulation of flagellar assembly. Curr. Opin. Microbiol. 5: 160-165.

Hughes, K.T., Gillen, K.L., Semon, M.J. and Karlinsey, J.E. 1993. Sensing structural intermediates in bacterial flagellar assembly by export of a negative regulator. Science. 262: 1277-1280.

Iriarte, M., Stainier, I., Mikulskis, A.V. and Cornelis, G.R. 1995. The *fliA* gene encoding σ28 in *Yersinia enterocolitica*. J Bacteriol. 177: 2299-2304.

Isberg, R.R. and Leong, J.M. 1990. Multiple β1 chain integrins are receptors for invasin, a protein that promotes bacterial penetration into mammalian cells. Cell. 60: 861-871.

Isberg, R.R., Voorhis, D.L. and Falkow, S. 1987. Identification of invasin: a protein that allows enteric bacteria to penetrate cultured mammalian cells. Cell. 50: 769-778.

Kapatral, V. and Minnich, S.A. 1995. Co-ordinate, temperature-sensitive regulation of the three *Yersinia enterocolitica* flagellin genes. Mol. Microbiol. 17: 49-56.

Kapatral, V., Olson, J.W., Pepe, J.C., Miller, V.L. and Minnich, S.A. 1996. Temperature-dependent regulation of *Yersinia enterocolitica* class III flagellar genes. Mol. Microbiol. 19: 1061-1071.

Khoramian-Falsafi, T., Harayama, S., Kutsukake, K. and Pechere, J.C. 1990. Effect of motility and chemotaxis on the invasion of *Salmonella typhimurium* into HeLa cells. Microb Pathog. 9: 47-53.

Kubori, T., Matsushima, Y., Nakamura, D., Uralil, J., Lara-Tejero, M., Sukhan, A., Galán, J.E. and Aizawa, S.-I. 1998. Supramolecular structure of the *Salmonella typhimurium* type III protein secretion system. Science. 280: 602-605.

Liu, S.L., Ezaki, T., Miura, H., Matsui, K. and Yabuuchi, E. 1988. Intact motility as a *Salmonella typhi* invasion-related factor. Infect. Immun. 56: 1967-1973.

Lucas, R.L., Lostroh, C.P., DiRusso, C.C., Spector, M.P., Wanner, B.L. and Lee, C.A. 2000. Multiple factors independently regulate *hilA* and invasion gene expression in *Salmonella enterica* Serovar *Typhimurium*. J. Bacteriol. 182: 1872-1882.

Macnab, R.M. (1996) Flagella and motility. In Neidhardt, F.C. (ed.) *Escherichia coli and Salmonella typhimurium; cellular and molecular biology*. ASM Press, Washington DC, Vol. 1, pp. 123-145.

Macnab, R.M. 1999. The bacterial flagellum: reversible rotary propellor and type III export apparatus. J. Bacteriol. 181: 7149-7153.

Marra, A. and Isberg, R.R. 1997. Invasin-dependent and invasin-independent pathways for translocation of *Yersinia pseudotuberculosis* across the Peyer's patch intestinal epithelium. Infect Immun. 65: 3412-3421.

Miller, V.L. and Falkow, S. 1988. Evidence for two genetic loci in *Yersinia enterocolitica* that can promote invasion of epithelial cells. Infect. Immun. 56: 1242-1248.

Minamino, T. and Macnab, R.M. 1999. Components of the *Salmonella* flagellar export apparatus and classification of export substrates. J. Bacteriol. 181: 1388-1394.

Parkhill, J., Wren, B.W., Thomson, N.R., Titball, R.W., Holden, M.T., Prentice, M.B., Sebaihia, M., James, K.D., Churcher, C., Mungall, K.L., Baker, S., Basham, D., Bentley, S.D., Brooks, K., Cerdeno-Tarraga, A.M., Chillingworth, T., Cronin, A., Davies, R.M., Davis, P., Dougan, G., Feltwell, T., Hamlin, N., Holroyd, S., Jagels, K., Karlyshev, A.V., Leather, S., Moule, S., Oyston, P.C., Quail, M., Rutherford, K., Simmonds, M., Skelton, J., Stevens, K., Whitehead, S. and Barrell, B.G. 2001. Genome sequence of *Yersinia pestis*, the causative agent of plague. Nature. 413: 523-527.

Petersen, S. and Young, G.M. 2002. An essential role for cAMP and its receptor protein in *Yersinia enterocolitica* virulence. Infect. Immun. 70: 3665-3672.

Rohde, J.R., Fox, J.M. and Minnich, S.A. 1994. Thermoregulation in *Yersinia enterocolitica* is coincident with changes in DNA supercoiling. Mol. Microbiol. 12: 187-199.

Schmiel, D.H., Wagar, E., Karamanou, L., Weeks, D. and Miller, V.L. 1998. Phospholipase A of *Yersinia enterocolitica* contributes to pathogenesis in a mouse model. Infect. Immun. 66: 3941-3951.

Schmiel, D.S., Young, G.M. and Miller, V.L. 2000. The *Yersinia enterocolitica* phospholipase gene *yplA* is part of the flagellar regulon. J. Bacteriol. 182: 2314-2320.

Smith, K.D. and Ozinsky, A. 2002. Toll-like receptor-5 and the innate immune response to bacterial flagellin. Curr. Top. Microbiol. Immunol. 270: 93-108.

Snellings, N.J., Popek, M. and Linder, L.E. 2001. Complete DNA sequence of *Yersinia enterocolitica* Serotype O:8 low-calcium response plasmid reveals a new virulence plasmid-associated replicon. Infect. Immun. 69: 4627-4638.

Tsubokura, M., Otsuki, K., Shimohira, I. and Yamamoto, H. 1979. Production of indirect hemolysin by *Yersinia enterocolitica* and its properties. Infect. Immun. 25: 939-942.

Venediktov, V.S., Timchenko, N.F., Antonenko, F.F. and Stepanenko, V.I. 1988. Chemotaxis of *Yersinia pseudotuberculosis* as a mechanism in its search for target tissues of the host organism. Zh. Mikrobiol. Epidemiol. Immunobiol. 5: 77-81.

Young, G.M., Badger, J.L. and Miller, V.L. 2000. Motility is required to initiate host cell invasion by *Yersinia enterocolitica*. Infect. Immun. 68: 4323-4326.

Young, G.M., Schmiel, D.H. and Miller, V.L. 1999. A new pathway for the secretion of virulence factors by bacteria: The flagellar export apparatus functions as a protein-secretion system. Proc. Natl. Acad. Sci. USA. 96: 6456-6461.

Young, G.M., Smith, M., Minnich, S.A. and Miller, V.L. 1998. The *Yersinia enterocolitica* motility master regulatory operon, *flhDC*, is required for flagellin production, swimming motility and swarming motility. J. Bacteriol. 181: 2823-2833.

Young, B.M. and Young, G.M. 2002. YplA is exported by the Ysc, Ysa and flagellar type III secretion systems of *Yersinia enterocolitica*. J. Bacteriol. 184: 1324-1334.

Chapter 13

Iron and Heme Uptake Systems

Robert D. Perry and Jacqueline D. Fetherston

ABSTRACT

The pathogenic yersiniae contain a number of different iron transport systems. The ability to use exogenous siderophores probably allows the enteropathogens to more effectively compete with other microbes in the intestinal lumen and in the environment outside the host. In *Y. pestis*, there appears to be a hierarchy of iron transport systems with some being more important than others during different stages of mammalian infection. Thus, the Ybt siderophore-dependent system is absolutely required during the initial stages of bubonic plague while the Yfe transporter is important during the later stages of disease. The other inorganic and heme transport systems do not seem to play a role in disease in mice by a subcutaneous route of infection. Perhaps one or more of these systems is relevant in pneumonic plague or is important for disease in other animals. Alternatively, some of the iron transport systems (the aerobactin/Fhu system, for example) may simply reflect the *Enterobacteriaceae* lineage of *Y. pestis*.

INTRODUCTION

Included among the defined virulence properties of the human pathogenic *Yersinia* species are iron acquisition systems. The connection between iron metabolism and disease outcome was established by Jackson and Burrows (Jackson and Burrows, 1956) who observed that non-pigmented (Pgm$^-$) mutants of *Yersinia pestis* were avirulent unless injected into mice with iron or hemin. These Pgm$^-$ isolates, which failed to bind large quantities of exogenous hemin and form greenish-brown colonies at 26°C, likely arose from deletion of a 102-kb chromosomal region now termed the *pgm* locus (Fetherston et al., 1992; Lucier and Brubaker, 1992). The *pgm* locus contains the hemin-storage (*hms*) locus that is required for the characteristic hemin-adsorption and is essential for the blockage of the flea proventricular valve (Hinnebusch et al., 1996; see also Chapter 4 of this book). However, the Hms system is not involved in iron or heme acquisition or storage for nutritional use by the bacterium (Lillard et al., 1999). The loss of mammalian virulence results from the deletion of the yersiniabactin (Ybt) siderophore-dependent iron transport system encoded by a high pathogenicity island (HPI; see Chapter 14 of this book) within the *pgm* locus. The Ybt system is critical for the pathogenesis of all three pathogenic *Yersinia* - *Y. pestis, Yersinia enterocolitica,* and *Yersinia pseudotuberculosis*. Extensive analysis of iron and heme transport systems has been performed in the first two organisms. *Y. pseudotuberculosis* and *Y. pestis* are closely related, thus systems encoded by *Y. pestis* are most likely also present in *Y. pseudotuberculosis* but have not been experimentally examined. The genomes of *Y. pestis* KIM10+ (biotype mediaevalis) and CO92 (biotype orientalis) have been sequenced and analysis has revealed 12 potential inorganic iron or heme transport systems.

INORGANIC IRON TRANSPORT SYSTEMS

SIDEROPHORE-DEPENDENT IRON TRANSPORT SYSTEMS

THE Ybt IRON TRANSPORT SYSTEM

Siderophores are low molecular mass compounds with a high binding affinity for ferric iron that are enzymatically synthesized and secreted from the bacterial cell. The genes of the Ybt siderophore-dependent systems of *Y. pestis* and *Y. enterocolitica* are nearly identical and have been extensively characterized (Gehring et al., 1998a; Pelludat et al., 1998; Buchrieser et al., 1999; Rakin et al., 1999). Although not as well studied, the *Y. pseudotuberculosis* Ybt system is almost certainly identical to the Ybt systems of the two other human pathogenic *Yersinia* species (Carniel et al., 1992; Chambers and Sokol, 1994; Buchrieser et al., 1998; Perry et al., 1999).

Production of the Ybt siderophore proceeds by a mixed nonribosomal peptide synthetase (NRPS)/polyketide synthase (PKS) mechanism that assembles salicylate, three cysteines, a malonyl linker group and three methyl groups into a four-ring structure composed of salicylate, one thiazolidine, and two thiazoline rings (Figure 1). The formation constant of Ybt with ferric iron is 4 x 10^{36}, higher than several other siderophores; ferrous iron is not bound by Ybt. The structure of Ybt is similar to that of pyochelin and anguibactin, produced by *Pseudomonas aeruginosa* and *Vibrio anguillarum,* respectively (Drechsel et al., 1995; Chambers et al., 1996; Perry et al., 1999; Crosa and Walsh, 2002). Assembly of Ybt begins with YbtD (a proposed phosphopantetheinyl transferase encoded

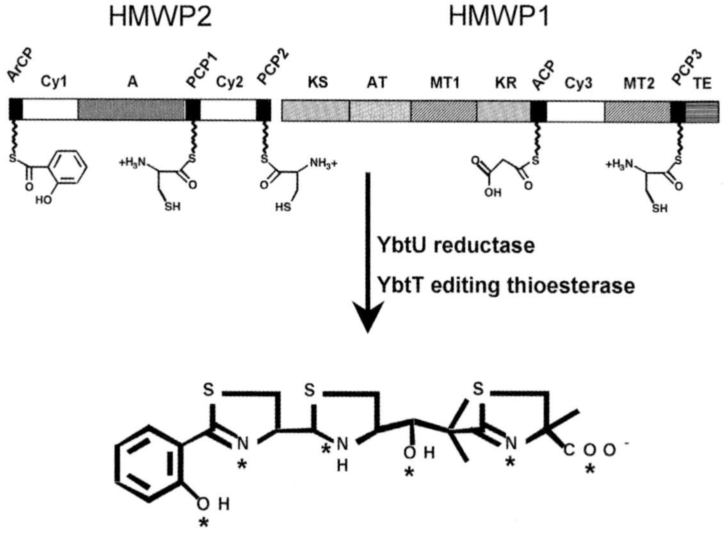

Figure 1. Biosynthesis of yersiniabactin. The HMWP1 and HMWP2 attachment sites for salicylate, malonate, and the three cysteine residues that are precursors for the thiazoline and two thiazolidine rings are indicated. Reactions prior to this stage involving YbtS (salicylate biosynthesis), YbtE (adenylates salicylate), and YbtD (P-pant transfer to HMWP1 and HMWP2) are not shown. Enzymatic domains: ArCP, aryl carrier protein; Cy, condensation/cyclization; A, adenylation; PCP, peptidyl carrier protein; KS, ketoacyl synthase; AT, acyltransferase; MT, methyltransferase; KR, β-ketoreductase; ACP, acyl carrier protein; TE, thioesterase. The predicted ferric ion coordination sites of the completed Ybt siderophore are indicated by asterisks.

outside the HPI) which activates HMWP1 and HMWP2 presumably by transferring the 4'-phosphopantethein (P-pant) moiety of coenzyme A to acyl, aryl, and peptidyl carrier domains (ACP, ArCP, PCP) on these enzymes (Bobrov et al., 2002; Figure 1). The P-pant moieties serve as tethers on which these groups are attached and then assembled into the siderophore (Quadri, 2000; Crosa and Walsh, 2002) . YbtS is essential for Ybt production and is likely involved in the synthesis of salicylate from chorismate. YbtE, a salicyl-AMP ligase, adenylates salicylate and transfers the activated compound to the ArCP domain of HMWP2. In addition, HMWP2 possesses adenylation (A), cyclization/condensation (Cy1 and Cy2), and two PCP domains required for the attachment, incorporation and cyclization of two cysteine molecules and the incorporation of salicylate. A previously proposed HMWP2 methyltransferase domain is likely nonfunctional (Gehring et al., 1998b; Suo et al., 1999; Keating et al., 2000a; Keating et al., 2000b; Suo et al., 2000; Miller and Walsh, 2001; Patel and Walsh, 2001). Following transfer of this partial structure to HWMP1, synthesis switches from an NRPS to a PK/fatty acid synthase mode (KS, AT, MT2, KR, and ACP domains in Figure 1), with the addition of a malonyl group linker. The end thiazoline ring is reduced by YbtU to form a thiazolidine ring. A methylation (MT1) and a reduction (KR) are accomplished before switching back to the final NRPS modules (Cy3, MT2, PCP3, and TE) which catalyze the condensation, cyclization, and methylation of the final thiazoline ring. A terminal thioesterase domain in HMWP1 likely releases the completed siderophore from the enzyme complex (Suo et al., 2000; Miller et al., 2001; Suo et al., 2001; Crosa and Walsh, 2002; Miller et al., 2002). The external thioesterase, YbtT, may remove aberrant or mischarged structures from the enzyme complex (Geoffroy et al., 2000; Bobrov et al., 2002). In vitro, HMWP1, HMWP2, YbtE, and YbtU plus reactants synthesize authentic Ybt (Miller et al., 2002). In vivo, YbtT, YbtS, and YbtD are required, in addition to these four Ybt enzymes, for siderophore production (Bearden et al., 1992; Fetherston et al, 1995; Gehring et al., 1998a; Geoffroy et al., 2000; Bobrov and Perry, 2002).

The mechanism by which Ybt is secreted from the bacterial cell is undetermined (Figure 2). Although YbtX has some similarities to EntS, which is required for the export of enterobactin (Furrer et al., 2002), a Δ*ybtX* mutant still secreted Ybt and had no observed in vitro defects. In addition, YbtP, YbtQ, and Psn, components of the uptake system, are not required for Ybt export (Fetherston et al., 1999). In vitro, Ybt removes iron from transferrin and lactoferrin; strains with a defective Ybt system are unable to use these compounds as sources of inorganic iron (unpublished results). Once the Ybt-Fe complex is formed, the outer membrane (OM) receptor Psn (termed FyuA in *Y. enterocolitica*) and the TonB system are required for translocation across the OM. This receptor and TonB are also required for transport of the bacteriocin, pesticin, across the OM. The *tonB* genes of *Y. enterocolitica* and now *Y. pestis* have been characterized and encode typical enteric TonB proteins. In *Y. pestis, exbB* and *exbD* likely form an operon that is not adjacent to *tonB* (Ferber et al., 1981; Koebnik et al., 1993a; Rakin et al., 1994; Fetherston et al., 1995; Rakin and Heesemann, 1995; Perry et al., 2003b).

Passage through the inner membrane (IM) involves an unusual ABC transporter (Figure 2) composed of YbtP and YbtQ which have both IM permease and ATP hydrolase domains. Such fused function ABC transporters are normally components of export systems. Both YbtP and YbtQ are essential for iron uptake via the Ybt system (Fetherston et al., 1999; Brem et al., 2001). There is evidence that additional components are required for Ybt utilization (Brem et al., 2001), possibly a periplasmic (or substrate) binding

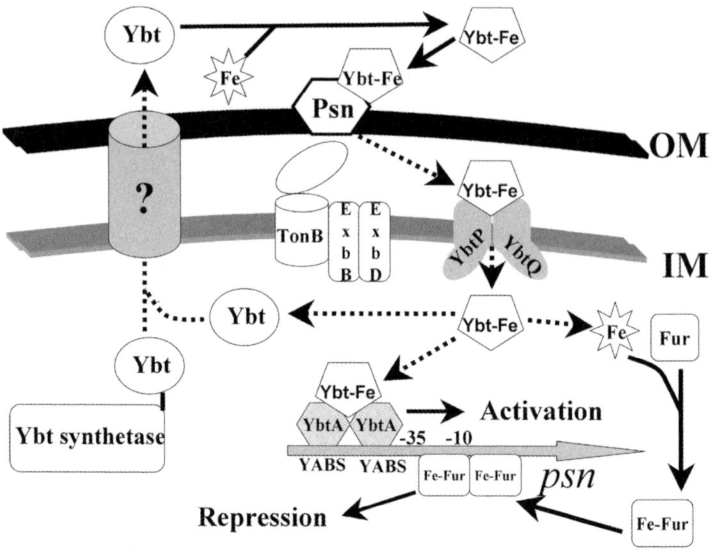

Figure 2. Model of the Ybt system. Dashed arrows indicate predicted mechanisms, steps, or transported substrates that have not been experimentally determined. It is unclear whether or not a SBP is required for Ybt uptake; none is shown in this model. For simplicity, only *Y. pestis* designations are shown. Figure is reproduced from Perry, 2004 with permission of ASM.

protein (SBP) and/or a mechanism for removing iron from the siderophore. If an SBP is part of the Ybt uptake system, its gene is located outside of the HPI that encodes most of the Ybt system. While Ybt is involved in the accumulation of iron, translocation of the siderophore into the bacterial cell has not been experimentally demonstrated. Finally, the mechanism for the removal of iron from the siderophore is unresolved; ferric iron reduction or degradation of Ybt is the most likely mechanism.

In vitro, mutations in the Ybt system reduce the ability of *Y. pestis* strains to accumulate iron and to grow at 37°C on Fe-chelated agar. In siderophore biosynthetic mutants of all three pathogenic species, addition of exogenous Ybt overcomes these defects (Fetherston et al., 1995; Bearden et al., 1997; Gehring et al., 1998a; Pelludat et al., 1998; Fetherston et al., 1999; Perry et al., 1999; Geoffroy et al., 2000; Bobrov et al., 2002). Ybt transport mutants still produce Ybt and display more severe iron uptake defects than do strains unable to synthesize Ybt (Fetherston et al., 1995; Fetherston et al., 1996; Perry et al., 1999; Brem et al., 2001; Perry et al., 2003b), probably because Ybt binds residual iron in the medium making it unavailable to other *Yersinia* iron transport systems. In mice, plague infection from peripheral sites (subcutaneous or intraperitoneal injection) requires a functional Ybt system. Thus the Ybt system is essential during the early stages of bubonic plague - spread via the lymphatics and growth in the lymph nodes. It is notable that the Ybt system is completely dispensable after this stage of the disease; Ybt$^-$ mutants are fully virulent if injected intravenously. This indicates that the Ybt system is not important for the later stages of plague (Une and Brubaker, 1984; Bearden et al., 1997; Bearden and Perry, 1999; Gong et al., 2001; Table 1). Ybt$^-$ strains of *Y. enterocolitica* and *Y. pseudotuberculosis* each show a loss of virulence by subcutaneous, intraperitoneal, or intravenous infection routes in mice (Une and Brubaker, 1984; Heesemann, 1987; Carniel et al., 1992; Rakin et al., 1994; Table 1)

The genetic organization of the Ybt locus is the same in *Y. pestis* KIM10+, CO92, and *Y. enterocolitica* 8081. These strains differ in the distance between the coding sequence for the Ybt receptor (*psn* in *Y. pestis* and *fyuA* in *Y. enterocolitica*) and YbtD. In CO92, *psn* and *ybtD* are separated by over 337 kb while the distance between these 2 genes is approximately 170.5 kb in KIM10+ and 17.7 kb in *Y. enterocolitica*. There are four *ybt* operons within the HPI – *psn, irp2-irp1-ybtU-ybtT-ybtE, ybtA,* and *ybtP-ybtQ-ybtX-ybtS* in *Y. pestis* (designated *fyuA, irp2-irp1-irp3-5,* and *irp6-9* in *Y. enterocolitica*) (Gehring et al., 1998a; Carniel, 2001; Pelludat et al., 2002; see Chapter 15). All three human pathogenic species of *Yersinia* use similar mechanisms to regulate *ybt* gene expression (Fetherston et al., 1996; Pelludat et al., 1998; Perry et al., 1999; Perry et al., 2001). In *Y. pestis*, each of these four operons is repressed by the Fur-Fe complex (Staggs et al., 1994; Fetherston et al., 1996; Fetherston et al., 1999; Perry et al., 2003a). In addition, YbtA and Ybt stimulate transcription from the *psn, irp2*, and *ybtP* promoters while repressing transcription from the *ybtA* promoter (Fetherston et al., 1996; Fetherston et al., 1999; Perry et al., 2003a). In contrast, transcription of *Y. pestis ybtD*, located outside the HPI and *pgm* locus, is not repressed by iron or activated by YbtA (Bobrov et al., 2002). These results suggest that *ybtD* may be a recently recruited gene of the Ybt system.

YbtA is an AraC-type transcriptional regulator that most likely binds to an inverted repeat sequence within each of the four *ybt* promoters. In the *psn* promoter region, altering one arm of this inverted repeat sequence greatly reduced the promoter activity under Fe-deficient conditions. Our working hypothesis is that YbtA binds the Ybt-Fe complex and activates transcription from the *psn*, *irp2*, and *ybtP* promoters (Fetherston et al., 1996; Perry et al., 2003a; Figure 2). If a siderophore-Fe complex were required for maximal expression this would prevent activation by apo-Ybt before it was secreted from the bacterium. In an *irp2::kan* mutant of *Y. pestis*, addition of exogenous Ybt stimulated β-galactosidase activity from a transcriptional *ybtP::lacZ* fusion 4-fold within 10 min with a maximal 7.5-fold activation by 30 min, compared to a culture with no Ybt added. A 500-fold lower concentration of exogenous Ybt stimulates transcription from the *ybtP* promoter 2-fold in 20 min than is required to promote the growth of the *irp2::kan* mutant. Thus Ybt is an extremely effective signal molecule for regulation of Ybt biosynthesis and transport (Perry et al., 2003a). We have also used the *ybtP::lacZ* reporter to determine the effects of 13 different *ybt* mutations on the expression of *ybt* genes. Our results indicate that these strains can be divided into 5 different regulatory classes (Table 2). The Class I strain, Δ*ybtX*, shows the same level of expression as the Ybt^+ strain KIM6+ and displays no in vitro defect in the Ybt system. Mutants lacking YbtA comprise the Class II group and show negligible transcriptional activity from the *ybtP::lacZ* reporter. Class III contains all Ybt transport mutants and exhibits β-galactosidase activity moderately higher than KIM6+ cells (Table 2). Transport mutants may be more iron-deprived from secreted Ybt binding the trace levels of iron in the medium and preventing iron acquisition from other *Y. pestis* transport systems. Class IV mutants produce no detectable Ybt by bioassay and exhibit ~18-fold lower transcriptional activity than Ybt^+ KIM6+ (Table 2). In contrast, Class V strains, which also include Ybt biosynthetic mutants, have ~25-fold higher β-galactosidase activity than Class IV mutants, ~1.4-fold higher than wild-type cells, and ~71% of the activity displayed by Class II mutants (Table 2) (Perry et al., 2003a).

The regulatory properties of Class II and Class IV mutants conform to our model (i.e., both YbtA and the siderophore are required to activate transcription of the *ybt* genes.). In the Class III transport mutants of *Y. pestis*, including Psn, minute quantities

Table 1. The effects of iron and heme transporters on the pathogenesis of *Yersinia*.

Strain[a]	Relevant Traits[a]	LD_{50} values by indicated infection route[b]	
		SC injection	IV injection
Y. pestis			
KIM5-2053.11 (pCD1::*yopJ*::MudI1734)+	Ybt^+ Yfe^+ Yfu^+ Hmu^+ Has^+ ($YopJ^-$ Psa^-)	120	NT
KIM5-2046.41 (pCD1::*yopJ*::Mu*d*I1734)	Ybt^- (Δ*irp2*) Yfe^+ Yfu^+ Hmu^+ Has^+ ($YopJ^-$ Psa^-)	> 1.3 x 10^7	NT
KIM5(pCD1Ap)+	Ybt^+ Yfe^+ Yfu^+ Hmu^+ Has^+ ($YopJ^+$ Psa^+)	< 8	NT
KIM5-2045.7 (pCD1::*yopJ*::Mu*d*I1734)	Ybt^- (Δ*psn*) Yfe^+ Yfu^+ Hmu^+ Has^+ ($YopJ^-$ Psa^-)	> 5.4 x 10^6	NT
KIM5-3173 (pCD1::*yopJ*::Mu*d*I1734)	Ybt^- (Δ*pgm*) Yfe^+ Yfu^+ Hmu^+ Has^+ ($YopJ^-$ Psa^-)	NT	< 12
KIM5-2031.12 (pCD1::*yopJ*::Mu*d*I1734)	Ybt^- (Δ*pgm*) Yfe^- (Δ*yfeAB*) Yfu^+ Hmu^+ Has^+ ($YopJ^-$ Psa^-)	NT	> 1.7 x 10^7
KIM5-2031.11 (pCD1::*yopJ*::Mu*d*I1734)+	Ybt^+ Yfe^- (Δ*yfeAB*) Yfu^+ Hmu^+ Has^+ ($YopJ^-$ Psa^-)	8.7 x 10^3	NT
KIM5-2031.12 (pCD1Ap)+	Ybt^+ Yfe^- (Δ*yfeAB*) Yfu^+ Hmu^+ Has^+ ($YopJ^+$ Psa^+)	74.3	NT
KIM5-2082.11 (pCD1Ap)+	Ybt^+ Yfe^- (Δ*yfeAB*) Yfu^- Hmu^+ Has^+ ($YopJ^+$ Psa^+)	< 82	NT
KIM5-2044.21 (pCD1::*yopJ*::Mu*d*I1734)+	Ybt^+ Yfe^+ Yfu^+ Hmu^- (Δ*hmuP'-V*) Has^+ ($YopJ^-$ Psa^-)	42	NT
KIM5-2081.1 (pCD1Ap)+	Ybt^+ Yfe^+ Yfu^+ Hmu^- (Δ*hmuP'-V*) Has^- (Δ*hasR-E*) ($YopJ^+$ Psa^+)	< 4.2	NT
Y. pseudotuberculosis			
IP 2790Sm	Ybt^+	3.6 x 10^6	6.9
IP 2790 H^-	Ybt^- (*irp2*::pJMA13)	> 2.1 x 10^9	1.9 x 10^3
PB1 Pst^s	Ybt^+	1.6 x 10^5	20
PB1 Pst^r	Ybt^- (*psn*)	>10^7	10^3
Y. enterocolitica			
WA-314	Ybt^+ Yfu^+	NT	5 x 10^2
WA *fyuA*(pYV08)	Ybt^- (*fyuA*) Yfu^+	NT	> 5 x 10^6
WA *yfuB*(pYV08)	Ybt^+ Yfu^- (*yfuB::kan*)	NT	7 x 10^2
WA Pst^s	Ybt^+	2.3 x 10^3	10^2
WA Pst^r	Ybt^- (*psn*)	1.1 x 10^5	2.3 x 10^3

[a] Some *Y. pestis* strains have mutations in *yopJ* and *psa* encoding YopJ and an adhesin, respectively, in addition to differences in iron transporters. Mutations in these two genes decreased virulence >15-fold compared to the wild-type strain. *Y. pseudotuberculosis* and *Y. enterocolitica* Pst^r strains are resistant to pesticin due to mutations in the Ybt OM receptor.

[b] SC - subcutaneous; IV - intravenous; NT - not tested; results for *Y. pestis* were compiled from Bearden et al. (1997), Bearden and Perry (1999), Fetherston et al. (1999), Thompson et al. (1999), Gong et al. (2001), and Rossi et al. (2001); results for *Y. pseudotuberculosis* are from Une and Brubaker (1984) and Carniel et al. (1992); results for *Y. enterocolitica* are from Une and Brubaker (1984), Rakin et al. (1994), and Saken et al. (2000).

of Ybt may nonspecifically pass through the OM and IM to activate the Ybt system. In unpublished studies, we found that exogenous Ybt siderophore activated transcription of the *ybtP::lacZ* reporter in a *psn irp2* mutant of *Y. pestis*. For Class V strains to fit our model, these mutants should produce either an aberrant molecule with signaling activity (i.e., *ybtS::kan* mutation) or small amounts of authentic Ybt that activate transcription (Δ*ybtT* and *irp1-2086* mutations). In an *ybtT* mutant, mischarged products may accumulate on the Ybt synthetase complex with just enough authentic Ybt produced to stimulate transcription. The *irp1-2086* mutation inactivates the terminal thioesterase domain of HMWP1 which likely releases the completed siderophore from the enzyme complex. Chemical cleavage of the ester bond may produce low levels of secreted Ybt that are sufficient to activate transcription but not to stimulate bacterial growth. Finally, an alternate phenolate ring structure may be inefficiently substituted for salicylate in a *ybtS* mutant; this could yield Ybt-like molecules that retain only signaling activity (Bobrov et al., 2002; Perry et al., 2003a).

Although exogenous Ybt activates transcription of *ybt* operons in both *Y. enterocolitica* and *Y. pseudotuberculosis*, similar to *Y. pestis* regulation (Fetherston et al., 1995; Fetherston et al., 1996; Bearden et al., 1997; Perry et al., 1999; Brem et al., 2001; Perry et al., 2003a), there are significant differences. *Y. enterocolitica* has an ERIC sequence inserted in the divergent promoter region between *ybtA* and *irp6-9* (*ybtPQXS* in *Y. pestis*) that might affect regulation of these genes; however, it is clear that both operons are expressed during iron starvation (Pelludat et al., 1998; Brem et al., 2001). Unlike *Y. pestis psn* mutants, a mutation in the Ybt OM receptor causes loss of siderophore production in *Y. enterocolitica* and *Y. pseudotuberculosis*. In *Y. enterocolitica*, a *fyuA* mutant does not express GFP from a *fyuA::gfp* reporter (Fetherston et al., 1996; Perry et al., 1999; Brem et al., 2001; Perry et al., 2003a). This is similar to the *Pseudomonas* pyochelin system where interaction of the OM receptor with the siderophore transmits a signal for the activation of pyochelin gene expression (Heinrichs and Poole, 1996). This difference among the Yersiniae may be explained by differing degrees of OM permeability in these bacterial species. Compared to the OM of *Y. pestis*, the OMs of *Y. enterocolitica* and *Y. pseudotuberculosis* are relatively

Table 2. β-galactosidase activity from a *ybtP::lacZ* reporter in *Y. pestis* derivatives[a] grown in Fe-deficient medium.

Relevant Characteristics	β-Galactosidase activity[b]	Relevant Characteristics	β-Galactosidase activity
Class I		Class IV (Ybt biosynthesis mutants)	
Ybt⁺	23,778 ± 2,658	*irp1::kan*	1,088 ±144
Δ*ybtX* (Ybt⁺)	24,693 ± 4,900	*irp2::kan*	1,173 ± 96
Class II		Δ*ybtE*	1,130 ± 294
Δ*pgm* (Ybt⁻)	363 ± 178	Δ*ybtU*	1,657 ± 57
ybtA::kan (Ybt⁻)	313 ± 14	Δ*ybtD*	1,468 ± 330
Class III (Ybt transport mutants)		Class V (Ybt biosynthesis mutants)	
Δ*psn*	43,205 ± 2,674	*ybtS::kan*	30,010 ± 2,291
Δ*ybtP*	44,794 ± 3,486	Δ*ybtT*	31,239 ± 2,012
tonB::kan	38,493 ± 3,064	*irp1-2086* (HMWP1 with defective thioesterase domain)	36,518 ± 1,049

[a] All strains carried the reporter plasmid pEUYbtP.
[b] Averages of Miller units from four or more samples ± standard deviation.

impermeable to small hydrophobic molecules (Bengoechea et al., 1998). Interestingly, mutations in the Ybt IM permeases in both *Y. pestis* (*ybtPQ*) and *Y. enterocolitica* (*irp6, 7*) upregulate expression of YbtA-dependent reporters. A *Y. enterocolitica irp8* (*ybtX* in *Y. pestis*) mutant was similar to wild-type cells while an *irp6,7,8* triple-mutant also displayed upregulated expression. This suggests that Irp8/YbtX does not serve as an alternate IM Ybt-Fe transporter that is sufficient for signal transduction but not for utilization of the siderophore (Fetherston et al., 1999; Brem et al., 2001; Perry et al., 2003a; Perry et al., 2003b). If our model of a cytoplasmic YbtA-Ybt-Fe complex activating transcription of *ybt* genes is correct, sufficient amounts of Ybt-Fe must be transported through the IM of both *Y. pestis* and *Y. enterocolitica* permease mutants.

Jacobi et al. (2001) found that *Y. enterocolitica* cells isolated from the spleens of infected mice expressed moderate levels of an *yfuA* reporter gene while cells from the peritoneal cavity exhibited strong expression. The lowest level of expression occurred in *Y. enterocolitica* cells isolated from the liver and intestinal lumen. The *Y. enterocolitica fyuA* reporter was expressed at 3-fold higher levels in cells isolated from the peritoneal cavity than in those cultured under Fe-deprived conditions in vitro. These results clearly demonstrate in vivo expression of the Ybt system (Jacobi et al., 2001) and suggest either different iron availability in various organ systems or unidentified in vivo regulatory signals that affect expression.

FERRIC HYDROXAMATE IRON TRANSPORT SYSTEMS

The Yersiniae species vary in their ability to produce and utilize various hydroxamate siderophores. Both *Y. pestis* (unpublished results) and *Y. enterocolitica* (Perry and Brubaker, 1979; Rutz et al., 1991; Bäumler et al., 1993; Chambers and Sokol, 1994) can use ferrichrome as a source of iron. The *Y. enterocolitica* ferrichrome receptor (FcuA) has been cloned and sequenced (Koebnik et al., 1993b). Curiously, FcuA more closely resembles FatA, the OM receptor for anguibactin, a siderophore produced by *V. anguillarum*, than FhuA, the ferrichrome receptor from *E. coli* (Coulton et al., 1983; Fecker and Braun, 1983; Walter et al., 1983; Actis et al., 1988; Koebnik et al., 1993b). The *Y. enterocolitica* FcuA receptor is TonB-dependent and recognizes some but not all of the same substrates as FhuA (Koebnik et al., 1993a; Koebnik et al., 1993b). A homologue to *fcuA* is present in the *Y. pestis* genomes (Parkhill et al., 2001; Deng et al., 2002). It is not known if *Y. pseudotuberculosis* or the environmental strains of *Yersinia* can use ferrichrome.

Y. enterocolitica and the environmental strains, *Yersinia kristensenii*, *Yersinia frederiksenii*, and *Yersinia intermedia,* can all use ferrioxamine as an iron source (Perry and Brubaker, 1979; Bäumler et al., 1993; Chambers and Sokol, 1994). However, ferrioxamine inhibits the growth of *Y. pestis* cells (Perry and Brubaker, 1979; unpublished results). There are conflicting reports on the ability of *Y. pseudotuberculosis* to use ferrioxamine which may be due to differences in the strains tested and/or the methods used (Perry and Brubaker, 1979; Chambers and Sokol, 1994). The *Y. enterocolitica* OM receptor (FoxA) for ferrioxamine has been cloned and characterized (Bäumler and K. Hantke, 1992). Uptake of ferrioxamine requires TonB (Koebnik et al., 1993a) and FoxA likely forms a β-barrel similar to the structures formed by FepA, FecA and FhuA (Bäumler and K. Hantke, 1992; Ferguson et al., 1998; Locher et al., 1998; Ferguson et al., 2002). A second *Y. enterocolitica* gene (*pcpY*) was identified that enabled an *E. coli* FhuE mutant to use ferrioxamine. However, additional experiments indicated that overexpression of the

PCP$_{Ye}$ lipoprotein probably resulted in OM changes that allowed ferrioxamine to leak into the cell (Bäumler and Hantke, 1992).

Early studies suggested that *Y. enterocolitica* and *Y. pseudotuberculosis* but not *Y. pestis* could use aerobactin as an iron source (Perry and Brubaker, 1979). However, later experiments indicated that aerobactin could not supply iron to *Y. enterocolitica* (Bäumler et al., 1993) while *Y. pestis* can obtain iron from aerobactin (unpublished results). Both *Y. pestis* KIM10+ and CO92 as well as *Y. pseudotuberculosis* but not *Y. enterocolitica* 8081 contain genes homologous to the *E. coli* aerobactin receptor, IutA. In *Y. pestis* the use of aerobactin is TonB-dependent (unpublished results). *Y. pestis* also contains the aerobactin biosynthetic genes, *iucA-D*, organized in a similar manner as the *E. coli iuc* operon with one difference, a frameshift mutation splits *iucA* into two separate Orfs and is likely responsible for the inability of *Y. pestis* to produce aerobactin (unpublished results). This mutation may also affect the expression of the downstream genes as the cloned *Y. pestis* locus was unable to complement an *E. coli* strain with an insertion in *iucB* (unpublished results). In *Y. pseudotuberculosis* IP32953, the aerobactin biosynthetic operon appears to be intact (http://bbrp.llnl.gov/bbrp/html/microbe.html) but whether or not a functional siderophore is produced is unknown. However, aerobactin is synthesized by some but not all environmental strains of *Yersinia* (Stuart et al., 1986; Reissbrodt and Rabsch, 1988; Romalde et al., 1991). The aerobactin biosynthetic genes are absent from the *Y. enterocolitica* 8081 genome.

The ABC transport system FhuCDB (Figure 3) is used for the uptake of a number of hydroxamate siderophores including aerobactin, ferrichrome, and ferrioxamine B (Braun and Hantke, 1991; Köster, 1991; Braun et al., 1998). Both *Y. enterocolitica* and *Y. pestis* contain *fhuCDB* but, unlike *E. coli*, these genes are not located adjacent to a gene encoding an OM receptor (Fecker and Braun, 1983; Bäumler et al., 1993; Koebnik et al., 1993b). The predicted amino acid sequence of *Y. enterocolitica* 8081 FhuB and FhuC are 86% and 95% identical, respectively, to their *Y. pestis* counterparts. However, *Y. pestis* KIM10+ and CO92 FhuD contains 53 additional amino acids in the amino-terminal portion of the predicted protein that are not present in *Y. enterocolitica* FhuD. In *Y. pestis*, mutants in *fhuCDB* are unable to use either ferrichrome or aerobactin (unpublished results).

THE ENTEROBACTIN-DEPENDENT IRON TRANSPORT SYSTEM

Y. enterocolitica biotypes IA, IB, and II, but not IV, contain genes orthologous to *E. coli fepBDGC* and *fes*, which encode ABC transport components and enterobactin esterase (Figure 3) and can use exogenous enterobactin (Rutz et al., 1991; Schubert et al., 1999). *E. coli fepD, fepG,* and *fes* mutants are complemented by the corresponding *Y. enterocolitica* genes and *Y. enterocolitica fepD* or *fes* mutants no longer use exogenous enterobactin. Although the *Y. enterocolitica* locus has a gene order similar to that in *E. coli*, it is missing *entS* (formerly designated P43) and the enterobactin-biosynthetic genes. Southern blot hybridization, using *Y. enterocolitica fepDGC* and *fes* sequences as probes, did not detect similar genes in *Y. pestis* and *Y. pseudotuberculosis* (Schubert et al., 1999). A growth stimulation assay suggested that *Y. pseudotuberculosis* but not *Y. pestis* can utilize exogenous enterobactin (Perry and Brubaker, 1979). Further experiments will be required to resolve the apparent discrepancy between DNA hybridization and enterobactin use by *Y. pseudotuberculosis*. Although *Y. pestis* CO92 and KIM10+ do not encode *fepD* or *fes* homologues, both *Y. pestis* genomes contain a *fepB* homolog that is isolated from other genes encoding potential iron transport components (Parkhill et al., 2001; Deng

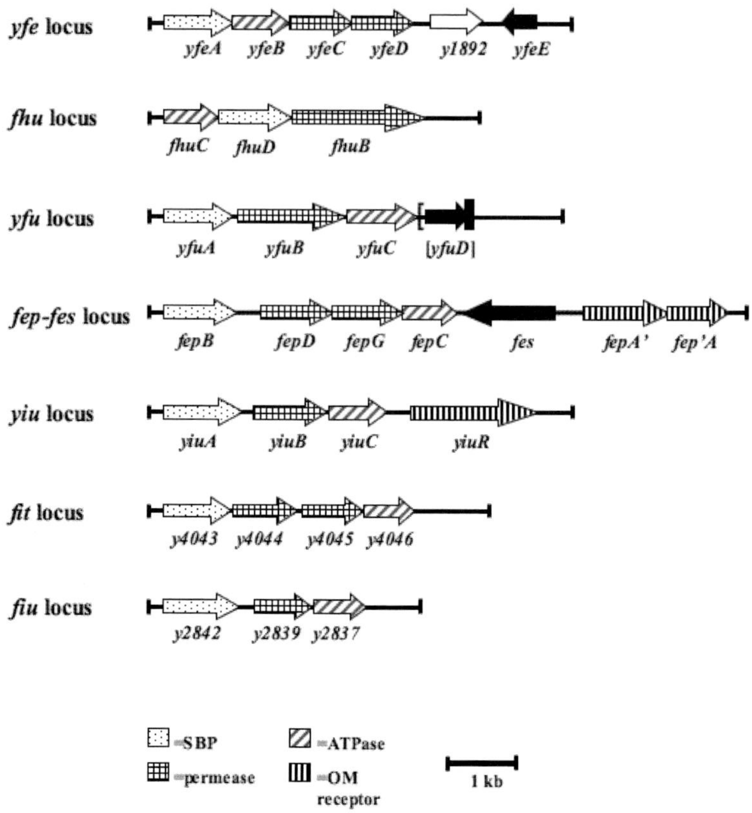

Figure 3. Genetic organization of proven and putative *Yersinia* iron ABC transporters. Arrows indicate the Orfs and direction of transcription. Orfs are drawn to scale. In the *yfe* locus, *y1892* (open arrow) is not required for transport function while *yfeE* (filled arrow) is required by the Yfe transporter in *E. coli* but not in *Y. pestis*. *yfuD* (filled arrow in parentheses) is present in *Y. enterocolitica* but not in *Y. pestis*. The *fep-fes* locus is present only in *Y. enterocolitica*.

et al., 2002). To test the hypothesis that this *Y. pestis* "FepB" is an SBP for the Ybt transporter, we constructed a *Y. pestis* KIM *fepB* mutant. This mutant was not defective in iron acquisition via the Ybt system (unpublished results). A gene homologous to the OM receptor for enterobactin, *fepA*, is located downstream of *fes* in *Y. enterocolitica*. However, the *Y. enterocolitica* WA-C gene has a frameshift mutation that yields a truncated 40 kDa product (Figure 3). CccA has been suggested as a receptor for catecholate and dihydroxybenzoylserine (DHBS); CccA or other iron-repressible OM proteins may substitute for a truncated FepA OM receptor which is likely nonfunctional (Bäumler et al., 1993; Schubert et al., 1999). In contrast, Rutz et al (1991) detected an iron-regulated OM protein similar in size to *E. coli* FepA that reacted with monoclonal antibodies against *E. coli* FepA. This suggests that *fepA* may be intact in some strains of *Y. enterocolitica*.

ABC IRON TRANSPORTERS

In addition to the systems described above, *Y. pestis* has other ABC transporters with significant similarities to iron or siderophore transport systems. None of the three systems in this section are associated with genes encoding biosynthetic enzymes for siderophores. Moreover, they all function in *E. coli* strains unable to synthesize enterobactin or aerobactin. Consequently, it is likely that these are siderophore-independent iron transporters.

THE Yfe IRON AND MANGANESE ABC TRANSPORTER

The *yfe* locus of *Y. pestis* (Figure 3) was initially identified by its ability to restore growth, under Fe-deficient conditions, to an enterobactin-negative (*ent*$^-$) strain of *E. coli* (Bearden et al., 1998). The locus consists of 2 separate operons, *yfeA*-D and *yfeE*, which are divergently transcribed and separated by an Orf, originally called ORF33 but designated Y1892 in the *Y. pestis* KIM10+ genome nomenclature (Bearden et al., 1998; Deng et al., 2002). All five *yfe* genes are required for a functional system in *ent*$^-$ *E. coli*; mutations in *y1892* have no effect (Bearden et al., 1998). In *Y. pestis*, growth and iron transport defects due to mutations in the Yfe system are only observed in strains missing the Ybt system. While mutations in the *yfeA-D* operon profoundly affect the growth of *Y. pestis* on iron-chelated gradient plates, a defect in *yfeE* is less severe. A Δ*yfeE* strain takes 24 h longer than the Yfe$^+$ strain to achieve full growth on gradient plates containing conalbumin (Bearden and Perry, 1999).

The Yfe system resembles a typical ABC transporter with YfeA being the putative SBP, YfeC and YfeD forming the inner membrane permease (IMP), and YfeB, a presumptive ATP hydrolase (ATPase) (Figure 4). YfeE is a putative IM protein with no homology to proteins of known function. Early publications speculated that YfeE might modulate expression of the YfeABCD system (Bearden and Perry, 1999). However, subsequent experiments have shown that a mutation in *yfeE* has no effect on expression from the *yfeA* promoter (Perry et al., 2003a). Consequently, YfeE does not regulate expression of the Yfe system and its function remains unknown. Curiously, no OM receptor or porin has been identified for the Yfe system in *Y. pestis* or any of the other bacteria that possess related systems (Bearden et al., 1998; Zhou et al., 1999; Runyen-Janecky et al., 2003).

Transport experiments revealed that both iron and manganese (Mn) are taken up by the Yfe system (Figure 4). Both Mn and zinc (Zn) competed with iron for uptake through the Yfe system; however only iron competed for Mn uptake (Bearden and Perry, 1999). In vitro, iron uptake via the Yfe system appears to be TonB independent (Perry et al., 2003b). Finally, the defects exhibited by Yfe mutants on iron-chelated gradient plates are due to iron and not Mn insufficiency since supplementation with iron but not Mn or Zn restored growth (Bearden and Perry, 1999).

Studies with transcriptional and translational reporter gene fusions have demonstrated that the *yfeA-D* operon is regulated by Fur and repressed by iron and Mn but unaffected by the addition of Zn (Figure 4). Excess iron repressed expression of a *yfeA* reporter approximately 5-fold while Mn resulted in an ~2-fold repression (Bearden et al., 1998). The SitABCD system of *Shigella flexneri,* which is highly similar to YfeABCD, is also regulated by iron and Mn (Runyen-Janecky et al., 2003). Although repression of the *yfeA* promoter by Mn required Fur, other Fur- and iron-repressible *Y. pestis* and *E. coli* promoters were unaffected by Mn (Bearden et al., 1998; Perry et al., 2003a). Recent studies with a *yfeE::lacZ* fusion have shown that this promoter is not regulated by the Fe-, Mn-, or Zn-status of *Y. pestis* cells (Perry et al., 2003a).

Strains of *Y. pestis* with defects in the yersiniabactin iron transport system are fully virulent by an intravenous route of infection in mice and avirulent when injected subcutaneously. By a subcutaneous route of infection, an Ybt$^+$ Yfe$^-$ strain exhibited an

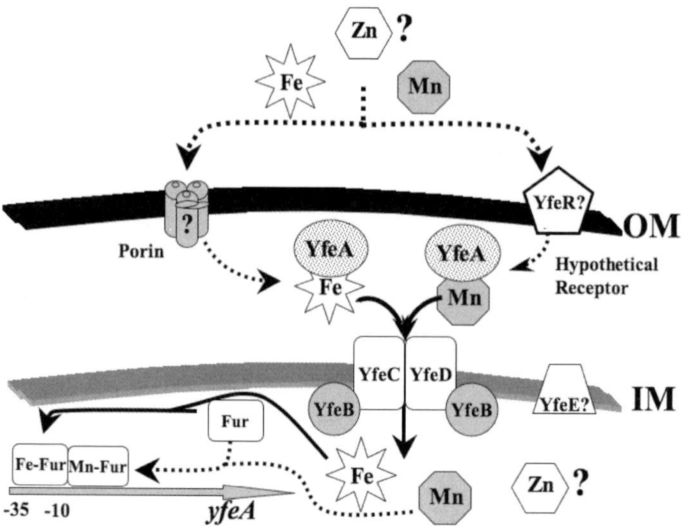

Figure 4. Model of the *Y. pestis* Yfe Fe and Mn transport system and regulation. The "YfeR receptor" and "porin" are alternative, completely speculative, channels through the OM. An OM component for this system has not been discovered. Dashed arrows identify steps that have not been experimentally proven. It is unclear whether Zn is accumulated by the Yfe system; the question mark underscores this uncertainty. Figure is reproduced from Perry, 2004 with permission of ASM.

~70-fold loss of virulence compared to its fully virulent parent. An Ybt$^-$ Yfe$^-$ mutant is completely avirulent by an intravenous route of infection (Bearden and Perry, 1999; Table 1). The fact that a Ybt$^+$ Yfe$^-$ strain can still cause disease by a subcutaneous route of infection suggests that the Ybt system can, to some extent, compensate for the absence of the Yfe system. However, the Yfe transport system may be more effective than the Ybt system in the later stages of bubonic plague. Thus one explanation for multiple transport systems in pathogens is that some systems may function more efficiently than others in specific organ systems or environments.

Based on sequence similarities, the YfeA-D ABC transporter is a member of the Mn/Zn/Iron chelate uptake transporter family (TC 3.A.1.15). Other members include MntCAB of *Synechocystis* (Bartsevich and Pakrasi, 1995), AdcCBA from *Streptococcus pneumoniae*, (Dintilhac and Claverys, 1997) and SitABCD of *Salmonella typhimurium* (Zhou et al., 1999). Sequences homologous to *yfeA* have also been detected in *Actinobacillus acetinomycetemcomitans* and *Haemophilus influenzae* (Graber et al., 1998; Smoot et al., 1999). PCR analysis and BLAST searches revealed the presence of sequences homologous to the *Y. pestis yfe* locus in the genomes of *Y. enterocolitica* and *Y. pseudotuberculosis* (Bearden et al., 1998). Both the *yfeA-D* and *yfeE* operons are present in the *Y. enterocolitica* 8081 genome. While all of the *yfe* genes appear to be highly conserved, it has not been experimentally determined if this system functions in other *Yersinia* species.

THE Yfu IRON TRANSPORTER

The *Yersinia* Yfu system is a member of the FeT cluster (TC 3.A.1.10) of ABC transporters that includes *Serratia marcescens* Sfu (Saier, 1999). Yfu has been characterized in *Y.*

enterocolitica (Saken et al., 2000) and subsequently in *Y. pestis*. *Y. pestis yfuABC* enhanced the growth of an *E. coli* enterobactin-biosynthetic mutant under Fe-deficient conditions, partially compensating for the loss of siderophore production. DNA hybridization analysis suggests that *Y. pseudotuberculosis* encodes a similar system (Gong et al., 2001). The proteins encoded by the *Y. enterocolitica yfu* locus are YfuA, YfuB, and YfuC, an SBP, IM permease, and ATP hydrolase, respectively (Figure 3), with >75% identity to the *Y. pestis* YfuABC proteins (Saken et al., 2000; Gong et al., 2001). YfuD, present in *Y. enterocolitica* but not in *Y. pestis*, has significant similarity to a hypothetical protein (YahN) in *E. coli* and relatively low similarities to several transporter proteins and membrane channels. It also possesses a LysE-type translocator domain. One member of the LysE-family effluxes lysine. YfuD enhances the function of the *Y. enterocolitica* YfuABC system in *E. coli* but is not essential. The *Y. enterocolitica* Ybt system masks Fe-deficient growth defects due to mutation of the Yfu system (Saken et al., 2000). Recent studies using Fe-chelator gradient plates have demonstrated growth defects in *Y. pestis yfu* mutants but only in a Ybt^- Yfe^- background (unpublished results).

Iron ABC transporters related to Yfu are often associated with OM receptors that bind transferrin or lactoferrin. Among the different bacteria, the transferrin and lactoferrin receptors have highly similar amino acid sequences. Our searches of the *Y. pestis* KIM10+ and CO92 genomes did not identify any putative OM receptors for these host iron-binding proteins. Whatever the in vitro substrate for the *Y. enterocolitica* and *Y. pestis* Yfu systems, translocation through the OM is TonB independent (Saken et al., 2000; Perry et al., 2003b).

In *Y. enterocolitica* and *Y. pestis,* Ybt^+ Yfu^- mutants are fully virulent; any defect may be masked by the Ybt system. In addition, a *Y. pestis* Ybt^+ Yfe^- Yfu^- mutant is as pathogenic as its Ybt^+ Yfe^- Yfu^+ parent (Table 1). Consequently, an in vivo role for the Yfu system has yet to be detected (Saken et al., 2000; Gong et al., 2001) .

In *Y. enterocolitica*, *yfuA-D* likely form an operon with a putative Fur-binding site (FBS) in the promoter region upstream of *yfuA* (Saken et al., 2000). Similarly, in *Y. pestis, yfuA-C* appear to constitute an operon with demonstrated regulation by iron and Fur (Gong et al., 2001; Figure 3).

THE Yiu IRON TRANSPORTER

The *yiu* locus (Figure 3), encoding four Orfs, is present in both the *Y. pestis* KIM10+ and CO92 genomes. The *Y. enterocolitica* 8081 genome encodes three of the Orfs but is missing *yiuA*. YiuA, YiuB, and YiuC have motifs that classify them, respectively, as an SBP, IM permease, and ATP hydrolase of an ABC transporter. YiuR is a proposed TonB-dependent OM receptor with sequence similarity to IrgA of *Vibrio cholerae* (Goldberg et al., 1992). YiuA-C show similarities to *Corynebacterium diphtheriae* Irp6A-C proteins essential for corynebactin-dependent iron uptake (Qian et al., 2002). YiuC also shows similarity to FepC, the ATP hydrolase of the Fep enterobactin ABC transporter (unpublished observations). In an *E. coli* enterobactin-deficient mutant, the *yiu* locus enhances Fe-deficient growth without restoring enterobactin production; this growth stimulation is not dependent upon YiuR. Using Fe-chelator gradient plates, we have demonstrated that a Δ*yiuABCR* mutation affects the growth of *Y. pestis*, but only in strains defective in the Ybt, Yfe and Yfu systems. Any one of these three iron transport systems will mask growth defects due to mutations in the Yiu system. A *Y. pestis* Δ*yiuR* mutant is also defective in growth under Fe-deficient conditions but again only in a Ybt^-,

Yfe⁻ and Yfu⁻ background. A *yiuR* mutant does grow better than a strain missing the entire *yiu* locus suggesting that the Yiu ABC transporter can function without the YiuR receptor (unpublished results).

The *yiuR* gene is located ~250 bp downstream of the predicted stop codon for *yiuC* and is probably transcribed separately from *yiuABC*. The predicted promoter regions of both *yiuABC* and *yiuR* possess putative FBSs. We have experimentally demonstrated Fur and iron regulation of the *yiuABC* promoter (unpublished results).

THE Feo FERROUS IRON TRANSPORTER

A ferrous iron uptake system (Feo) was initially identified and characterized in *E. coli* (Hantke, 1987; Kammler et al., 1993; Stojiljkovic et al., 1993) and subsequently found in a number of other bacteria (Tsolis et al., 1996; Velayudhan et al., 2000; Katoh et al., 2001; Leipe et al., 2002; Robey and Cianciotto, 2002; Runyen-Janecky and Payne, 2002). The *feo* operon encodes anywhere from 1 to 3 Orfs, depending upon the bacterial species. FeoB is the largest Orf and is always present. Initially, FeoB was thought to function as a unique type of transport ATPase (Kammler et al., 1993) but more recent studies indicate that this protein may be a primordial precursor to eukaryotic G proteins (Marlovits et al., 2002). Thus, FeoB actually binds and hydrolyzes GTP and not ATP (Marlovits et al., 2002). However, studies with inhibitors clearly indicated a role for ATPase activity in ferrous iron transport (Velayudhan et al., 2000). This raises the question of whether FeoB serves a direct or indirect function in Fe(II) uptake (Marlovits et al., 2002). That it does play some role is clear from numerous studies showing that *feoB* mutations affect iron transport and/or growth under iron-deficient conditions (Kammler et al., 1993; Velayudhan et al., 2000; Katoh et al., 2001; Marlovits et al., 2002; Robey and Cianciotto, 2002).

The genetic organization of the Feo system in *Y. pestis* KIM10+ and CO92 as well as in *Y. enterocolitica* 8081 is the same as that of *E. coli*, with three Orfs (*feoABC*) comprising a putative operon (Parkhill et al., 2001; Deng et al., 2002). *Y. pestis feoB* mutants with a functional Ybt iron transport system did not exhibit growth defects under any of the tested conditions (unpublished results). Strains missing both the Ybt and Feo systems grew normally in J774.1 cells and in iron-deficient media under aerobic conditions. However, *ybt⁻ feo⁻* strains incubated in candle jars to reduce the oxygen content, exhibited growth defects on gradients plates containing nitrilotriacetic acid. In addition, the growth of *ybt⁻ feo⁻* strains under static conditions in liquid media containing 5uM EDDA was significantly reduced compared to the Ybt⁻ FeoB⁺ parent strain (unpublished results). Thus *Y. pestis* contains a functional Feo system whose relevance can be masked by other iron transport systems.

HEME TRANSPORT SYSTEMS

THE Hmu/Hem HEME TRANSPORTER

The pathogenic *Yersinia* species can use hemin as a source of iron (Perry and Brubaker, 1979). In addition, *Y. pestis* and *Y. enterocolitica* can acquire iron from a variety of heme-containing proteins (Sikkema and Brubaker, 1989; Staggs and Perry, 1991; Bracken et al., 1999) via a periplasmic SBP-dependent transport system (Figure 5) designated Hmu and Hem, respectively. The Hem system of *Y. enterocolitica* was the first ABC transporter for heme to be characterized (Stojiljkovic and Hantke, 1992, 1994). Subsequently, the nearly identical *hmu* locus was identified in *Y. pestis* (Hornung et al., 1996; Thompson et al., 1999). The loci consist of 6 genes, *P, R, S, T, U,* and *V*. In *Y. pestis, hmuP'* is truncated relative to *hemP*, however, this difference is insignificant since HemP is not an essential component of the Hem system (Stojiljkovic and Hantke, 1994). HemR/HmuR is a TonB-dependent OM receptor that binds hemin and a variety of hemoproteins (Figure 5) including hemoglobin, myoglobin, hemoglobin-haptoglobin, heme-hemopexin, and heme-albumin (Sikkema and Brubaker, 1989; Koebnik et al., 1993a; Bracken et al., 1999; Thompson et al., 1999; Perry et al., 2003b). Precisely how the receptor interacts with these different proteins is unknown. While it is possible that the receptor recognizes the common heme moiety in these hemoproteins, in hemoglobin most of the porphyrin structure is not surface exposed (Stojiljkovic and Perkins-Balding, 2002). An analysis of bacterial heme and hemoglobin receptors identified two highly conserved amino acid motifs, FRAP and NPNL, with a conserved histidine residue (corresponding to H461 of HemR) located between the two motifs (Bracken et al., 1999). Mutational studies revealed that two histidine residues in HemR, H128 and H461, are important in the utilization of heme and hemoproteins as sources of heme and iron. Alteration of H461 abolished the ability to use hemoproteins as a source of heme and iron. In addition, H461 mutants were unable to use heme as an iron source and were severely impaired in the use of hemin as a source of heme (Bracken et al., 1999). A mutation at H128 abolished the ability to use heme and hemoproteins as heme sources and to use myoglobin and heme/albumin as iron sources. H128 mutants could still utilize heme and hemoglobin as sources of iron but not as efficiently as the wild-type parent strain. Both histidine residues are surface-accessible and are likely involved in heme binding (Bracken et al., 1999)

Presumably only the heme moiety is translocated into the periplasm where it is bound by the periplasmic binding protein, HemT/HmuT. In *Y. enterocolitica*, HemT is needed for the efficient transport of hemin into the cells but is not required to use hemin as an iron

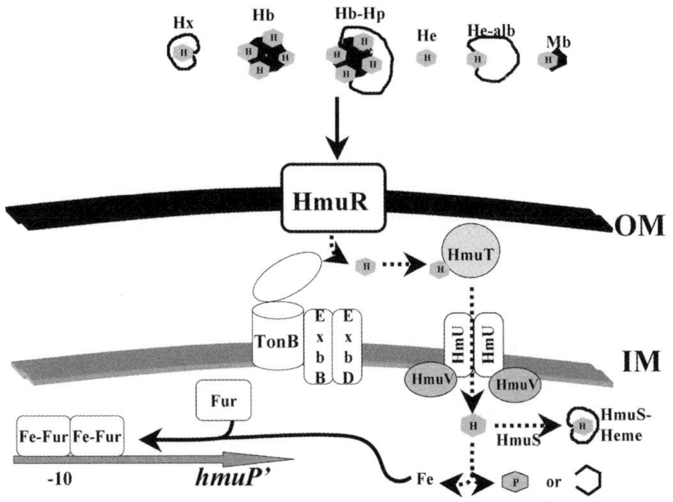

Figure 5. Model of the Hmu/Hem heme transport system and regulation. For simplicity, only Hmu designations are used. Dashed arrows designate predicted functions or mechanisms that have not been experimentally determined. Abbreviations: Hx, heme-hemopexin; Hb, hemoglobin; Hb-Hp, hemoglobin-haptoglobin; He, heme; He-albumin, heme-albumin; Mb, myoglobin.. Figure is reproduced from Perry, 2004 with permission of ASM.

source (Stojiljkovic and Hantke, 1994). The heme moiety is transported across the IM by the permease-ATP hydrolase complex of HemU/HmuU and HemV/HmuV (Figure 5). The growth of *hemU* and *hemV* mutants on heme was severely reduced (Stojiljkovic and Hantke, 1994). Similarly, *hmuTUV* mutants were unable to use heme but could use hemoglobin as an iron source (Thompson et al., 1999).

hemS/*hmuS* encodes a cytoplasmic protein and appears to be an essential gene in *Y. enterocolitica* but not in *Y. pestis* (Stojiljkovic and Hantke, 1994; Thompson et al., 1999). Based on studies with the related protein, ShuS, of *Shigella dysenteriae*, HemS/HmuS may form a large oligomeric complex which binds and sequesters heme thereby preventing heme toxicity (Wilks, 2001; Stojiljkovic and Perkins-Balding, 2002). An *E. coli hemA aroB* mutant strain carrying *hemR* on a low copy number plasmid could grow on hemin as an iron source. However strains expressing HemR from a high-copy plasmid could not, unless HemS was also present - presumably due to heme toxicity (Stojiljkovic and Hantke, 1994). Initial studies showing that *Y. pestis* HmuR alone would not allow *E. coli* mutants to grow on hemin were probably also the result of toxicity due to over expressing *hmuR* as low-copy number vectors encoding *hmuR* are functional (Thompson et al., 1999; unpublished results).

In other bacteria, additional genes have been found associated with heme utilization systems (Torres and Payne, 1997; Wyckoff et al., 1998; O'Malley et al., 1999; Ochsner et al., 2000; Henderson et al., 2001). Three Orfs (designated Y0545-Y0547 in KIM10+ genomic nomenclature), that are homologous to sequences present in other heme uptake systems, are found upstream of *hmuP'* in *Y. pestis.* Y0547 contains a domain found in oxygen-independent coproporphyrinogen III oxidases and is homologous to HugW of *Plesiomonas shigelloides* and PhuW of *Vibrio parahemolyticus* (O'Malley et al., 1999; Henderson et al., 2001). Y0546 and Y0545, previously designated OrfX and OrfY, are related to HugX of *Plesiomonas* and ShuY of *S. dysenteriae* (Wyckoff et al., 1998; Henderson et al., 2001). A mutation in *y0547*, alone or in combination with a *hmuS* deletion, had no effect on the ability of *Y. pestis* cells to grow in a medium containing hemoglobin and EDDA (unpublished results).

The *hem*/*hmuPRSTUV* locus as well as the three upstream ORFs found in other heme uptake systems are highly conserved in *Y. pestis* KIM10+, CO92, and *Y. enterocolitica* 8081. HemV may be slightly larger than HmuV due to a duplication of 15 base pairs near the predicted start site of HemV. In addition, *hemR* is predicted to encode 11 additional amino acids that are not found in the predicted coding sequence for *hmuR*. A second locus encoding sequences similar to HemR (52% identity/67% similarity), HemS (53% identity/69% similarity), HemT (60% identity/75% similarity) and part of HemP is present in *Y. enterocolitica* 8081 but not in the *Y. pestis* genomes. The *hemR* and *hemS*-related sequences apparently do not compensate for mutations in *hemR* and *hemS* making it unlikely that they function in heme uptake and utilization. However, it would be interesting to determine if the *hemT*-related sequence is responsible for the ability of a *hemT* mutant to use hemin as an iron source (Stojiljkovic and Hantke, 1992, 1994)

It is not clear whether the yersiniae extract inorganic iron from heme. Searches of the *Y. pestis* genomic sequences fail to identify Orfs with significant similarities to heme oxygenases (Schmitt, 1997; Zhu et al., 2000; Ratliff et al., 2001) or other proteins implicated in the extraction of inorganic iron from heme. In yersiniae, acquired heme might be used directly in cytochromes and other heme-requiring enzymes, thus reducing the need for inorganic iron in heme biosynthesis.

The *hem*/*hmu* loci contain 2 promoters, one for *hem*/*hmuPR* and a separate promoter for *hem*/*hmuSTUV* (Stojiljkovic and Hantke, 1992, 1994; Thompson et al., 1999). Expression from the *PR* promoter is regulated by Fur and iron (Stojiljkovic and Hantke, 1992; Thompson et al., 1999). Although a putative FBS is located in the *hmuSTUV* promoter region, transcriptional reporter gene fusions indicate that the *hmuS* promoter is not iron regulated but rather has low constitutive activity. Potential transcriptional start sites upstream of *hmuP'* and *hmuS* were identified by primer extension analysis. The *hmuP'* promoter likely initiates transcription of the entire *hmu* operon while transcription from the *hmuS* promoter yields a *hmuSTUV* mRNA. Western blot analysis confirms that HmuR is strongly repressed by iron while HmuS displays only modest iron repression possibly due to read-through from the iron-regulated *hmuP'* promoter (Thompson et al., 1999).

HemR mutants were as virulent as wild-type by an intravenous route of infection (Stojiljkovic and Hantke, 1992). Similarly, *Y. pestis* Δ*hmuP'RSTUV* cells injected subcutaneously or intravenously into mice retained full virulence (Thompson et al., 1999; Table 1). A *hemR-gfp* translational fusion was expressed at low levels in the liver as well as in the intestinal lumen and at moderate levels in Peyer's patches and the spleen following infection. Bacterial cells recovered from the peritoneal cavity of infected mice exhibited the highest level of expression. Similar expression patterns were observed for an *yfuA* reporter construct and may simply reflect the relative levels of bacterial iron stress in these diverse organ systems (Jacobi et al., 2001). Although not required for disease in mice, the Hmu hemoprotein transport system is important for growth in J774 macrophage-like cells. After infection of a J774 monolayer with *Y. pestis* cells grown at 37°C under Fe-deficient conditions, there is an initial decline in intracellular bacteria followed by a growth phase and eventual destruction of the J774 monolayer. In general, bacterial cells with mutations in the Ybt and Yfe inorganic iron transport systems do not affect the growth of *Y. pestis* in J774 cells under these conditions. However, mutants in *hmu* or *tonB* fail to fully recover from the initial decline in cell numbers (unpublished observations). Whether the Hmu system is required for intracellular growth in vivo or in non-phagocytic cells remains to be determined.

THE Has HEMOPHORE SYSTEM

The Has system was initially characterized in *S. marcescens* and consists of a small protein (HasA) that is secreted from the cell through a Type I secretion system consisting of HasD, HasE, and HasF (Létoffé et al., 1994a, 1994b; Binet and Wandersman, 1996). In *E. coli* TolC can replace the function of HasF (Binet and Wandersman, 1996). HasA is a hemophore that binds heme; the HasA/heme complex then interacts with an OM receptor, HasR (Létoffé et al., 1999). Heme utilization in *Serratia* is regulated by iron and requires SecB as well as either $TonB_{sm}$ or HasB, a TonB-like protein that is part of the *has* operon in *Serratia* (Létoffé et al., 1994a; Delepelaire and Wandersman, 1998; Paquelin et al., 2001).

Y. pestis possesses sequences that are related to the *S. marcescens has* genes and are organized in a similar fashion. $HasF_{yp}$ (TolC) is encoded by a locus that is separate from *hasRADEB*$_{yp}$. Expression from the *hasRADEB*$_{yp}$ promoter is regulated by iron and Fur. $HasA_{yp}$ is secreted from *E. coli* cells expressing a heterologous system consisting of $HasDE_{sm}$ and TolC. Furthermore $HasA_{yp}$ bound heme and could acquire heme from hemoglobin. However, $HasA_{yp}$ secretion could not be detected in a number of different

Y. pestis strains grown under a variety of conditions or in *E. coli* strains carrying the cloned *Y. pestis hasRADEB* locus. The deduced amino acid sequences of HasD, HasE, and HasF from *Y. pestis* do not contain any obvious defects to explain this result (Rossi et al., 2001).

An *E. coli hemA* strain expressing $HasR_{sm}$ can use hemoglobin as a source of porphyrins (Ghigo et al., 1997). Although in vitro transcription/translation experiments indicated that both HasR and HasA from *Y. pestis* were expressed, *hemA* mutants carrying either $hasRADEB_{yp}$ or $hasRA_{yp}$ were unable to grow on media containing hemoglobin as a porphyrin source. In addition, $HasA_{yp}$ did not interact with HasR from either *Y. pestis* or *S. marcescens*. Perhaps not surprisingly, the growth of a *Y. pestis hmu has* mutant in iron-deficient medium supplemented with hemoglobin was not significantly different from its *hmu* has^{+} parent (Rossi et al., 2001). Finally, a *Y. pestis hmu has* mutant retained full virulence in mice infected subcutaneously (Table 1). Thus the *Y. pestis* Has system is apparently not functional under the experimental conditions tested.

A second *tonB*-like gene, *hasB*, is encoded by the *has* locus. Our studies indicate that $HasB_{Yp}$ cannot substitute for TonB in either the Ybt or Hmu transport systems (Perry et al., 2003b).

Sequences related to *hasRADEB* are also present in the *Y. enterocolitica* 8081 genome. The genes are organized in the same manner as in *Y. pestis* KIM10+ and CO92 except that *Y. enterocolitica* 8081 possesses 4 tandem copies of *hasA*-related sequences. These *hasA*-like genes have diverged from each other with "*hasA1*" and "*hasA4*" being similar to each other (42% identity/52% similarity) and "*hasA2*" being related to "*hasA3*" (42% identity/51% similarity). All four *Y.enterocolitica hasA* homologues are more closely related to *Y. pestis* than to *S. marcescens hasA*.

PUTATIVE IRON TRANSPORTERS

THE Ysu SYSTEM

The *ysu* locus of *Y. pestis* (Figure 6) contains a number of Orfs that have their highest similarities to siderophore biosynthetic enzymes and transport components. *ysuA-D* encode a SBP, two permeases, and an ATP hydrolase, respectively. *ysuR* potentially encodes an OM receptor with the most homology to FauA, the receptor for alcaligin, the siderophore produced by *Bordetella* (Brickman and Armstrong, 1999). YsuE and YsuG have the most similarity to AlcC while YsuH and YsuI have similarities to AlcB, and AlcA, respectively. AlcA-C are biosynthetic enzymes required for the production of the siderophore alcaligin from *Bordetella* (Giardina et al., 1997). Although intact in *Y. pestis* CO92, *ysuE* in strain KIM10+ is most likely a pseudogene due to a frameshift mutation that yields two Orfs of 318 and 287 amino acids. Decarboxylase domains present in YsuJ are found in a number of proteins including some involved in siderophore biosynthesis. Finally, YsuF is related to FhuF, an *E. coli* reductase implicated in the removal of iron from hydroxamate siderophores (Müller et al., 1998). Currently, there is no experimental evidence that this system either produces a siderophore or is involved in iron acquisition. Blast searches with 4 of the *ysu* genes failed to identify homologues in *Y. enterocolitica* 8081.

THE Ynp NRPS SYSTEM

A locus tentatively designated *ynp* for *Yersinia* non-ribosomal peptide synthesis (Figure 6), spans ~30 kb. It includes genes for a potential OM receptor (KIM Y3404), an ABC transporter, as well as Orfs (Y3421 and Y3419) with similarities to YbtX and YbtU, respectively. In addition to two putative reductases (Y3419 and Y3418), there are five Orfs (Y3406, Y3410-Y3412, Y3416) with NRPS and PKS domains found in Ybt, pyochelin, and/or anguibactin biosynthetic enzymes. The putative Ynp biosynthetic system is complex and contains enzymatic modules for condensation/cyclization and adenylation reactions as well as methyltransferase and thioesterase activities. The Orfs also possess putative peptidyl and acyl carrier protein domains and a ketoacyl synthase domain. However, it is unclear whether this system can produce a final or partial product. Some of the Orfs appear to be disrupted or degenerate. For example, in *Y. pestis* KIM10+, Y3416 has an internal 6 amino acid deletion and a 50 amino acid region that is not 100% conserved in its counterpart in CO92 (YPO0776). In addition, there is an IS*100* insertion between *y3406* and *y3410* that may prevent synthesis of a product (Figure 6). Y3411 (only 12.5 kDa) with similarities to regions of HMWP2 of the Ybt system and AngR of the *V. anguillarum* anguibactin system, may be the result of a frameshift mutation. Finally, the *ynp* locus does not encode a P-pant transferase. P-pant transferases initiate the assembly of siderophores, polyketides, and fatty acids by modifying acyl, aryl, or peptidyl carrier protein domains. A BLAST search of the *Y. pestis* genome identified only two potential P-pant transferases – YbtD (used by the Ybt system) and a homologue of the *E. coli* acyl

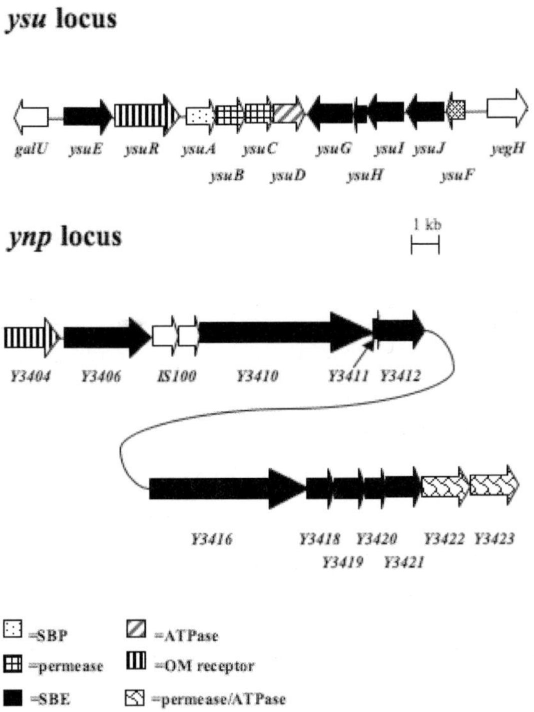

Figure 6. Genetic organization of the *ysu* and *ynp* loci of *Y. pestis*. Arrows indicate the Orfs and direction of transcription. Orfs are drawn to scale. SBE: sideropore biosynthetic enzyme.

carrier protein (ACPS), which is essential for fatty acid synthesis. If the Ynp system synthesizes a product, one of these P-pant transferases must catalytically activate the Ynp biosynthetic enzymes.

The putative OM receptor (Y3404) contains a TonB box and has similarities to putative or proven siderophore receptors, including Psn/FyuA. The two ABC transporter components (Y3422 and Y3423) have fused permease and ATPase domains similar to YbtP and YbtQ. These transport functions, encoded by the *ynp* locus, may be functional even if a product is not produced.

The similarities of the Ynp system to the Ybt system are numerous. Both systems possess 3 Cy domains that would yield 3 cyclized cysteines, as well as external and internal TE domains, an unusual two-component ABC transporter and potential siderophore exporters (YbtX and Y3421). As in the *ybt* locus, the putative OM receptor gene (*y3404*) is separated from the ABC transporter alleles by NRPS/PKS genes and neither system contains an Orf corresponding to a SBP. These similarities suggest that there may be an ancestral relationship between Ybt and Ynp.

The *ynp* locus has been rearranged in *Y. pestis* CO92 with *y3404* (*YPO1011* in CO92) and *y3406* (*YPO1012*) over 285 kb away from the remainder of the *ynp* genes. Blast searches with 3 of the *ynp* genes indicate that this system is probably absent from *Y. enterocolitica* 8081.

THE Fit TRANSPORTER

We have tentatively designated a locus encoding four Orfs in *Y. pestis* CO92 and KIM10+ as *fit* (Figure 3). The *Y. pestis* KIM10+ Orfs designated Y4043-Y4046 encode putative IM permeases, a SBP, and an ATP hydrolase with the highest similarities to ABC transporters of proven or putative siderophore-iron transporters. Except for differences in the assigned start sites, the deduced protein sequences of Y4043-Y4046 from KIM10+ are identical to those in *Y. pestis* CO92 (Parkhill et al., 2001; Deng et al., 2002) and are also highly conserved in *Y. enterocolitica* 8081. Whether or not the Fit system is functional or involved in iron acquisition remains to be determined.

THE Fiu TRANSPORTER

Another potential iron ABC transporter is encoded in *Y. pestis* KIM10+ by *y2837, y2839,* and *y2842* and tentatively designated the *fiu* locus (Figure 3). These genes are present in *Y. pestis* CO92 but not in *Y. enterocolitica* 8081. Y2842 is a member of the periplasmic SBP family and has similarity to a number of hypothetical SBPs, some annotated as iron or siderophore transporters. Of proven iron transporters, *Y. pestis* YiuA (see above) and *C. diphtheriae* Irp6A (required for uptake of the siderophore, corynebactin) show the best similarities to Y2842. Y2839 is a putative IM permease that possesses a FecCD transport sub-family domain. Of transporter components with a proven function, YiuB and *C. diphtheriae* Irp6B exhibit the highest similarities to Y2839. The ATP hydrolase component, Y2837, has significant similarity to the FhuC ATP hydrolase of *E. coli* (unpublished observations). While the bioinformatics analysis suggests that these genes may encode an iron ABC transporter, there is no supporting experimental evidence.

THE PUTATIVE 26°C IRON TRANSPORTER

Lucier et al. (1996) showed that higher concentrations of a variety of iron chelators were required to prevent growth of *Y. pestis* at 26°C than at 37°C and that the enhanced growth at the lower temperature was not Ybt-dependent. From these results it was hypothesized that *Y. pestis* contained an iron transport system that was functional at 26°C but not at mammalian temperatures. Since then, experimental studies have eliminated the Yfe (Bearden and Perry, 1999), Yfu (Gong et al., 2001), and Yiu (unpublished results) systems as the putative 26°C iron transporter. Preliminary studies with Ynp⁻ and Ysu⁻ derivatives of *Y. pestis* suggest that these systems are also not responsible for the enhanced growth at 26°C under iron-chelated conditions. Although several more putative iron transport systems remain to be tested, it is possible that a number of different iron transporters contribute to this growth response. Alternatively, enhanced growth during iron deprivation at 26°C compared to 37°C may be due to differences in intermediary metabolism that substantially decrease the iron requirements at the lower temperature.

IRON STORAGE

The *Y. pestis* genome encodes genes for bacterioferritin (*bfrA*), bacterioferritin-associated ferridoxin (*bfd*), and ferritin (*ftnA*) (Parkhill et al., 2001; Deng et al., 2002). Whether any of these Orfs have a storage function in *Y. pestis* has not yet been experimentally determined. In *Y. pestis* KIM6+, a high molecular mass complex that bound iron was composed of a 19-kDa cytoplasmic polypeptide that could be either BfrA or FtnA. HemS/HmuS, like ShuS from *Shigella,* may prevent heme toxicity by sequestering heme (Stojiljkovic and Hantke, 1994; Wilks, 2001; Stojiljkovic and Perkins-Balding, 2002). The Hms system was first proposed to have a hemin storage function (Perry et al., 1990); however, it has now been clearly shown that the Hms phenotype is not involved in storage but is important in the blockage of fleas (Lillard et al., 1999; Hinnebusch et al., 1996; see Chapter 4).

IRON REGULATION

In Gram-negative bacteria, the ferric uptake regulator (Fur) represses expression of numerous genes, including iron uptake systems. Under conditions of surplus iron, a Fur-Fe^{2+} complex binds to the FBS of regulated promoters and limits their transcription. Under iron starvation conditions, free Fur does not bind and the regulated genes are fully expressed (Hantke, 2001). A Pgm^+ *Y. pestis fur* mutant does not grow at all under Fe-surplus conditions and even a Δ*pgm* (Δ*ybt*) *fur* mutant grows poorly. We have speculated that constitutive overexpression of the Ybt system leads to the accumulation of iron to toxic levels (Staggs et al., 1994). In *Y. pestis* a minimum of 38 iron-repressible and 6 iron-inducible proteins have been noted. Most but not all of these are regulated through Fur (Staggs et al., 1994; Lucier et al., 1996). *Y. pestis* Fur is highly similar to Fur proteins in *E. coli* and other enterics (Staggs and Perry, 1991, 1992; Staggs et al., 1994). Fur_{Yp} represses expression of the Ybt, Yfe, Yfu, Yiu, Has and Hmu systems (Bearden et al., 1998; Thompson et al., 1999; Gong et al., 2001; Perry et al., 2001; Rossi et al., 2001; Perry et al., 2003a; unpublished results). The Ybt, Hem, and enterobactin systems of *Y. enterocolitica* are also regulated by Fur (Carniel et al., 1992; Staggs and Perry, 1992; Stojiljkovic and Hantke, 1992; Heesemann et al., 1993; Schubert et al., 1999) and a putative FBS overlaps the promoter region of the *yfu* locus (Saken et al., 2000). In *Y. pestis,* putative FBSs are upstream of *yfeE, hmuS* and *feoAB*. However FBS-like sequences may not always be functional as neither *yfeE* or *hmuS* promoters are Fur or iron regulated (Thompson et al., 1999; Perry et al., 2003a; unpublished observations). Iron repression and Fur regulation of the *Y. pestis* Feo, Fhu, Fit, Fiu, Ynp, and Ysu systems has not been examined.

In *Y. pestis*, Mn-repression of the *yfeA-D* promoter is Fur-dependent. In vivo Mn-repression via Fur is not a typical property of Fur-regulated promoters from *Y. pestis* or other Gram-negative bacteria (Bearden et al., 1998; Perry et al., 2003a). This unique regulation may be due to characteristics of the *Y. pestis* Fur protein, the *yfeA-D* promoter, or both.

Two-dimensional gel electrophoresis suggests that at least nine Fe-repressible polypeptides of *Y. pestis* are regulated independently of Fur (Staggs et al., 1994). Searches of the *Y. pestis* genome have failed to identify any known alternative Fe-responsible regulators other than PmrA and PmrB (unpublished observations). In *Salmonella*, this two component regulatory system induces the expression of their target genes in high-iron environments (Chamnongpol et al., 2002). The *Y. pestis* PmrA/PmrB system may regulate the expression of two Fe-inducible polypeptides that are Fur independent. Alternatively these proteins could be induced by regulators responsive to the oxidative stress that would be caused by the accumulation of excess iron in *fur* mutants. Four polypeptides are upregulated during growth with surplus iron in a Fur-dependent mechanism (Staggs et al., 1994). In *E. coli*, Fur-Fe represses the transcription of *ryhB* which encodes a small RNA that inhibits the synthesis of several proteins. Consequently these proteins are highly expressed only during growth with excess iron (Massé and Gottesman, 2002). *Y. pestis* contains a *ryhB* gene (unpublished observations) but its expression or role in iron regulation has not been determined.

ACKNOWLEDGEMENTS

In our laboratory Jennifer Abney, Bill Baker, Ildefonso Mier, Jr., Mike Nagiec as well as Drs. Alexander Bobrov, Valérie Geoffroy, and Olga Kirillina contributed unpublished observations and results described in this chapter. We thank Dr. Luis Quadri for his assistance in analyzing the enzymatic domains encoded by Orfs of the *ynp* locus. Our research program on iron transport systems of *Y. pestis* has been supported by Public Health Service grants AI25098, AI33481, and AI42738. The sequence data for *Y.enterocolitica* 8081 was produced by the *Yersinia enterocolitica* Sequencing Group at the Sanger Institute and can be obtained from ftp://ftp.sanger.ac.uk/pub/pathogens/ye.

The contributions of Igor Stojiljkovic to our understanding of bacterial heme transport systems is difficult to overestimate. He will be missed as a scientist and friend.

REFERENCES

Actis, L.A., Tolmasky, M.E., Farrell, D.H., and Crosa, J.H. 1988. Genetic and molecular characterization of essential components of the *Vibrio anguillarum* plasmid-mediated iron-transport system. J. Biol. Chem. 263: 2853-2860.

Bartsevich, V.V., and Pakrasi, H.B. 1995. Molecular identification of an ABC transporter complex for manganese: analysis of a cyanobacterial mutant strain impaired in the photosynthetic oxygen evolution process. EMBO J. 14: 1845-1853.

Bäumler, A., Koebnik, R., Stojiljkovic, I., Heesemann, J., Braun, V., and Hantke, K. 1993. Survey on newly characterized iron uptake systems of *Yersinia enterocolitica*. Zbl. Bakt. 278: 416-424.

Bäumler, A.J., and Hantke, K. 1992. A lipoprotein of *Yersinia enterocolitica* facilitates ferrioxamine uptake in *Escherichia coli*. J. Bacteriol. 174: 1029-1035.

Bäumler, A.J., and Hantke, K. 1992. Ferrioxamine uptake in *Yersinia enterocolitica*: characterization of the receptor protein FoxA. Mol. Microbiol. 6: 1309-1321.

Bearden, S.W., Fetherston, J.D., and Perry, R.D. 1997. Genetic organization of the yersiniabactin biosynthetic region and construction of avirulent mutants in *Yersinia pestis*. Infect. Immun. 65: 1659-1668.

Bearden, S.W., Staggs, T.M., and Perry, R.D. 1998. An ABC transporter system of *Yersinia pestis* allows utilization of chelated iron by *Escherichia coli* SAB11. J. Bacteriol. 180: 1135-1147.

Bearden, S.W., and Perry, R.D. 1999. The Yfe system of *Yersinia pestis* transports iron and manganese and is required for full virulence of plague. Mol. Microbiol. 32: 403-414.

Bengoechea, J.-A., Brandenburg, K., Seydel, U., Díaz, R., and Moriyón, I. 1998. *Yersinia pseudotuberculosis* and *Yersinia pestis* show increased outer membrane permeability to hydrophobic agents which correlates with lipopolysaccharide acyl-chain fluidity. Microbiology 144: 1517-1526.

Binet, R., and Wandersman, C. 1996. Cloning of the *Serratia marcescens hasF* gene encoding the Has ABC exporter outer membrane component: a TolC analogue. Mol. Microbiol. 22: 265-273.
Bobrov, A.G., Geoffroy, V.A., and Perry, R.D. 2002. Yersiniabactin production requires the thioesterase domain of HMWP2 and YbtD, a putative phosphopantetheinylate transferase. Infect. Immun. 70: 4204-4214.
Bracken, C.S., Baer, M.T., Abdur-Rashid, A., Helms, W., and Stojiljkovic, I. 1999. Use of heme-protein complexes by the *Yersinia enterocolitica* HemR receptor: histidine residues are essential for receptor function. J. Bacteriol. 181: 6063-6072.
Braun, V., and Hantke, K. 1991. Genetics of bacterial iron transport. In: CRC Handbook of Microbial Iron Chelates. G. Winkelmann, ed. CRC Press, Boca Raton, Florida. p. 107-138.
Braun, V., Hantke, K., and Köster, W. 1998. Bacterial iron transport: mechanisms, genetics, and regulation. In: Metal Ions in Biological Systems. Vol. 35. A. Sigel and Sigel, H., eds. Marcel Dekker, Inc., New York. p. 67-145.
Brem, D., Pelludat, C., Rakin, A., Jacobi, C.A., and Heesemann, J. 2001. Functional analysis of yersiniabactin transport genes of *Yersinia enterocolitica*. Microbiology 147: 1115-1127.
Brickman, T.J., and Armstrong, S.K. 1999. Essential role of the iron-regulated outer membrane receptor FauA in alcaligin siderophore-mediated iron uptake in *Bordetella* species. J. Bacteriol. 181: 5958-5966.
Buchrieser, C., Brosch, R., Bach, S., Guiyoule, A., and Carniel, E. 1998. The high-pathogenicity island of *Yersinia pseudotuberculosis* can be inserted into any of the three chromosomal *asn* tRNA genes. Mol. Microbiol. 30: 965-978.
Buchrieser, C., Rusniok, C., Frangeul, L., Couve, E., Billault, A., Kunst, F., Carniel, E., and Glaser, P. 1999. The 102-kilobase *pgm* locus of *Yersinia pestis*: sequence analysis and comparison of selected regions among different *Yersinia pestis* and *Yersinia pseudotuberculosis* strains. Infect. Immun. 67: 4851-4861.
Carniel, E., Guiyoule, A., Guilvout, I., and Mercereau-Puijalon, O. 1992. Molecular cloning, iron-regulation and mutagenesis of the *irp2* gene encoding HMWP2, a protein specific for the highly pathogenic *Yersinia*. Mol. Microbiol. 6: 379-388.
Carniel, E. 2001. The *Yersinia* high-pathogenicity island: an iron-uptake island. Microbes Infect. 3: 561-569.
Chambers, C.E., and Sokol, P.A. 1994. Comparison of siderophore production and utilization in pathogenic and environmental isolates of *Yersinia enterocolitica*. J. Clin. Microbiol. 32: 32-39.
Chambers, C.E., McIntyre, D.D., Mouck, M., and Sokol, P.A. 1996. Physical and structural characterization of yersiniophore, a siderophore produced by clinical isolates of *Yersinia enterocolitica*. BioMetals 9: 157-167.
Chamnongpol, S., Dodson, W., Cromie, M.J., Harris, Z.L., and Groisman, E.A. 2002. Fe(III)-mediated cellular toxicity. Mol. Microbiol. 45: 711-719.
Coulton, J.W., Mason, P., and DuBow, M.S. 1983. Molecular cloning of the ferrichrome-iron receptor of *Escherichia coli* K-12. J. Bacteriol. 156: 1315-1321.
Crosa, J.H., and Walsh, C.T. 2002. Genetics and assembly line enzymology of siderophore biosynthesis in bacteria. Microbiol. Mol. Biol. Rev. 66: 223-249.
Delepelaire, P., and Wandersman, C. 1998. The SecB chaperone is involved in the secretion of the *Serratia marcescens* HasA protein through an ABC transporter. EMBO J. 17: 936-944.
Deng, W., Burland, V., Plunkett, G., III, Boutin, A., Mayhew, G.F., Liss, P., Perna, N.T., Rose, D.J., Mau, B., Zhou, S., Schwartz, D.C., Fetherston, J.D., Lindler, L.E., Brubaker, R.R., Plano, G.V., Straley, S.C., McDonough, K.A., Nilles, M.L., Matson, J.S., Blattner, F.R., and Perry, R.D. 2002. Genome Sequence of *Yersinia pestis* KIM. J. Bacteriol. 184: 4601-4611.
Dintilhac, A., and Claverys, J.-P. 1997. The *adc* locus, which affects competence for genetic transformation in *Streptococcus pneumoniae*, encodes an ABC transporter with a putative lipoprotein homologous to a family of streptococcal adhesins. Res. Microbiol. 148: 119-131.
Drechsel, H., Stephan, H., Lotz, R., Haag, H., Zähner, H., Hantke, K., and Jung, G. 1995. Structure elucidation of yersiniabactin, a siderophore from highly virulent *Yersinia* strains. Liebigs Ann. 1995: 1727-1733.
Fecker, L., and Braun, V. 1983. Cloning and expression of the *fhu* genes involved in iron(III) hydroxamate uptake by *Escherichia coli*. J. Bacteriol. 156: 1301-1314.
Ferber, D.M., Fowler, J.M., and Brubaker, R.R. 1981. Mutations to tolerance and resistance to pesticin and colicins in *Escherichia coli*. J. Bacteriol. 146: 506-511.
Ferguson, A.D., Hofmann, E., Coulton, J.W., Diederichs, K., and Welte, W. 1998. Siderophore-mediated iron transport: crystal structure of FhuA with bound lipopolysaccharide. Science 282: 2215-2220.
Ferguson, A.D., Chakraborty, R., Smith, B.S., Esser, L., van der Helm, D., and Deisenhofer, J. 2002. Structural basis of gating by the outer membrane transporter FecA. Science 295: 1715-1719.
Fetherston, J.D., Schuetze, P., and Perry, R.D. 1992. Loss of the pigmentation phenotype in *Yersinia pestis* is due to the spontaneous deletion of 102 kb of chromosomal DNA which is flanked by a repetitive element. Mol. Microbiol. 6: 2693-2704.

Fetherston, J.D., Lillard, J.W., Jr., and Perry, R.D. 1995. Analysis of the pesticin receptor from *Yersinia pestis*: role in iron-deficient growth and possible regulation by its siderophore. J. Bacteriol. 177: 1824-1833.

Fetherston, J.D., Bearden, S.W., and Perry, R.D. 1996. YbtA, an AraC-type regulator of the *Yersinia pestis* pesticin/yersiniabactin receptor. Mol. Microbiol. 22: 315-325.

Fetherston, J.D., Bertolino, V.J., and Perry, R.D. 1999. YbtP and YbtQ: two ABC transporters required for iron uptake in *Yersinia pestis*. Mol. Microbiol. 32: 289-299.

Furrer, J.L., Sanders, D.N., Hook-Barnard, I.G., and McIntosh, M.A. 2002. Export of the siderophore enterobactin in *Escherichia coli*: involvement of a 43 kDa membrane exporter. Mol. Microbiol. 44: 1225-1234.

Gehring, A.M., DeMoll, E., Fetherston, J.D., Mori, I., Mayhew, G.F., Blattner, F.R., Walsh, C.T., and Perry, R.D. 1998a. Iron acquisition in plague: modular logic in enzymatic biogenesis of yersiniabactin by *Yersinia pestis*. Chem. Biol. 5: 573-586.

Gehring, A.M., Mori, I., Perry, R.D., and Walsh, C.T. 1998b. The nonribosomal peptide synthetase HMWP2 forms a thiazoline ring during biogenesis of yersiniabactin, an iron-chelating virulence factor of *Yersinia pestis*. Biochemistry 37: 11637-11650.

Geoffroy, V.A., Fetherston, J.D., and Perry, R.D. 2000. *Yersinia pestis* YbtU and YbtT are involved in synthesis of the siderophore yersiniabactin but have different effects on regulation. Infect. Immun. 68: 4452-4461.

Ghigo, J.-M., Létoffé, S., and Wandersman, C. 1997. A new type of hemophore-dependent heme acquisition system of *Serratia marcescens* reconstituted in *Escherichia coli*. J. Bacteriol. 179: 3572-3579.

Giardina, P.C., Foster, L.A., Toth, S.I., Roe, B.A., and Dyer, D.W. 1997. Analysis of the *alcABC* operon encoding alcaligin biosynthesis enzymes in *Bordetella bronchiseptica*. Gene 194: 19-24.

Goldberg, M.B., Boyko, S.A., Butterton, J.R., Stoebner, J.A., Payne, S.M., and Calderwood, S.B. 1992. Characterization of a *Vibrio cholerae* virulence factor homologous to the family of TonB-dependent proteins. Mol. Microbiol. 6: 2407-2418.

Gong, S., Bearden, S.W., Geoffroy, V.A., Fetherston, J.D., and Perry, R.D. 2001. Characterization of the *Yersinia pestis* Yfu ABC iron transport system. Infect. Immun. 67: 2829-2837.

Graber, K.R., Smoot, L.M., and Actis, L.A. 1998. Expression of iron binding proteins and hemin binding activity in the dental pathogen *Actinobacillus actinomycetemcomitans*. FEMS Microbiol. Lett. 163: 135-142.

Hantke, K. 1987. Ferrous iron transport mutants in *Escherichia coli* K12. FEMS Microbiol. Lett. 44: 53-57.

Hantke, K. 2001. Iron and metal regulation in bacteria. Curr. Opin. Microbiol. 4: 172-177.

Heesemann, J. 1987. Chromosomal-encoded siderophores are required for mouse virulence of enteropathogenic *Yersinia* species. FEMS Microbiol. Lett. 48: 229-233.

Heesemann, J., Hantke, K., Vocke, T., Saken, E., Rakin, A., Stojiljkovic, I., and Berner, R. 1993. Virulence of *Yersinia enterocolitica* is closely associated with siderophore production, expression of an iron-repressible outer membrane polypeptide of 65 000 Da and pesticin sensitivity. Mol. Microbiol. 8: 397-408.

Heinrichs, D.E., and Poole, K. 1996. PchR, a regulator of ferripyochelin receptor gene (*fptA*) expression in *Pseudomonas aeruginosa*, functions both as an activator and as a repressor. J. Bacteriol. 178: 2586-2592.

Henderson, D.P., Wyckoff, E.E., Rashidi, C.E., Verlei, H., and Oldham, A.L. 2001. Characterization of the *Plesiomonas shigelloides* genes encoding the heme iron utilization system. J. Bacteriol. 183: 2715-2723.

Hinnebusch, B.J., Perry, R.D., and Schwan, T.G. 1996. Role of the *Yersinia pestis* hemin storage (*hms*) locus in the transmission of plague by fleas. Science 273: 367-370.

Hornung, J.M., Jones, H.A., and Perry, R.D. 1996. The *hmu* locus of *Yersinia pestis* is essential for utilization of free haemin and haem-protein complexes as iron sources. Mol. Microbiol. 20: 725-739.

Jackson, S., and Burrows, T.W. 1956. The virulence-enhancing effect of iron on non-pigmented mutants of virulent strains of *Pasteurella pestis*. Br. J. Exp. Pathol. 37: 577-583.

Jacobi, C.A., Gregor, S., Rakin, A., and Heesemann, J. 2001. Expression analysis of the yersiniabactin receptor gene *fyuA* and the heme receptor *hemR* of *Yersinia enterocolitica* in vitro and in vivo using the reporter genes for green fluorescent protein and luciferase. Infect. Immun. 69: 7772-7782.

Kammler, M., Schön, C., and Hantke, K. 1993. Characterization of the ferrous iron uptake system of *Escherichia coli*. J. Bacteriol. 175: 6212-6219.

Katoh, H., Hagino, N., Grossman, A.R., and Ogawa, T. 2001. Genes essential to iron transport in the cyanobacterium *Synechocystis* sp. strain PCC 6803. J. Bacteriol. 183: 2779-2784.

Keating, T.A., Miller, D.A., and Walsh, C.T. 2000a. Expression, purification, and characterization of HMWP2, a 229 kDa, six domain protein subunit of yersiniabactin synthetase. Biochemistry 39: 4729-4739.

Keating, T.A., Suo, Z., Ehmann, D.E., and Walsh, C.T. 2000b. Selectivity of the yersiniabactin synthetase adenylation domain in the two-step process of amino acid activation and transfer to a holo-carrier protein domain. Biochemistry 39: 2297-2306.

Koebnik, R., Bäumler, A.J., Heesemann, J., Braun, V., and Hantke, K. 1993a. The TonB protein of *Yersinia enterocoltiica* and its interactions with TonB-box proteins. Mol. Gen. Genet. 237: 152-160.

Koebnik, R., Hantke, K., and Braun, V. 1993b. The TonB-dependent ferrichrome receptor FcuA of *Yersinia enterocolitica*: evidence against a strict coevolution of receptor structure and substrate specificity. Mol. Microbiol. 7: 383-393.

Köster, W. 1991. Iron(III) hydroxamate transport across the cytoplasmic membrane of Escherichia coli. Biol. Metals 4: 23-32.

Leipe, D.D., Wolf, Y.I., Koonin, E.V., and Aravind, L. 2002. Classification and evolution of P-loop GTPases and related ATPases. J. Mol. Biol. 317: 41-72.

Létoffé, S., Ghigo, J.M., and Wandersman, C. 1994a. Iron acquisition from heme and hemoglobin by a *Serratia marcescens* extracellular protein. Proc. Natl. Acad. Sci. USA 91: 9876-9880.

Létoffé, S., Ghigo, J.M., and Wandersman, C. 1994b. Secretion of the *Serratia marcescens* HasA protein by an ABC transporter. J. Bacteriol. 176: 5372-5377.

Létoffé, S., Nato, F., Goldberg, M.E., and Wandersman, C. 1999. Interactions of HasA, a bacterial haemophore, with haemoglobin and with its outer membrane receptor HasR. Mol. Microbiol. 33: 546-555.

Lillard, J.W., Jr., Bearden, S.W., Fetherston, J.D., and Perry, R.D. 1999. The haemin storage (Hms^+) phenotype of *Yersinia pestis* is not essential for the pathogenesis of bubonic plague in mammals. Microbiology 145: 197-209.

Locher, K.P., Rees, B., Koebnik, R., Mitschler, A., Moulinier, L., Rosenbusch, J.P., and Moras, D. 1998. Transmembrane signaling across the ligand-gated FhuA receptor: crystal structures of free and ferrichrome-bound states reveal allosteric changes. Cell 95: 771-778.

Lucier, T.S., and Brubaker, R.R. 1992. Determination of genome size, macrorestriction pattern polymorphism, and nonpigmentation-specific deletion in *Yersinia pestis* by pulsed-field gel electrophoresis. J. Bacteriol. 174: 2078-2086.

Lucier, T.S., Fetherston, J.D., Brubaker, R.R., and Perry, R.D. 1996. Iron uptake and iron-repressible polypeptides in *Yersinia pestis*. Infect. Immun. 64: 3023-3031.

Marlovits, T.C., Haase, W., Herrmann, C., Aller, S.G., and Unger, V.M. 2002. The membrane protein FeoB contains an intramolecular G protein essential for Fe(II) uptake in bacteria. PNAS 99: 16243-16248.

Massé, E., and Gottesman, S. 2002. A small RNA regulates the expression of genes involved in iron metabolism in *Escherichiacoli*. PNAS 99: 4620-4625.

Miller, D.A., and Walsh, C.T. 2001. Yersiniabactin synthetase: probing the recognition of carrier protein domains by the catalytic heterocyclization domains, Cy1 and Cy2, in the chain-initiating HMWP2 subunit. Biochemistry 40: 5313-5321.

Miller, D.A., Walsh, C.T., and Luo, L. 2001. C-methyltransferase and cyclization domain activity at the intraprotein PK/NRP switch point of yersiniabactin synthetase. J. Am. Chem. Soc. 123: 8434-8435.

Miller, D.A., Luo, L., Hillson, N., Keating, T.A., and Walsh, C.T. 2002. Yersiniabactin synthetase: a four-protein assembly line producing the nonribosomal peptide/polyketide hybrid siderophore of *Yersinia pestis*. Chem. Biol. 9: 333-344.

Müller, K., Matzanke, B.F., Schünemann, V., Trautwein, A.X., and Hantke, K. 1998. FhuF, an iron-regulated protein of *Escherichia coli* with a new type of [2Fe-2S] center. Eur. J. Biochem. 258: 1001-1008.

Ochsner, U.A., Johnson, Z., and Vasil, M.L. 2000. Genetics and regulation of two distinct haem-uptake systems, *phu* and *has*, in *Pseudomonas aeruginosa*. Microbiology 146: 185-198.

O'Malley, S.M., Mouton, S.L., Occhino, D.A., Deanda, M.T., Rashidi, J.R., Fuson, K.L., Rashidi, C.E., Mora, M.Y., Payne, S.M., and Henderson, D.P. 1999. Comparison of the heme iron utilization systems of pathogenic Vibrios. J. Bacteriol. 181: 3594-3598.

Paquelin, A., Ghigo, J.M., Bertin, S., and Wandersman, C. 2001. Characterization of HasB, a *Serratia marcescens* TonB-like protein specifically involved in the haemophore-dependent haem acquisition system. Mol. Microbiol. 42: 995-1005.

Parkhill, J., Wren, B.W., Thomson, N.R., Titball, R.W., Holden, M.T., Prentice, M.B., Sebaihia, M., James, K.D., Churcher, C., Mungall, K.L., Baker, S., Basham, D., Bentley, S.D., Brooks, K., Cerdeno-Tarraga, A.M., Chillingworth, T., Cronin, A., Davies, R.M., Davis, P., Dougan, G., Feltwell, T., Hamlin, N., Holroyd, S., Jagels, K., Karlyshev, A.V., Leather, S., Moule, S., Oyston, P.C., Quail, M., Rutherford, K., Simmonds, M., Skelton, J., Stevens, K., Whitehead, S., and Barrell, B.G. 2001. Genome sequence of *Yersinia pestis*, the causative agent of plague. Nature 413: 523-527.

Patel, H.M., and Walsh, C.T. 2001. In vitro reconstitution of the *Pseudomonas aeruginosa* nonribosomal peptide synthesis of pyochelin: characterization of backbone tailoring thiazoline reductase and N-methyltransferase activities. Biochemistry 40: 9023-9031.

Pelludat, C., Rakin, A., Jacobi, C.A., Schubert, S., and Heesemann, J. 1998. The yersiniabactin biosynthetic gene cluster of *Yersinia enterocolitica*: organization and siderophore-dependent regulation. J. Bacteriol. 180: 538-546.

Pelludat, C., Hogardt, M., and Heesemann, J. 2002. Transfer of the core region genes of the *Yersinia enterocolitica* WA-C serotype O:8 high-pathogenicity island to *Y. enterocolitica* MRS40, a strain with low levels of pathogenicity, confers a yersiniabactin biosynthesis phenotype and enhanced mouse virulence. Infect. Immun. 70: 1832-1841.

Perry, R.D. 2004. *Yersinia.* In: Iron Transport in Bacteria: Molecular Genetics, Biochemistry, Microbial Pathogenesis and Ecology. J.H. Crosa, A.R. Mey and S.M. Payne, eds. ASM Press, Washington, D.C. p. 219-240.

Perry, R.D., and Brubaker, R.R. 1979. Accumulation of iron by yersiniae. J. Bacteriol. 137: 1290-1298.

Perry, R.D., Pendrak, M.L., and Schuetze, P. 1990. Identification and cloning of a hemin storage locus involved in the pigmentation phenotype of *Yersinia pestis*. J. Bacteriol. 172: 5929-5937.

Perry, R.D., Balbo, P.B., Jones, H.A., Fetherston, J.D., and DeMoll, E. 1999. Yersiniabactin from *Yersinia pestis*: biochemical characterization of the siderophore and its role in iron transport and regulation. Microbiology 145: 1181-1190.

Perry, R.D., Bearden, S.W., and Fetherston, J.D. 2001. Iron and heme acquisition and storage systems of *Yersinia pestis*. Recent Res. Devel. Microbiol. 5: 13-27.

Perry, R.D., Abney, J., Mier, I., Jr., Lee, Y., Bearden, S.W., and Fetherston, J.D. 2003a. Regulation of the *Yersinia pestis* Yfe and Ybt iron transport systems. In: The Genus *Yersinia*: entering the functional genomic era. M. Skurnik, Bengoechea, J.A., and Granfors, K., eds. Kluwer Academic/Plenum Publishers, New York. p. 275-283.

Perry, R.D., Shah, J., Bearden, S.W., Thompson, J.M., and Fetherston, J.D. 2003b. *Yersinia pestis* TonB: role in iron, heme and hemoprotein utilization. Infect. Immun. 71: 4159-4162.

Qian, Y., Lee, J.H., and Holmes, R.K. 2002. Identification of a DtxR-regulated operon that is essential for siderophore-dependent iron uptake in *Corynebacterium diphtheriae*. J. Bacteriol. 184: 4846-4856.

Quadri, L.E.N. 2000. Assembly of aryl-capped siderophores by modular peptide synthetases and polyketide synthases. Mol. Microbiol. 37: 1-12.

Rakin, A., Saken, E., Harmsen, D., and Heesemann, J. 1994. The pesticin receptor of *Yersinia enterocolitica*: a novel virulence factor with dual function. Mol. Microbiol. 13: 253-263.

Rakin, A., and Heesemann, J. 1995. Yersiniabactin/pesticin receptor: a component of an iron uptake system of highly pathogenic *Yersinia*. Contrib. Microbiol. Immunol. 13: 244-247.

Rakin, A., Noelting, C., Schubert, S., and Heesemann, J. 1999. Common and specific characteristics of the high-pathogenicity island of *Yersinia enterocolitica*. Infect. Immun. 67: 5265-5274.

Ratliff, M., Zhu, W., Deshmukh, R., Wilks, A., and Stojiljkovic, I. 2001. Homologues of Neisserial heme oxygenase in gram-negative bacteria: degradation of heme by the product of the *pigA* gene of *Pseudomonas aeruginosa*. J. Bacteriol. 183: 6394-6403.

Reissbrodt, R., and Rabsch, W. 1988. Further differentiation of Enterobacteriaceae by means of siderophore-pattern analysis. Zentralbl. Bakteriol. Mikrobiol. Hyg. [A] 268: 306-317.

Robey, M., and Cianciotto, N.P. 2002. *Legionella pneumophila feoAB* promotes ferrous iron uptake and intracellular infection. Infect. Immun. 70: 5659-5669.

Romalde, J.L., Conchas, R.F., and Toranzo, A.E. 1991. Evidence that *Yersinia ruckeri* possesses a high affinity iron uptake system. FEMS Microbiol. Lett. 64: 121-125.

Rossi, M.-S., Fetherston, J.D., Létoffé, S., Carniel, E., Perry, R.D., and Ghigo, J.-M. 2001. Identification and characterization of the hemophore-dependent heme acquisition system of *Yersinia pestis*. Infect. Immun. 69: 6707-6717.

Runyen-Janecky, L.J., and Payne, S.M. 2002. Identification of chromosomal *Shigella flexneri* genes induced by the eukaryotic intracellular environment. Infect. Immun. 70: 4379-4388.

Runyen-Janecky, L.J., Reeves, S.A., Gonzales, E.G., and Payne, S.M. 2003. Contribution of the *Shigella flexneri* Sit, Iuc, and Feo iron acquisition systems to iron acquisition in vitro and in cultured cells. Infect. Immun. 71: 1919-1928.

Rutz, J.M., Abdullah, T., Singh, S.P., Kalve, V.I., and Klebba, P.E. 1991. Evolution of the ferric enterobactin receptor in gram-negative bacteria. J. Bacteriol. 173: 5964-5974.

Saier, M.H., Jr. 1999. A functional-phylogenetic system for the classification of transport proteins. J. Cell. Biochem. Suppl. 32/33: 84-94.

Saken, E., Rakin, A., and Heesemann, J. 2000. Molecular characterization of a novel siderophore-independent iron transport system in *Yersinia*. Intl. J. Med. Microbiol. 290: 51-60.

Schmitt, M.P. 1997. Utilization of host iron sources by *Corynebacterium diphtheriae*: identification of a gene whose product is homologous to eukaryotic heme oxygenases and is required for acquisition of iron from heme and hemoglobin. J. Bacteriol. 179: 838-845.

Schubert, S., Fischer, D., and Heesemann, J. 1999. Ferric enterochelin transport in *Yersinia enterocolitica*: molecular and evolutionary aspects. J. Bacteriol. 181: 6387-6395.

Sikkema, D.J., and Brubaker, R.R. 1989. Outer membrane peptides of *Yersinia pestis* mediating siderophore-independent assimilation of iron. Biol. Metals 2: 174-184.

Smoot, L.M., Bell, E.C., Crosa, J.H., and Actis, L.A. 1999. Fur and iron transport proteins in the Brazilian purpuric fever clone of *Haemophilus influenzae* biogroup aegyptius. J. Med. Microbiol. 48: 629-636.

Staggs, T.M., and Perry, R.D. 1991. Identification and cloning of a *fur* regulatory gene in *Yersinia pestis*. J. Bacteriol. 173: 417-425.

Staggs, T.M., and Perry, R.D. 1992. Fur regulation in *Yersinia* species. Mol. Microbiol. 6: 2507-2515.

Staggs, T.M., Fetherston, J.D., and Perry, R.D. 1994. Pleiotropic effects of a *Yersinia pestis fur* mutation. J. Bacteriol. 176: 7614-7624.

Stojiljkovic, I., and Hantke, K. 1992. Hemin uptake system of *Yersinia enterocolitica*: similarities with other TonB-dependent systems in Gram-negative bacteria. EMBO J. 11: 4359-4367.

Stojiljkovic, I., Cobeljic, M., and Hantke, K. 1993. *Escherichia coli* K-12 ferrous iron uptake mutants are impaired in their ability to colonize the mouse intestine. FEMS Microbiol. Lett. 108: 111-116.

Stojiljkovic, I., and Hantke, K. 1994. Transport of haemin across the cytoplasmic membrane through a haemin-specific periplasmic binding-protein-dependent transport system in *Yersinia enterocolitica*. Mol. Microbiol. 13: 719-732.

Stojiljkovic, I., and Perkins-Balding, D. 2002. Processing of heme and heme-containing proteins by bacteria. DNA Cell. Biol. 21: 281-295.

Stuart, S.J., Prpic, J.K., and Robins-Browne, R.M. 1986. Production of aerobactin by some species of the genus *Yersinia*. J. Bacteriol. 166: 1131-1133.

Suo, Z., Walsh, C.T., and Miller, D.A. 1999. Tandem heterocyclization activity of the multidomain 230 kDa HMWP2 subunit of *Yersinia pestis* yersiniabactin synthetase: interaction of the 1-1382 and 1383-2035 fragments. Biochemistry 38: 14023-14035.

Suo, Z., Chen, H., and Walsh, C.T. 2000. Acyl-CoA hydrolysis by the high molecular weight protein 1 subunit of yersiniabactin synthetase: Mutational evidence for a cascade of four acyl-enzyme intermediates during hydrolytic editing. PNAS 97: 14188-14193.

Suo, Z., Tseng, C.C., and Walsh, C.T. 2001. Purification, priming, and catalytic acylation of carrier protein domains in the polyketide synthase and nonribosomal peptidyl synthetase modules of the HMWP1 subunit of yersiniabactin synthetase. PNAS 98: 99-104.

Thompson, J.M., Jones, H.A., and Perry, R.D. 1999. Molecular characterization of the hemin uptake locus (*hmu*) from *Yersinia pestis* and analysis of *hmu* mutants for hemin and hemoprotein utilization. Infect. Immun. 67: 3879-3892.

Torres, A.G., and Payne, S.M. 1997. Haem iron-transport system in enterohaemorrhagic *Escherichia coli* O157: H7. Mol. Microbiol. 23: 825-833.

Tsolis, R.M., Bäumler, A.J., Heffron, F., and Stojiljkovic, I. 1996. Contribution of TonB- and Feo-mediated iron uptake to growth of *Salmonella typhimurium* in the mouse. Infect. Immun. 64: 4549-4556.

Une, T., and Brubaker, R.R. 1984. In vivo comparison of avirulent Vwa$^-$ and Pgm$^-$ or Pst^r phenotypes of yersiniae. Infect. Immun. 43: 895-900.

Velayudhan, J., Hughes, N.J., McColm, A.A., Bagshaw, J., Clayton, C.L., Andrews, S.C., and Kelly, D.J. 2000. Iron acquisition and virulence in *Helicobacter pylori*: a major role for FeoB, a high-affinity ferrous iron transporter. Mol. Microbiol. 37: 274-286.

Walter, M.A., Potter, S.A., and Crosa, J.H. 1983. Iron uptake system mediated by *Vibrio anguillarum* plasmid pJM1. J. Bacteriol. 156: 880-887.

Wilks, A. 2001. The ShuS Protein of *Shigella dysenteriae* is a heme-sequestering protein that also binds DNA. Arch. Biochem. Biophys. 387: 137-142.

Wyckoff, E.E., Duncan, D., Torres, A.G., Mills, M., Maase, K., and Payne, S.M. 1998. Structure of the *Shigella dysenteriae* haem transport locus and its phylogenetic distribution in enteric bacteria. Mol. Microbiol. 28: 1139-1152.

Zhou, D., Hardt, W.-D., and Galán, J.E. 1999. *Salmonella typhimurium* encodes a putative iron transport system within the centisome 63 pathogenicity island. Infect. Immun. 67: 1974-1981.

Zhu, W., Wilks, A., and Stojiljkovic, I. 2000. Degradation of heme in gram-negative bacteria: the product of the *hemO* Gene of Neisseriae is a heme oxygenase. J. Bacteriol. 182: 6783-6790.

Chapter 14

The High-Pathogenicity Island: A Broad-Host-Range Pathogenicity Island

Biliana Lesic and Elisabeth Carniel

ABSTRACT

Highly pathogenic *Yersinia* (*Y. enterocolitica* 1B, *Y. pseudotuberculosis* and *Y. pestis*) harbor a pathogenicity island termed the High-Pathogenicity Island (HPI). The *Yersinia* HPI carries genes involved in the biosynthesis, transport and regulation of the siderophore yersiniabactin and can thus be regarded as an iron-uptake island. Its presence confers to the bacterium the ability to disseminate in its host and cause a systemic infection. The *Yersinia* HPI has kept the potential to be mobile within the chromosome of *Y. pseudotuberculosis*. A unique characteristic of this island is its wide distribution among different enterobacteria genera such as *E. coli*, *Klebsiella*, *Citrobacter*, *Enterobacter*, *Serratia* and *Salmonella*. In *E. coli*, the island is prevalent in extraintestinal isolates and is associated with a higher virulence. A second and degenerate HPI is present on the chromosomes of *Y. pestis* and *Y. pseudotuberculosis*. HPI-like elements have also been recently identified in the insect pathogen *Photorhabdus luminescens* and the Gram-positive species *Corynebacterium diphtheriae*. These HPI-like elements potentially encode iron-uptake systems, but whether they are still functional remains to be determined.

INTRODUCTION

Among the 11 species included in the genus *Yersinia*, three are pathogenic for humans and other animals: *Y. pestis*, *Y. pseudotuberculosis,* and *Y. enterocolitica* (except biotype 1A). All pathogenic species harbor a conserved 70 kb plasmid designated pYV (for plasmid associated with *Yersinia* virulence) which is essential for their virulence phenotype (see chapter 16 for more details about this plasmid). Pathogenic *Yersinia* can be further subdivided into low-pathogenicity strains, i.e. strains that induce a mild intestinal infection in humans and are non-lethal for mice at low doses, and high-pathogenicity strains which cause severe systemic infections in humans and are mouse-lethal at low doses.

One of the major differences between low- and high-pathogenicity *Yersinia* lies in their ability to capture the iron molecules necessary for their systemic dissemination in the host. This capacity is mediated by a high-affinity iron-chelating system (i.e. siderophore) called yersiniabactin (Heesemann, 1987). The machinery of yersiniabactin biosynthesis, transport and regulation is located on a large chromosomal region which has the characteristics of a pathogenicity island (PAI). The concept of PAI was coined by J. Hacker (Hacker et al., 1997) to describe a large chromosomal segment which carries pathogenicity as well as mobility genes and is characteristically bordered at one extremity by a *tRNA* gene. PAIs are sometimes unstable and their GC content is usually different from that of the host chromosome, suggesting a horizontal mode of acquisition and transfer. The chromosomal segment which carries the yersiniabactin locus fulfills these criteria: (i) it is a large chromosomal DNA fragment (36 to 43 kb); (ii) it carries genes essential for bacterial dissemination in the host; (iii) it incorporates several mobility genes such as insertion sequences (IS) and a bacteriophage P4-like integrase gene; (iv) it is bordered on one side by an *asn* tRNA locus; (v) its GC content (56.4%) is much higher than that of the core genome (47.6%); and (vi) this region is able in some strains to excise from the host chromosome. Since the presence of this specific entity correlates with the expression of a high-pathogenicity phenotype, this PAI was termed "High-Pathogenicity Island" or HPI (Carniel et al., 1996).

DISTRIBUTION OF THE HPI WITHIN THE GENUS *YERSINIA*

The HPI is never present in the avirulent species *Y. intermedia, Y. kristensenii, Y. frederiksenii, Y. mollaretii, Y. bercovieri,* and *Y. aldovae* (De Almeida et al., 1993).

In *Y. enterocolitica,* the island is never found in the non-virulent biotype 1A or in the low-pathogenicity strains of biotypes 2 to 5. In contrast, it is systematically present in all highly pathogenic biotype 1B strains (De Almeida et al., 1993).

In *Y. pseudotuberculosis*, an initial study performed on European isolates indicated that the HPI is always absent from strains of serotypes II, IV, and V; while a complete island is found in some (but not all) strains of serotype I, and a HPI truncated of its 9-kb left-hand part is present in some serotype III strains (De Almeida et al., 1993; Rakin et al., 1995; Buchrieser et al., 1998b). A recent study confirmed the distribution of the HPI among European isolates and further showed that the island may also be present in Asian isolates of serotypes O:5, O:13 and O:15 (Fukushima et al., 2001).

The HPI is systematically present in freshly isolated strains of *Y. pestis* (De Almeida et al., 1994) but it may be lost at high frequencies upon culture *in vitro* (Fetherston et al., 1992; De Almeida et al., 1993; Buchrieser et al., 1998a; Hare and McDonough, 1999).

GENETIC ORGANIZATION OF THE HPI IN THE THREE PATHOGENIC *YERSINIA* SPECIES

The size of the HPI varies depending on the species. The island is 36.1 kb long in *Y. pseudotuberculosis* I, 36.4 kb in *Y. pestis,* and 43.1 kb in *Y. enterocolitica* 1B. The island may be divided into three distinct parts: the highly conserved ≈29 kb right-hand part (termed the yersiniabactin (*ybt*) locus), the variable 5.6-12.5 kb left-hand part, and the extremities of the HPI (Figure 1). The main characteristics of the HPI in the three highly pathogenic yersiniae are summarized in Table 1.

THE YERSINIABACTIN LOCUS

The locus involved in yersiniabactin-mediated iron-uptake is composed of 11 genes organized in four operons (Figure 1). Their functions are not yet entirely elucidated but recent biochemical and biological investigations have provided a better understanding of the respective role of each gene. Since a detailed description of these functions is presented in chapter 13 of this book, only a brief overview of the function of the yersiniabactin system will be given here. The *ybt* locus can be roughly divided into three functional parts involved in yersiniabactin biosynthesis, uptake, and regulation. The *Y. pestis* nomenclature for the HPI-borne genes is used here and the name of the *Y. enterocolitica* 1B homologs is given in Figure 1.

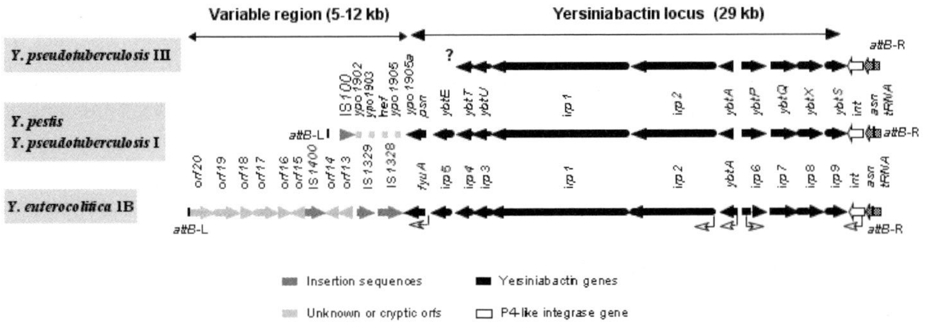

Figure 1. Genetic organizations of the HPI in highly pathogenic *Yersinia*. Arrows below the genetic maps indicate the start and direction of transcription.

Yersiniabactin Biosynthesis. Formation of yersiniabactin is governed by at least 6 HPI-borne genes (the operon *irp21ybtUTE* and *ybtS*), and one gene (*ybtD*) located outside the HPI. The siderophore is formed by a mechanism of non-ribosomal peptide synthesis which involves multifunctional and multidomain proteins, following a thiotemplating process (Chapter 13 of this volume). Yersiniabactin is a 482 dal molecule that belongs to a small sub-group of phenolate siderophores and has a much higher affinity for ferric iron (KD = 4 x 10-36) than for ferrous iron (Gehring et al., 1998; Perry et al., 1999). Its structure is identical in the three *Yersinia* species and is closely related to those of pyochelin and anguibactin produced by *Pseudomonas aeruginosa* and *Vibrio anguillarum*, respectively (Perry et al., 1999).

Table 1. Main characteristics of the HPI of the three pathogenic *Yersinia* species.

	Y. pestis	*Y. pseudotuberculosis* I	*Y. enterocolitica*
Size of the HPI	36.38 kb	36.10 kb	43.15 kb
Yersiniabactin locus			
Size	28.83 kb	28.83 kb	28.90 kb
Range of nt identity[1]	Pst[2]: 100% Ent[4]: 97-98%	Pestis[3]: 99-100% Ent: 97-98%	Pestis: 97-98% PST: 97-98%
ERIC sequence[5]	No	No	Yes
Left-hand part			
Size	5.91 kb	5.65 kb	12.56 kb
IS elements	IS*100*	IS*100*	IS*1328*, IS*1329*, IS400, IS*1222*'
ypo1905	Truncated	Intact	Absent
Extremities			
attB-R	Intact	Intact	Intact
attB-L	Intact	Intact	Intact
int	Intact	Intact or mutated	Mutated
Insertion site	*asn1*	*asn1*, *asn2*, or *asn3*	*asn2*

[1]nucleotide identity for each orf constituting the *ybt* locus.
[2]*Y. pseudotuberculosis*
[3]*Y. pestis*
[4]*Y. enterocolitica*
[5]125 nt sequence located upstream of *ybtA*

Yersiniabactin Uptake. Yersiniabactin-mediated internalization of iron involves three genes (*psn, ybtP* and *ybtQ*). Psn is the outer membrane receptor for both the siderophore-iron complex and the bacteriocin pesticin (Heesemann et al., 1993; Rakin et al., 1994; Fetherston et al., 1995). This receptor acts in concert with TonB to transport the Ybt-Fe complex across the outer membrane. No obvious HPI-encoded candidate for a periplasmic-binding protein has yet been identified. YbtP and YbtQ act as inner-membrane permeases to translocate either the Ybt-Fe complex or Fe alone into the bacterial cytosol (Fetherston et al., 1999).

Yersiniabactin Regulation. One regulator, YbtA, is encoded by the HPI (Fetherston et al., 1996). This regulator, which belongs to the AraC family of transcriptional regulators, activates expression from the *psn, irp2* and *ybtP* promoters but represses expression of its own promoter. In addition to YbtA, yersiniabactin biosynthesis is controlled by the Fur repressor and is upregulated by yersiniabactin itself (Fetherston et al., 1995; Pelludat et al., 1998; Perry et al., 1999).

The Yersiniabactin locus is extremely conserved among highly pathogenic *Yersinia* (Table 1). In the two strains of *Y. pestis* (CO92 and KIM10) whose genomes have been sequenced (Parkhill et al., 2001; Deng et al., 2002), the nucleotide identity of the *ybt* genes is 100%, with one exception, *ybtP*, which displays one nucleotide (nt) difference. The HPI of *Y. pestis* and *Y. pseudotuberculosis* IP32953 (sequence available at greengenes.llnl.gov/bbrp/html/microbe.html) are also highly conserved. Four out of the 11 genes composing the *ybt* locus are 100% identical (*ybtQ, X, T, S*), 5 genes display 1 nt difference (*psn, irp1, ybtE, U, A*), and two genes differ by 2 nt (*irp2, ybtP*). As expected, the *ybt* locus of *Y. enterocolitica* Ye8081 (sequence available at www.sanger.ac.uk/Projects/Y_enterocolitica/) is the most divergent with nt identities of 97 - 98%. In addition to minor nt variations in the *ybt* genes, the yersiniabactin locus of *Y. enterocolitica* differs from those of *Y. pestis* and *Y. pseudotuberculosis* by the presence of a 125 bp ERIC sequence upstream of the *ybtA* promoter (Rakin et al., 1999). This additional sequence may have some regulatory function on yersiniabactin production.

THE VARIABLE LEFT-HAND PART OF THE HPI

In contrast to the highly conserved yersiniabactin locus, the left-hand portion of the HPI is much more polymorphic among pathogenic *Yersinia* spp.

In *Y. pestis*, the segment located downstream of *psn* is 5.9 kb-long (Figure 1). It carries several short open reading frames (orfs) which share homologies with phage genes (Buchrieser et al., 1999; Parkhill et al., 2001; Deng et al., 2002), and an IS*100* insertion sequence (Fetherston and Perry, 1994; Podladchikova et al., 1994; Prentice and Carniel, 1995; McDonough and Hare, 1997).

In *Y. pseudotuberculosis* serotype I, the region between *psn* and the IS*100* locus (Figure 1) is almost 100% identical to that of *Y. pestis,* except for the presence of one additional nt in the ypo1905 orf of *Y. pestis* which induces a frame shift in the coding sequence. Another difference between the two islands is noticeable in the small chromosomal segment located between the left-hand border of the HPI and IS*100*. This region is 266 bp longer in *Y. pestis* because of the presence of the remnants of an IS*630*-like element (Hare et al., 1999).

In *Y. pseudotuberculosis* serotype III (Figure 1), the entire left-hand region of the HPI as well as the adjacent *psn* and *ybtE* genes are missing (Buchrieser et al., 1998b). The nature and organization of the genes located at this part of the HPI in serotype III strains are unknown.

In *Y. enterocolitica* 1B (Figure 1), the left-hand part of the HPI extends 12.5 kb downstream of the *psn* gene (Carniel et al., 1996; Rakin et al., 1999). It contains a cluster of four IS elements (IS*1328*, IS*1329*, a portion of IS*1222* interrupted by IS*1329*, and IS*1400*) that all belong to the IS*3* family of insertion sequences (Rakin and Heesemann, 1995; Carniel et al., 1996; Rakin et al., 2000). Interestingly, these IS are inserted in an AT rich region of the HPI which might represent a hot spot of integration for this class of elements. In addition, the left-hand part of the *Y. enterocolitica* HPI contains 7 orfs, of which 4 (orf15-18) have the capacity to code for products with some similarity to hypothetical proteins (YfjK and YfjL) of *E. coli* (Rakin et al., 1999). Since this region is not well conserved among various *Y. enterocolitica* 1B isolates (Carniel et al., 1996), these genes may vary depending on the isolates.

THE EXTREMITIES OF THE HPI

The borders of the HPI are defined by a 17-bp sequence (5'-CCAGTCAGAGGAGCCAA-3') located at each extremity of the island. These flanking repeats are homologous to the *attP* site of bacteriophage P4 (Buchrieser et al., 1998b), and may thus be considered as *attB*-right *(attB*-R) and *attB*-left *(attB*-L) sites.

attB-R is entirely contained within the *asn* tRNA locus sequence (Buchrieser et al., 1998b; Hare et al., 1999; Rakin et al., 1999). tRNA loci are known to be preferential sites for phage integration. Adjacent to *attB*-R, and inside the HPI (Figure 1), there is a gene (*int*) that is homologous to the integrase gene of bacteriophage P4 (Buchrieser et al., 1998b; Bach et al., 1999; Hare et al., 1999; Rakin et al., 1999). This gene is highly conserved (≥98% identity) among the three pathogenic *Yersinia* species.

attB-L is present and perfectly conserved in *Y. pestis* and *Y. pseudotuberculosis* I. Although this site was initially found to be degenerate in *Y. enterocolitica* Ye8081 (Bach et al., 1999), an intact *attB* site was reported in strain WA (Rakin et al., 1999), and the availability of the entire genome sequence of strain Ye8081 revealed that the degenerate *attB-L* site initially identified is actually 1394 bp downstream of an intact *attB*-L site.

The chromosomal regions flanking the HPI are conserved among *Y. enterocolitica* 1B isolates because the island is always inserted into the same *asn* tRNA locus in this species (Carniel et al., 1996; Bach et al., 1999; Rakin et al., 1999). This also holds true for the *Y. pestis* HPI which is systematically inserted into a single *asn* tRNA locus, which is different from that of the *Y. enterocolitica* HPI (Hare et al., 1999). In contrast, the *Y. pseudotuberculosis* HPI can be found inserted into any of the three copies of the *asn* tRNA locus present on the *Yersinia* chromosome (Buchrieser et al., 1998b).

HPI AND VIRULENCE

Iron acquisition is an essential requirement for the majority of microorganisms. In mammals, this metal is bound to eukaryotic proteins (hemoglobin, ferritin, transferrin and lactoferrin) which maintain a level of free iron that is far too low (10-18 M) to sustain bacterial growth. Numerous clinical reports have shown that low-pathogenicity strains of *Y. enterocolitica* (biotypes 2 and 4), which are usually responsible for moderate intestinal symptoms, can cause systemic infections in patients with iron-overload (thalassemia, hemochromatosis, etc.) (Robins-Browne et al., 1979; Boelaert et al., 1987; Vadillo et al., 1994). Moreover, these strains can become lethal for mice at low doses if an exogenous siderophore (Desferal) is administered to the animals (Robins-Browne and Kaya Prpic,

1985). Thus, the availability of free iron, provided exogenously or by Desferal treatment, enables low-pathogenicity *Y. enterocolitica* to multiply in the host and to cause systemic infections.

High-pathogenicity strains of *Yersinia* possess the HPI-borne yersiniabactin system which is an intrinsic and potent iron-chelating system (Heesemann, 1987). Because of its high affinity for iron, this siderophore can solubilize the metal bound to host proteins and transport it back to the bacteria. Several pieces of information suggest that the yersiniabactin system is important for the systemic dissemination of the bacteria in their host: (i) strains of *Y. pestis* spontaneously deleted of a 102 kb chromosomal segment (designated the *pgm* locus) which encompasses the entire HPI had a reduced virulence for mice (Jackson and Burrows, 1956; Une and Brubaker, 1984; Carniel et al., 1991; Iteman et al., 1993); (ii) these attenuated strains could regain virulence if iron was injected into animals prior to bacterial challenge (Jackson and Burrows, 1956); (iii) the presence of HPI-specific genes or products perfectly correlated with the level of pathogenicity of natural isolates of *Yersinia* spp. (Carniel et al., 1987; Heesemann, 1987; Carniel et al., 1989; De Almeida et al., 1993; Heesemann et al., 1993; Chambers and Sokol, 1994; Rakin et al., 1995); and (iv) abolition of yersiniabactin production by random Tn*5* insertions altered the pathogenicity of *Y. enterocolitica* 1B strains (Heesemann, 1987).

Definitive proof for a role of the HPI in *Yersinia* virulence was obtained by specifically mutating HPI-borne genes (*irp2*, *fyuA/psn* or *ybtP*) in highly pathogenic *Yersinia* (Carniel et al., 1992; Rakin et al., 1994; Bearden et al., 1997; Bearden and Perry, 1999; Fetherston et al., 1999; Gong et al., 2001). The 50% lethal dose (LD_{50}) for mice of the various mutants was systematically increased (by a factor usually ≥100) in the three pathogenic *Yersinia* species, whatever the HPI-borne gene mutated (Table 2). Furthermore, the number of bacteria recovered two days post-infection from the spleens of mice infected intravenously with an *irp6-7* mutant of *Y. enterocolitica* WA-C (a strain cured of the virulence plasmid) was 10^5-fold lower than that of the parental strain. In contrast, an *irp8* mutant was able to grow as well as the parental WA-C strain (Brem et al., 2001). The role of the HPI in *Yersinia* pathogenicity is thus now clearly established.

However, this did not prove that the presence of the island is sufficient to account for the difference in the level of virulence between low- and high-pathogenicity *Y. enterocolitica* strains. Recently, the entire cloned *ybt* locus of *Y. enterocolitica* WA was introduced into a non-pathogenic biotype 1A and a low-pathogenicity biotype 2 *Y. enterocolitica* strains

Table 2. Virulence (LD_{50}) for mice of *Yersinia* strains mutated in various HPI-borne genes.

Species	Strain	Mutated gene	Route of infection	LD 50		Reference
				Wild type	Mutant	
Y. pseudotuberculosis	IP2790	*irp2*	sc	3.5×10^6	$> 2.1 \times 10^9$	Carniel et al., 1992
		irp2	iv	6.9×10^0	1.9×10^3	Carniel et al., 1992
		irp2	ig	7.6×10^7	2.7×10^9	Carniel et al., 1992
Y. enterocolitica	WA	*fyuA*	iv	5×10^4	$> 1 \times 10^6$	Rakin et al., 1994
Y. pestis	KIM5-2053.11+ (Δ*psa*, *yopJ*-)	*psn*	sc	1.3×10^2	$> 2.9 \times 10^5$	Bearden et al., 1997
		irp2	sc	1.3×10^2	$> 1.3 \times 10^6$	Bearden et al., 1997
		irp2	sc	1.1×10^2	$> 1.3 \times 10^7$	Bearden and Perry, 1999
		ybtP	sc	1.1×10^2	$> 7.6 \times 10^4$	Fetherston et al., 1999

sc: subcutaneous, iv: intravenous, ig: intragastric.

(Pelludat et al., 2002). The recombinant non-pathogenic strain produced undetectable amounts of yersiniabactin, suggesting that factors outside the HPI are necessary for the synthesis of the siderophore and are missing in this group of strains. Upon intraperitoneal infection, this biotype 1A strain had no enhanced capacity to multiply in the spleen or liver of the infected mice. This result was expected since biotype 1A strains do not harbor the pYV virulence plasmid which is essential for *Yersinia* virulence. In contrast, when the Ybt locus was introduced into the chromosome of a biotype 2 low-pathogenicity *Y. enterocolitica*, this strain became able to produce yersiniabactin and to grow in the organs of infected mice, leading to a number of bacteria 1000-fold higher than that of the parental strain (Pelludat et al., 2002). These data demonstrate that the presence of the HPI in a low-pathogenicity genetic background is sufficient to increase the virulence of the recipient strain.

Using translational gene fusion, the expression of *fyuA* was monitored in various organs of mice 24h post-infection with a *Y. enterocolitica* 1B strain (Jacobi et al., 2001). Curiously, *fyuA* expression was found to be high in the peritoneal cavity, moderate in the spleen, and low in the Peyer's patches and liver of infected animals. Therefore, although the role of the HPI in promoting a higher level of virulence in *Yersinia* is now clearly established, the mechanisms underlying the enhanced capacity to cause bacterial dissemination in the host remain to be elucidated.

Table 3. Putative products of the *Y. pestis* CO92 HPI-like element.

HPI-like product	Size (aa)	aa identity (length)	Homologs	Organisms	Putative function
YPO0770	586	30.8% (582 aa) 33.7% (594 aa)	Irp6 YbtQ	*Y. enterocolitica* *Y. pestis*	ABC-transporter transmembrane protein
YPO0771	591	32.4% (574 aa) 32.6% (574 aa)	Irp6 YbtP	*Y. enterocolitica* *Y. pestis*	ABC-transporter transmembrane protein
YPO0772	435	27.9% (416 aa) 27.1% (421 aa)	Irp8 YbtX	*Y. enterocolitica* *Y. pestis*	Putative membrane protein
YPO0773	255	38.6% (236 aa) 36.4% (225 aa)	PchC Orf1 (C-terminus)	*Pseudomonas aeruginosa* *Streptomyces griseus*	Pyochelin biosynthetic protein Putative thioesterase
YPO0774	376	33.7% (362 aa) 33.8% (364 aa)	Irp3 YbtU	*Y. enterocolitica* *Y. pestis*	Conserved hypothetical protein
YPO0775	359	23.3% (339 aa) 27.8% (209 aa)	Dtxr/Irp2 PA4362 (N-terminus)	*Corynebacterium diphtheriae* *Pseudomonas aeruginosa*	Conserved hypothetical protein
YPO0776	1940	31.1% (1659 aa) 34.1% (1957 aa)	HMWP2 PchF	*Y. pestis* *Pseudomonas*	Ybt biosynthesis protein Pyochelin synthetase
YPO0777	551	45.1% (439 aa) 40.7% (553 aa)	HMWP2 (internal region) AngR (C-terminus)	*Y. pestis* *Vibrio anguillarum*	Ybt biosynthesis protein Anguibactin biosynthesis protein
YPO0778	2202	36.8% (2031 aa) 32% (1435 aa)	HMWP2 MtaD (C-terminus)	*Y. pestis* *Stigmatella aurantiaca*	Ybt biosynthesis protein Myxothiazol biosynthetic protein

MOBILITY OF THE HPI

INTEGRATION OF THE HPI INTO THE *YERSINIA* CHROMOSOME

The 3' end of each chromosomal *asn* tRNA locus, present in three copies on the *Yersinia* chromosome, contains a 17-bp sequence homologous to the *attP* site of bacteriophage P4 (Buchrieser et al., 1998b; Rakin et al., 2001). Integration of the HPI into the bacterial chromosome occurs at this 17-bp *attB* site and results in a duplication of *attB* at each extremity of the island. In *Y. pseudotuberculosis,* the HPI can insert into any of the three chromosomal copies (Buchrieser et al., 1998b), while it is inserted specifically into the *asn2* locus in *Y. enterocolitica* (Carniel et al., 1996; Rakin et al., 1999), and into the *asn1* locus in *Y. pestis* (Hare et al., 1999) (Table 1). Adjacent to the *asn* locus, the HPI carries an integrase gene (*int*) which is also homologous to that of bacteriophage P4 (Figure 1). Recently, it was shown that a suicide plasmid containing the cloned *int* gene and a 266-bp DNA segment containing *attP* and the sequences surrounding the HPI, could insert into any of the chromosomal *asn* tRNA loci of *E. coli* or *Y. pestis* (Rakin et al., 2001). This mechanism is RecA-independent. Thus the presence of intact *int* and *att* loci is sufficient to promote the integration of the HPI into the bacterial chromosome.

EXCISION OF THE HPI FROM THE YERSINIA CHROMOSOME

The *int* and *att* loci are also involved in the precise excision of the HPI from the bacterial chromosome. Upon HPI excision by site specific recombination between *attB*-L and *attB*-R, a unique *attB* sequence is generated at the junction site and an intact *asn* tRNA locus is restored (Buchrieser et al., 1998b). The role of *int* in HPI excision was recently demonstrated by showing that inactivation of *int* abolishes the excision of the island (Lesic et al., 2004). However, *int* and *att* are not sufficient for HPI excision. A third HPI-borne gene, designated *hef* (for HPI excision factor), is also required (Lesic et al., 2004). Hef belongs to a family of Recombination Directionality Factors (RDF; Lewis and Hatfull, 2001), and likely plays an architectural rather than a catalytic role. This factor is predicted to promote HPI excision from the chromosome by driving the function of the integrase towards an excisionase activity. The ability of the HPI to excise from the bacterial chromosome differs among the three pathogenic *Yersinia* species.

In *Y. enterocolitica* 1B, no strain precisely deleted of the island has ever been detected (De Almeida et al., 1993). Absence of HPI excision has at least two causes: (i) a nt substitution at position 415 in the HPI-borne *int* gene leads to a product truncated of the last 282 amino acids (Bach et al., 1999; Rakin et al., 1999); and (ii) no *hef* homolog is present on the HPI of this species (Lesic et al., 2004). These mutations may be considered as a process of HPI stabilization in the chromosome of *Y. enterocolitica.* HPI-deleted *Y. enterocolitica* mutants were observed in vitro, with a frequency of approximately $5x10^{-8}$, but the deletion was not limited to the HPI and encompassed a much larger chromosomal fragment of ≈140 kb (Bach et al., 1999).

Similarly, the excision of the *Y. pestis* HPI is not precise but occurs as part of a much larger chromosomal deletion of a 102 kb DNA region designated "*pgm* locus", which encompasses most of the HPI and extends further rightward over a ≈68 kb region (Fetherston et al., 1992). Spontaneous excision of the *pgm* locus occurs at very high frequencies ($2x10^{-3}$) by homologous recombination between two IS*100* flanking copies (Fetherston and Perry, 1994; Hare and McDonough, 1999). The HPI of *Y. pestis* has the virtual potential to excise precisely from the chromosome since it possesses intact copies of *attB* at each border, and potentially functional *int* and *hef* genes (Buchrieser et al., 1998b; Hare et al., 1999). Most likely, excision of the *Y. pestis* HPI may occur but is masked by the much higher deletion frequency of the 102 kb *pgm* locus.

In *Y. pseudotuberculosis* serotype I, spontaneous and precise excision of the HPI is observed in some strains, with a frequency between $6x10^{-5}$ and $8x10^{-7}$ (Lesic et al., 2004). These strains possess functional *int, attB* and *hef* loci. Most of the *Y. pseudotuberculosis* isolates with a non-excisable HPI exhibit a 5-bp deletion at position 602 in *int*, leading to a product truncated of the last 211 amino acids and to stabilization of the island (Lesic et al., 2004). In one strain, lack of HPI excision was due to a defect in *hef* expression (Lesic et al., 2004). Upon excision from the chromosome, the HPI is able to form a circular episome (Lesic et al., 2004). Remarkably, in individual colonies of the same *Y. pseudotuberculosis* isolate, the HPI is found inserted into different *asn* tRNA loci (Buchrieser et al., 1998a). This indicates that this element is able to excise from the *Y. pseudotuberculosis* chromosome, to form an episomal molecule, and to reinsert into either the same or another location on the bacterial chromosome. The *Y. pseudotuberculosis* HPI has thus retained intragenomic mobility.

AN HPI-LIKE ELEMENT IS PRESENT ON THE *Y. PESTIS* AND *Y. PSEUDOTUBERCULOSIS* CHROMOSOMES

The availability of the entire genome sequence of two *Y. pestis* strains (Parkhill et al., 2001; Deng et al., 2002) has allowed the identification of a second locus (designated the HPI-like element here) which contains paralogs of several HPI-borne genes belonging to the *ybt* locus (*ybtU, irp2, ybtP, ybtQ* and *ybtX*). However, not all HPI-borne *ybt* genes are present on the HPI-like element, and its genetic organization is quite different from that of the HPI (Figure 2). Portions of *irp2* are found in three adjacent orfs (*ypo0776-0778*) with different sizes. The product of *ypo0776* is homologous to HMWP2 (the *irp2* product) for most of its length, except at the C-terminus of the protein. The short YPO0777 product corresponds to an internal portion of HMWP2, while the YPO0778 protein is homologous for most of its length to HMWP2. The amino acid identity between the products of the HPI and HPI-like paralogs is around 30% (Table 3), but no identity is detectable at the nucleotide level. The HPI-like element also contains three genes (*ypo0773, ypo0775* and *ypo0776*) that are

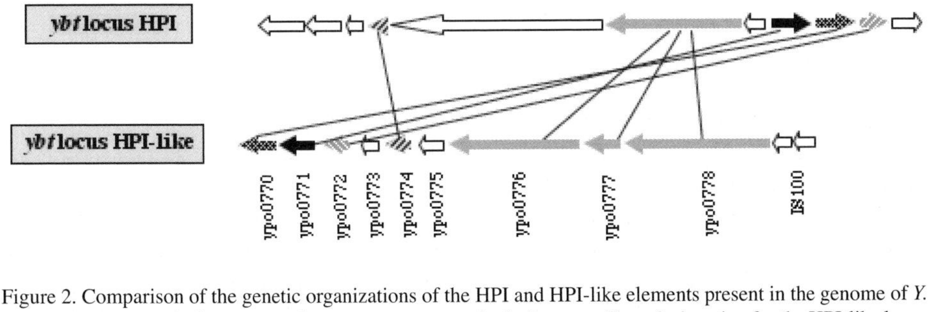

Figure 2. Comparison of the genetic organizations of the HPI and HPI-like elements present in the genome of *Y. pestis*. Blank arrows indicate genes that are not present on both elements. Gene designation for the HPI-like locus are from the *Y. pestis* CO92 genome sequence.

not homologous to HPI genes but are homologous to other iron-uptake genes (*pchC* and *pchF* from *Pseudomonas aeruginosa* involved in pyochelin biosynthesis, and *dtXR/irp2,* an iron-regulated gene from *Corynebacterium diphtheriae*). A more detailed analysis of their putative function is provided in chapter 13 of this book. Based on the similarity data, it may be predicted that the HPI-like locus encodes an iron-acquisition system. However, whether this system is functional remains to be determined. A similar locus is present in the genome of *Y. pseudotuberculosis* but contains a longer *ypo0778* homolog. The shorter size of the *Y. pestis ypo0778* gene appears to be the result of the insertion of an IS*100* element, probably followed by genetic rearrangements. No HPI-like element is present in the genome of *Y. enterocolitica* Ye8081.

A search in the vicinity of the HPI-like element for tRNA loci, *att* sites, integrase genes, or phage-like loci, as well as for genes homologous to those present on the left-hand part of the HPI, was negative. Moreover, the GC% of the HPI-like element (49.2%) is closer to that of the *Y. pestis* chromosome (47.6%) than to that of the HPI (56.4%). Therefore, this element has none of the characteristics that define a PAI. The possibility that the HPI-like element corresponds to a duplication and subsequent rearrangement of a portion of the HPI-borne *ybt* locus is highly unlikely. The comparative analysis of the two loci rather suggests that the HPI and the HPI-like element have been acquired independently, after the branching of the *Y. pestis-Y. pseudotuberculosis* lineage from *Y. enterocolitica,* probably from an element ancestral to the *ybt* loci whose function was related to iron-uptake.

PRESENCE OF THE HPI AMONG VARIOUS NON-*YERSINIA* BACTERIA

The HPI was first identified in the three pathogenic *Yersinia* species but its presence was subsequently detected in *E. coli,* and then in various members of the family *Enterobacteriaceae*. To date, the HPI is the only pathogenicity island identified in a large number of bacterial genera.

GENETIC ORGANIZATION AND CHARACTERISTICS OF THE *E. COLI* HPI

In *E. coli*, the HPI has been identified in a wide variety of pathotypes: EAEC, EPEC, EIEC, ETEC, EHEC/STEC (serotypes O26:H11/H- and O128:H2/H-), UTIEC, SEPEC, MNEC, ExPEC, and non-typable *E. coli* (Schubert et al., 1998; Karch et al., 1999; Johnson and Stell, 2000; Schubert et al., 2000; Xu et al., 2000; Clermont et al., 2001; Johnson et al., 2003; Koczura and Kaznowski, 2003b).

The *E. coli* HPI is genetically more closely related to the islands of *Y. pestis* and *Y. pseudotuberculosis* than to that of *Y. enterocolitica* (Rakin et al., 1995). The *E. coli* HPI is classically integrated into an *asnT* locus which contains a perfectly conserved *attB*-R site (Karch et al., 1999; Schubert et al., 1999; Dobrindt et al., 2002; Girardeau et al., 2003; Koczura and Kaznowski, 2003b). In contrast, the *attB*-L site, located at the other extremity of the island, is absent from *E. coli* (Schubert et al., 1999; Bach et al., 2000; Dobrindt et al., 2002). Next to the right-hand extremity of the HPI, a P4-like *int* gene is commonly present and shares 94% (STEC) to 99% (UTIEC) nt identity with the *Y. pestis* homolog (Karch et al., 1999; Schubert et al., 1999). However, this gene is commonly mutated because of either a 347-bp deletion (Karch et al., 1999; Gophna et al., 2001; Girardeau et al., 2003), or a premature stop codon (Schubert et al., 1999), leading to a truncated *int* product. The left-hand part of the *E. coli* HPI is different from that of the *Y. pestis* and *Y. pseudotuberculosis* island. Some HPI-positive *E. coli* strains lack the *psn* gene (Schubert et al., 1998; Janssen et al., 2001) and/or the IS*100* element (Schubert et al., 1999; Koczura and Kaznowski, 2003b), a situation reminiscent of the HPI of *Y. pseudotuberculosis* serotype III. In the UTIEC strain 536, the HPI terminates approximately 1 kb downstream of *psn* (Schubert et al., 1999; Dobrindt et al., 2002). In contrast to the left-hand part, the Ybt part of the HPI is highly conserved in *E. coli* and *Yersinia*, as demonstrated by the 97-100% nt identity of the *ybt* genes of the two genera (Rakin et al., 1995; Schubert et al., 1998; Dobrindt et al., 2002).

The recent availability of the genome of the HPI-positive *E. coli* strain CFT073 (Welch et al., 2002) allows a comparison of the complete genetic maps of the *E. coli* and *Yersinia* HPI. The organization of the two islands is highly conserved in a region encompassing 13 genes and extending from the *asn* tRNA locus to *ypo1905a* (Figure 3). The identity between the products of each paralog is usually extremely high (between 98 and 100%, Figure 3). However, several differences can be noted. One concerns the absence of the *attB*-L and left-hand part of the HPI, as already reported in other *E. coli* strains. Another difference is the shortening of the integrase gene because of a G to A replacement which destroys the ATG start codon and leads to a shorter coding sequence in *E. coli*. The other major differences are the truncations of *ybtE, irp1* and *irp2* into two or three segments. The truncation of *irp2* may be explained by the presence of a remnant of an IS*1541*-like element at the junction between the two truncated portions of the gene. In *irp1*, the three gene segments are the consequences of the deletion of 5 non-contiguous nt (site A on Figure 3), and the insertion of one nt (site B on Figure 3), which introduce premature stop codons. The truncation of *ybtE* results from a frame shift caused by the deletion of 1 nt in the 3' portion of the gene. The sequences of other *E. coli* HPIs would be needed to determine whether the differences observed between the *E. coli* and the *Yersinia* HPIs are restricted to strain CFT073, or whether they are a common characteristic of the former species. The functionality of the Ybt system in CFT073 would also deserve investigation.

Most HPI-positive *E. coli* isolates produce the iron-regulated proteins HMWP1 and HMWP2 encoded by *irp1* and *irp2,* respectively, but they do not always synthesize Psn, even when they harbor the *psn* locus (Schubert et al., 1998; Karch et al., 1999). In *Yersinia*, Psn acts not only as a yersiniabactin receptor but also as the receptor for the bacteriocin pesticin (Rakin et al., 1994). In contrast, the majority of HPI-positive and Psn-positive *E. coli* are insensitive to pesticin (Schubert et al., 1998). Nonetheless, the HPI seems to be functional in most *E. coli* since the majority of the pathogenic strains produce yersiniabactin (Schubert et al., 2002; Koczura and Kaznowski, 2003b).

One particular case is that of the HPI of *E. coli* strain ECOR31. In this strain, the HPI extends 35 kb leftward from the *psn* locus. This additional segment carries three distinct regions: (i) a segment coding for a functional mating pair formation system; (ii) a DNA-processing region related to that of plasmid CloDF13; and (iii) a segment sharing some homology with *Vibrio cholerae* genes (Schubert et al., 2004). In addition to this large additional left-hand portion, the ECOR31 HPI differs from the island of other *E. coli* by its insertion site in an *asnV* tRNA, the presence of an intact 17-bp *attB* site flanking each extremity, and a functional integrase gene. A circular extrachromosomal form of the HPI is observed upon induction of *int* expression. This element resembles a subgroup of integrative and conjugative elements designated ICE, and the ECOR31 HPI has been proposed to be the progenitor of the HPI of *E. coli* and *Yersinia* (Schubert et al., 2004).

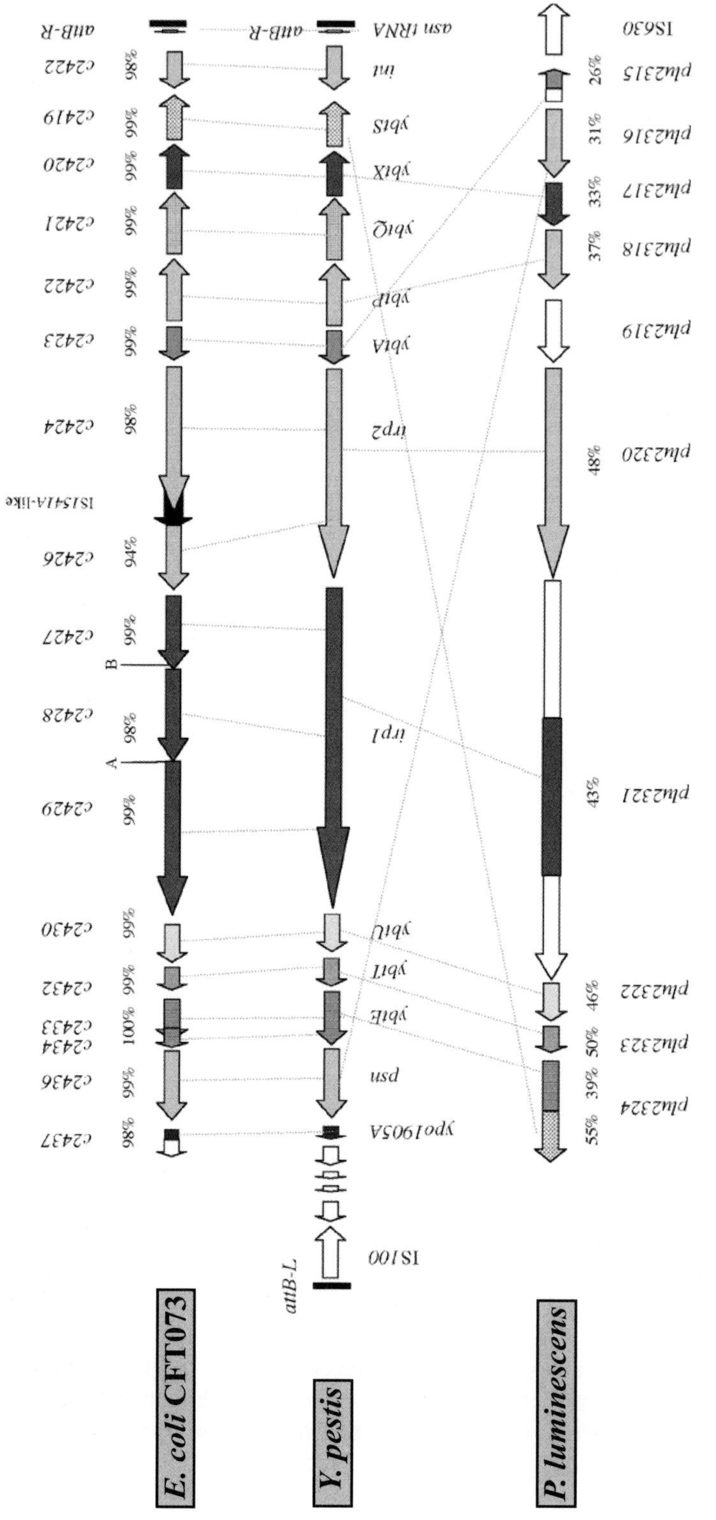

Figure 3. Comparison of the genetic maps of the HPI of *E. coli* CFT073, *Y. pestis* and *Photorhabdus luminescens*. Blank arrows represent genes with no homolog on the HPI counterpart. Numbers above or below the arrows indicate the percentages of amino acid identity with the *Y. pestis* homolog products.

IMPORTANCE OF THE HPI FOR *E. COLI* VIRULENCE

One of the most remarkable features that emerges from the numerous recent studies on the distribution of the HPI in various *E. coli* isolates is the association between the presence of the island and the severity of disease. In the ECOR reference collection, the HPI is much more prevalent in pathogenic strains belonging to the phylogenetic groups B2 (92-100%) and D (83%), than in isolates belonging to the non-pathogenic groups A (32-52%) and B1 (25-48%) (Clermont et al., 2001; Johnson et al., 2001; Schubert et al., 2002). Furthermore, among HPI-positive strains belonging to the various phylogenetic groups, synthesis of yersiniabactin is more common in B2 (94%) than in A (71%) or B1 (45%) strains (Schubert et al., 2002).

The frequency of HPI-positive strains causing septicemia is particularly high (Table 4). The island has been detected in 93% of blood culture isolates from patients with urosepsis in Seattle (Johnson and Stell, 2000), and in 75-83% of isolates (SEPEC) from septicemic patients in Germany (Schubert et al., 1998; Schubert et al., 2000), suggesting that the presence of the HPI is a common characteristic of bacteremic *E. coli*. Carriage of the HPI is also extremely frequent in *E. coli* strains causing severe forms of infection in poultry. In one study, 68% of the *E. coli* strains isolated from visceral organs of poultry with fatal colibacillosis carried the HPI (Janssen et al., 2001). In another study, the island was present in 89% of serotype 078, and 100% of serotype 02 *E. coli* strains isolated from poultry with acute septicemia (Gophna et al., 2001). The prevalence of the HPI is also remarkably high (71-93%) in *E. coli* strains responsible for urinary tract infections and pyelonephritis (Johnson and Stell, 2000; Schubert et al., 2000; Dobrindt et al., 2002; Girardeau et al., 2003; Johnson et al., 2003). During an analysis of *E. coli* strains isolated from dogs with urinary tract infections, the presence of the HPI emerged as one of the two most predictive factors of a risk of urinary infection in multivariate analyses (Johnson et al., 2003). Even more importantly, the HPI appears to be a hallmark of strains causing neonatal meningitis (MNEC). Two independent studies reported that the HPI is systematically present (100%, Table 4) in *E. coli* strains causing neonatal meningitis in humans (Clermont et al., 2001; Gophna et al., 2001). While the HPI is one of the most widely distributed virulence factors among extraintestinal *E. coli*, it is much less frequently found in fecal isolates (Table 4) (Schubert et al., 1998; Xu et al., 2000; Girardeau et al., 2003; Johnson et al., 2003).

Indirect experimental evidence for a role of the HPI in *E. coli* virulence came from the infection of mice with various *E. coli* isolates from the ECOR collection. Of all the virulence factors investigated in these various isolates, the presence of a functional HPI was statistically the most significant factor associated with lethality (Schubert et al., 2002). In another study where 77 *E. coli* isolates were tested for serum resistance, the HPI was one of the three virulence factors whose presence was significantly correlated with serum resistance (Girardeau et al., 2003). A direct demonstration for the role of the HPI in *E. coli* pathogenicity was recently obtained by inactivating the *irp1* gene. While 53-93% of mice subcutaneously infected with 10^8 cfu of the wild-type strains died of the infection, only 0-3% of the animals infected with the same number of *irp1* mutants succumbed (Schubert et al., 2002).

The HPI is thus clearly associated with extraintestinal *E. coli* and with the most severe forms of infection. As in yersiniae, this island appears to be one of the key factors that determines the potential of *E. coli* strains to disseminate in their host and cause systemic infections.

Table 4. Distribution of the HPI among various pathotypes of *E. coli*.

Study	Pathotype	Frequency	Genes detected	*E. coli* strains analyzed
Schubert et al., 1998	EAEC EIEC ETEC EPEC EHEC SEPEC Fecal isolates	93% 27% 5% 5% 0% 83% 32%	*fyuA*, *irp1*, *irp2*	220 strains from strain collections, human patients and healthy individuals in Germany
Karch et al., 1999	STEC	27.2%	*fyuA*, *irp2*	206 STEC from patients
Johnson and Stell, 2000	Urosepsis	93%	*fyuA*	75 blood-culture strains from patients with urosepsis in Seattle (USA)
Xu et al., 2000	Fecal isolates	23.9%	*fyuA*, *irp2*	176 isolates from patients with diarrhea in Shandong province (China)
Schubert et al., 2000	UTIEC SEPEC	71% 75%	*fyuA*, *irp1*, *irp2*	100 strains from blood and 185 from urine of inpatients in Munich (Germany)
Janssen et al., 2001	APEC	68%	*fyuA*, *irp2*	150 strains from visceral organs of dead poultry in various farms in Germany
Gophna et al., 2001	O78 (human) O78 (avian) O78 (ovine) O2 (avian) O35, O11 (avian)	100% 89% 0% 80% 25%	*fyuA*, *irp2*, *int*	Strains from septicemic poultry and sheep, and from human newborns with meningitis
Clermont et al., 2001	MNEC	100%	*fyuA*, *irp2*	124 strains from human cases of neonatal meningitis in France
Dobrindt et al., 2002	UTIEC MNEC+SPEC Diarrheagenic EC	85.5% 89.3% 27.8%	*int*, *irp1*, *irp2*, *ybtE-psn*	132 strains from a strain collection and from women with chronic urinary tract infections in Germany
Girardeau et al., 2003	UTIEC (human) SEPEC (human) SEPEC (animal) Non-STEC (animal) STEC	72% 78% 84% 47% 57%	*fyuA*, *irp1*, *irp2*	77 *afa-8* strains from humans (32 strains) with septicemia or UTI, and from calves and piglets (45 strains) with intestinal or extrainestinal infections
Koczura and Kaznowski, 2003	ExPEC	77%	*fyuA*, *irp1*, *irp2*	35 strains from human specimen of urine, blood, cervical canal, cerebrospinal fluid, conjunctiva, semen, or wound
Johnson et al., 2003	UTIEC Rectal EC	84% 35%	*fyuA*	54 paired rectal and urinary isolates from dogs in California (USA)

EAEC: enteroaggregative *E. coli*, EIEC: enteroinvasive *E. coli*, ETEC: enterotoxinogenic *E. coli*, EPEC: enteropathogenic *E. coli*, EHEC: enterohemorrhagic *E. coli*, SEPEC: sepsis-causing *E. coli*, STEC: Shiga toxin-producing *E. coli*, APEC: avian pathogenic *E. coli*, UTIEC: urinary tract infection *E. coli*, MNEC: meningitis-causing *E. coli*, ExPEC: extra intestinal *E. coli*.

DISTRIBUTION OF THE HPI AMONG OTHER MEMBERS OF THE ENTEROBACTERIACEAE FAMILY

After its first description in *Yersinia* spp., and its detection in *E. coli*, the HPI has been identified in several other members of the *Enterobacteriaceae* family (Table 5) such as *Enterobacter* (*E. cloacae* and *E. aerogenes*), *Klebsiella* (*K. pneumoniae, K. rhinoscleromatis, K. ozaenae, K. planticola*, and *K. oxytoca*), *Citrobacter* (*C. diversus* and *C. koseri*), *Salmonella enterica, Photorhabdus luminescens* and *Serratia liquefaciens* (Bach et al., 2000; Schubert et al., 2000; Duchaud et al., 2003; Koczura and Kaznowski, 2003a; Mokracka et al., 2003; Oelschlaeger et al., 2003; Olsson et al., 2003).

Studies using several sets of primers to encompass most of the HPI region indicated that the overall organization of the island is usually well conserved among the different HPI-positive enterobacteria (Bach et al., 2000; Koczura and Kaznowski, 2003a; Oelschlaeger et al., 2003; Olsson et al., 2003). With very few exceptions (Bach et al., 2000; Oelschlaeger et al., 2003), the enterobacteria HPI is integrated into an *asn* tRNA locus (Koczura and Kaznowski, 2003a; Oelschlaeger et al., 2003; Olsson et al., 2003). In contrast, the IS*100* element located at the left-hand extremity of the *Y. pestis*-*Y. pseudotuberculosis* island is classically absent from these HPI-positive enterobacteria (Bach et al., 2000). Some differences with the *Yersinia* HPI could also be identified in

Table 5. Members of the *Enterobacteriaceae* family (excluding *Yersinia* and *E. coli*) harboring the HPI.

Study	Species	HPI+/ total	Genes detected	Characteristics of the strains analyzed
Bach et al., 2000	*Klebsiella pneumoniae* *Klebsiella rhinoscleromatis* *Klebsiella ozaenae* *Klebsiella planticola* *Klebsiella oxytoca* *Citrobacter diversus*	1/3 1/2 1/2 1/1 1/1 1/1	*irp2*	67 isolates belonging to 18 genera and 52 species of enterobacteria from a strain collection.
Schubert et al., 2000	*Klebsiella oxytoca* *Klebsiella pneumoniae* *Enterobacter cloacae* *Citrobacter koseri* *Citrobacter freundii*	14/24 14/79 5/41 6/6 2/26	*fyuA, irp1, irp2*	Strains isolated from blood or urine of inpatients in Munich (Germany).
Koczura and Kaznowski, 2003	*Klebsiella pneumoniae*	6/34	*fyuA, irp1, irp2*	34 *K. pneumoniae* strains from human specimen of urine, blood, cervical canal, cerebrospinal fluid, conjunctiva, or wound.
Olsson et al., 2003	*Serratia liquefaciens*	1	*irp2*	103 colonies from unknown species isolated from meat from local shops in Sweden.
Mokracka et al., 2003	*Enterobacter cloacae* *Enterobacter aerogenes*	1 1	*fyuA, irp1, irp2*	89 *Enterobacter* strains belonging to the species *E. cloacae, E. aerogenes* and *E. sakazakii.*
Oelschlaeger et al., 2003	*Salmonella enterica* Group IIIa Group IIIb Group VI	 1/9 11/11 2/8	*ybtS-X, irp1-2, irp4-fyuA*	74 *Salmonella* strains belonging to the species *S. enterica* and *S. bongori.*
Duchaud et al., 2003	*Photorhabdus luminescens*	1	Genome sequence	Strain TT01

individual isolates or groups of strains. Partial or total deletion of *int* was detected in one strain of *C. diversus* (Bach et al., 2000). In one *K. oxytoca* isolate, the right-hand part of the island extending from *ybtP* to *attB*-R border was deleted (Bach et al., 2000). Deletions of some portions of the HPI and a shorter *fyuA* product were also detected in some *K. pneumoniae* isolates (Koczura and Kaznowski, 2003a). In *S. enterica*, strains of group VI have an HPI identical to that of *Yersinia*, including the presence of an integrase gene downstream of *ybtS*, and an insertion site into an *asn* tRNA gene. In contrast, *S. enterica* strains of groups IIIa and IIIb harbor several genes (*int, ybtS, ybtP-ybtA* intergenic region, *irp1)* containing insertions or deletions, and their HPI is not inserted into an *asn* tRNA gene but is downstream of *ych* (Oelschlaeger et al., 2003).

Most of the HPI-positive enterobacteria tested are able to synthesize the HPI-encoded HMWP1 and HMWP2 in an iron-dependent manner (Bach et al., 2000; Oelschlaeger et al., 2003), and to produce yersiniabactin (Schubert et al., 2000; Koczura and Kaznowski, 2003a; Oelschlaeger et al., 2003). Therefore, in most enterobacteria, the core region of the island is well conserved and functional.

The recent sequencing of the genome of the enterobacterium *P. luminescens* strain TT01 has revealed the presence of a region homologous to the *Yersinia* HPI on the chromosome of this organism, which is a symbiont of nematodes and is pathogenic for a wide range of insects (Duchaud et al., 2003). Ten of the *Y. pestis ybt* genes have homologs in *P. luminescens* (Figure 3). The genetic organization of the cluster of 5 genes extending from *irp2* to *ybtE* is conserved between the two species, while the order and location of the 5 other genes is different. The amino acid identity between the products of the *Y. pestis* and *P. luminescens* homologs is between 26% and 55% (Figure 3). Genetic markers of pathogenicity islands, such as the P4-like *int, hef, asn* tRNA, attB-L and *attB*-R, were not detected in the vicinity of the *ybt*-like locus on the *P. luminescens* chromosome. This suggests that, as for the HPI-like element of *Y. pestis*, the *ybt* locus is not carried by a PAI in *P. luminescens*. However, the GC% of the *ybt*-like locus (≈48%) is significantly higher than that of the core genome (42.8%), still suggesting that this locus has been horizontally acquired.

PRESENCE OF AN ELEMENT RESEMBLING THE HPI IN *CORYNEBACTERIUM DIPHTHERIAE*

The carriage of an HPI-like element is not restricted to Gram-negative bacteria. Recently, the presence of a chromosomal region homologous to a portion of the *ybt* locus has been identified in *C. diphtheriae*, a Gram-positive bacterium causing diphtheria (Kunkle and Schmitt, 2003). This region, designated the *sid* operon, contains two putative siderophore biosynthesis genes homologous to *irp1* and *irp2,* and two genes predicted to code for siderophore transport functions and homologous to *ybtP* and *ybtQ* (Figure 4). The products of the *Y. pestis* and *C. diphtheriae* homologs display approximately 30% amino acid identity (Figure 4). Although no siderophore biosynthesis genes other than *sidA* and *sidB* are located on the *sid* operon, homologs to *ybtE* and *ybtD* are present on the *C. diphtheriae* chromosome (Kunkle and Schmitt, 2003). The *sid* operon is negatively regulated by the global iron-dependent DtxR repressor, in a manner similar to the regulation of the *ybt* operons by Fur (Kunkle and Schmitt, 2003). Interestingly, although no integrase gene or *attB* sites have been identified in the vicinity of the *sid* operon on the *C. diphtheriae* chromosome, phage genes and an IS*30* element are present on the left-hand part of the operon, an organization similar to that found on the left-hand part of the *Y. pestis* HPI

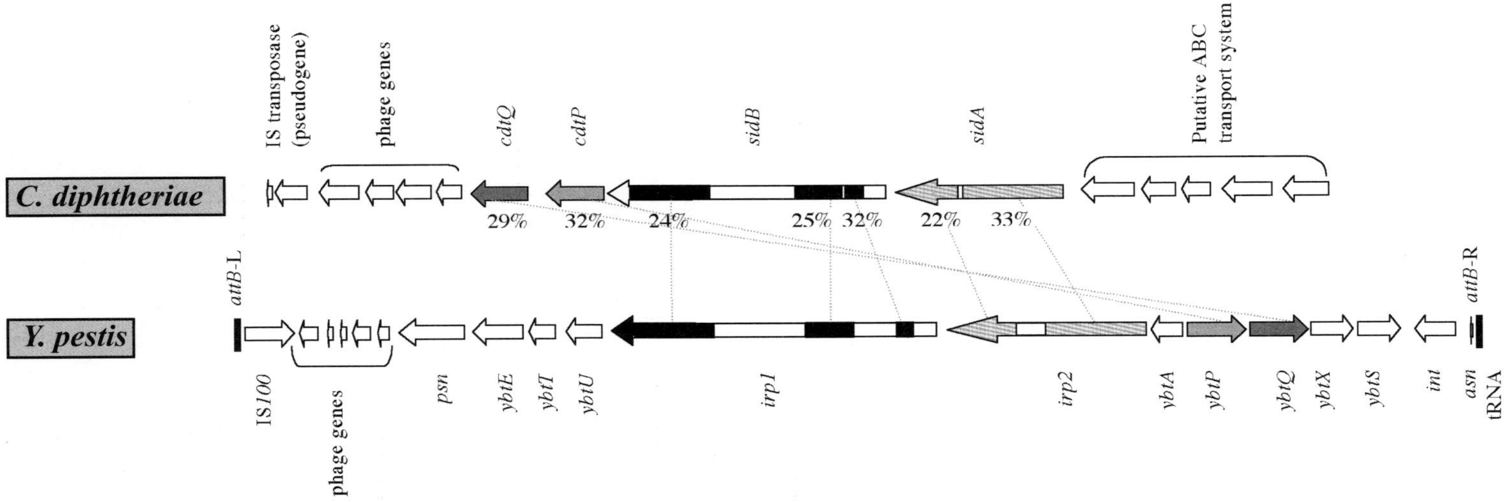

Figure 4. Comparison of the genetic maps of the HPI of *C. diphtheriae* and *Y. pestis*. Blank arrows indicate genes that are not present on both elements. Numbers above the arrows indicate the percentages of amino acid identity with the *Y. pestis* homolog products.

(Figure 4). The function of the *sid* operon is not yet known but it may be predicted that it encodes an iron-uptake system. Mutations in *sidA* or *sidB* affected neither siderophore production nor growth of *C. diphtheriae* in a low-iron medium (Kunkle and Schmitt, 2003). However, since these strains produce the siderophore corynebactin, which is not encoded by the *sid* operon, the presence of this siderophore may mask the inactivation of the *sid*-encoded iron-uptake system.

CONCLUSION

Acquisition of the HPI by pathogenic *Yersinia* has certainly been an essential step in the emergence of a subgroup of highly pathogenic strains. Since the identity between HPI-borne genes of *Y. enterocolitica* and *Y. pseudotuberculosis* is higher than that of other chromosomal genes, and since the two HPIs have a different genetic organization, it may be estimated that the HPI was acquired independently and horizontally by the two species, after their divergence from the *Yersinia* ancestor about one million years ago (Achtman et al., 1999). However, due to the high conservation of the *Y. pestis* and *Y. pseudotuberculosis* HPI, and to the very recent emergence of *Y. pestis* less than 20,000 years ago, it is reasonable to consider that the HPI was vertically acquired by *Y. pestis* from its *Y. pseudotuberculosis* progenitor.

The high degree of conservation between HPI-borne genes of *Yersinia* and *E. coli* also argues for a recent and horizontal acquisition of the island in these two genera, long after their divergence from a common ancestor. The bacterial species that may have been the donor of the HPI remains to be identified.

A degenerate *ybt* locus (HPI-like element) is present on the chromosome of *Y. pestis*, *Y. pseudotuberculosis*, *P. luminescens* and *C. diphtheriae*. The homology between these loci and the HPI is limited to certain genes of the *ybt* system. Several *ybt* genes are missing from these loci, and their genetic organization is clearly different from that of the HPI and varies depending on the bacterial species. All these HPI-like elements, however, seem to encode (or to have encoded) iron-uptake systems. No remnants of a horizontal acquisition are found in their vicinity, suggesting that these elements were acquired much earlier than the HPI and have gradually eliminated the non-essential mobility genes, while keeping the functional locus which was subjected to rearrangements.

In *Yersinia* spp. as well as in *E. coli*, the presence of the HPI is clearly a marker of the high-pathogenicity potential of the strains. The island provides the recipient bacterium with the ability to disseminate in its host and to cause severe and systemic infections. However, some HPI-positive enterobacteria are isolated as commensal microorganisms from asymptomatic carriers, or are known to persist as saprophytic bacteria in healthy humans (*Citrobacter, Enterobacter*, *Klebsiella*, etc.). One hypothesis, which would deserve investigation, is that HPI-positive commensal enterobacteria which are normally non-pathogenic in a normal host may cause a nosocomial infection in immuno-compromised patients because of the presence of the island. In contrast, HPI-negative commensals would remain in their initial niche and would not produce an infection in debilitated hosts. The presence of the HPI would then be a key factor in determining whether the bacterium will remain localized at the site of infection or will invade its host. This capability would of course also depend on the genetic background of the HPI-bearing bacterium. Another, and non-exclusive hypothesis is that the HPI was initially acquired by non-pathogenic bacteria and allowed them to better adapt to the iron-limiting conditions encountered in the environment. In these bacteria, the HPI would thus originally have been a "Metabolic island" (Hacker and Carniel, 2001). Subsequent transfer of the island to a bacterium which harbored some virulence features provided the organism with a better capacity to disseminate in vivo, and the HPI then became a "Pathogenicity island".

The *Yersinia* HPI has kept the capacity to excise from the chromosome, form an episomal molecule, and re-insert in the chromosome of the host bacterium. Whether this island has retained the potential to be transmitted to new bacterial hosts is certainly one of the most challenging and important questions to address.

REFERENCES

Achtman, M., Zurth, K., Morelli, C., Torrea, G., Guiyoule, A. and Carniel, E. 1999. *Yersinia pestis*, the cause of plague, is a recently emerged clone of *Yersinia pseudotuberculosis*. Proc. Natl. Acad. Sci. USA. 96: 14043-14048.

Bach, S., de Almeida, A. and Carniel, E. 2000. The *Yersinia* high-pathogenicity island is present in different members of the family *Enterobacteriaceae*. FEMS Microbiol. Lett. 183: 289-294.

Bach, S., Buchrieser, C., Prentice, M., Guiyoule, A., Msadek, T. and Carniel, E. 1999. The high-pathogenicity island of *Yersinia enterocolitica* Ye8081 undergoes low-frequency deletion but not precise excision, suggesting recent stabilization in the genome. Infect. Immun. 67: 5091-5099.

Bearden, S.W. and Perry, R.D. 1999. The Yfe system of *Yersinia pestis* transports iron and manganese and is required for full virulence of plague. Mol. Microbiol. 32: 403-414.

Bearden, S.W., Fetherston, J.D. and Perry, R.D. 1997. Genetic organization of the yersiniabactin biosynthetic region and construction of avirulent mutants in *Yersinia pestis*. Infect. Immun. 65: 1659-1668.

Boelaert, J.R., Van Landuyt, H.W., Valcke, Y.J., Cantinieaux, B., Lornoy, W.F., Vanherweghem, J.-L., Moreillon, P. and Vandepitte, J.M. 1987. The role of iron overload in *Yersinia enterocolitica* and *Yersinia pseudotubercu losis*bacteremia in hemodialysis patients. J. Infect. Dis. 156: 384-387.

Brem, D., Pelludat, C., Rakin, A., Jacobi, C.A. and Heesemann, J. 2001. Functional analysis of yersiniabactin transport genes of *Yersinia enterocolitica*. Microbiology. 147: 1115-1127.

Buchrieser, C., Prentice, M. and Carniel, E. 1998a. The 102-kilobase unstable region of *Yersinia pestis* comprises a high-pathogenicity island linked to a pigmentation segment which undergoes internal rearrangement. J. Bacteriol. 180: 2321-2329.

Buchrieser, C., Brosch, R., Bach, S., Guiyoule, A. and Carniel, E. 1998b. The high-pathogenicity island of *Yersinia pseudotuberculosis* can be inserted into any of the three chromosomal *asn tRNA* genes. Mol. Microbiol. 30: 965-978.

Buchrieser, C., Rusniok, C., Frangeul, L., Couve, E., Billault, A., Kunst, F., Carniel, E. and Glaser, P. 1999. The 102-kb *pgm* locus of *Yersinia pestis*: sequence analysis and comparison of selected regions among different *Yersinia pestis* and *Yersinia pseudotuberculosis* strains. Infect. Immun. 67: 4851-4861.

Carniel, E., Mazigh, D. and Mollaret, H.H. 1987. Expression of iron-regulated proteins in *Yersinia* species and their relation to virulence. Infect. Immun. 55: 277-280.

Carniel, E., Mercereau-Puijalon, O. and Bonnefoy, S. 1989. The gene coding for the 190,000-dalton iron-regulated protein of *Yersinia* species is present only in the highly pathogenic strains. Infect. Immun. 57: 1211-1217.

Carniel, E., Guilvout, I. and Prentice, M. 1996. Characterization of a large chromosomal "high-pathogenicity island" in biotype 1B *Yersinia enterocolitica*. J. Bacteriol. 178: 6743-6751.

Carniel, E., Guiyoule, A., Mercereau-Puijalon, O. and Mollaret, H.H. 1991. Chromosomal marker for the high pathogenicity phenotype in *Yersinia*. Contrib. Microbiol. Immunol. 12: 192-197.

Carniel, E., Guiyoule, A., Guilvout, I. and Mercereau-Puijalon, O. 1992. Molecular cloning, iron-regulation and mutagenesis of the *irp2* gene encoding HMWP2, a protein specific for the highly pathogenic *Yersinia*. Mol. Microbiol. 6: 379-388.

Chambers, C.E. and Sokol, P.A. 1994. Comparison of siderophore production and utilization in pathogenic and environmental isolates of *Yersinia enterocolitica*. J. Clin. Microbiol. 32: 32-39.

Clermont, O., Bonacorsi, S. and Bingen, E. 2001. The *Yersinia* high-pathogenicity island is highly predominant in virulence-associated phylogenetic groups of *Escherichia coli*. FEMS Microbiol. Lett. 196: 153-157.

De Almeida, A.M.P., Guiyoule, A., Leal, N.C. and Carniel, E. 1994. Survey of the *irp2* gene among *Yersinia pestis* strains isolated during several plague outbreaks in Northeast Brazil. Mem. Inst. Oswaldo Cruz. 89: 87-92.

De Almeida, A.M.P., Guiyoule, A., Guilvout, I., Iteman, I., Baranton, G. and Carniel, E. 1993. Chromosomal *irp2* gene in *Yersinia*: distribution, expression, deletion and impact on virulence. Microb. Pathog. 14: 9-21.

Deng, W., Burland, V., Plunkett, G., Boutin, A., Mayhew, G.F., Liss, P., Perna, N.T., Rose, D.J., Mau, B., Zhou, S.G., Schwartz, D.C., Fetherston, J.D., Lindler, L.E., Brubaker, R.R., Plano, G.V., Straley, S.C., McDonough, K.A., Nilles, M.L., Matson, J.S., Blattner, F.R. and Perry, R.D. 2002. Genome sequence of *Yersinia pestis* KIM. J. Bacteriol. 184: 4601-4611.

Dobrindt, U., Blum-Oehler, G., Nagy, G., Schneider, G., Johann, A., Gottschalk, G. and Hacker, J. 2002. Genetic structure and distribution of four pathogenicity islands (PAI I(536) to PAI IV(536)) of uropathogenic *Escherichia coli* strain 536. Infect. Immun. 70: 6365-6372.

Duchaud, E., Rusniok, C., Frangeul, L., Buchrieser, C., Givaudan, A., Taourit, S., Bocs, S., Boursaux-Eude, C., Chandler, M., Charles, J.F., Dassa, E., Derose, R., Derzelle, S., Freyssinet, G., Gaudriault, S., Medigue, C., Lanois, A., Powell, K., Siguier, P., Vincent, R., Wingate, V., Zouine, M., Glaser, P., Boemare, N., Danchin, A. and Kunst, F. 2003. The genome sequence of the entomopathogenic bacterium *Photorhabdus luminescens*. Nat. Biotechnol. 21: 1307-1313.

Fetherston, J.D. and Perry, R.D. 1994. The pigmentation locus of *Yersinia pestis* KIM6+ is flanked by an insertion sequence and includes the structural genes for pesticin sensitivity and HMWP2. Mol. Microbiol. 13: 697-708.

Fetherston, J.D., Schuetze, P. and Perry, R.D. 1992. Loss of the pigmentation phenotype in *Yersinia pestis* is due to the spontaneous deletion of 102 kb of chromosomal DNA which is flanked by a repetitive element. Mol. Microbiol. 6: 2693-2704.

Fetherston, J.D., Lillard, J., J.W. and Perry, R.D. 1995. Analysis of the pesticin receptor from *Yersinia pestis*: role in iron-deficient growth and possible regulation by its siderophore. J. Bacteriol. 177: 1824-1833.

Fetherston, J.D., Bearden, S.W. and Perry, R.D. 1996. YbtA, an AraC-type regulator of the *Yersinia pestis* pesticin/yersiniabactin receptor. Mol. Microbiol. 22: 315-325.

Fetherston, J.D., Bertolino, V.J. and Perry, R.D. 1999. YbtP and YbtQ: two ABC transporters required for iron uptake in *Yersinia pestis*. Mol. Microbiol. 32: 289-299.

Fukushima, H., Matsuda, Y., Seki, R., Tsubokura, M., Takeda, N., Shubin, F.N., Paik, I.K. and Zheng, X.B. 2001. Geographical heterogeneity between Far Eastern and Western countries in prevalence of the virulence plasmid, the superantigen *Yersinia pseudotuberculosis*-derived mitogen, and the high-pathogenicity island among *Yersinia pseudotuberculosis* strains. J. Clin. Microbiol. 39: 3541-3547.

Gehring, A.M., Demoll, E., Fetherston, J.D., Mori, I., Mayhew, G.F., Blattner, F.R., Walsh, C.T. and Perry, R.D. 1998. Iron acquisition in plague - modular logic in enzymatic biogenesis of yersiniabactin by *Yersinia pestis*. Chem. Biol. 5: 573-586.

Girardeau, J.P., Lalioui, L., Said, A.M.O., De Champs, C. and Le Bouguenec, C. 2003. Extended virulence genotype of pathogenic *Escherichia coli* isolates carrying the *afa-8* operon: Evidence of similarities between isolates from humans and animals with extraintestinal infections. J. Clin. Microbiol. 41: 218-226.

Gong, S., Bearden, S.W., Geoffroy, V.A., Fetherston, J.D. and Perry, R.D. 2001. Characterization of the *Yersinia pestis* Yfu ABC inorganic iron transport system. Infect. Immun. 69: 2829-2837.

Gophna, U., Oelschlaeger, T.A., Hacker, J. and Ron, E.Z. 2001. *Yersinia* HPI in septicemic *Escherichia coli* strains isolated from diverse hosts. FEMS Microbiol. Lett. 196: 57-60.

Hacker, J. and Carniel, E. 2001. Ecological fitness, genomic islands and bacterial pathogenicity - A Darwinian view of the evolution of microbes. EMBO Reports. 2: 376-381.

Hacker, J., Blum-Oehler, G., Mühldorfer, I. and Tschäpe, H. 1997. Pathogenicity islands of virulent bacteria: structure, function and impact on microbial evolution. Mol. Microbiol. 23: 1089-1097.

Hare, J.M. and McDonough, K.A. 1999. High-frequency RecA-dependent and -independent mechanisms of Congo red binding mutations in *Yersinia pestis*. J. Bacteriol. 181: 4896-4904.

Hare, J.M., Wagner, A.K. and McDonough, K.A. 1999. Independent acquisition and insertion into different chromosomal locations of the same pathogenicity island in *Yersinia pestis* and *Yersinia pseudotuberculosis*. Mol. Microbiol. 31: 291-303.

Heesemann, J. 1987. Chromosomal-encoded siderophores are required for mouse virulence of enteropathogenic *Yersinia* species. FEMS Microbiol. Lett. 48: 229-233.

Heesemann, J., Hantke, K., Vocke, T., Saken, E., Rakin, A., Stojiljkovic, I. and Berner, R. 1993. Virulence of *Yersinia enterocolitica* is closely associated with siderophore production, expression of an iron-repressible outer membrane polypeptide of 65000 Da and pesticin sensitivity. Mol. Microbiol. 8: 397-408.

Iteman, I., Guiyoule, A., de Almeida, A.M.P., Guilvout, I., Baranton, G. and Carniel, E. 1993. Relationship between loss of pigmentation and deletion of the chromosomal iron-regulated *irp2* gene in *Yersinia pestis*: evidence for separate but related events. Infect. Immun. 61: 2717-2722.

Jackson, S. and Burrows, T.W. 1956. The virulence enhancing effect of iron on non-pigmented mutants of virulent strains of *Pasteurella pestis*. Br. J. Exp. Pathol. 37: 577-583.

Jacobi, C.A., Gregor, S., Rakin, A. and Heesemann, J. 2001. Expression analysis of the yersiniabactin receptor gene *fyuA* and the heme receptor *hemR* of *Yersinia enterocolitica* in vitro and in vivo using the reporter genes for green fluorescent protein and luciferase. Infect. Immun. 69: 7772-7782.

Janssen, T., Schwarz, C., Preikschat, P., Voss, M., Philipp, H.C. and Wieler, L.H. 2001. Virulence-associated genes in avian pathogenic *Escherichia coli* (APEC) isolated from internal organs of poultry having died from colibacillosis. Int. J. Med. Microbiol. 291: 371-378.

Johnson, J.R. and Stell, A.L. 2000. Extended virulence genotypes of *Escherichia coli* strains from patients with urosepsis in relation to phylogeny and host compromise. J. Infect. Dis. 181: 261-272.

Johnson, J.R., Delavari, P., Kuskowski, M. and Stell, A.L. 2001. Phylogenetic distribution of extraintestinal virulence-associated traits in *Escherichia coli*. J. Infect. Dis. 183: 78-88.

Johnson, J.R., Kaster, N., Kuskowski, M.A. and Ling, G.V. 2003. Identification of urovirulence traits in *Escherichia coli* by comparison of urinary and rectal *E. coli* isolates from dogs with urinary tract infection. J. Clin. Microbiol. 41: 337-345.

Karch, H., Schubert, S., Zhang, D., Zhang, W., Schmidt, H., Ölschläger, T. and Hacker, J. 1999. A genomic island, termed high-pathogenicity island, is present in certain non-O157 shiga toxin-producing *Escherichia coli* clonal lineages. Infect. Immun. 67: 5994-6001.

Koczura, R. and Kaznowski, A. 2003a. Occurrence of the *Yersinia* high-pathogenicity island and iron uptake systems in clinical isolates of *Klebsiella pneumoniae*. Microb. Pathog. 35: 197-202.

Koczura, R. and Kaznowski, A. 2003b. The *Yersinia* high-pathogenicity island and iron-uptake systems in clinical isolates of *Escherichia coli*. J. Med. Microbiol. 52: 637-642.

Kunkle, C.A. and Schmitt, M.P. 2003. Analysis of the *Corynebacterium diphtheriae* DtxR regulon: identification of a putative siderophore synthesis and transport system that is similar to the *Yersinia* high-pathogenicity island-encoded yersiniabactin synthesis and uptake system. J. Bacteriol. 185: 6826-6840.

Lesic, B., Bach, S., Ghigo, J.M., Dobrindt, U., Hacker, J., and Carniel, E. 2004. Excision of the high-pathogenicity island of *Yersinia pseudotuberculosis* requires the combined actions of its cognate integrase and Hef, a new recombination directionality factor. Mol. Microbiol. 52: 1337-1348.

Lewis, J.A. and Hatfull, G.F. 2001. Control of directionality in integrase-mediated recombination: examination of recombination directionality factors (RDFs) including Xis and Cox proteins. Nucleic Acids Res. 29: 2205-2216.

McDonough, K.A. and Hare, J.M. 1997. Homology with a repeated *Yersinia pestis* DNA sequence IS*100* correlates with pesticin sensitivity in *Yersinia pseudotuberculosis*. J. Bacteriol. 179: 2081-2085.

Mokracka, J., Kaznowski, A., Szarata, M. and Kaczmarek, E. 2003. Siderophore-mediated strategies of iron acquisition by extraintestinal isolates of *Enterobacter* spp. Acta Microbiol. Pol. 52: 81-86.

Oelschlaeger, T.A., Zhang, D., Schubert, S., Carniel, E., Rabsch, W., Karch, H. and Hacker, J. 2003. The high-pathogenicity island is absent in human pathogens of *Salmonella enterica* subspecies I but present in isolates of subspecies III and VI. J. Bacteriol. 185: 1107-1111.

Olsson, C., Olofsson, T., Ahrne, S. and Molin, G. 2003. The *Yersinia* HPI is present in *Serratia liquefaciens* isolated from meat. Lett. Appl. Microbiol. 37: 275-280.

Parkhill, J., Wren, B.W., Thomson, N.R., Titball, R.W., Holden, M.T.G., Prentice, M.B., Sebaihia, M., James, K.D., Churcher, C., Mungall, K.L., Baker, S., Basham, D., Bentley, S.D., Brooks, K., Cerdeño-Tárraga, A.M., Chillingworth, T., Cronin, A., Davies, R.M., Davis, P., Dougan, G., Feltwell, T., Hamlin, N., Holroyd, S., Jagels, K., Karlyshev, A.V., Leather, S., Moule, S., Oyston, P.C.F., Quail, M., Rutherford, K., Simmonds, M., Skelton, J., Stevens, K., Whitehead, S. and Barrell, B.G. 2001. Genome sequence of *Yersinia pestis*, the causative agent of plague. Nature. 413: 523-527.

Pelludat, C., Hogardt, M. and Heesemann, J. 2002. Transfer of the Core Region Genes of the *Yersinia enterocolitica* WA-C Serotype O:8 High-Pathogenicity Island to *Y. enterocolitica* MRS40, a Strain with Low Levels of Pathogenicity, Confers a Yersiniabactin Biosynthesis Phenotype and Enhanced Mouse Virulence. Infect. Immun. 70: 1832-1841.

Pelludat, C., Rakin, A., Jacobi, C.A., Schubert, S. and Heesemann, J. 1998. The yersiniabactin biosynthetic gene cluster of *Yersinia enterocolitica*: organization and siderophore-dependent regulation. J. Bacteriol. 180: 538-546.

Perry, R.D., Balbo, P.B., Jones, H.A., Fetherston, J.D. and DeMoll, E. 1999. Yersiniabactin from *Yersinia pestis*: biochemical characterization of the siderophore and its role in iron transport and regulation. Microbiology. 145: 1181-1190.

Podladchikova, O.N., Dikhanov, G.G., Rakin, A.V. and Heesemann, J. 1994. Nucleotide sequence and structural organization of *Yersinia pestis* insertion sequence IS*100*. FEMS Microbiol. Lett. 121: 269-274.

Prentice, M.B. and Carniel, E. 1995. Sequence analysis of a *Yersinia pestis* insertion sequence associated with an unstable region of the chromosome. Contr. Microbiol. Immunol. 13: 294-298.

Rakin, A. and Heesemann, J. 1995. Virulence-associated *fyuA* /*irp2* gene cluster of *Yersinia enterocolitica* biotype 1B carries a novel insertion sequence IS*1328*. FEMS Microbiol. Lett. 129: 287-292.

Rakin, A., Urbitsch, P. and Heesemann, J. 1995. Evidence for two evolutionary lineages of highly pathogenic *Yersinia* species. J. Bacteriol. 177: 2292-2298.

Rakin, A., Saken, E., Harmsen, D. and Heesemann, J. 1994. The pesticin receptor of *Yersinia enterocolitica*: a novel virulence factor with dual function. Mol. Microbiol. 13: 253-263.

Rakin, A., Noelting, C., Schubert, S. and Heesemann, J. 1999. Common and specific characteristics of the high-pathogenicity island of *Yersinia enterocolitica*. Infect. Immun. 67: 5265-5274.

Rakin, A., Noelting, C., Schropp, P. and Heesemann, J. 2001. Integrative module of the high-pathogenicity island of *Yersinia*. Mol. Microbiol. 39: 407-415.

Rakin, A., Schubert, S., Guilvout, I., Carniel, E. and Heesemann, J. 2000. Local hopping of IS*3* elements into the A+T-rich part of the high-pathogenicity island in *Yersinia enterocolitica* 1B, O : 8. FEMS Microbiol. Lett. 182: 225-229.

Robins-Browne, R.M. and Kaya Prpic, J. 1985. Effects of iron and desferrioxamine on infections with *Yersinia enterocolitica*. Infect. Immun. 47: 774-779.

Robins-Browne, R.M., Rabson, A.R. and Koornhof, H.J. 1979. Generalized Infection with *Yersinia enterocolitica* and the Role of Iron. Contrib. Microbiol. Immunol. 5: 277-282.

Schubert, S., Cuenca, S., Fischer, D. and Heesemann, J. 2000. High-pathogenicity island of *Yersinia pestis* in enterobacteriaceae isolated from blood cultures and urine samples: Prevalence and functional expression. J. Infect. Dis. 182: 1268-1271.

Schubert, S., Dufke, S., Sorsa, J. and Heesemann, J. 2004. A novel integrative and conjugative element (ICE) of *Escherichia coli*: the putative progenitor of the *Yersinia* high-pathogenicity island. Mol. Microbiol. 51: 837-848.

Schubert, S., Rakin, A., Karch, H., Carniel, E. and Heesemann, J. 1998. Prevalence of the "high-pathogenicity island" of *Yersinia* species among *Escherichia coli* strains that are pathogenic to humans. Infect. Immun. 66: 480-485.

Schubert, S., Rakin, A., Fischer, D., Sorsa, J. and Heesemann, J. 1999. Characterization of the integration site of *Yersinia* high-pathogenicity island in *Escherichia coli*. FEMS Microbiol. Lett. 179: 409-414.

Schubert, S., Picard, B., Gouriou, S., Heesemann, J. and Denamur, E. 2002. *Yersinia* high-pathogenicity island contributes to virulence in *Escherichia coli* causing extraintestinal infections. Infect. Immun. 70: 5335-5337.

Une, T. and Brubaker, R.R. 1984. In vivo comparison of avirulent Vwa^- and Pgm^- or Pst^r phenotypes of yersiniae. Infect. Immun. 43: 895-900.

Vadillo, M., Corbella, X., Pac, V., Fernandez-Viladrich, P. and Pujol, R. 1994. Multiple liver abscesses due to *Yersinia enterocolitica* discloses primary hemochromatosis: Three case reports and review. Clin. Infect. Dis. 18: 938-941.

Welch, R.A., Burland, V., Plunkett, G., 3rd, Redford, P., Roesch, P., Rasko, D., Buckles, E.L., Liou, S.R., Boutin, A., Hackett, J., Stroud, D., Mayhew, G.F., Rose, D.J., Zhou, S., Schwartz, D.C., Perna, N.T., Mobley, H.L., Donnenberg, M.S. and Blattner, F.R. 2002. Extensive mosaic structure revealed by the complete genome sequence of uropathogenic *Escherichia coli*. Proc. Natl. Acad. Sci. USA. 99: 17020-17024.

Xu, J.G., Cheng, B., Wen, X., Cui, S. and Ye, C. 2000. High-pathogenicity island of *Yersinia* spp. in *Escherichia coli* strains isolated from diarrhea patients in China. J. Clin. Microbiol. 38: 4672-4675.

Chapter 15

YAPI, a New Pathogenicity Island in Enteropathogenic Yersiniae

François Collyn, Michaël Marceau and Michel Simonet

ABSTRACT

We describe a large (98 kilobases) DNA segment found in enteropathogenic species *Y. pseudotuberculosis*. Baptised "YAPI" for *Yersinia* Adhesion Pathogenicity Island because it codes for a type IV pilus that promotes bacterial adherence to the intestinal mucosa, this chromosomal fragment fulfills all the pathogenicity island (PAI) criteria: (i) it bears at least one gene cluster involved in bacterial pathogenicity (the *pil* locus, encoding a type IV pilus); (ii) it is associated with a tRNA locus (*phe-tRNA*); (iii) is flanked by directly-repeated sequences; (iv) it comprises a variety of mobility genes (whether intact or not), especially an ORF encoding a prophage integrase and situated upstream of the flanking tRNA gene; and finally (v) its sequence has a heterogenous GC content, reflecting a genetic mosaic. An homologous YAPI is present in the *Y. enterocolitica* chromosome while the PAI is absent in *Y. pestis*.

INTRODUCTION

To date, only one PAI has been well characterized on the chromosome of the three pathogenic *Yersinia*, *Y. pestis*, *Y. pseudotuberculosis*, and *Y. enterocolitica* (1): the so-called High-Pathogenicity Island (HPI) ranges from 36 to 43 kb in size (according to the species in question) and bears genes involved in the biosynthesis, transport and regulation of the yersiniabactin siderophore (see Chapter 14). We discovered a 98,058-base pair (bp) segment that is inserted in the chromosome of the HPI-positive *Y. pseudotuberculosis* O:1 strain 32777 between the *Y. pestis* CO92 *YPO0339* and *YPO0341* homolog genes (Parkhill et al., 2001). The segment is associated with a phenylalanine (*phe*)-specific tRNA gene, a frequent target for PAI integration (Hacker and Kaper, 2000). The 3' end of the *phe* tRNA-encoding gene contains a 54-nucleotide motif that is directly repeated (DR) at the opposite extremity of the 98-kb segment, DR sequences being another feature common to PAIs (Hacker and Kaper, 2000). This large chromosomal segment (called YAPI, for *Yersinia* Adhesion Pathogenicity Island) is missing in the region separating *YPO0339* and *YPO0341* in the fully-sequenced genome of serotype O:1 strain 32953 (HPI^+).

GENETIC ORGANIZATION OF YAPI

The 98,058-bp sequence contains ninety-five putative open reading frames (ORFs) (Collyn et al., 2004). The overall GC content of the fully-sequenced DNA segment is 47.8% but the base composition was found to be heterogenous (between 31.2 and 60.6 %) along the length of the 98-kb region, reflecting a genetic mosaic (Figure 1). Of these ORFs, eight encoded putative proteins that lack significantly similar counterparts in public databases,

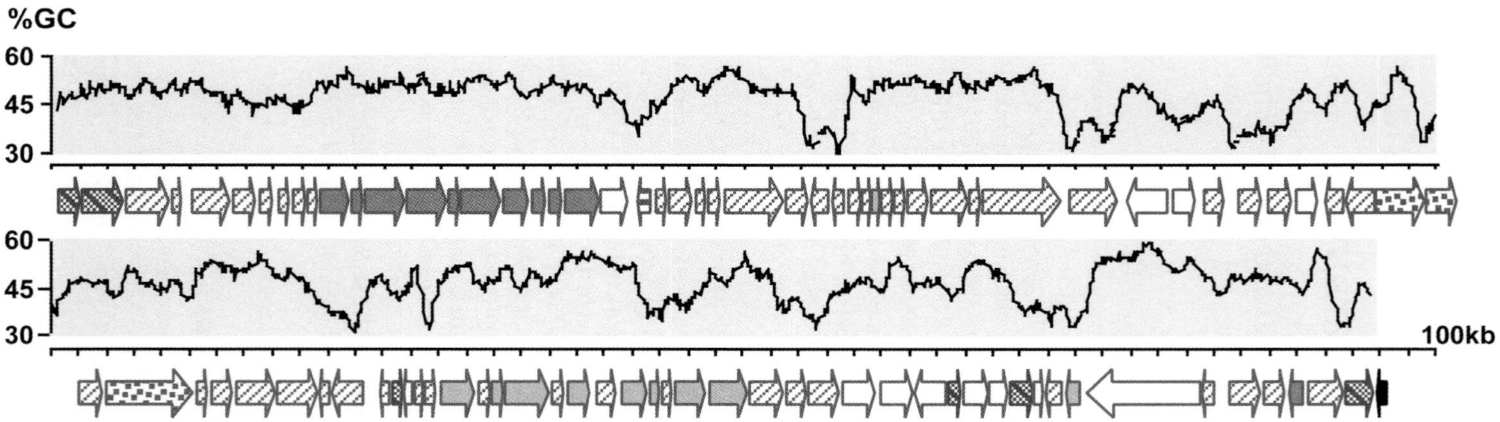

Figure 1. Genetic organization of YAPI from *Y. pseudotuberculosis* 32777. ORFs related to virulence ( ), metabolism ( ), regulation ( ), phage and plasmid function ( ), insertion elements ( ), type I restriction-modification systems ( ) and unknown functions ( ) are positioned and orientated along the 98-kilobase (kb) segment. The *phe-tRNA* gene ( ) at the 3' extremity flanks a putative integrase-encoding gene, i.e. the last gene of the PAI. Numbers on the right hand side of the map indicate the scale in kb. The DNA sequence's guanine and cytosine (G + C) percentage (determined using Artemis software with a 500-base window size) is indicated above the genetic map. The nucleotide sequence data reported here have been deposited in the EMBL nucleotide sequence data bases (accession no. AJ 627388).

whereas forty-three are similar to known protein sequences of unknown function. The latter are specified by coding sequences (CDSs) found principally on the 134-kb PAI (SPI-7, previously called the Major Pathogenicity Island) of *S. enterica* serovars Typhi and Dublin (Zhang et al., 2000), and, to a lesser extent, on the *Ralstonia solanacearum* megaplasmid (Salanoubat et al., 2002). Forty-four ORFs code for proteins with putative functions (Table 1), a marked fraction of which are derived from mobile, accessory genetic elements such as IS elements, bacteriophages, and plasmids. The IS elements include different subtypes (IS*Sod13*-like, IS*100*, IS*110*-like, IS*285*, IS*630*-like, IS*911*-like and IS*1353*-like) and can be complete or partial. One of these elements (IS*100)* disrupts an ORF which, at the nucleotide level, is identical to a gene previously identified in the *Y. pestis* genome (*YPO1092),* where it encodes a DNA-binding protein of phage origin. An intact ORF (*api95*) adjacent to the *phe-tRNA* gene specifies a 326-amino acid product displaying a high degree of homology to recombinases from the Cre family, which includes various bacteriophage integrases. Furthermore, for two CDSs at the 5' extremity of this large DNA segment, the deduced proteins are homologous to phage and plasmid DNA helicases (for *api2*), ATPases involved in plasmid partitioning (for *api1*). Finally, one ORF (*api88*) code for product similar to rearrangement hot spot (Rhs)-related protein (Hill, 1999).

The 98-kb fragment harbors eleven genes, *pilLMNOPQRSUVW*, that are organized in a polycistronic unit, and which products (except that of *pilW*) display 30-55 % identity and 46-70 % similarity to PilLMNOPQRSUV proteins involved in the biogenesis of type IV pili in *Salmonella* (Collyn et al., 2002). These appendages, which may be peritrichous or polar on the bacterial cell surface sometimes form bundles, and have been implicated in a variety of microbial functions including cell adhesion, bacteriophage adsorption, plasmid transfer and twitching motility, a form of flagellum-independent locomotion. The assembly machinery involved in their formation consists of a set of proteins encoded by genes either scattered throughout the bacterial genome (IVA subclass), or organized into operons consisting of 11 to 14 genes (IVB subclass) (Manning and Meyer, 1997). Expression of *Y. pseudotuberculosis pilLMNOPQRSUV* genes in *Escherichia coli* K-12 reconstitutes a bundle-forming pilus (Figure 2). Pili are composed of pilin subunits and pilins -the primary structures of which are conserved- are produced from prepilin molecules through cleavage of the leader sequence by a prepilin peptidase. Type IV pili are classified into two subclasses, IVA and IVB, on the basis of similarities in the deduced amino acid sequences of prepilins. Type IVA prepilins have a very short leader sequence (5 to 6 amino acids) and the N-terminal amino acid of the mature pilin is a methylated phenylalanine. Type IVB prepilins tend to have longer signal sequences (13 to 30 amino acids) and the N-terminal amino acid of the mature protein may be a methionine (methylated in some cases), leucine, tryptophan or serine (Manning and Meyer, 1997). The products of *Y. pseudotuberculosis pilS* and *pilV* are thought to be structural prepilins, and a putative signal sequence cleavage site has been identified between the 14 and 15th amino acid of prePilS and the 11 and 12th amino acid of prePilV, i.e. a glycine and a tryptophan: therefore, *Y. pseudotuberculosis* type IV pili belong to the IVB subclass. Type IV pili contribute to *Y. pseudotuberculosis* virulence, since deletion of almost every element of the gene cluster is associated with a decrease of bacterial pathogenicity *per os* in a murine model (Figure 3). A striking finding is the presence of an ORF homologous to the *Salmonella pilK* gene (*api 93*) near the 3' end of the 98-kb segment (in *Salmonella, pilK* is upstream of *pilL*). *In vitro* transcription of the *Y. pseudotuberculosis* type IV pilus gene cluster is regulated by oxygen tension,

Table 1. Characteristics of the ninety-five CDS harbored on YAPI from *Y. pseudotuberculosis* 32777.				
CDS	Size (aa[a])	Similarities (determined by BLAST)	% Identity/ Homology	Accession number
api1	293	plasmid partitioning ATPase STY4521 (*Salmonella enterica* Typhi)	61/77	NP_458616.1
api2	453	DNA helicase STY4522 (*S. enterica* Typhi)	56/73	NP_458617.1
api3	560	hypothetical protein STY4523 (*S. enterica* Typhi)	47/62	NP_458618.1
api4	193	hypothetical protein STY4526 (*S. enterica* Typhi)	50/72	CAD09309.1
api5	399	hypothetical protein STY4528 (*S. enterica* Typhi)	46/62	NP_458622.1
api6	252	hypothetical protein STY4529 (*S. enterica* Typhi)	45/60	NP_458623.1
api7	180	hypothetical protein STY4534 (*S. enterica* Typhi)	64/83	NP_458625.1
api8	143	ORFC8 (*Ralstonia solanacearum*)	43/59	NP_052313.1
api9	77	hypothetical protein NMB1666 (*Neisseria meningitidis*)	45/58	NP_274671.1
api10	128	hypothetical protein NMB1665 (*N. meningitidis*)	45/65	NP_274670.1
pilL	356	PilL (*S. enterica* Typhi)	44/55	NP_458628.1
pilM	145	PilM (*S. enterica* Typhi)	42/57	NP_458629.1
pilN	530	PilN (*S. enterica* Typhimurium)	48/65	BAA77974.1
pilO	445	PilO (*S. enterica* Typhi)	33/51	NP_458631.1
pilP	164	PilP (*S. enterica* Typhi)	36/55	NP_458632.1
pilQ	517	PilQ (*S. enterica* Typhi)	55/71	NP_458633.1
pilR	369	PilR (*S. enterica* Typhimurium)	47/67	BAA77978.1
pilS	195	PilS (*S. enterica* Typhimurium)	54/70	BAA77979.1
pilU	222	PilU (*S. enterica* Typhi)	34/51	NP_458637.1
pilV	425	Shufflon PilV A' (*S. enterica* Typhimurium)	54/68	P09746
pilW	319	YPO0076 transposase (*Yersinia pestis*)	73/83	AD0010
api22	118	putative transcriptional regulator (*Serratia* spp.)	56/70	AAM47490.1
api23	110	unknown		
api24	272	hypothetical protein STY4558 (*S. enterica* Typhi)	54/72	AE1029
api25	212	hypothetical protein STY4559 (*S. enterica* Typhi)	59/73	NP_458645.1
api26	173	hypothetical protein STY4560 (*S. enterica* Typhi)	45/64	NP_458646.1
api27	706	hypothetical protein STY4562 (*S. enterica* Typhi)	72/85	NP_458648.1
api28	250	hypothetical protein STY4563 (*S. enterica* Typhi)	51/68	NP_458649.1
api29	230	unknown		
api30	156	unknown		
api31	102	hypothetical protein STY4564 (*S. enterica* Typhi)	62/76	NP_458650.1
api32	81	hypothetical protein STY4565 (*S. enterica* Typhi)	38/64	NP_458651.1
api33	116	putative oxydase STY4566 (*S. enterica* Typhi)	41/58	NP_458652.1
api34	124	hypothetical protein (*Pseudomonas fluorescens*)	37/56	ZP_00087885.1
api35	217	hypothetical protein STY4568 (*S. enterica* Typhi)	70/82	NP_458653.1
api36	277	hypothetical protein STY4569 (*S. enterica* Typhi)	58/69	NP_458654.1
api37	505	hypothetical protein STY4570 (*S. enterica* Typhi)	51/66	NP_458655.1
api38	134	hypothetical protein STY4571 (*S. enterica* Typhi)	53/70	NP_458656.1
api39	948	hypothetical proteins STY4572 (*S. enterica* Typhi) and STY4573 (*S. enterica* Typhi)	55/70 65/77	NP_458657.1 NP_458658.1
api40	571	putative abortive infection phage resistance protein AbiR (*Nitrosomas europaea*)	20/39	CAD84400.1
api41	402	IS*285* transposase (*Y. pestis*)	99/100	NP_403677.1
api42	309	IS*630*-like transposase (*S. enterica* Choleraesuis)	70/81	BAB20556.1
api43	293	unknown		
api44	381	unknown		

api45	245	putative poly(A)-specific ribonuclease subunit, C end (*Schizosaccharomyces pombe*)	24/44	CAA91128.1
api46	289	ISSod*13*-like transposase, N end (*Shewanella oneidensis*)	80/89	AAN56855.1
api47	226	hypothetical protein (*Streptomyces avermitilis*)	33/50	NP_824450.1
api48	239	hypothectical protein (*Mesorhizobium loti*) and hypothetical protein (*S. avermitilis*)	40/58 25/43	NP_824451.1 NP_104999.1
api49	568	type I restriction-modification system, methylation subunit HsdM (*Methanosarcina mazei*)	74/83	NP_632453.1
api50	449	type I restriction-modification system, specificity subunit HsdS (*Klebsiella pneumoniae*)	40/55	AAB70708.1
api51	395	hypothetical protein (*Haemophilus somnus*)	41/61	ZP_00122800.1
api52	1043	type I restriction-modification system, restriction subunit HsdR (*M. mazei*)	69/82	NP_632455.1
api53	134	hypothetical protein STY4575 (*S. enterica* Typhi)	49/65	NP_458660.1
api54	317	hypothetical protein STY4576 (*S. enterica* Typhi)	63/78	NP_458661.1
api55	468	hypothetical protein STY4577 (*S. enterica* Typhi)	60/78	NP_458662.1
api56	507	hypothetical protein STY4579 (*S. enterica* Typhi)	45/65	NP_458664.1
api57	98	unknown		
api58	419	hypothetical protein BPP0992 (*Bordetella parapertussis*) and hypothetical protein BPP0991 (*B. parapertussis*)	54/76 48/71	NP_883312.1 NP_883311.1
api59	65	unknown		
api60	55	DNA-binding phagic protein YPO1092, N end (*Y. pestis*)	51/76	AE0134
api61	80	IS*1353*-like transposase, C end (*Shigella flexneri*)	55/63	AAL72503.1
api62	63	unknown		
api63	83	hypothetical protein c4523, C end (*E. coli* CFT073)	60/71	NP_756383.1
api64	486	probable tRNA synthetase RSp1438 (*R. solanacearum*)	48/64	NP_522997.1
api65	154	hypothetical protein RSp1437 (*R. solanacearum*)	35/46	NP_522996.1
api66	157	putative acetyltransferase RSp1436 (*R. solanacearum*)	47/61	NP_522995.1
api67	542	putative AMP-binding enzyme RSp1434 (*R. solanacearum*)	50/62	NP_522993.1
api68	219	hypothetical proteins RSp 1433 (*R. solanacearum*) and RSp 1432 (*R. solanacearum*)	54/70 42/60	CAD18584.1 CAD18583.1
api69	348	putative oxydoreductase signal peptide RSp1431 (*R. solanacearum*)	39/54	NP_522990.1
api70	264	hypothetical protein RSp1430 (*R. solanacearum*)[b]	42/58	NP_522989.1
api71	298	putative esterase RSp1429 (*R. solanacearum*)[b]	36/49	NP_522988.1
api72	132	rhodanese-like protein RSp1428 (*R. solanacearum*)[b]	66/84	NP_522987.1
api73	121	hypothetical protein RSp1426 (*R. solanacearum*)	45/62	NP_522985.1
api74	443	probable L-ornithine 5-monooxygenase oxidoreductase protein PvdA (*R. solanacearum*)	57/71	NP_522984.1
api75	443	probable diaminobutyrate--pyruvate aminotransferase protein EctB (*R. solanacearum*)	60/73	NP_522983.1
api76	410	putative transmembrane protein RSp1423 (*R. solanacearum*)	44/60	NP_522982.1
api77	319	putative transmembrane protein RSp1427 (*R. solanacearum*)	28/51	NP_522986.1
api78	305	hypothetical protein BtrH (*Bacillus circulans*)	25/39	BAC41215.1
api79	344	IS*110*-like transposase (*S. oneidensis*)	47/66	NP_719476.1
api80	88	IS*911*-like inactive transposase (*E. coli*)	95/98	CAD48417.1
api81	344	IS*110*-like transposase (*S. oneidensis*)	45/68	NP_719476.1
api82	102	DNA-binding phagic protein YPO1092, N end (*Y. pestis*)	50/72	NP_404706.1
api83	340	IS*100* transposase (*Y. pestis*)	100	NP_403697.1
api84	259	orfB of IS*100* (*Y. pestis*)	100	NP_395139.1
api85	236	DNA-binding phagic protein YPO1092, C end (*Y. pestis*)	47/67	NP_404706.1
api86	17	Transposase, fragment (*S. oneidensis*)	77/94	NP_858158.1
api87	206	hypothetical protein (*Photorhabdus luminescens*)	66/82	NP_931453.1

api88	157	hypothetical protein SciY (*S. enterica* Typhimurium)	40/57	NP_459292.1
api89	1423	hypothetical protein similar to Rhs family Plu4280 (*P. luminescens*)	54/67	NP_931456.1
api90	143	hypothetical protein STY0320 (*S. enterica* Typhi)	33/53	AF0538
api91	323	hypothetical protein (*P. fluorescens*)	47/60	ZP_00087873.1
api92	301	putative membrane protein YccB (*S. enterica* Typhimurium)	32/52	NP_052478.1
api93	215	PilK (*S. enterica* Typhimurium)	24/40	BAA77971.1
api94	398	hypothetical protein STY 4665 (*S. enterica* Typhi)	26/40	NP_928477.1
api95	326	probable phage integrase STY 4666 (*S. enterica* Typhi)	54/70	NP_928478.1

[a] amino acids
[b] CDS (entire or not) present in the *Y. pestis* chromosome

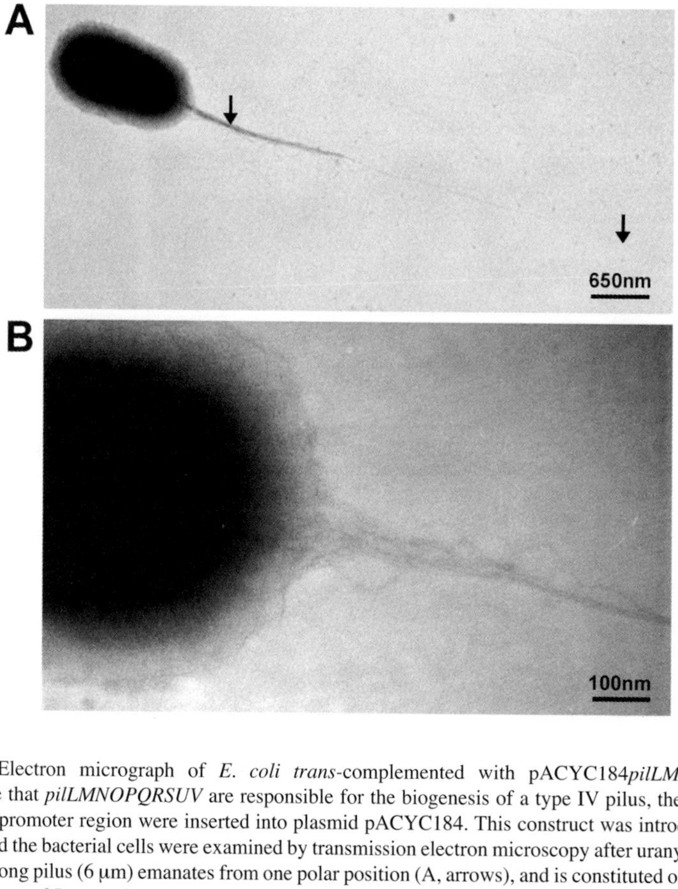

Figure 2. Electron micrograph of *E. coli trans*-complemented with pACYC184*pilLMNOPQRSUV*. To demonstrate that *pilLMNOPQRSUV* are responsible for the biogenesis of a type IV pilus, the gene cluster and its putative promoter region were inserted into plasmid pACYC184. This construct was introduced into *E. coli* MC1061 and the bacterial cells were examined by transmission electron microscopy after uranyl acetate negative staining. A long pilus (6 μm) emanates from one polar position (A, arrows), and is constituted of bundles of fibres with a diameter of 5 nm, a characteristic feature of type IV pili. Control strain MC1061 carrying only pACYC184 was non-piliated (not shown).

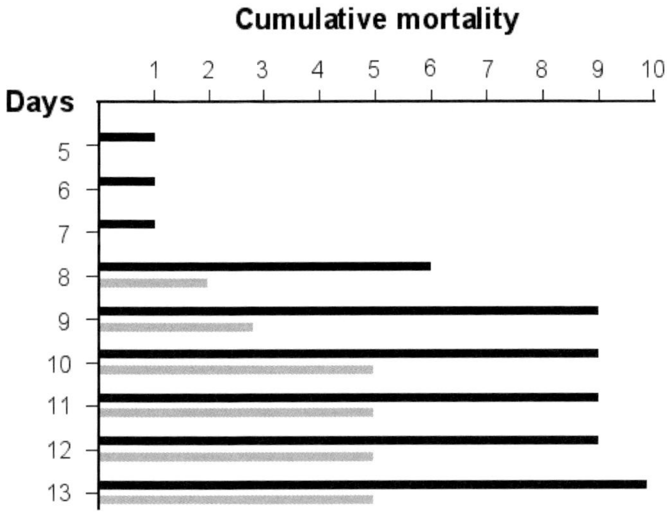

Figure 3. Role of the type IV pilus in *Y. pseudotuberculosis* 32777 pathogenicity. BALB/c mice (groups of ten) were inoculated intragastrically with 10^8 viable bacteria (wild type or type IV pilus-deficient mutant), and mortality was recorded over the two-week period following bacterial challenge. No mice infected with the type IV pilus-deficient mutant died after day 14. Black, wild type; hatched, type IV pilus-deficient mutant.

temperature, cellular density and osmolarity (Collyn et al., 2002), and indeed one can identify a putative transcriptional regulator gene (*api22*) contiguous to *pilW* (the last gene of the type IV pilus gene cluster) but with opposite polarity.

Besides the type IV pilus genes, another notable feature is the presence of a gene cluster (*api49*, *50* and *52*) that is predicted to encode enzymatic subunits of a type I restriction modification system. Finally, the downstream region of the 98-kb chromosomal fragment comprises a 13-kb region (*api64* to *api77*), the gene organization and products of which are similar to a *R. solanacerum* megaplasmid segment containing fifteen CDSs (*rsp1423* to *rsp1438*) mostly involved in general metabolism.

PAIs have a tendency to spontaneously delete specific sequences or even the whole genetic element. Screening of spontaneous deletants after insertion of reporter genes into *pilS* showed that deletion of YAPI (i) does not differ in frequency (c.a. 1 x 10^{-7}) from that of the urease locus, a chromosomal region known to be stable, (ii) results in perfect excision of the 98,058-bp segment, and finally (iii) does not result in a significant decrease of bacterial virulence for the BALB/c mouse when compared to inactivation of the *pil* gene cluster alone. Bearing in mind the limitations of the experimental model of infection, this latter finding strongly suggests that no virulence genes other than *pil* are present within YAPI (Collyn et al., 2004).

DISTRIBUTION OF YAPI IN PATHOGENIC YERSINIAE

By screening with PCR for the presence of YAPI in randomly selected *Y. pseudotuberculosis* isolates collected from various areas throughout the world, we showed that the PAI is not distributed in all strains of the species: it is noteworthy that the genome-sequenced strain 32953 lacks YAPI. Its presence was found to be independent of the O-serotype.

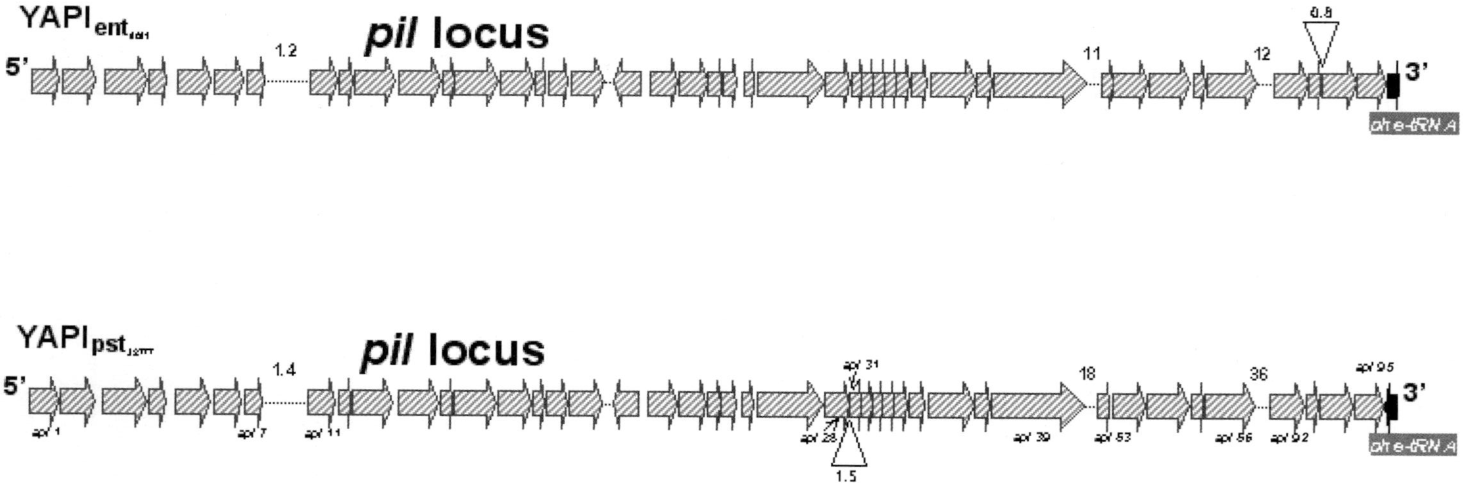

Figure 4. Genetic organization of YAPI from enteropathogenic *Yersinia* species. Only homologous CDSs present on YAPI from *Y. pseudotuberculosis* 32777 ($YAPI_{pst}$) and *Y. enterocolitica* 8081 ($YAPI_{ent}$) are shown. Target *phe-tRNA* genes are represented by black arrows. The size (in kb) of non-homologous regions on the two PAIs are indicated. The nucleotide sequence of $YAPI_{ent}$ was downloaded from the Sanger web site (www.sanger.ac.uk/Projects/Y_enterocolitica/).

Table 2. Characteristics of the sixty-one CDS of YAPI from *Y. enterocolitica* 8081.			
CDS	Size (aa[a])	Homologue as determined by BLAST	% Identity/ Homology
1	293	*Y. pseudotuberculosis* 32777 Api1	86/92
2	455	*Y. pseudotuberculosis* 32777 Api2	86/92
3	528	*Y. pseudotuberculosis* 32777 Api3	65/73
4	172	*Y. pseudotuberculosis* 32777 Api4	83/91
5	400	*Y. pseudotuberculosis* 32777 Api5	78/84
6	252	*Y. pseudotuberculosis* 32777 Api6	83/89
7	180	*Y. pseudotuberculosis* 32777 Api7	83/83
8	82	hypothetical protein Rpal4228 (*Rhodopseudomonas palustris*)	67/75
9	149	hypothetical protein Plu4118 (*Photorhabdus luminescens*)	67/82
10	324	*Y. pseudotuberculosis* 32777 PilL	74/79
11	145	*Y. pseudotuberculosis* 32777 PilM	79/88
12	562	*Y. pseudotuberculosis* 32777 PilN	74/77
13	435	*Y. pseudotuberculosis* 32777 PilO	69/77
14	159	*Y. pseudotuberculosis* 32777 PilP	67/73
15	517	*Y. pseudotuberculosis* 32777 PilQ	90/93
16	369	*Y. pseudotuberculosis* 32777 PilR	81/85
17	195	*Y. pseudotuberculosis* 32777 PilS	65/69
18	222	*Y. pseudotuberculosis* 32777 PilU	71/76
19	425	*Y. pseudotuberculosis* 32777 PilV	86/91
20	306	*Y. pseudotuberculosis* 32777 PilW	97/98
21	118	*Y. pseudotuberculosis* 32777 Api22	88/96
22	272	*Y. pseudotuberculosis* 32777 Api24	86/90
23	212	*Y. pseudotuberculosis* 32777 Api25	86/91
24	173	*Y. pseudotuberculosis* 32777 Api26	92/93
25	706	*Y. pseudotuberculosis* 32777 Api27	94/94
26	250	*Y. pseudotuberculosis* 32777 Api28	78/80
27	102	*Y. pseudotuberculosis* 32777 Api31	85/85
28	81	*Y. pseudotuberculosis* 32777 Api32	71/71
29	116	*Y. pseudotuberculosis* 32777 Api33	64/64
30	124	*Y. pseudotuberculosis* 32777 Api34	84/86
31	217	*Y. pseudotuberculosis* 32777 Api35	95/96
32	277	*Y. pseudotuberculosis* 32777 Api36	96/96
33	502	*Y. pseudotuberculosis* 32777 Api37	94/95
34	134	*Y. pseudotuberculosis* 32777 Api38	81/81
35	934	*Y. pseudotuberculosis* 32777 Api39	91/92
36	78	putative antitoxin CcdA of gyrase inhibiting toxin (*E. coli* O157:H7)	66/84
37	104	putative gyrase inhibiting toxin CcdB (*E. coli* O157:H7)	54/74
38	378	probable membrane protein YdaA (*S. enterica* Typhimurium)	81/86
39	97	hypothetical protein YcjA (*S. enterica* Typhimurium)	96/98
40	203	arsenic resistance protein ArsH (*S. enterica* Typhimurium)	90/91
41	397	IS *1330* transposase (*Y. enterocolitica*)	95/95
42	141	arsenate reductase ArsC (*S. enterica* Typhimurium)	93/95

43	429	transmembranar arsenical pump ArsB (*S. enterica* Typhimurium)	89/91
44	588	ATPase catalytic subunit ArsA (*E. coli*)	91/93
45	120	arsenic resistance protein ArsD (*S. enterica* Typhimurium)	95/96
46	117	inducible repressor ArsR (*S. enterica* Typhimurium)	94/96
47	133	*Y. pseudotuberculosis* 32777 ApiI53	92/98
48	317	*Y. pseudotuberculosis* 32777 ApiI54	86/89
49	468	*Y. pseudotuberculosis* 32777 ApiI55	89/89
50	506	*Y. pseudotuberculosis* 32777 ApiI56	91/92
51	327	conserved hypothetical protein ORF SG78 (*Pseudomonas aeruginosa*)	27/46
52	365	conserved hypothetical protein ORF SG77 (*P. aeruginosa*)	27/50
53	295	putative DNA-binding phagic protein YPO1092 (*Y. pestis*)	48/68
54	333	anti-restriction protein ArdC (*Y. enterocolitica*)	43/60
55	300	*Y. pseudotuberculosis* 32777 Api92	85/92
56	194	*Y. pseudotuberculosis* 32777 PilK	61/68
57	800	putative hemagglutinin-like secreted protein Y1701 (*Y. pestis*)	68/74
58	113	hypothetical protein Plu3583 (*P. luminescens*)	41/58
59	114	hypothetical protein Plu3707 (*P. luminescens*)	32/47
60	398	*Y. pseudotuberculosis* 32777 Api94	83/89
61	327	*Y. pseudotuberculosis* 32777 Api95	92/95

[a] amino acids.

Additionally, we found that YAPI was inserted into either of the two copies of *phe-tRNA* gene. An homologous DNA segment (although shorter in length, 66 kb) was detected *in silico* in the genome-sequenced strain 8081 of *Y. enterocolitica*. Also associated with a *phe*-specific tRNA locus, the region shared 41 of its 61 predicted ORFs (64 to 97% identity at the protein level) with the YAPI from *Y. pseudotuberculosis* 32777 ($YAPI_{pst}$), and homologous CDSs (including *pil* genes) were arranged in the same manner in the two chromosomal segments (Figure 4). The first 36 kb of $YAPI_{pst}$ and $YAPI_{ent}$ comprise thirty-one homologous ORFs whereas the downstream regions of both PAIs were found to be unrelated, except for the 3' boundary where a putative integrase-encoding ORF terminates the island (Table 2). Differences in size and gene composition between $YAPI_{pst}$ and $YAPI_{ent}$ may have resulted from DNA deletion, insertion or recombination events during the evolution of the ancestral genetic unit in the two *Yersinia* species. This last point is evidenced by the presence of a DNA fragment on $YAPI_{pst}$ that contains ORFs similar to *Ralstonia* megaplasmid-borne metabolic genes, and an arsenic resistance operon on $YAPI_{ent}$. In contrast, *in silico* analysis of the genomes of *Y. pestis* biovars Medievalis KIM (Deng et al., 2002) and Orientalis CO92 (Parkhill et al., 2001) failed to detect YAPI-like sequences. PCRs performed on several other strains of the Medievalis and Orientalis biovars (as well as in biovar Antiqua strains) confirmed the absence of YAPI in the plague agent. At this point in time, we do not know whether YAPI is present in non-pathogenic Yersiniae and, like the HPI, is harboured by other Enterobacteriaceae (Bach et al., 2000). Analysis of YAPI distribution and polymorphism in a collection of enteropathogenic *Yersinia* strains from various animal and environmental sources (in progress in our laboratory) should provide insight into the pathogenesis and ecological fitness of these microorganisms.

ACKNOWLEDGEMENTS

Y. enterocolitica sequence data were produced by the *Yersinia enterocolitica* Sequencing Group at the Sanger Institute and were obtained from ftp://ftp.sanger.ac.uk/pub/pathogens/ye/

REFERENCES

Bach, S., de Almeida, A., and Carniel E. 2000. The *Yersinia* high-pathogenicity island is present in different members of the family *Enterobacteriaceae*. FEMS Microbiol. Lett. 183: 289-294.

Carniel, E. 2001. The *Yersinia* high-pathogenicity island: an iron-uptake island. Microbes Infect. 3: 561-569.

Collyn, F., Lety, M.A., Nair, S., Escuyer, V., Ben Younes, A., Simonet, M., and Marceau, M. 2002. *Yersinia pseudotuberculosis* harbors a type IV pilus gene cluster that contributes to pathogenicity. Infect. Immun. 70: 6196-6205.

Collyn, F., Billault, A., Mullet, C., Simonet, M., and Marceau, M. 2004. YAPI, a new *Yersinia pseudotuberculosis* pathogenicity island. Infect. Immun. 72: 4784-4790.

Deng, W., Burland, V., Plunkett, G. 3rd, Boutin, A., Mayhew, G.F., Liss, P., Perna, N.T., Rose, D.J., Mau, B., Zhou, S., Schwartz, D.C., Fetherston, J.D., Lindler, L.E., Brubaker, R.R., Plano, G.V., Straley, S.C., McDonough, K.A., Nilles, M.L., Matson, J.S., Blattner, F.R., and Perry, R.D. 2002. Genome sequence of *Yersinia pestis* KIM. J. Bacteriol. 184: 4601-4611.

Hacker, J., and Kaper, J. B. 2000. Pathogenicity islands and the evolution of microbes. Annu. Rev. Microbiol. 54: 641-679.

Hill, C. W. 1999. Large genomic sequence repetitions in bacteria: lessons from rRNA operons and Rhs elements. Res. Microbiol. 150: 665-674.

Manning, P.A., and Meyer, T.F. 1997. Type-4 pili: biogenesis, adhesins, protein export and DNA import. Proceedings of a workshop. Gene 192: 1-198.

Parkhill, J., Wren, B.W., Thomson, N.R., Titball, R.W., Holden, M.T., Prentice, M.B., Sebaihia, M., James, K.D., Churcher, C., Mungall, K.L., Baker, S., Basham, D., Bentley, S.D., Brooks, K., Cerdeno-Tarraga, A.M., Chillingworth, T., Cronin, A., Davies, R.M., Davis, P., Dougan, G., Feltwell, T., Hamlin, N., Holroyd, S., Jagels, K., Karlyshev, A.V., Leather, S., Moule, S., Oyston, P.C., Quail, M., Rutherford, K., Simmonds, M., Skelton, J., Stevens, K., Whitehead, S., and Barrell, B.G. 2001. Genome sequence of *Yersinia pestis*, the causative agent of plague. Nature 413: 523-527.

Salanoubat, M., Genin, S., Artiguenave, F., Gouzy, J., Mangenot, S., Arlat, M., Billault, A., Brottier, P., Camus, J.C., Cattolico, L., Chandler, M., Choisne, N., Claudel-Renard, C., Cunnac, S., Demange, N., Gaspin, C., Lavie, M., Moisan, A., Robert, C., Saurin, W., Schiex, T., Siguier, P., Thebault. P., Whalen, M., Wincker, P., Levy, M., Weissenbach, J., and Boucher, C.A. 2002. Genome sequence of the plant pathogen *Ralstonia solanacearum*. Nature 415: 497-502.

Zhang, X.L., Tsui, I.S., Yip, C.M., Fung, A.W., Wong, D.K., Dai, X., Yang, Y., Hackett, J., and Morris, C. 2000. *Salmonella enterica* serovar Typhi uses type IVB pili to enter human intestinal epithelial cells. Infect. Immun. 68: 3067-3073.

Chapter 16

The pYV Plasmid and the Ysc-Yop Type III Secretion System

Marie-Noëlle Marenne, Luís Jaime Mota and Guy R. Cornelis

ABSTRACT

The three pathogenic *Yersinia* spp. (*Y. pestis*, *Y. pseudotuberculosis* and *Y. enterocolitica*) harbor a 70-kb pYV plasmid that is essential for their virulence. The pYV plasmid encodes the Yop virulon consisting of a complete type III secretion system, called Ysc-Yop. This highly sophisticated virulence system allows extracellular *Yersinia* bacteria to inject "effector" Yop proteins directly into the cytosol of the eukaryotic host cells. The proteins are secreted across the two bacterial membranes and are also translocated across the eukaryotic cell membrane. This is achieved in a tightly regulated manner by a complex protein secretion machinery, the Ysc injectisome, and by "translocator" Yop proteins that are secreted by the Ysc machinery and presumably insert in the eukaryotic cell membrane. The proper functioning of the system also requires the assistance in the bacterial cytoplasm of the Syc chaperones, which bind and assist secretion of Yop proteins. Once inside the eukaryotic cell, the Yop effector proteins will subvert and disrupt host cell signaling pathways, incapacitating the host innate immune system, in particular inhibiting phagocytosis and downregulating the anti-inflammatory response.

INTRODUCTION

The *Yersinia* genus contains three pathogenic species: *Y. pestis*, the agent of plague, and two enteropathogens, *Y. pseudotuberculosis* and *Y. enterocolitica*, which are generally acquired after ingestion of contaminated food. Although these three pathogens have different entry routes and cause different diseases, they share a common tropism for the lymphoid tissues where they resist the host innate immune system. This capacity is essentially conferred by a 70-kb plasmid that is conserved among the three species and is required for virulence. This plasmid, called pYV, encodes the Yop virulon that consists of a complete type III secretion (TTS) system, called Ysc-Yop (reviewed by Cornelis, 2002a; Cornelis, 2002b; Cornelis et al., 1998) (Figure 1). TTS is a highly sophisticated virulence mechanism used by several Gram-negative bacteria that are pathogenic for animals or plants, or symbionts for plants and insects (reviewed by Cornelis and Van Gijsegem, 2000). By this mechanism, bacteria that are either extracellular or localized in phagosomes communicate with eukaryotic cells by injecting bacterial proteins into the cytosol of these cells. The proteins not only are secreted across the two bacterial membranes but are also translocated across the eukaryotic cell membrane. Inside the eukaryotic cell, these bacterial proteins will subvert and disrupt host cell signaling pathways to the benefit of the pathogenic bacteria.

In *Yersinia*, TTS can be triggered by incubating the bacteria at 37°C in the absence of calcium ions. In these conditions *Yersinia* bacteria no longer grow but instead release large

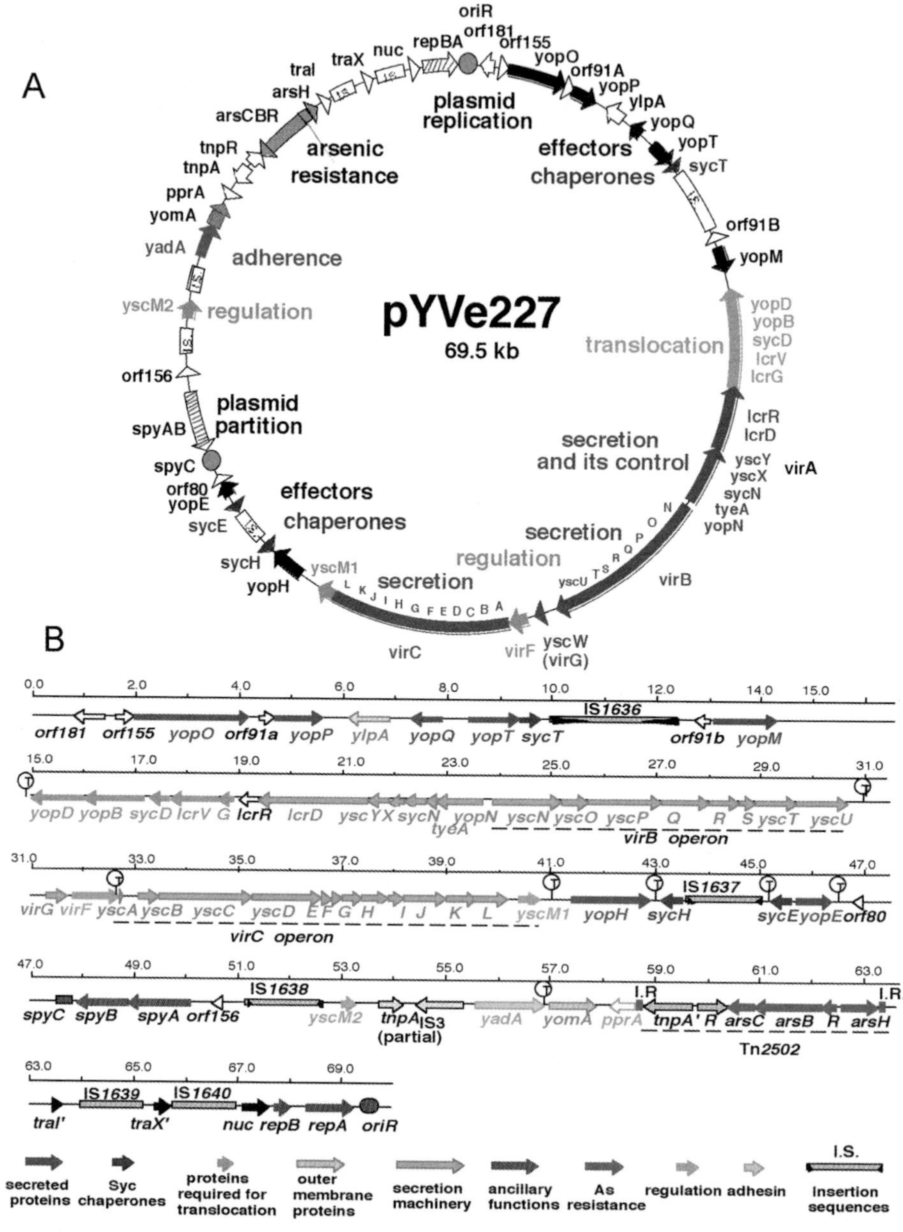

Figure 1. Detailed genetic map of the pYVe227 plasmid of *Y. enterocolitica* W22703. Reprinted from Iriarte and Cornelis (1999a) with permission of the publisher.

amounts of proteins called Yops in the culture supernatant (Michiels et al., 1990). This is probably an artifactual leakage, but this observation allowed the identification of the *ysc* (Yop secretion) genes, involved in the process of Yops release. As a result of these studies, the Ysc-Yop TTS system has become one of the best understood. As mentioned above, the system consists of secreted proteins called Yops and their dedicated TTS apparatus, called the Ysc injectisome. Secretion of some Yops requires the assistance, in the bacterial cytosol, of small individual chaperones called the Syc (specific Yop chaperone) proteins that bind specifically to their cognate Yop. The Yop proteins include intracellular "effectors" (YopE, YopH, YopM, YpkA/YopO, YopJ/P, YopT) and "translocators" (YopB, YopD, LcrV), needed to deliver the effectors across the plasma membrane into the cytosol of eukaryotic target cells. The system also secretes proteins that seem to have an exclusive regulatory role (YopN, YopQ, YscM/LcrQ), components of the Ysc injectisome (YscP, YscF) and one protein with unknown function (YopR). Physiological secretion of Yops is triggered by intimate contact between an invading bacterium and a target cell. The whole system is regulated at several levels to ensure that the cocktail of six Yop effectors is delivered inside the eukaryotic cells at the right time and in the right place. Even if the mechanistic details of how the pYV-encoded *Yersinia* TTS system works are far from being totally understood, astonishing progress has been achieved in the last ten-fifteen years and a clear overall picture has emerged (Figure 2).

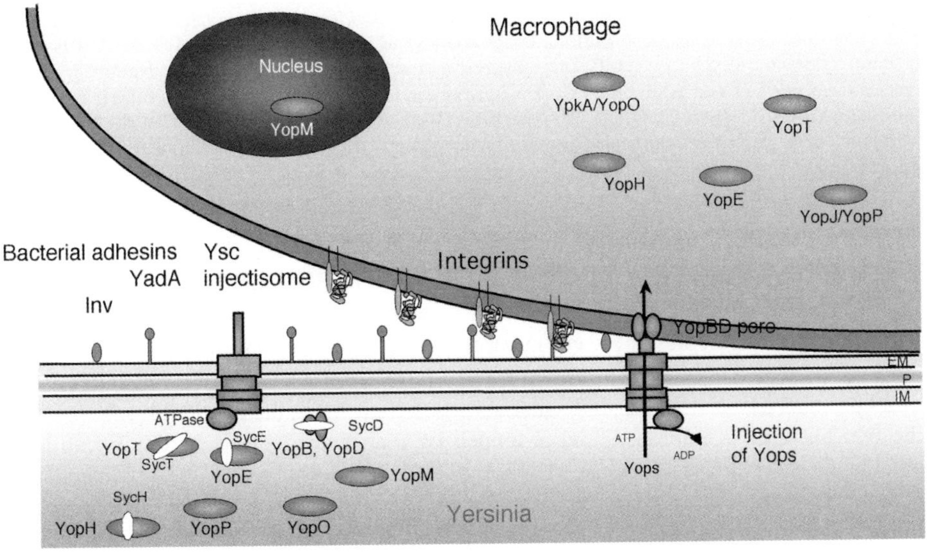

Figure 2. Secretion of Yops by the Ysc injectisome and translocation across the target cell membrane. When *Yersinia* are placed at 37°C, a needle-like structure called the Ysc injectisome is assembled and a stock of Yop proteins is produced. Some Yops are kept in the cytoplasm bound to their specific Syc chaperone. In the absence of contact with a eukaryotic cell, the secretion channel is closed. Upon contact with a eukaryotic cell, the bacterial adhesins Invasin and YadA interact with integrins at the surface of the eukaryotic cell, which docks the bacterium at the cell's surface and promotes opening of the secretion channel. YopB and YopD form a pore in the target cell plasma membrane, and the Yop effectors are delivered into the eukaryotic cell cytosol through this pore. Among the effectors, YopM is further translocated into the cell nucleus. EM, outer membrane. P, peptidoglycan; IM, plasma membrane. Reprinted from Cornelis (2002a) with permission of the publisher.

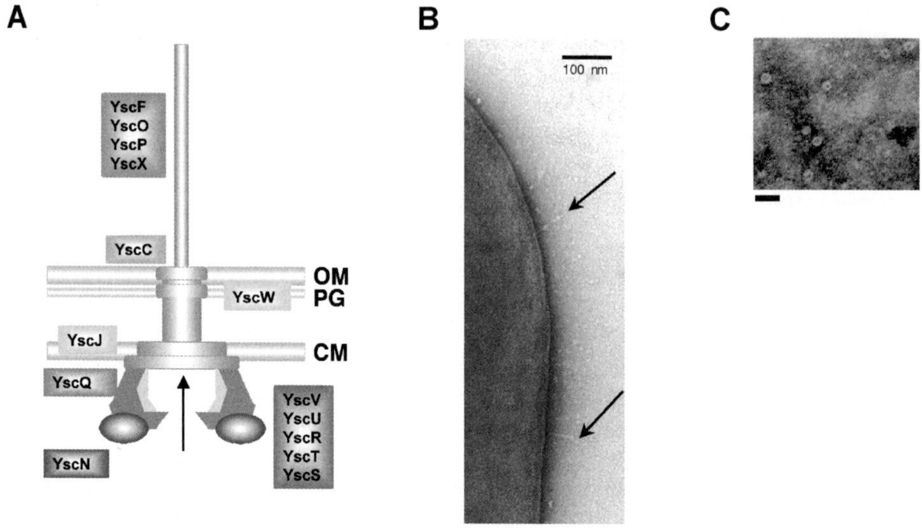

Figure 3. The Ysc injectisome. A. Schematic representation of the Ysc injectisome spanning the outer membrane (OM), the peptidoglycan layer (PG) and the cytosplamic membrane (CM) of the bacterium. The ring spanning the OM is made of the secretin YscC, assisted by lipoprotein YscW. YscJ is another lipoproptein. YscF, YscO, YscP and YscX are external parts of the injectisome. YscF is the main constituent of the needle. YscV, YscU, YscR, YscT and YscS are proteins of the basal body that are in contact with the CM. YscN is the ATPase of the pump. YscQ is probably localized to the large inner cylinder. B. An electron micrograph of injectisome needles protruding from *Yersinia enterocolitica* E40. Image courtesy of L. Journet, P. Broz and G. R. C., University of Basel (Biozentrum), Switzerland. C. Macromolecular structure of the YscC complex. Negatively stained preparation of purified YscC analyzed by electron microscopy. Circular structures with stain-filled pores that are sometimes located off-centre are visible. The scale bar is 50 nm. A. and B. Reprinted from Cornelis (2002b) with permission of the publisher; C. reprinted from Koster et al. (1997) with permission of the publisher.

THE Ysc INJECTISOME, A MOLECULAR WEAPON TO INJECT BACTERIAL PROTEINS INTO EUKARYOTIC CELLS

Prior to contact with eukaryotic target cells, *Yersinia* incubated at the temperature of their host build multiple copies of a secretion apparatus at their surface (Michiels and Cornelis, 1991) (Figure 3). This protein secretion apparatus, called the Ysc injectisome, is composed of 29 proteins associated in a molecular syringe which spans the peptidoglycan layer and the two bacterial membranes, and is topped by a stiff needle-like structure protruding outside the bacterium (Hoiczyk and Blobel, 2001). The internal part of the Ysc injectisome contains proteins that have counterparts in the basal body of the flagellum, indicating that the two structures have a common evolutionary origin. An essential protein of this internal part is an ATPase resembling the α and β subunits of F_0F_1 proton translocase (Woestyn et al., 1994). The external part of the Ysc injectisome, which spans the bacterial outer membrane, is a homomultimeric ring-shaped structure with an external diameter of about 200 Å and an apparent central pore of 50 Å (Koster et al., 1997) (Figure 3C). YscC, the monomer of this complex, belongs to the family of secretins. The Ysc injectisome ends with a 6-7 nm-wide needle formed by the polymerization of monomers of the 6-kDa YscF protein that are secreted by the Ysc apparatus itself (Hoiczyk and Blobel, 2001). The needle has a hollow center of about 2 nm. The length of the needle varies between different

Yersinia species and presumably also between different serotypes of *Y. enterocolitica*. It is 58 ± 10 nm in *Y. enterocolitica* O:9 but only 41 ± 8 nm in *Y. pestis* ((Journet et al., 2003). This length is controlled by the protein YscP acting as a molecular ruler. Indeed, there is a linear correlation between the needle length and the number of residues in YscP, with an increment of 1.9 Å per residue (Journet et al., 2003). *yscP* null mutant bacteria display needles of indefinite length and do not secrete Yops. This suggests that, in addition to its molecular ruler function, YscP also acts as a switch controlling which substrate is exported. When the needle reaches its normal length, YscP signals the export apparatus to stop secreting YscF subunits and to be ready to secrete Yops upon cell contact (Journet et al., 2003; Edqvist et al., 2003).

The complete Ysc injectisome of *Yersinia* has not been isolated yet. However, the entire TTS complex of *Salmonella enterica* serovar Typhimurium (Kimbrough and Miller, 2000; Kubori et al., 1998; Kubori et al., 2000) and *Shigella flexneri* (Blocker et al., 2001) have been purified and visualized by electron microscopy. These structures are composed of three domains: the export apparatus, the basal body and the needle. The sequence similarity between proteins composing the *Salmonella*/*Shigella* injectisome and the *Yersinia* injectisome suggests that the *Yersinia* injectisome is similar to the ones visualized so far. The export apparatus, installed in the basal body, exports the components of the needle as well as the Yop proteins. The basal body consists of two pairs of rings that are anchored to the inner and outer membranes of the bacterial envelope and joined by a central rod.

THE TYPE III SECRETION SIGNAL CONTROVERSY

One of the still unsolved main questions about the TTS process is how proteins apparently not structurally related are recognized and secreted by the same machinery. At the time of the discovery of type III secretion, Michiels et al. (1990) noticed that the Yop proteins are recognized via their amino terminus, and that no classical signal sequence is cleaved off during Yop secretion. By successive deletions of hybrid proteins composed of a Yop and a reporter, the minimal region sufficient for secretion was determined to be only about 10-15 residues (Anderson and Schneewind, 1999; Sory et al., 1995). However, the idea that the secretion signal resides in the secreted protein itself was challenged by Anderson and Schneewind (1997, 1999), who performed systematic mutagenesis experiments of the amino-terminal secretion signal of hybrid proteins composed of a Yop protein and the neomycin phosphotransferase (Npt) protein. These authors did not observe any point mutation that specifically abolished secretion. Some frameshift mutations that altered the peptide sequences of the whole secretion signal also failed to prevent secretion of hybrid proteins. These results suggested that the secretion signal is encoded in the mRNA rather than in the peptide sequence (Anderson and Schneewind, 1997; Anderson and Schneewind, 1999).

The mRNA signal hypothesis was challenged by Lloyd et al. (2001). These authors observed that mutations in the first 11 codons of YopE that modify the mRNA but not the amino-acid sequence did not impair secretion. This observation indicates that the amino-terminus of YopE, and not the 5' end of *yopE* mRNA, serves as a secretion signal. The same authors also observed that the amino terminus of Yops contain an amphipathic sequence composed of alternating hydrophobic and polar residues (Lloyd et al., 2001; Lloyd et al., 2002). By systematic replacement of all the amino acids of the N-terminus region of YopE, Lloyd et al. (2002) observed that amino terminal amphipathic sequences have a higher probability to allow YopE secretion than do hydrophobic or hydrophilic sequences. These results suggested that the amino terminal amphipatic sequence could be the secretion signal of the Yop proteins.

Cheng et al. (1997) showed that YopE possesses not one but two distinct secretion signals. In agreement with the work of Sory et al. (1995), these authors observed that a fusion between the first 15 residues of YopE and Npt is secreted independently of the presence of the chaperone. The secretion of the hybrid protein is abolished when a +1 frameshift is introduced in the first secretion signal (Anderson and Schneewind, 1997; Cheng et al., 1997). When the +1 frameshift is introduced in a fusion protein composed of the entire YopE and Npt, the hybrid is weakly secreted, but only in presence of the SycE chaperone. This observation suggested the presence of a second, chaperone-dependent secretion signal located downstream of the first one. This signal was localized between residues 15 and 100 of YopE. However, experiments by Lloyd et al. (2001) performed with the native YopE protein are in contradiction with the observations of Cheng et al. (1997). Lloyd et al. (2001) confirmed that YopE could be secreted in the absence of SycE. A frameshift mutation in the amino terminus secretion signal did not abolish YopE secretion in presence of SycE. However, the deletion of the first 15 amino terminal residues completely prevented the secretion of YopE in the presence or absence of SycE. This contradicts the model of Cheng et al. (1997) proposing the existence of a second secretion signal which is independent of the first one and is sufficient for secretion.

Recently, Birtalan et al. (2002) hypothesized that the three-dimensional structure of the complex effector/chaperone could interact with the type III secretion machinery, probably with the ATPase YscN, and function as a secretion signal. However, no experimental observation confirms this hypothesis.

THE Syc CHAPERONES

COMMON PROPERTIES OF THE Syc CHAPERONES

The chaperones in the *Yersinia* Ysc-Yop type III secretion system are called Syc proteins (for "specific Yop chaperone"). Chaperones have been described for three of the effector Yops, SycE for YopE, SycH for YopH, and SycT for YopT. The translocator Yops, YopB, and YopD share the same chaperone, SycD, that due to its specific characteristics will be discussed later. Finally, the regulatory YopN has two chaperones, SycN and YscB, and YscY may be a chaperone of YscX, an element of the Ysc injectisome. Generally, the Syc chaperone genes are located next to the gene encoding their cognate protein. There are no obvious sequence similarities between the different chaperones but they are characteristically small (less than 20 kDa), acidic (pI ~4-5), bear a C-terminal amphiphilic α-helix, and specifically bind with high affinity to their partner (Wattiau et al., 1996).

ROLES OF THE Syc CHAPERONES

The exact function(s) of the Syc chaperones remain mysterious. The YopE-SycE pair has been the subject of much research. Amino-acid residues 15-75 of YopE (219 residues) are sufficient to allow the binding of a SycE homodimer (Birtalan et al., 2002; Wattiau and Cornelis, 1993; Woestyn et al., 1996). In good agreement with this, the three-dimensional structure of SycE bound to the SycE-binding domain of YopE indicates that SycE covers the first 78 residues of YopE (Birtalan et al., 2002). Thus, the binding site of SycE on YopE is located in the N-terminal part of the effector, just after the secretion signal and before the GTPase activating protein (GAP) domain of YopE (residues 96-219) (see below) (Birtalan et al., 2002).

The Syc chaperones were first thought to be necessary mainly for the secretion of their cognate Yop, and Wattiau and Cornelis hypothesized that they could act as a kind of secretion pilot to drive nascent Yops to the secretion machinery (Wattiau et al., 1994; Wattiau and Cornelis, 1993). This hypothesis was questioned by Woestyn et al. (1996), who analyzed the secretion of different hybrid YopE-Cya proteins. In fact, the $YopE_{1-40}$-Cya hybrid devoid of the SycE chaperone-binding site is secreted normally in the presence or absence of SycE. However, the $YopE_{1-130}$-Cya hybrid that possesses the SycE chaperone-binding site is secreted only when SycE is present. These results suggested that the chaperone is not indispensable for the secretion of its cognate Yop, but becomes necessary when its binding site is present. Later on, Boyd et al. (2000) localized the region of YopE which creates the need for the SycE chaperone between residues 50 to 77. Interestingly, Krall et al. (2004) showed recently that residues 54-75 are necessary and sufficient for the correct intracellular localization of YopE into mammalian cells. This observation explains why YopE possesses this "troublesome" sequence and hence the need for SycE.

In parallel, Lloyd et al. (2001) suggested that the Syc chaperones could also play an anti-folding role. These authors showed that YopE is rapidly secreted in conditions permissive for secretion. The immediate secretion of stored YopE does not require *de novo* protein synthesis but is completely dependent on SycE, whereas long-term co-translational delivery of YopE is not. This observation suggests that SycE may maintain preformed, stored YopE in a secretion-competent (presumably unfolded or partially folded) state. Indeed, the observed diameter of the needle (2 nm) is too small to allow folded, globular proteins to travel through it (Hoiczyk and Blobel, 2001). Moreover, the molecular diameter of the catalytic domain of YopE is nearly 2.5 nm, a size that should not allow folded YopE to travel through the needle (Evdokimov et al., 2002). If the Yops travel through the needle, then they have to be at least partially unfolded. Therefore, SycE might prevent YopE from folding prematurely. However, Birtalan et al. (2002) showed that the folding state of YopE is not affected by the binding of SycE. These authors observed that the purified SycE-YopE complex produced in *E. coli* displays GAP activity, which indicates a perfect folding of the C-terminal domain of YopE.

To investigate further a possible role of the chaperones as anti-folding factors, Feldman et al. (2002) studied the secretion of mouse dihydrofolate reductase (DHFR), a cytosolic globular protein, by the *Yersinia* Ysc injectisome as a YopE-DHFR hybrid protein. A fusion between the first 52 residues of YopE and the wild type DHFR was secreted, and this secretion was dependent on SycE. However, a fusion between the first 81 residues of YopE and the wild type DHFR was not secreted, even in the presence of SycE. Since SycE bound to YopE covers the first 78 residues of the effector protein, Feldman et al. hypothesized that in the case of the Yop_{52}-DHFR hybrid protein, SycE covers a part of DHFR and prevents its folding. Although SycE seems to play an anti-folding role with the Yop_{52}-DHFR hybrid protein, Feldman et al. did not suggest that this function could be applicable to the entire YopE. However, the observations with the different YopE-DHFR hybrids can be taken as an evidence that Yops travel through the injectisome in an unfolded form. Another role for the TTS chaperones, suggested by observations of Boyd et al., (2000) is that the chaperones could orchestrate a defined order of secretion. In a wild type *Yersinia* strain, the Yop_{1-15}-Cya hybrid protein devoid of the

SycE chaperone-binding site is not translocated. However, in a polymutant *Yersinia* strain deleted of all the Yop effectors, the same Yop_{1-15}-Cya hybrid protein is translocated at the same level as the Yop_{1-130}-Cya hybrid protein, which carries the chaperone-binding site. These results suggest that SycE is required for the secretion of YopE when the other Yop proteins are present. The SycE chaperone could be some kind of hierarchy factor organizing the secretion of the Yop proteins in a defined order. In good agreement with this hypothesis, the results of Birtalan et al. (2002) indicate that the SycE-YopE complex could provide a three-dimensional secretion signal helping secretion that is initiated by the amino-terminal signal. These authors speculate that the SycE-YopE complex could interact with the Ysc machinery, probably with the ATPase YscN.

The majority of the TTS chaperones contribute to the cytosolic stability of their cognate Yops. For example, SycE plays an antidegradation role since the half-life of YopE is longer in wild type *Yersinia* than in *sycE* mutant bacteria (Frithz-Lindsten et al., 1995). There are some exceptions, however. SycH is the chaperone required for the secretion not only of YopH, but also of LcrQ in *Y. pseudotuberculosis* and YscM1-2 in *Y. enterocolitica*, two negative regulators of the system (Cambronne et al., 2000; Wattiau et al., 1994). SycH does not play a clear antidegradation role since YopH and YscM1 can be detected in the cytosol of *sycH* mutant bacteria (Cambronne et al., 2000; Persson et al., 1995; Wattiau et al., 1994). In spite of the absence of the chaperone, YopH is stable and enzymatically active. Similarly, LcrQ/YscM1-2 maintains its negative regulatory role in the absence of SycH since the production of the Yop proteins is reduced in *sycH* mutant bacteria. Instead of stabilizing its partners, SycH induces the inactivation of the negative regulators and allows an increase in the transcription of the *yop* genes. This idea is consistent with the observations that increased quantities of SycH induce expression of the *yop* virulon and stimulate the type III secretion machinery. These observations suggest that SycH has an indirect regulatory role by means of its effect on the regulatory protein LcrQ/YscM1-2. This role of the TTS chaperones in the regulation of the transcription of the virulence factors was first observed for SycD (LcrH in *Y. pestis* and *Y. pseudotuberculosis*) (Bergman et al., 1991).

The situation of the SycN and YscB chaperones is special because these two chaperones possess the same substrate (YopN). Cross-linking experiments followed by co-immunoprecipitation allowed Day and Plano (1998) to show that SycN and YscB bind together to YopN. Some two- and three-hybrid experiments in yeast demonstrate that SycN and YscB can interact together in the absence of YopN, but that either chaperone individually does not bind YopN (Day and Plano, 1998). It seems then that the interaction between YscN and YscB induces conformational modifications of the chaperones allowing them to bind to YopN. One obtains a complex of three proteins: two molecules of chaperones for one molecule of secreted protein. This stoichiometry is the same for the SycE-YopE complex, in which the SycN-YscB complex replaces the SycE homodimer. According to Day et al. (2003), SycN and YscB could facilitate the secretion and the subsequent translocation of YopN.

YscY is a protein that shows the same properties as the chaperones. This protein interacts with YscX, a secreted element of the type III secretion machinery (Day and Plano, 2000; Iriarte and Cornelis, 1999b).

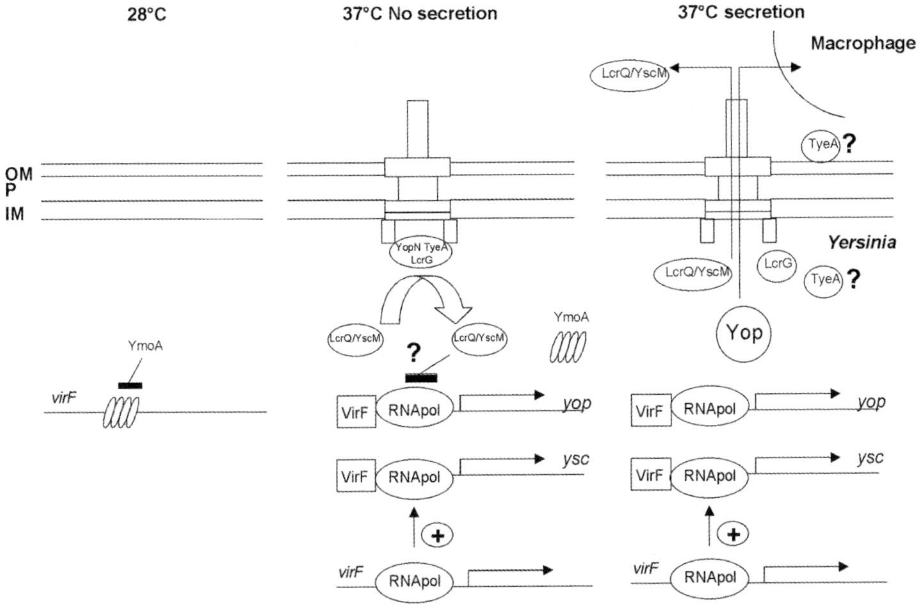

Figure 4. Schematic representation of transcriptional regulation of the type III genes of *Yersinia*. At 28°C, YmoA repress the transcription of *virF*. At 37°C, and in absence of contact between *Yersinia* and a eukaryotic cell, *virF* is transcribed. Transcription of all *yop* genes and some *ysc* genes requires the activator VirF. The plug, composed of the YopN, TyeA and LcrG proteins, prevents the release of LcrQ/YscM and of Yops. LcrQ/YscM accumulates inside the bacteria cytosol and indirectly represses *yop* transcription. At 37°C, upon contact between *Yersinia* and the eukaryotic cell, the *yop* and *ysc* genes are transcribed allowing for the injection of Yops into the cytosol of the target cell. The release of LcrQ/YscM amplifies *yop* transcription. The subcellular localization of TyeA is not known with certainty.

REGULATION OF THE TYPE III SECRETION TRIGGER MECHANISM

Transcription of *yop* genes is increased at 37 °C through the concerted action of VirF and YmoA, a histone-like protein. However, in the absence of contact with a target cell the secretion channel remains closed, through the action of YopN, TyeA, and LcrG proteins, and a mechanism of feedback inhibition prevents the accumulation of Yops (Figure 4). Upon close contact with a target cell, secretion starts and active transcription of the *yop* genes occurs (Cornelis et al., 1998).

REGULATION OF Yop SECRETION BY CONTACT WITH EUKARYOTIC CELLS

In vivo, the expression of the type III secretion system is activated by contact between the bacterium and the target eukaryotic cell (Pettersson et al., 1996; Rosqvist et al., 1994). Rosqvist et al. (1994) have shown that *Yersinia* do not secrete Yops when they are incubated in a cell-free eukaryotic cell culture medium. However, in the presence of target cells, they inject their Yop effectors, indicating that physical contact with cells triggers this process. This hypothesis was elegantly confirmed by placing a luciferase gene under the control of the *yopE* promoter and infecting cells with the recombinant *Yersinia* (Pettersson et al., 1996). Only adhering bacteria became luminescent, demonstrating clearly the necessity of contact in triggering the system. The different proteins involved in the trigger mechanism will be described below in more detail.

REGULATION OF Yop SECRETION BY YopN, TyeA AND LcrG

In vitro, *Yersinia* secretes the Yop proteins when placed at 37°C in a rich medium devoid of Ca^{2+} ions. The isolation of Ca^{2+}-blind mutants allowed the identification of three genes whose products are involved in the control of Yop release: *yopN*, *tyeA* and *lcrG* (Boland et al., 1996; Cheng and Schneewind, 2000; Forsberg et al., 1991; Iriarte et al., 1998; Skryzpek and Straley, 1993; Yother and Goguen, 1985).

YopN

YopN, also known as LcrE, is a 32.6-kDa protein with two coiled-coil domains. It is secreted at 37°C in the absence of Ca^{2+} ions while in the presence of Ca^{2+} ions, the protein is not released but it is exposed at the bacterial surface (Iriarte et al., 1998). A *yopN* mutant secretes Yops at 37°C, even in presence of Ca^{2+}. Upon contact with eukaryotic cells, the *yopN* mutant can still deliver Yops into the cytosol of the target cell, but it secretes more Yops into the eukaryotic cell medium than do wild type *Yersinia* strains (Boland et al., 1996). Translocation of YopN into the cytosol of the eukaryotic cell was not detected using the Cya reporter method (Boland et al., 1996). However, fractionation experiments and a new reporter method have shown that YopN is translocated into the cytosol of eukaryotic cells (Day et al., 2003). It has been suggested that YopN could function as a sensor and a plug controlling Yop secretion. After contact with the eukaryotic cell, the YopN sensor could interact with a ligand on the target cell surface, be removed, and allow Yop secretion and delivery into the target cell (Rosqvist et al., 1994). However, YopN has never been shown to interact with a cell receptor.

TyeA

TyeA is a protein of 10.8-kDa. It was named TyeA because it plays a role in translocation of the YopE and YopH but not the YopM, YopO, YopP and YopT effectors (Iriarte et al., 1998). However, Cheng and Schneewind (2000) have observed by fractionation experiments that TyeA is involved in the translocation of all Yop effectors. TyeA is detected in the bacterial cytosolic fraction but not in the culture supernatant, irrespective of the presence of Ca^{2+} in the culture medium. Like YopN, TyeA is loosely associated with the membrane (Iriarte et al., 1998). This localization of TyeA was refuted by Cheng and Schneewind (2000), who consider that TyeA is located solely in the bacterial cytosol. TyeA has the capacity to bind to the second coiled coil of YopN and to interact with YopD (Cheng and Schneewind, 2000; Iriarte et al., 1998).

LcrG

LcrG a protein of 11 kDa that controls the release of Yops *in vitro* (Skryzpek and Straley, 1993; Sarker et al., 1998b) but is also required for efficient translocation of the Yop effectors (Sarker et al., 1998b). This protein has been shown to be primarily cytosolic, but it has also been detected in the membrane and in the extracellular medium (Nilles et al., 1997; Skryzpek and Straley, 1993). The fact that *tyeA* and *lcrG* mutants, like the *yopN* mutant, are deregulated for Yop secretion in the presence of Ca^{2+} ions or depolarized in the presence of eukaryotic cells (Iriarte et al., 1998; Sarker et al., 1998b; Skryzpek and Straley, 1993) suggests that the control of the delivery of the effectors requires not simply YopN but, rather, a complex system comprised of at least YopN, TyeA and LcrG. The control of the delivery of the effectors requires also that the complex of these three proteins be located in the bacterial cytosol. In fact, how could the *lcrG* mutant secrete Yop proteins in the presence of Ca^{2+} if YopN acts as a surface-localized stop valve of the type III machinery?

Transcriptional Regulation of *yop* Gene Expression

In vitro, Yop secretion occurs only at 37°C in the absence of Ca^{2+}. This secretion correlates with growth arrest of the bacterium, a phenomenon called LCR for "low-calcium response". At 26°C (or at 37°C in the presence Ca^{2+} ions), Yop expression is repressed. The activation of Yop expression by temperature and repression by Ca^{2+} are two distinct phenomena. Temperature control acts directly at the level of the transcription of the *yop* genes (Cornelis et al., 1989; Lambert de Rouvroit et al., 1992), while Ca^{2+} ions inhibit Yop secretion by the type III machinery (Forsberg et al., 1987). However, when secretion is inhibited by the presence of Ca^{2+} ions, a feedback inhibition mechanism blocks transcription of *yop* genes (Cornelis et al., 1987; Forsberg and Wolf-Watz, 1988; Straley et al., 1993). This feedback mechanism allows a strict control of the amount of intrabacterial Yops.

Positive Control of Yop synthesis by Temperature

VirF

VirF is a 30.9-kDa protein that belongs to the AraC family of regulators (Cornelis et al., 1989). Transcription of many pYV genes including all the *yop* genes, *sycE*, *ylpA*, *yadA*, and the *virC* operon, is dependent on VirF (China et al., 1990; Cornelis et al., 1989; Michiels and Cornelis, 1991; Skurnik and Toivanen, 1992). VirF seems to be dispensable or less important for the transcription of the *virA* and *virB* operons, which encode the Ysc secretion apparatus, and some other genes such as *sycH* (Lambert de Rouvroit et al., 1992; Wattiau and Cornelis, 1994). All these genes, whether dependent or independent of VirF, are silent at low temperature but strongly expressed at 37°C. DNase I footprinting experiments carried out by Wattiau and Cornelis (1994) on four promoters (*yopE*, *yopH*, *virC*, and *lcrGVsycDyopBD*) showed that VirF binds to a 40-bp region localized immediately upstream from the RNA polymerase binding site.

In *Y. enterocolitica*, the *virF* gene itself is strongly thermoregulated (Cornelis et al., 1987). This thermoinduction occurs when VirF is encoded on the pYV plasmid or when the *virF* gene transcribed from its own promoter is cloned on an independent plasmid in a *Yersinia* strain devoid of the pYV plasmid. This observation suggests that the *virF* regulation is achieved through a protein encoded by the chromosome. Moreover, thermoinduction of VirF can also occur in *E. coli* containing the plasmid just described, suggesting that VirF self-activates its transcription or that a ubiquitous factor is involved. The fact that *virF* is itself thermoregulated can explain why Yops are produced only at 37°C. However, it does not prove that the control of the *ysc-yop* genes expression by temperature occurs only through VirF. In fact, when *virF* is transcribed at low temperature from a *tac* promoter, the *yop* genes are poorly transcribed. At 37°C, transcription again reaches normal levels, suggesting that the optimal expression of the *yop* genes requires a temperature of 37°C as well as VirF (Lambert de Rouvroit et al., 1992).

YmoA

The *ymoA* (for *Yersinia* modulator) gene is a chromosomal gene that encodes an 8.1-kDa protein extremely rich in positively and negatively charged residues, a feature also found in the *E. coli* histone-like protein H-NS (Dorman et al., 1999). H-NS is involved in the temperature regulation of virulence gene expression through modifications in DNA supercoiling (Dorman et al., 1999). Although there is no sequence similarity between YmoA and H-NS, it is very likely that YmoA is a histone-like protein. This idea is reinforced by the fact that the level of supercoiling is higher in *ymoA* mutant bacteria that in wild type bacteria, as shown by chloroquine agarose gel electrophoresis of plasmid DNA (Cornelis et al., 1991). To determine whether the chromatin structure influences expression of *yop* genes, Lambert de Rouvroit et al. (1992) measured the expression of a *yopH-cat* operon fusion in the *ymoA* mutant. They observed that the expression of *yopH* becomes VirF-independent in a *ymoA* mutant but still remains thermoinducible. This observation suggests that temperature and chromatin structure strongly affect the *yopH* promoter. If this hypothesis is true, VirF would reinforce the activity of the *yopH* promoter in the *ymoA* mutant even at low temperature. Indeed, expression of *virF* under the control of the *tac* promoter in the *ymoA* mutant induces the same level of expression of a *yopH-lacZ* fusion at 25°C and at 37°C. These results lead to the hypothesis that temperature could modify the structure of chromatin, making the promoters more accessible to VirF. Rhode et al. (1999) have demonstrated that temperature indeed alters the level of DNA supercoiling and they hypothesized that raising the temperature dislodges a repressor, perhaps YmoA, bound on promoter regions of VirF-sensitive genes.

NEGATIVE CONTROL BY Ca^{2+} IONS THROUGH A FEEDBACK REGULATION MECHANISM

Ca^{2+} ions inhibit secretion of the Yop proteins at 37°C but they also have an indirect role on the production of the Yop proteins. Transcription of the *ysc* genes is only weakly affected by the presence of Ca^{2+} but the expression of the *yop* genes is strongly reduced (Cornelis et al., 1987; Forsberg and Wolf-Watz, 1988; Goguen et al., 1984; Mulder et al., 1989). It is difficult to imagine how an ion like Ca^{2+} could penetrate into the bacterial cytosol to block transcription. It is more plausible that the abundance of Ca^{2+} ions at the bacterial surface could influence Yop secretion, and that a feedback inhibition mechanism reduces transcription of the *yop* genes to avoid a cytosolic accumulation of the Yop proteins that could be toxic for the bacterium (Cornelis et al., 1987).

ROLE OF LcrQ/YscM1-2

The mechanism of feedback inhibition has been better understood by the identification of a factor in *Y. pseudotuberculosis* that acts negatively on Yop secretion (Pettersson et al., 1996; Rimpilainen et al., 1992). This negative regulator, called LcrQ in *Y. pseudotuberculosis*, is a 12.4-kDa secreted protein. Rimpiläinen et al. (1992) have observed that the overproduction of this protein abolishes Yop production. In contrast, an *lcrQ* mutant synthesizes more Yops than the wild type strain and accumulates Yops in the cytosol of the bacteria when secretion is prevented by Ca^{2+} or by mutation in the genes coding for the type III secretion machinery. In the presence of Ca^{2+}, this mutant secretes YopD and LcrV. LcrQ is rapidly secreted when bacteria are shifted from a medium containing Ca^{2+} (nonpermissive conditions for Yop secretion) to a medium containing a Ca^{2+}-chelator (permissive conditions for Yop secretion). LcrQ seems to be a negative regulator that accumulates in the bacterial cytosol in conditions that are nonpermissive for Yop secretion and that acts negatively on *yop* transcription. The LcrQ level in the bacterium is inversely proportional to the expression level of the Yop proteins: Yop expression is maximal when the LcrQ concentration in the bacterial cytosol is low, and a high concentration of LcrQ leads to a decrease of Yop expression. In *Y. enterocolitica*, the situation appears to be slightly different: YscM, the homologue of LcrQ, is also a secreted protein (Stainier et al., 1997) but a *Y. enterocolitica yscM* mutant does not show the same phenotype as the *Y. pseudotuberculosis lcrQ* mutant, although overexpression of YscM blocks Yop synthesis (Allaoui et al., 1995). The reason for this discrepancy was elucidated by the discovery on the pYV virulence plasmid of a second gene related to *yscM*. The original *yscM* gene was renamed *yscM1*, and the new gene has been called *yscM2* (Stainier et al., 1997). The *yscM1yscM2* double mutant in *Y. enterocolitica* shows the same phenotype as the *lcrQ* mutant of *Y. pseudotuberculosis*. Thus, two different YscM proteins in *Y. enterocolitica* behave like LcrQ in *Y. pseudotuberculosis*.

The experimental observations suggest that LcrQ in *Y. pseudotuberculosis* and YscM1-2 in *Y. enterocolitica* are negative regulators of *yop* transcription. However, some results seem to contradict this view. First, the *lcrQ* and *yscM1-2* mutants secrete YopD and LcrV in the presence of Ca^{2+} (Rimpilainen et al., 1992; Stainier et al., 1997). How to explain that in the absence of these negative regulators, the secretion channel is open, allowing the passage of YopD and LcrV but not the secretion of the other Yop proteins? Second, Stainier et al. (1997) have observed that overexpression of YscM1-2 in a simplified system consisting only of a *yopH-cat* reporter gene and *virF* had no effect on *yopH* transcription, but in the presence of the pYV plasmid it leads to decreased *yopH* transcription. These results do not contradict the negative role of LcrQ and YscM1-2, but they do suggest that these factors require one or more pYV-encoded proteins to act in the feedback inhibition mechanism.

ROLE OF YopD

Williams and Straley (1998) have suggested that YopD acts with LcrQ/YscM1-2 in the feedback inhibition mechanism. This will be discussed below.

TRANSLOCATION OF THE YOP EFFECTOR PROTEINS ACROSS THE CELL MEMBRANE

Purified secreted Yops have no cytotoxic effect on cultured cells, although live extracellular *Yersinia* have such an activity. Cytotoxicity nevertheless depends on the capacity of the bacterium to secrete YopE and YopD, and YopE alone is cytotoxic when microinjected into the cells (Rosqvist et al., 1991). This observation led to the hypothesis that YopE is a cytotoxin that needs to be injected into the eukaryotic cell's cytosol by a mechanism involving YopD in order to exert its effect (Rosqvist et al., 1991). This hypothesis was demonstrated by confocal laser scanning microscopy (Rosqvist et al., 1994), and by the adenylate cyclase reporter enzyme strategy, an approach that was introduced by Sory and Cornelis and is now widely used in "type III secretion" research (Sory and Cornelis, 1994): infection of eukaryotic cells with a recombinant *Y. enterocolitica* producing hybrid proteins consisting of the N-terminus various of Yops (other than YopB and YopD) fused to the catalytic domain of a calmodulin-dependant adenylate cyclase (Yop-Cya proteins), leads to an accumulation of cyclic AMP (cAMP) in the cells. Since there is no calmodulin in the bacterial cell or culture medium, this accumulation of cAMP signifies the internalization of Yop-Cya into the cytosol of eukaryotic cells (Sory and Cornelis, 1994). The phenomenon is strictly dependent on the presence of YopD and YopB. Thus, extracellular *Yersinia* inject Yops into the cytosol of eukaryotic cells by a mechanism that involves at least YopD and YopB (Boland et al., 1996; Hakansson et al., 1996b). Yops are thus a collection of intracellular "effectors" (YopE, YopH, YopM, YpkAYopO, YopP/J, YopT) and "translocators" which are required for the translocation of the effectors across the plasma membrane of eukaryotic cells (Cornelis and Wolf-Watz, 1997).

YopB and YopD are encoded by the translocation operon *lcrGVsycDYopBD* (Bergman et al., 1991). This operon encodes the "translocator" Yops, YopB, YopD and LcrV; the chaperone of YopB and YopD, SycD; and a regulatory molecule, LcrG. Their action in translocation of the Yop effectors and in regulation is described below.

YopD AND YopB

The 33.3-kDa YopD was the first protein to be shown to be involved in the translocation of the Yop effector proteins into the cytosol of the eukaryotic cell (Boland et al., 1996; Persson et al., 1995; Rosqvist et al., 1994; Sory et al., 1995; Sory and Cornelis, 1994). Translocation of the hybrid protein YopE-Cya is dependent on YopB as well as on YopD (Boland et al., 1996). Moreover, a nonpolar *yopB* mutant is unable to induce cytotoxicity in HeLa cells and does not inhibit phagocytosis by macrophages (Hakansson et al., 1996b). This mutant is not pathogenic for mice (Hakansson et al., 1996b). The YopE and YopH effector proteins are not recovered in the cytosol of eukaryotic cells infected by the nonpolar *yopB* mutant (Hakansson et al., 1996b). These various results indicate that YopB is individually required for the translocation of the effector proteins into the cytosol of the target cells.

YopB is a 41.8-kDa protein with a moderate level of similarity to members of the RTX family of alpha-hemolysins and leukotoxins. The homology between YopB and the RTX proteins is limited to the hydrophobic regions. Since, in the RTX proteins, these hydrophobic regions are believed to be involved in disrupting the target cell membrane, this suggested that YopB could form a pore in the membrane of the eukaryotic cell. To confirm this hypothesis, Håkansson et al. (1996b) studied the hemolytic activity of *Y. pseudotuberculosis* when contact between the bacteria and the erythrocytes is achieved by centrifugation. These authors observed that hemolysis is dependent on the type III secretion apparatus and on YopB. This YopB-dependent lytic activity is higher when the effector *yop* genes are deleted, suggesting that the pore is normally filled with effector proteins during contact. The presence of sugar molecules whose size exceeds the size of the translocation pore prevented the YopB-dependent hemolysis owing to osmotic pressure equilibration between the intra- and extracellular compartments of erythrocytes. Using this approach, the inner diameter of the pore was estimated to be 1.2 to 3.5 nm. Subsequently, Neyt and Cornelis (1999a) demonstrated that polymutant *Y. enterocolitica* also generate pores in the membranes of macrophages. These pores allow the release of small membrane-permeant dyes such as BCECF (623 Da) but not large molecules such as lactate dehydrogenase (Neyt and Cornelis, 1999a). These pores also allow the entry of a small membrane-impermeant dye such as lucifer yellow CH (443 Da) but not of the larger Texas red-phalloidin (1490 Da). The use of these different dyes led to the conclusion that translocation pore formation by *Y. enterocolitica* in the macrophage cell membrane requires YopB and YopD but not LcrG. Based on the dye exclusion experiments, the translocation pore diameter was estimated to be 1.6 to 2.3 nm. This value coincides with the one obtained with the osmoprotection experiments (Hakansson et al., 1996b).

The fact that YopB and YopD are both needed for translocation suggests that they could interact at some stage to induce pore formation. This idea is reinforced by the presence of hypothetical coiled coil structures in both proteins. In agreement with this hypothesis, YopB and YopD appear to associate in the bacterium prior to their secretion and could form a trimeric complex with their chaperone SycD (Neyt and Cornelis, 1999b). In an attempt to localize the domain of YopB involved in the interaction with YopD, these authors demonstrated that the binding between the two translocators does not occur at a specific site on YopB but at different sites along the protein. These results suggest that YopB and YopD could insert together in the eukaryotic membrane to form the translocation pore. To study the components of the translocation pores, Tardy et al. (1999) infected liposomes with *Y. enterocolitica*. In this strategy, liposomes are mixed with bacteria and Yop secretion is induced by incubation in a rich medium devoid of Ca^{2+}. Secreted YopB and YopD proteins were found to be inserted into the liposomes. To characterize the putative pore-forming properties of the lipid-bound Yops, the proteoliposomes were fused to a planar lipid bilayer, and ectophysiological experiments revealed the presence of channels. Channels appeared with wild type *Y. enterocolitica*, and with *lcrG* mutant bacteria devoid of the Yop effectors. In contrast, mutant bacteria devoid of YopB did not generate channels, and mutants devoid of YopD led to current fluctuations that were different from those observed with wild type *Y. enterocolitica*. These results are in perfect agreement with the experiments performed on macrophages and erythrocytes and suggest that YopB and YopD do indeed form channels (Hakansson et al., 1996b; Neyt and Cornelis, 1999a).

LcrV

LcrV is a 37-kDa Yop protein that was known as the V antigen long before Yops were discovered. Not only is it a protective antigen, but it also has strong immunomodulatory effects on the host (Motin et al., 1994; Roggenkamp et al., 1997; Sing et al., 2002). This protein is able to interact with LcrG (Matson and Nilles, 2002; Nilles et al., 1997; Sarker et al., 1998b) as well as with the translocators YopB and YopD (Sarker et al., 1998a) but this last point has been contradicted (Lee et al., 2000). LcrV plays a role as a positive regulatory protein of the *yop* regulon by interacting with LcrG (Nilles et al., 1997). This regulatory role will be discussed later.

The analysis of the phenotype of a polar *lcrG* mutant led Nilles et al. (1997) to suggest that LcrV has an indirect role in the Yop translocation process through its effect on the secretion and the deployment of the translocators YopB and YopD. In the same vein, Pettersson et al. (1999) have observed that antibodies directed again LcrV are able to block the injection of effector Yop proteins into the cytosol of eukaryotic cells, suggesting that LcrV plays a key function in the translocation of the Yop effectors. Moreover, these authors constructed a non-polar *lcrV* mutant, and analysis of the phenotype of this mutant suggested that LcrV has a direct role in the translocation of the Yop effectors (Pettersson et al., 1999). However, demonstration of the role of LcrV in Yop effector translocation is hampered by the fact that LcrV also plays a positive regulatory role on the synthesis of YopB and YopD.

Holmström et al. observed that a mutation in the *lcrQ* negative regulatory gene can suppress the effect on regulation of *lcrV* mutation. They then tested an *lcrQ*, *lcrV* double mutant for its effect on pore formation and observed that it has no hemolytic activity (Holmstrom et al., 2001). This indicates that LcrV, along with YopB and YopD, is required for pore formation. Surprisingly, these authors also showed that a pure preparation of LcrV could form pores in artificial lipid bilayers. The authors made the hypothesis that LcrV inserts in the eukaryotic cell membrane to initiate channel formation, and that YopB and YopD are then inserted into the membrane to stabilize the channel and form a functional translocation pore. LcrV would be the size-determining structural component of the translocation pore. A few years ago, YopQ was shown to influence the size of the translocation pore, as deduced from osmoprotection experiments (Holmstrom et al., 1997). The mechanism involved in the regulation of the translocation pore size by YopQ is unknown.

REGULATION OF SECRETION BY LcrG

LcrG has been shown to exert a negative regulatory effect on Yop secretion and this effect is counteracted by LcrV. One hypothesis, commonly called the Nilles hypothesis, (Matson and Nilles, 2001; Nilles et al., 1998; Nilles et al., 1997), suggests the following: under conditions that are nonpermissive for secretion (presence of Ca^{2+} or absence of eukaryotic cell contact), the Ysc machinery is blocked by LcrG and by YopN. In secretion-permissive conditions, YopN is secreted and there is an increase of the intrabacterial content of LcrV. LcrV titrates LcrG by formation of a stable LcrG-LcrV complex involving a hydrophobic domain located in the N-terminus of LcrG and a predicted coiled-coil motif in LcrV (Matson and Nilles, 2001; Sarker et al., 1998b). Lawton et al. (2002) have suggested that upon induction of type III secretion, high levels of LcrV are produced predominantly as dimers. The high affinity of LcrV for LcrG would disrupt LcrV dimers and the formation of a LcrV-LcrG complex would result in the removal of the internal gating of the Ysc apparatus, thus enabling Yop secretion to occur. Since the *lcrG* and *lcrV* genes are located on the same operon and transcribed from the same promoter, an increase of *lcrV* transcription implies also an increase of *lcrG* transcription. LcrV expression must therefore be regulated at a post-transcriptional level in order to titrate LcrG.

SycD

SycD is a 19-kDa protein that is encoded by the translocation operon and that serves two Yops, namely YopB and YopD (Wattiau et al., 1994). In *E. coli*, this protein protects the host against the cytotoxicity of YopB and YopD (Neyt and Cornelis, 1999a). Certain observations suggest that SycD could prevent association between LcrV and the YopB/D complex in the bacterial cytosol (Neyt and Cornelis, 1999b). SycD also acts as a negative regulator of *yop* gene transcription (Bergman et al., 1991): a *sycD* mutant shows increased transcription of *yop* genes in the presence of Ca^{2+} ions or in a secretion mutant background (Bergman et al., 1991). According to Williams and Straley (1998), the role of SycD on the *yop* transcription could be indirect, resulting from the destabilization of YopD. However, this view is inconsistent with the fact that SycD has a repressor effect when overexpressed in a *yopD* mutant (Bergman et al., 1991). Francis et al. (2001) have identified some recombinant SycD proteins that can bind and stabilize YopD without affecting regulation. Their results suggest that the SycD-YopD complex could act as a repressor of Yop protein synthesis. After the secretion of YopD, SycD could interact with YscY, a partner protein identified by the two hybrid system (Francis et al., 2001). The SycD-YscY complex could derepress the expression of the type III secretion genes. This hypothesis is supported by the observation that YopD and SycD regulate expression of YopQ by a posttranscriptional mechanism and bind to *yopQ* mRNA (Anderson et al., 2002; Francis et al., 2001).

THE ACTION OF THE Yop EFFECTORS INSIDE THE EUKARYOTIC CELL

Yersinia spp. survive within the host by remaining extracellular and avoiding the innate immune system. In particular, they inhibit phagocytosis and they downregulate the anti-inflammatory response. The six Yop effectors (YopE, YopH, YpkA/YopO, YopT, YopJ/P, YopM), which are delivered inside eukaryotic cells by the complex TTS system described above, are crucial for the anti-innate immune system capacity of *Yersinia*. Four of the Yop effectors, YopE, YopH, YpkA/YopO, and YopT, have been shown to have a negative impact on host cell cytoskeletal dynamics (Figure 5), thus conferring resistance to phagocytosis by macrophages and polymorphonuclear leukocytes (PMNs) (Grosdent et al., 2002).

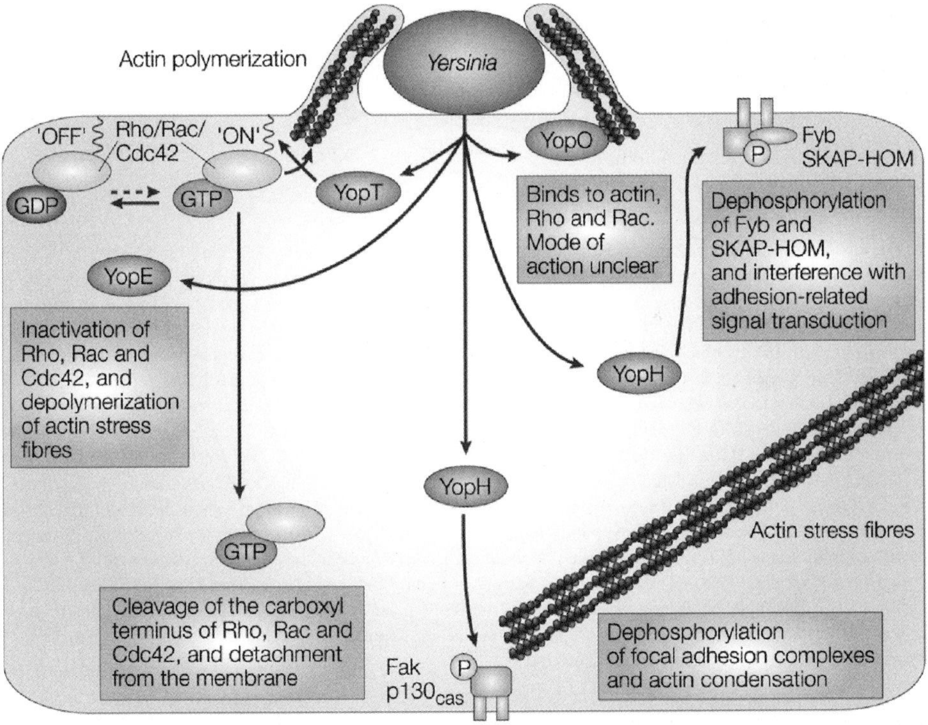

Figure 5. The anti-phagocytic action of YopE, YopH, YopT and YopO/YpkA. YopH is a powerful protein tyrosine phosphatase that dephosphorylates Fak and $p130_{cas}$ in focal adhesion complexes, as well as Fyb and SKAP-HOM in an adhesion-signalling complex. YopE exerts a GTPase-activacting protein (GAP) activity on monomeric GTPases of the Rho family (Rho, Rac, Cdc42), and YopT cleaves the same monomeric GTPases close to their carboxyl terminus, which releases them from the membrane anchor. The actions of YopE and YopT both result from the inactivation of Rho proteins and the depolimerization of actin stress fibres. YopO is a kinase with homology to eukaryotic serine/threonine kinases that is activacted by actin binding and binds RhoA and Rac1. Its mode of action is unclaer at present. Reprinted from Cornelis (2002b) with permission of the publisher.

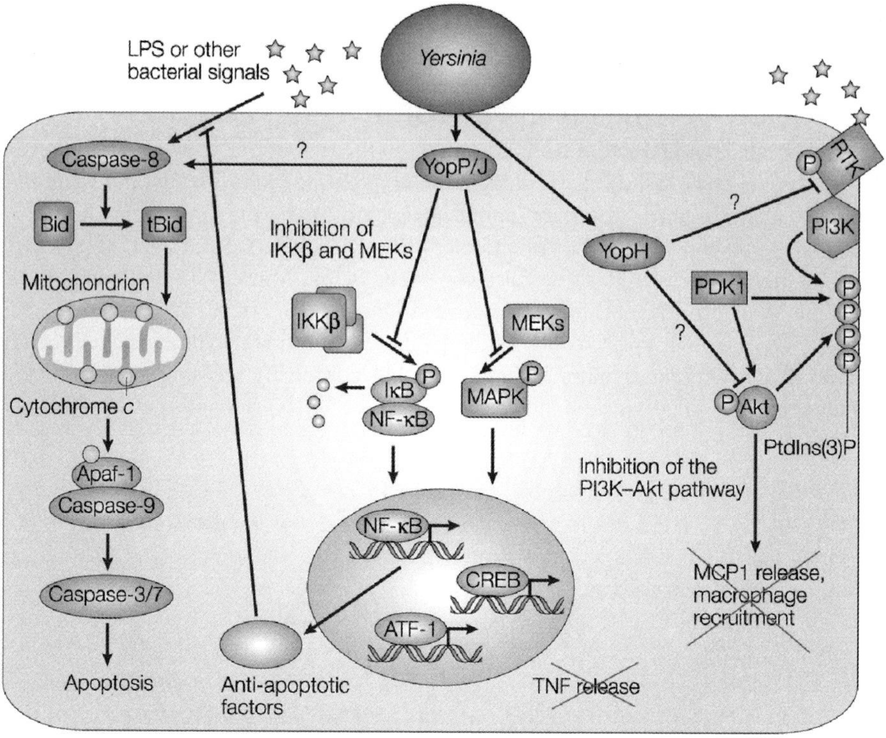

Figure 6. The anti-inflammatory action of Yops. YopJ/P binds to IKKβ and MAPK and blocks their activation through phosphorylation by upstream MAPK kinases (MEKs). This inhibits the activity of the NF-kB, CREB and ATF-1 transcription factors, thus preventing transcription of anti-apoptotic genes and of pro-inflammatory cytokines, such as TNF-α. YopJ/P also induces apoptosis in macrophages, either by directly activating a death pathway or indirectly by blocking the synthesis of anti-apoptotic factors. In this second hypothesis, bacterial LPS or lipoproteins would act as the pro-apoptotic signal. The apoptotic cascade is probably triggered by caspase-8 activation, leading to Bid cleavage. The subsequent translocation of tBid to the mitochondria induces the release of cytochrome c, which binds to the apoptotic protease activating factor-1 (Apaf-1) and leads to recruitment and activation of caspase-9 that, in turn, activates executioner caspases-3 and –7. The YopJ/P protease may act on SUMOylated and/or ubiquitinated proteins, but the relevance of these activities for YopJ/P function is unclear. The right hand part of the figure shows that, on contact of LPS or other bacterial signals with an unidentified receptor tyrosine kinase (RTK), phosphatidylinositol 3-kinase (PI3K) is recruited to the membrane. Activated PI3K phosphorylates lipid phosphoinositides, leading to the appearance in the membrane of phophatidylinositol-3-phosphates (Ptdins(3)Ps). These, is turn, recruit proteins such as phosphoinositide-dependent-kinase-1 (PDK-1) that phosphorylates and activates Akt (also known as Protein Kinase B). Activated Akt, in turn, can phosphorylate different classes of proteins, which leads, among many other events, to the production of monocyte chemotactic protein 1 (MCP-1).

In cell culture systems, YopJ/P displays a strong anti-inflammatory activity (Figure 6) (Orth, 2002). However, a microarray analysis of transcription in *Yersinia*-infected macrophages clearly indicated that pYV-encoded factors other than YopJ/P can modulate the host inflammatory response (Sauvonnet et al., 2002b). Accordingly, two recent reports describe the anti-inflammatory actions of YopE and YopH (Sauvonnet et al., 2002a; Viboud et al., 2003). Exactly how YopM contributes to the anti-host action of *Yersinia* remains unknown.

YopE

Delivery of YopE into epithelial cells and macrophages leads to a cytotoxic response characterized by cell rounding and detachment from the extracellular matrix, resulting from the disruption of the actin microfilament network (Rosqvist et al., 1991). YopE was the first Yop effector shown to have an inhibitory effect on the actin cytoskeleton by inactivation of small GTP-binding proteins of the Rho family (Black and Bliska, 2000; Von Pawel-Rammingen et al., 2000).

The so-called Rho GTPases are master regulators of the actin cytoskeleton dynamics (Hall, 1998), and thus play an important role in phagocytic processes (Caron and Hall, 1998; Massol et al., 1998). The most extensively characterized Rho GTPases are Cdc42, Rac1, and RhoA. These proteins cycle between two interconvertible forms, an inactive GDP-bound state and an activated GTP-bound state. Activation occurs by GDP and GTP exchange, a process promoted by guanine nucleotide exchange factors (GEFs). Inactivation occurs by GTP hydrolysis through the action of their intrinsic GTPase activity, a process facilitated by GTPase activating proteins (GAPs). When activated, Rho GTPases interact with a wide variety of effector proteins to mediate downstream signaling. These proteins have sequences at their carboxyl terminus that undergo post-translational modification with lipid, such as farnesyl or geranygeranyl. This lipid modification is required for their binding to membranes and activation of downstream effectors.

YopE bears an arginine-finger motif similar to those used by mammalian GAP proteins for catalysis and displays in vitro GAP activity towards RhoA, Rac1 and Cdc42 (Figure 5) (Black and Bliska, 2000; Von Pawel-Rammingen et al., 2000). The mechanism of inhibition of phagocytocis by YopE results from its GAP activity (Black and Bliska, 2000). The actual preferred substrate of YopE under physiological conditions seems to be Rac1 (Andor et al., 2001).

Recently, it has been shown that the GAP activity of YopE may also play a role in minimizing plasma membrane damage caused by the YopB-YopD translocation pore (Viboud and Bliska, 2001), and in counteracting YopB-stimulated pro-inflammatory signaling in infected epithelial cells (Viboud et al., 2003). These data suggest that apart from its established anti-phagocytic action, YopE may also exert an anti-inflammatory role.

The crystal structure of YopE has been recently solved (Evdokimov et al., 2002). YopE shares a high degree of structural similarity with the GAP domains of Exoenzyme S (ExoS) of *Pseudomonas aeruginosa* and SptP from *Salmonella typhimurium*, but has no obvious structural similarity with known mammalian functional GAP homologues, suggesting that convergent evolution has generated the bacterial GAPs (Stebbins and Galan, 2001).

YopH

YopH is among the most powerful protein tyrosine phosphatases (PTPases) known and it has long been known to contribute to ability of *Yersinia* to resist phagocytosis by macrophages (Rosqvist et al., 1988). The carboxy-terminal domain resembles eukaryotic phosphatases, and contains a phosphate-binding loop including a critical cysteine residue (Guan and Dixon, 1990), but the phosphotyrosylpeptide-binding activity also localizes to an amino-terminal domain (Black et al., 1998; Montagna et al., 2001). When injected into J774 macrophages, YopH dephosphorylates $p130^{Cas}$ and disrupts focal adhesions (Figure 5) (Hamid et al., 1999). Other YopH targets in J774 macrophages are the Fyn-binding protein

Fyb (Hamid et al., 1999) and the scaffolding protein SKAP-HOM (Figure 5), which interact with each other and become tyrosine-phosphorylated in response to macrophage adhesion (Black et al., 2000). It is likely that the action of YopH against focal adhesions, p130Cas, and Fyb, is relevant to the antiphagocytic action (Deleuil et al., 2003; Persson et al., 1999). YopH also suppresses the oxidative burst in macrophages (Bliska and Black, 1995), and protects against phagocytosis by polymorphonuclear leukocytes (PMNs) (Grosdent et al., 2002; Visser et al., 1999).

Recent observations have shown that YopH may also contribute to the downregulation of the inflammatory response (Sauvonnet et al., 2002a; Viboud et al., 2003). Upon infection, macrophages release the monocyte chemoattractant protein 1 (MCP-1), a chemokine involved in the recruitment of other macrophages to the sites of infection. In fact, MCP-1 mRNA levels are downregulated in macrophages infected with *Y. enterocolitica,* and this inhibition is dependent upon YopH. MCP-1 synthesis is known to be under the control of the PI 3-kinase pathway, which is involved in the control of multiple cellular processes. A wide variety of stimulated receptors can recruit PI 3-kinase to the cell membrane. Activated PI 3-kinase phosphorylates inositol phospholipids, leading to the appearance in the membrane of phosphatidylinositol 3-phosphate [PtdIns3P], phosphatidylinositol 3,4-bisphosphate [PtdIns(3,4)P_2] and phosphatidylinositol 3,4,5-triphosphate [PtdIns(3,4,5)P_3]. These, in turn, recruit proteins that control actin polymerization and other proteins such as phosphoinositide-dependent-kinase-1 (PDK-1) that phosphorylate and activate the protein kinase B (PKB, also called Akt) (Vanhaesebroeck and Alessi, 2000). Consistent with its negative action on MCP-1 mRNA levels, YopH was shown to abrogate the PI 3-kinase-dependent activation of PKB (Figure 6) (Sauvonnet et al., 2002a). The site of action of YopH in this cascade is still unknown but it is likely to be a tyrosine-phosphorylated receptor upstream from PI 3-kinase. Furthermore, as observed with YopE, YopH also counteracts YopB-stimulated pro-inflammatory signaling in infected epithelial cells, and it is likely that this also depends on its action on the PI 3-kinase pathway (Viboud et al., 2003).

In addition, in cell culture models, YopH also seems to incapacitate the host adaptive immune response. T and B cells transiently exposed to *Y. pseudotuberculosis* are impaired in their ability to be activated through their antigen receptors. YopH appears to be the main effector involved in blocking this activation (Yao et al., 1999), which most likely also involves the PI 3-kinase/PKB pathway (Sauvonnet et al., 2002a). In agreement with these observations, recombinant YopH introduced into human T lymphocytes was shown to dephosphorylate the primary signal transducer for the T cell receptor, the Lck tyrosine kinase (Alonso et al., 2004)

YpkA/YopO

YpkA (for *Yersinia* protein kinase A; called YopO in *Y. enterocolitica*) is an effector that also modulates the dynamics of the cytoskeleton (Hakansson et al., 1996a), and that contributes to *Yersinia* resistance to phagocytosis (Grosdent et al., 2002). It is an autophosphorylating serine-threonine kinase (Galyov et al., 1993), which shows some sequence and structural similarity to RhoA-binding kinases (Dukuzumuremyi et al., 2000), and only becomes active by interacting with actin (Figure 5) (Juris et al., 2000). In addition to being an activator of YpkA, actin can also function as an in vitro substrate of the kinase. YpkA/YopO interacts with, but does not phosphorylate, RhoA and Rac1 irrespective of the nucleotide bound, and apparently without affecting the guanine-nucleotide exchange capacity (Barz et al., 2000; Dukuzumuremyi et al., 2000). The N-terminal part of YopO/YpkA contains the kinase domain, while the C-terminal part of the kinase binds to actin. The C-terminal part also contains sequences that bear similarity to several eukaryotic RhoA-binding kinases and it binds to RhoA and Rac, but not to Cdc42 (Barz et al., 2000; Dukuzumuremyi et al., 2000). The kinase domain is required to localize YopO/YpkA to the plasma membrane, while the C-terminal part is responsible for its effect on the actin cytoskeleton in *Yersinia*-infected cells (Dukuzumuremyi et al., 2000; Juris et al., 2000). Although binding of YpkA/YopO to actin, RhoA, and Rac-1 appears to be relevant in the context of cytoskeleton dynamics and phagocytosis inhibition, the kinase target and the exact mode of action of YpkA/YopO remain unknown.

YopT

Infection of mammalian cells with a *Y. enterocolitica* strain that only expresses the YopT effector leads to rounding up of the cells and disruption of the cytoskeleton, which contributes to the anti-phagocytic activity of YopT (Grosdent et al., 2002; Iriarte and Cornelis, 1998). YopT releases RhoA from the cell membrane to the cytosol and modifies its isoelectric point (Figure 5) (Sorg et al., 2001; Zumbihl et al., 1999). YopT is a cysteine protease that releases RhoA from the membrane i by cleavage of isoprenylated RhoA near its carboxyl terminus. A polybasic sequence in the COOH terminus of RhoA is essential for recognition and YopT cleavage (Shao et al., 2003). YopT also releases Rac and Cdc42 from membranes by an identical mechanism (Shao et al., 2002), cleaving N-terminal to the prenylated cysteine in RhoA, Rac, and Cdc42 (Shao et al., 2003), but YopT injected by *Yersinia* seems to act only on RhoA (Aepfelbacher et al., 2003) This removes the prenyl group of the Rho GTPases and results in the irreversible inactivation of the targeted Rho GTPases, in contrast to the GAP activity of YopE, which can be reverted by the GEFs within the cell.

YopJ/P

YopJ (YopP in *Y. enterocolitica*) counteracts the normal pro-inflammatory response in various cell culture systems (Boland and Cornelis, 1998; Denecker et al., 2002; Schesser et al., 1998). This capacity results from the ability of YopJ/P to inhibit the NF-κB signaling pathway and also the mitogen-activated protein kinase (MAPK) pathways, c-Jun NH_2-terminal kinase (JNK), p38, extracellular signal regulatory kinase 1 (ERK1) and ERK2. Inhibition of the MAPK pathways abrogates phosphorylation of CREB, a transcription factor involved in the immune response (reviewed by Orth, 2002). Furthermore, the NF-κB transcription factor plays a central role in the modulation of cell survival, immune and inflammatory responses (Karin and Lin, 2002; Mercurio and Manning, 1999). NF-κB exists in the cytoplasm of resting cells in an inactive form associated with inhibitory proteins termed inhibitor-kappa B (IκB). After cell stimulation, NF-κB activation is achieved through the phosphorylation of IκB by the IκB kinase (IKK) complex. The components of the IKK complex that regulate the NFκB pathway include IKKα and IKKβ, which are activated by morphogenic and proinflammatory signals, respectively. Phosphorylated IκB is then selectively ubiquitinated, which targets its degradation by the proteasome. This unmasks the nuclear localization signal (NLS) of NF-κB, freeing it to translocate to the nucleus and activate transcription of genes involved in immune responses, including several cytokines.

YopJ/P binds directly to members of the superfamily of MAPK kinases (MEKs), blocking their phosphorylation and consequent activation (Figure 6) (Orth et al., 1999). Similarly, YopJ/P interacts with IKKβ, presumably also preventing its phosphorylation and activation (Orth et al., 1999). Thus, YopJ/P acts by preventing activation of MEKs and IKK in the MAPK and NF-κB signaling pathways, respectively (Orth et al., 1999). The exact molecular mechanism by which YopJ/P acts is still elusive, but its secondary structure resembles that of an adenovirus cysteine protease. The presumed protease catalytic triad of YopJ/P is required for the capacity of YopJ/P to inhibit the MAPK and NF-κB signaling pathways (Orth et al., 2000). Based on the observations that overexpression of YopJ/P results in a decrease of SUMOylated and ubiquitinylated proteins, it was proposed that YopJ/P could be an ubiquitin or a SUMO protease (SUMO are ubiquitin-like modifiers that are involved in stabilization or destabilization of proteins) (Orth, 2002; Orth et al., 2000). It is unclear how a de-SUMOylating activity could contribute to the ability of YopJ/P to disrupt MAPK and NF-κB signaling, but it is interesting to note that the NF-κB pathway is regulated at two distinct points by ubiquitinylation (Deng et al., 2000; Mercurio and Manning, 1999; Wang et al., 2001). Whatever the exact mechanism, it is clear that its elucidation will also provide valuable insights on how ubiquitinylation (or SUMOylation) contribute to for MAPK and NF-κB signaling, and also on how ubiquitinylation is involved in processes other than targeting proteins for proteasome degradation.

YopJ/P induces apoptosis of macrophages, but not of other cell types (Mills et al., 1997; Monack et al., 1997). This apoptosis is accompanied by cleavage of the cytosolic protein Bid, the release of cytochrome c from the mitochondria, and the activation of caspase-3, -7 and –9 (Figure 6) (Denecker et al., 2001). The apoptosis-inducing activity of YopJ/P is lost when the cysteine residue forming the catalytic triad responsible for its putative protease activity is replaced by a threonine (Denecker et al., 2001). It is very tempting to speculate that YopJ/P could induce apoptosis by cleaving a pro- or anti-apoptotic factor. However, this has never been demonstrated and it is unclear how a de-SUMOylating or de-ubiquitinylation activity could be related to the induction of a death pathway. It is even unclear whether apoptosis results from a YopP-induced early cell death signal or from YopP blockage of the NF-κB signaling, coupled to cellular activation by bacterial lipopolysaccharide (LPS) (Denecker et al., 2001; Ruckdeschel et al., 2001). On one side, it seems clear that a significant percentage of YopJ/P transfected macrophages undergo apoptosis. On the other side, this percentage of cell death is increased considerably by LPS treatment (Ruckdeschel et al., 2001).

Remarkably, despite its ability to disrupt several host signaling pathways and to induce apoptosis, YopJ/P seems to be dispensable for virulence of *Y. pseudotuberculosis* in the mouse model (Galyov et al., 1994). Nevertheless, YopJ/P-induced apoptosis helps in the establishment of a systemic infection in vivo (Monack et al., 1998).

YopM

YopM belongs to a family of type III effectors that has several representatives in *Shigella* (IpaH family) and *Salmonella* (SspH) (Kobe and Kajava, 2001). It is a strongly acidic protein composed almost entirely of 20-22 residue leucine-rich repeats (LRR). The repeating LRR unit of YopM is the shortest among all LRRs known to date, and depending on the *Yersinia* species, the number of copies can vary between 13 and 20 repeats. The crystal structure has revealed that the LRRs, consisting of parallel β-sheets, form a crescent shape, which is preceded by a α-helical hairpin at the N-terminus (Evdokimov et al., 2001). The latter domain has been shown to be part of the signal necessary to target YopM for translocation into eukaryotic cells (Boland et al., 1996). Intriguingly, individual YopM molecules form a tetramer in the crystal, creating a hollow cylinder with an inner diameter of 35 Å (Evdokimov et al., 2001). YopM has been shown to traffic to the nucleus via a vesicle-associated pathway, but its action in the nucleus remains unknown (Skrzypek et al., 1998).

The function of this Yop effector, which is essential for *Yersinia* virulence in vivo (Leung et al., 1990; Mulder et al., 1989), has been a mystery for many years. However, new insights about the role of YopM came from two recent studies. A transcriptome analysis of *Yersinia*-infected macrophages revealed that YopM might control the expression of genes involved in the eukaryotic cell cycle and cell growth (Sauvonnet et al., 2002b). Also, two kinases, protein kinase C-like 2 (PRK2) and ribosomal S6 protein kinase 1 (RSK1), have been shown to interact directly with YopM, by co-immunoprecipitation experiments using transfected HEK293 cells. Furthermore, YopM also interacts with RSK1 in infected macrophages (McDonald et al., 2003). The two kinases associate only when YopM is present, and their enzymatic activity is stimulated in a YopM-dependent manner (McDonald et al., 2003). This suggests that YopM may interfere with cellular signaling by modulating the phosphorylation status of yet to be identified host protein(s). Despite this remarkable recent progress, no phenotypic effect of YopM on eukaryotic cells has been identified so far, and the cellular basis of its pathogenic action remains an enigma.

CONCLUSIONS

Pathogenic *Yersinia* have in common a 70-kb virulence plasmid that encodes the Ysc-Yop TTS system, which is essential for virulence. Upon close contact with eukaryotic target cells, this system enables *Yersinia* to deliver six different effector Yops into the cytosol of contacted cells. As a result of this injection, phagocytosis is inhibited and the onset of the pro-inflammatory response is delayed. An overall picture of how the *Yersinia* Ysc-Yop system works now exists, as a result of intensive studies by several groups during the last fifteen years. When *Yersinia* are placed at the temperature of their host, they build multiple copies of the Ysc secretion apparatus at their surface. This protein-pumping machine is a complex assembly of 29 pYV-encoded Ysc proteins that spans the peptidoglycan layer and the two bacterial membranes, and is topped by a needle-like structure that has been visualized by electron microscopy. As long as there is no contact with a target cell, the secretion channel remains closed by the action of the YopN, TyeA and LcrG proteins, and a feedback inhibition mechanism limits transcription of the *yop* genes. Upon contact between the bacterium and the target cell, the feedback inhibition mechanism is relieved, the secretion channel opens, and the six Yop effectors (YopE, YopH, YpkA/YopO, YopT, YopJ/P and YopM) are translocated through the cell membrane and delivered into the eukaryotic cells. The translocation process is accomplished by the action of the Ysc-secreted YopB, YopD and LcrV that are thought to form a pore in the eukaryotic cell membrane. Inside of the eukaryotic cells, four Yop effectors (YopE, YopH, YpkA/YopO and YopT) contribute to the anti-phagocytic action of *Yersinia*, as their concerted action leads to the complete destruction of the actin cytoskeleton. YopE (GAP) and YopT (a cysteine protease) target the Rho family of GTPases directly and inhibit their activation. YpkA/YopO (a serine/threonine kinase) also interacts with the Rho family of GTPases. However, although this leads to a partial destruction of the cytoskeleton, no direct target has yet been identified. The anti-phagocytic function of YopH (a PTPase) is to disassemble adhesion complexes at the cell membrane. YopJ/P (a cysteine protease) efficiently shuts down multiple kinase cascades, and in this way may be responsible for the downregulation of the inflammatory response, an important host response to a bacterial infection. YopJ/P is also responsible for the induction of apoptosis of macrophages. Recently it is becoming clear that at least YopH and YopE may also have a role in the downregulation of the inflammatory response. Thus, the action of the different Yops may converge on one key physiological process, but a single Yop may have more than one effect. Despite recent progress, the role of the YopM effector remains a mystery and is a topic for further research.

But many other questions remain. Electron microscopy visualization of the entire Ysc machinery is still a challenge, as is the understanding of how the Ysc machinery associates, or not, with the translocator Yops in the process of effector Yop delivery. The exciting idea that there is a defined hierarchy of Yop secretion has been raised in the last few years, but experimental evidence to support it is lacking, and exactly how this hierarchy is determined is not known at all. The chaperones have been proposed to participate in setting this hierarchy, but the function of the chaperones themselves is still enigmatic, as many roles have been proposed. Also, the question of the recognition of substrates is not solved, as is not the question of how the system is unplugged upon cell contact. Regarding the function of the Yop effectors, in addition to the enigmatic YopM, the actual cellular targets and exact mode of action of YopJ/P and YpkA/YopO are unknown. As for YopH, it is becoming clear that its powerful phosphatase activity can have effects on the host cell beyond its well known anti-phagocytic action. The apparent connection between the YopE GAP activity and the inhibition of the inflammatory response is also an open field for future research. Clearly, future studies on the *Yersinia* Ysc-Yop TTS system promise not only an even a better understanding of *Yersinia* pathogenicity, but also should give rise to new concepts at the molecular biology and cellular biology levels.

ACKNOWLEDGEMENTS

The laboratory of G. R. C. in Brussels was supported by the Belgian Fonds National de la Recherche Scientifique Médicale (Convention 3.4595.97) and the Direction générale de la Recherche Scientifique-Communauté Française de Belgique (Action de Recherche Concertée 94/99-172). In Switzerland, the laboratory of G. R. C. is supported by the Swiss National Science Foundation (contract Nr 32-65393.01). The laboratory is also a member of a European Union network (HPRN-CT-2000-00075). M-N. M. is funded by the Belgian "Fonds pour la formation à la recherche dans l'Industrie et dans l'Agriculture" (FRIA). L.J.M. is supported by a post-doctoral fellowship (SFRH/BPD/3582/2000) from Fundação para a Ciência e Tecnologia (Portugal).

REFERENCES

Aepfelbacher, M., Trasak, C., Wilharm, G., Wiedemann, A., Trulzsch, K., Krauss, K., Gierschik, P, and Heesemann, J. Characterization of YopT effects on Rho GTPases in *Yersinia enterocolitica*-infected cells. J. Biol. Chem. 278:33217-33223.

Allaoui, A., Schulte, R., and Cornelis, G.R. 1995. Mutational analysis of the *Yersinia enterocolitica virC* operon: characterization of *yscE, F, G, I, J, K* required for Yop secretion and *yscH* encoding YopR. Mol. Microbiol. 18:343-55.

Alonso, A., Bottini, N., Bruckner, S., Rahmouni, S., Williams, S., Schoenberger, S.P., and Mustelin, T. (2004). Lck dephosphorylation at Y394 and inhibition of T cell antigen receptor signaling by *Yersinia* phosphatase YopH. J. Biol. Chem. 279: 4922-4928.

Anderson, D.M., Ramamurthi, K.S., Tam, C., and Schneewind, O. 2002. YopD and LcrH regulate expression of *Yersinia enterocolitica* YopQ by a posttranscriptional mechanism and bind to *yopQ* RNA. J. Bacteriol. 184: 1287-1295.

Anderson, D.M., and Schneewind, O. 1997. A mRNA signal for the type III secretion of Yop proteins by *Yersinia enterocolitica*. Science. 278:1140-1143.

Anderson, D.M., and Schneewind, O. 1999. *Yersinia enterocolitica* type III secretion: an mRNA signal that couples translation and and secretion of YopQ. Mol. Microbiol. 31:1139-1148.

Andor, A., Trulzsch, K., Essler, M., Roggenkamp, A., Wiedemann, A., Heesemann, J. and Aepfelbacher, M. 2001. YopE of *Yersinia*, a GAP for Rho GTPases, selectively modulates Rac- dependent actin structures in endothelial cells. Cell. Microbiol. 3:301-310.

Barz, C., Abahji, T.N., Trulzsch, K., and Heesemann, J. 2000. The *Yersinia* Ser/Thr protein kinase YpkA/YopO directly interacts with the small GTPases RhoA and Rac-1. FEBS Lett. 482:139-143.

Bergman, T., Hakansson, S., Forsberg, A., Norlander, L., Macellaro, A., Backman, A., Bolin, I., and Wolf-Watz, H. 1991. Analysis of the V antigen *lcrGVH-yopBD* operon of *Yersinia pseudotuberculosis*: evidence for a regulatory role of LcrH and LcrV. J. Bacteriol. 173:1607-1616.

Birtalan, S.C., Phillips, R.M., and Ghosh, P. 2002. Three-dimensional secretion signals in chaperone-effector complexes of bacterial pathogens. Mol. Cell. 9:971-980.

Black, D.S., and Bliska, J.B. 2000. The RhoGAP activity of the *Yersinia pseudotuberculosis* cytotoxin YopE is required for antiphagocytic function and virulence. Mol. Microbiol. 37:515-527.

Black, D.S., Marie-Cardine, A., Schraven, B., and Bliska, J.B. 2000. The *Yersinia* tyrosine phosphatase YopH targets a novel adhesion- regulated signalling complex in macrophages. Cell. Microbiol. 2:401-414.

Black, D.S., Montagna, L.G., Zitsmann, S., and Bliska. J.B. 1998. Identification of an amino-terminal substrate-binding domain in the *Yersinia* tyrosine phosphatase that is required for efficient recognition of focal adhesion targets. Mol. Microbiol. 29:1263-1274.

Bliska, J.B., and Black, D.S. 1995. Inhibition of the Fc receptor-mediated oxidative burst in macrophages by the *Yersinia pseudotuberculosis* tyrosine phosphatase. Infect. Immun. 63:681-685.

Blocker, A., Jouihri, N., Larquet, E., Gounon, P., Ebel, F., Parsot, C., Sansonetti, P., and Allaoui, A. 2001. Structure and composition of the *Shigella flexneri* "needle complex", a part of its type III secreton. Mol. Microbiol. 39:652-663.

Boland, A., and Cornelis, G.R. 1998. Role of YopP in suppression of tumor necrosis factor alpha release by macrophages during *Yersinia* infection. Infect. Immun. 66:1878-1884.

Boland, A., Sory, M.P., Iriarte, M., Kerbourch, C., Wattiau, P., and Cornelis, G.R. 1996. Status of YopM and YopN in the *Yersinia* Yop virulon: YopM of *Y.enterocolitica* is internalized inside the cytosol of PU5-1.8 macrophages by the YopB, D, N delivery apparatus. EMBO J. 15:5191-5201.

Boyd, A.P., Lambermont, I., and Cornelis, G.R. 2000. Competition between the Yops of *Yersinia enterocolitica* for delivery into eukaryotic cells: role of the SycE chaperone binding domain of YopE. J. Bacteriol. 182: 4811-21.

Cambronne, E.D., Cheng, L.W., and Schneewind, O. 2000. LcrQ/YscM1, regulators of the *Yersinia yop* virulon, are injected into host cells by a chaperone-dependent mechanism. Mol. Microbiol. 37:263-273.

Caron, E., and Hall, A. 1998. Identification of two distinct mechanisms of phagocytosis controlled by different Rho GTPases. Science. 282:1717-1721.

Cheng, L.W., Anderson, D.M., and Schneewind, O. 1997. Two independent type III secretion mechanisms for YopE in *Yersinia enterocolitica*. Mol. Microbiol. 24:757-765.

Cheng, L.W., and Schneewind, O. 2000. *Yersinia enterocolitica* TyeA, an intracellular regulator of the type III machinery, is required for specific targeting of YopE, YopH, YopM, and YopN into the cytosol of eukaryotic cells. J. Bacteriol. 182:3183-3190.

China, B., Michiels, T., and Cornelis, G.R. 1990. The pYV plasmid of *Yersinia* encodes a lipoprotein, YlpA, related to TraT. Mol. Microbiol. 4:1585-1593.

Cornelis, G., Sluiters, C., de Rouvroit, C.L., and Michiels, T. 1989. Homology between VirF, the transcriptional activator of the *Yersinia* virulence regulon, and AraC, the *Escherichia coli* arabinose operon regulator. J. Bacteriol. 171:254-262.

Cornelis, G., Vanootegem, J.C., and Sluiters, C. 1987. Transcription of the yop regulon from *Y. enterocolitica* requires trans acting pYV and chromosomal genes. Microb. Pathog. 2:367-379.

Cornelis, G.R. 2002a. *Yersinia* type III secretion: send in the effectors. J. Cell. Biol. 158:401-408.

Cornelis, G.R. 2002b. The *Yersinia* Ysc-Yop 'Type III' weaponry. Nat. Mol. Cell Biol. Rev. 3:742-752.

Cornelis, G.R., Boland, A., Boyd, A.P., Geuijen, C., Iriarte, M., Neyt, C., Sory, M.P., and Stainier, I. 1998. The virulence plasmid of *Yersinia*, an Antihost Genome. Microbiol. Mol. Biol. Rev. 62:1315-1352.

Cornelis, G.R., Sluiters, C., Delor, I., Geib, D., Kaniga, K., Lambert de Rouvroit, C., Sory, M.P., Vanooteghem, J.C., and Michiels, T. 1991. *ymoA*, a *Yersinia enterocolitica* chromosomal gene modulating the expression of virulence functions. Mol. Microbiol. 5:1023-1034.

Cornelis, G.R., and Van Gijsegem, F. 2000. Assembly and function of type III secretory systems. Ann. Rev. Microbiol. 54:735-774.

Cornelis, G.R., and Wolf-Watz, H. 1997. The *Yersinia* Yop virulon: a bacterial system for subverting eukaryotic cells. Mol. Microbiol. 23:861-867.

Day, J.B., Ferracci, F. and Plano, G.V. 2003. Translocation of YopE and YopN into eukaryotic cells by *Yersinia pestis yopN, tyeA, sycN, yscB* and *lcrG* deletion mutants measured using a phosphorylatable peptide tag and phosphospecific antibodies. Mol. Microbiol. 47:807-823.

Day, J.B., and Plano, G.V. 1998. A complex composed of SycN and YscB functions as a specific chaperone for YopN in *Yersinia pestis*. Mol. Microbiol. 30:777-788.

Day, J.B., and Plano, G.V. 2000. The *Yersinia pestis* YscY protein directly binds YscX, a secreted component of the type III secretion machinery. J. Bacteriol. 182:1834-1843.

Deleuil, F., Mogemark, L., Francis, M.S., Wolf-Watz, H., and Fallman, M. (2003). Interaction between the *Yersinia* protein tyrosine phosphatase YopH and eukaryotic Cas/Fyb is an important virulence mechanism. Cell. Microbiol. 5: 53-64.

Denecker, G., Declercq, W., Geuijen, C.A., Boland, A., Benabdillah, R., van Gurp, M., Sory, M.P., Vandenabeele, P., and Cornelis, G.R. 2001. *Yersinia enterocolitica* YopP-induced apoptosis of macrophages involves the apoptotic signaling cascade upstream of Bid. J. Biol. Chem. 276:19706-19714.

Denecker, G., Tötemeyer, S., Mota, L.J., Troisfontaines, P., Lambermont, I., Youta, C., Stainier, I., Ackermann, M., and Cornelis, G.R. 2002. Effect of Low- and High-Virulence *Yersinia enterocolitica* Strains on the Inflammatory Response of Human Umbilical Vein Endothelial Cells. Infect. Immun. 70:3510-3520.

Deng, L., Wang, C., Spencer, E., Yang, L., Braun, A., You, J., Slaughter, C., Pickart, C., and Chen, Z.J. 2000. Activation of the IkappaB kinase complex by TRAF6 requires a dimeric ubiquitin-conjugating enzyme complex and a unique polyubiquitin chain. Cell. 103:351-361.

Dorman, C.J., Hinton, J.C., and Free, A. 1999. Domain organization and oligomerization among H-NS-like nucleoid-associated proteins in bacteria. Trends Microbiol. 7:124-128.

Dukuzumuremyi, J.M., Rosqvist, R., Hallberg, B., Akerstrom, B., Wolf-Watz, H., and Schesser, K. 2000. The Yersinia protein kinase A is a host factor inducible RhoA/Rac- binding virulence factor. J. Biol. Chem. 275: 35281-35290.

Edqvist, P.J., Olsson, J., Lavander, M., Sundberg, L., Forsberg, A., Wolf-Watz, H., and Lloyd, S.A. 2003. YscP and YscU regulate substrate specificity of the *Yersinia* type III secretion system. J. Bacteriol. 185:2259-2266.

Evdokimov, A.G., Anderson, D.E., Routzahn, K.M., and Waugh, D.S. 2001. Unusual molecular architecture of the *Yersinia pestis* cytotoxin YopM: a leucine-rich repeat protein with the shortest repeating unit. J. Mol. Biol. 312:807-821.

Evdokimov, A.G., Tropea, J.E., Routzahn, K.M., and Waugh, D.S. 2002. Crystal structure of the *Yersinia pestis* GTPase activator YopE. Protein Sci. 11:401-408.

Feldman, M.F., Muller, S., Wuest, E., and Cornelis, G.R. 2002. SycE allows secretion of YopE-DHFR hybrids by the *Yersinia enterocolitica* type III Ysc system. Mol. Microbiol. 46:1183-1197.

Forsberg, A., Bolin, I., Norlander, L., and Wolf-Watz, H. 1987. Molecular cloning and expression of calcium-regulated, plasmid-coded proteins of *Y. pseudotuberculosis*. Microb. Pathog. 2:123-37.

Forsberg, A., Viitanen, A.M., Skurnik, M., and Wolf-Watz, H. 1991. The surface-located YopN protein is involved in calcium signal transduction in *Yersinia pseudotuberculosis*. Mol Microbiol. 5:977-986.

Forsberg, A., and Wolf-Watz, H. 1988. The virulence protein Yop5 of *Yersinia pseudotuberculosis* is regulated at transcriptional level by plasmid-plB1-encoded trans-acting elements controlled by temperature and calcium. Mol. Microbiol. 2:121-133.

Francis, M.S., Lloyd, S.A., and Wolf-Watz, H. 2001. The type III secretion chaperone LcrH co-operates with YopD to establish a negative, regulatory loop for control of Yop synthesis in *Yersinia pseudotuberculosis*. Mol. Microbiol. 42:1075-1093.

Frithz-Lindsen, E., Rosqvist, R., Johansson, L., Forsberg, A. 1995. The chaperone-like protein YerA of *Yersinia pseudotuberculosis* stabilizes YopE in the cytoplasm but is dispensable for targeting to the secretion loci. Mol. Microbiol. 16: 635-647.

Galyov, E.E., Hakansson, S., Forsberg, A., and Wolf-Watz, H. 1993. A secreted protein kinase of *Yersinia pseudotuberculosis* is an indispensable virulence determinant. Nature. 361:730-732.

Galyov, E.E., Hakansson, S., and Wolf-Watz, H. 1994. Characterization of the operon encoding the YpkA Ser/Thr protein kinase and the YopJ protein of *Yersinia pseudotuberculosis*. J. Bacteriol. 176:4543-4548.

Goguen, J.D., Yother, J., and Straley, S.C. 1984. Genetic analysis of the low calcium response in *Yersinia pestis* mu d1(Ap lac) insertion mutants. J. Bacteriol. 160:842-848.

Grosdent, N., Maridonneau-Parini, I., Sory, M.P. and Cornelis, G.R. 2002. Role of Yops and adhesins in resistance of *Yersinia enterocolitica* to phagocytosis. Infect. Immun. 70:4165-4176.

Guan, K.L., and Dixon, J.E. 1990. Protein tyrosine phosphatase activity of an essential virulence determinant in *Yersinia*. Science. 249:553-556.

Hakansson, S., Galyov, E.E., Rosqvist, R., and Wolf-Watz, H. 1996a. The *Yersinia* YpkA Ser/Thr kinase is translocated and subsequently targeted to the inner surface of the HeLa cell plasma membrane. Mol. Microbiol. 20:593-603.

Hakansson, S., Schesser, K., Persson, C., Galyov, E.E., Rosqvist, R., Homble, F., and Wolf-Watz, H. 1996b. The YopB protein of *Yersinia pseudotuberculosis* is essential for the translocation of Yop effector proteins across the target cell plasma membrane and displays a contact-dependent membrane disrupting activity. EMBO J. 15:5812-5823.

Hall, A. 1998. Rho GTPases and the actin cytoskeleton. *Science*. 279:509-514.

Hamid, N., Gustavsson, A., Andersson, K., McGee, K., Persson, C., Rudd, C.E., and Fallman, M. 1999. YopH dephosphorylates Cas and Fyn-binding protein in macrophages. Microb. Pathog. 27:231-242.

Hoiczyk, E., and Blobel, G. 2001. Polymerization of a single protein of the pathogen *Yersinia enterocolitica* into needles punctures eukaryotic cells. Proc. Natl. Acad. Sci. USA. 98:4669-4674.

Holmstrom, A., Olsson, J., Cherepanov, P., Maier, E., Nordfelth, R., Pettersson, J., Benz, R., Wolf-Watz, H., and Forsberg, A. 2001. LcrV is a channel size-determining component of the Yop effector translocon of *Yersinia*. Mol. Microbiol. 39:620-632.

Holmstrom, A., Petterson, J., Rosqvist, R., Hakansson, S., Tafazoli, F., Fallman, M., Magnusson, K.E., Wolf-Watz, H., and Forsberg, A. 1997. YopK of *Yersinia pseudotuberculosis* controls translocation of Yop effectors across the eukaryotic cell membrane. Mol. Microbiol. 24:73-91.

Iriarte, M., and Cornelis, G.R.. 1998. YopT, a new *Yersinia* Yop effector protein, affects the cytoskeleton of host cells. Mol. Microbiol. 29:915-929.

Iriarte, M. and Cornelis, G.R. 1999a. The 70-Kilobase Virulence Plasmid of Yersiniae. In: Pathogenicity Islands and Other Mobile Virulence Elements. J.B. Kapper and J. Hacker ed. American Society for Microbiology, Washington, D.C. p. 91-126.

Iriarte, M., and Cornelis, G.R.. 1999b. Identification of SycN, YscX and YscY, three new elements of the *Yersinia* Yop virulon. J. Bacteriol. 181:675-680.

Iriarte, M., Sory, M.P., Boland, A., Boyd, A.P., Mills, S.D., Lambermont, I., and Cornelis, G.R.. 1998. TyeA, a protein involved in control of Yop release and in translocation of *Yersinia* Yop effectors. EMBO J. 17:1907-1918.

Journet, L., Agrain, C, Broz, P. and Cornelis, G.R. 2003. The needle length of bacterial injectisomes is determined by a molecular ruler. Science 302: 1757-1760.

Juris, S.J., Rudolph, A.E., Huddler, D., Orth, K., and Dixon, J.E. 2000. A distinctive role for the *Yersinia* protein kinase: actin binding, kinase activation, and cytoskeleton disruption. Proc. Natl. Acad. Sci. USA. 97:9431-9436.

Karin, M., and Lin, A. 2002. NF-kappaB at the crossroads of life and death. Nat. Immunol. 3:221-227.

Kimbrough, T.G., and Miller, S.I. 2000. Contribution of *Salmonella typhimurium* type III secretion components to needle complex formation. Proc. Natl. Acad. Sci. USA. 97:11008-11013.

Kobe, B., and Kajava, A.V. 2001. The leucine-rich repeat as a protein recognition motif. Curr. Opin. Struct. Biol. 11:725-732.

Koster, M., Bitter, W., de Cock, H., Alloui, A., Cornelis, G.R., and J. Tommassen. 1997. The outer membrane component, YscC, of the Yop secretion machinery of *Yersinia enterocolitica* forms a ring-shaped multimeric complex. Mol. Microbiol. 26:789-798.

Krall, R., Zhang, Y. and Barbieri J. 2004. Intracellular membrane localization of *Pseudomonas* ExoS and *Yersinia* YopE in mammalian cells. J. Biol. Chem. 279: 2747-2753.

Kubori, T., Matsushima, Y., Nakamura, D., Uralil, J., Lara-Tejero, M., Sukhan, A., Galan, J.E., and Aizawa, S.I. 1998. Supramolecular structure of the *Salmonella typhimurium* type III protein secretion system. Science. 280:602-605.

Kubori, T., Sukhan, A., Aizawa, S.I., and Galan, J.E. 2000. Molecular characterization and assembly of the needle complex of the *Salmonella typhimurium* type III protein secretion system. Proc. Natl. Acad. Sci. USA. 97:10225-10230.

Lambert de Rouvroit, C., Sluiters, C., and Cornelis, G.R.. 1992. Role of the transcriptional activator, VirF, and temperature in the expression of the pYV plasmid genes of *Yersinia enterocolitica*. Mol. Microbiol. 6:395-409.

Lawton, D.G., Longstaff, C., Wallace, B.A., Hill, J., Leary, S.E., Titball, R.W., and Brown, K.A. 2002. Interactions of the type III secretion pathway proteins LcrV and LcrG from *Yersinia pestis* are mediated by coiled-coil domains. J. Biol. Chem. 277:38714-38722.

Lee, V.T., Tam, C., and Schneewind, O. 2000. LcrV, a substrate for *Yersinia enterocolitica* type III secretion, is required for toxin targeting into the cytosol of HeLa cells. J. Biol. Chem. 275:36869-36875.

Leung, K.Y., Reisner, B.S., and Straley, S.C. 1990. YopM inhibits platelet aggregation and is necessary for virulence of *Yersinia pestis* in mice. Infect. Immun. 58:3262-3271.

Lloyd, S.A., Norman, M., Rosqvist, R., and Wolf-Watz, H. 2001. *Yersinia* YopE is targeted for type III secretion by N-terminal, not mRNA, signals. Mol. Microbiol. 39:520-531.

Lloyd, S.A., Sjostrom, M., Andersson, S., and Wolf-Watz, H. 2002. Molecular characterization of type III secretion signals via analysis of synthetic N-terminal amino acid sequences. Mol. Microbiol. 43:51-59.

Massol, P., Montcourrier, P., Guillermot, J.C., and Chavrier, P. 1998. Fc receptor phagocytosis reqiured Cdc42 and Rac1. EMBO *J.* 17:6219-6229.

Matson, J.S., and Nilles, M.L. 2001. LcrG-LcrV interaction is required for control of Yops secretion in *Yersinia pestis*. J. Bacteriol. 183:5082-5091.

Matson, J.S., and Nilles, M.L. 2002. Interaction of the *Yersinia pestis* type III regulatory proteins LcrG and LcrV occurs at a hydrophobic interface. BMC Microbiol. 28:16.

McDonald, C., Vacratsis, P.O., Bliska, J.B., and, Dixon, J.E. 2003. The *Yersinia* virulence factor YopM forms a novel protein complex with two cellular kinases. J. Biol. Chem. 278:18514-18523

Mercurio, F., and Manning, A.M. 1999. Multiple signals converging on NF-kB. Curr. Opin. Cell Biol. 11:226-232.

Michiels, T., and Cornelis, G.R. 1991. Secretion of hybrid proteins by the *Yersinia* Yop export system. J. Bacteriol. 173:1677-85.

Michiels, T., Wattiau, P., Brasseur, R., Ruysschaert, J.M., and Cornelis, G. 1990. Secretion of Yop proteins by Yersiniae. Infect. Immun. 58:2840-2849.

Mills, S.D., Boland, A., Sory, M.P., van der Smissen, P., Kerbourch, C., Finlay, B.B., and Cornelis, G.R. 1997. *Yersinia enterocolitica* induces apoptosis in macrophages by a process requiring functional type III secretion and translocation mechanisms and involving YopP, presumably acting as an effector protein. Proc. Natl. Acad. Sci. USA. 94:12638-12643.

Monack, D.M., Mecsas, J., Bouley, D., and Falkow, S. 1998. *Yersinia*-induced apoptosis in vivo aids in the establishment of a systemic infection of mice. J. Exp. Med. 188:2127-2137.

Monack, D.M., Mecsas, J., Ghori, N., and Falkow, S. 1997. *Yersinia* signals macrophages to undergo apoptosis and YopJ is necessary for this cell death. Proc. Natl. Acad. Sci. USA. 94:10385-10390.

Montagna, L.G., Ivanov, M.I., and Bliska, J.B. 2001. Identification of residues in the N-terminal domain of the Yersinia tyrosine phosphatase that are critical for substrate recognition. J. Biol. Chem. 276:5005-5011.

Motin, V.L., Nakajima, R., Smirnov, G.B., and Brubaker, R.R. 1994. Passive immunity to yersiniae mediated by anti-recombinant V antigen and protein A-V antigen fusion peptide. Infect. Immun. 10:4192-4201.

Mulder, B., Michiels, T., Simonet, M., Sory, M.P., and Cornelis, G. 1989. Identification of additional virulence determinants on the pYV plasmid of *Yersinia enterocolitica* W227. Infect. Immun. 57:2534-2541.

Neyt, C., and Cornelis, G.R. 1999a. Insertion of a Yop translocation pore into the macrophage plasma membrane by *Yersinia enterocolitica*: requirement for translocators YopB and YopD, but not LcrG. Mol. Microbiol. 33: 971-981.

Neyt, C., and Cornelis, G.R. 1999b. Role of SycD, the chaperone of the *Yersinia* translocators YopB and YopD. Mol. Microbiol. 31:143-156.

Nilles, M.L., Fields, K.A., and Straley, S.C. 1998. The V antigen of *Yersinia pestis* regulates Yop vectorial targeting as well as Yop secretion through effects on YopB and LcrG. J. Bacteriol. 180:3410-20.

Nilles, M.L., Williams, A.W., Skrzypek, E., and Straley, S.C. 1997. *Yersinia pestis* LcrV forms a stable complex with LcrG and may have a secretion-related regulatory role in the low-Ca2+ response. J. Bacteriol. 179: 1307-1316.

Orth, K. 2002. Function of the *Yersinia* effector YopJ. Curr. Opin. Microbiol. 5:38-43.

Orth, K., Palmer, L.E., Bao, Z.Q., Stewart, S., Rudolph, A.E., Bliska, J.B., and Dixon, J.E. 1999. Inhibition of the mitogen-activated protein kinase kinase superfamily by a *Yersinia* effector. Science. 285:1920-1923.

Orth, K., Xu, Z., Mudgett, M.B., Bao, Z.Q., Palmer, L.E., Bliska, J.B., Mangel, W.F., Staskawicz, B., and Dixon, J.E. 2000. Disruption of signaling by *Yersinia* effector YopJ, a ubiquitin-like protein protease. Science. 290: 1594-1597.

Persson, C., Nordfelth, R., Andersson, K., Forsberg, A., Wolf-Watz, H., and Fallman, M. 1999. Localization of the *Yersinia* PTPase to focal complexes is an important virulence mechanism. Mol. Microbiol. 33:828-838.

Persson, C., Nordfelth, R., Holmstrom, A., Hakansson, S., Rosqvist, R., and Wolf-Watz, H. 1995. Cell-surface-bound *Yersinia* translocate the protein tyrosine phosphatase YopH by a polarized mechanism into the target cell. Mol. Microbiol. 18:135-50.

Pettersson, J., Holmstrom, A., Hill, J., Leary, S., Frithz-Lindsten, E., von Euler-Matell, A., Carlsson, E., Titball, R., Forsberg, A. and Wolf-Watz, H. 1999. The V-antigen of *Yersinia* is surface exposed before target cell contact and involved in virulence protein translocation. Mol. Microbiol. 32:961-76.

Pettersson, J., Nordfelth, R., Dubinina, E., Bergman, T., Gustafsson, M., Magnusson, K.E., and Wolf-Watz, H. 1996. Modulation of virulence factor expression by pathogen target cell contact. Science. 273:1231-1233.

Rimpilainen, M., Forsberg, A., and Wolf-Watz, H. 1992. A novel protein, LcrQ, involved in the low-calcium response of *Yersinia pseudotuberculosis* shows extensive homology to YopH. J. Bacteriol. 174:3355-3363.

Roggenkamp, A., Geiger, A.M., Leitritz, L., Kessler, A., and Heesemann, J. 1997. Passive immunity to infection with *Yersinia* spp. mediated by anti-recombinant V antigen is dependent on polymorphism of V antigen. Infect. Immun. 65:446-451.

Rohde, J.R., Luan, X.S., Rohde, H., Fox, J.M., and Minnich, S.A. 1999. The *Yersinia enterocolitica* pYV virulence plasmid contains multiple intrinsic DNA bends which melt at 37 degrees C. J. Bacteriol. 181:4198-204.

Rosqvist, R., Bolin, I., and Wolf-Watz, H. 1988. Inhibition of phagocytosis in *Yersinia pseudotuberculosis*: a virulence plasmid-encoded ability involving the Yop2b protein. Infect. Immun. 56:2139-2143.

Rosqvist, R., Forsberg, A., and Wolf-Watz, H. 1991. Intracellular targeting of the *Yersinia* YopE cytotoxin in mammalian cells induces actin microfilament disruption. Infect. Immun. 59:4562-4569.

Rosqvist, R., Magnusson, K.E., and Wolf-Watz, H. 1994. Target cell contact triggers expression and polarized transfer of *Yersinia* YopE cytotoxin into mammalian cells. EMBO J. 13:964-72.

Ruckdeschel, K., Mannel, O., Richter, K., Jacobi, C.A., Trulzsch, K., Rouot, B., and Heesemann, J. 2001. Yersinia outer protein P of *Yersinia enterocolitica* simultaneously blocks the nuclear factor-kappa B pathway and exploits lipopolysaccharide signaling to trigger apoptosis in macrophages. J. Immunol. 166:1823-1831.

Sarker, M.R., Neyt, C., Stainier, I., and Cornelis, G.R. 1998a. The *Yersinia* Yop virulon: LcrV is required for extrusion of the translocators YopB and YopD. J. Bacteriol. 180:1207-1214.

Sarker, M.R., Sory, M.P., Boyd, A.P., Iriarte, M., and Cornelis, G.R. 1998b. LcrG is required for efficient translocation of *Yersinia* Yop effector proteins into eukaryotic cells. Infect. Immun. 66:2976-2979.

Sauvonnet, N., Lambermont, I., van der Bruggen, P., and Cornelis, G.R. 2002a. YopH prevents monocyte chemoattractant protein 1 expression in macrophages and T-cell proliferation through inactivation of the phosphatidylinositol 3-kinase pathway. Mol. Microbiol.45:805-815.

Sauvonnet, N., Pradet-Balade, B., Garcia-Sanz, J.A., and Cornelis, G.R. 2002b. Regulation of mRNA expression in macrophages following *Yersinia enterocolitica* infection: role of different Yop effectors. J. Biol. Chem. 2: 25133-25142.

Schesser, K., Spiik, A.K., Dukuzumuremyi, J.M., Neurath, M.F., Pettersson, S., and Wolf-Watz, H. 1998. The *yopJ* locus is required for *Yersinia*-mediated inhibition of NF- kappaB activation and cytokine expression: YopJ contains a eukaryotic SH2-like domain that is essential for its repressive activity. Mol. Microbiol. 28: 1067-1079.

Shao, F., Merritt, P.M., Bao, Z., Innes, R.W., and Dixon, J.E. 2002. A *Yersinia* effector and a *Pseudomonas* avirulence protein define a family of cysteine proteases functioning in bacterial pathogenesis. Cell. 109:575-588.

Shao, F., Vacratsis, P.O., Bao, Z., Bowers, K.E., Fierke, C.A., and Dixon, J.E. 2003. Biochemical characterization of the *Yersinia* YopT protease: cleavage site and recognition elements in Rho GTPases. Proc. Natl. Acad. Sci. USA. 100:904-909.

Sing, A., Roggenkamp, A., Geiger, A.M., and Heesemann, J. 2002. *Yersinia enterocolitica* evasion of the host innate immune response by V antigen-induced IL-10 production of macrophages is abrogated in IL-10-deficient mice. J. Immunol. 168:1315-1321.

Skryzpek, E., and Straley, S.C. 1993. LcrG, a secreted protein involved in negative regulation of the low-calcium response in *Yersinia pestis*. J. Bacteriol. 175:3520-3528.

Skrzypek, E., Cowan, C., and Straley, S.C. 1998. Targeting of the *Yersinia pestis* YopM protein into HeLa cells and intracellular trafficking to the nucleus. Mol. Microbiol. 30:1051-1065.

Skurnik, M., and Toivanen, P. 1992. LcrF is the temperature-regulated activator of the *yadA* gene of *Yersinia enterocolitica* and *Yersinia pseudotuberculosis*. J. Bacteriol. 174:2047-2051.

Sorg, I., Goehring, U.M., Aktories, K., and Schmidt, G. 2001. Recombinant *Yersinia* YopT leads to uncoupling of RhoA-effector interaction. Infect. Immun. 69:7535-7543.

Sory, M.P., Boland, A., Lambermont, I., and Cornelis, G.R. 1995. Identification of the YopE and YopH domains required for secretion and internalization into the cytosol of macrophages, using the *cyaA* gene fusion approach. Proc. Natl. Acad. Sci. USA. 92:11998-2002.

Sory, M.P., and Cornelis, G.R. 1994. Translocation of a hybrid YopE-adenylate cyclase from *Yersinia enterocolitica* into HeLa cells. Mol. Microbiol. 14:583-594.

Stainier, I., Iriarte, M., and Cornelis, G.R. 1997. YscM1 and YscM2, two *Yersinia enterocolitica* proteins causing down regulation of *yop* transcription. Mol. Microbiol. 26:833-843.

Stebbins, C.E., and Galan, J.E. 2001. Maintenance of an unfolded polypeptide by a cognate chaperone in bacterial type III secretion. Nature. 414:77-81.

Straley, S.C., Plano, G.V., Skrzypek, E., Haddix, P.L., and Fields, K.A. 1993. Regulation by Ca2+ in the *Yersinia* low-Ca2+ response. Mol. Microbiol. 8:1005-1010.

Tardy, F., Homble, F., Neyt, C., Wattiez, R., Cornelis, G.R., Ruysschaert, J.M., and Cabiaux, V. 1999. *Yersinia enterocolitica* type III secretion-translocation system: channel formation by secreted Yops. EMBO J. 18: 6793-6799.

Vanhaesebroeck, B., and Alessi, D.R. 2000. The PI3K-PDK1 connection: more than just a road to PKB. Biochem J. 346:561-576.

Viboud, G.I., and Bliska, J.B. 2001. A bacterial type III secretion system inhibits actin polymerization to prevent pore formation in host cell membranes. EMBO J. 20:5373-5382.

Viboud, G.I., So, S.S.K., Ryndak, M.B., and Bliska, J.B. 2003. Proinflammatory signalling stimulated by the type III translocation factor YopB is counteracted by multiple effectors in epithelial cells infected with *Yersinia pseudotuberculosis*. Mol. Microbiol. 47:1305-1315.

Visser, L.G., Seijmonsbergen, E., Nibbering, P.H., van den Broek, P.J., and van Furth, R. 1999. Yops of *Yersinia enterocolitica* inhibit receptor-dependent superoxide anion production by human granulocytes. Infect. Immun. 67:1245-1250.

Von Pawel-Rammingen, U., Telepnev, M.V., Schmidt, G., Aktories, K., Wolf-Watz, H., and Rosqvist, R. 2000. GAP activity of the *Yersinia* YopE cytotoxin specifically targets the Rho pathway: a mechanism for disruption of actin microfilament structure. Mol. Microbiol. 36:737-748.

Wang, C., Deng, L., Hong, M., Akkaraju, G.R., Inoue, J.I., and Chen, J.C. 2001. TAK1 is a ubiquitin-dependent kinase of MKK and IKK. Nature. 412:346-351.

Wattiau, P., Bernier, B., Deslee, P., Michiels, T., and Cornelis, G.R. 1994. Individual chaperones required for Yop secretion by *Yersinia*. Proc Natl Acad Sci USA. 91:10493-10497.

Wattiau, P., and Cornelis, G.R. 1993. SycE, a chaperone-like protein of *Yersinia enterocolitica* involved in the secretion of YopE. Mol. Microbiol. 8:123-131.

Wattiau, P., and Cornelis, G.R. 1994. Identification of DNA sequences recognized by VirF, the transcriptional activator of the *Yersinia* yop regulon. J Bacteriol. 176:3878-3884.

Wattiau, P., Woestyn, S., and Cornelis, G.R. 1996. Customized secretion chaperones in pathogenic bacteria. Mol. Microbiol. 20:255-262.

Williams, A.W., and Straley, S.C. 1998. YopD of *Yersinia pestis* plays a role in negative regulation of the low-calcium response in addition to its role in translocation of Yops. J. Bacteriol. 180:350-358.

Woestyn, S., Allaoui, A., Wattiau, P., and Cornelis, G.R.. 1994. YscN, the putative energizer of the *Yersinia* Yop secretion machinery. J. Bacteriol. 176:1561-1569.

Woestyn, S., Sory, M.P., Boland, A., Lequenne, O., and Cornelis, G.R. 1996. The cytosolic SycE and SycH chaperones of *Yersinia* protect the region of YopE and YopH involved in translocation across eukaryotic cell membranes. Mol. Microbiol. 20:1261-1271.

Yao, T., Mecsas, J., Healy, J.I., Falkow, S., and Chien, Y. 1999. Suppression of T and B lymphocyte activation by a *Yersinia pseudotuberculosis* virulence factor, yopH. J. Exp. Med. 190:1343-1350.

Yother, J., and Goguen, J.D. 1985. Isolation and characterization of Ca2+-blind mutants of *Yersinia pestis*. J Bacteriol. 164:704-711.

Zumbihl, R., Aepfelbacher, M., Andor, A., Jacobi, C.A., Ruckdeschel, K., Rouot, B., and Heesemann, J. 1999. The cytotoxin YopT of *Yersinia enterocolitica* induces modification and cellular redistribution of the small GTP-binding protein RhoA. J. Biol. Chem. 274:29289-29293.

Chapter 17

The Plasminogen Activator Pla of *Yersinia pestis*: Localized Proteolysis and Systemic Spread

Timo K. Korhonen, Maini Kukkonen, Ritva Virkola, Hannu Lang, Marjo Suomalainen, Päivi Kyllönen and Kaarina Lähteenmäki

ABSTRACT

The plasminogen activator (Pla) of the plague bacterium *Yersinia pestis* is a cell-surface protease that belongs to the omptin family of enterobacterial aspartic proteases. Pla is a critical virulence factor in the pathogenesis of plague and specifically enables the spread of the bacterium from the subcutaneous site of a flea bite. Pla may enhance the invasiveness of *Y. pestis* by multiple mechanisms. It proteolytically activates human plasminogen to plasmin and inactivates α2-antiplasmin, the major plasmin inhibitor in human serum, which would be predicted to result in uncontrolled, highly potent proteolysis at the infection site. Pla also is an adhesin with affinity for laminin of basement membranes, and we hypothesize that bacterial adhesiveness localizes the formed plasmin activity onto susceptible targets, such as laminin and basement membranes. The combined activities of Pla are known to damage basement membrane and extracellular matrix and impair their barrier function in vitro, a behaviour that is reminiscent of metastatic tumor cell migration. Pla also degrades the complement component C3, which may increase resistance of *Y. pestis* to killing by phagocytes. Finally, Pla is an invasin that promotes the in-vitro invasion of *Y. pestis* into human epithelial and endothelial cells. The adhesive and the invasive functions of Pla can be genetically dissected from the proteolytic activity. Pla is predicted to have a 10-stranded antiparallel β-barrel conformation, with five short surface-exposed loops. The active site is located in a groove at the outermost surface of the barrel. Residues and regions in the surface loops that are important for the functions of Pla have been identified. Pla has a dual interaction with lipopolysaccharide: it requires rough lipopolysaccharide to be functionally active but its functions are sterically hindered by an O-side chain. *Y. pestis* lacks an O-antigen, and an advantage for *Y. pestis* of having a rough lipopolysaccharide may be its ability to fully utilize Pla functions.

THE OMPTIN FAMILY OF SURFACE PROTEASES

The *Yersinia pestis* Pla belongs to the omptin family of outer membrane proteases that have been detected in enterobacterial species pathogenic to humans or plants (Table 1). The name of the family is derived from the first well-characterized member, OmpT of *Escherichia coli*. The omptins are highly related in structure: they share about 50% sequence identity, their mature forms are 292-298 amino acid residues long, and

Table 1. Members of the omptin family of aspartic proteases			
Bacterium	Omptin	Gene location	Reference
Yersinia pestis	Pla	plasmid pPCP1	Sodeinde and Goguen, 1989
Salmonella enterica	PgtE	chromosome	Guina et al., 2000
Escherichia coli	OmpT	chromosome	Grodberg et al., 1988
Escherichia coli	OmpP	F plasmid	Matsuo et al., 1999
Shigella flexneri	SopA	virulence plasmid pWR100	Egile et al., 1997
Erwinia pyrifoliae	Pla endopeptidase A	plasmid pEP36	McGhee et al., 2002

they lack or have a low content of cysteines. The omptin genes are located either on the chromosome or on a plasmid, which in *Y. pestis* and in *Shigella flexneri* is associated with virulence. OmpT has been crystallized and serves as the prototype of an omptin molecule. The structure of OmpT shows a 10-stranded antiparallel β-barrel that traverses the outer membrane and protrudes far from the outer membrane into the extracellular space (Vandeputte-Rutten et al., 2001). The OmpT β-barrel has five surface loops, which are located at the distance of about 40 Å from the outer membrane. The high sequence similarity of the omptins, as well as topology and structure modeling have led to the conclusion that the β-barrel structure is shared by the other omptin molecules as well. It is not presently known whether the omptins occur in the outer membrane as monomers or as polymeric complexes. For sequence comparison and structural data on the omptin family, the reader is referred to the MEROPS data base (http://merops.sanger.ac.uk/).

The omptins are classified as aspartic proteases and cleave substrates next to basic residues, preferentially between two consecutive basic amino acids (Dekker et al., 2001). The omptins do not contain any conserved active site sequences of other known protease families. In an analogy to eukaryotic plasminogen activators, the omptins were originally considered to be serine proteases, but commonly used serine protease inhibitors have little or no effect on their activity. The crystal structure of OmpT showed that the proteolytic site is located in a groove at the extracellular top of the β-barrel (Vandeputte-Rutten et al., 2001). The catalytic site is formed by a His-Asp dyad and an Asp-Asp couple that is fully conserved within the omptin family and present in Pla as well.

ASSOCIATION OF Pla WITH VIRULENCE

Plague is transmitted by fleas that infect both rodents and humans. *Y. pestis* exhibits exceptionally efficient organ invasion during plague infection and spreads from the subcutaneous site of the flea bite to lymph nodes and the circulation (reviewed by Perry and Fetherston, 1997). The bacterium multiplies to large numbers and causes swelling in the lymph nodes (bubonic plague); systemic infection leads to septicemia and bacterial colonization of the liver, the spleen and sometimes the lungs (pneumonic plague), which facilitates direct transmission from human to human. Virulent isolates of *Y. pestis* harbour the 9.5-kb plasmid pPCP1 (pPla, pPst) which is needed for the invasive character of plague. Loss of pPCP1 increases the median lethal dose of *Y. pestis* by a million-fold, but only when the infection initiates subcutaneously (Brubaker et al., 1965; Ferber and Brubaker, 1981). pPCP1 is specific to *Y. pestis* and encodes at least three proteins: the bacteriocin pesticin, the protein conferring immunity to pesticin, and the outer membrane protease Pla (Sodeinde and Goguen, 1988; Hu et al., 1998). (For additional information on the *Y. pestis* pPla plasmid, see Chapter 3). Sodeinde et al (1992) showed by mutagenesis

Table 2. Proposed functions of Pla of *Y. pestis*	
Function	References
Cleavage and activation of Plg to active plasmin	Sodeinde et al., 1992; Goguen et al., 2000
Inactivation of the serpin α2-antiplasmin	Kukkonen et al., 2001
Degradation of the complement component C3	Sodeinde et al., 1992
Coagulase activity	Beesley et al., 1967; McDonough and Falkow., 1989
Degradation of Yersinia outer membrane proteins	Sodeinde et al., 1988
Adhesion to eukaryotic cells	Kienle et al., 1992
Adhesion to laminin and basement membrane	Lähteenmäki et al., 1998
Invasion into human endothelial and epithelial cells	Cowan et al., 2000; Lähteenmäki et al. 2001a

that the virulence-associated property in pPCP1 is encoded by *pla*. When bacteria were injected subcutaneously into mice, the LD_{50} value of a Pla-negative mutant strain was close to 10^7 bacteria, as compared to an LD_{50} value of less than 50 bacteria for the isogenic Pla-positive strain. However, when bacteria were injected intravenously, there was no difference between the virulence of the strains. These data indicate that Pla specifically enables the dissemination of *Y. pestis* from the subcutaneous infection site.

Much less is known about the virulence associations of the other omptins, but overall, our current knowledge indicates that the different omptins affect bacterial virulence by various mechanisms and to different degrees. For example, SopA increases virulence of *Shigella* indirectly by processing the IcsA surface protein needed for actin-based motility within eukaryotic cells (Shere et al., 1997; Egile et al., 1997). Interestingly, expression of *ompT* in *Shigella* leads to degradation, rather than processing, of IcsA and to loss of spreading ability (Nakata et al., 1993), which indicates that SopA and OmpT cleave IcsA differently and probably differ in substrate recognition. The presence of *ompT* is not correlated with invasiveness of *E. coli*; instead OmpT seems to be a housekeeping protease that degrades denaturated proteins (White et al., 1995). However, a role in bacterial resistance to innate immunological defense is suggested by the finding that OmpT cleaves protamine (Stumpe et al., 1998). PgtE cleaves cationic antimicrobial peptides and has been proposed to increase survival of *Salmonella* within macrophages (Guina et al., 2000), but its other possible roles in salmonellosis have not been addressed.

FUNCTIONS OF Pla

Pla appears to be an exceptionally multifunctional surface protein that enhances virulence of *Y. pestis* by several mechanisms (Table 2). As will be discussed in the following sections, it is an efficient surface protease that creates uncontrolled proteolysis in humans by activating the plasminogen (Plg) proteolytic cascade and inactivating the antiprotease α2-antiplasmin (α2AP). In addition, Pla interferes with the complement system by degrading the complement component C3. Pla is an adhesin with affinity for basement membranes (BMs) and epithelial cells, as well as an invasin that potentiates bacterial intake into endothelial and epithelial cells. The adhesive and the invasive functions of Pla can be genetically dissected from the proteolytic activity; however, the relative roles of these functions in plague virulence have not yet been estimated.

PROTEOLYTIC FUNCTIONS

A major virulence-associated function of Pla is its proteolytic activation of Plg. Plg is a 90-kDa circulating proenzyme that is converted to the active serine protease plasmin by a proteolytic activation. The mammalian Plg activators cleave the peptide bond between Arg_{560} and Val_{561} in Plg, leaving the two peptide chains of the formed plasmin molecule joined via two disulphide bonds. Plasmin is a broad-spectrum protease that has an important role in several physiological processes in mammals: it is a key enzyme in fibrinolysis, degrades components of extracellular matrices (ECMs) and BMs, and is involved in activation of certain prohormones and growth factors as well as in tumor metastasis (for reviews on plasminogen and its functions, see Saksela, 1985; Stephens and Vaheri, 1993 Lijnen and Collen, 1995; Plow et al., 1999). Plasmin cleaves procollagenases to active collagenases and thus indirectly enhances degradation of collagen fibers or networks, which is needed for cellular metastasis. Plg circulates in the body in large amounts: in adult human plasma the concentration is 180-200 μg per ml (ca. 2 μM), and it is understandable that Plg activation and plasmin proteolysis must be tightly regulated. This is achieved by specific plasminogen activators, by inhibitors that control the Plg system both at the level of Plg activation and plasmin, and by immobilization of Plg and plasmin to cellular receptors or to target molecules.

The Plg system offers a high-potential proteolytic system to be exploited by pathogenic bacteria for tissue invasion or nutritional requirements (reviewed by Lähteenmäki et al., 2001b). The central roles of Plg activation and Pla in the systemic spread of *Y. pestis* in mice infected subcutaneously have been well established. Pla activates Plg by cleaving the same Arg_{560}-Val_{561} bond as do human plasminogen activators, and it was estimated that the efficiency of Plg activation per cell surface area is essentially the same for Pla on the surface of *Y. pestis* and for urokinase on mammalian cells (Sodeinde et al., 1992). In recombinant *E. coli* expressing *ompT* or *pla*, Plg activation by Pla is dramatically more rapid and efficient than activation by OmpT (Kukkonen et al., 2001). Plg-deficient mice are a hundred-fold more resistant to *Y. pestis* infection than normal mice (Goguen et al., 2000), which underlines the pathogenetic role of Plg activation. However, the difference is clearly less dramatic than the million-fold difference between normal mice infected subcutaneously with either Pla-positive or Pla-negative bacteria, which suggests that other functions of Pla also influence virulence.

The formation and activity of proteases in the human circulation are tightly controlled, and approximately 10% of the protein mass in human plasma consists of proteinase inhibitors or antiproteases (reviewed by Travis and Salvesen, 1983). Free plasmin in circulation is rapidly inactivated by the serpin (serine protease inhibitor) α2AP, which forms an equimolar complex with plasmin in solution but is unable to bind plasmin immobilized on a cellular Plg/plasmin receptor. Plg/plasmin receptors are common in pathogenic bacteria and function to enhance Plg activation and protect plasmin from inactivation by α2AP (Lähteenmäki et al., 2001b). Pla-positive *Y. pestis* forms active plasmin in human plasma (Sodeinde et al., 1992). No Plg/plasmin receptors have so far been detected in *Y. pestis,* but we found that Pla-positive *Y. pestis* and recombinant *E. coli* expressing *pla* cleave and inactivate α2AP (Kukkonen et al., 2001). The inactivation was not observed with bacteria expressing proteolytically deficient mutants of Pla, which indicates that Pla degrades α2AP. Inactivation of α2AP most likely is very important in the pathogenesis of plague, as it allows full utilization of the Pla-generated plasmin to create uncontrolled proteolysis at the infection site.

Pla may enhance the spread of *Y. pestis* in the circulatory system also. A major physiological function of plasmin is fibrinolysis, and it was proposed that Pla-generated plasmin may clear fibrin deposits that could hinder bacterial migration in the circulation (Beesley et al., 1967). This does not appear to be required for disease because Pla-minus strains are fully virulent after i.v. inoculation. Sodeinde et al. (1992) found that Pla cleaves the complement component C3 and that infected lesions in mice infected with wild-type *Y. pestis* contained fewer inflammatory cells than the lesions in mice infected with a *pla* mutant. Sodeinde et al. (1992) proposed that Pla-mediated cleavage of complement components leads to suppression of the migration of inflammatory cells to the infection site. On the other hand, degradation of C3 might also reduce opsonophagocytosis of Pla-expressing bacteria via the C3b receptors and thereby promote serum resistance of *Y. pestis*. Degradation of complement component(s) may be particularly important for *Y. pestis,* which lacks the O-antigen of lipopolysaccharide (LPS) that is known to increase serum resistance of other enterobacterial species (reviewed by Rautemaa and Meri, 1999). Pla also exhibits a weak, low-temperature dependent coagulase activity (Beesley et al., 1967; Sodeinde and Goguen, 1988) that is detectable with rabbit but not with human or mouse plasma. Its role in human infections has therefore remained uncertain; a possible role in the flea host also seems unlikely as *Y. pestis pla* is not required for infection or virulence in the flea vector (Hinnebusch et al., 1998).

Expression of *pla* in *Yersinia* leads to degradation of Yops, *Yersinia* outer membrane proteins that are essential virulence factors in yersiniosis (Sodeinde et al., 1988). The degradation seems specific to Yops as other *Yersinia* proteins or *E. coli* outer membrane proteins are not significantly degraded by expression of Pla. However, Pla-positive and Pla-negative *Y. pestis* target Yops equally well into human HeLa cells (Skrzypek et al., 1998), and the biological significance of Yop degradation in the pathogenesis of plague remains unclear.

NONPROTEOLYTIC FUNCTIONS

The first indications of a nonproteolytic function for Pla was provided by Kienle et al. (1992), who reported that expression of Pla enhanced bacterial adherence to human epithelial cell lines, extracted glycolipids, and type IV collagen of BMs. Bacterial adhesiveness was reduced by treating target cells with β-galactosidase, and adhesion was detected to globoside that contains the terminal tetrasaccharide GalNAc(β1-3) Gal(α1-4)Gal(β1-4)Glc. Adhesion to collagen was of low affinity, and later studies showed that Pla binds to laminin and, to a lesser extent, to heparan sulfate proteoglycan (Lähteenmäki et al., 1998). Both are components of the BM, and Pla-expressing *Y. pestis* as well as recombinant *E. coli* exhibit specific adherence to the mouse BM preparation Matrigel, as well as to ECMs from the human lung cell line NCI-H292 and the human endothelial-like cell line ECV304 (Lähteenmäki et al., 1998, 2001a). Adherence via the Pla protein has not been directly demonstrated, and the possibility remains that Pla induces bacterial adhesiveness by degrading outer membrane proteins to uncover cryptic adhesion sites. This seems unlikely, however, because adhesiveness to ECM is also exhibited by proteolytically inactive mutants of Pla (Lähteenmäki et al., 2001a). Although laminin supports adhesion, it is not directly degraded by Pla-positive bacteria, which indicates that laminin is not a substrate for Pla-mediated proteolysis. However, laminin is a well-characterized target for plasmin, and degradation of laminin was seen when Pla-expressing bacteria and plasminogen were incubated on a laminin-coated surface (Lähteenmäki et al., 1998).

BMs form important tissue barriers that invading bacteria (and metastatic tumor cells on the other hand) must penetrate in order to get into circulation. The tissue barrier function is mainly provided by the type IV collagen and the laminin networks. BMs are also considered to be reservoirs of the Plg system and contain Plg activators as well as Plg which can be activated to functional plasmin (Farina et al., 1996). BMs also contain procollagenases that are proteolytically activated by plasmin and function to disrupt the ECM barrier. Activation of procollagenases and adherence to BM are also involved in metastasis of tumor cells into circulation (for reviews, see Liotta et al., 1986; Mignatti and Rifkin, 1993; Plow et al., 1999), and it is striking that the highly invasive *Y. pestis* shares these properties. Actual penetration through BM or activation of procollagenases by Pla-positive *Y. pestis* has not been demonstrated yet, but we observed that the matrix from ^{35}S-labelled NCI-H292 lung cells is degraded by Pla-positive bacteria in the presence of exogeneous plasminogen (Lähteenmäki et al., 1998). Our hypothesis is that the binding of Pla to laminin localizes the generated plasmin onto a susceptible target, laminin, in BMs and that the integrity of the BM is further weakened by plasmin-induced procollagenase activation.

Y. pestis is regarded as an extracellular pathogen and has lost by frameshift mutation or transposon insertion the major invasins and adhesins expressed by enteropathogenic *Yersinia* species (see references in Parkhill et al., 2001). However, *Y. pestis* can also cause systemic infection from aerogenic or oral-fecal routes of infection (Butler et al., 1982; Rust et al., 1972), which often involve intracellular invasion. Cowan et al. (2000) observed an efficient *in vitro* invasion of *Y. pestis* into the human epithelial HeLa cell line and that loss of the plasmid pPCP1 reduced invasiveness by more than 90%. Expression of Pla in recombinant *E. coli* enhanced bacterial invasiveness into the human endothelial-like cell line ECV304 (Lähteenmäki et al., 2001a). Invasiveness was seen also with proteolytically inactive mutant forms of Pla and was not enhanced by adding exogenous Plg, which indicate that the invasive function does not involve the proteolytic activity of Pla, or Plg activation. Pla-expressing bacteria also adhered to the matrix of ECV304 cells, but the possible role of laminin-binding or the basal/apical direction of the invasion were not studied. Interestingly, serum at a low concentration inhibits the invasion to HeLa and endothelial cells, which indicates that *Y. pestis* should not be invasive once it has reached the circulation. The biological role and the molecular mechanism of Pla-mediated invasion as well as the nature of the serum inhibitory factor(s) remain to be established.

Figure 1 summarizes the mechanisms by which Pla may enhance invasiveness of *Y. pestis*. *Y. pestis* is genetically very close to and has evolved from *Y. pseudotuberculosis* (Achtman et al., 1999). Although the central role of Pla in the invasiveness of plague is well established, it is interesting to note that the introduction of *pla* into *Y. pseudotuberculosis* does not confer increased virulence to the organism, even though it expresses functional Pla (Kutyrev et al., 1999). This indicates that Pla alone is not sufficient to explain the high invasiveness of *Y. pestis* and that the nature of the additional invasion factors remain to be identified.

STRUCTURE-FUNCTION RELATIONSHIPS IN Pla

The *pla* gene encodes a polypeptide of 312 amino acids containing a signal sequence of 20 amino acids (Sodeinde and Goguen, 1989). The predicted sequences of mature Pla polypeptides from the three *Y. pestis* biotypes are identical for all 292 residues, and the partially sequenced Pla associated with some medieval black death victims differed at one

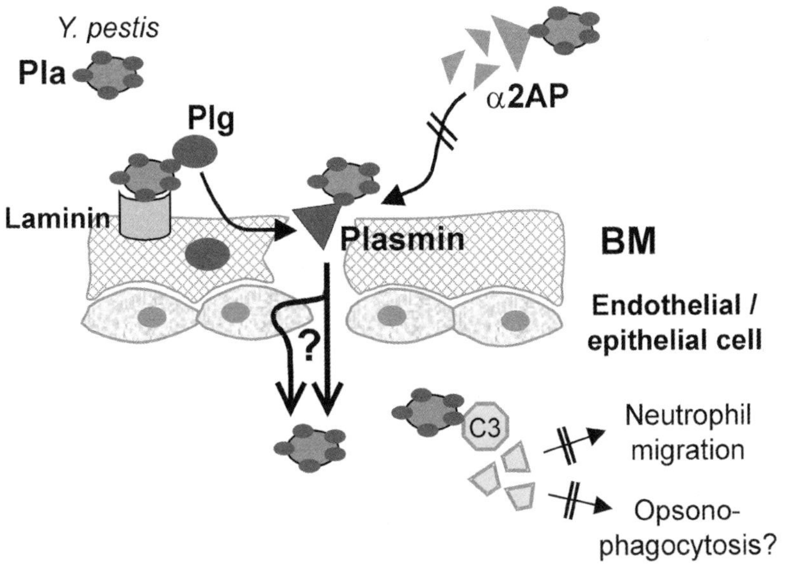

Figure 1. Mechanisms of how Pla may enhance invasiveness of *Y. pestis* through tissue barriers. *Y. pestis* is infected subcutaneously by a flea bite, and Pla binds to laminin, a major glycoprotein of the basement membrane (BM), which also contains components of the plasminogen (Plg) system as well as latent procollagenases. Activation of Plg within BM and/or in circulation directly damages the laminin network in BM and will indirectly lead to damage of the collagen network via plasmin-induced procollagenase activation. Degradation of α_2-antiplasmin (α2AP) by Pla will lead to uncontrolled proteolysis at the infection site, and eventually the barrier function of BM is lost. Pla enhances invasion of the bacteria into human cells, but it is not known whether *Y. pestis* utilizes this property to migrate through cell layers or whether the bacterium translocates via the pericellular route. Once the bacteria have entered the circulation, Pla-mediated degradation of the complement component C3 may decrease migration of neutrophils to the infection site and/or protect the bacteria from opsonophagocytosis via the C3b receptors. Question marks indicate functions that have not been demonstrated experimentally.

residue only (Raoult et al., 2000). Topology modeling suggested that OmpT and Pla are similar transmembrane, antiparallel β-barrel structures (Kramer et al., 2000; Kukkonen et al., 2001), and the β-barrel conformation was confirmed by the crystal structure of OmpT (Vandeputte et al., 2001). The mature forms of Pla and OmpT share 51% sequence identity, and the catalytic residues are conserved. Figure 2 shows the putative structure of the β-barrel and the active site groove in Pla modeled according to the OmpT crystal structure. Pla contains short turns at the periplasmic side of the barrel, and the position of the molecule within the outer membrane is determined by the girdle of aromatic residues that flank a hydrophobic region of about 25 Å (Figure 2A). Pla protrudes about 40 Å from the lipid bilayer with the outermost loops L1-L5 located just above the rim of the LPS core region.

The β-barrel fold is common in bacterial outer membrane proteins and forms a rigid membrane-embedded structure that tolerates large changes in the mobile loops on the extracellular side (Koebnik et al., 2000). As will be discussed below, sequence differences at critical sites in these surface loops can have profound effects on the specificities of the individual omptins. The omptin family offers an easily manipulated model system to study the molecular and functional features that convert a housekeeping protease (OmpT) into

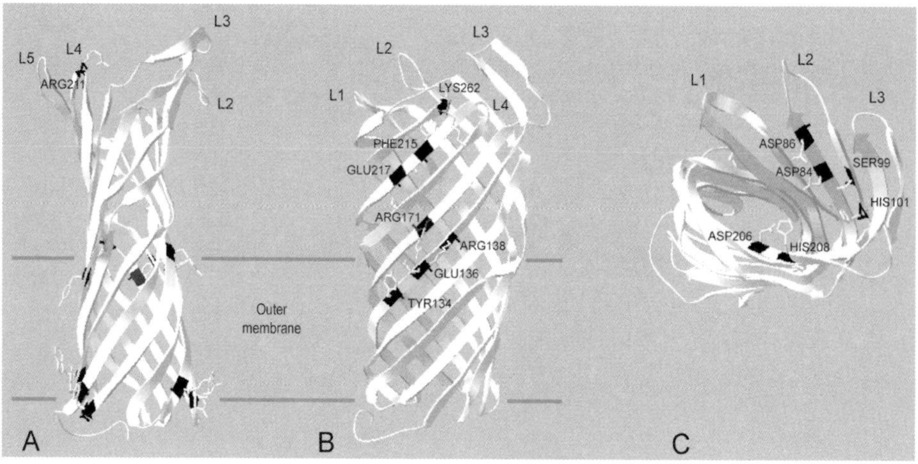

Figure 2. The β-barrel structure of Pla modeled with the help of Swiss-Model homology modelling server (http://www.expasy.org/swissmod/SWISS-MODEL.html) using the 3D model of OmpT (ExPDB 1I78) as a template. The figures were drawn with the Swiss-Pdb viewer 3.7 (SP4) programme. A and B show sideviews rotated 90° along the barrel axis, the outer membrane is positioned on the basis of the two rims of aromatic residues shown in A. L1 through L5 indicate the surface loops. The residue Arg_{211} is important in recognition of α2-antiplasmin, the autoprocessing site is Lys_{262}, and Phe_{215} and Glu_{217} are important for self-recognition by Pla. Arg_{171}, Arg_{138}, Glu_{136} and Tyr_{134} form the putative LPS-binding site. C shows a top view of the active-site groove: Asp_{86}, Asp_{84}, Asp_{206} and His_{208} are the catalytic residues, and Ser_{99} and His_{101} are important for substrate binding.

an efficient invasion factor (Pla). On the other hand, the omptin sequences share conserved structural features in the β-strands that influence function by mediating contacts between the barrel and the lipid A of LPS. We have begun a systematic mapping of functionally important residues and domains in the Pla molecule. Loop structures or single amino acid residues were exchanged between Pla and OmpT, and the resulting hybrid proteins were analyzed for surface expression and for functionality. The basis of these studies was the observation that the putative virulence functions of Pla, i.e. Plg activation, α2AP inactivation, and invasiveness into human cells, were lacking or were only poorly expressed by OmpT, as analyzed by comparing *E. coli* hosts expressing either *pla* or *ompT* cloned into the pSE380 vector (Kukkonen et al., 2001, 2004). Both OmpT-positive and –negative *E. coli* strains transformed with the recombinant plasmids gave essentially the same results, which probably results from poor expression of endogenous OmpT under the culture conditions we used for induction of the cloned omptins (Kukkonen et al., 2001).

AUTOPROCESSING OF Pla

In an electrophoretic analysis of outer membrane proteins from *Y. pestis* or recombinant *E. coli*, Pla migrates in four molecular forms: the unprocessed pre-Pla with an apparent molecular weight of 36 kDa, α-Pla (35 kDa), β-Pla (33 kDa), and γ-Pla (31 kDa; Sodeinde and Goguen, 1988; Sodeinde et al., 1988; Kutyrev et al., 1999; Kukkonen et al., 2001). In cell wall preparations from recombinant *E. coli*, the amounts of α-Pla, β-Pla, and γ-Pla are roughly in a ratio of 2:2:1 (Kukkonen et al., 2001). This ratio of the Pla molecular forms seems to be very similar to the ratio in cell walls from *Y. pestis* and from recombinant *Y. pseudotuberculosis* (Sodeinde and Goguen, 1988; Kutyrev et al., 1999; Kukkonen et al.,

2004). Truncation of α-Pla to β-Pla takes place at the C-terminus of the Pla molecule and is not seen with catalytic-site mutants of Pla, indicating that it is formed by autoprocessing. Substitution analyses of basic residues at the C-terminus showed that the cleavage is at a single site, Lys_{262} at the end of the β-strand leading to L5 (see Figure 2B), and the cut does not destroy the overall β-barrel structure. In addition to substitution of the catalytic residues, the autoprocessing is abolished by substitution of His_{101} and Ser_{99} which are located close to the active-site residues and have an influence on all proteolytic functions of Pla, most likely by participating in substrate binding (Figure 2C). Autoprocessing is also specifically affected by the residues Glu_{217} and Phe_{215}, (Figure 2B), which are oriented inwards in the β-barrel and probably are involved in self-recognition by Pla (Kukkonen et al., 2001). γ-Pla, on the other hand, is seen with proteolytically inactive Pla constructs as well and may represent full-size mature Pla that has folded differently from α-Pla.

Several of the proteases involved in eukaryotic proteolytic cascades (including the Plg system) are activated by a proteolytic cleavage, and it was therefore suspected that the cleavage of α-Pla that yields β-Pla is an activation process. However, we did not detect any effect on Plg activation or α2AP inactivation by preventing the autoprocessing by substitutions of the residues Lys_{262}, Glu_{217} or Phe_{215}, which indicates that the proteolytic activity of Pla is not enhanced by autoprocessing. The autoprocessing of Pla is incomplete and could also result from accidental cleavage of neighboring Pla molecules, or possibly depend on the formation of multimeric complexes in the membrane.

PLASMINOGEN ACTIVATION AND α2AP INACTIVATION

Our hypothesis is that sequence differences of surface-exposed loops are responsible for the differing proteolysis specificities and nonproteolytic functions observed with omptins from different enterobacterial species (Kukkonen et al., 2001, 2004; Lähteenmäki et al., 2001a). Indeed, substitutions of amino acids at or close to the active site groove have profound effects on substrate recognition by Pla and OmpT, and we were able to convert OmpT into a Pla-like protease by stepwise, cumulative substitutions at surface loops (Kukkonen et al., 2001). It appears that the size of the surface loops as well as the nature of the amino acids are important for the ability of Pla to recognize Plg and α2AP. The key elements in conversion of OmpT into a Pla-like protease (Figure 3) were the shortening of L4 by two residues, Asp_{214} and Pro_{215} (OmpT numbering), close to the catalytic Asp-His pair (residues 206 and 208 in Pla; see Figure 2C), substituting Lys_{217} to Arg in L4 (residue Arg_{211} in Pla; Figure 2A), and substitution of OmpT L3 to the Pla L3 sequence. The resulting hybrid protein OmpT/ΔDP/K217R/L3 exhibited a remarkably efficient Plg activation (Figure 3) and a low level of α2AP degradation not seen with OmpT. More efficient α2AP degradation was seen when other loop structures of the OmpT hybrid were substituted to match the corresponding loop structures of Pla. This study confirmed that sequence differences in the surface loops of OmpT and Pla are responsible for their differential substrate recognition. OmpT has longer L3 and L4 loops than Pla, and the active-site groove in Pla appears to be more open than in OmpT and thus able to interact more readily with large physiologically important substrates, such as Plg and α2AP. OmpT, on the other hand, cleaves small-molecular-weight substrates (Kramer et al., 2000; Kukkonen et al., 2001), which may reflect its function in degradation of cationic antimicrobial peptides (Stumpe et al., 1998) and denaturated proteins (White et al., 1995). The in vitro conversion of OmpT- to Pla-activity provides an example of how a bacterial

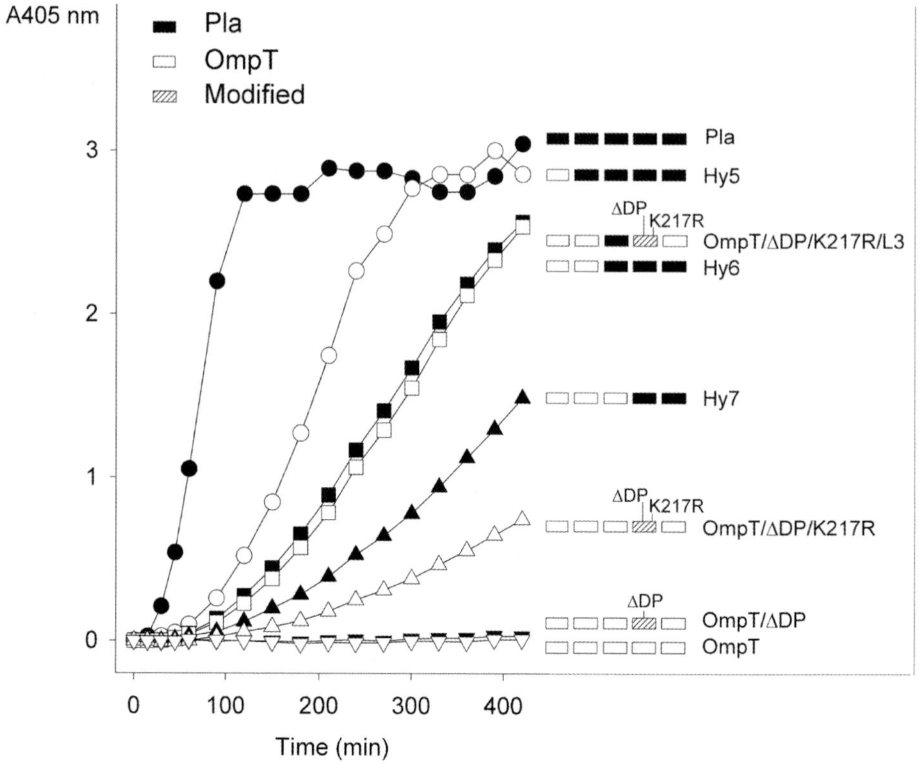

Figure 3. Conversion of the substate specificity of OmpT into a Pla-like protease. Enhancement of plasmin formation was assessed by incubating *E. coli* expressing different omptin constructs and the chromogenic plasmin substrate Val-Leu-Lys-p-nitroaniline. The recombinant *E. coli* expressed either OmpT, Pla-OmpT hybrid proteins (Hy), or modifed OmpT. The surface loop structures of the proteins are shown: Pla (closed boxes), OmpT (open boxes), and the genetically modified L4 of OmpT (hatched box). Reproduced from Kukkonen et al. (2001) with the permission of Blackwell Science Ltd.

protein can evolve into a powerful virulence factor by accumulating subtle mutations at critical sites without changing the overall architecture of the protein.

DEPENDENCE OF Pla FUNCTION ON ROUGH LIPOPOLYSACCHARIDE

The O-antigens are important virulence factors of a range of enterobacterial pathogens and mediate resistance to complement-mediated and phagocytic killing of bacteria (Rautemaa and Meri, 1999). The genes encoding the O-antigen synthesis are inactivated in *Y. pestis*, and the nature of the selective advantage of the loss of the O-antigen has remained unexplained (Parkhill et al., 2001). LPS affects folding and assembly of several outer membrane proteins. Kramer et al. (2002) were able to reconstitute purified OmpT into an enzymatically active form in the presence of rough LPS and concluded that OmpT requires LPS for activity. Kramer et al. (2002) did not test smooth LPS in the reconstitution. Several lines of evidence indicate that Pla requires rough LPS for activity but is sterically inhibited by the O antigen repeats in smooth LPS (Kukkonen et al., 2004). First, purified His_6-Pla was successfully refolded into an enzymatically active form with rough LPS of the Re chemotype, whereas reconstitution with smooth LPS resulted in poor proteolytic activity. *Y. pestis* has evolved from *Y. pseudotuberculosis* O:1b which shows a temperature-sensitive O-antigen synthesis (Bengoechea et al., 1998; Skurnik et al., 2000). Expression of the Pla-encoding plasmid pMRK1 in the *Y. pseudotuberculosis* O:1b strain PB1 gave efficient Plg activation only in cells cultivated at 37°C, the temperature at which O:1b-antigen is not made; similarly, Plg activation was seen in rough but not in smooth recombinant *E. coli* expressing *pla*. Furthermore, expression of pla in the rough Δ*wb* derivative of *Y. pseudotuberculosis* PB1 gave high Plg activation levels regardless of the cultivation temperature. The presence of an O-antigen prevented adhesive and invasive functions of Pla as well, and it was concluded that the O-antigen prevents Pla function by steric hindrance, i.e. by covering the surface loops(Kukkonen et al., 2004). This hypothesis is in agreement with the lack of O-antigen in *Y. pestis* and with the structure and membrane position of Pla (Figure 2). Thus, Pla has a dual interaction with LPS: it requires rough LPS to be enzymatically active but is sterically inhibited by long O-side chains which cover the active-site pocket. A pathogenic advantage of the loss of the O-antigen in *Y. pestis* may be the possibility of rough bacteria to fully make use of the various functions of Pla (Kukkonen et al., 2004).

The physical mechanisms of the Pla-LPS interaction are currently being studied. A structural model for a β-barrel-LPS interaction was provided by the crystal structure of FhuA of *E. coli* with bound LPS (Ferguson et al., 1998). Eukaryotic and prokaryotic LPS-binding proteins share a putative LPS-binding motif of four basic amino acids, which bind to phosphate groups in lipid A (Ferguson et al., 2000). This three-dimensional motif is present on one side of the FhuA β-barrel, which also contains residues that bind to acyl groups of lipid A within the outer membrane. Vandeputte-Rutten et al. (2001) identified the lipid A phosphate-binding residues in the correct spatial constellation in the OmpT crystal structure, and the motif is partially conserved in all omptin sequences. The OmpT and the predicted Pla structures have four amino acids oriented outwards in the correct positions to bind LPS: Arg_{138} and Arg_{171} that are predicted to bind to the 4'-phosphate of lipid A, and Tyr_{134} and Glu_{136} whose counterparts in FhuA bind to other regions of lipid A (Vandeputte-Rutten et al., 2001) (see Figure 2B). The hypothesis has been presented that the binding of the two arginines to the 4'-phosphate is needed to obtain a native active-site geometry in OmpT (Kramer et al., 2002). Also, the sequence and predicted structure of PgtE contains the putative LPS-binding amino acids indicated in Figure 2B, and recent evidence shows that substitution of Arg_{138} and Arg_{171} abolishes enzymatic activity but not surface localization or overall folding of PgtE (Kukkonen et al., 2004), which supports the hypothesis of Kramer et al. (2002). For Pla, however, the mechanisms of LPS-binding have not been studied in detail.

EVOLUTION OF Pla

The omptins offer an intriguing model for a lateral transfer of a common protein fold, the β-barrel, and its divergence through substitutions at critical surface-exposed sites. On the basis of nucleotide sequence comparison, Sodeinde and Goguen (1989) proposed that pla originated from pgtE of *Salmonella*. Only two omptin genes, *pgtE* and *ompT* are chromosomal (Table 1), and of these, *ompT* is located on a lambdoid prophage that is lacking in some pathogenic *E. coli* strains. Hence it seems likely that lateral transfer of omptin genes has taken place during evolution of enterobacterial species. The omptin family can be divided into two subgroups on the basis of the predicted omptin sequences.

Pla belongs to the same subgroup as PgtE of *Salmonella* and PlaA of the plant pathogen *E. pyrifoliae*. At the DNA level, *pla* has 71% identity with *plaA* and 69% identity with *pgtE*, and the mature proteins have 76% and 71% identity, respectively, to Pla. In contrast, the amino acid sequence of Pla is only 51% identical to OmpT. We have observed that Pla and PgtE are also functionally more close to each other than to OmpT (Kukkonen et al., 2004), which is in accordance with the hypothesis by Sodeinde and Goguen (1989). However, our on-going work indicates that PlaA is functionally unrelated to Pla or PgtE; thus, it seems that Pla and PlaA have diverged from PgtE, or a close ancestor of PgtE, to serve different bacterial functions in the mammalian and the plant hosts. The functions of PlaA remain unknown, but the existence of an omptin homolog in a plant-associated bacterium is intriguing and suggests that lateral transfer of the omptin β-barrel fold may have been widespread and important in evolution of surface-associated functions of enteric bacteria.

ACKNOWLEDGEMENTS

We have been supported by the Academy of Finland (project numbers 42103, 45162, 80666, 201967, and the Microbes and Man Programme).

REFERENCES

Achtman, M., Zurth, K., Morelli, G., Torrea, G., Guiyoule, A., and Carniel, E. 1999. *Yersinia pestis*, the cause of plague, is a recently emerged clone of *Yersinia pseudotuberculosis*. Proc. Natl. Acad. Sci. USA. 96: 14043-14048.

Beesley, E.D., Brubaker, R.R., Janssen, W.A., and Surgalla, M.J. 1967. Pesticins. III. Expression of coagulase and mechanisms of fibrinolysis. J. Bacteriol. 94: 19-26.

Bengoechea, J.A., Brandenburg, K., Seydel, U., Diaz, R., and Moriyon, I. 1998. *Yersinia pseudotuberculosis* and *Yersinia pestis* show increased outer membrane permeability to hydrophobic agents which correlates with lipopolysaccharide acyl-chain fluidity. Microbiology 144: 1517-1526.

Brubaker, R.R., Beesley, E.D., and Surgalla, M.J. 1965. *Pastereulla pestis*. Role of pesticin I and iron in experimental plague. Science 149: 422-424.

Butler, T., Fu, Y., Furman, L., Almeida, C., and Almeida, A. 1982. Experimental *Yersinia pestis* infection in rodents after intragastric inoculation and ingestion of bacteria. Infect. Immun. 36: 1160-1167.

Cowan, C., Jones, H.A., Kaya, Y.H., Perry, R.D., and Straley, S.C. 2000. Invasion of epithelial cells by *Yersinia pestis*: evidence for a *Y. pestis*-specific invasin. Infect. Immun. 68: 4523-4530.

Dekker, N., Cox, R.C., Kramer, R.A., and Egmond, M.R. 2001. Substrate specificity of the integral membrane protease OmpT determined by spatially addressed peptide libraries. Biochemistry 40: 1694-1701.

Egile, C., d'Hauteville, H., Parsot, C., and Sansonetti, P.J. 1997. SopA, the outer membrane protease responsible for polar localization of IcsA in *Shigella flexneri*. Mol. Microbiol. 23: 1063-1073.

Farina, A.R., Tiberio, A., Tacconelli, A., Cappabianca, L., Gulino, A., and Mackay, A.R. 1996. Identification of plasminogen in Matrigel and its activation by reconstitution of this basement membrane extract. Biotechniques 21: 904-909.

Ferber, D.M. and Brubaker, R.R. 1981. Plasmids in *Yersinia pestis*. Infect. Immun. 31: 839-841.

Ferguson, A.D., Hofmann, E., Coulton, J.W., Diedrichs, K., and Welte, W. 1998. Siderophore-mediated iron transport: crystal structure of FhuA with bound lipopolysaccharide. Science 282: 2215-2220.

Ferguson, A.D., Welte, W., Hofmann, E., Lindner, B., Holst, O., Coulton, J.W., and Diederichs, K. 2000. A conserved structural motif for lipopolysaccharide recognition by procaryotic and eucaryotic proteins. Structure Fold Des. 8: 585-592.

Goguen, J.D., Bugge, T., and Degen, J.L. 2000. Role of the pleiotropic effects of plasminogen deficiency in infection experiments with plasminogen-deficient mice. Methods 21: 179-183.

Grodberg, J., Lundrigan, M.D., Toledo, D.L., Mangel, W.F., and Dunn, J.J. 1988. Complete nucleotide sequence and deduced amino acid of the ompT gene of *Escherichia coli* K12. Nucleic Acids Res. 16: 1209.

Guina, T., Yi, E.C., Wang, H., Hackett, M., and Miller, S.I. 2000. A PhoP-regulated outer membrane protease of *Salmonella enterica* serovar Typhimurium promotes resistance to alpha-helical antimicrobial peptides. J. Bacteriol. 182: 4077-4086.

Hinnebusch, J.B., Fischer, E.R., and Schwan, T.G. 1998. Evaluation of the role of the *Yersinia pestis* plasminogen activator and other plasmid-encoded factors in temperature-dependent blockage of the flea. J. Inf. Dis. 178: 1406-1415.

Hu, P., Elliott, J., McCready, P., Skowronski, E., Garne, J., Kobayashi, A., Carrano, A.V., Brubaker, R.R., and Garcia, E. 1998. Structural organization of virulence-associated plasmids in *Yersinia pestis*. J. Bacteriol. 180: 5192-5202.

Kienle, Z., Emödy, L., Svanborg, C., and O'Toole, P.W. 1992. Adhesive properties conferred by the plasminogen activator of *Yersinia pestis*. J. Gen. Microbiol. 138: 1679-1687.

Koebnik, R., Locher, K.P., and Van Gelder, P. 2000. Structure and function of bacterial outer membrane proteins: barrels in a nutshell. Mol. Microbiol. 37: 239-253.

Kramer, R.A., Zandwijken, D., Egmond, M.R., and Dekker, N. 2000. In vitro folding, purification and characterization of *Escherichia coli* outer membrane protease OmpT. Eur. J. Biochem. 267: 885-893.

Kramer, R.A., Brandenburg, K., Vandeputte-Rutten, L., Werkhoven, M., Gros, P., Dekker, N., and Egmond, M.R. 2002. Lipopolysaccharide regions involved in the activation of *Escherichia coli* outer membrane protease OmpT. Eur. J. Biochem. 269: 1746-1752.

Kukkonen, M., Lähteenmäki, K., Suomalainen, M., Kalkkinen, N., Emödy, L., Lang, H., and Korhonen, T.K. 2001. Protein regions important for plasminogen activation and inactivation of α_2-antiplasmin in the surface protease Pla of *Yersinia pestis*. Mol. Microbiol. 40: 1097-1111.

Kukkonen, M., Suomalainen, M., Kyllönen, P., Lähteenmäki, K., Lang, H., Virkola, R., Helander, I.M., Holst, O., and Korhonen, T.K. 2004. Lack of O-antigen is essential for plasminogen activation by *Yersinia pestis* and *Salmonella* enterica. Mol.Microbiol. 51: 215-225.

Kutyrev, V., Mehigh, R.J., Motin, V.L., Pokrovskaya, M.S., Smirnov, G.B., and Brubaker, R.R. 1999. Expression of the plague plasminogen activator in *Yersinia pseudotuberculosis* and *Escherichia coli*. Infect. Immun. 67: 1359-1367.

Lähteenmäki, K., Virkola, R., Saren, A., Emödy, L., and Korhonen, T.K. 1998. Expression of plasminogen activator Pla of *Yersinia pestis* enhances bacterial attachment to the mammalian extracellular matrix. Infect. Immun. 66: 5755-5762.

Lähteenmäki, K., Kukkonen, M., and Korhonen, T.K. 2001a. The Pla surface protease/adhesin of *Yersinia pestis* mediates bacterial invasion into human endothelial cells. FEBS Lett. 504: 69-72.

Lähteenmäki, K., Kuusela, P., and Korhonen, T.K. 2001b. Bacterial plasminogen activators and receptors. FEMS Microbiol. Rev. 25: 531-532.

Lijnen, H.R. and Collen, D. 1995. Mechanisms of physiological fibrinolysis. Baillieres Clin. Haematol. 8: 277-290.

Liotta, L.A., Rao, C.N., and Wewer, U.M. 1986. Biochemical interactions of tumor cells with the basement membrane. Ann. Rev. Biochem. 55: 1037-1057.

Matsuo, E., Sampei, G., Mizabuchi, K., and Ito, K. 1999. The plasmid F OmpP protease, a homologue of OmpT, as a potential obstacle to *E. coli*-based protein production. FEBS Lett. 461: 6-8.

McDonough, K.A. and Falkow, S. 1989. A *Yersinia pestis* DNA fragment encodes temperature-dependent coagulase and fibrinolysin-associated phenotypes. Mol. Microbiol. 3: 767-775.

McGhee, C.C., Schnabel, E.L., Maxson-Stein, K., Jones, B., Stromberg, V.K., Lacy, G.H., and Jones, A.L. 2002. Relatedness of chromosomal and plasmid DNAs of *Erwinia pyrifoliae* and *Erwinia amylovora*. Appl. Environ. Microbiol. 68: 6182-6192.

Mignatti, P. and Rifkin, D.B. 1993. Biology and biochemistry of proteinases in tumor invasion. Physiol. Rev. 73: 161-195.

Nakata, N., Tobe, T., Fukuda, I., Suzuki, T, Komatsu, K., Yoshikawa, M., and Sasakawa, C. 1993. The absence of a surface protease, OmpT, determines the intercellular spreading ability of *Shigella*: the relationship between the *ompT* and *kcpA* loci. Mol. Microbiol. 9: 459-468.

Parkhill, J., Wren, B.W., Thomson, N.R., Titball, R.W., Holden, M.T.G., Prentice, M.B., Sebaihia, M., James, K.D., Churcher, C., Mungall, K.L., Baker, S., Basham, D., Bentley, S.D., Brooks, K., CerdeZo-Tárraga, A.M., Chillingworth, T., Cronin, A., Davies, R.M., Davis, P., Dougan, G., Feltwell, T., Hamlin, N., Holroyd, S., Jagels, K., Karlyshev, A.V., Leather, S., Moule, S., Oyston, P.C.F., Quail, M., Rutherford, K., Simmonds, M., Skelton, J., Stevens, K., Whitehead, S., and Barrell, B.G. 2001. Genome sequence of *Yersinia pestis*, the causative agent of plague. Nature 413: 523-527.

Perry, R.D., and Fetherston, J.D. 1997. *Yersinia pestis* - etiologic agent of plague. Clin. Microbiol. Rev. 10: 35-66.

Plow, E.F., Ploplis, V.A., Carmeliet, P., and Collen, D. 1999. Plasminogen and cell migration in vivo.Fibrinol. Proteol. 13: 49-53.

Raoult, D., Abouharam, G., Crubézy, E., Larrouy, G., Ludes, B., and Drancourt, M. 2000. Molecular identification by "suicide PCR" of *Yersinia pestis* as the agent of medieval black death. Proc. Natl. Acad. Sci. USA 97: 12800-12803.

Rautemaa, R., and Meri, S. 1999. Complement-resistance mechanisms of bacteria. Microbes Infect. 1: 785-794.

Rust, J.H., Harrison, D.N., Marshall, J.D., and Cavanaugh, D.C. 1972. Susceptibility of rodents to oral plague infection: a mechanism for the persistence of plague in inter-epidemic periods. J. Wildlife Dis. 8: 127-133.

Saksela, O. 1985. Plasminogen activation and regulation of pericellular proteolysis. Biochim. Biophys. Acta. 823: 35-65.

Shere, K.D., Sallustio, S., Manessis, A., D'Aversa, T.G., and Goldberg, M.B. 1997. Disruption of IcsP, the major *Shigella* protease that cleaves IcsA, acelerates actin-based motility. Mol. Microbiol. 25: 451-462.

Skrzypek, E., Cowan, C., and Straley, S.C. 1998. Targeting of *Yersinia pestis* YopM protein into HeLa cells and intracellular trafficking to the nucleus. Mol. Microbiol. 30: 1051-1065.

Skurnik, M., Peippo, A., and Ervelä, E. (2000) Characterization of the O-antigen clusters of Yersinia pseudotuberculosis and the cryptic gene cluster of *Yersinia pestis* shows that the plague bacillus is most closely related to and has evolved from *Y. pseudotuberculosis* O:1b. Mol. Microbiol. 37: 316-330.

Sodeinde, O.A. and Goguen, J.D. 1988. Genetic analysis of the 9.5-kilobase virulence plasmid of *Yersinia pestis*. Infect. Immun. 56: 2743-2748.

Sodeinde, O.A., and Goguen, J.D. 1989. Nucleotide sequence of the plasminogen activator gene of *Yersinia pestis*: relationship to ompT of *Escherichia coli* and gene E of *Salmonella* typhimurium. Infect. Immun. 57: 1517-1523.

Sodeinde, O.A., Sample, A.K., Brubaker, R.R., and Goguen, J.D. 1988. Plasminogen activator/coagulase gene of *Yersinia pestis* is responsible for degradation of plasmid-encoded outer membrane proteins. Infect. Immun. 56: 2749-2752.

Sodeinde, O.A., Subrahmanyam, Y.V.B.K., Stark, K., Quan, T., Bao, Y., and Goguen, J. D. 1992. A surface protease and the invasive character of plague. Science 258: 1004-1007.

Stephens, R.W. and Vaheri, A. 1993. Plasminogen. Plasminogen activator inhibitor 1 (PAI-1). Plasminogen activator inhibitor 2 (PAI-2). In: Guidebook to the Extracellular Matrix and Adhesion Proteins. T. Kreis and R. Vale, eds. Oxford University Press, New York. p. 81-85.

Stumpe, S., Schmid, R., Stephens, D.L., Georgiou, G., and Bakker, E.P. 1998. Identification of OmpT as the protease that hydrolyzes the antimicrobial peptide protamine before it enters growing cells of *Escherichia coli*. J. Bacteriol. 180: 4002-4006.

Travis, J. and Salvesen, G.S. 1982. Human plasma proteinase inhibitors. Ann. Rev. Biochem. 52: 655-709.

Vandeputte-Rutten, L., Kramer, R.A., Kroon, J., Dekker, N., Egmond, M.R. and Gros, P. 2001. Crystal structure of the outer membrane protease OmpT from *Escherichia coli* suggests a novel catalytic site. EMBO J. 20: 5033-5039.

White, C.B., Chen Q., Kenyon, G.L., and Babbitt, P.C. 1995. A novel activity of OmpT proteolysis under extreme denaturating conditions. J. Biol. Chem. 270: 12990-12994.

Chapter 18

Structure, Assembly and Applications of the Polymeric F1 Antigen of *Yersinia pestis*

Sheila MacIntyre, Stefan D. Knight and Laura J. Fooks

ABSTRACT

F1 polymer is a fine, fibrillar, surface structure with a capsule-like appearance. It is unique to *Yersinia pestis*, is produced in large amounts early in infection and is a powerful immunogen. It stimulates production of protective antibodies which complement V antigen induced protection. Thus, despite the fact that it is not essential for virulence in rodent models, it remains a major component of newer generation subunit and DNA vaccines and a target of diagnostic kits. The polymer is assembled by the periplasmic chaperone: outer membrane usher pathway. The polymer is a series of Ig-like modules formed via donor strand complementation between neighbouring Caf1 subunits. The high resolution crystal structure of two assembly intermediates has revealed the structure of F1 and provided the basis for a novel model of chaperone driven polymer assembly. As the nature of the assembly of F1 unfolds, information critical to recombinant vaccine development is being revealed.

INTRODUCTION

The third pandemic of plague started just before the 'Golden Era' of bacteriology when Robert Koch and Louis Pasteur formulated the basic principles and methods for identification of the causative organism of infectious diseases. The time was ripe for isolation of the plague bacterium and when the disease struck Hong Kong in May 1894, Alexandre Yersin and Shibasaburo Kitasato both actively sought to identify 'the bacterium of bubonic plague'. By late June/early July of the same year both investigators announced isolation and identification of the plague bacillus (Kitasato, 1894; Yersin, 1894). On this first correlation of *Yersinia pestis* (then named *Bacterium pestis*) with plague, production of capsule by the bacterium was already recognised. While there is some doubt as to the identity of Kitasato's final isolate (Bibel and Chen, 1976; Butler, 1983) both he and Yersin undoubtedly repeatedly observed the characteristic bilobed staining, encapsulated *Y. pestis* (Figure 1a) from buboes of plague victims.

The capsular-like structure, named F1 antigen, is formed by a high molecular weight (HMW) polymer of a single polypeptide. Since early in the 20th century F1 antigen has attracted interest primarily due to its ability to induce a remarkably strong protective humoral response (Baker et al., 1947; 1952). Early heat killed whole cell vaccines were developed shortly following initial identification of *Y. pestis* (Haffkine, 1897). Killed whole cell vaccine preparations are still in use today. However, whole cell plague vaccines,

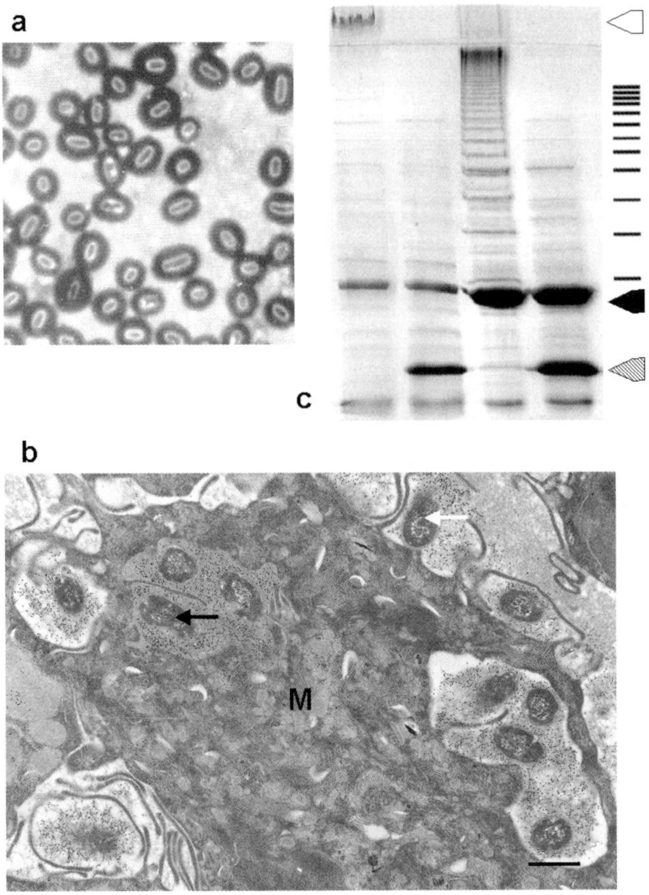

Figure 1. Visualisation of bacteria-associated or purified Fraction 1 Antigen. (a) *Y. pestis* smear stained with India Ink revealing capsular appearance of F1 polymer (Amies, 1951) (b) Electron micrograph of an immunogold labelled thin section of *Y. pestis* strain CO92 infected lung from an African Green Monkey that died from pneumonic plague. M, alveolar macrophage; dark arrow, phagocytosed bacterium; white arrow, extracellular bacterium. F1 polymer extending from all bacteria is labelled with monospecific rabbit anti-F1 serum and 10 nm-gold conjugate. (Reproduced with kind permission of (Davis et al., 1996) (c) SDS-polyacrylamide gel showing polymeric nature of F1 antigen. Lanes 1 and 2, rF1 (HMW polymer) isolated from the surface of *E. coli*/ pFMA1 (expresses complete *caf* locus) (MacIntyre et al., 2001); lanes 3 and 4, Caf1M chaperone:Caf1 subunit complexes isolated from the periplasm of *E. coli*, following expression in the absence of usher. Samples incubated at RT (lanes 1 and 3) or at 100°C for 5 min (lanes 2 and 4) prior to electrophoresis. Open arrowhead, very high molecular weight surface polymer of F1; bars, oligomeric periplasmic polymer increasing in size by one Caf1 subunit; cross-hatched arrow, denatured Caf1 subunit; filled arrowhead, Caf1M chaperone.

whether live attenuated or killed, have several disadvantages. These include short-lived protection, relatively severe side-effects and in addition, concern over reversion for attenuated ones and poor protection against the pneumonic form of plague for killed vaccines (Titball and Williamson, 2001). Indeed the disadvantages outweigh the benefits to such an extent that there is currently no plague vaccine available in the USA. Plague is often considered a disease of the past. There are, however, a number of enzootic foci of plague around the world particularly in Asia, Africa and the American continent, which have recently led to human outbreaks, and plague is now recognised as a re-emerging disease (Tikhomirow, 1999). Worldwide reports have increased particularly in Africa (an average of approximately 2,500 cases per year in the 1990s) (Tikhomirow, 1999; WHO, 2002). Also a number of areas that had been free from plague for several decades have recently seen a resurgence of the disease. Examples include the epidemic in Surat, India in 1994 (WHO, 1994) and in Madagascar where 200-1600 cases were reported per year during the 1990's (Chanteau et al., 2000b). In addition to the natural foci of plague, *Y. pestis* has been considered as a weapon of mass destruction and there is increasing concern that it could be used as a bioterrorist weapon, culminating in an outbreak of primary pneumonic plague (Inglesby et al., 2000).

Antibiotic treatment, assuming rapid diagnosis and prompt treatment, is relatively effective against plague with mortality reduced from greater than 50 % to around the 5-14 % range (Butler, 1983). In cases of pneumonic plague, due to the rapid progression of disease, antibiotics must be administered within 18-24 h after the first sign of symptoms. Antibiotic resistance in *Y. pestis* is rare, but of considerable concern were the reports of two clinical antibiotic resistant isolates in Madagascar. Both carried antibiotic resistance genes on conjugative plasmids and one isolate carried a multi-drug resistance plasmid encoding resistance to all antibiotics of choice for treatment of plague (Galimand et al., 1997; Guiyoule et al., 2001). Equally alarming is the demonstration of high frequency transfer of antibiotic resistance genes in the flea gut (Hinnebusch et al., 2002). In the absence of effective antibiotics, this first line of defence would be lost increasing the demand for effective vaccines. For these reasons interest in the polymeric F1 antigen and the potential of this antigen as a component of newer generation vaccines and diagnostic processes was rejuvenated in the last decade of the 20th century.

There have been numerous excellent reviews on plague, which generally include some detail on the significance of F1 antigen to virulence and immunity (Butler, 1983; Brubaker, 1991; 2000; Perry and Fetherston, 1997; Titball and Williamson, 2001). Until recently, however, structural detail of this polymer remained elusive. The high resolution crystal structure of F1 subunit has recently been solved (Zavialov et al., 2003b) and fundamental questions on the mechanism of polymerisation and surface localisation of F1 polymer are now beginning to be answered. F1 is assembled by a chaperone: usher pathway, analogous to that of the well-characterised P and type 1 pili chaperone: usher systems of *Escherichia coli* (Sauer et al., 2000a; Knight, 2000). While the new structural data is of fundamental relevance to chaperone:usher systems in general, it also provides insight into problems and potential answers relating to the application of F1 in newer generation vaccines and diagnostic tools. This account addresses past and current information relating to the composition, biological function and applications of the F1 capsular antigen, but in particular highlights newer aspects relating to the structure and assembly of the F1 polymer and the relevance of this information to newer generation vaccine and diagnostics design.

THE CAPSULE OF *YERSINIA PESTIS*

The *Y. pestis* 'capsular' material is comprised primarily of a high molecular weight (HMW) polymer of a single 15.56 kDa polypeptide, the Caf1 subunit. In the absence of expression of this gene product, no capsule is formed (Davis et al., 1996) and recombinant expression of the *caf* operon results in a capsular phenotype of the host bacterium (Andrews et al.,

1996; Titball et al., 1997). At 37 °C F1 envelops the cell forming a gelatinous mass that can be visualised by staining with India ink (Figure 1a) (Amies, 1951), hence the designation capsule. This polymer is soluble in water (Vorontsov et al., 1990) and large amounts are also shed from the cell surface into the surrounding medium (Englesberg and Levy, 1954; Vorontsov et al., 1990; Andrews et al., 1996) or host fluid during infection (Davis et al., 1996; Chanteau et al., 2003). No defined structure of this polymer has been visualised by electron microscopy. Analysis by scanning electron microscopy revealed irregular layers of granular material covering the surface of the bacteria and extending into the surrounding medium (Chen and Erlberg, 1977). In one transmission electron microscopy (TEM) study, using immunogold labelled antibody, negative staining and *in vitro* grown bacteria, no structure for F1 was detected in contrast to pH6 antigen, which was visualised as wiry fibrillae of roughly 4 nm diameter and bundles of strands (Lindler and Tall, 1993). In another study of sections of infected lungs from African Green monkeys, immunogold labelled F1 was visualised surrounding the bacterial cell and also extending in long strands from the bacterial surface in a form consistent with labelling of fibrillar polymers (Figure 1b; Davis et al., 1996).

Both cell surface material and shed polymer have been used as a source of material for analysis of chemical composition and antigenic nature of the capsular material. Baker and coworkers (Baker et al., 1947;1952) developed a procedure involving sequential ammonium sulphate precipitation of material washed from the surface of acetone dried cells using physiological saline. Antigenic activity correlated with fraction I, giving rise to the acronym Caf1, Capsular Fraction 1 antigen. This fraction could be subfractionated into a protein-polysaccharide complex (fraction IA) and pure protein (fraction IB). Both fractions appeared immunologically identical and hence antigenic activity was attributed to the protein. F1 antigen has frequently been referred to as a glycoprotein or glycolipoprotein complex with enrichment in particular of galactose and in some reports also of fucose (Baker et al., 1952; Bennett, 1974; Glosnicka, 1980; Simpson et al., 1990). F1 has been isolated, from *Y. pestis* EV76, with <0.12% associated carbohydrate (Vorontsov et al., 1990) and several studies have reported the lack of influence of associated carbohydrate on the antigenic properties of the protein (Baker et al., 1952; Bennett, 1974). On comparison, by SDS-polyacrylamide gel electrophoresis, of denatured *Y. pestis* F1 with recombinant F1 from *E. coli*, no difference in apparent molecular weight was observed (Simpson et al., 1990; Andrews et al., 1996). This was the case irrespective of whether the F1 was cell-derived or isolated from the spent culture media. Also, the absence of covalently linked carbohydrate in recombinant F1 is apparent from the results of Q-ToF (time-of -flight) mass spectrometry analysis of recombinant F1 polymer isolated from *E. coli* (Tito et al., 2001). Only monomer of the anticipated 15.56 kDa protein was observed. Thus, based on current evidence, the enrichment in galactose associated with some F1 preparations would appear to be from tightly associated glycolipid, rather than a covalently modified subunit. Gram negative bacteria release fragments of outer membrane material during growth (Li et al., 1998). Therefore, it is not surprising that some fractions of concentrated F1 material, either from the cell surface or the spent culture media, have lipid, carbohydrate and even other proteins associated with them. Interestingly, addition of terminal galactose to lipooligosaccharide of *Y. pestis* is variable and subject to PhoP control (Hitchen et al., 2002). Perhaps this activity, or one of the other putative galactosyltransferases identified in the *Y. pestis* genome sequence, is responsible for production of a galactose enriched glycolipid that might co-purify with F1 antigen.

As recombinant F1 can form a polymer on the surface of *E. coli, E. coli* has provided a safe source of material for analysis (Simpson et al., 1990). Both recombinant and native F1 isolated from either the cell surface or medium (with or without ammonium sulphate) is primarily a very HMW polymer greater than 0.5 -1 x 10^6 Da (Figure 1c, Andrews et al., 1996; MacIntyre et al., 2001). Caf1 has an immunoglobulin-like fold with 67% β-structure (Zavialov et al., 2003b). This is reflected in the Fourier transform infrared spectra of F1 polymer, but far-UV-circular dichroism analysis does not give a spectra characteristic of β-secondary structure (Miller et al., 1998; Abramov et al., 2001). Both subunit and polymer have an isoelectric point of 4.4 to 4.5 and are negatively charged at physiological pH (Zavialov et al., 2002). The polymer is resistant to proteases (trypsin, proteinase K). It is extremely stable and remains associated in 0.5 % (w/v) SDS at 75 °C. Heating above 75 °C in SDS is required to dissociate the polymer and fully denature the subunit. A consequence of this is that polymers can be readily visualised by SDS-PAGE (Figure 1c; MacIntyre et al., 2001; Zavialov et al., 2002). Polymer can also be dissociated by boiling for 10 min in a non-denaturing buffer such as Tris or sodium phosphate at neutral pH or at room temperature in 7M urea (Vorontsov et al., 1990; Miller et al., 1998; MacIntyre et al., 2001). Analysis by gel filtration chromatography has suggested formation of stable dimeric and tetrameric species with reformation of some form of HMW species at 4°C following removal of denaturant. The nature of both denatured and reassociated species requires further investigation.

THE *caf* LOCUS

The *caf* locus, which is required for surface expression of capsular fraction 1 polymer, was first cloned and sequenced from *Y. pestis* pEV76 (Galyov et al., 1990; 1991; Karlyshev et al., 1992a; 1992b). It is located on the largest *Y. pestis* plasmid, called pFra after capsular Fraction 1 or pMTI denoting 'mouse murine toxin' (Lindler et al., 1998; Hu et al., 1998). The plasmid is around 100 kb in size but shows small differences in strains of different origin, apparently a consequence of biovar specific deletions (Prentice et al., 2001). The organisation of genes (Figure 2B) and complete 4.206 kb nucleotide sequence of the *caf* locus is identical in the two sequenced *Y. pestis* strains CO92 biovar Orientalis and KIM5 biovar Medievalis (Lindler et al., 1998; Hu et al., 1998; Parkhill et al., 2001). The recombinant locus mediates production and surface assembly of high levels of Caf1 polymer in both *E. coli* (Karlyshev et al., 1992b; Andrews et al., 1996) and *Salmonella typhimurium* (Titball et al., 1997).

The locus encodes four genes: *caf1R*, encoding a 36.83 kDa regulator which shares homology with the AraC family of regulators (Karlyshev et al., 1992a); *caf1* encoding the sole structural subunit (17.6 kDa precursor, 15.56 kDa mature subunit) of F1 polymer (Galyov et al., 1990; Zavialov et al., 2002) and two assembly related genes, *caf1M* and *caf1A*. Caf1M (for Caf1 mediator), is synthesised as a 28.76 kDa precursor of a 26.3 kDa periplasmic chaperone (Galyov et al., 1991; Chapman et al., 1999) and Caf1A (for Caf1 assembly) is synthesised as a 93.0 kDa precursor of a 90.52 kDa outer membrane usher (Karlyshev et al., 1992b). Homology of both Caf1M and Caf1A to components of the *pap* locus involved in assembly of P pili of uropathogenic *E. coli* (Thanassi et al., 1998a) readily identified the *caf* locus as belonging to the family of chaperone:usher systems involved in assembly of adhesive pili and surface fibrillae. Recently exciting advances have been made in understanding the role of Caf1M chaperone in capping interactive sites of incompletely folded Caf1 subunit, stabilising folded subunit and facilitating

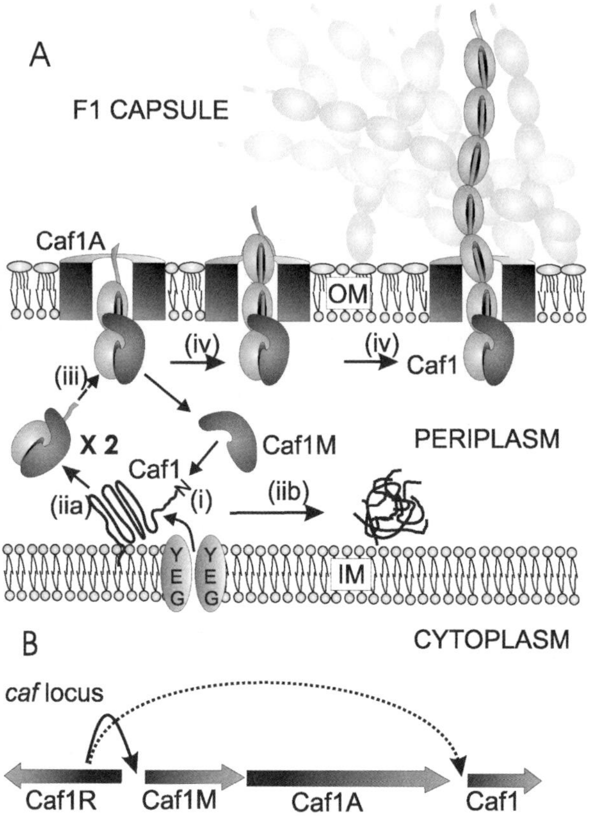

Figure 2. Diagrams of chaperone mediated assembly of F1 and of the *caf* locus. (A) Assembly of F1 polymer. Following translocation across the inner membrane presumably via the general Sec pathway (i), an incompletely folded Caf1 subunit interacts with Caf1M chaperone at the periplasmic surface of the inner membrane (iia). The mature Caf1 polypeptide has potential to form a marginally stable six stranded β-sandwich with an immunoglobulin (Ig) like topology. This is stabilised initially by interaction with the chaperone. In the absence of this interaction, subunit is unstable, forms aggregates and is degraded (iib). On interaction of two chaperone: subunit complexes, the N-terminal extension on the subunit in one complex replaces the chaperone of the second complex leading to a ternary chaperone:subunit:subunit complex (iii) (See Fig 3 below for details). Interaction of additional chaperone-subunit complex(es) with the outer membrane translocation usher, Caf1A, further catalyses insertion of subunit into the growing polymer and leads to surface localisation of long flexible fibres of Caf1 polymer (> 30 subunits) (iv). Surface fibres interact and collapse into a capsule-like structure which accumulates on the surface of the bacteria and is also shed into the surrounding milieu (MacIntyre et al., 2001; Zavialov et al., 2002; 2003b). (B) Genetic organisation of the *caf* locus. The *caf* locus encodes: Caf1R, regulator; Caf1M, periplasmic chaperone; Caf1A, outer membrane usher; Caf1, subunit. Arrow and dotted arrow indicate proposed positive control of transcription of *caf1M-caf1A* and *caf1* by Caf1R.

correct polymerisation (Chapman et al., 1999; MacIntyre et al., 2001; Zavialov et al., 2002; Zavialov et al., 2003b). An overview of the current understanding of F1 assembly is depicted in Figure 2A and is described in detail below under '*Assembly of F1 polymer*'.

The F1 capsule is produced at high levels at 35-37 °C, but not at lower temperatures of 26 °C and below (Burrows and Bacon, 1956; Chen and Erlberg, 1977; Simpson et al., 1990). A study using a *luxAB* transcriptional fusion in *Y. pestis* EV76 estimated a 20-40 fold higher level of transcription of *caf1* at 37 °C compared to 26 °C (Du et al., 1995). As a consequence, F1 is expressed in the mammalian host but not in the flea vector. Temperature dependent control of capsule expression has also been noted with recombinant *caf* locus in both *E. coli* (Simpson et al., 1990) and *S. typhimurium* (Titball et al., 1997). During infection, Caf1 must be one of the most highly expressed polypeptides in the bacterial cell. Capsule can extend to twice the diameter of the cell and large amounts of secreted polymer are present in the blood and fluids of infected hosts (Chanteau et al., 2003; Figure 1b) as well as in the culture media during *in vitro* growth (Andrews et al., 1996). Expression of the *caf* locus is positively controlled by the transcriptional activator, Caf1R (Karlyshev et al., 1994). Divergent promoter elements and five direct repeats (AT/TA GCT AA/TT) exist within the 331 bp region of DNA located between *caf1R* and *caf1M*. There has been no detailed analysis of either the regulator or the promoter elements, but requirement of this region for F1 expression has been verified by analysis of two spontaneous F1$^-$ strains which each possessed a deletion between the same direct repeats (Welkos et al., 1995; Friedlander et al., 1995). Existence of an additional promoter within the 80 bp upstream of *caf1* has been suggested (Galyov et al., 1990). Current evidence indicates that following injection of *Y. pestis*, via a flea bite, bacteria which are phagocytosed by macrophage can survive and multiply within the phagocytic cell (Cavanaugh and Randall, 1959; Du et al., 2002). These studies provide evidence that F1 expression is induced during this early intracellular stage of disease, as bacteria released by the macrophage are coated in capsule. The mechanism of temperature mediated regulation of the *caf* locus has attracted little attention to date. Undoubtedly, over the next few years a wealth of information on the relative expression of F1 and other virulence determinants will be gleaned from analysis of the complete transcriptome of *Y. pestis* following growth at different temperatures as well as under a variety of environmental conditions.

STRUCTURE OF F1 ANTIGEN

STRUCTURE OF THE CAF1 SUBUNIT: DONOR STRAND COMPLEMENTATION BETWEEN ADJACENT SUBUNITS COMPLETES AN IMMUNOGLOBULIN-LIKE FOLD

Elucidation of the structure of Caf1 subunit has been plagued by the fact that the subunit naturally exists in a HMW polymeric mass and progressively reassociates to oligomers and polymers of varying size following dissociation of the polymer. Caf1 subunit requires the periplasmic chaperone, Caf1M, to stabilise subunit monomer (Figures 2A, 3 and 4). Expression of Caf1M together with Caf1 subunit in the absence of the outer membrane usher leads to periplasmic accumulation of short polymers complexed to a single chaperone molecule (Figure 1c) (MacIntyre et al., 2001; Zavialov et al., 2002). By dissecting this polymerisation pathway in the periplasm of *E. coli*, two mutant forms of Caf1 subunit have been trapped, purified and crystallised: (i) a single subunit (lacking the G_d donor strand) bound to chaperone (see below and Figure 4 for definition of G_d strand) and (ii) a chaperone:subunit:subunit complex in which Ala9 of each subunit G_d strand is replaced by Arg (Zavialov et al., 2003a; 2003b). Thus, the high resolution crystal structure of the shortest polymeric unit of F1, originally cloned from *Y. pestis* EV76, has been solved in the form of a chaperone:subunit$_2$ ternary complex. A ribbon diagram of this complex is shown in Figure 3.

The dimeric F1 unit in the Caf1M:Caf1$_2$ ternary complex shows that polymeric F1 consists of Ig-like modules separated by a short linker. Remarkably, formation of each

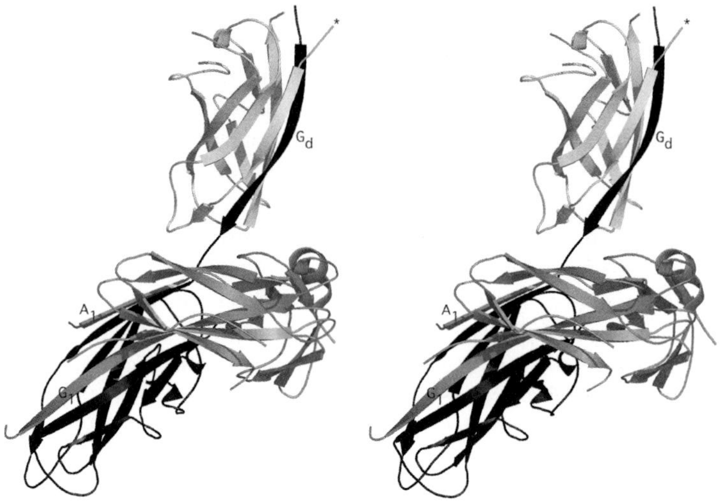

Figure 3. Ribbon stereo diagram of the crystal structure of Caf1M:Caf1:Caf1 complex (PDB 1P5U). Caf1M is shown in dark grey, the chaperone bound subunit (Caf1′) in black, and the second subunit (Caf1″) in light grey. The chaperone A_1 and G_1 donor strand, and the N-terminal G_d donor strand of Caf1′ are labelled, the disordered N-terminus of Caf1″ is indicated by an asterisk (*) next to the first visible residue of the Caf1″ subunit. The G_d donor strand of polymeric Caf1 (Caf1″) remains disordered and is not visible in the electron density map. In both subunits, Ala9 is replaced by Arg.

Ig-like module is dependent on two consecutive Caf1 subunits in the fibrous polymer. Each Caf1 subunit is only capable of forming an incomplete Ig-like fold. In each subunit six of the β-strands (strands A-F) are provided by a single subunit polypeptide chain (Figures 3 and 5). The seventh (G) strand, required to produce a complete Ig fold, is provided by the N-terminus of the adjacent subunit in the polymer. This strand has been named G_d to emphasize the fact that the G strand of each Ig-like module in the polymer is donated by a neighbouring subunit. The G_d strand hydrogen bonds in an anti-parallel orientation to the C-terminal F-strand, giving each module in the polymer the topology of a canonical seven-stranded Ig domain. It shields what would otherwise be a hydrophobic cleft along the entire length of the subunit, and in so doing mediates subunit-subunit binding and provides the key to polymer formation. The side chains of the alternating hydrophobic residues Leu13, Ala11, Ala9, Thr7 and Ala5 of the G_d strand point towards the cleft. Proof that the same principle of donor strand complementation is involved in F1 assembly at the cell surface and extends throughout the surface polymer was obtained from a series of double cysteine mutagenesis studies in which a donor strand Cys10 is covalently bound to an F strand Cys139 (Zavialov et al., 2003b). Importantly, this study not only demonstrated that each subunit in surface F1 polymer is linked to the adjacent subunit via donor strand complementation as outlined in Figure 4, but also confirmed the register of donor strand alignment seen in the mutant ternary complex crystal structure (Figure 3). This mode of binding between neighbouring subunits in the polymer, termed donor strand complementation, was initially predicted to occur during pili formation (Choudhury et al., 1999; Sauer et al., 1999), but has not yet been directly proven for these structures.

The Ig-like modules of polymeric Caf1 consist of two sheets packed against each other in a β-sandwich. Sheet 1 comprises strands ABED and sheet 2 comprises strands

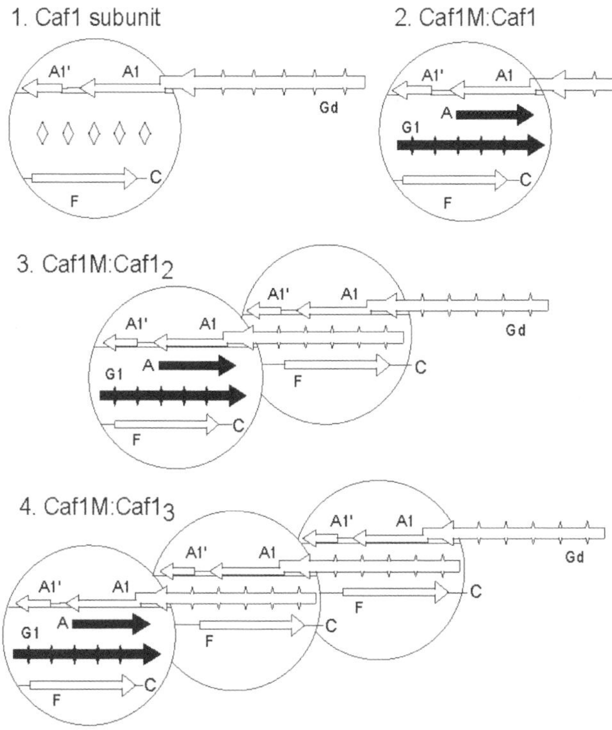

Figure 4. Schematic diagram of donor strand complementation. Caf1 subunits have an immunoglobulin-like β-sandwich fold but lack the 7th G strand, resulting in a groove along the surface of the molecule, which exposes the hydrophobic core (◊) between the two β-sheets of the subunit β-sandwich (unfilled arrows, edge β-strands) (1). Nascent Caf1 subunits are "capped" by the G_1 and A strands (black arrows) of the Caf1M chaperone to form a "super barrel" binary complex (2) by donor strand complementation (DSC). This interaction compensates for the missing G strand thereby stabilising the subunit and preventing subunit aggregation and degradation. The G_1 and A_1 strands of the chaperone lie parallel with the subunit F strand. Large hydrophobic side chains from the conserved motif of alternating hydrophobic residues in the G_1 donor strand (■) are deeply inserted between the two sheets of the subunit β-sandwich and form an integral part of the subunit hydrophobic core. This interaction forces the subunit β-sheets apart and effectively arrests subunit folding at an intermediate state. Polymerisation of Caf1 subunits occurs by a donor strand exchange (DSE) mechanism (3) and (4). During DSE the G_d strand of a second (Caf1′) subunit substitutes the chaperone mediated G_1/A_1 strand interaction in the hydrophobic cleft of Caf1″. This process removes the large chaperone G_1 donor residues inserted in the groove between the two β-sheets of the subunit β-sandwich, allowing subunit folding to proceed. The two β-sheets move closer and pack tightly against each other to form a close packed hydrophobic core (indicated by a narrow groove in polymeric subunits in (3) and (4)). Thus, although polymerised subunits interact with a smaller surface area of the binding cleft than in the chaperone:subunit interface, assembly can proceed because the smaller binding area is compensated by the repacking of the subunit core to form the final compact folded structure.

G_dFCC′. Sheet 2 is thus a composite sheet of β-strands from two neighbouring subunits in the F1 fibre. The C strand is divided in two parts by a loop near the middle. The C′/D strand runs diagonally across the base of the sandwich, opposite to the G_d-binding hydrophobic acceptor cleft. The first half of this strand (C′) is hydrogen bonded to sheet 2, the second half (D) is part of sheet 1. The D and E strands are connected *via* a long loop containing an irregular β hairpin that protrudes away from the body of the subunit sandwich. This loop

```
1   ADLTASTTAT ATLVEPARIT LTYKEGAPIT IMDNGNIDTE LLVGTLTLGG
          Gd                 A          A'              B

51  YKTGTTSTSV NFTDAAGDPM YLTFTSQDGN NHQFTTKVIG KDSRDFDISP
                C1           C2               C'/D       D'

101 KVNGENLVGD DVVLATGSQD FFVRSIGSKG GKLAAGKYTD AVTVTVSNQ
    D''                         E                      F
```

Figure 5. Primary amino acid sequence of mature Caf1 subunit (149 residues). Residues in β-strand secondary structure (67.3% of residues) are underlined with strand assignment given below each strand. The N-terminal G_d donor strand (which is flexible prior to polymerisation) is shaded grey; the sequence identified as crucial for polymerisation (Zavialov et al., 2002) is in bold.

packs against the C-strand loop, and together they close off this end of the sandwich. In contrast, the opposite end of the subunit sandwich, defined by the A and G_d edge strands, remains open and could potentially mediate secondary subunit-subunit interactions that might occur between individual fibres in surface expressed F1 capsule.

ULTRASTRUCTURE OF F1 ANTIGEN

F1 polymer is formed according to the same basic principle used in pili assembly with head-to-tail polymerisation mediated by donor strand complementation. The six shortest periplasmic Caf1 polymer: Caf1M chaperone complexes, isolated following expression in the absence of usher (Figure 1c), were analysed for theoretical versus experimental isoelectric point, stoichiometry of subunit and chaperone in each purified complex and also their elution profile on analytical gel filtration chromatography (Zavialov et al., 2002). All the evidence supported a model where the polymer grows by addition of a single Caf1 subunit at a time to form a linear rod or coil-shaped structure.

The two subunits in the crystal structure of the ternary Caf1M:$Caf1_2$ complex are separated by a short four residue linker (15-Glu Pro Ala 18-Arg) between the end of the G_d donor strand and the beginning of the A strand. Additional interaction between the two subunits is limited. Two of the linker residues, Glu15′ and Arg18′, stabilize the interaction between the subunits by hydrogen bonds and electrostatic interactions with residues Gly50′, Lys52′, Ile31″, Asp33″, and Ala135″ (Zavialov et al., 2003b). The angle of rotation between the two Caf1 subunits in the complex is 132° with a tilt of 30°.

An F1 fibre has been modelled, on the assumption that the subunit: subunit conformation observed in the crystal structure of the chaperone: $subunit_2$ complex is maintained throughout the polymer (Zavialov et al., 2003b). According to this model F1 polymer would form a thin extended linear fibre of approximately 20 Å diameter. The model fibre is wound into an open right-handed helical structure with a rise per subunit of ~47 Å (corresponding roughly to the length of a single Caf1 subunit) and ~2.7 subunits/turn, giving it a wavy appearance in projection. This F1 model fibre has several parameters that fit with those observed for pilus tip fibrillae. The P pilus tip fibrillum is a ~20 Å thin fibre of PapE subunits. High-resolution micrographs of the fibrillum show a wavy outline with a repeat of 15 nm (Holmgren et al., 1992), similar to the pitch of 2.7 x 47 Å in the F1 model fibre. Thus, both pilus tip fibrillae and fibrillar non-pilus surface structures such as F1 polymer may share the same basic architecture with subunits arranged more or less linearly end-to-end in a thin open helical conformation. F1 antigen appears not as fibres but as an amorphous mass, or capsule, in the electron microscope. High-resolution micrographs show "clumps" of capsule material covering and also surrounding *Y. pestis* expressing F1 antigen (Chen and Erlberg, 1977). The interface between Caf1′ and Caf1″ in the ternary complex (270 Å^2 buried area) is relatively small and might not be sufficient to support a rigid orientation between successive subunits. This suggests that F1 fibres might be quite flexible and that rather than extending laterally as individual fibres from the surface of the bacterium they collapse into a tangle of thin fibres of varying lengths that might be difficult if not impossible to resolve using standard electron microscopy techniques.

An earlier proposal suggested that F1 may form a bilayer type structure around the cell (Vorontsov et al., 1990). This model was based on the observation that upon dissociating polymer by heat or denaturants, a dimer appeared to be the most stable form. The nature of this heat induced dimer is not known and formation of a dimer based bilayer is difficult to model based on linear assembly via donor strand complementation. Heat or denaturant induced depolymerisation presumably involves at least partial unfolding of the Caf1 Ig-modules in the F1 fibre. Individual subunits released by such treatment would not be expected to be very stable due to the absence of a seventh β-strand to complete the fold. It is possible that following cooling or removal of denaturant, Caf1 subunits reassociate aberrantly to form dimers, possibly by interacting via the A and F strands to form some sort of superbarrel similar to that observed between Caf1M chaperone and Caf1 subunit prior to polymer assembly (see below). Another more recent suggestion is based on electrospray ionization time-of-flight mass spectrometry (ESI-TOF MS) analysis of purified surface F1 polymer (Tito et al., 2001). This study identified primarily tetradecameric and heptameric subunit complexes together with monomeric Caf1. Proposed interpretations of this have included linear helices with 7 subunits per turn (Tito et al., 2001) and formation of a closed seven membered ring by insertion of the donor strand of the leading subunit into the cleft of a free base subunit (Zavialov et al., 2002). The former model would lead to very thick (~15 nm) pilus-like structures that should be easy to detect by conventional electron microscopy. The latter model would require a controlled break in subunit insertion in order to create a free base cleft every seventh subunit. As yet multiples of seven have not been identified by other techniques such as gel filtration or acrylamide gel electrophoresis.

ASSEMBLY OF F1 POLYMER

The last four years has seen dramatic progress in the understanding of the mechanism of assembly of surface organelles via the chaperone:usher pathway (Choudhury et al., 1999; Sauer et al., 1999; 2002; Zavialov et al., 2003b) and in particular in the assembly of Caf1 polymeric 'capsule' (MacIntyre et al., 2001; Zavialov et al., 2002; 2003b). The series of directed mutagenesis studies (Chapman et al., 1999; MacIntyre et al., 2001; Zavialov et al., 2002; 2003b), which ultimately allowed determination of the Caf1 subunit structure ('*Structure of F1 antigen*' above), were instrumental in understanding the F1 assembly pathway, outlined in Figure 2A. As well as confirming the basic principles of chaperone interaction and subunit polymerisation, the studies furnished two key assembly intermediates, the high resolution crystal structures of which provide snapshots of the assembly process (Zavialov et al., 2003b). The key principle of F1 assembly lies in the fact that the encoded Caf1 subunit is incapable of folding into a stable structure on its own and requires one additional β-strand to complete the Ig-like fold. During *in vivo* assembly this additional β-strand is provided in a series of two donor strand complementation reactions (Figure 4). The first is between the periplasmic chaperone and subunit, and the second is the donor strand exchange reaction where an incoming subunit replaces the chaperone and complements the fold of a neighbouring subunit, thereby linking subunits in a linear polymer.

THE PERIPLASMIC CHAPERONE, CAF1M, AND SUBUNIT STABILISATION

Caf1M belongs to the family of immunoglobulin-like periplasmic chaperones, the prototype of which has been the PapD chaperone (Hung et al., 1996; Sauer et al., 2000b) involved in P pili assembly in uropathogenic *E. coli*. The crystal structure of Caf1M in complex with an N-terminal His-tagged subunit was solved to 1.8 Å resolution (PDB, 1P5V; Zavialov et al., 2003b). The same chaperone: subunit conformation exists in the ternary chaperone:subunit:subunit complex shown in Figure 3. As expected, Caf1M has the typical PapD-like bilobed structure consisting of two Ig-like domains. The A_1 and G_1 edge strands in domain 1 contribute almost all of the residues involved in subunit binding. The chaperone G_1 β-strand occupies the same cleft between the subunit A and F edge strands that harbours the G_d strand in polymeric F1 (see '*Structure of F1 antigen*' above), and thus blocks this essential assembly surface (Figure 4). In contrast to the subunit G_d strand, the chaperone G_1 strand is oriented parallel to the subunit F strand rather than antiparallel. The four hydrophobic residues Val126, Val128, Val130, and Phe132, which are significantly larger than the corresponding G_d residues, face into the cleft and penetrate the hydrophobic core of the subunit by inserting between the two sheets of the subunit β-sandwich. Arg20, which is absolutely conserved among all PapD-like chaperones (Hung et al., 1996), and Lys139 anchor the subunit C-terminus to the chaperone via hydrogen bonds. The key requirement of chaperone residues Arg20 and Val126 to subunit folding and stabilisation has been highlighted by mutagenesis (MacIntyre et al., 2001).

A second chaperone β-strand, the A_1 strand, is also directly involved in subunit interaction and stabilisation. It interacts, at the opposite edge of the cleft, via an antiparallel β-strand interaction with the subunit A strand. Hydrophobic residues of the chaperone A_1 strand lie perpendicular to the subunit cleft and contribute to a layer of hydrophobic residues which lies above the cleft and which is continuous with the core of domain 1 of the chaperone. The chaperone and the subunit thus bind via edge strands (A_1 and G_1 in Caf1M; A and F in Caf1) to form a closed "super-barrel" with a common hydrophobic core. Interaction with both chaperone strands is essential for formation of stable subunit:chaperone complex (Chapman et al., 1999; MacIntyre et al., 2001; Moslehi and MacIntyre, unpublished data). In the absence of functional chaperone, Caf1 subunit cannot fold correctly in the bacterial periplasm. This is recognised by a low recovery of subunit together with an 'irregular' banding pattern on PAGE (Chapman et al., 1999; MacIntyre et al., 2001). Exposure of hydrophobic surface area presumably leads to incorrect interaction of subunits, aggregate formation and degradation.

In Caf1M, the G_1 strand is linked to the F_1 strand via a disulphide bond between Cys98 and Cys137. The presence of a disulphide bond in both free and bound Caf1M was demonstrated biochemically (Zav'yalov et al., 1997; Chapman et al., 1999). Disulphide bond isomerase (DsbA) is essential for formation of the disulphide bond and folding of Caf1M *in vivo* (Zav'yalov et al., 1997). Microcalorimetry studies indicate that on reduction of the disulphide bond, the N-terminal domain of Caf1M is more readily denatured (Tishchenko, 2002). This is consistent with the low level of Caf1M when either Cys residue alone or together are mutated (MacIntyre, unpublished data). Interestingly, in the *E. coli* model system used this very low level of chaperone functioned as efficiently as wt Caf1M to assemble surface F1 polymer. Thus, while formation of the disulphide bond is not required for activity per se, it is critical to folding of the chaperone. Attenuated mutants with deletions in *dsbA* have been considered as vaccine candidates. In the Caf system, the Caf1A usher also contains Cys residues, with at least one disulphide bond that would appear to be required for activity. In an *E. coli dsbA* strain expressing the complete Caf assembly system, there is no production of surface F1, as monitored by both quantitative immunofluorescence and EM following immunogold labelling (Zav'yalov et al., 1997; MacIntyre, unpublished data). An attenuated *Y. pestis dsbA* mutant would therefore presumably lack F1 and quite likely be a poorer vaccine candidate than an other attenuated strain expressing F1.

SUBUNIT FOLDING IN THE ABSENCE OF CHAPERONE

Caf1A is not essential for polymerisation of F1. Polymerisation can occur directly between chaperone:subunit complexes although apparently with less efficiency (Figure 1c; Zavialov et al., 2002). Could F1 polymer also form in the absence of chaperone? It had been considered that a consequence of the donor strand complementation requirement for completion of the β-sheet structure of *E. coli* P and type 1 pilin subunits was that each individual subunit would be incapable of folding in the absence of a donor strand, which in the folding pathway is initially supplied by the chaperone (Choudhury et al., 1999; Sauer et al., 1999). This appeared to be borne out by the ability to refold the *E. coli* type 1 adhesin (FimH) *in vitro* in the presence of chaperone, but the inability to do so in the absence of chaperone (Barnhart et al., 2000). However, detailed control of refolding conditions has revealed that the six β-stranded pilin domain of FimH can indeed refold autonomously in the absence of chaperone, albeit to a form with very low stability (Vetsch et al., 2002). Hence, there is not an absolute requirement for a donor strand for autonomous refolding. Successful refolding was achieved only at very low protein and salt concentrations. Increase of either resulted in irreversible non-specific aggregation rather than refolding.

In the bacterial periplasm, in the absence of chaperone, Caf1 subunit is degraded and any accumulated undegraded subunit is in some form of misfolded aggregate (Chapman et al., 1999; MacIntyre et al., 2001). This phenomenon has been studied in some detail for the Pap system in *E. coli* (Jones et al., 1997; Hung et al., 2001). Incompletely folded subunits and aggregates remain associated with the inner membrane and are susceptible to DegP protease. Moreover, accumulation of this state of pilin subunit at the inner membrane is sensed by the Cpx two component sensor regulator system which in turn increases DegP production and proteolysis. Thus, the periplasmic chaperones undoubtedly have a role in preventing aggregation of structural subunits, as well as stabilising the subunit fold and targeting the subunits to the outer membrane. Another consequence of chaperone interaction is that the reactive subunit cleft is masked and thus, under normal conditions, prevents polymerisation in the periplasm.

It remains to be established whether, *in vivo*, this family of chaperones also act as a folding platform binding partially folded subunits and catalysing the folding reaction or whether they bind the folded six β-stranded sheet. The solution structure of an isolated self-folded subunit such as the FimH pilin domain obtained by Vetsch and co-workers (Vetsch et al., 2002) is not known. However, they demonstrated that the folded structure obtained in the absence of chaperone was only marginally stable, and that chaperone binding of pre-folded subunit was significantly less efficient than binding of unfolded subunit. The expanded subunit conformation observed in the crystal structures of chaperone-bound subunit is expected to be highly unstable in solution, and binding of a more collapsed conformation as observed for polymeric Caf1 would require reopening of the subunit β-sandwich to allow insertion of chaperone donor residues between the two sheets of the sandwich. It was therefore suggested that the periplasmic chaperones bind to a relatively unfolded state of the subunit, and that binding to chaperone and subunit folding are coupled events (Zavialov et al., 2003b). As current development of newer generation anti-plague vaccines includes cytosolic expression of Caf1 in both eukaryotic and prokaryotic cells (Oyston et al., 1995; Heath et al., 1998; Osorio et al., 2003; Grosfeld et al., 2003), understanding the details of Caf1 folding and polymerisation could have substantial benefits in this field.

THE SUBUNIT DONOR STRAND AND POLYMERISATION

The significance of the N-terminal region of Caf1 to subunit polymerisation was first noted following creation of a series of deletion mutants (Zavialov et al., 2002). The sequence 7-ThrThrAlaThrAlaThrLeu-13 was found to be essential for polymerisation while loss of it had no effect on subunit folding or stability. It was replacement of this polymerisation signal with a tag of six histidine residues that permitted ready purification of the chaperone:subunit complex free from contaminating polymers and suitable for crystallisation (Zavialov et al., 2003a; 2003b). During subunit:subunit polymerisation this flexible N-terminal sequence somehow exchanges with the chaperone G_1 strand of an adjacent complex to form the ordered antiparallel β-strand interaction with the subunit F strand, seen in Figure 3. The exact mechanism by which this occurs is not known, but in the final construct this G_d strand replaces the chaperone G_1 donor strand residue for residue. The G_d residues are much smaller than the corresponding chaperone G_1 residues and only Leu13 of the G_d strand is deeply inserted into the core of the subunit. The other donor residues (Ala11, Ala9, Thr7, and Ala5) shield much smaller hydrophobic pockets.

F1 polymer elongation requires 1) dissociation of Caf1 subunit from the chaperone and 2) association of subunits by binding of the G_d donor strand from a second subunit into the exposed acceptor cleft of the released subunit (Figure 4). This process can occur spontaneously *in vitro* without input of external energy and without the aid of the outer membrane usher (Zavialov et al., 2002). The same appears to hold for *in vivo* assembly. There is no direct source of energy in the bacterial periplasm. Some other outer membrane transport systems utilise either cytosolic ATP e.g. haemolysin export (Koronakis et al., 2001) or the inner membrane proton motif force e.g. TonB mediated uptake systems (Braun, 1995) via inner membrane anchored proteins. The proton motif force and probably also ATP were excluded as sources of energy for P pili assembly (Jacob-Dubuisson et al., 1994) and this is most likely also true for all chaperone/usher systems. Hence, both assembly and secretion of F1 *in vivo* appear to be spontaneous processes. Yet, the interaction between chaperone and subunit is much more extensive than that between two subunits, going from a two stranded interaction in the chaperone:subunit super-barrel to a single stranded subunit:subunit interaction in the polymer (Figures 3 and 4). So what is the driving force for release of the chaperone-bound subunit and polymerisation? Comparison of the crystal structure of polymeric subunit (Caf1″, Figure 3) with that of chaperone bound subunit (Caf1′, Figure 3) revealed a crucial difference in the extent of condensation of the hydrophobic core (Zavialov et al., 2003b). The observed difference provided an explanation for the mechanism by which periplasmic chaperones mediate subunit folding and polymerization in the absence of cellular energy donors. The chaperone appears to arrest subunit folding by blocking condensation of the subunit hydrophobic core, thereby trapping subunits in an expanded molten-globule-like high-energy conformation as seen in the crystal structures of Caf1 bound to Caf1M. On release from the chaperone, folding continues and there is a remarkable conformational change involving the whole hydrophobic core of the subunit leading to the final fully condensed core. A model, applicable to all chaperone-usher systems, was proposed whereby the energy trapped in the chaperone-bound intermediate drives polymerisation of Caf1 subunits.

To date, the actual translocation step has not been investigated in any detail. In analogy with PapC (Thanassi et al., 1998b) and outer membrane secretins (Thanassi et al., 2002) one can assume that Caf1A is also an oligomer. From the comparative short length of polymers which accumulate in the periplasm in the absence of Caf1A and those much longer polymers which assemble on the cell surface in the presence of Caf1A (Figure 1c), one can conclude that Caf1A catalyses and improves efficiency of F1 polymer assembly. *In vitro* studies with purified complexes found the chaperone:subunit complex to be the most highly reactive and the ternary chaperone:subunit:subunit complex to be more stable (Zavialov et al., 2002). For this reason and because the overall shape of the ternary complex mirrors that of the adhesin:chaperone complex which has been shown to interact with the usher in type 1 pili assembly, the ternary complex is a candidate for targeting of Caf1 to the outer membrane translocation site (Caf1A) (Zavialov et al., 2003b).

OTHER CHAPERONE: USHER SYSTEMS OF *YERSINIA SPP.*

Chaperone:usher systems are common throughout the *Enterobacteriaceae* and a single bacterium generally carries a number of different loci belonging to this superfamily. Analyses of the genome sequence of *Y. pestis* CO92 has revealed ten loci in this bacterium (Parkhill et al., 2001). Two of these loci were previously known and characterised: the *caf* locus described in this chapter, and the *psa* locus responsible for expression of pH 6 antigen (Lindler and Tall, 1993). The *psa* locus, like the *caf* locus, encodes a one-subunit FGL-type (Hung et al., 1996) chaperone/usher system. Expression of the *psa* locus is induced at temperatures above 35 °C and at a pH below 6.7 allowing biogenesis and surface expression of the thin fibrillar pH 6 antigen. Unlike F1 antigen, the pH 6 antigen is a known adhesin with specificity for β1-linked galactosyl glycosphingolipids (Payne et al., 1998). The structural basis for this specificity is not known. It has been suggested that pH 6 antigen-mediated binding to human plasma lipoproteins containing apolipoprotein B could be of importance for the establishment of *Y. pestis* infections by preventing recognition of the pathogen by the host defence systems (Makoveichuk et al., 2003).

In addition to these two known and characterised loci, there are eight previously unknown chaperone/usher loci in *Y. pestis* CO92, two of which are presumably nonfunctional due to the presence of insertion elements or frameshift mutations within the usher sequences (Table 1; Parkhill et al., 2001). The other six could potentially express surface adhesins and be of significance to pathogenesis and/or epidemiology of plague. Three of these loci encode chaperones of the FGS type (Hung et al., 1996) normally associated with the assembly of complex multi-subunit pili, together with an FGS pilin-like subunit (YPO1707, YPO1922, YPO3877) and a two-domain adhesin-like subunit (YPO1710, YPO1919, YPO3880) with the same basic organisation as typical FGS adhesins such as FimH or PapG. Such adhesins consist of an N-terminal receptor-binding domain fused to a C-terminal pilin domain and are located at the tip of pili where they mediate binding to carbohydrate receptors on target cell surfaces. Interestingly, each of

Table 1. Putative chaperone-usher loci identified in *Y. pestis* genome.				
Surface structure	Subunit	Chaperone/Usher	Comment	Reference[1]
F1 capsular antigen	Caf1	Caf1M/ Caf1A	Antiphagocytic, immunogen	(Karlyshev *et al.*, 1994)
pH6 fibrillar antigen	PsaA	PsaB/ PsaC	Low pH induced adhesin	(Lindler *et al.*, 1993)
?	YPO0301	YPO0303/ YPO0302		(Parkhill *et al.*, 2001)
?	YPO0700[2]	YPO0699/YPO0698	Homology to F17G adhesin	(Parkhill *et al.*, 2001)
?	YPO1698, 1699, 1700	YPO1697/YPO1696		(Parkhill *et al.*, 2001)
?	YPO1707, 1710[2]	YPO1708, 1711/YPO1709		(Parkhill *et al.*, 2001)
?	YPO1922, 1919[2]	YPO1921, 1918/YPO1920		(Parkhill *et al.*, 2001)
?	YPO2945, 2940	YPO2944,/ (YPO2943)-IS		(Parkhill *et al.*, 2001)
?	YPO3877, 3880[2]	YPO3878, 3881/ YPO3897		(Parkhill *et al.*, 2001)
?	YPO4044	YPO4041/YPO(4042)-IS		(Parkhill *et al.*, 2001)

[1] All sequences present in *Y. pestis* CO92 (Parkhill *et al.*, 2001) and *Y. pestis* KIM5 sequence (Lindler et al., 1998).
[2] Putative two-domain subunit characteristic of terminal adhesins.

the three loci, which are similarly organized, appears to encode two FGS chaperones, and might potentially represent cases where a separate chaperone is used for different subunits. Another locus encodes one FGS chaperone together with a potential adhesin (YPO0700) that is similar to the *N*-acetyl-*D*-glucosamine-binding F17G adhesin (26 % identity over the entire sequence) of enteropathogenic *E. coli* (Lintermans et al., 1991). Of the remaining two loci, one encodes a single chaperone (YPO0303) together with a single subunit (YPO0301). Both of these appear to belong to an uncharacterised family of vaguely FGS-like, chaperone/usher systems with hypothetical homologues in *Pseudomonas aeruginosa*, *Salmonella* species, and *E. coli*. The second locus potentially encodes three subunits (YPO1698-1700) and one chaperone (YPO1697). Together with a number of recently discovered loci in the genome sequences of soil and plant associated bacteria, this final system appears to define a new subfamily of the "alternative" or non-classical chaperone/usher systems involved in assembly of for example, CS1 pili in enterotoxigenic *E. coli* (Sakellaris and Scott, 1998). The structural detail, and the methods used to isolate F1 antigen assembly intermediates and to characterise F1 structure and assembly, will undoubtedly be applicable to future investigations of the plethora of newly identified chaperone:usher system being identified by genome sequencing.

BIOLOGICAL SIGNIFICANCE OF F1 ANTIGEN

While the power of F1 polymer as an antigen is indisputable, there remains a question mark over the significance of this plasmid encoded 'virulence factor' to *Y. pestis* induced disease and to its life-style.

ANTIPHAGOCYTIC ACTIVITY

It has been estimated that on feeding, a *Y. pestis* infected flea can inject as many as 11,000 to 24,000 bacterial cells intradermally into the mammalian host (Perry and Fetherston, 1997). Knowledge regarding the immediate fate of these bacteria, *in vivo*, and the consequential pathogenesis is limited. Current understanding is that these non-encapsulated *Y. pestis* cells are readily phagocytosed and that those taken up by neutrophils are mostly killed; whereas those internalised by monocytes and nonactivated macrophage can survive and multiply (Cavanaugh and Randall, 1959; Perry and Fetherston, 1997). The intraphagocytic stage is clearly important. Production of a number of virulence determinants, including F1 and the antiphagocytic Type III secreted Yop products, are induced during this early stage of infection (Cavanaugh and Randall, 1959; Du et al., 2002). On release from the phagocytic cells, these factors function in concert to drastically reduce phagocytosis and promote extracellular multiplication of the bacterium.

The capsular nature of F1 led to early correlation of F1 production with resistance to phagocytosis (Amies, 1951; Burrows and Bacon, 1956; Cavanaugh and Randall, 1959). Burrows and coworkers (1956) identified three forms of virulent *Y. pestis*. Following growth in vitro at 28°C, the bacteria were non-encapsulated and phagocytosis-sensitive (S-type), a property shared with the bacteria obtained from infected flea stomachs (Bacot and Martin, 1914). After 3 h growth at 37°C *in vitro* or *in vivo* in mice, *Y. pestis* (then called *Pasteurella pestis*) became phagocytosis-resistant although there was no readily visible capsule and V-antigen could be detected (R-type). We now know that this must correlate with induction of the antiphagocytic YOP type III secretion system that is rapidly induced, after about 30 min, inside macrophages (Burrows, 1957; Du et al., 2002). Only after 9-16 h infection of mice was the third form, with a large clearly visible F1 capsule and high level of resistance to phagocytosis (M-type), recovered. These findings were extended to show that (i) on ingestion of the phagosensitive S-type, *Y. pestis* multiplied within free phagocytes and (ii) upon release from the damaged cells, bacteria were of the M-type and thereby resistant to phagocytosis by both polymorphs and monocytes (Cavanaugh and Randall, 1959). The ability of recombinant F1 to confer upon *E. coli* resistance to phagocytosis, by mouse peritoneal macrophages, further supported the antiphagocytic role of F1 capsule (Friedlander et al., 1995).

The antiphagocytic activity of the pYV encoded Type III secretion system (TTSS) has been intensively studied primarily in *Y. pseudotuberculosis* and *Y. enterocolitica* (Cornelis, 2002). Due to the effect of some secreted Yops on signal processing within the phagocytic cell, this TTSS leads to general impairment of phagocytic function. The consequence of this is inhibition of phagocytosis of unrelated micro-organisms, as well as *Y. pseudotuberculosis* (Fallman et al., 1995). Analysis, using defined deletion mutants, to assess the relative contributions of the *Y. pestis* TTSS and F1 to inhibition of phagocytosis during simulated plague infection has only recently been performed (Du et al., 2002). In this study, phagocytosis of *Y. pestis* EV76, *Y. pestis* EV76Δ*caf1M* (an F1 negative mutant), and *Y. pestis* EV76C (cured of the 70 kb pCD plasmid, but still expressing F1 capsule) by the macrophage-like cell line, J774, were compared. The F1 negative mutant, *Y. pestis* EV76Δ*caf1M*, has an in frame deletion within the Caf1M chaperone and therefore is incapable of assembling F1 polymer on the cell surface (see Figure 2A for overview of chaperone function). *Y. pestis* EV76, pregrown at 37°C for 5 h to fully induce capsule formation, was highly resistant to uptake, with only about 5% being phagocytosed. In contrast, with the EV76Δ*caf1M* and EV76C strains ca. 30 % and 70 % of the bacteria were phagocytosed, respectively. A combined mutant EV76CΔ*caf1M* was virtually unable to resist uptake, only 5% of bacteria remained extracellular. Thus, at least in this *in vitro* system, F1 capsule and TTSS Yops had a cumulative negative effect on the phagocytic process.

Both the above study (Du et al., 2002) and an earlier study (Williams et al., 1972) highlighted the inability of F1 to generally impair phagocytic function. The earlier study used isolated F1, and although there was no effect on opsonisation of most micro-organisms tested, F1 was found to interfere specifically with opsonisation of *E. coli.* This was attributed to activation of the complement system and binding of C′4 and C′2 components of complement (Williams et al., 1972). There is circumstantial evidence that F1 capsule may reduce the efficiency of invasion of epithelial cells by *Y. pestis* (Cowan et al., 2000) as well as the number of extracellular *Y. pestis* cells associating with the surface of macrophage (Du et al., 2002). Hence, the main antiphagocytic effect of large amounts of F1 may simply be to physically decrease cell: cell interaction and the efficiency of receptor mediated interaction during the early stages of phagocytosis. Accordingly, although some antibody can still reach and interact with other surface components of *Y. pestis* in the presence of F1 the efficiency of this antibody-mediated opsonisation was found to be lower for cells expressing F1 than for those devoid of F1 (Du et al., 2002).

F1, AN ADHESIN?

The *caf* locus belongs to a large family of loci involved in assembly of surface pili, fibrillar and afimbrial structures. Virtually all characterised members of this family are known adhesins and were initially identified as such. To date F1 antigen, identified as an antiphagocytic capsule, is a rare exception. Does this structure have any role as an adhesin?

The one documented role of F1 as an adhesin is controversial. High affinity binding of ^{125}I-labelled Caf1 to mouse fibroblasts of the cell line (NIH 3T3), known to express large numbers of interleukin-1 (IL-1) receptors, human thymic epithelial cells (TEC2.HS) and the macrophage (U-937) cell line has been reported (Abramov, 2001; 2002). ^{125}I-labelled Caf1 binding was inhibited by either IL-1α or IL-1β and vice-versa, indicating specificity of binding to IL-1 receptors. Based on these results it was suggested that such Caf1 - IL-1 receptor interactions might play an important role in the early stages of plague development, by mediating adhesion to immunocompetent cells or by interfering with cytokine activity. In a separate study, IL-1 receptor antagonist (IL-1ra) activity of polymeric F1 was assessed by measuring the ability of F1 to suppress the IL-1β mediated induction of the adhesion molecules ICAM and ELAM (Krakauer and Heath, 1998). The ability to activate peripheral blood monocytes to produce IL-4 and IL-10 was also tested. None of these IL-1ra activities were detected. In the former study (Abramov, 2001), binding was not observed with HMW F1 polymer, but only with low molecular weight oligomers of Caf1 (dimers or tetramers) obtained either by dilution of the polymer or by heating F1 polymer at 100°C for 10 min. While this might explain why such F1 binding has not been observed in other assays (Du et al., 2002; Krakauer and Heath, 1998), it presents another concern relating to the nature of the Caf1 dimers and tetramers. Bearing in mind the known structure of F1 (Figure 3), it is hard to envisage how specific fractionation of polymer into dimeric or tetrameric species would occur. Furthermore, a Caf1 dimer produced by specific breaking of F1 fibres into units of two would be expected to consist of two subunits held together by donor strand complementation but with an exposed hydrophobic cleft in one subunit. This would be unlikely to be stable and no exposed hydrophobic surface in heat induced dimers was detected with ANS (8-anilino-1-naphthalene sulphonate binding) (Abramov et al., 2002). An alternate explanation might be an aberrant interaction between two subunits induced by heat treatment and conceivably also by some *in vivo* condition. To date there is no evidence of dimers on the surface of the cell.

Human macrophages express the chemokine receptor CCR5, which is also one of the receptors required for infection by HIV (Horuk, 1999). A mutation, Δ32, results in a frame shift that generates a non-functional receptor (Samson et al., 1996) which is not expressed at the cell surface (Liu et al., 1996). Individuals homozygous for the CCR5-Δ32 mutation are highly protected from infection by HIV (Wilkinson et al, 1998). It is found solely in European populations (Samson et al., 1996) with the incidence being characteristic of mutations having arisen due to a selective pressure rather than natural variation in the population (Stephens et al., 1998). Calculations suggest that the selection of CCR5-Δ32 receptor arose at a time when outbreaks of bubonic plague were frequent and widespread throughout Europe (Stephens et al., 1998). Therefore, it was hypothesised that a surface molecule of *Y. pestis* may interact with the CCR5 receptor, that the Δ32 mutation interfered with this interaction, and that this led to either prevention of establishment of infection or to milder symptoms and decreased mortality in individuals carrying this mutation. F1 antigen fits the criteria of being a major surface protein and a potential adhesin. However, preliminary studies which monitored interaction between F1 antigen and the CCR5 receptor, using the HeLa cell-line HJC24 (overexpresses the CCR5 receptor), purified F1 polymer and *E. coli* cells expressing recombinant surface F1, provided no indication of binding of F1 by CCR5 receptor (Kersley, McKeating and MacIntyre, unpublished data).

The pFra plasmid is unique to *Y. pestis*. It is a mosaic of insertion elements and much of the pFra plasmid is closely related to the *Salmonella enterica* serovar Typhi pHMC2 plasmid (Prentice et al., 2001). The low GC content of the *caf* locus and flanking bacteriophage homologues indicate that this locus has been acquired independently (Lindler et al., 1998). The closest known homologues of both Caf1M chaperone, AfaB (42% identical), and Caf1A usher, AfaC (43% identical), are involved in assembly of the afimbrial adhesin, AfaE – VII and VIII, of diarrhoegenic *E. coli* strains (Lalioui et al., 1999). The Caf1R regulator is closely related (42% identical) to the AfrR regulator of the AF/R1 pilus locus of rabbit enteroadherent strains of *E. coli* (gi: 3372510). Thus, the *caf* locus would appear to have arisen from a common ancestor of one or both of these *E. coli* loci; a scenario consistent with the hypothesis that the pFra plasmid was acquired by a *Y. pseudotuberculosis* progenitor of *Y. pestis* by horizontal transfer of genes in the mammalian gut (Carniel, 2003). Subunit encoding genes of chaperone:usher systems are generally very poorly conserved. As yet, there is no close homologue of Caf1 subunit itself in the DNA database to give any further clues as to the origin of this surface structure.

The *caf* locus encodes one of the simplest chaperone:usher systems. It belongs to the FGL subfamily (Hung et al., 1996) most of which assemble structures composed of a single subunit. Unlike the complex type 1 and P pili, these structures do not incorporate a dedicated adhesin. With other members of this subfamily, such as the *Yersinia* pH 6.0 antigen (Lindler and Tall, 1993), and *E. coli* Dr family of adhesins (Carnoy and Moseley, 1997), a single structural subunit also doubles as the receptor binding molecule. Comparison of the structures of Caf1 and pili subunits shows that the six strands ABCC′/DEF in the subunit β-sandwich define a common structural framework that is likely to be shared by all subunits assembled into filamentous structures via a chaperone/usher pathway. In contrast, the D-E and C-strand loops vary considerably between the different structures. They may be considered as different "decorations" of the basic structural framework that might be used to confer functionality, such as receptor binding onto the basic β-sandwich. For example, modelling of the Dr haemagglutinin adhesin, DraE, based on the Caf1 crystal structure suggests that residues involved in receptor binding in this system are mainly located in the C-strand loop region (Berglund and Knight, 2003). Caf1 may have an, as yet unrecognised, binding activity or, on the other hand, it may have no adhesin function. It may represent a progenitor of simple adhesins prior to evolution of a C-strand loop binding site or Caf1 may be a more highly evolved afimbrial adhesin, which has arisen to optimise antiphagocytic activity by forming a densely packed polymer. In this latter scenario, adhesin activity might be lost or retained as a vestigial function.

F1 ANTIGEN AS AN IMMUNOGEN

That F1 polymer is a major immunogen of *Y. pestis* is indisputable. As reflected in the name, initial interest in F1 polymer (antigen) was its ability to induce a strong protective antibody response (Baker et al., 1952). The strong humoral response to this antigen has been demonstrated in rodents, non-human primates and humans (Meyer, 1974b; 1974c; Chen, 1974; Andrews, 1996; Williamson, 1997) and has been reiterated in the numerous studies dealing with immunisation with isolated F1 or killed whole cells, and also in the routine diagnosis of plague (see below '*The role of F1 in anti-plague vaccines*' and '*Serology:quantitation of anti-F1 antibodies*'). The relative strength of F1 as an antigen during infection has recently been highlighted in a study examining survivors in a mouse pneumonic plague model (Benner et al., 1999). The serum IgG response to F1 antigen was at least ten times higher compared to that of twelve other surface virulence factors of *Y. pestis*, including V antigen, other TTSS components, LPS, Pla and pH6 antigen. In addition, in this study, all mice responded to F1, compared to 8-90 % non-responding mice to other antigens. A similar trend reflecting a dramatically higher response to F1 compared to other surface antigens has been observed in human victims recovering from bubonic plague (Friedlander, personal communication). In a study of patients surviving bubonic plague in Madagascar (Rasoamanana et al., 1997), seroconversion (accumulation of anti-F1 IgG) occurred on average between 6-7 days post onset of disease and plateaued around day 15. IgM is prominent in the first 5-9 days post infection (Williams et al., 1986). High responders (82% of individuals) initiated an early IgG response and possessed a high, prolonged final antibody titre; in contrast low responders (18%) produced much lower levels of IgG with a short half-life. It is not uncommon to have a small percent of a population apparently not responding serologically to F1 antigen (Marshall et al., 1974; Williams et al., 1986; Rasoamanana et al., 1997).

The response to F1 polymer administered to mice intramuscularly is primarily of the IgG_1 subclass of immunoglobulin, reflecting a Th_2-type response (Williamson et al., 1999). The response can be modulated by different routes of delivery of F1. Expression of F1 in an attenuated *Salmonella* oral delivery system was shown to stimulate mucosal IgA production as well as production of the IgG_{2a} subclass of serum antibody, indicating primarily a Th_1 response (Leary et al., 1997; Bullifent et al., 2000). A similar Th_1 response was reported following expression of *caf1* from an injected DNA vaccine (Brandler et al., 1998; Grosfeld et al., 2003). There is a dearth of information regarding the early, cell-mediated and innate immune responses during the course of infection in animals, largely as a consequence of the restrictions imposed and facilities required when working with *Y. pestis* infected animals. Studies with attenuated strains and the closely related *Y. pseudotuberculosis*, as well as the increased focus and funding on studies with dangerous pathogens, should alleviate this. In addition, *in vitro* DNA microarray analysis with *Y. pestis* infected macrophages are already producing interesting results on signalling events during the early intracellular stage of *Y. pestis* infection (Ng et al., 2003).

Various correlations have been made between serum anti-F1 IgG and protection against plague. High levels of serum anti-F1 antibody, of both human (Marshall et al., 1974; Meyer et al., 1974b) and animal origin (Chen, 1974; Meyer et al., 1974d; Williamson et al., 1999) generally correlate positively with passive protection in the standard mouse challenge model. Efficient passive protection against systemic infection by *Y. pestis* was also conferred on SCID/Beige mice, which lack mature B- and T- lymphocytes and have defective natural killer (NK) cells (Green et al., 1999). The mAb F1-04-A-G1 confers passive protection in murine bubonic and pneumonic plague model systems (Anderson et al., 1997). Also, individuals surviving bubonic plague infection generally possess high anti-F1 titres (Butler and Hudson, 1977; Rasoamanana et al., 1997). Hence F1 antigen became and remains a primary focus in vaccine design. Furthermore, many surface molecules of pathogenic bacteria are under high selective pressure to vary antigenically, making them less useful as vaccines. *Y. pestis* clearly does not have a mechanism to mediate rapid antigenic variation of F1. A screen of over 500 isolates in Madagascar with anti-F1 mAb revealed that all were F1 producers (Rasoamanana et al., 1997; Chanteau et al., 2003) and the encoded amino acid sequence is identical in the *caf1* gene from *Y. pestis* strain CO92 (Parkhill et al., 2001), strain KIM5 (Lindler et al., 1998), strain EV76 (Galyov et al., 1990), six Orientalis biovars isolated from rodents (Moslehi and MacIntyre, unpublished data), an Iranian isolate (accession no. AF542378) and the *caf1* fragment (93 % of the gene) sequenced from each of eight Indian isolates (EMBL accession nos. AF528530-37).

High levels of production, surface location and release of F1 into the body fluids may contribute to the strong anti-F1 response during *Y. pestis* infection. As yet the relative contribution of the polymeric nature of F1 and specific epitopes to this strong humoral response remains virtually unexplored. The one study which has used synthetic peptides in an attempt to identify immunodominant epitopes of F1 produced conflicting results (Sabhnani and Rao, 2000). Availability of the three dimensional structure and knowledge of the polymerisation pathway of F1 will now pave the way for a detailed, directed analysis of linear and conformational epitopes of this key antigen. Since F1 is induced during an intracellular stage early in *Y. pestis* infection, details of the T cell response to F1 as well as the humoral response should be of interest.

ROLE OF F1 IN VIRULENCE AND LIFESTYLE OF *YERSINIA PESTIS*

Despite the fact that F1 capsule contributes to the antiphagocytic arsenal of *Y. pestis*, it is not an essential virulence determinant in laboratory animals. This was first highlighted in 1957 by Burrows (Burrows, 1957) when he isolated an $F1^-$ mutant (strain M23) from vaccinated mice. The isolate was fully virulent for mice and exhibited only slightly decreased virulence in guinea pigs. A number of conflicting reports followed this. Some recorded decreased virulence and others host species dependent sensitivity to $F1^-$ mutants (Meyer, 1974c; Williams and Cavanaugh, 1984; Kutyrev et al., 1989). The one human isolate of $F1^-$ *Y. pestis* was reportedly less virulent for guinea pigs (Winter,

1960). Interpretation of these results, however, is difficult due to the undefined nature and stability of the mutations. In addition, to the potential of unlinked mutations, it has been demonstrated that the pFra plasmid can insert reversibly into the chromosome leading to an $F1^-$ phenotype in the integrated state (Protsenko et al., 1991). A number of different defined $F1^-$ mutants have since been created and have been used to confirm the minimal role of F1 in virulence in laboratory mice. These mutants include: *Y. pestis* strains, 231 and 358 pFra/pFS23, carrying an allelic replacement of a *caf1A-caf1* deletion with a *kan* cassette (Drozdov et al., 1995); *Y. pestis* CO92-C12, encoding two stop codons in place of Lys2 and Lys3 of the Caf1 signal sequence (Worsham et al., 1995; Friedlander et al., 1995); and *Y. pestis* EV76 *caf1::luxAB* (Du et al., 1995). Following subcutaneous challenge of mice, *Y. pestis* CO92-C12 had an increase in mean time to death (10.6 days vs 5.1 days) compared to the isogenic parent. This resulted in a small apparent increase in LD_{50} shortly after infection (at 14 days), but the same LD_{50} for parent and mutant strain after 56 days. Strain 358 pFra/pFS23 exhibited a similar increase in time to death in both mice and guinea pigs, while no effect was observed with the same mutation in strain 231, or with *Y. pestis* EV76 *caf1::luxAB* in mice. Another study (Davis et al., 1996), compared the virulence of inhaled *Y. pestis* CO92-C12 and the parent strain *Y. pestis* CO92 in the African Green monkey. All monkeys died within 4 - 10 days of exposure and there was no significant difference between the mean survival times of monkeys infected with F1 positive and negative strains. Interestingly, all monkeys infected with the $F1^-$ strain had large aggregates of intra-alveolar bacteria, a phenomenon that was absent from the lungs of monkeys infected with the wild-type parent strain. *Y. pestis* expresses no O antigen and has only lipooligosaccharide rather than lipopolysaccharide (Skurnik et al., 2000). Hence, in the absence of F1, the surface of *Y. pestis* is likely to have reduced hydrophilicity. It was suggested that the aggregation of $F1^-$ strains may be a consequence of loss of the negatively charged capsule.

F1 must have been acquired by a recent ancestor of *Y. pestis*, since it is absent from the closely related *Y. pseudotuberculosis* strains (Achtman et al., 2000). It is plasmid encoded and carried in a bacterium with a propensity to convert superfluous genes to pseudogenes (Parkhill et al., 2001). Moreover, it must be one of the most highly expressed genes in *Y. pestis*. There has been only one recorded human plague case caused by an F1-deficient strain (Winter, 1960). An analyses of over 500 isolates in Madagascar revealed that all expressed F1 capsule (Rasoamanana et al., 1997; Chanteau et al., 2003). A similar Kazakhstan study (Meka-Mechenko, 2003) on isolates from mammals and fleas identified five non-F1 producers, four of these however had lost the pFra plasmid; a potential problem during laboratory storage of cultures (Protsenko et al., 1991; Chanteau et al., 2003). Reported mechanisms by which $F1^-$ mutants spontaneously occur include deletion of the regulatory region of DNA between two direct repeats (Welkos et al., 1995; Friedlander et al., 1995) and insertion of the pFra plasmid into the chromosome (Protsenko et al., 1991; Friedlander et al., 1995). While a thorough screen of rodent and flea isolates for ability to express F1 is required, thus far it would appear that natural occurrence of such mutants, even from vaccinated animals infected with $F1^+$ strains of *Y. pestis* (Friedlander et al., 1995), is a rare event. So what advantage does the F1 capsule confer on *Y. pestis*? Humans are not part of the normal lifecycle of *Y. pestis* which is a zoonotic disease affecting primarily rodents. While little difference was observed between $F1^-$ and $F1^+$ strains in the laboratory models of virulence using either mice or non-human primates, there could be natural rodent species that exhibit a stronger requirement for F1 for virulence. It is also possible that F1 has very little to do with virulence in any individual animal, perhaps it is more important to the epidemiology of plague. Transmission via fleas ensures the spread of *Y. pestis* and the persistence of plague in the animal reservoir (Perry and Fetherston, 1997). Capsule may simply give the bacteria 'an edge' to enhance multiplication in the rodent host and spread to appropriate sites. Properties of F1 capsule that might enhance multiplication and dissemination include: the anti-phagocytic activity; surface charge of the 'capsule', which appears to prevent aggregation of *Y. pestis* in the body (Davis et al., 1996); and decoy activity of F1 shed into the surrounding milieu removing the low levels of antibody produced early in infection. Death of the animal is important for dissemination of *Y. pestis* (Perry and Fetherston, 1997). This ensures that the flea carrying *Y. pestis* leaves one host, as the body temperature drops, and moves on to a new host. An important element of this is that sufficient bacteria are taken up with the blood by the flea. Levels of *Y. pestis* in the blood of an infected animal vary during the course of infection. Using a quantitative PCR assay, it has been estimated that $>10^6$ bacteria ml^{-1} in the host blood is required for detectable infection of the flea and $>10^7$ bacteria ml^{-1} for efficient flea mediated transmission from an infected to an uninfected animal (Engelthaler et al., 2000). Possession of F1 may ensure that these bacterial loads are reached in the host often enough to maintain productive transfer to the flea and a pool of infected animals in an enzootic area. Interestingly, *Y. pestis* appears to have evolved other strategies to ensure high bacterial numbers prior to death. For example, lipooligosaccharide synthesis of *Y. pestis* switches from the more toxic highly acylated forms (hexa-, penta-, tetra- and tri-acylated LPS) which all exist at 27° C to the less toxic tri- and tetra- acylated forms at 37° C (Kawahara et al., 2002).

THE ROLE OF F1 IN ANTI-PLAGUE VACCINES

Killed or live whole cell anti-plague vaccines were used quite extensively in the 20th century during the Asian epidemic and in endemic regions such as Madagascar and Vietnam. The efficacy of these vaccine preparations against bubonic plague, although indicated, has never been substantiated in controlled trials (Jefferson, 2002). Variations in immunogenicity and reactogenicity, together with the short-lived protection mean that protecting communities against exposure to enzootic disease by vaccination is not currently feasible. Thus the World Health Organisation recommend vaccination only for exceptionally high risk personnel such as clinical and laboratory workers continually exposed to *Y. pestis* and military personnel at high risk (Gage et al., 1996; Poland and Dennis, 1999b; WHO, 2002). Antibiotics are the mainstay of treatment and prevention of disease in contacts in endemic areas today (Poland and Dennis, 1999b). The successful development of effective and cheap, single-dose, anti-plague vaccines would therefore benefit populations in endemic regions, or during a sudden outbreak, as well high risk workers.

KILLED WHOLE CELL VACCINES

The first killed whole cell plague vaccine, the Haffkine vaccine, was developed and in production by 1897 in Bombay (Haffkine, 1897; Taylor, 1933), at the height of the plague epidemic in India. This vaccine was a combination of cells, killed by heating at 70°C for 1h, and the spent culture medium. Inoculum was of virulent strains isolated directly from buboes of plague patients and bacterial growth was at ambient temperature generally around 27°C, except in very hot weather, for an extended period of 3-6 weeks.

Table 2. Selected published mouse protection studies with different forms of F1 as vaccine.

Vaccine	Immunisation	Challenge[1]			Survivors	Reference
		Strain	Route	Dose (LD_{50})		
Whole cell	Porton outbred mice					(Russell et al., 1995)
KWC-USP[2]	10^8cfu i.m.; one booster	GB	s.c. aerosol	$10^3/10^4$ 15/150	100%/60% 80%/80%	
Attenuated *Y. pestis* EV76	10^7cfu i.m., single dose	GB	s.c. aerosol	$10^5/10^6$ 15/150	100%/100% 100%/100%	
Whole cell vs subunit	Swiss Webster mice					(Andrews et al., 1996; 1999; Heath et al., 1998)
KWC-USP	10^8cfu; {one} or two boosters	CO92 (F1$^+$)	s.c. aerosol	100 100	{50%}, 78% 10%	
KWC-USP		CO92-C12 (F1$^-$)	s.c. aerosol	100 590	22% 0%	
rF1p[3]	10 µg in Alhydrogel, s.c.; one booster	CO92	s.c. aerosol	100 100	100% 90%	
YpF1[4]		CO92	s.c. aerosol	100 100	90% 70%	
	BALB/c mice					(Reddin et al., 1998)
KWC-USP	10^8cfu i.m.; one booster	GB	s.c.	$10^3/10^5/10^7$	100%/60%/40%	
rF1p	10 µg in Alhydrogel, i.p.; two boosters	GB	s.c.	$10^3/10^5/10^7$	100%/90%/10%	
Polymer vs denatured	BALB/c mice					(Miller et al., 1998)
rF1p	10 µg in IFA i.m.; one booster	GB	s.c.	$10^5/10^6$	100%/72%	
rF1ds[5]		GB	s.c.	$10^5/10^6$	100%/16%	
Combined subunit						(Williamson et al., 2000; Jones et al., 2000)
rF1p+ rV	BALB/c mice, 10 µg each in Alhydrogel, i.m.; one booster	GB	s.c. aerosol	$10^5/10^7$ $10^2/10^4$	100%/100% 100%/100%	
KWC-USP	10^8cfu; one booster	GB	aerosol	10^4	16%	
Heterologous expression -Salmonella	BALB/c					(Titball et al., 1997)
-rF1p, surface	10^8 cfu, i.g.	GB	s.c.	$10^5/10^7$	100%/100%	
-rF1s,[6] cytosolic		GB	s.c.	$10^5/10^7$	50%/33%	
DNA vaccines *caf1*DNA[7] (rF1s)	BALB/c mice 3 x 10 µg, i.m.	CO92	aerosol	400	0%	(Brandler et al., 1998)
*caf1*DNA (-SS) (rF1s)	BALB/c mice 3 x 0.5 µg, gene gun	KIM53	s.c.	$10^2/10^3$	100%/100%	(Grosfeld et al., 2003)
Viral vaccine Poxvirus-tPA-F1 (rF1s)	AJ mice 10^7pfu x 2, footpad	CO92	s.c.	$10^2/10^3/10^4$	100%/75%/25%	(Osorio et al., 2003)
Passive immunity mAb F1-04-A-G1	Swiss Webster mice, 250 µg, i.p., 6h pre-challenge	CO92	s.c. aerosol	48 74	100% 40%	(Anderson et al., 1997)

mAb F1-04-A-G1	BALB/c mice 100μg, i.p. 4h pre-challenge	GB	s.c. aerosol	91 88	20% 90%	(Hill et al., 2003).
mAb F1-04-A-G1 + anti-V mAb[8]	i.p. 4h pre-challenge	GB	s.c. aerosol	$10^4/10^5$ 88	88%/100% 90%	
mAb F1-04-A-G1 + anti-V mAb[8]	i.p. 48h postchallenge	GB	s.c.	91	80%	

[1]Challenge was generally 4-6 weeks after the final boost, except for (Russell, 1995 #1707) and DNA vaccination, when challenge was 8 days and 18 weeks after the final booster, respectively. LD_{50} for s.c. challenge was generally about 1 cfu; for aerosol challenge it ranged from 10^2 for GB strain to 10^4 for strain CO92. [2]KWC, killed whole cell-USP vaccine. [3]rF1p recombinant F1 polymer isolated from culture supernatant of *E. coli*. [4]YpF1 recombinant F1 polymer isolated from culture supernatant of *Y. pestis* CO92. [5]rF1ds, recombinant Caf1 polymer denatured in SDS. [6]rF1s recombinant Caf1 subunit, expressed in the absence of chaperone, conformation not established – likely to be largely in a non-native unpolymerised form. [7]*caf1* DNA (-SS), *caf1* gene minus a signal sequence. [8]mAb7.3 (anti-V) alone conferred 70% and 100% protection against 46 LD_{50} s.c. and 88 LD50 aerosol challenge and 20% protection when administered 48h postchallenge (91 LD_{50} s.c.).

Systemic reaction in vaccinated individuals to this first vaccine was considered essential and was often severe with a temperature of 39°C and several days incapacitation. Over 30 million doses of this vaccine were administered in India over the following years. The few cases where records were kept indicate possible efficacy of this vaccine in reducing the incidence and the severity of disease. For example, in the initial, non-randomised trial in 1897 in the House of Correction, Byculla, India, of 147 immunised prisoners 2 contracted plague and both survived; whereas of 172 non-immunised inmates 12 contracted plague and six died (Taylor, 1933). The highly virulent 195/P strain of *Y. pestis,* originally isolated from a plague patient in India, eventually became the standard strain for killed whole cell vaccine (KWC vaccine) production at the Haffkine Institute and later worldwide. Heat (Commonwealth Serum Laboratories (CSL), Australia) or formaldehyde (second generation Haffkine vaccine, and Plague vaccine, USP) to kill the bacteria, a growth temperature of 37°C (enhancing F1 production) and a shorter incubation period of 48h (dramatically reducing systemic reactions) was subsequently adopted. The agar-grown, formalin-killed Plague vaccine USP was developed in the USA and became the major plague vaccine in the Western world (Meyer, 1970; Meyer et al., 1974a; Gage et al., 1996) until production ceased in 1999 (Inglesby et al., 2000). The CSL vaccine remains in production.

Both early protection trials, with a variety of animal species (Meyer, 1970; Meyer et al., 1974d), and more recent studies, with mice (Table 2), demonstrated efficacy of the USP KWC vaccine in protecting against subcutaneous (s.c.) challenge (the model for bubonic plague) and delaying time to death in those animals which eventually succumbed. Protection from aerosol challenge (the model for pneumonic plague) was much lower and less consistent (Table 2; Russell et al., 1995; Andrews et al., 1996). There is some indication that KWC induced protection could be improved by priming the mucosal immune system using an appropriate delivery route and formulation of killed whole cell vaccine (Baca-Estrada et al., 2000).

There is a substantial body of evidence identifying F1 as the major protective immunogen of these killed vaccines. Most convincingly, Plague vaccine USP does not protect mice against low dose s.c. challenge with an $F1^-$ strain of *Y. pestis* (Table 2); survival rates were the same as that in mice injected with the adjuvant control (Heath et al., 1998). In addition, data from several studies reveal a trend where high titres of anti-F1 antibody in immunised rodents and non-human primates correlate with protection against subcutaneous challenge (Meyer, 1970; Chen et al., 1974; Williams and Cavanaugh, 1979; Gage et al., 1996). A similar positive correlation has been reported between anti-F1 titres in serum from immunised animals and humans with ability of the serum to confer passive protection in mice (Meyer, 1970; Bartelloni et al., 1973; Williams and Cavanaugh, 1979). In contrast, antibody against the TTSS component, V antigen (the second proven protective antigen), was undetectable in mice immunised with plague vaccine USP (Heath et al., 1998; Williamson et al., 1999). Hence serological conversion, specifically of anti-F1 IgG, is taken as a presumptive measure of vaccine induced protective immunity in humans - with a titre of 128, assessed by passive haemagglutination or ELISA, considered to be a protective level (Bartelloni et al., 1973; Gage et al., 1996). Primary immunisation with killed whole cell vaccines requires a series of at least 2 and often 3 or more doses to induce seroconversion and additional booster injections every 6 months to 2 years are required to maintain adequate antibody levels (Gage et al., 1996). A small percentage of individuals do not respond to the vaccine. While adverse reaction to the USP vaccine was minimised by production procedures and decreasing the number of killed organisms in each dose (2.0 x 10^9 in the first followed by 4 x 10^8 in subsequent doses), reports indicate that local side-effects were not uncommon and systemic side-effects (headache, elevated temperature, malaise) occurred in about 4-10% of recipients. Severe side-effects were rare (Marshall, 1974; Gage et al., 1996).

Evidence supporting the efficacy of KWC vaccines in humans is indirect. From World War II to the end of the twentieth century, the USP vaccine was administered to USA armed force personnel under threat of exposure to plague (Meyer et al., 1974a). There were no reported cases of plague in vaccinated soldiers during World War II and the incidence of plague in American soldiers serving in Vietnam was negligible, despite higher prevalence in the local population and exposure of soldiers to *Y. pestis* infected fleas and rats (Cavanaugh, 1974). Moreover, 12% of soldiers hospitalised with typhus, in Vietnam in 1969, had higher anti-F1 antibody levels in their convalescent serum compared to acute serum, indicating probable exposure to *Y. pestis* coincident with *Rickettsia typhi* from infected fleas. There have been two confirmed cases of pneumonic plague in vaccinated individuals (Meyer, 1970). Taking this information together with the animal protection studies, it is generally accepted that this KWC vaccine could protect against or decrease the severity of bubonic disease, but was unlikely to be effective against pneumonic plague (Gage et al., 1996; Titball and Williamson, 2001). Safety considerations during production made this an expensive vaccine to produce, further compounding the drawbacks of short-lived immunity and apparent limited protection.

LIVE ATTENUATED VACCINES

Early vaccines also included live attenuated plague vaccines that were first tested in humans in 1908 (Strong, 1908). The EV vaccine (Girard, 1933) and the Tjiwidej vaccine (Otten, 1941) were then developed in Madagascar and Indonesia, respectively. The EV 76 vaccine has been widely used in Madagascar, Asia and was adopted by the former Soviet Union. It was isolated by consecutive subculture over 5 years. Attenuation markers are not fully defined, but EV76, which is F1 positive, is known to have a disrupted *pgm* locus and is phenotypically pigmentation negative (does not bind haemin) (Meyer, 1970; Galyov et al., 1990; Podladchikova et al., 2002). Animal studies revealed a high level of protective immunity induced by the EV 76 vaccine (Meyer, 1970), and in one direct comparison with the USP killed vaccine, EV 76 proved to be more effective in protecting against both subcutaneous and intranasal challenge (Table 2) (Meyer, 1974a; Russell et al., 1995). As with KWC vaccines, anti-F1 titres generally correlated with the passive mouse protection index. Pre-growth of the attenuated vaccine strain at 28°C or 37°C gave the same results, indicating that prior production of F1 was not essential. Some reports are more varied with respect to immunogenicity and protective properties of EV strains. A fact that has been attributed, in part, to strain variation on storage (Meyer, 1970; Butler, 1983). Immunisation with EV 76 was fairly extensive in endemic areas during the 20th century and appears to have had a beneficial effect on the severity and spread of disease (Meyer et al., 1974a). For example, when first used in Madagascar 815,000 individuals (80% of the population) were immunised between 1933-1936, incidence of both bubonic and pneumonic plague dropped dramatically and in 1937 only 31.5% of the total plague cases occurred in vaccinated individuals (Meyer, 1970; Meyer et al., 1974c). Nevertheless, as with the KWC vaccines, the absence of randomised trials both here and in other anti-plague programmes makes it difficult to define the true efficacy of the vaccines (Jefferson, 2002). Major drawbacks with EV76 include recurring reports of residual virulence, severe adverse reaction and even death in animals (Hallett et al., 1973; Russell et al., 1995; Welkos et al., 2002) and humans, (Meyer, 1970) in addition to concern about enhanced virulence following immunisation.

The apparent better protection against pneumonic and bubonic plague afforded by the EV76 vaccine, has stimulated a search for alternate attenuated vaccines. Enhanced protection may relate to *in vivo* expression of other key protective antigen(s) in addition to F1. Also, continued exposure to the immune system during several rounds of bacterial multiplication and possible stimulation of a cell mediated response may contribute to the enhanced protective properties of an attenuated vaccine. The availability of two *Y. pestis* genome sequences (Parkhill et al., 2001; Deng et al. 2002) should dramatically accelerate the identification of additional major protective antigens, induced at different stages of infection. Ideally an attenuated strain would still express F1 and V antigens as well as other major protective antigens identified. Attempts have been made to construct defined attenuated mutants of *Y. pestis*, using targets identified as attenuation markers in *Salmonella*. Targets have included *aroA*, *phoP* and *degP*, but in *Y. pestis* inactivation of these genes led to only marginal attenuation (Titball and Williamson, 2001). Proven attenuating mutations in *Y. pestis* include mutations in the TTSS genes (Perry and Fetherston, 1997), plasminogen activator (Sodeinde et al., 1992; Welkos et al., 2002) and an enzyme required for nucleotide metabolism, adenylate kinase (Munier-Lehmann et al., 2003). Conversion of any attenuated strain into a stable, safe, effective vaccine acceptable today for human administration would be a major task, particularly in light of the known plasticity of the *Y. pestis* genome (Parkhill et al., 2001; Radnedge et al., 2002).

SUBUNIT VACCINES

With the focus on development of safe, effective protection against pneumonic plague, the search for improved vaccines turned primarily to subunit vaccines with F1 as an obvious first candidate. Since Baker identified the F1 antigen as being immunogenic (Baker et al., 1947), evidence has accumulated to support its immunoprotective properties (see above '*F1 as an Immunogen*'). Highly purified, endotoxin-free, polymeric F1 derived from the avirulent *Y. pestis* CO92 Pgm^- Lcr^- and from recombinant *E. coli* were found to induce similar high titres of antibody and to confer similar levels of protection in mice (Simpson et al., 1990; Andrews et al., 1996). Both preparations produced significantly better protection than USP vaccine against both s.c. and aerosol challenge (Table 2). Results were independent of whether F1 was cell surface or culture medium derived.

As early as 1957, Burrows reported the isolation of fully virulent F1 negative strains, from immunised animals and suggested that protection studies focus on the second major protective antigen 'V (virulence) antigen' as well as F1 (Burrows, 1957). The report of a human F1 negative isolate (Winter, 1960) underscored the requirement for inclusion of a second protective antigen in a combined subunit vaccine. V antigen is part of the Type Three Secretion System (TTSS), is secreted to the cell surface and the medium (Pettersson et al., 1999), is essential for TTSS mediated resistance to phagocytosis (Du et al., 1995) and has immunosuppressive properties (Nakajima and Brubaker, 1993; Brubaker, 2003). Immunisation with rV antigen induces high levels of protection in mice from subcutaneous or aerosol challenge (Leary et al., 1995; Andrews et al., 1999) and protects against challenge with the F1 negative strain (*Y. pestis* CO92-C12) (Anderson et al., 1996). V antigen is the only TTSS component thus far proven to confer high levels of protection in mice, against wild type *Y. pestis* (Andrews et al.,1999; Titball and Williamson, 2001).

Two approaches have been used for production of combined rF1 and rV subunit vaccines. One approach attempted to minimise purification of vaccine components by creating a gene fusion encoding V-antigen fused to the C-terminus of a Caf1 polypeptide carrying either an N-terminal His-tag (Heath et al., 1998) or lacking a signal sequence (Leary et al., 1997; Titball et al., 1997). This product was purified primarily from cytosolic inclusion bodies but also from a small proportion of soluble protein. The purified rF1-V fusion induced very good protection in mice against subcutaneous and aerosol challenge but was subject to proteolytic degradation which led to lower recoveries of F1 (Leary et al., 1997; Heath et al., 1998). Susceptibility to proteolysis and inclusion body formation are indicative of incomplete protein folding and consistent with the concept that cytosolic expression, in the absence of chaperone, is not conducive to F1 folding and polymer formation. Indeed, the efficiency with which Caf1 subunit can correctly fold and polymerise, if at all, in the absence of chaperone remains to be established (see '*Polymer assembly*' above). Sodium dodecyl sulphate denatured F1 (rF1ds) still protects against subcutaneous challenge of *Y. pestis*, but was found to be much less effective than F1 polymer at a high challenge dose (10^6 cfu, Table 2) (Miller et al., 1998). A similar result was obtained with cytosolically expressed Caf1 (rF1s) compared to the surface polymer (rF1p) in recombinant *Salmonella* (Table 2, Titball et al., 1997). Hence, although both of these non-native forms of F1 still induce some protective immune response, it would seem judicious to use native folded F1 polymer in subunit vaccine development, at least until more is understood about the protective epitopes and folding requirements of Caf1.

The second approach to subunit vaccine development combined independently purified rF1 polymer from culture supernatants with rV antigen, using an optimised molar ratio of two Caf1 subunits to one V antigen subunit (Williamson et al., 1997; 1999). While both rF1 and rV antigens separately are good immunogens, co-immunisation with rF1 and rV had the desired improved effect inducing solid protective immunity in mice against remarkably high challenge doses (10^7 LD_{50} s.c. challenge and 10^4 LD_{50} aerosol challenge, Table 2) (Jones et al., 2000; Williamson et al., 2000). Protection against both 'pneumonic' and 'bubonic' plague was attributed to high levels of circulating IgG to both antigens, titres of which remained high for at least one year. The combined titre of the anti-F1 and anti-V IgG1 subclass of antibody was found to correlate directly with the level of protection against subcutaneous infection. A 1 μg dose level was sufficient to fully protect against a

10^5 cfu challenge, whereas a 5 μg dose of each subunit was required to protect against an increased challenge of 10^7 (Williamson et al., 1999). These two antigens appear to have a complementary effect in protection. IgG directed against F1 most likely neutralises the antiphagocytic effect of F1 and protects by opsonising the bacteria and enhancing phagocytosis; whilst anti-V antibody appears to specifically block the activities of LcrV (V antigen). Anti-V antibody inhibits *Y. pestis* induced phagocyte toxicity, promotes phagocytosis and may also block the immunomodulatory activity of LcrV (Weeks et al., 2002). A similar approach of enhancing opsonisation together with neutralisation of a key virulence determinant has been proposed for other bacterial vaccines (Schneerson et al., 2003). While the *Y. pestis* proteome may reveal additional promising vaccine candidates, currently F1 and LcrV are the only two validated vaccine candidates identified.

An alhydrogel formulation of the combined rF1 + rV vaccine for parenteral immunisation (two doses 20 days apart) is now at the developmental stage of clinical trials (Titball and Williamson, 2003). The oral or nasal route of administration would obviously be cheaper and more efficient. Encapsulation of rF1 and rV antigen in microspheres, for example poly-L-lactide, is under investigation to enhance mucosal immunity in mice (Williamson et al., 1996; Eyles et al., 1998; 2000; Titball and Williamson, 2001). Ideally, formulation in microspheres would be optimised to induce protection from both bubonic and pneumonic plague following administration of a single dose .

HETEROLOGOUS EXPRESSION OF F1 IN ATTENUATED BACTERIA

Attenuated strains of *Salmonella spp* have received considerable attention regarding their potential as carriers for heterologous expression of foreign antigens of significance to human disease (Cardenas and Clements, 1992). A consequence of the limited invasion of the host and prolonged survival of these strains is the potentially enhanced stimulation of long term cell and antibody mediated immunity through continued presentation of the antigen to the immune system. Other benefits include stimulation of mucosal and systemic immunity as well as cheap production. As members of the *Enterobacteriaceae*, *Salmonella* species have the advantage that the recombinant F1 assembly apparatus functions efficiently to assemble surface capsule (Titball et al., 1997). An *aroA* mutant of *Salmonella enterica* serovar Typhimurium, the mouse model of typhoid, has been investigated to assess the potential of *Salmonella* species as carriers of *Y. pestis* F1 and V antigens. Surface polymer afforded a high level of protection against s.c. challenge, a level markedly better than that of the cytosolically expressed and presumably unpolymerised subunit (Table 2) (Titball et al., 1997). Expression was optimised and maximum serum and mucosal antibody obtained using the *in vivo* induced *phoP* promoter (Bullifent et al., 2000). rV expressed in *S. enterica* serovar Typhimurium *aroA* has also recently been shown to induce a protective response (Garmory et al., 2003). These results provide the basis for development of attenuated strains of *S. typhi* expressing one or both of these antigens.

PASSIVE IMMUNISATION AGAINST PLAGUE

Passive immunisation, the first attempted route of immunisation against plague (Yersin, 1894), is now being revisited. In endemic areas treatment of plague is dependent on antibiotic therapy and serious problems would arise in the event of infection with a multi-antibiotic resistant strain. Some vaccinees do not respond to existing vaccines and consistently have anti-F1 titres below the recommended level (Gage et al., 1996).

Also, even with improved vaccines (fewer side-effects and prolonged immunity) only groups at recognised risk would be vaccinated. Hence, if *Y. pestis* was ever used as a bioweapon, unprotected individuals would undoubtedly be exposed. As discussed above, levels of anti-F1 antibody have been shown to correlate with protection. The ability of such serum to passively protect mice from s.c. infection has routinely been demonstrated with the mouse protection assay used to monitor vaccinee serum samples (Williams and Cavanaugh, 1979). Monoclonal antibodies are now being assessed for their ability to prevent disease. In one study (Anderson et al., 1997), the mAb F1-04-A-G1 (Navy Medical Research Institute, Bethesda, USA) was found to be highly protective in mice. A dose of 250 μg given intraperitoneally (ip) 6 h prior to infection, afforded 100% protection against s.c. challenge (48 LD_{50}) and 500 μg given ip 24 h prior to infection, afforded good protection (90% survival) against an aerosol challenge (74 LD_{50}). In cases of pneumonic plague, antibiotic treatment must be extremely rapid. The great advantage of mAbs would be to treat already infected individuals and improve survival rates, particularly those recognised at a later stage of infection. Another study has addressed this problem and found that the same anti-F1 mAb together with anti-V mAb (Mab 7.3) protected mice when administered up to 48 h post – aerosol or s.c. infection, but not at 60h postinfection (Hill et al., 2003). As with the subunit vaccination, combined anti-F1 and anti-V mAbs had an additive effect when compared to administration of each mAb alone (see Table 2 for details). Clinical use of mAbs to treat infected individuals would require either 'humanisation' of identified mouse monoclonal antibodies or a search for new high affinity mAbs from human antibody libraries.

VIRAL AND DNA VACCINES

Third generation anti-plague vaccines are also being investigated and include naked DNA and viral vaccines. Both of these entail antigen expression in the eukaryotic host cell. The native F1 assembly system cannot be readily reproduced within the eukaryotic cell and therefore consideration of the targeting of the recombinant Caf1 molecule is important. The nature of the rCaf1 can modulate not only the extent but also the type of the immune response.

DNA vaccines have been shown to elicit primarily cell-mediated immune responses but can also stimulate a good humoral response and offer the advantages of extended antigen expression over a long period of time, cheap production and a thermostable vaccine product (Reyes-Sandoval and Ertl, 2001). Initial attempts with the complete *caf1* gene as a DNA vaccine conferred no protection against aerosol challenge, but led to a strong humoral response following priming with *caf1* DNA and a single booster injection of purified protein (Brandler et al., 1998; Williamson et al., 2002). Another report (Grosfeld et al., 2003) compared the efficacy of three forms of *caf1* DNA derivatives encoding - full-length Caf1 (which appeared to be rapidly degraded, possibly following secretion), Caf1 devoid of a signal sequence (deF1 which was expressed in the cytosol) and mature Caf1 carrying the signal sequence of the E3 polypeptide of Semliki forest virus (E3/F1) (which was targeted to the secretory cisternae). The vector expressing the signal-sequenceless Caf1 was most effective in eliciting anti-F1 IgG antibodies and IgG1 levels, in particular, were enhanced by immunisation with a gene gun in outbred as well as inbred mice. Complete protection against subcutaneous infection with the *Y. pestis* strain KIM53, at relatively low challenges of 4,000 cfu, was obtained following a schedule of 3 immunisations (0.5 μg deF1 DNA each) (Table 2). Interestingly, in some preparations of cells expressing the cytosolic deF1, immunoblots revealed a ladder of bands reminiscent of the low molecular weight polymer observed in periplasmic preparations of F1 polymer (Figure 1c, lane 3). Although the nature of this 'polymer' remains to be established, should it prove to be correctly folded and polymerised, this would be an indication of the ability of the signal sequenceless Caf1 subunit to form at least low levels of cytosolic F1 polymer in the absence of Caf1M chaperone. Enhanced stability as a consequence of folding of the subunit and polymerisation could contribute to the improved immune response observed with this construct. It remains to be established how effective such an immunisation schedule would be against higher challenge doses s.c. and against aerosol challenge.

Viral delivery systems are also being considered as vehicles for plague vaccines. The raccoon poxvirus (RCN) expression system, has been assessed for ultimate potential use as a wild-life animal vaccine to reduce environmental reservoirs of plague (Osorio et al., 2003). This study compared the efficacy of native Caf1 to Caf1 with the secretory signal of tissue plasminogen activator (tPA) with and without a membrane anchor. As with the DNA vaccine trial, Caf1 with the native bacterial signal sequence was less effective. Protection and antibody levels were lower and incorrect folding may be a factor as the recombinant Caf1, which appeared to be largely processed, was insoluble and apparently less stable. Another indication that folding is important to protection was noted from the fact that, when anchored to the membrane, Caf1 induced high levels of antibody but poor protection. Although much lower than protection conferred by purified rF1 polymer, targeting of Caf1 to the secretory vesicles via the tPA signal elicited some protection against s.c. challenge (Table 2).

APPLICATION OF F1 TO DIAGNOSIS OF *YERSINIA PESTIS* INFECTION

Plague is a Class I notifiable disease; all cases must be reported to the World Health Organisation, Geneva. Accurate diagnosis is central to establishing the epidemiology of plague. In light of the rapid decline in health of patients with plague, prompt and accurate diagnosis is also of paramount importance for effective treatment and control of spread of the disease. F1 antigen has been in the past and remains the principle biological marker for plague infection. Its uniqueness to *Y. pestis* (yet conservation among different strains), stability, abundant levels on the bacterial cell surface and surrounding milieu and its strong immunogenicity has made it an excellent target for diagnostic procedures. The recommended diagnosis of confirmed plague is either isolation and identification of *Y. pestis* from clinical material or demonstration of seroconversion (to F1 antigen) in convalescent patients. Tests for the presumptive diagnosis of plague include identification of F1 antigen, presence of anti-F1 antibody (in non-immunised individuals) and a positive polymerase chain reaction (PCR) (Poland and Dennis, 1999a; Chu, 2000) .

The level of diagnosis is dependent on resources available and experience of clinical and laboratory personnel. In endemic areas, facilities for even basic diagnosis are generally only available at specialised regional centres, entailing long delays and problems in transport and storage of samples. Hence, unfortunately in many endemic regions primary diagnosis is often based on clinical symptoms with or without direct microscopic examination, followed by retrospective confirmation where possible. For this reason and also to ensure a rapid response in the event of any bioterrorist incident, development of newer rapid diagnostic tests that are sensitive, accurate, cheap and simple to use is viewed as a priority. The outbreak of pneumonic plague in Surat, India in 1994, where within 5 days 452 suspected cases were reported and 41 died highlights the potential requirement of high throughput as well as rapid and accurate tests (WHO, 1994; Campbell and Hughes, 1995; Ramalingaswami, 1995).

DIRECT AND INDIRECT BACTERIAL DIAGNOSIS

Y. pestis, a member of the *Enterobacteriaceae,* is a Gram negative, oxidase negative rod which ferments glucose. With respect to biochemical tests *Y. pestis* is relatively nonreactive and commercial tests can lead to misdiagnosis (Chu, 2000). Characteristic features which aid in identification include its bipolar "closed safety pin" appearance on staining with Giemsa or Waysons stain, "fried-egg" colony morphology with a "hammered copper" shiny surface on Sheep blood agar, clumping stalactite-like growth in early unshaken broth cultures leading to settled 'cotton wool fluff' in older cultures, and optimum growth at 28°C with poorer growth at 37°C (Perry and Fetherston, 1997; Brubaker, 2000). Clinical samples, for subculture and direct staining, are collected from blood, aspirates from suspected buboes, autopsy pieces or sputum and bronchial wash from patients with suspected pneumonic plague. Samples are preferably taken prior to initiation of antibiotic treatment and, due to the intermittent appearance of *Y. pestis* in the blood, blood samples are ideally taken as a set of three over 45 min. Recovery of *Y. pestis* from contaminated clinical specimens can be optimised by inoculation of mice (Poland and Dennis, 1999a).

In addition to staining, smears or isolated cultures can also be tested for F1 using a direct fluorescent anti-F1 antibody (DFA) test. Due to the amount of surface F1 antigen, a positive reaction gives an intense 'starry night' appearance (Ramalingaswami, 1995; Chu, 2000). Assuming access to facilities, this can provide a rapid presumptive test for plague in contrast to the several days required to isolate the bacterium. Following death of a plague patient, immunostaining for F1 capsule has also been used to detect *Y. pestis* directly in autopsy tissue (Guarner et al., 2002; Gabastou et al., 2000). For isolated bacteria, the DFA test is an additional presumptive identification of *Y. pestis* with the limitation that false negatives can occur if the *in vitro* cultured bacteria have been grown below the recommended temperature rather than at 35-37°C to ensure F1 induction. Bacteriophage typing, which is independent of growth temperature, is the recommended confirmatory test for isolated *Y. pestis* (Poland and Dennis, 1999a; Chu, 2000).

SEROLOGY: QUANTITATION OF ANTI-F1 ANTIBODIES

Isolation of *Y. pestis* from clinical samples taken from a suspected plague patient is often unsuccessful, particularly if antibiotic treatment has been initiated. Serological diagnosis, for anti-F1 antibodies, has therefore been routine and can substantially increase the number of presumptive or confirmed cases (Williams et al., 1982). For example, in the 1994 outbreak of plague in India the presumptive 876 cases were primarily identified following positive serological analyses for F1 antibodies (WHO, 1995). In a plague outbreak in Namibia in 1982, of 38 confirmed cases, 50% were identified by isolation of *Y. pestis* and a further 34% by anti-F1 serological response (Williams et al., 1986) and in Madagascar in 1995, 101 of 342 suspected cases were identified by bacteriology and an additional 35 by anti-F1 ELISA (Rasoamanana et al., 1997).

The simplest method of detecting anti-F1 antibody is the passive haemagglutination assay (PHA), which was used worldwide throughout the second half of the 20th century and is still a primary rapid diagnostic test today (Chen and Meyer, 1954; Butler and

Hudson, 1977; Butler, 1983; Chu, 2000). The assay depends on anti-F1 antibody mediated agglutination of red blood cells sensitised by coating them with polymeric F1, which is washed from the surface of bacterial cells and isolated by a series of differential ammonium sulphate precipitation steps. Although preparation of antigen and cells has been standardised by WHO (Chu, 2000), a problem with non-specific agglutination remains. Hence, positive agglutination requires confirmation by inhibition of haemagglutination with F1 antigen (PHI assay). While the PHA/PHI assay is simple and cheap with no requirement for specialised equipment, enzyme-linked immunosorbent assays (ELISAs) are more sensitive and specific and have, where the equipment is available, superseded the PHA assay (Cavanaugh et al., 1979; Williams et al., 1982; 1986; Rasoamanana et al., 1997; Chanteau et al., 2000a).

Direct ELISA assays (with F1 capsular antigen bound directly to the plate) (Cavanaugh et al., 1979; Williams et al., 1982; Rasoamanana et al., 1997; Chanteau et al., 2000a) or sandwich ELISA (with F1 capsular antigen attached to the plate via mAb (3G8)) (Williams et al., 1986; Arntzen and Frean, 1997) have been used to quantitate serum anti-F1 antibody. In the direct assay, retention of F1 on the plate was reportedly improved by crosslinking F1 polymer with glutaraldehyde to polyalanine on a Dacron disk (PANI) (Coêlho et al., 2001). With both the combined PHA/PHI and the anti-F1 IgG ELISA a titre of ≥1:16 in unvaccinated individuals is considered a presumptive diagnosis and an increase in titre (≥4 fold) between acute and convalescent serum (taken after 3 -4 weeks or longer) is considered to be a confirmed case of plague (Chu, 2000). A competitive blocking assay in which serum anti-F1 is quantitated via its ability to block binding of a labelled anti-F1 antibody decreases the time and reagents required (Chu, 2000). By using a specific second antibody, ELISA also has the advantage that it can differentiate between relative levels of serum anti-F1 IgG and IgM, which can in turn provide an indicator of recent infections (Williams et al., 1986). An important application of this procedure is as a retrospective tool to more accurately estimate the extent and epidemiology of infection.

One method which is both rapid and sensitive and has been assessed for detection of anti-F1 antibody is the fibre-optic biosensor system (Anderson et al., 1998). This was most successful with a competitive fluoroimmunoassay in which F1 polymer is bound to the probe via polyclonal antibodies. Serum antibodies are then bound to the F1-probe followed by an excess of fluorescently labelled anti-F1 mAb6H3-IgG. Levels of specific serum antibody are quantitated from the decrease in fluorescence returning to the probe. Laboratory tests compared favourably with ELISA and detected anti-F1 antibodies in the range of 2-15 μg/ml. It has been anticipated that as this technology develops it will become available for use in the field, providing cheap, rapid and accurate analysis.

It is estimated that approximately 5-12 % of an infected population might be negative in a standard serological test (Marshall, 1974; Butler and Hudson, 1977). The two most likely explanations for this are that a small fraction of the population will not respond adequately to F1 antigen and that specific anti-F1 IgG levels may be below the level of sensitivity of the assay if blood samples are taken very early in infection. In early infections detection of F1 antigen may be more successful. This reinforces the importance of diagnosis by more than one route (Rasoamanana et al., 1997; Poland and Dennis, 1999a).

DETECTION OF F1 ANTIGEN IN CLINICAL SAMPLES

Due to the high levels of F1 that accumulate in tissue and blood surrounding *Y. pestis* during infection (see Figure 1b as example), F1 antigen can be directly monitored in clinical samples. Moreover its detection can identify plague infection at an early stage prior to a detectable antibody response. The standard assay is an ELISA immunocapture assay. Both a mAb (3G8) coated plate to capture F1 antigen with anti-*Y. pestis* polyclonal serum for quantitation of F1 (Williams et al., 1986) and an anti-*Y. pestis* polyclonal antibody to capture F1 with a specific anti-F1 mAb (F1-04-A-G1) for quantitation (Chanteau et al., 1998; 2000a; 2003) have been used. This type of assay can detect approximately 1-2 ng of F1 polymer ml^{-1}(Williams et al., 1986; Chanteau et al., 2000a). It has been successfully used to detect F1 in serum (Williams et al., 1986; Chanteau et al., 2000a), buboes and urine (Chanteau et al., 2000a) as well as in sputum in cases of pneumonic plague (Ratsitorahina et al., 2000). In one report (Chanteau et al., 2000a) estimated mean concentrations of F1 detected in patients with bubonic plague were 70.8 ng ml^{-1} (buboes), 42 ng ml^{-1} (serum) and 13.5 ng ml^{-1} (urine). Sensitivity of the assay, when compared to culture positive confirmed cases, was found to be excellent for bubo aspirates (100 %), but less effective in serum (52 %) and urine specimens (58 %). As with the F1 antibody ELISA, F1 ELISA will identify further presumptive cases by identifying F1 antigen in clinical samples from culture negative patients.

There is evidence that the concentration of F1 in serum can serve as a marker for the gravity of disease and success of antibiotic treatment (Chanteau et al., 1998; 2000a). As might be expected, in patients who responded to antibiotic treatment and survived, F1 antigen levels dropped 48 h after initiation of antibiotic treatment and were undetectable by 24-48 h later. In contrast, patients who died within 24 h of admission to hospital had remarkably high levels of serum F1 (10-30 μg ml^{-1}), a reflection of high bacterial numbers in the blood. A variation of the ELISA assay, the highly sensitive dissociation-linked lanthanide fluorescent immunoassay (DELPHIA), is also being investigated for detection of F1 in clinical samples (Smith et al., 2001). This one report detected 24 pg F1 ml^{-1} but with considerable problems with background.

F1 ELISA, while reliable in analyses of bubo specimens, is not appropriate for local use in endemic areas due to the requirement of specialised reagents and equipment. Undoubtedly the most significant recent advance in the diagnosis of plague has been development of the rapid, immunochromatography dipstick assay (Chanteau et al., 2000a; 2003). This is a handheld assay, the original prototype of which was developed by the Naval Medical Institute, Bethesda. The original version (Chanteau et al., 2000a) is based on colloidal gold-labelled anti-F1 mAb (F1-04-A-G1) and a fine line of anti-*Y. pestis* polyclonal antibody as the capture antibody. The strip is dipped into a small volume of clinical specimen. F1 antigen, if present, is bound by the mobile phase gold-labelled mAb as it moves through the matrix and 10-15 min later a pink test line develops in positive samples, as the gold-labelled Ab-F1 complex is sequestered by the capture antibody. In a field trial with 213 suspect plague patients in Madagascar in 1997, the original dipstick assay could detect 5 ng ml^{-1} F1 and was shown to be as sensitive as F1 ELISA in the early diagnosis of plague. An improved immunogold dipstick assay, based on two high affinity anti-F1 mAbs with different binding epitopes (B18-1, $kD<5x10^{-11}$ mol, gold conjugate mobile phase and G6-18, $kD=1x10^{-10}$ mol, capture antibody), has since been developed (Chanteau et al., 2003). The test is highly specific for *Y. pestis* and has a detection range of 0.5 ng ml^{-1} to 50 μg ml^{-1}. Its sensitivity was 100 % with all clinically isolated bacterial strains and in a screen of a bank of *Y. pestis* isolates from different countries it only failed to detect two. The two negative results were due to loss of the pFra plasmid, which was attributed to laboratory storage. This new rapid diagnostic test (RDT) was assessed in an extensive field trial in Madagascar during 2001. Of 691 suspected cases, 197 were confirmed by bacteriology, 212 were presumptive by F1 ELISA and 279 were positive by plague RDT analysis of clinical samples (bubo aspirate, sputum or post-mortem organ puncture). The high success rate of the RDT is attributed to its very low detection threshold. However, 31 samples that were negative by RDT were positive for bacteriology. The potential of the dipstick RDT for general bedside use in endemic regions was validated by a comparison of results from the Central Reference laboratory, Antananarivo, with those from 26 remote sites using the local staff. There was 89.9 % agreement between tests done centrally and those performed at remote sites in detection of bubonic and pneumonic plague. The simplicity, reliability, sensitivity and low cost of this kit make it ideal for use in rural endemic areas. The test kit has now been distributed to all health centres in endemic regions of Madagascar and it is recommended that they be used together with bacteriological tests for surveillance of plague (Chanteau et al., 2003). The successful transfer of this test to the endemic regions of Madagascar provides a solid basis for use of the test worldwide. Success of the kit in rapid diagnosis early in infection and its continued applicability, for at least a short time, following initiation of antibiotic treatment are additional key attributes that would be a distinct advantage in monitoring any threatened bioterrorist attack.

Biosensors are also being developed for the rapid detection of F1 antigen. In response to the threat of misuse of a range of different biological agents as weapons, the target may be either F1 antigen alone or combined with marker antigens of other dangerous pathogens. Published reports are of immunosensor systems with F1 antigen trapped by the mAb 6H3 and detected by the same mAb, labelled with a fluorescent probe (Cao et al., 1995; Wadkins et al., 1998; Rowe et al., 1999). A fibre optic biosensor accurately quantitated levels of F1 between 50 and 400 ng ml^{-1} in 15 min and had a detection limit of 5 ng ml^{-1} (Cao et al., 1995). A slide array system has also been reported for simultaneous detection of F1, staphylococcal enterotoxin and ricin using a CCD camera for detection and quantitation (Wadkins et al., 1998; Rowe et al., 1999). The utility of any of these systems for detection in a clinical setting remains to be evaluated.

POLYMERASE CHAIN REACTION (PCR)

PCR is a sensitive and specific technique for the rapid diagnosis of bacterial pathogens. It is considered a presumptive diagnostic test for *Y. pestis* infection and requires complementation by another standard test for confirmatory diagnosis (Chu, 2000). Targets of the *caf1* gene have varied from the entire gene (Engelthaler et al., 1999) to a short 171 bp fragment (Tsukano et al., 1996). Unlike the protein and antibody detection assays, there is no *a priori* reason for targeting the *caf1* gene in PCR. Despite the very close relationship to *Y. pseudotuberculosis* there are other unique genes, including the *pla* (plasminogen activator) gene, which is carried on a higher copy number plasmid and which has been frequently used as a target (Hinnebusch and Schwan, 1993; Norkina et al., 1994; Tsukano et al., 1996; Engelthaler et al., 1999; Iqbal et al., 2000; Radnedge et al., 2001). Inclusion of *caf1* as a target along with other genes, such as *pla* and *lcrV* or *yop* genes, ensures identification should PCR fail for one gene and concomitantly monitors the presence of key virulence determinants (Tsukano et al., 1996; Engelthaler et al., 1999; Neubauer et al., 2000). Multiplex PCR, where several genes are targeted in one reaction, is now being applied to identification of several virulence factors within *Y. pestis* (Engelthaler et al., 1999; Tsukano et al., 1996), to monitoring association of *Y. pestis* and other flea-borne pathogens with fleas and their host (Stevenson et al., 2003), and to the identification of *Y. pestis* along with a number of other potential BW agents (McDonald et al., 2001). Due to the absence of expression of *caf1* (and many other virulence determinants) in the flea, screening of fleas for carriage of *Y. pestis* has in the past been very expensive and time consuming. It required inoculation of mice with flea material and analysis of tissue from moribund mice (Engelthaler et al., 1999). Hence, PCR is potentially an economical and time-efficient replacement for detecting *Y. pestis* infected fleas (Hinnebusch and Schwan, 1993; Engelthaler et al., 1999). PCR has also been successfully applied to the identification of *Y. pestis* in preserved autopsy tissue (Ramalingaswami, 1995; Panda et al., 1996; Gabastou et al., 2000).

Whilst PCR in the laboratory has been highly successful, this was not the case for a clinical trial in Madagascar involving transport of samples to a regional laboratory for testing (Rahalison et al., 2000). It was concluded that, with the conditions and equipment available at that time, PCR was not sufficiently reliable for routine clinical analysis. Basic techniques are now being adapted to improve PCR in the field. With the application of real-time PCR to *Y. pestis* detection (Higgins et al., 1998; Lindler et al., 2001) and development of a hand-held real-time PCR machine (Idaho Technologies Ltd, Idaho, USA), there is real promise of adapting this rapid, high-throughput technology to *Y. pestis* detection in the field.

CONCLUSIONS

By studying the assembly of F1 antigen the subunit structure and polymeric nature of F1 have been resolved (Zavialov et al., 2003b). The structure confirms the predicted assembly of the polymer via donor strand interaction between neighbouring subunits and has revealed the high resolution detail of this interaction. Knowledge of the structure and assembly of F1 will undoubtedly be a valuable model in the study of the many uncharacterised chaperone:usher systems identified in the genomes of *Yersinia* species. The solved structures have also raised many new and exciting questions about polymer assembly at the cell surface of Gram negative bacteria. For example, is the chaperone induced formation of a high-energy folding intermediate prior to polymerisation a more general phenomenon among chaperones with no obvious energy source? How is donor strand exchange linked to translocation via the outer membrane usher? The possibility also exists that some principles identified with F1 polymer assembly may be extended to other β-structural polymers of medical importance.

Knowledge of the F1 structure now provides a solid platform for identification of the major epitopes of this key antigen. For the majority of the 20th century, F1 antigen almost exclusively attracted attention as the primary protective antigen in vaccines and as the key target for diagnostic procedures. As we enter the post-genomic era, a wealth of information will undoubtedly accrue regarding other antigens induced during specific stages of *Y. pestis* infection. An F1 + V antigen subunit vaccine is the most promising anti-plague vaccine to date (Titball and Williamson, 2001). In light of the concern over the possibility of occurrence of $F1^-$ strains, and the high level of protective immunity conferred by V antigen in animal models, development of a vaccine dependent solely on an anti-F1 response would now seem less ideal. However, the ability of anti-F1 antibodies to complement protection conferred by anti-V antibodies, presumably by promoting opsonisation relatively early in infection, the detailed knowledge of the molecular biology of F1 polymer, and the actual, apparently ubiquitous, possession of F1 by *Y. pestis* strains, together, are likely to ensure that F1 retains a place of priority in vaccine development for the foreseeable future. Detailed understanding of the folding of Caf1 with and without chaperone will be critical to optimisation of the folding of Caf1 in eukaryotic cells, a prerequisite to effective DNA and viral vector based vaccines. Similarly, the uniqueness of F1 to *Y. pestis*, its stability and high level of production resulting in highly sensitive assays, have ensured that F1 remains a prime target for the newer rapid, detection assays being developed. Alternate (or complementary), specific markers for *Y. pestis* may be identified, permitting rapid detection of the bacterium in the unlikely event of a plague outbreak with an F1 negative strain. Identification of alternate, suitable secreted antigens as targets for screening clinical samples would be much more challenging.

ACKNOWLEDGEMENTS

We would like to thank Alex Pudney and Elham Moslehi for preparation of figures and Table 1, Kelly Davis for kindly providing the immunogold labelled image of *in vivo* F1 and Arthur Friedlander for unpublished information. The Biotechnology and Biological Sciences Research Council/ Ministry of Defense (UK) and the Swedish Research Council provided financial support.

REFERENCES

Abramov, V.M., Vasiliev, A.M., Vasilenko, R.N., Kulikova, N.L., Kosarev, I.V., Khlebnikov, V.S., Ishchenko, A.T., MacIntyre, S., Gillespie, J.R., Khurana, R., Korpela, T., Fink, A.L., and Uversky, V.N. 2001. Structural and functional similarity between *Yersinia pestis* capsular protein Caf1 and human interleukin-1 β. Biochem. 40: 6076-6084.

Abramov, V.M., Vasiliev, A.M., Khlebnikov, V.S., Vasilenko, R.N., Kulikova, N.L., Kosarev, I.V., Ishchenko, A.T., Gillespie, J.R., Millett, I.S., Fink, A.L., and Uversky, V.N. 2002. Structural and functional properties of *Yersinia pestis* Caf1 capsular antigen and their possible role in fulminant development of primary pneumonic plague. J. Proteome Res. 1: 307-315.

Achtman, M., Zurth, K., Morelli, G., Torrea, G., Guiyoule, A., and Carniel, E. 1999. *Yersinia pestis*, the cause of plague, is a recently emerged clone of *Yersinia pseudotuberculosis*. Proc. Natl. Acad. Sci. USA. 96: 14043-14048

Amies, C.R. 1951. The envelope substance of *Pasteurella pestis*. Brit. J. Exp. Pathol. 32: 259-273.

Anderson, G.P., King, K.D., Cao, L.K., Jacoby, M., Ligler, F.S., and Ezzell, J. 1998. Quantifying serum antiplague antibody with a fiber-optic biosensor. Clin. Diag. Lab. Immunol. 5: 609-612.

Anderson, G.W., Worsham, P.L., Bolt, C.R., Andrews, G.P., Welkos, S.L., Friedlander, A.M., and Burans, J.P. 1997. Protection of mice from fatal bubonic and pneumonic plague by passive immunization with monoclonal antibodies against the F1 protein of *Yersinia pestis*. Am. J. Trop. Med. Hyg. 56: 471-473.

Anderson, G.W., Jr., Leary, S.E., Williamson, E.D., Titball, R.W., Welkos, S.L., Worsham, P.L., and Friedlander, A.M. 1996. Recombinant V antigen protects mice against pneumonic and bubonic plague caused by F1-capsule-positive and -negative strains of *Yersinia pestis*. Infect. Immun. 64: 4580-4585.

Andrews, G.P., Heath, D.G., Anderson, G.W., Welkos, S.L., and Friedlander, A.M. 1996. Fraction 1 capsular antigen (F1) purification from *Yersinia pestis* CO92 and from an *Escherichia coli* recombinant strain and efficacy against lethal plague challenge. Infect. Immun. 64: 2180-2187.

Andrews, G.P., Strachan, S.T., Benner, G.E., Sample, A.K., Anderson, G.W., Jr., Adamovicz, J.J., Welkos, S.L., Pullen, J.K., and Friedlander, A.M. 1999. Protective efficacy of recombinant *Yersinia* outer proteins against bubonic plague caused by encapsulated and nonencapsulated *Yersinia pestis*. Infect. Immun. 67: 1533-1537.

Arntzen, L., and Frean, J.A. 1997. The laboratory diagnosis of plague. Belgian J. Zool. 127: 91-96.

Baca-Estrada, M.E., Foldvari, M., Snider, M., Harding, K., Kournikakis, B., Babiuk, L.A., and Griebel, P. 2000. Intranasal immunization with liposome-formulated *Yersinia pestis* vaccine enhances mucosal immune responses. Vaccine 18: 2203-2211.

Bacot, A.W., and Martin, C.J. 1914. Observation on the mechanism of the transmission of plague by fleas. J. Hyg.; plague suppl. 3: 423-439.

Baker, B.W., Sommer, H., Foster, L.E., Meyer, E., and Meyer, K.F. 1952. Studies on immunization against plague. I. The isolation and characterisation of the soluble antigen of *Pasteurella pestis*. J. Immunol. 68: 131-145.

Baker, E.E., Sommer, H., Foster, L.E., and Meyer, K.F. 1947. Antigenic structure of *Pasteurella pestis* and the isolation of a crystalline antigen. Proc. Soc. Exp. Biol. 64: 139-141.

Barnhart, M.M., Pinkner, J.S., Soto, G.E., Sauer, F.G., Langermann, S., Waksman, G., Frieden, C., and Hultgren, S.J. 2000. PapD-like chaperones provide the missing information for folding of pilin proteins. Proc. Natl. Acad. Sci. USA 97: 7709-7714.

Bartelloni, P.J., Marshall, J.D.J., and Cavanaugh, D.C. 1973. Clinical and serological responses to plague vaccine USP. Milit. Med. 138: 720-722.

Benner, G.E., Andrews, G.P., Byrne, W.R., Strachan, S.D., Sample, A.K., Heath, D.G., and Friedlander, A.M. 1999. Immune response to *Yersinia* outer proteins and other *Yersinia pestis* antigens after experimental plague infection in mice. Infect. Immun. 67: 1922-1928.

Bennett, L.G.a.T., T.G. 1974. Characterization of the antigenic subunits of the envelope protein of *Yersinia pestis*. J. Bacteriol. 117: 48-55.

Berglund, J., and Knight, S.D. 2003. Structural basis of bacterial adhesion in the urinary tract. In Glycoimmunology 3. Axford, J.S. (ed).: Kluwer Academic/Plenum Publishers, New York. p. 33-52.

Bibel, D.J., and Chen, T.H. 1976. Diagnosis of Plague: an Analysis of the Yersin-Kitasato Controversy. Bacteriol. Rev. 40: 633-651.

Brandler, P., Saikh, K.U., Heath, D., Friedlander, A., and Ulrich, R.G. 1998. Weak anamnestic responses of inbred mice to *Yersinia* F1 genetic vaccine are overcome by boosting with F1 polypeptide while outbred mice remain nonresponsive. J. Immunol. 161: 4195-4200.

Braun, V. 1995. Energy coupled transport and signal transduction through the gram negative outer membrane via TonB ExbB ExbD dependent receptor proteins. FEMS Microbiol. Rev. 16: 295-307.

Brubaker, B. 2000. *Yersinia pestis* and bubonic plague. In The Prokaryotes: An evolving electronic resource for the microbiology community. Dworkin, M. (ed).Springer-Verlag, New York.

Brubaker, R.R. 1991. Factors promoting acute and chronic diseases caused by *Yersiniae*. Clin. Microbiol. Rev. 4: 309-324.

Brubaker, R.R. 2003. Interleukin-10 and inhibition of innate immunity to yersiniae: Roles of Yops and LcrV (V antigen). Infect. Immun. 71: 3673-3681.

Bullifent, H.L., Griffin, K.F., Jones, S.M., Yates, A., Harrington, L., and Titball, R.W. 2000. Antibody responses to *Yersinia pestis* F1 antigen expressed in *Salmonella typhimurium aroA* from in vivo inducible promoters. Vaccine. 18: 2668-2676.

Burrows, T.W., and Bacon, G.A. 1956. The basis of virulence in *Pasteurella pestis*: An antigen determining virulence. Brit. J. Exp. Path. 37: 481-493.

Burrows, T.W. 1957. Virulence of *Pasteurella pestis*. Nature. 179: 1246-1247.

Butler, T., and Hudson, B.W. 1977. The serological response to *Yersinia pestis* infection. Bulletin World Health Organisation 55: 39-42.

Butler, T. (ed) 1983. Plague and other *Yersinia* infections. New York: Plenum Press.

Campbell, G.L., and Hughes, J.M. 1995. Plague in India: A new warning from an Old Nemesis. Annal. Int. Med. 122: 151-153.

Cao, L.K., Anderson, G.P., Ligler, F.S., and Ezzell, J. 1995. Detection of *Yersinia pestis* fraction-1 antigen with a fiber optic biosensor. J. Clin. Microbiol. 33: 336-341.

Cardenas, L., and Clements, J.D. 1992. Oral immunization using live attenuated *Salmonella* spp. as carriers of foreign antigens. Clin. Microbiol. Rev. 5: 328-342.

Carniel, E. 2003. Evolution of pathogenic *Yersinia*, some lights in the dark. In Genus Yersinia: Entering the Functional Genomic Era. Vol. 529, pp. 3-12.

Carnoy, C., and Moseley, S.L. 1997. Mutational analysis of receptor binding mediated by the Dr family of *Escherichia coli* adhesins. Mol. Microbiol. 23: 365-379.

Cavanaugh, D.C., and Randall, R. 1959. The role of multiplication of *Pasteurella pestis* in mononuclear phagocytes in the pathogenesis of flea-borne plague. J. Immunol. 83: 348-363.

Cavanaugh, D.C., Fortier, M.K., Robinson, D.M., Williams, J.E., and Rust, J., J.H., 1979. Application of the ELISA technique to problems in the serologic diagnosis of plague. Bull. Pan. Am. Health Org. 13: 399-402.

Cavanaugh, D.C., Elisberg, B.L., Llewellyn, C.H. Marshall, Jr., J.D., Rust, Jr., J.H., Williams, J.E. and Meyer, K.F. 1974. Plague Immunization. V. Indirect Evidence for the Efficacy of Plague Vaccines. J. Infect. Dis.129: S37-S40.

Chanteau, S., Ratsifasoamanana, L., Rasoamanana, B., Rahalison, L., Randriambelosoa, J., Roux, J., and Rabeson, D. 1998. Plague, a reemerging disease in Madagascar. Emerg. Infect. Dis. 4: 101-104.

Chanteau, S., Rahalison, L., Ratsitorahina, M., Mahafaly, Rasolomaharo, M., Boisier, P., O'Brien, T., Aldrich, J., Keleher, A., Morgan, C., and Burans, J. 2000a. Early diagnosis of bubonic plague using F1 antigen capture ELISA assay and rapid immunogold dipstick. Int. J. Med. Microbiol. 290: 279-283.

Chanteau, S., Ratsitorahina, M., Rahalison, L., Rasoamanana, B., Chan, F., Boisier, P., Rabeson, D., and Roux, J. 2000b. Current epidemiology of human plague in Madagascar. Microbes Infect. 2: 25-31.
Chanteau, S., Rahalison, L., Ralafiarisoa, L., Foulon, J., Ratsitorahina, M., Ratsifasoamanana, L., Carniel, E., and Nato, F. 2003. Development and testing of a rapid diagnostic test for bubonic and pneumonic plague. Lancet. 361: 211-216.
Chapman, D.A.G., Zavialov, A.V., Chernovskaya, T.V., Karlyshev, A.V., Zav'yalova, G.A., Vasiliev, A.M., Dudich, I.V., Abramov, V.M., Zav'yalov, V.P., and MacIntyre, S. 1999. Structural and functional significance of the FGL sequence of the periplasmic chaperone Caf1M of *Yersinia pestis*. J. Bacteriol. 181: 2422-2429.
Chen, T.H., and Meyer, K.F. 1954. Studies on immunisation against plague VII A haemagglutination test with the protein fraction of *Pasteurella pestis*. J. Immunol. 72: 282-298.
Chen, T.H., Foster, L.E., and Meyer, K.F. 1974. Comparison of the immune response to three different *Yersinia pestis* vaccines in guinea pigs and langurs. J. Infect. Dis. 129: S53-S61.
Chen, T.H., and Erlberg, S.S. 1977. Scanning electron microscopy study of virulent *Yersinia pestis* and *Yersinia pseudotuberculosis* type I. Infect. Immun. 15: 972-977.
Chen, T.H. and Meyer., K.F. 1974. Susceptibility and Immune Response to Experimental Plague in Two Species of Langurs and in African Green (Grivet) Monkeys. J. Infect. Dis.129: S46-S52.
Choudhury, D., Thompson, A., Stojanoff, V., Langermann, S., Pinkner, J., Hultgren, S.J., and Knight, S.D. 1999. X-ray structure of the FimC-FimH chaperone-adhesin complex from uropathogenic *Escherichia coli*. Science. 285: 1061-1066.
Chu, M.C. 2000. Laboratory manual of plague diagnostic tests. World Health Organisation, Geneva.
Coêlho, R.A.L., Santos, G.M.P., Azevedo, P.H.S., Jaques, G.D., Azevedo, W.M., and Carvalho, L.B. 2001. Polyaniline-dacron composite as solid phase in enzyme linked immunosorbent assay for *Yersinia pestis* antibody detection. J. Biomed. Mat. Res. 56: 257-260.
Cornelis, G.R. 2002. *Yersinia* type III secretion: send in the effectors. J. Cell Biol. 158: 401-408.
Cowan, C., Jones, H.A., Kaya, Y.H., Perry, R.D., and Straley, S.C. 2000. Invasion of epithelial cells by *Yersinia pestis:* evidence for a *Y. pestis* specific invasin. Infect. Immun. 68: 4523-4530.
Davis, K.J., Fritz, D.L., Pitt, M.L., Welkos, S.L., Worsham, P.L., and Friedlander, A.M. 1996. Pathology of experimental pneumonic plague produced by Fraction 1 positive and Fraction 1 negative *Yersinia pestis* in African Green Monkeys (*Cercopithecus aethiops*). Arch. Pathol. Lab. Med. 120: 156-163.
Deng, W., Burland, V., Plunkett, G., 3rd, Boutin, A., Mayhew, G.F., Liss, P., Perna, N.T., Rose, D.J., Mau, B., Zhou, S., Schwartz, D.C., Fetherston, J.D., Lindler, L.E., Brubaker, R.R., Plano, G.V., Straley, S.C., McDonough, K.A., Nilles, M.L., Matson, J.S., Blattner, F.R. and Perry, R.D. 2002. Genome sequence of *Yersinia pestis* KIM. J. Bacteriol. 184: 4601-4611.
Drozdov, I.G., Anisimov, A.P., Samoilova, S.V., Yezhov, I.N., Yeremin, S.A., Karlyshev, A.V., Krasilnikova, V.M., and Kravchenko, V.I. 1995. Virulent non-capsulate *Yersinia pestis* variants constructed by insertion mutagenesis. J. Med. Microbiol. 42: 264-268.
Du, Y., Galyov, E., and Forsberg, A. 1995. Genetic analysis of virulence determinants unique to *Yersinia pestis*. In Yersiniosis: Present and future. Vol. 13. Ravagnan, G. and Chiesa, C. (eds). Karger, Basel. p. 321-324.
Du, Y.D., Rosqvist, R., and Forsberg, A. 2002. Role of Fraction 1 antigen of *Yersinia pestis* in inhibition of phagocytosis. Infect. Immun. 70: 1453-1460.
Engelthaler, D.M., Gage, K.L., Montenieri, J.A., Chu, M., and Carter, L.G. 1999. PCR detection of *Yersinia pestis* in fleas: Comparison with mouse inoculation. J. Clin. Microbiol. 37: 1980-1984.
Engelthaler, D.M., Hinnebusch, B.J., Rittner, C.M., and Gage, K.L. 2000. Quantitative competitive PCR as a technique for exploring flea - *Yersinia pestis* dynamics. Am. J. Trop. Med. Hyg. 62: 552-560.
Englesberg, E., and Levy, J.B. 1954. Studies on the immunisation against plague. VI. Growth of *Pasteurella pestis* and the production of the envelope and other soluble antigens in a casein hydrolysate mineral glucose medium. J. Bacteriol. 67: 438-449.
Eyles, J.E., Spiers, I.D., Williamson, E.D., and Alpar, H.O. 1998. Analysis of local and systemic immunological responses after intra-tracheal, intra-nasal and intra-muscular administration of microsphere co-encapsulated *Yersinia pestis* sub-unit vaccines. Vaccine. 16: 2000-2009.
Eyles, J.E., Williamson, E.D., Spiers, I.D., and Alpar, H.O. 2000. Protection studies following bronchopulmonary and intramuscular immunisation with *Yersinia pestis* F1 and V subunit vaccines coencapsulated in biodegradable microspheres: a comparison of efficacy. Vaccine. 18: 3266-3271.
Fallman, M., Andersson, K., Hakansson, S., Magnusson, K.E., Stendahl, O., and Wolfwatz, H. 1995. *Yersinia pseudotuberculosis* inhibits Fc receptor mediated phagocytosis in J774 cells. Infect. Immun. 63: 3117-3124.
Friedlander, A.M., Welkos, S.L., Worsham, P.L., Andrews, G.P., Heath, D.G., Anderson, G.W., Pitt, M.L.M., Estep, J., and Davis, K. 1995. Relationship between virulence and immunity as revealed in recent studies of the F1 capsule of *Yersinia pestis*. Clin. Infect. Dis. 21: S178-S181.

Gabastou, J.M., Proano, J., Vimos, A., Jaramillo, G., Hayes, E., Gage, K., Chu, M., Guarner, J., Zaki, S., Bowers, J., Guillemard, C., Tamayo, H., and Ruiz, A. 2000. An outbreak of plague including cases with probable pneumonic infection, Ecuador, 1998. Trans. Royal Soc. Trop. Med. Hyg. 94: 387-391.

Gage, K.L., Dennis, D.T., and Tsai, T.F. 1996. Prevention of Plague: Recommendations of the Advisory Committee on immunisation Practices (ACIP). Morbid. Mortal. Week. Rep. 45: 1-15.

Galimand, M., Guiyoule, A., Gerbaud, G., Rasoamanana, B., Chanteau, S., Carniel, E., and Courvalin, P. 1997. Multidrug resistance in *Yersinia pestis* mediated by a transferable plasmid. N. Engl. J. Med. 337: 677-680.

Galyov, E.E., Smirnov, O.Y., Karlishev, A.V., Volkovoy, K.I., Denesyuk, A.J., Nazimov, I.V., Rubtsov, K.S., Abramov, V.M., Dalvadyanz, S.M., and Zav'yalov, V.P. 1990. Nucleotide sequence of the *Yersinia pestis* gene encoding F1 antigen and the primary structure of the protein putative T-cell and B-cell epitopes. FEBS Lett. 277: 230-232.

Galyov, E.E., Karlishev, A.V., Chernovskaya, T.V., Dolgikh, D.A., Smirnov, O.Y., Volkovoy, K.I., Abramov, V.M., and Zav'yalov, V.P. 1991. Expression of the envelope antigen F1 of *Yersinia pestis* is mediated by the product of *caf1M* gene having homology with the chaperone protein PapD of *Escherichia coli*. FEBS Lett. 286: 79-82.

Garmory, H.S., Griffin, K.F., Brown, K.A., and Titball, R.W. 2003. Oral immunisation with live *aroA* attenuated *Salmonella enterica* serovar Typhimurium expressing the *Yersinia pestis* V antigen protects mice against plague. Vaccine. 21: 3051-3057.

Girard, G. 1933. L'immunité dans l'infection pesteuse. Biol. Med. 631.

Glosnicka, R. and Gruszkiewicz, E. 1980. Chemical composition and biological activity of the *Yersinia pestis* envelope substance. Infect. Immun. 30: 506-512.

Green, M., Rogers, D., Russell, P., Stagg, A.J., Bell, D.L., Eley, S.M., Titball, R.W., and Williamson, E.D. 1999. The SCID/Beige mouse as a model to investigate protection against *Yersinia pestis*. FEMS Immunol. Med. Microbiol. 23: 107-113.

Grosfeld, H., Cohen, S., Bino, T., Flashner, Y., Ber, R., Mamroud, E., Kronman, C., Shafferman, A., and Velan, B. 2003. Effective protective immunity to *Yersinia pestis* infection conferred by DNA vaccine coding for derivatives of the F1 capsular antigen. Infect. Immun. 71: 374-383.

Guarner, J., Shieh, W.J., Greer, P.W., Gabastou, J.M., Chu, M., Hayes, E., Nolte, K.B., and Zaki, S.R. 2002. Immunohistochemical detection of *Yersinia pestis* in formalin- fixed, paraffin-embedded tissue. Am. J. Clin. Pathol. 117: 205-209.

Guiyoule, A., Gerbaud, G., Buchrieser, C., Galimand, M., Rahalison, L., Chanteau, S., Courvalin, P., and Carniel, E. 2001. Transferable plasmid-mediated resistance to streptomycin in a clinical isolate of *Yersinia pestis*. Emerg. Infect. Dis. 7: 43-48.

Haffkine, W.M. 1897. The plague prophylactic fluid. Br. Med. J. 1:1461-1462.

Hallett, A.F., Isaacson, M., and Meyer, K.F. 1973. Pathogenicity and immunogenic efficacy of a live attenuated plague vaccine in Vervet monkeys. Infect. Immun. 8: 876-881.

Heath, D.G., Anderson, G.W., Mauro, J.M., Welkos, S.L., Andrews, G.P., Adamovicz, J., and Friedlander, A.M. 1998. Protection against experimental bubonic and pneumonic plague by a recombinant capsular F1-V antigen fusion protein vaccine. Vaccine. 16: 1131-1137.

Higgins, J.A., Ezzell, J., Hinnebusch, B.J., Shipley, M., Henchal, E.A., and Ibrahim, M.S. 1998. 5' nuclease PCR assay to detect *Yersinia pestis*. J. Clin. Microbiol. 36: 2284-2288.

Hill, J., Copse, C., Leary, S., Stagg, A.J., Williamson, E.D., and Titball, R.W. 2003. Synergistic protection of mice against plague with monoclonal antibodies specific for the F1 and V antigens of *Yersinia pestis*. Infect. Immun. 71: 2234-2238.

Hinnebusch, B.J., Rosso, M.L., Schwan, T.G., and Carniel, E. 2002. High-frequency conjugative transfer of antibiotic resistance genes to *Yersinia pestis* in the flea midgut. Mol. Microbiol. 46: 349-354.

Hinnebusch, J., and Schwan, T.G. 1993. New method for plague surveillance using polymerase chain reaction to detect *Yersinia pestis* in fleas. J. Clin. Microbiol. 31: 1511-1514.

Hitchen, P.G., Prior, J.L., Oyston, P.C.F., Panico, M., Wren, B.W., Titball, R.W., Morris, H.R., and Dell, A. 2002. Structural characterization of lipo-oligosaccharide (LOS) from *Yersinia pestis*: regulation of LOS structure by the PhoPQ system. Mol. Microbiol. 44: 1637-1650.

Holmgren, A., Kuehn, M.J., Branden, C.I., and Hultgren, S.J. 1992. Conserved Immunoglobulin-Like Features in a Family of Periplasmic Pilus Chaperones in Bacteria. EMBO J. 11: 1617-1622.

Horuk, R. 1999. Chemokine receptors and HIV-1: the fusion of two major research fields. Immunol. Today. 20: 89-94.

Hu, P., Elliott, J., McCready, P., Skowronski, E., Garnes, J., Kobayashi, A., Brubaker, R.R., and Garcia, E. 1998. Structural organization of virulence-associated plasmids of *Yersinia pestis*. J. Bacteriol. 180: 5192-5202.

Hung, D.L., Knight, S.D., Woods, R.M., Pinkner, J.S., and Hultgren, S.J. 1996. Molecular basis of two subfamilies of immunoglobulin-like chaperones. EMBO J. 15: 3792-3805.

Hung, D.L., Raivio, T.L., Jones, C.H., Silhavy, T.J., and Hultgren, S.J. 2001. Cpx signaling pathway monitors biogenesis and affects assembly and expression of P pili. EMBO J. 20: 1508-1518.

Inglesby, T.V., Dennis, D.T., Henderson, D.A., Bartlett, J.G., Ascher, M.S., Eitzen, E., Fine, A.D., Friedlander, A.M., Hauer, J., Koerner, J.F., Layton, M., McDade, J., Osterholm, M.T., O'Toole, T., Parker, G., Perl, T.M., Russell, P.K., Schoch-Spana, M., and Tonat, K. 2000. Plague as a biological weapon - Medical and public health management. J. Am. Med. Ass. 283: 2281-2290.

Iqbal, S.S., Chambers, J.P., Goode, M.T., Valdes, J.J., and Brubaker, R.R. 2000. Detection of *Yersinia pestis* by pesticin fluorogenic probe- coupled PCR. Mol. Cell. Probes 14: 109-114.

Jacob-Dubuisson, F., Striker, R., and Hultgren, S.J. 1994. Chaperone assisted self assembly of pili independent of cellular energy. J. Biol. Chem. 269: 12447-12455.

Jefferson, T., Demicheli, V. and Pratt, M. 2002. Vaccines for preventing plague. Cochrane Review.

Jones, C.H., Danese, P.N., Pinkner, J.S., Silhavy, T.J., and Hultgren, S.J. 1997. The chaperone-assisted membrane release and folding pathway is sensed by two signal transduction systems. EMBO J. 16: 6394-6406.

Jones, S.M., Day, F., Stagg, A.J., and Williamson, E.D. 2000. Protection conferred by a fully recombinant sub-unit vaccine against *Yersinia pestis* in male and female mice of four inbred strains. Vaccine. 19: 358-366.

Karlyshev, A.V., Galyov, E.E., Abramov, V.M., and Zav'yalov, V.P. 1992a. Caf1R gene and its role in the regulation of capsule formation of *Y. pestis*. FEBS Lett. 305: 37-40.

Karlyshev, A.V., Galyov, E.E., Smirnov, O.Y., Guzayev, A.P., Abramov, V.M., and Zav'yalov, V.P. 1992b. A new gene of the Fl operon of *Y. pestis* involved in the capsule biogenesis. FEBS Lett. 297: 77-80.

Karlyshev, A.V., Galyov, E.E., Smirnov, O., Abramov, V., and Zav'yalov, V.P. 1994. Structure and regulation of a gene cluster involved in capsule formation of *Yersinia pestis*. In Biological Membranes: Structure, Biogenesis and Dynamics. NATO ASI Series. Vol. H 82. Op den Kamp, J.A.F. (ed). Springer-Verlag, Berlin-Heidelberg. p. 321-330.

Kawahara, K., Tsukano, H., Watanabe, H., Lindner, B., and Matsuura, M. 2002. Modification of the structure and activity of lipid A in *Yersinia pestis* lipopolysaccharide by growth temperature. Infect. Immun. 70: 4092-4098.

Kitasato, S. 1894. The bacillus of bubonic plague. The Lancet: 428-431.

Knight, S.D., Berglund, J., and Choudhury, D. 2000. Bacterial adhesins: structural studies reveal chaperone function and pilus biogenesis. Curr. Opin. Chem. Biol. 4: 653-660.

Koronakis, V., Andersen, C., and Hughes, C. 2001. Channel tunnels. Curr. Opin. Struct. Biol. 11: 403-407.

Krakauer, T., and Heath, D. 1998. Lack of IL-1 receptor antagonistic activity of the capsular F1 antigen of *Yersinia pestis*. Immunol. Lett. 60: 137-142.

Kutyrev, V.V., Filippov, Y.A., Shavina, N.I., and Protsenko, O.A. 1989. Genetic analysis and modelling of the virulence of *Yersinia pestis* [in Russian]. Mol. Gen. Mikrobiol. Virusol. 8: 42-47.

Lalioui, L., Jouve, M., Gounon, P., and Le Bouguenec, C. 1999. Molecular cloning and characterization of the *afa*-7 and *afa*-8 gene clusters encoding afimbrial adhesins in *Escherichia coli* strains associated with diarrhea or septicemia in calves. Infect. Immun. 67: 5048-5059.

Leary, S.E.C., Williamson, E.D., Griffin, K.F., Russell, P., Eley, S.M., and Titball, R.W. 1995. Active immunization with recombinant V antigen from *Yersinia pestis* protects mice against plague. Infect. Immun. 63: 2854-2858.

Leary, S.E.C., Griffin, K.F., Garmory, H.S., Williamson, E.D., and Titball, R.W. 1997. Expression of an F1/V fusion protein in attenuated *Salmonella typhimurium* and protection of mice against plague. Microb. Pathog. 23: 167-179.

Li, Z.S., Clarke, A.J., and Beveridge, T.J. 1998. Gram-negative bacteria produce membrane vesicles which are capable of killing other bacteria. J. Bacteriol. 180: 5478-5483.

Lindler, L.E., and Tall, B.D. 1993. *Yersinia pestis* pH-6 antigen forms fimbriae and is induced by intracellular association with macrophages. Mol. Microbiol. 8: 311-324.

Lindler, L.E., Plano, G.V., Burland, V., Mayhew, G.F., and Blattner, F.R. 1998. Complete DNA sequence and detailed analysis of the *Yersinia pestis* KIM5 plasmid encoding murine toxin and capsular antigen. Infect. Immun. 66: 5731-5742.

Lindler, L.E., Fan, W., and Jahan, N. 2001. Detection of ciprofloxacin-resistant *Yersinia pestis* by fluorogenic PCR using the lightcycler. J. Clin. Microbiol. 39: 3649-3655.

Lintermans, P.F., Bertels, A., Schlicker, C., Deboeck, F., Charlier, G., Pohl, P., Norgren, M., Normark, S., van Montagu, M., and De Greve, H. 1991. Identification, characterization, and nucleotide sequence of the F17-G gene, which determines receptor binding of *Escherichia coli* F17 fimbriae. J. Bacteriol. 173: 3366-3373.

Liu, R., Paxton, W.A., Choe, S., Ceradini, D., Martin, S.R., Horuk, R., MacDonald, M.E., Stuhlmann, H., Koup, R.A., and Landau, N.R. 1996. Homozygous defect in HIV-1 coreceptor accounts for resistance of some multiply-exposed individuals to HIV-1 infection. Cell. 86: 367-377.

MacIntyre, S., Zyrianova, I.M., Chernovskaya, T.V., Leonard, M., Rudenko, E.G., Zav'yalov, V.P., and Chapman, D.A.G. 2001. An extended hydrophobic interactive surface of *Yersinia pestis* Caf1M chaperone is essential for subunit binding and F1 capsule assembly. Mol. Microbiol. 39: 12-25.

Makoveichuk, E., Cherepanov, P., Lundberg, S., Forsberg, A., and Olivecrona, G. 2003. pH6 antigen of *Yersinia pestis* interacts with plasma lipoproteins and cell membranes. J. Lipid Res. 44: 320-330.

Marshall, J.D., Bartelloni, P.J., Cavanaugh, D.C., Kadhull, P.J. and Meyer, K.F. 1974. Plague Immunization. II. Relation of adverse clinical reactions to multiple immunizations with killed vaccine. J. Infect. Dis. 129: S19-S25.

Marshall, J.D., Cavanaugh, D.C., Bartelloni, P.J., and Meyer, K.F. 1974. Plague immunization. III. Serologic response to multiple inoculations of vaccine. J. Infect. Dis. 129: S26-S29.

McDonald, R., Cao, T., and Borschel, R. 2001. Multiplexing for the detection of multiple biowarfare agents shows promise in the field. Milit. Med. 166: 237-239.

Meka-Mechenko, T.V. 2003. F1-negative natural *Y. pestis* strains. Adv. Exp. Med. Biol. 529: 379-381.

Meyer, K.F. 1970. Effectiveness of live or killed plague vaccines in man. Bull. World Health Org. 42: 653-666.

Meyer, K.F., Cavanaugh, D.C., Bartelloni, P.J., and Marshall, J.D. 1974a. Plague immunization. I. Past and present trends. J. Infect. Dis. 129: S13-S18.

Meyer, K.F., Hightower, J.A., and McCrumb, F.R. 1974b. Plague immunisation. VI. Vaccination with the F1 antigen of *Yersinia pestis*. J. Infect. Dis. 129: S41-S45.

Meyer, K.F., Smith, G., Foster, L., Brookman, M., and Sung, M. 1974c. Live, attenuated *Yersinia pestis* vaccine: virulent in nonhuman primates, harmless to guinea pigs. J. Infect. Dis. 129: S85-S120.

Meyer, K.F., Smith, G., Foster, L.E., Marshall, J., J.D. , and Cavanaugh, D.C. 1974d. Plague immunization. IV. Clinical reactions and serologic response to inoculations of Haffkine and freeze-dried plague vaccine. J. Infect. Dis. 129: S30-S36.

Miller, J., Williamson, E.D., Lakey, J.H., Pearce, M.J., Jones, S.M., and Titball, R.W. 1998. Macromolecular organisation of recombinant *Yersinia pestis* F1 antigen and the effect of structure on immunogenicity. FEMS Immunol. Med. Microbiol. 21: 213-221.

Munier-Lehmann, H., Chenal-Francisque, V., Ionescu, M., Christova, P., Foulon, J., Carniel, E. and Barzu, O. 2003. Relationship between bacterial virulence and nucleotide metabolism: a mutation in the adenylate kinase gene renders *Yersinia pestis* avirulent. Biochem. J. 373: 515-522.

Nakajima, R., and Brubaker, R.R. 1993. Association between virulence of *Yersinia pestis* and suppression of gamma interferon and tumour necrosis factor alpha. Infect. Immun. 61: 21-31.

Neubauer, H., Meyer, H., Prior, J., Aleksic, S., Hensel, A., and Splettstosser, W. 2000. A combination of different polymerase chain reaction (PCR) assays for the presumptive identification of *Yersinia pestis*. J. Vet. Med. 47: 573-580.

Ng, L.C., Forslund, O., Koh, S., Kuoppa, K., and Sjosted, A. 2003. The response of murine macrophages to infection with *Yersinia pestis* as revealed by DNA microarray analysis. Adv. Exp. Med Biol. 529: 155-160.

Norkina, O.V., Kulichenko, A.N., Gintsburg, A.L., Tuchkov, I.V., Popov, Y.A., Aksenov, M.U., and Drosdov, I.G. 1994. Development of a diagnostic test for *Yersinia pestis* by the polymerase chain reaction. J. Appl. Bacteriol. 76: 240-245.

Osorio, J.E., Powell, T.D., Frank, R.S., Moss, K., Haanes, E.J., Smith, S.R., Rocke, T.E., and Stinchcomb, D.T. 2003. Recombinant raccoon pox vaccine protects mice against lethal plague. Vaccine. 21: 1232-1238.

Otten, L. 1941. Live plague vaccine and results. Mededeeliagen van den dienst der Volksgezondheid in Nederlandsch-Indië 30: 61-110.

Oyston, P.C.F., Williamson, E.D., Leary, S.E.C., Eley, S.M., Griffin, K.F., and Titball, R.W. 1995. Immunization with live recombinant *Salmonella typhimurium aroA* producing F1 antigen protects against plague. Infect. Immun. 63: 563-568.

Panda, S.K., Nanda, S.K., Ghosh, A., Sharma, C., Shivaji, S., Kumar, G.S., Kannan, K., Batra, H.V., Tuteja, U., Ganguly, N.K., Chakrabarty, A., and Chandra, H.S. 1996. The 1994 plague epidemic of India: Molecular diagnosis and characterization of *Yersinia pestis* isolates from Surat and Beed. Curr. Science. 71: 794-799.

Parkhill, J., Wren, B.W., Thomson, N.R., Titball, R.W., Holden, M.T.G., Prentice, M.B., Sebaihia, M., James, K.D., Churcher, C., Mungall, K.L., Baker, S., Basham, D., Bentley, S.D., Brooks, K., Cerdeno-Tarraga, A.M., Chillingworth, T., Cronin, A., Davies, R.M., Davis, P., Dougan, G., Feltwell, T., Hamlin, N., Holroyd, S., Jagels, K., Karlyshev, A.V., Leather, S., Moule, S., Oyston, P.C.F., Quail, M., Rutherford, K., Simmonds, M., Skelton, J., Stevens, K., Whitehead, S., and Barrell, B.G. 2001. Genome sequence of *Yersinia pestis*, the causative agent of plague. Nature. 413: 523-527.

Payne, D., Tatham, D., Williamson, E.D., and Titball, R.W. 1998. The pH 6 antigen of *Yersinia pestis* binds to beta 1-linked galactosyl residues in glycosphingolipids. Infect. Immun. 66: 4545-4548.

Perry, R.D., and Fetherston, J.D. 1997. *Yersinia pestis* - Etiologic agent of plague. Clin. Microbiol. Rev. 10: 35-66.

Pettersson, J., Holmstrom, A., Hill, J., Leary, S., Frithz-Lindsten, E., von Euler-Matell, A., Carlsson, E., Titball, R., Forsberg, A., and Wolf-Watz, H. 1999. The V-antigen of *Yersinia* is surface exposed before target cell contact and involved in virulence protein translocation. Mol. Microbiol. 32: 961-976.

Podladchikova, O.N., Rykova, V.A., Ivanova, V.S., Eremenko, N.S., and Lebedeva, S.A. 2002. Study of Pgm mutation mechanism in *Yersinia pestis* (plague pathogen) vaccine strain EV76 (Article in Russian). Mol. Gen. Mikrobiol. Virusol. 2: 14-19.

Poland, J.D. and Dennis, D.T. 1999a. Diagnosis and clinical manifestations. In Plague Manual: Epidemiology, Distribution, Surveillance and Control. World Health Organisation, Geneva. p. 43-53.

Poland, J.D. and Dennis, D.T. 1999b. Treatment of plague. In: Plague Manual: Epidemiology, Distribution, Surveillance and Control. World Health Organisation: Geneva. p. 55-62.

Prentice, M.B., James, K.D., Parkhill, J., Baker, S.G., Stevens, K., Simmonds, M.N., Mungall, K.L., Churcher, C., Oyston, P.C.F., Titball, R.W., Wren, B.W., Wain, J., Pickard, D., Hien, T.T., Farrar, J.J., and Dougan, G. 2001. *Yersinia pestis* pFra shows biovar-specific differences and recent common ancestry with a *Salmonella enterica* serovar Typhi plasmid. J. Bacteriol. 183: 2586-2594.

Protsenko, O.A., Filippov, A.A., and Kutyrev, V.V. 1991. Integration of the plasmid encoding the synthesis of capsular antigen and murine toxin into *Yersinia pestis* chromosome. Microb. Path. 11: 123-128.

Radnedge, L., Gamez-Chin, S., McCready, P.M., Worsham, P.L., and Andersen, G.L. 2001. Identification of nucleotide sequences for the specific and rapid detection of *Yersinia pestis*. Appl. Environ. Microbiol. 67: 3759-3762.

Radnedge, L., Agron, P.G., Worsham, P.L., and Andersen, G.L. 2002. Genome plasticity in *Yersinia pestis*. Microbiol. 148: 1687-1698.

Rahalison, L., Vololonirina, E., Ratsitorahina, M., and Chanteau, S. 2000. Diagnosis of bubonic plague by PCR in Madagascar under field conditions. J. Clin. Microbiol. 38: 260-263.

Ramalingaswami, V. 1995. Plague in India. Nat. Med. 1: 1237-1239.

Rasoamanana, B., Leroy, F., Boisier, P., Rasolomaharo, M., Buchy, P., Carniel, E., and Chanteau, S. 1997. Field evaluation of an immunoglobulin G anti-F1 enzyme-linked immunosorbent assay for serodiagnosis of human plague in Madagascar. Clin. Diag. Lab. Immunol. 4: 587-591.

Ratsitorahina, M., Chanteau, S., Rahalison, L., Ratsifasoamanana, L., and Boisier, P. 2000. Epidemiological and diagnostic aspects of the outbreak of pneumonic plague in Madagascar. Lancet 355: 111-113.

Reddin, K.M., Easterbrook, T.J., Eley, S.M., Russell, P., Mobsby, V.A., Jones, D.H., Farrar, G.H., Williamson, E.D., and Robinson, A. 1998. Comparison of the immunological and protective responses elicited by microencapsulated formulations of the F1 antigen from *Yersinia pestis*. Vaccine. 16: 761-767.

Reyes-Sandoval, A., and Ertl, H.C.J. 2001. DNA Vaccines. Curr. Mol. Med. 1: 217-243.

Rowe, C.A., Scruggs, S.B., Feldstein, M.J., Golden, J.P., and Ligler, F.S. 1999. An array immunosensor for simultaneous detection of clinical analytes. Anal. Chem. 71: 433-439.

Russell, P., Eley, S.M., Hibbs, S.E., Manchee, R.J., Stagg, A.J., and Titball, R.W. 1995. A comparison of plague vaccine, USP and EV76 vaccine-induced protection against *Yersinia pestis* in a murine model. Vaccine. 13: 1551-1556.

Sabhnani, L., and Rao, D.N. 2000. Identification of immunodominant epitope of F1 antigen of *Yersinia pestis*. FEMS Immunol. Med. Microbiol. 27: 155-162.

Sakellaris, H., and Scott, J.R. 1998. New tools in an old trade: CS1 pilus morphogenesis. Mol. Microbiol. 30: 681-687.

Samson, M., Libert, F., Doranz, B.J., Rucker, J., Liesnard, C., Farber, C.M., Saragosti, S., Lapoumeroulie, C., Cognaux, J., Forceille, C., Muyldermans, G., Verhofstede, C., Burtonboy, G., Georges, M., Imai, T., Rana, S., Yi, Y.J., Smyth, R.J., Collman, R.G., Doms, R.W., Vassart, G., and Parmentier, M. 1996. Resistance to HIV-1 infection in Caucasian individuals bearing mutant alleles of the CCR-5 chemokine receptor gene. Nature. 382: 722-725.

Sauer, F.G., Futterer, K., Pinkner, J.S., Dodson, K.W., Hultgren, S.J., and Waksman, G. 1999. Structural basis of chaperone function and pilus biogenesis. Science. 285: 1058-1061.

Sauer, F.G., Barnhart, M., Choudhury, D., Knights, S.D., Waksman, G., and Hultgren, S.J. 2000a. Chaperone-assisted pilus assembly and bacterial attachment. Curr. Opin. Struct. Biol. 10: 548-556.

Sauer, F.G., Knight, S.D., Waksman, G., and Hultgren, S.J. 2000b. PapD-like chaperones and pilus biogenesis. Semin. Cell Dev. Biol. 11: 27-34.

Sauer, F.G., Pinkner, J.S., Waksman, G., and Hultgren, S.J. 2002. Chaperone priming of pilus subunits facilitates a topological transition that drives fiber formation. Cell. 111: 543-551.

Schneerson, R., Kubler-Kielb, J., Liu, T.Y., Dai, Z.D., Leppla, S.H., Yergey, A., Backlund, P., Shiloach, J., Majadly, F., and Robbins, J.B. 2003. Poly(gamma-D-glutamic acid) protein conjugates induce IgG antibodies in mice to the capsule of *Bacillus anthracis*: A potential addition to the anthrax vaccine. Proc. Natl. Acad. Sci. USA. 100: 8945-8950.

Simpson, W.J., Thomas, R.E., and Schwan, T.G. 1990. Recombinant capsular antigen (Fraction 1) from *Yersinia pestis* induces a protective antibody response in BALB/c mice. Am. J. Trop. Med. Hyg. 43: 389-396.

Skurnik, M., Peippo, A., and Ervela, E. 2000. Characterization of the O-antigen gene clusters of *Yersinia pseudotuberculosis* and the cryptic O-antigen gene cluster of *Yersinia pestis* shows that the plague bacillus is most closely related to and has evolved from *Y. pseudotuberculosis* serotype O : 1b. Mol. Microbiol. 37: 316-330.

Smith, D.R., Rossi, C.A., Kijek, T.M., Henchal, E.A., and Ludwig, G.V. 2001. Comparison of dissociation-enhanced lanthanide fluorescent immunoassays to enzyme-linked immunosorbent assays for detection of staphylococcal enterotoxin B, *Yersinia pestis*- specific F1 antigen, and Venezuelan equine encephalitis virus. Clin. Diag. Lab. Immunol. 8: 1070-1075.

Sodeinde, O.A., Subrahmanyam, Y., Stark, K., Quan, T., Bao, Y.D., and Goguen, J.D. 1992. A surface protease and the invasive character of plague. Science. 258: 1004-1007.

Stephens, J.C., Reich, D.E., Goldstein, D.B., Shin, H.D., Smith, M.W., Carrington, M., Winkler, C., Huttley, G.A., Allikmets, R., Schriml, L., Gerrard, B., Malasky, M., Ramos, M.D., Morlot, S., Tzetis, M., Oddoux, C., di Giovine, F.S., Nasioulas, G., Chandler, D., Aseev, M., Hanson, M., Kalaydjieva, L., Glavac, D., Gasparini, P., Kanavakis, E., Claustres, M., Kambouris, M., Ostrer, H., Duff, G., Baranov, V., Sibul, H., Metspalu, A., Goldman, D., Martin, N., Duffy, D., Schmidtke, J., Estivill, X., O'Brien, S.J., and Dean, M. 1998. Dating the origin of the CCR5-Delta 32 AIDS-resistance allele by the coalescence of haplotypes. Am. J. Human Genet. 62: 1507-1515.

Stevenson, H.L., Bai, Y., Kosoy, M.Y., Montenieri, J.A., Lowell, J.L., Chu, M.C., and Gage, K.L. 2003. Detection of novel *Bartonella* strains and *Yersinia pestis* in prairie dogs and their fleas (*Siphonaptera*: *Ceratophyllidae* and *Pulicidae*) using multiplex polymerase chain reaction. J. Med. Entomol. 40: 329-337.

Strong, R.P. 1908. Protective inoculation against plague. J. Med. Res. 18: 325-346.

Taylor, J. 1933. Haffkine's plague vaccine. Indian Med. Res. Memoirs. 27: 1-125.

Thanassi, D.G., Saulino, E.T., and Hultgren, S.J. 1998a. The chaperone/usher pathway: a major terminal branch of the general secretory pathway. Curr. Opin. Microbiol. 1: 223-231.

Thanassi, D.G., Saulino, E.T., Lombardo, M.J., Roth, R., Heuser, J., and Hultgren, S.J. 1998b. The PapC usher forms an oligomeric channel: Implications for pilus biogenesis across the outer membrane. Proc. Natl. Acad. Sci. USA. 95: 3146-3151.

Thanassi, D.G. 2002. Ushers and secretins: Channels for the secretion of folded proteins across the bacterial outer membrane. J. Mol. Microbiol. Biotechnol. 4: 11-20.

Tikhomirow, E. 1999. Epidemiology and distribution of plague. In Plague Manual: Epidemiology, Distribution, Surveillance and Control. World Health Organisation, Geneva. p. 11-41.

Tishchenko, V.M. 2002. Investigation of the domain structure of Caf1M from *Yersinia pestis*. Structure and role of some domains. Biofizika 47: 228-235.

Titball, R.W., Howells, A.M., Oyston, P.C.F., and Williamson, E.D. 1997. Expression of the *Yersinia pestis* capsular antigen F1 antigen on the surface of an aroA mutant of *Salmonella typhimurium* induces high levels of protection against plague. Infect. Immun. 65: 1926-1930.

Titball, R.W., and Williamson, E.D. 2001. Vaccination against bubonic and pneumonic plague. Vaccine. 19: 4175-4184.

Titball, R.W., and Williamson, E.D. 2003. Second and third generation plague vaccines. Adv. Exp. Med Biol. 529: 397-406.

Tito, M.A., Miller, J., Griffin, K.F., Williamson, E.D., Titball, R.W., and Robinson, C.V. 2001. Macromolecular organization of the *Yersinia pestis* capsular F1 antigen: insights from time-of-flight mass spectrometry. Protein Sci. 10: 2408-2413.

Tsukano, H., Itoh, K., Suzuki, S., and Watanabe, H. 1996. Detection and identification of *Yersinia pestis* by polymerase chain reaction (PCR) using multiplex primers. Microbiol. Immunol. 40: 773-775.

Vetsch, M., Sebbel, P., and Glockshuber, R. 2002. Chaperone-independent folding of type 1 pilus domains. J. Mol. Biol. 322: 827-840.

Vorontsov, E.D., Dubichev, A.G., Serdobintsev, L.N., and Naumov, A.V. 1990. Association-dissociation processes and supermolecular organisation of the capsule antigen (protein F1) of *Yersinia pestis*. Biomed. Sci. 1: 391-396.

Wadkins, R.M., Golden, J.P., Pritsiolas, L.M., and Ligler, F.S. 1998. Detection of multiple toxic agents using a planar array immunosensor. Biosens. Bioelect. 13: 407-415.

Weeks, S., Hill, J., Friedlander, A., and Welkos, S. 2002. Anti-V antigen antibody protects macrophages from *Yersinia pestis* induced cell death and promotes phagocytosis. Microb. Pathog. 32: 227-237.

Welkos, S., Pitt, M.L.M., Martinez, M., Friedlander, A., Vogel, P., and Tammariello, R. 2002. Determination of the virulence of the pigmentation-deficient and pigmentation-/plasminogen activator-deficient strains of *Yersinia pestis* in non-human primate and mouse models of pneumonic plague. Vaccine. 20: 2206-2214.

Welkos, S.L., Davis, K.M., Pitt, L.M., Worsham, P.L., and Friedlander, A.M. 1995. Studies on the contribution of the F1 capsule-associated plasmid pFra to the virulence of *Yersinia pestis*. In Yersiniosis: present and future. Vol. 13. Basel: Karger, pp. 299-305.

WHO 1994. Weekly Epidemiological Record. World Health Organisation, Geneva. p. 289-291.

WHO 1995. Weekly Epidemiological Record. World Health Organisation, Geneva. p. 35.

WHO 2002. Plague: Fact sheet. World Health Organisation, Geneva.

Williams, E.W., Arntzen, L., Tyndal, G.L., and Isaacson, M. 1986. Application of enzyme immunoassays for the confirmation of clinically suspect plague in Namibia, 1982. Bull. World Health Org. 64: 745-752.

Williams, J.E., and Cavanaugh, D.C. 1979. Measuring the efficacy of vaccination in affording protection against plague. Bull. World Health Org. 57: 309-313.

Williams, J.E., Arntzen, L., Robinson, D.M., Cavanaugh, D.C., and Isaacson, M. 1982. Comparison of passive haemagglutination and enzyme-linked immunosorbent assay for serodiagnosis of plague. Bull. World Health Org. 60(5): 777-781.

Williams, J.E., and Cavanaugh, D.C. 1984. Potential for rat plague from nonencapsulated variants of the plague bacillus (*Yersinia pestis*). Experientia 40: 739-740.

Williams, R.C., Gewurz, H., and Quie, P.G. 1972. Effects of Fraction I from *Yersinia pestis* on phagocytosis in vitro. J. Infect. Dis. 126: 235-241.

Williamson, E.D., Sharp, G.J.E., Eley, S.M., Vesey, P.M., Pepper, T.C., Titball, R.W., and Alpar, H.O. 1996. Local and systemic immune response to a microencapsulated sub-unit vaccine for plague. Vaccine. 14: 1613-1619.

Williamson, E.D., Eley, S.M., Stagg, A.J., Green, M., Russell, P., and Titball, R.W. 1997. A sub-unit vaccine elicits IgG in serum, spleen cell cultures and bronchial washings and protects immunized animals against pneumonic plague. Vaccine. 15: 1079-1084.

Williamson, E.D., Vesey, P.M., Gillhespy, K.J., Eley, S.M., Green, M., and Titball, R.W. 1999. An IgG1 titre to the F1 and V antigens correlates with protection against plague in the mouse model. Clin. Exp. Immunol. 116: 107-114.

Williamson, E.D., Eley, S.M., Stagg, A.J., Green, M., Russell, P., and Titball, R.W. 2000. A single dose sub-unit vaccine protects against pneumonic plague. Vaccine. 19: 566-571.

Williamson, E.D., Bennett, A.M., Perkins, S.D., Beedham, R.J., Miller, J., and Baillie, L.W.J. 2002. Co-immunisation with a plasmid DNA cocktail primes mice against anthrax and plague. Vaccine. 20: 2933-2941.

Winter, C.C., Cherry, W.B. and Moody, M.D. 1960. An unusual strain of *Pasteurella pestis* isolated from a fatal human case of plague. Bull. World Health Org. 23: 408-409.

Worsham, P.L., Stein, M.-P., and Welkos, S.L. 1995. Construction of defined F1 negative mutants of virulent *Yersinia pestis*. In Yersiniosis: Present and future. Vol. 13. Ravagnan, G. and Chiesa, C. (eds). Basel: Karger, pp. 325-328.

Yersin, A. 1894. La peste bubonique a Hong-Kong. Ann. Inst. Pasteur Paris 8: 662-667.

Zavialov, A.V., Kersley, J., Korpela, T., Zav'yalov, V.P., MacIntyre, S., and Knight, S.D. 2002. Donor strand complementation mechanism in the biogenesis of non-pilus systems. Mol. Microbiol. 45: 983-995.

Zavialov, A.V., Berglund, J., and Knight, S.D. 2003a. Overexpression, purification, crystallization and preliminary X-ray diffraction analysis of the F1 antigen Caf1M-Caf1 chaperone-subunit pre-assembly complex from *Yersinia pestis*. Acta Crystall. Section D-Biol. Crystallography. 59: 359-362.

Zavialov, A.V., Berglund, J., Pudney, A.F., Fooks, L.J., Ibrahim, T.M., MacIntyre, S., and Knight, S.D. 2003b. Structure and biogenesis of the capsular F1 antigen from *Yersinia pestis*: preserved folding energy drives fiber formation. Cell. 113: 587-596.

Zav'yalov, V.P., Chernovskaya, T.V., Chapman, D.A.G., Karlyshev, A.V., MacIntyre, S., Zavialov, A.V., Vasiliev, A.M., Denesyuk, A.I., Zav'yalova, G.A., Dudich, I.V., Korpela, T., and Abramov, V.M. 1997. Influence of the conserved disulphide bond, exposed to the putative binding pocket, on the structure and function of the immunoglobulin- like molecular chaperone Caf1M of *Yersinia pestis*. Biochem. J. 324: 571-578.

Chapter 19

pVM82: A Conjugative Plasmid of the Enteric *Yersinia* that Contributes to Pathogenicity

George B. Smirnov

ABSTRACT

The 123-kb conjugative plasmid pVM82 is often present in *Yersinia pseudotuberculosis* serotype I strains. Strains possessing pVM82 have been isolated from patients during large-scale outbreaks of the disease, but not from sporadic cases. *Y. pseudotuberculosis* strains that contain pVM82 cause more acute, severe, and generalized infections; and survive longer in infected experimental animals than do pVM82-free strains. The presence of pVM82 correlated with suppression of the antibody response of rabbits against specific *Y. pseudotuberculosis* antigens and with suppression of the cellular immune response in a mouse model of pseudotuberculosis infection. pVM82 also conferred increased resistance to the bactericidal effect of normal rabbit serum and certain basic dyes.

INTRODUCTION

The genus *Yersinia* contains three species pathogenic for humans. *Yersinia enterocolitica* and *Yersinia pseudotuberculosis* cause enteric diseases, while *Yersinia pestis* is the causative agent of plague, the most dangerous of known bacterial diseases. The phenotypic properties of *Yersinia* are determined by plasmid as well as chromosomal genes. In the early 1980s, several groups of researchers showed that all three *Yersinia* species pathogenic for humans possess plasmids. Each of the three species carried a closely related 70 to 75-kb plasmid associated with virulence (Gemski et al., 1980; Zink, et al., 1980; Ben-Gurion and Shafferman, 1981; Ferber and Brubaker, 1981; Portnoy and Falkow, 1981; Kutyrev et al., 1986). This virulence plasmid was later called pIB in *Y. pseudotuberculosis*, pYV in *Y. enterocolitica*, pCD in *Y. pestis*, or referred to generically as pCad, pYV, or the Lcr plasmid; and is responsible for Ca^{2+}-dependence of *Yersinia* growth at 37°C (see Chapter 16 on the *Yersinia* virulence plasmid).

In addition to the pCD virulence plasmid, the majority of *Y. pestis* strains possess two other plasmids, which contribute to virulence: pPCP (9.5 kb), and a 100-110-kb plasmid called pFra, pTox or pMT. Most isolates of *Y. enterocolitica* and *Y. pseudotuberculosis* possess only a virulence plasmid, which varies in size between 65-75 kb. However, some isolates contain extra plasmids. For example, 4, 5.2, 85, and 123-kb plasmids were isolated from different strains of *Y. pseudotuberculosis* by Shubin et al. (1985). The subject of this chapter is the 123-kb plasmid of *Y. pseudotuberculosis*.

The 123-kb plasmid was originally isolated from *Y. enterocolitica* by researchers from the Centers for Disease Control (CDC) in the United States (Kay et al., 1982). The authors proposed a role for this plasmid in *Yersinia* virulence, but did not further investigate this. We have worked with strains of *Y. pseudotuberculosis* from the Far East that possess this 123-kb plasmid, which was designated pVM82 (82-MDa plasmid isolated in Vladivostok and studied in Moscow). The work was halted in 1991, but sequencing and further characterization of pMV82 isolated from Far Eastern *Y. pseudotuberculosis* strains is planned.

EPIDEMIOLOGY

To study pVM82 prevalence and the possible relationship between plasmid content and the epidemiological features of disease outbreaks, 151 *Y. pseudotuberculosis* strains isolated over several years in Anadyr, a city in the Chukotka region of the Russian Far East, were analyzed (Shubin et al., 1985). Between 1971 and 1974, epidemic outbreaks of pseudotuberculosis were caused by serotype I strains that contained both the 72-kb pIB and the 123-kb pVM82 plasmids (pIB+, pVM82+). During 1975-1976, the incidence of the disease was much lower and the isolates contained only a 67.5-kb pIB plasmid. Similar observations were made in 1981, when two types of strains possessing a 67.5-kb pIB, or a 72-kb pIB and a 85.5-kb plasmid, were isolated. In 1977 and again in 1983-1985, high incidence of pseudotuberculosis returned and, as in 1971-1974, the isolates contained the 72-kb pIB and the 123-kb pVM82 plasmids. In 1986, a strain carrying four plasmids of 123, 72, 5.25, and 4.05 kb was prevalent, and the incidence of disease was again high. In 1987 and 1988, strains isolated from humans, animals, and vegetables were similar to those found in 1971-1974 and 1977, i.e., they contained 72 and 123-kb plasmids, but the morbidity associated with these strains was extremely high. All isolates possessing pVM82 or the 85.5-kb plasmid belonged to serotype I. During each period of surveillance, all isolates of the prevalent strain had identical plasmid content, whether isolated from stored vegetables, animals, or patients. Disease incidence associated with these strains differed significantly; low incidence years were characterized by less then 400 cases per 100,000; whereas high incidence years were characterized by more than 1,000 and sometimes more than 4,000 cases per 100,000 inhabitants.

According to Shubin et al. (1985), *Y. pseudotuberculosis* strains that carried the 72 and 123-kb plasmids were associated with higher disease incidence than strains carrying the 67.5-72- kb plasmid and the 85.5-kb plasmid. Because the 67-72 kb plasmid is the extensively studied virulence plasmid common to all the pathogenic yersiniae and known to confer anti-host functions, we hypothesized that the 123-kb plasmid, designated pVM82, encodes additional properties that contribute to virulence.

It is appropriate to mention here that during last 30 years the epidemic character of pseudotuberculosis has manifested itself in massive outbreaks that occurred in territories of the Russian Federation (Shubin, 1993). The first large outbreak of this disease was registered around 40 years ago. This outbreak, as well as many subsequent ones, was unusual in that the disease had peculiar clinical manifestations. This led physicians to consider it a new disease, which they called Far East scarlet-like fever (FESLF) (Somov, 1979).

CLINICAL MANIFESTATIONS OF FESLF

The first reported and studied outbreak of FESLF occurred in 1959 in the city of Vladivostok and involved more than 300 persons. 200 of them were hospitalized in a specially organized facility and were studied very carefully. It was concluded that the source of infection was contaminated cottage cheese. At disease onset the majority of patients presented with a punctate rash on hyperemic skin, tonsillitis, a pale nasolabial triangle, enlarged submandibular lymph nodes, and a crimson tongue tip. Given these symptoms, it is no surprise that the initial diagnosis was scarlet fever. A few days later, the symptoms of scarlet fever subsided, new ones appeared, and the initial diagnosis was rejected. The term FESLF was introduced by Grunin, Somov, and Zalmover in 1960 (see Somov, 1979). Initially, FESLF appeared in different regions of the huge territory of the Russian Far East, but after its clinical manifestations were described, it was subsequently reported in Leningrad, the Voronezsh, Lipetsk, and Stavropol regions; Ukraine, Sverdlovsk, Novosibirsk, Kemerovo, and other places, showing no bounds to any particular territory.

It is important to discuss the differences between FESLF and the typical pseudotuberculosis observed in Europe and America and why FESLF is not prevalent in the West. Based on clinical observations of 570 pseudotuberculosis patients, Zalmover (1969) identified six stages of FESLF: incubation period, initial onset, accrual, (increase of symptoms), remission, recurrence with exacerbation, and convalescence.

The incubation period usually lasts 7 to 10 days. The initial stage begins with the appearance of acute symptoms, including fever, shivering, and the standard symptoms of acute intoxication (headache, muscle, joint, and back pain, weakness, and loss of appetite). Hyperemia of the face and neck, pale nasolabial triangle, hyperemia of the conjunctiva and swelling of the scleral vessels may also be seen in this stage. Catarrh of the upper respiratory tract and gastrointestinal pain are also often observed.

In the accrual stage, symptoms of intoxication, gastrointestinal symptoms, and joint pain increase. Body temperature reaches its maximum on day 3 but then decreases by day 6. The major symptom of this stage is the scarlet fever-like rash, which appears on day 2-3. The rash was the reason for many misdiagnoses in which physicians initially diagnosed scarlet fever. It was later shown that the rash could have different presentations. It could be spotty (rubella- or measles-like) or confluent (erythematous-like). The alimentary tract pathology usually manifested as acute gastritis, gastroenteritis, mesenteric adenitis, terminal ileitis, and sometimes acute cholecystitis. Different combinations of all these symptoms were observed. Often, sharp pain in the ileocecal area prompted physicians to diagnose appendicitis, and 40% of such patients underwent appendectomies during the 1960s.

Infection of the gut was not the only manifestation of alimentary tract pathology during FESLF. In approximately 50% of cases, liver lesions exhibiting the features of acute parenchymatous hepatitis were seen. Jaundice in FESLF patients was maximal at the end of the first week and gradually decreased during the second week. These symptoms often led to a misdiagnosis of viral hepatitis. Liver pathology usually resolved and was subsiding by the end of the fever.

The central nervous system was also affected during the accrual stage. General weakness, hypotonia, severe headache, photophobia, vomiting, and insomnia were observed. Meningeal symptoms were also seen in some patients.

In some cases, lesions occurred in the cardiovascular system. Hypotonia characterized by a sudden drop in blood pressure was the most frequent symptom. Muffled heart sounds and, rarely, arrhythmia were observed.

Based on the frequency and combination of the various symptoms, the following forms of FESLF were proposed, listed in order of decreasing frequency: scarlet fever-like, abdominal, icteric, arthralgic, and generalized (Zalmover, 1969). In accord with this classification, incorrect initial diagnoses usually were scarlet fever, acute respiratory infection, tonsillitis, viral hepatitis, appendicitis, gastroenteritis, polyarthritis, and acute rheumatism.

Many years after these FESLF outbreaks, the relationship between the plasmid content of *Y. pseudotuberculosis* strains and clinical manifestations of pseudotuberculosis in humans was analyzed (Klimov et al., 1999). It was found that strains isolated from patients in the Northern Irkutsk region contained only the 70-kb virulence plasmid (pIB), whereas strains isolated from Southern Irkustsk contained two plasmids, pIB and the 123-kb pVM82. Statistical analysis revealed the clinical relationships indicated in Table 1.

As can be seen from Table 1, *Y. pseudotuberculosis* strains that contained pVM82 affected the alimentary tract, liver, spleen and musculoskeletal system much more frequently than pVM82-negative strains. Strains possessing both pIB and pVM82 were also more frequently associated with some (but not all) symptoms of FESLF.

What manifestations of pseudotuberculosis are most frequent in Western Europe, and therefore the most well-known and familiar? According to Knapp (1959) and Mollaret (1962), the most common symptom is mesenteric lymphadenitis, which is only one of many predominant symptoms of FESLF. Thus, in contrast to Western Europe, many more manifestations of pseudotuberculosis are encountered in the Russian Far East. The reasons for this disparity are unknown, but we anticipate that molecular analyses of the strains isolated from the two different geographic areas will provide the answer.

Others have also noted the correlation between the severity and diversity of pseudotuberculosis symptoms and the geographic origin of the responsible strains (Yoshino et al., 1995). These authors further observed a correlation between the expression of the superantigen YPM (*Y. pseudotuberculosis*-derived mitogen) and the severity of clinical manifestations of disease, and showed that the *ypm* gene is significantly more prevalent in Far Eastern than in European *Y. pseudotuberculosis* isolates. Later, Abe et al. (1997) reported the very high correlation between pseudotuberculosis systemic symptoms and the expression of YPM. They demonstrated that YPM is produced by the infecting microorganisms in vivo and induces IgG antibody response in patients. The highest titers of IgG and the expansion of YPM-responsive T-cells were seen in patients with

Table 1. Occurrence (percent of cases) of clinical symptoms in 146 patients infected with (pIB+, pVM82-) strains; and in 207 patients infected with (pIB+, pVM82+) strains of *Y. pseudotuberculosis* (from Klimov et al., 1999).

Clinical symptom	Infection caused by pIB(+) / pVM82(-) strains	Infection caused by pIB(+) / pVM82(+) strains
Rash	89	59
Upper respiratory tract lesions	70.5	69.5
Nausea, vomiting	11	47.4
Diarrhea	5.5	46.4
Abdominal pain	3	78
Hepatomegaly	14.4	55.5
Splenomegaly	0	17
Arthralgia	31	68.6

systemic symptoms of pseudotuberculosis similar to those described for FESLF, further implicating YPM in the pathogenesis of severe disease. (For further information on YPM, see Chapter 10). The possible relationship between the expression of the *ypm* gene and the effects of pVM82 and clinical manifestation of FESLF remain to be determined.

PATHOPHYSIOLOGY IN ANIMAL MODELS

Subcutaneous infection of guinea pigs with natural isolates or an isogenic series of (pIB+, pVM82+), (pIB+, pVM82-), and (pIB-, pVM82+) *Y. pseudotuberculosis* strains led to disease characterized primarily by a necrotic suppurative focus at the site of injection, macroscopic lesions in internal organs, and purulent septicemia (Shubin, 1993). Lethality occurred only after infection with the strains containing pIB or pIB and pVM82.

The following manifestations of infection were seen in guinea pigs infected with (pIB-, pVM82-) strains. The primary effect at the site of subcutaneous inoculation was miliary tubercles which appeared on day 10 and which progressed to necrotic foci by the third week in 50% of cases. Lesions in the lungs and liver developed 10 to 28 days after infection. Small foci of inflammation were seen in the liver, and the granulomas contained histiocytes and lymphocytes. Nodules with lymphocyte and histiocyte infiltration were also found in the lungs near the vessels and bronchi.

Infection caused by (pIB-, pVM82+) or (pIB+, pVM82+) strains looked different. The changes in the primary focus were more pronounced and developed more rapidly. Edema, hemorrhage, and infiltration were observed on day 3, miliary necrotic nodules on day 6, and confluent necrotic foci on day 10 after infection. In some cases, the necrotic foci gave rise to suppurating cutaneous fistulas 14 to 28 days after infection. A connective tissue capsule formed around the foci, and inguinal lymphadenitis often developed. Miliary as well as larger necrotic nodules were seen in the liver, spleen, and lungs, and the multiple foci of necrosis were associated with hepato- and splenomegaly. From day 21 to 28, necrotic foci disappeared, and small scars formed on the organs' surface. Histopathological changes were even more pronounced. Intramuscular edema at the infection site was replaced by increasing migration of leukocytes to the site, accompanied with cellular breakdown; rare bacteria could be isolated from the cell debris. Lesions in parenchymal organs were more serious than those produced by (pIB-, pVM82-) strains. Focal destruction of hepatocytes and proliferation of mononuclear phagocytes (von Kupffer cells) with mesenchymal infiltration of portal tracts were recorded. In some cases, subcapsular foci of inflammation with eosinophilic dystrophy of hepatocytes and pyknosis were observed. Granulomatous foci containing macrophages, lymphocytes and some granulocytes were seen in the lungs. In some animals, granulomas with central disintegration of cell nuclei were seen in the lungs and liver.

Finally, guinea pigs infected with (pIB+, pVM82+) and (pIB+, pVM82-) strains were studied. It was found that the overall dynamics of the disease caused by these two strains was very similar to that seen for (pIB-, pVM82+) strains; however, the pathological changes were more severe and extensive, and the presence of pVM82 did not make any difference. In summary, *Y. pseudotuberculosis* containing pVM82 alone produced more severe pathological changes at the site of primary infection and in tissues and parenchymal organs than did plasmidless bacteria. The strains containing both pIB and pVM82 were always the most virulent, however.

Different results were obtained with CBA mice (Klimov et al., 1999). It was shown that experimental infection with a (pIB+, pVM82+) *Y. pseudotuberculosis* strain was characterized by much greater involvement of gut, liver and spleen than was infection with a (pIB+, pVM82-) strain. More specifically, infiltrative and destructive changes developed sooner and were much more pronounced.

One may conclude that pVM82 contributed to the virulence of pIB(-) *Y. pseudotuberculosis* strains in certain animal models and conditions, and also contributed to the virulence of pIB(+) strains. The presence of pVM82 correlated with a more rapid and pronounced infection of gut, liver and spleen at the cellular level.

KINETICS OF DISEASE PROGRESSION AND PERSISTENCE

The LD_{50} of different *Y. pseudotuberculosis* strains was measured after intraperitoneal inoculation of white mice. No difference in LD_{50} was seen between (pIB-, pVM82+) and (pIB-, pVM82-) strains (Subin, 1993; Klimov et al., 1999). As expected, when the strains carried the pIB plasmid they were much more virulent.

Accumulation of *Y. pseudotuberculosis* in organs and tissues was studied in guinea pigs (Subin, 1993). Original patient or animal isolates as well as isogenic derivatives constructed in the laboratory were used. A gradual increase in the bacterial count of pIB-positive strains in the spleen and liver took place during the first 3 weeks after infection. pVM82 had no effect on the kinetics of accumulation. Bacterial counts for (pIB+, pVM82+) strains decreased significantly after 4 weeks. When strains lacking the virulence plasmid were used, the rate of bacterial accumulation in internal organs depended on the presence of pVM82 plasmid. The (pIB-, pVM82-) strain produced maximum bacterial load 7 days after infection and after 3 weeks no bacteria were found in the organs. In contrast, (pIB-, pVM82+) bacteria increased in number during the first 2 weeks after infection and could be isolated from the lungs, liver and spleen even after 4 weeks. The results were the same with isogenic strains and original isolates. These results indicate that pVM82 helps *Y. pseudotuberculosis* to persist longer in the infected host.

STRUCTURE OF pVM82

Four strains of *Y. pseudotuberculosis* isolated from patients in different areas of the Soviet Union were used as the sources of pVM82 for molecular studies. The strains were isolated in Anadyr in 1977, Anjero-Sudzshensk in 1975, Leningrad in 1978, and Artem in 1983. Here I shall refer to these strains as AN, AS, L, and AR, respectively. Before isolation of pVM82, all strains were cured of the 72-kb virulence plasmid pIB. Then, pVM82 plasmids were isolated from each strain and restriction enzyme analysis of the plasmid DNA was performed using *Hind*III, *Eco*R1, and *Bsp*R1. pMV82 restriction patterns were identical for all four strains studied, irrespective of their geographic origin and year of isolation. Moreover, the restriction pattern of pVM82 taken from a recent *Y. pseudotuberculosis* isolate was indistinguishable from that of pMV82 from strains AN, AS, L, and AR; and from the 123-kb plasmid isolated from *Y. enterocolitica* by Kay and coworkers (1982). This suggests that pVM82 is genetically stable and widespread in *Y. pseudotuberculosis* strains (Ginzburg et al., 1988; Klimov et al., 1999). Other than the original report by Kay et al. (1982), the structure and prevalence of the 123-kB plasmid in *Y. enterocolitica* populations has not been investigated.

It was mentioned above that in 1981 a *Y. pseudotuberculosis* strain possessing 72 and 85- kb plasmids was isolated. The source of this strain was a rat caught in a vegetable store in Anadyr, during a period when pseudotuberculosis cases were sporadic and morbidity

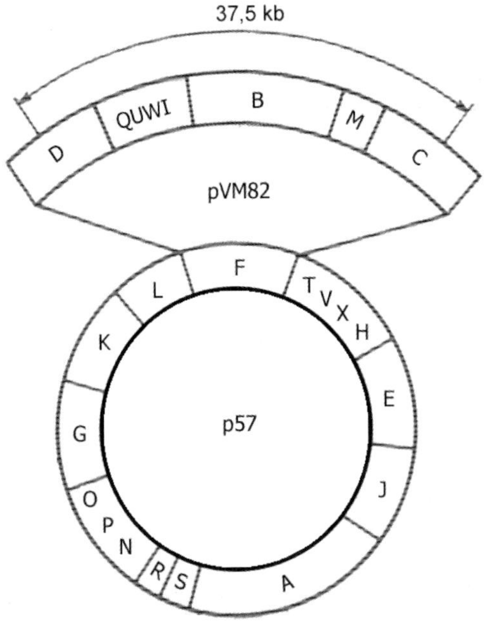

Figure 1. Restriction map of pVM82 and p57 plasmids. *Hin*dIII fragments are indicated by capital letters. Fragments with unknown order are shown in one segment.

was low. The strain was cured of the 72-kb pIB plasmid, and the 85-kb plasmid was isolated and analyzed. It turned out that all *Hin*dIII restriction fragments of the 85-kb (57-MDa) plasmid (named p57) except one were identical to pVM82 *Hin*dIII restriction fragments. Conversely, however, pVM82 possessed five *Hin*dIII fragments absent from p57. The homology of p57 to pVM82 was confirmed further by Southern hybridization and electron microscopy heteroduplex analyses (Ginzburg et al., 1988). We have assumed that p57 is a precursor of pVM82 formed by the insertion of a 37.5-kb DNA fragment into p57, and that the *Hin*dIII fragment (Figure 1, fragment F) present in p57 but absent in pVM82 contains the integration site of the 37.5-kb fragment. Alternatively, of course, p57 could be a deletion derivative of pVM82.

The restriction map of pVM82 was constructed using the "genome walking" method (Shovadaeva et al., 1990). A pVM82 genomic library was made using the vector phage λ EMBL3. Plasmid DNA was digested with *Sau*3A and ligated to the *Bam*H1-digested cohesive ends of the phage vector. Overlapping *Hin*dIII fragments were then identified by hybridization of recombinant phage DNAs with each *Hin*dIII fragment of pMV82, and by restriction analysis of every recombinant phage DNA (Shovadaeva et al., 1990). The resulting plasmid map is shown in Figure 1.

Comparison of restriction profiles of pVM82 and the *Y. pestis*-specific plasmid pFra/Tox that encodes the F1 antigen (Protsenko et al., 1982) showed no homology, indicating that pVM82 is probably unique to the enteric *Yersinia* (Ginzburg et al., 1988).

IN VITRO PHENOTYPE OF pVM82-CONTAINING STRAINS

CONJUGATIVE TRANSFER OF pVM82

Isogenic *Y. pseudotuberculosis* strains carrying pVM82 or p57 labeled with Tn5-341::25lac were used as donors in conjugative crosses with different *Y. pseudotuberculosis* or *E. coli* recipients. The results showed that, unlike p57-carrying strains, pVM82 donors were able to generate pVM82 transconjugants in the crosses with all recipients tested (Shovadaeva et al., 1990; Sever et al., 1991). Thus, pVM82 (but not p57) is a conjugative plasmid. Presumably the *tra* operon of pVM82 is located in its unique 37.5-kb fragment (Figure 1). Interestingly, the *Y. pseudotuberculosis* transconjugants obtained in these crosses were able to be active donors of pVM82 in subsequent conjugal transfers, whereas *E. coli* transconjugants were not, indicating that the Tra function of pVM82 was active only in the *Y. pseudotuberculosis* background.

Few other conjugative plasmids have been found in enteric *Yersinia*. Two of these (Kimura et al., 1975; Cornelis et al., 1976) were not cryptic but contained either drug-resistance or catabolite genes. Recently, two cryptic homologous conjugative plasmids from *Y. enterocolitica* were described (Hertwig et al., 2003).

RESISTANCE TO ANTIBACTERIAL COMPOUNDS

It is well known that increased resistance to serum complement and to phagocytosis by macrophages in many Gram-negative enteropathogenic bacteria correlates with reduced cell wall permeability for several hydrophobic compounds (for review see Sever et al., 1991). Therefore, the sensitivity of isogenic *Y. pseudotuberculosis* strains differing by the presence or absence of pVM82 and p57 to the hydrophobic antibacterial compounds lincomycin, novobiocin, erythromycin, fusidic acid, and gentian violet was studied. At the concentrations chosen, only pVM82-containing bacteria were able to form colonies on agar plates containing erythromycin, fusidic acid, or gentian violet. Sensitivity to lincomycin and novobiocin was independent of the plasmid content of the strains. Bacteria grown at 8°C exhibited plasmid-dependent differences in their sensitivity to much lower concentrations of the compounds than bacteria grown at 37°C. These results suggest that pVM82 encodes factors that modify the cell membrane, making the host strains more resistant to the hydrophobic agents.

Additional proof was obtained in experiments testing the ability of the strains to absorb Congo red dye during growth on nutrient agar. This test is usually used for differentiation of pIB-containing (virulent) and pIB-negative (avirulent) strains of *Y. pseudotuberculosis* (see Shubin, 1993 for review). Increased adsorption of the dye was observed with the (pIB+, pVM82+) strain. The (pIB-, pVM82+) strain reached the same level of pigmentation one day later than the (pIB+, pVM82+) strain, while the (pIB-, p57+) and the (pIB-, pVM82-, p57-) strains remained nonpigmented (Sever et al., 1991).

LOW TEMPERATURE-DEPENDENT SERUM RESISTANCE

Another characteristic phenotypic trait of some virulent bacteria is their increased resistance to normal serum. *Y. pseudotuberculosis* is serum resistant when grown at 37°C (Perry and Brubaker, 1983). This temperature-dependent resistance is encoded by chromosomal genes. The possible contribution of pVM82 to serum resistance was studied using bacteria grown at low (8°C) or at room temperature. The (pIB-, pVM82+)strain grown at 8°C was resistant to the bactericidal effect of up to 6% serum, whereas the

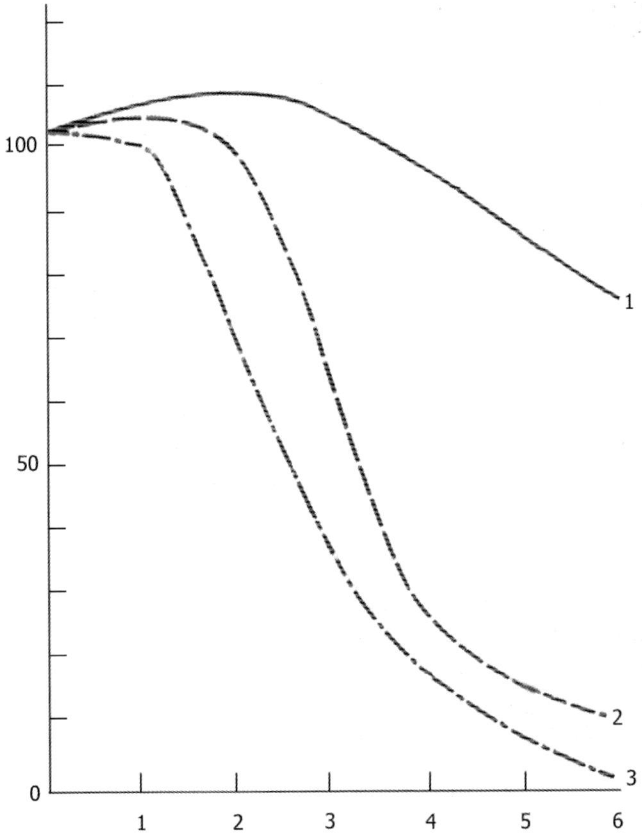

Figure 2. Effect of normal rabbit serum on the viability of isogenic *Y.pseudotuberculosis* containing pVM82 (curve 1); p57 (curve 2) or neither plasmid (curve 3). Abscissa axis: serum concentration (%); ordinate axis: viable bacteria (%).

isogenic (pIB+, pMV82-, p57+) and (pIB-, pVM82-, p57-) strains showed much greater sensitivity (Figure 2). Identical results were obtained when the strains were grown at room temperature (Sever et al., 1991).

One can conclude that pVM82 provides the cells with the ability to resist the bactericidal effect of normal serum at low temperatures. Since this effect was specific to low temperatures, its mechanism is different from serum resistance at 37°C described by Perry and Brubaker (1983).

It is well known that some conjugative plasmids of other enterobacteria also encode properties such as resistance to hydrophobic compounds, serum resistance, and the ability to transfer genetic material, similar to the properties that pVM82 confers to *Y. pseudotuberculosis*. For example, it was shown that the *traT* gene of the *E. coli* plasmids F, R6, R6-5, the *Salmonella typhimurium* virulence plasmid, and plasmids present in clinical isolates of *Shigella* and *Klebsiella* was responsible for the resistance phenotype (see Sever et al., 1991 for review). The effect was due to the fact that TraT is one of the major outer membrane proteins responsible for bacterial resistance to serum and hydrophobic compounds.

We attempted to find the *traT* homologue on the pVM82 plasmid. Southern blot hybridization between a *traT* gene probe from the *S. typhimurium* virulence plasmid and *Hin*dIII-restriction fragments of pVM82 and p57, and *Eco*RI-fragments of R6, was performed. Only the *Eco*RI restriction fragment of R6 containing *traT* gave a positive hybridization signal. Thus, pVM82 has no transfer gene exhibiting strong nucleotide sequence homology with the *traT* from the *S. typhimurium* virulence plasmid.

There are at least three possible explanations of the results. First, the putative *traT* homologue of pMV82 may differ significantly in nucleotide sequence from the *S. typhimurium traT*, since the TraT proteins encoded by different *E. coli* plasmids vary significantly in their amino acid sequences (see Sever et al., 1991 for review). Second, in addition to *traT*, another plasmid gene, *iss,* was also found to confer increased resistance of *E. coli* to serum complement. The product of the *iss* gene is also an outer membrane protein, and has a molecular mass similar to that of TraT but with different antigenic properties (see Sever et al., 1991 for review). It is possible that pVM82 contains a gene similar to *iss*. Therefore the genetic determinant of pVM82 conferring serum resistance may be a) a *traT*-like gene with low homology to the *traT* of *S. typhimurium*; b) an *iss* homologue; or c) an unknown gene with functional homology to *traT* and *iss.*

INFLUENCE OF pMV82 ON THE IMMUNE RESPONSE

ANTIBODY RESPONSE

Antibody induction was studied in rabbits infected with *Y. pseudotuberculosis* serotype I strains containing different plasmids. Animals were immunized 4 times at weekly intervals by i.v. injection of 10^9 bacteria that had been heated 5 min at 55^0 C. Sera from rabbits immunized with (pIB-, pVM82+, p57-), (pIB-, pVM82+, p57+), or (pIB-, pVM82-, p57-) strains were used in immunoblots to detect antigens in the lysates of the same strains (Figure 3). (All strains used in this set of experiments were isogenic and cured of pIB plasmids). The antisera raised against pVM82(+) or p57(+) strains grown at 37°C recognized four main antigens in the lysates of plasmid-containing (Figure 3A) and plasmidless (not shown) strains. These antigens (designated a, b, c, and d) had molecular masses 90, 69, 50, and 25 kDa, respectively (Figure 3). Seven or more additional minor antigens were detected upon longer incubation (more than 10 hours) of the immunoblots, but they were not studied.

The immunoblot results indicated that the pVM82(+) strains did not express additional immunogens compared to the pVM82(-) strains under the experimental conditions used, indicating that the four major antigens detected were encoded by chromosomal genes and not by pVM82 or p57. However, antiserum obtained after immunization of rabbits with the pVM82(+) strain grown at 14°C did not recognize antigens "a" and "b" in lysates of pVM82(+) or p57(+)-bacteria grown at 37°C (Figure 3A, lanes 3 and 4). Thus, antibodies against "a" and "b" were not produced by animals immunized with the pVM82(+) strain grown at 14°C. When rabbits were immunized with (pVM82-, p57-) or (pVM82-, p57+) bacteria grown at 14°or 37°C, or pVM82(+) bacteria grown at 37°C, antibodies against all four antigens were produced.

To study the ability of pVM82(+) strains to produce antigens "a" and "b" at low growth temperature, lysates prepared from 14°C cultures of (pVM82+, p57+) or (pVM82-, p57-),*Y. pseudotuberculosis* were used in immunoblots hybridized with antisera raised against the same strains grown at the same temperature. The anti-p57 serum (Figure3B,

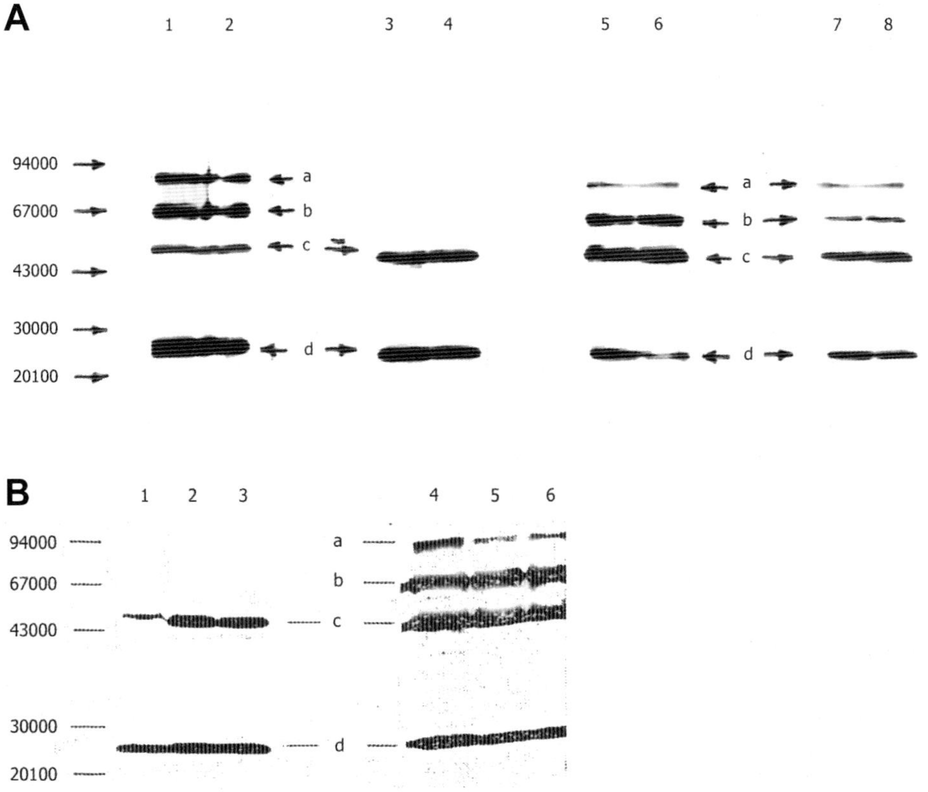

Figure 3. Immunoblots of plasmidless or plasmid-containing *Y. pseudotuberculosis* strains, hybridized with antisera raised against the same strains grown 14°or 37°C. (A) lanes 1, 3, 5, and 7– lysate of pVM82(+) strain; lanes 2, 4, 6, 8 – lysate of p57(+) (85 kb) strain. All lysates were prepared from bacteria grown at 37°C. Lanes 1 and 2 were hybridized with antiserum against the pVM82(+) strain grown at 37°C; lanes 3 and 4 with antiserum against the pVM82(+) strain grown at 14°C; lanes 5 and 6 with antiserum against the p57(+) strain grown at 37°C; and lanes 7 and 8 with antiserum against the p57(+) strain grown at 14°C. (B) Immunoblots of lysates prepared from bacteria grown at 14°C. Lanes 1 and 4 were hybridized with antiserum against the plasmidless strain; lanes 2 and 5 with antiserum against the pVM82(+) strain; and lanes 3 and 6 with antiserum against the p57(+) strain. All antisera used for (B) were raised against bacteria grown at 14°C.

lanes 4-6) and the antiserum against the plasmidless strain (not shown) revealed all four antigens in both strains, but the anti-pVM82 serum did not reveal antigens "a" and "b" (Figure 3B, lanes 1-3). These results indicate that the pVM82(+) strain produced antigens "a" and "b" at 14°C, but that the presence of pVM82 (but not p57) inhibited the formation of antibodies against these antigens in rabbits.

If the observed pVM82-mediated inhibition of immunogenicity was not due to lack of synthesis of antigens "a" and "b", then one may assume that some product encoded by the 37.5-kb fragment present in pVM82 but absent from p57 plasmid was responsible for the difference in antibody induction. The temperature-dependence of the observed immunosuppression mediated by pVM82 is in good agreement with the observation that growth of *Y. pseudotuberculosis* at low temperature increases its virulence (Varvashevich and Sidorova, 1985).

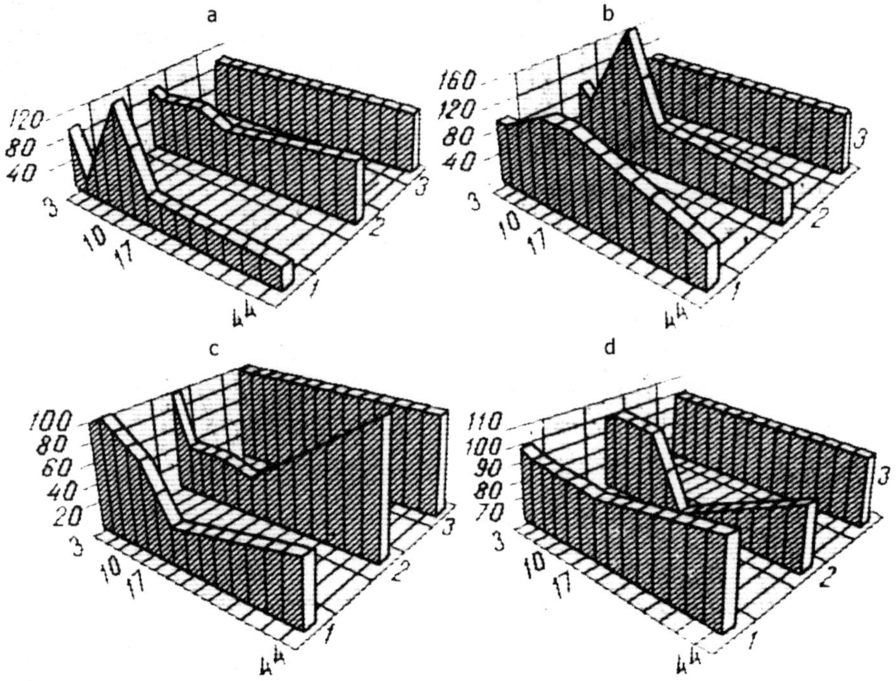

Figure 4. Proliferative response to ConA and LPS of splenocytes collected from CBA mice at different times after infection with *Y. pseudotuberculosis* strains containing different plasmids. Response to ConA after infection with bacteria grown at 10°C (A) or 37°C (B). Response to LPS after infection with bacteria grown at 10°C (C) or 37°C (D). Results with mice infected with the pVM82(+) strain (curve 1); the plasmidless strain (curve 2); and results for uninfected control mice (curve 3) are shown. Abscissa axis: days after infection; ordinate axis: proliferative response of splenocytes to mitogens (% of intact control).

CELL-MEDIATED IMMUNE RESPONSE

All *Yersinia* pathogenic for humans have very efficient mechanisms that provide resistance to elements of the innate immune response, such as phagocytosis and digestion by macrophages or polymorphonuclear leukocytes. It is now commonly accepted that yersiniae remain mostly extracellular during infection and have evolved the capacity to avoid uptake and killing by phagocytes. This implies that cellular immunity is a significant defense mechanism of the eukaryotic host, and that the yersiniae were unable to ignore it in the course of evolution. When the data suggesting that pVM82 might influence antibody induction were obtained, we went on to study its influence on cellular immunity. Lymphocyte proliferation in the course of experimental pseudotuberculosis and the kinetics of antigen-specific proliferative lymphocyte responses induced by isogenic (pIB-, pVM82+) and (pIB-, pVM82-) *Y. pseudotuberculosis* strains were studied (Volkov et al., 1991). (As in the experiments described in previous section, the strains used were isogenic and cured of the pIB plasmid, and, in addition, did not contain p57). Splenocytes taken from uninfected CBA mice or mice infected with plasmidless or pVM82-containing *Y. pseudotuberculosis* were incubated for 72 hours with the standard mitogens ConA or LPS (Figure 4). The dynamics of the proliferative response to the B cell-specific mitogen LPS were similar or identical irrespective of the growth temperature (10°or 37°C) or presence of pMV82 (Figure 4C, D). In all cases, the proliferative response was inhibited during day 3 – 17 after infection (40% of the control level of proliferative response for 10°C culture by day 17). By day 44, the tendency to recovery was seen (60% of the proliferative response of the 10°C control culture). The inhibition of proliferative response to LPS was greater when the animals were infected with bacteria grown at 10°C, consistent with previous data showing increased virulence of *Y. pseudotuberculosis* grown at low (8 – 14°C) temperatures (Varvashevich and Sidorova, 1985).

In experiments with the T cell-specific mitogen ConA, the following results were obtained (Figure 4A, B). On day 3 after infection, proliferative responses of splenocytes from mice infected with the pVM82(+) strain grown at 37°C, or the plasmidless strain grown at 10°C, were indistinguishable from the response of splenocytes taken from uninfected animals. On day 10 after infection, a slight increase in the response was noticed. On day 17, a slight decrease was observed, and on day 44 the response returned to normal in mice infected with the plasmidless strain grown at 10°C infection and was slightly reduced (60% of control) in mice infected with the pVM82(+) strain grown at 37°C. In contrast, the proliferative response of splenocytes from mice infected with the pVM82(+) strain grown at 10°C was significantly different from that of uninfected controls. On day 3 the response was 17% lower, and did not return to normal but remained at the low level on day 17 and even day 44 (Figure 4A).

Previously it was shown that antigen-specific proliferation of splenocytes in response to *Y. pseudotuberculosis* antigens in infected mice could not be studied using the standard scheme (estimation of the response after 5-6 days of contact of spleen cells with the antigen), because *Y. pseudotuberculosis* showed a delayed mitogenic response (Volkov et al., 1991). Therefore, longer terms of contact were applied. In keeping with the previous results, there was no response of intact mice splenocytes after 6 days of contact with *Y. pseudotuberculosis*. On day 17, stimulation of proliferation was noticed in splenocytes taken from mice infected with pVM82(+) or pVM82(-) strains grown at 37°C (stimulation indices of 2.3 and 1.9, respectively). On day 44, antigen-specific stimulation of splenocyte proliferation was detected in the animals infected with the plasmidless strain grown at 10°C (stimulation index 2.1). However, no proliferation response at all throughout the entire observation period was detected in splenocytes from mice infected with *Y. pseudotuberculosis* pVM82(+) grown at 10°C.

Thus, inhibition of the T-lymphocyte proliferative response 3 to 44 days after infection, and a dramatic fall of antigen-specific response during the entire period of the experiment occurred when the animals were infected with the *Y. pseudotuberculosis* pVM82(+) strain grown at 10°C. These findings are in good agreement with the results indicating that pVM82 –containing bacteria grown at 14°C inhibited formation of antibodies against at least two chromosomally encoded antigens (Ginzburg et al.,1988).

REFERENCES

Abe. J, Onimaru, M., Matsumoto, S. Noma, S., Baba, K., Ito, Y., Kohsaka, T., and Takeda, T. 1997. Clinical role for a superantigen in *Yersinia pseudotuberculosis* infection. J. Clin. Invest. 99: 8, 1823-1830.

Ben-Gurion, R., and Shafferman, A. 1981. Essential virulence determinants of different *Yersinia* species are carried on a common plasmid. Plasmid 5:183-187.

Cornelis, G., Bennet, P.M., and Grinsted, J. 1976. Properties of pGC1, a *lac* plasmid originating in *Y. enterocolitica* 842. J. Bacteriol. 127: 1058-1062.

Ferber, D.M., and Brubaker, R.R. 1981. Plasmids in *Yersinia pestis*. Infect. Immun. 31: 839-841.

Gemski, P., Lazere, J.R., Casey, T., and Wohlhieter, J.A. 1980. Presence of a virulence-associated plasmid in

Yersinia pseudotuberculosis. Infect. Immun. 28:1044-1047.
Ginzburg, A.L., Shubin, F.N., Shovadaeva, G.A., Kulichenko, A.N., Yanishevski, N.V., and Smirnov, G.B. 1988. [A novel pathogenic trait encoded by *Yersinia pseudotuberculosis* plasmid pVM82 Genetika] (in Russian) 24: 1562-1571.
Hertwig, S., Klein, I., Hammerl, J.A., and Appel, D. 2003. Characterization of two conjugative *Yersinia* plasmids mobilizing pYV. In: The Genus Yersinia. M.Skurnik et al., eds. Kluwer Academic/Plenum Publishers, New York. pp. 35-38.
Kay B.A., Wachsmuth, K., and Gemski, P. 1982. New virulence-associated plasmid in *Yersinia enterocolitica*. J. Clin. Microbiol. 15: 1161-1163.
Kimura, S., Eda, T., Ikeda, T., and Suzuki, M. 1975. Detection of conjugative R-plasmids in genus *Yersinia*. Japan. J. Microbiol. 19: 449-451.
Klimov, V.T., Maramovich, A.S., Malov, I.V., Shurygina, I.A., Chesnokova, M.V., Breneva, N.V., and Mik, O.V. 1999. [Epidemiological and morphoclinical features of pseudotuberculosis in Irkutsk region.] (in Russian) Epidemiology and Infectious Diseases. 4:29-34.
Knapp, W. 1959. *Pasteurella pseudotuberculosis* unter besondereberucksichtigung ihrer human medisinishen bedeutung. Erg. Mikrob. Exper. Therap. 32: 196-269.
Kutyrev, V.V., Popov, Yu, A., and Protsenko, O.A. 1986. Pathogenicity plasmids of the plague microbe (*Yersinia pestis*). Molec. Genet. Microbiol. Virusol. 6: 3-11.
Mollaret, H.H. 1962. Le bacille de Malassez et Vignal (caracteres culturaux et biochimiques). Pasteur Instutute. Paris. 131.
Perry, R.D. and Brubaker, R.R. 1983. Vwa^{+} phenotype of *Yersinia enterocolitica*. Infect. Immun. 40: 166-171.
Portnoy, D.A., and Falkow, S. 1981. Virulence associated plasmids from *Yersinia enterocolitica* and *Yersinia pestis*. J. Bacteriol. 148:877-883.
Protsenko, O.A., Anisimov, P.I., Mozharov, O.T., Konnov, N.P., Popov, Y.A., and Kokushkin, A.M. 1982. [Detection and characterization of *Yersinia pestis* plasmids determining pesticin I, fraction I antigen and "mouse" toxin synthesis.] (in Russian) Genetika 19: 1081-1090.
Sever, I.S., Ilyina, T.S., Motin, V.L., and Smirnov, G.B. 1991. [Phenotypic properties of *Yersinia pseudotuberculosis* strain harboring the pVM82 plasmid and isolated in the foci of outbreaks of pseudotuberculosis.] (in Russian) Molec. Genet. Microbiol. Virol. 12: 10-14.
Shovadaeva, G.A., Markov, A.P., Pokrovskaya, M.S., Chetina E.V., Yanishevsky, N.V., Shubin, F.N., Ginzburg, A.L., and Smirnov, G.B.1990. [Structural and functional characteristics of *Yersinia pseudotuberculosis* plasmid pVM82] (in Russian) Genetika 26: 621-629.
Shubin, F.N. 1993. [Ecological and molecular-genetic aspects of epidemiology of pseudotuberculosis.] Med. Dr. Theses. Vladivostok.
Shubin, F.N., Sibirtsev, Yu, T., and Rasskazov, V.A. 1985. [*Yersinia pseudotuberculosis* plasmids and their role in realization of pseudotuberculosis epidemic process.] (in Russian) Microbiologia. 12: 53-56.
Somov, G.P. 1979. [Far East scarlet-like fever.] Medicina, Moscow.
Varvashevich, T.N. and Sidorova, V.E. 1985. [Population changeability and biological properties of bacteria are the function of incubation conditions]. In: (Pathogenesis, microbiology, and laboratory diagnostics of yersinioses). Novosibirsk; SO AMS USSR, 3-8.
Volkov, L.V., Barteneva, N.S., Pronin, A.V., Markov, A.P., Filkova, S.L., Sanin, A.V., and Smirnov, G.B. 1991. The role of plasmid having molecular mass 82 MD on modulation of *Yersinia pseudotuberculosis* influence on proliferation activity of mice lymphoid cells. Molec. Genet. Microbiol. Virol. 12: 29-32.
Yoshino, K., Ramamurthy, T., Nair, G.B., Fukushima, H., Ohtomo, Y., Takeda, N., Kaneko, S., and Takeda, T. 1995. Geographical heterogeneity between Far East and Europe in prevalence of *ypm* gene encoding the novel superantigen among *Yersinia pseudotuberculosis* strains. J. Clin. Microbiol. 33: 3356-3358.
Zalmover, I.Y. 1969. [Clinical forms of Far East scarlet-like fever.] In: Proceedings of Vladivostok Res. Inst. Epidemiol. Vladivostok. 197-202.
Zink, D.L., Feeley, J.C., Wells, G., Vanderzant, C., Vickery, J.C., Roof, W.D., and O'Donovan, G.A. 1980. Plasmid mediated tissue invasiveness in *Yersinia enterocolitica*. Nature 283: 224-226.

Index

Notes

Notes

Notes

Notes

Notes